ANSWERS for

CARBON-13 NMR BASED
ORGANIC SPECTRAL PROBLEMS

by

Philip L. Fuchs
Charles A. Bunnell

John Wiley & Sons, Inc.
New York · Chichester · Brisbane · Toronto

Reproduction or translation of any part of
this work beyond that permitted by Sections
107 or 108 of the 1976 United States Copyright
Act without the permission of the copyright
owner is unlawful. Requests for permission
or further information should be addressed to
the Permissions Department, John Wiley & Sons, Inc.

ISBN 0 471 05138 1
Printed in the United States of America

10 9 8 7 6 5 4 3 2 1

This material may be reproduced for testing or
instructional purposes by people using the text.

Answer Book to
"C-13 NMR-Based Organic
Spectral Problems"

P. L. Fuchs, C. A. Bunnell

1. 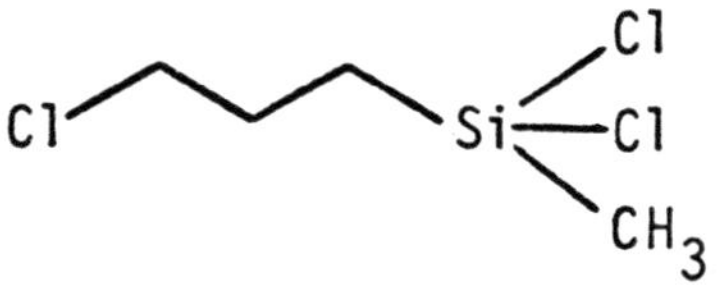 Cl-CH₂CH₂CH₂-Si(Cl)(Cl)(CH₃)

2. 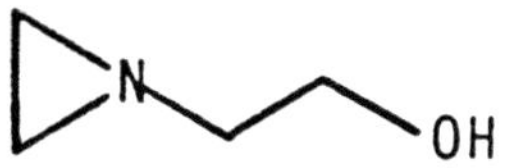 N-aziridinyl-CH₂CH₂-OH

3A. 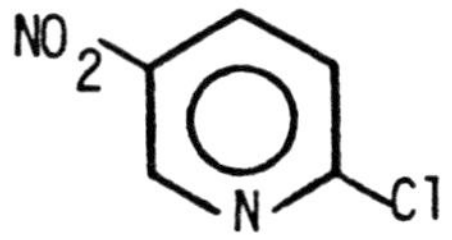 2-chloro-5-nitropyridine

3B. 2-chloro-3-nitropyridine

4. pyrrole-2-carboxaldehyde

5. Br-CH₂-CH=CH-CO₂CH₃ (trans)

6. 3,4-dihydro-2H-pyran

7. δ-valerolactone (tetrahydro-2H-pyran-2-one)

8. 1-cyclopropylethanol

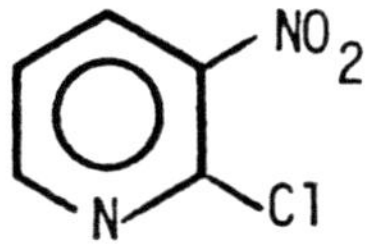

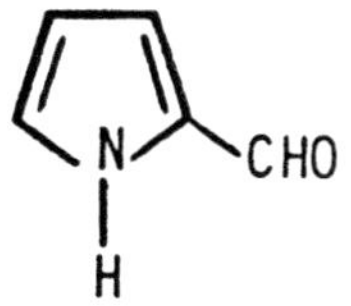

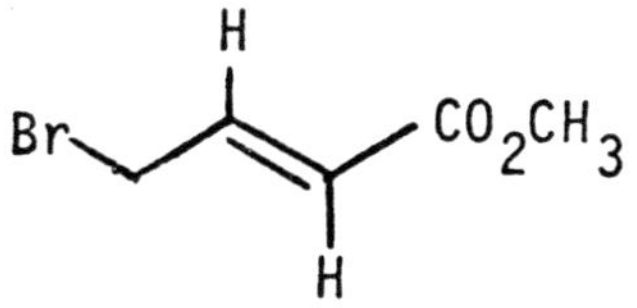

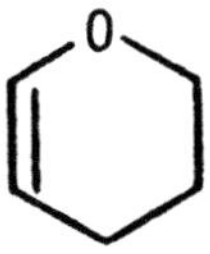

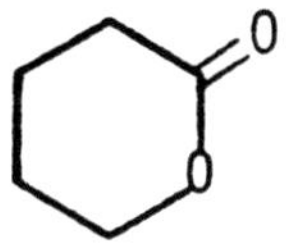

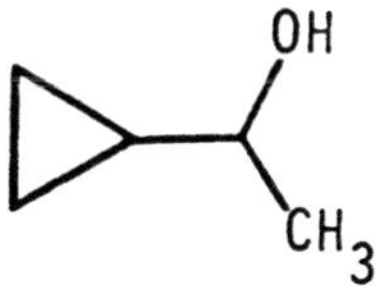

9. CH₃-CH(OCH₃)-CH₂-CH₂-OH

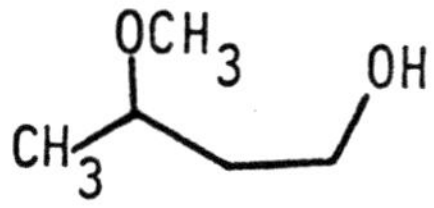

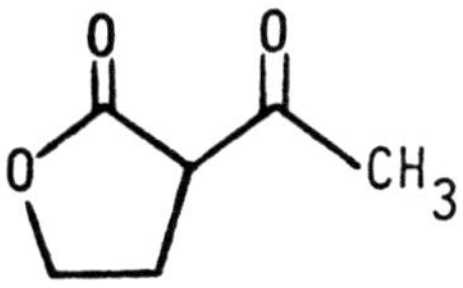

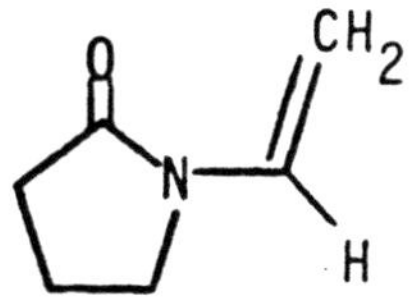

10. 2-acetyl-γ-butyrolactone

11. N-vinyl-2-pyrrolidinone

12A. cyclohexanone

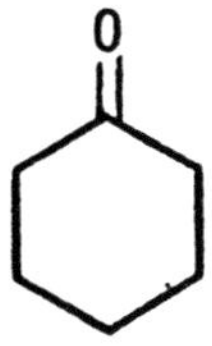

12B. HOCH₂CH₂C≡CCH₂CH₃

13A. 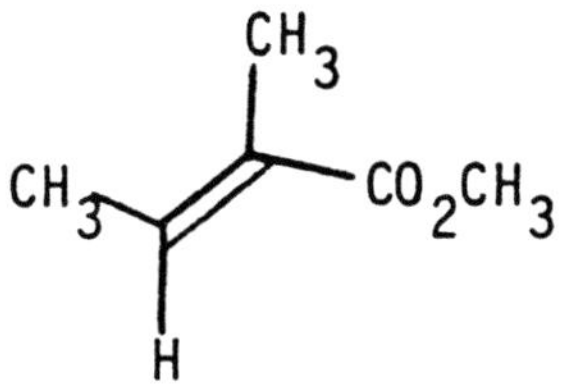 (E)-methyl 2-methyl-2-butenoate: CH₃-C(CH₃)=CH... CH₃C(=CHH)(CH₃)CO₂CH₃

13B. 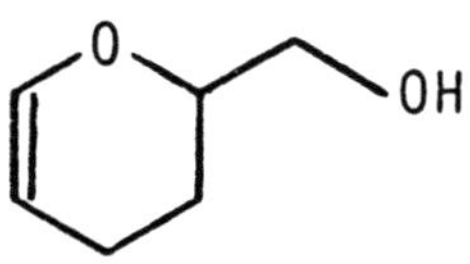 3,4-dihydro-2H-pyran-2-methanol

14A.  Br-CH₂CH₂CH₂-CO₂C₂H₅

14B. 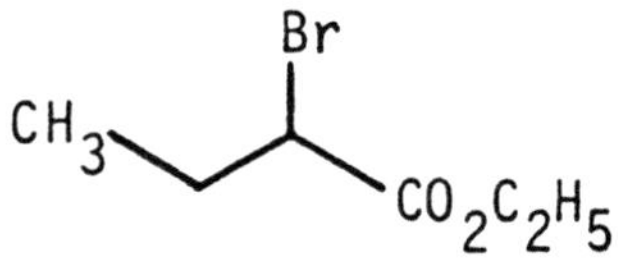 CH₃CH(Br)CH(Br)... CH₃CH₂CH(Br)CO₂C₂H₅

15. 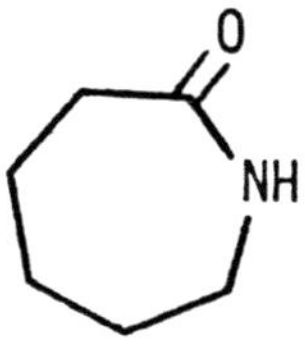 caprolactam (7-membered ring with C=O and NH)

16. 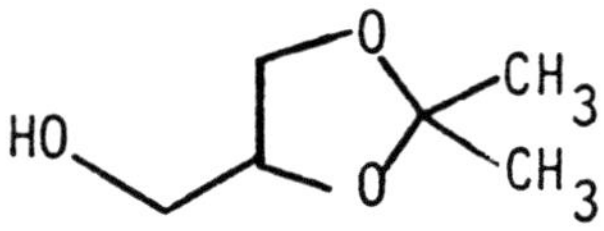 HOCH₂—(2,2-dimethyl-1,3-dioxolane)

17. 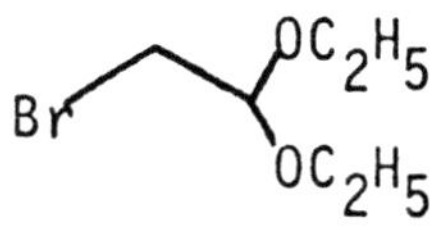 BrCH₂CH(OC₂H₅)₂

18. 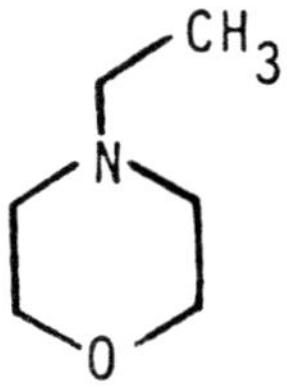 4-ethylmorpholine

19A. 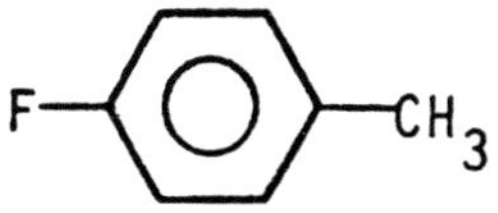 1-fluoro-4-methylbenzene (p-F, CH₃)

19B. 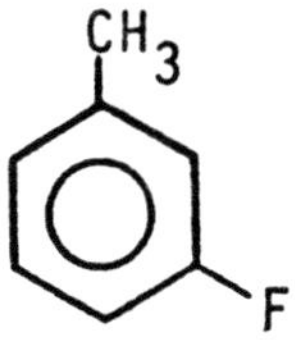 3-methyl-(2-fluoro...) — CH₃ and F on pyridine/benzene ring

19C. 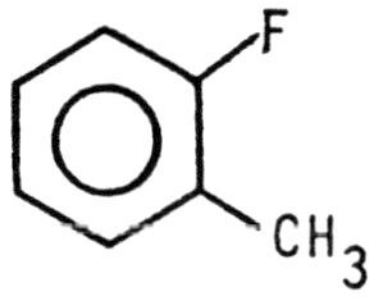 1-fluoro-2-methylbenzene (o-F, CH₃)

20A. 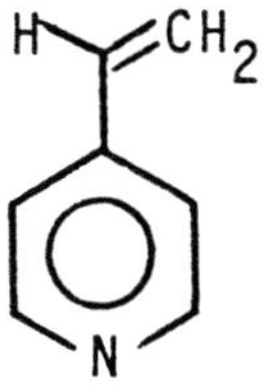 4-vinylpyridine

20B. 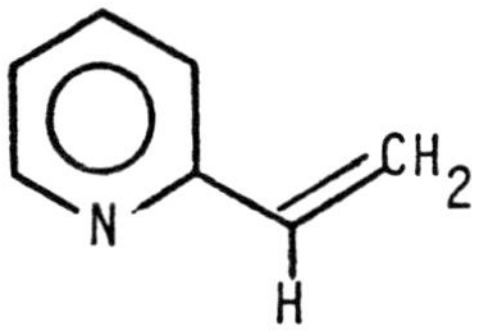 2-vinylpyridine

21. 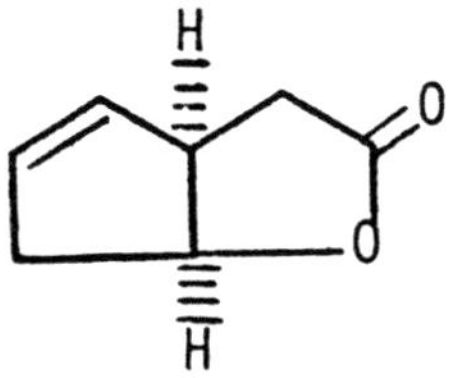 cis-bicyclic cyclopentene-fused γ-butyrolactone

22. 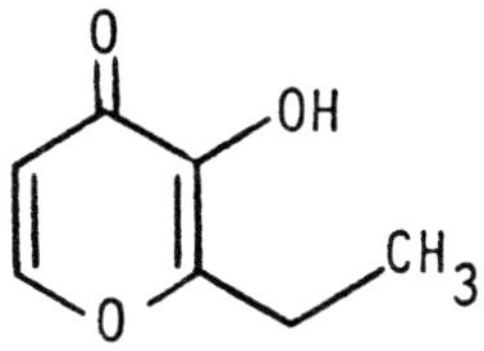 3-hydroxy-2-ethyl-4H-pyran-4-one

23. HOCH₂C≡CCH₂CH₂CH₂CN

24. norbornene (bicyclo[2.2.1]hept-2-ene)

25. 3-methoxycyclohex-2-en-1-one

26A. 2-methylcyclohexan-1-one

26B. 3-methylcyclohexan-1-one

27. ethyl cyclobutanecarboxylate (cyclobutyl-CO$_2$C$_2$H$_5$)

28. CH$_3$CO-O-(CH$_2$)$_4$-Cl

29. 7a-methyl-tetrahydro-oxazolo[2,3-b]oxazole (bicyclic N,O,O acetal with CH$_3$ at ring junction)

30. oxiranyl-CH(OC$_2$H$_5$)$_2$

31. CH$_3$O-CH(-)-CH$_2$-CH(-)-OCH$_3$ with OCH$_3$ substituents (1,1,3,3-tetramethoxy... 1,2,3,3-tetramethoxybutane arrangement): CH$_3$O, OCH$_3$, CH$_3$O, OCH$_3$

32. C$_6$H$_5$-CH(-O-CH$_2$) styrene oxide

33. piperonyl alcohol (1,3-benzodioxol-5-yl-CH$_2$OH)

34. C$_6$H$_5$-C(=O)-NHCH$_3$

35A. 4-amino-1,2-dimethylbenzene (3,4-dimethylaniline)

35B. 3-amino-1,4-dimethylbenzene (2,5-dimethylaniline)

35C. 2,3-dimethylaniline

35D. 2,6-dimethylaniline

35E. 4-amino-3,5-dimethyl... (2,4-dimethylaniline with NH₂ on ring)

35E. 2,4-dimethylaniline (NH₂ at position 1, CH₃ groups)

35F. 3,5-dimethylaniline (NH₂ with two CH₃ groups on ring)

36A. 1,5-cyclooctadiene

36B. 3-vinylcyclohexene (cyclohexene with CH=CH₂ substituent)

37. cyclooct-4-enone

38. 3,3-dimethyl-4-(propan-2-ylidene)oxetan-2-one (β-lactone with isopropylidene and gem-dimethyl)

39. 1-(ethoxycarbonyl)-4-piperidone (N-CO₂C₂H₅)

40. (CH₃)₂CHCH₂-C(OH)(CH₃)-C≡C-H (2,4-dimethyl... with OH and C≡CH)

41A. (Z)-CH₃-CH=CH-CH₂-CO₂C₂H₅ (cis)

41B. (E)-CH₃-CH=CH-CH₂-CO₂C₂H₅ (trans)

41C. bicyclic ether with OH (9-oxabicyclic alcohol)

42. CH₃CH₂CH₂-C(=O)-CH₂-CO₂C₂H₅

43. CH₃-C(=O)-CH(CH₃)-P(=O)(OC₂H₅)₂

44. (C₂H₅O)₂P(=O)-CH₂-CO₂C₂H₅

45. CH₃(CH₂)₆CH₂OH

46. (CH₃)₃C-C(OCH₃)... CH₃-C(CH₃)₂-C(OCH₃)(...)-CH₂-CH(OCH₃)₂

47A. 2-indanone

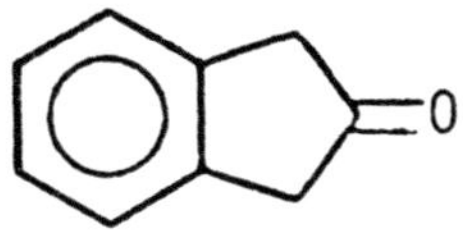

47B. 1-indanone

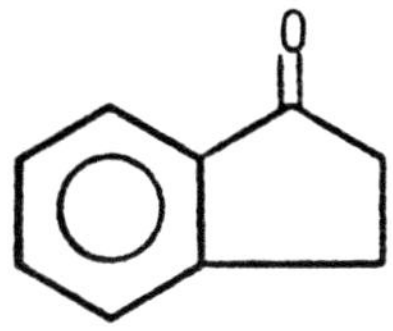

48. 3-chloro-2-methoxycarbonyl-norbornadiene (methyl 3-chlorobicyclo[2.2.1]hepta-2,5-diene-2-carboxylate)

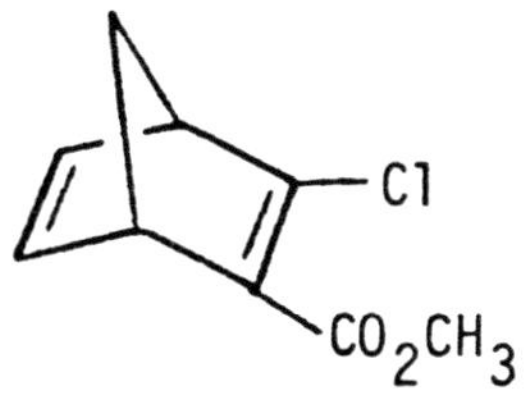

49. 2-indanol

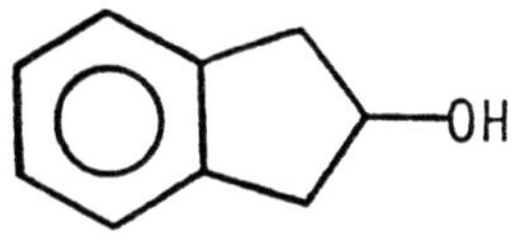

50A. 2,3-dimethylanisole

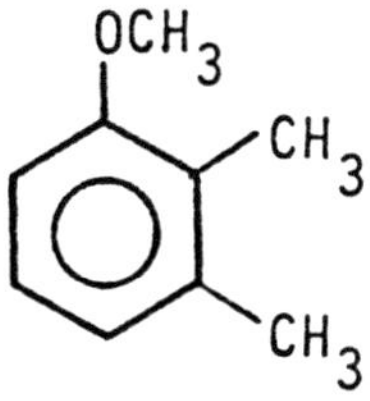

50B. 2,4-dimethylanisole

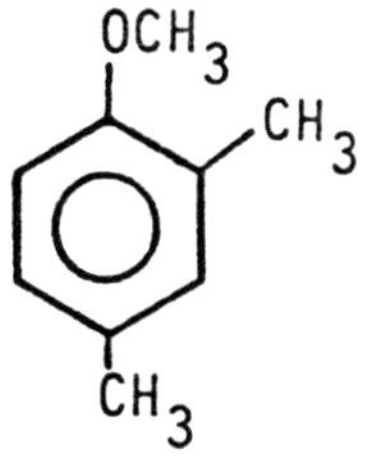

50C. 3,5-dimethylanisole

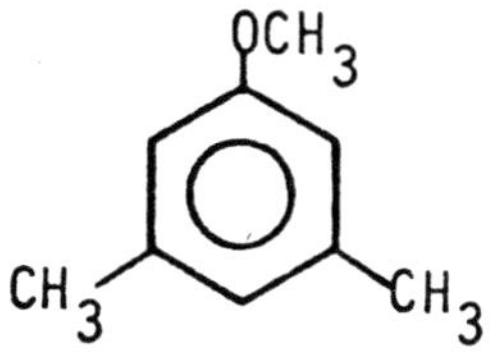

50D. 2,5-dimethylanisole

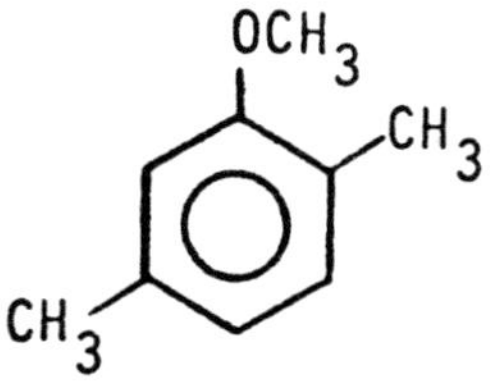

50E. 2,6-dimethylanisole

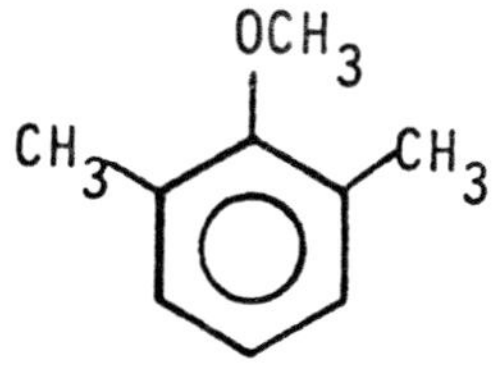

50F. 2,5-dimethyl-4-methoxy... (1-methoxy-2,5-dimethylbenzene derivative)

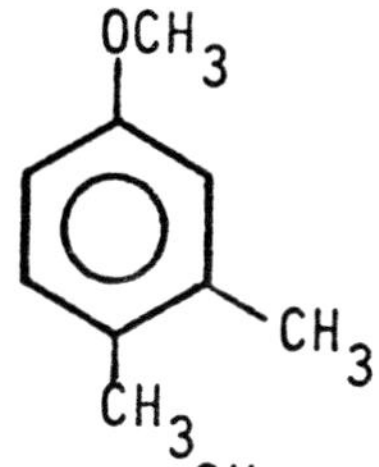

51. 3,5-dimethoxybenzyl alcohol

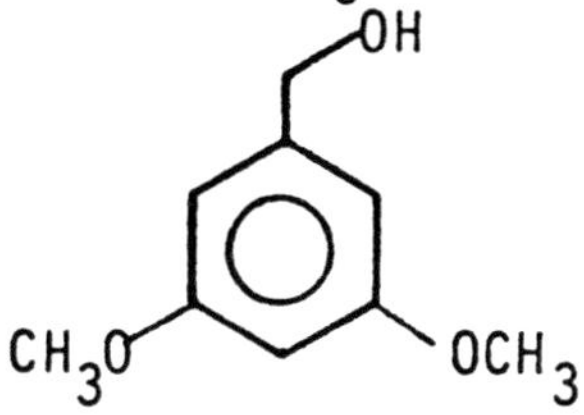

52. 4,4-dimethyl-1-cyclohexene-1-carbonitrile

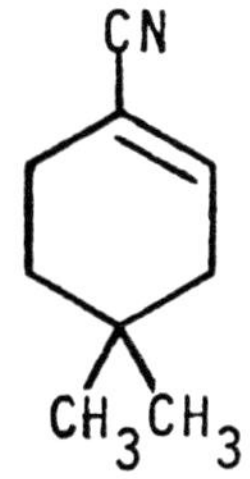

53. 2,3-epoxy-4,4,6-trimethylcyclohexanone (isophorone oxide)

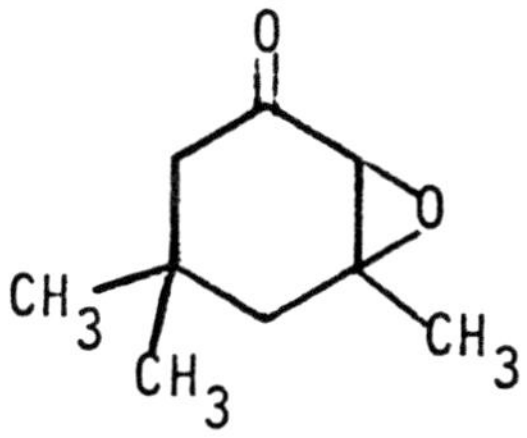

54. 1-(1-cyclopentenyl)pyrrolidine

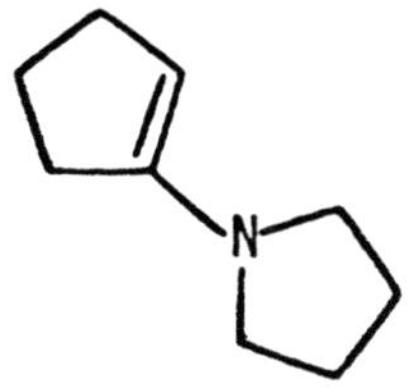

55. 1-(2-cyanoethyl)hexamethyleneimine

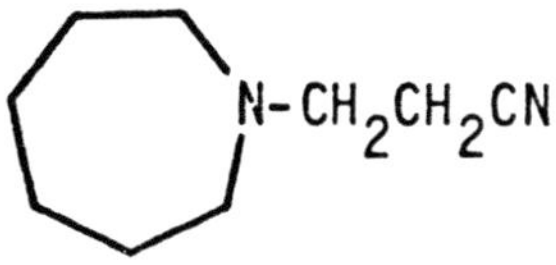

56. 3,5,5-trimethyl-3-... alkene (CH$_2$=CH-CH(CH$_3$)-CH$_2$-C(CH$_3$)$_3$)

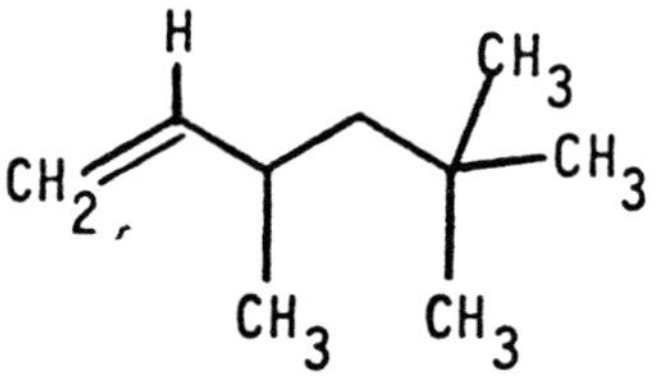

57. [structure: 2,2-diisopropyl-1,3-dioxolane]

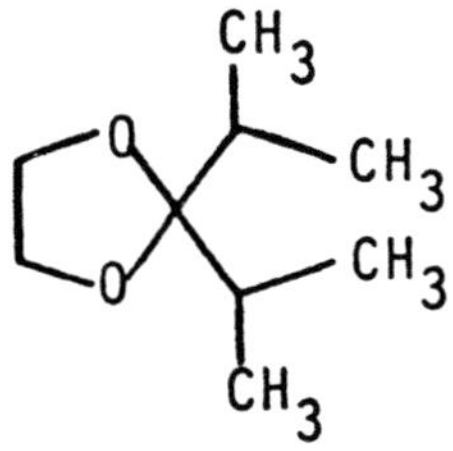

58. 2,2,6,6-tetramethylpiperidine

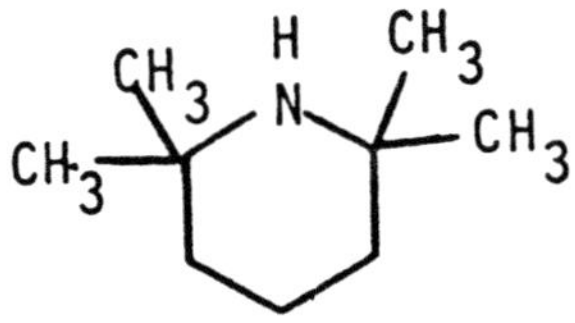

59. [(o-methoxyphenyl)Cr(CO)₃ complex]

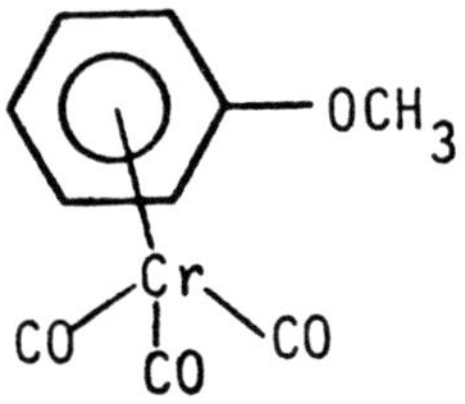

60A. naphthalene-2,3-diol

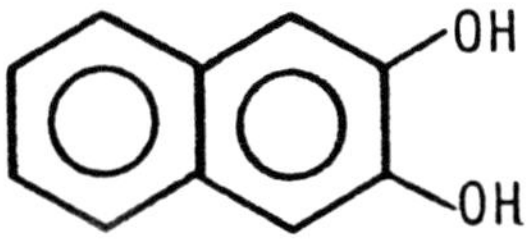

60B. naphthalene-2,7-diol

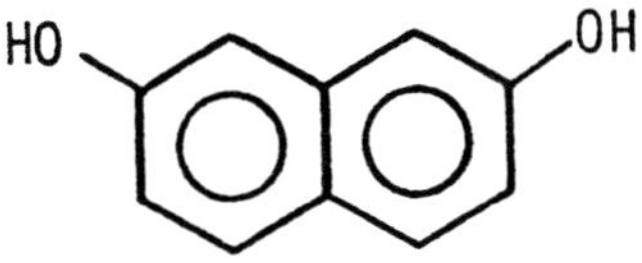

61A. CH₃-CO-CH(Cl)-CH₂-C₆H₄-Cl (para)

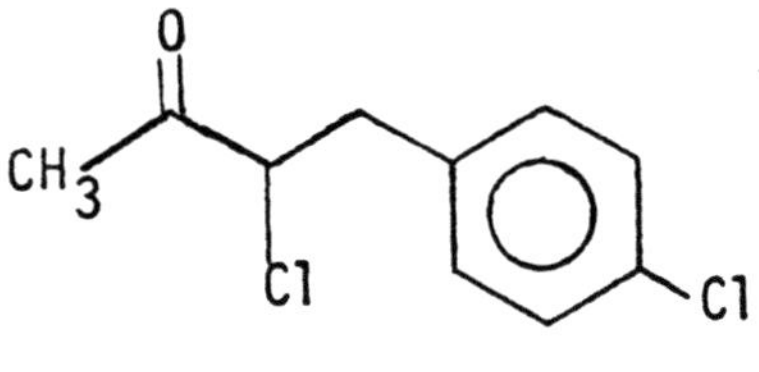

61B. CH₃-CO-CH(Cl)-CH₂-C₆H₄-Cl (ortho)

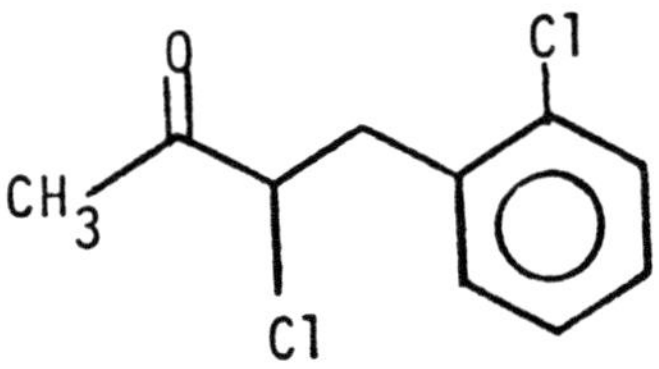

61C. CH₃-CO-CH(Cl)-CH₂-C₆H₄-Cl (meta)

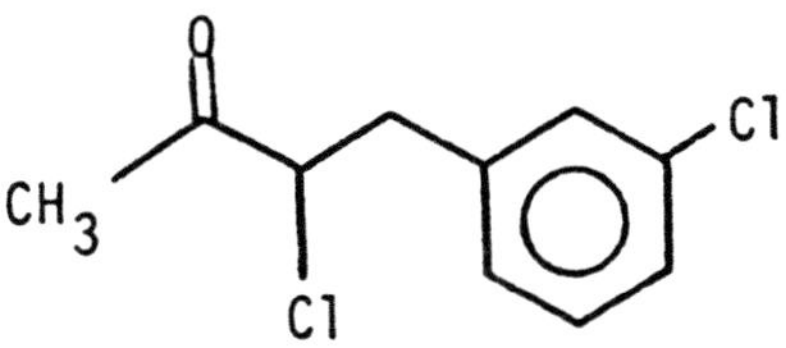

62. trans-epoxide: C₆H₅, H / H, COCH₃

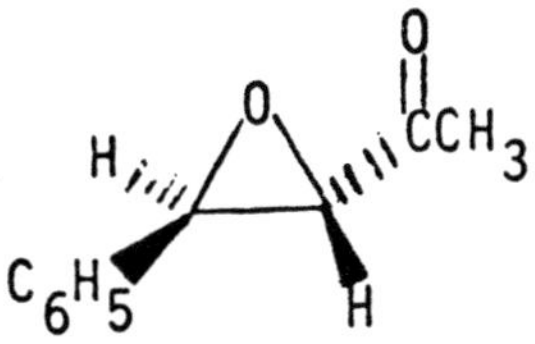

63A. 3-(phenylthio)-2,5-dihydrothiophene 1,1-dioxide

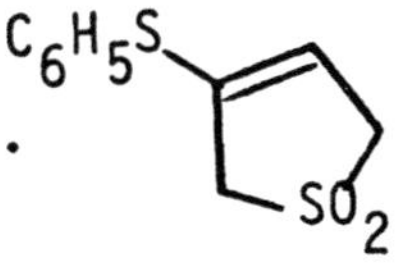

63B. 3-(phenylthio)-2,3-dihydrothiophene 1,1-dioxide

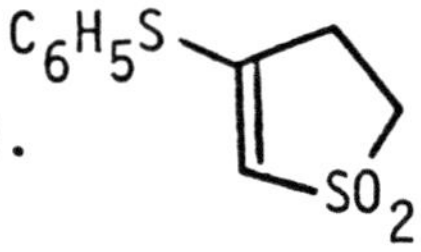

64. dicyclopentadiene

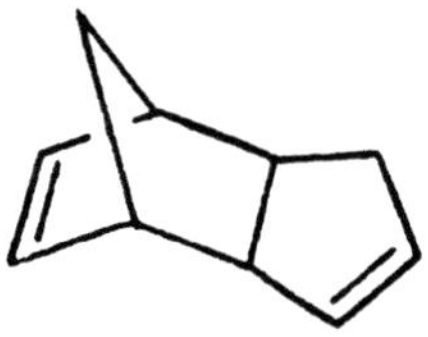

65. nicotine

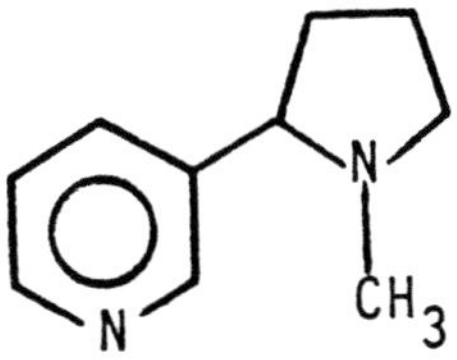

66. carvone

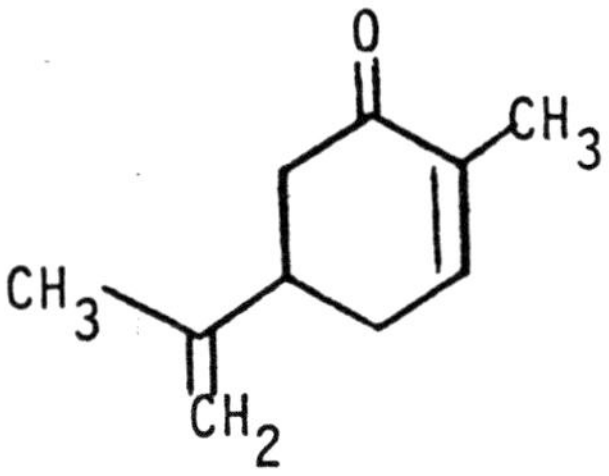

67A. myrtenol

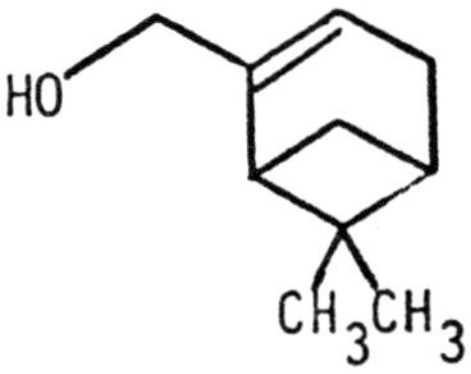

67B. [2-isopropyl-5-methyl-cyclohex-4-en-1-one]

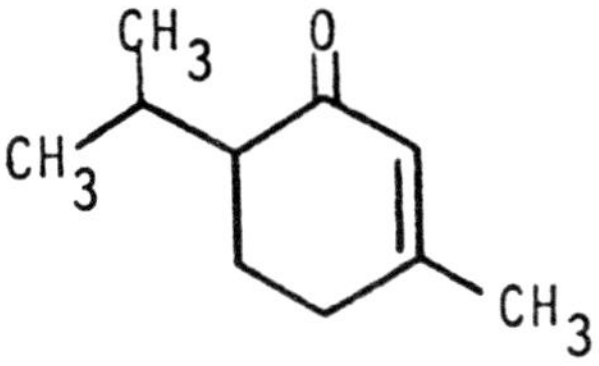

68. [structure]

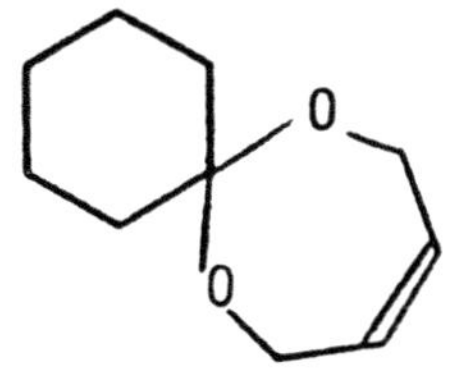

69. [structure]

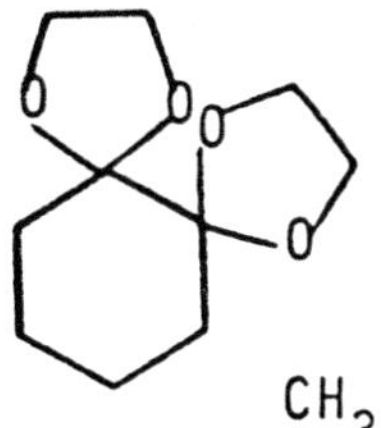

70. [structure]

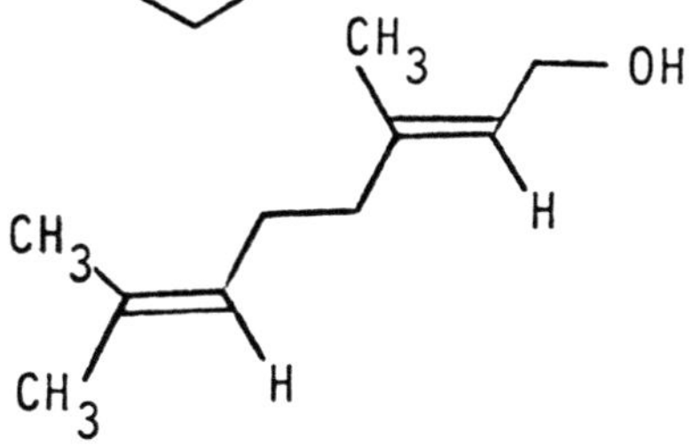

71. [structure]

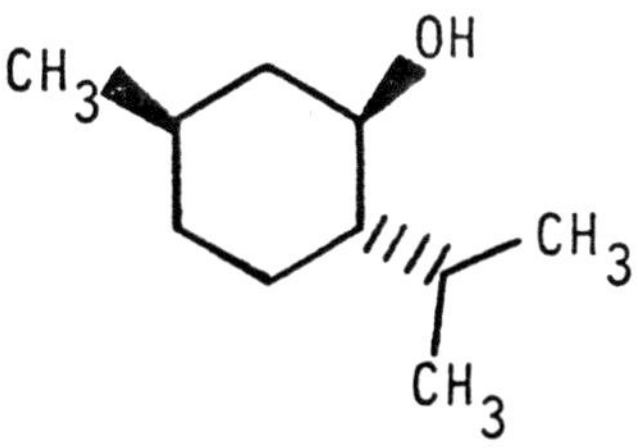

72. [structure]

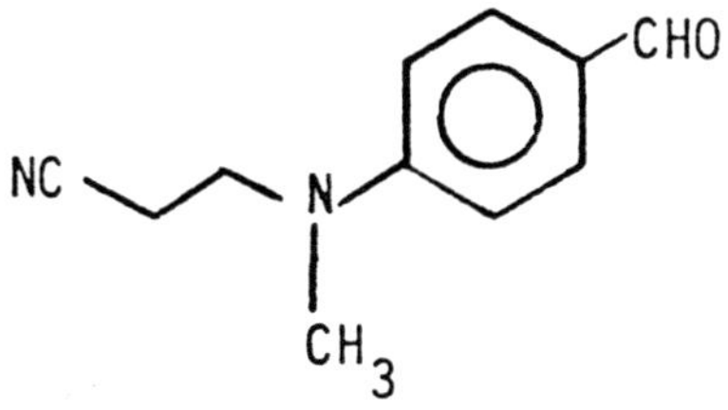

73. [structure]

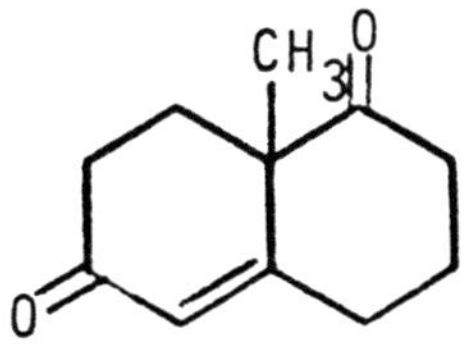

74. (C₂H₅)₂N—C₆H₄—CHO

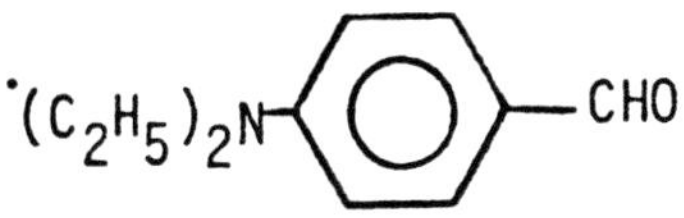

75. [structure: 1-phenethylpiperazine]

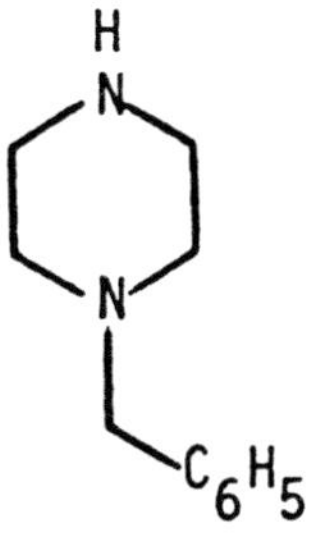

76. [structure]

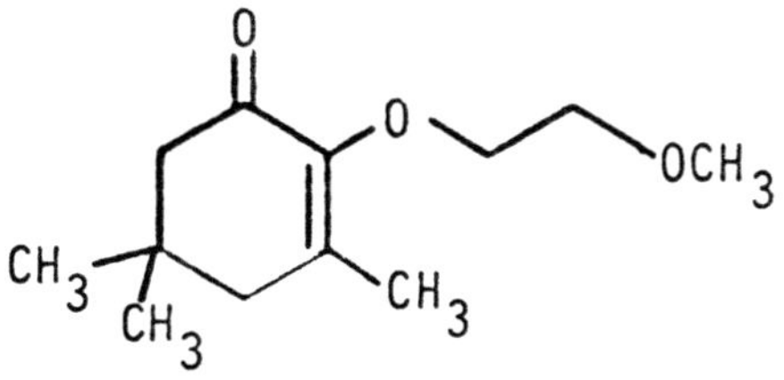

77. HO—C₆H₄—SO₂—C₆H₄—OH

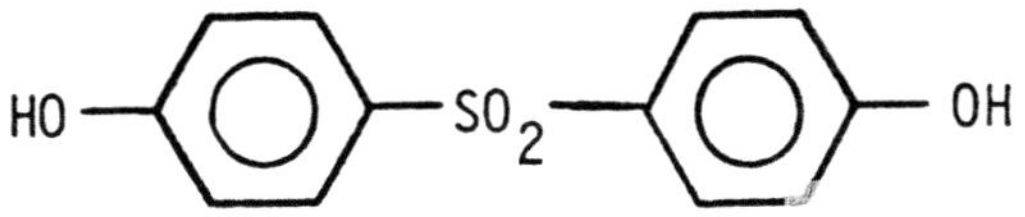

78. C₆H₅S—[cyclohexenyl]

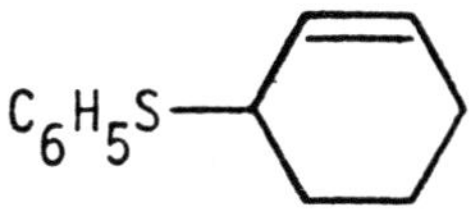

79. [structure: hexamethyl bicyclobutene]

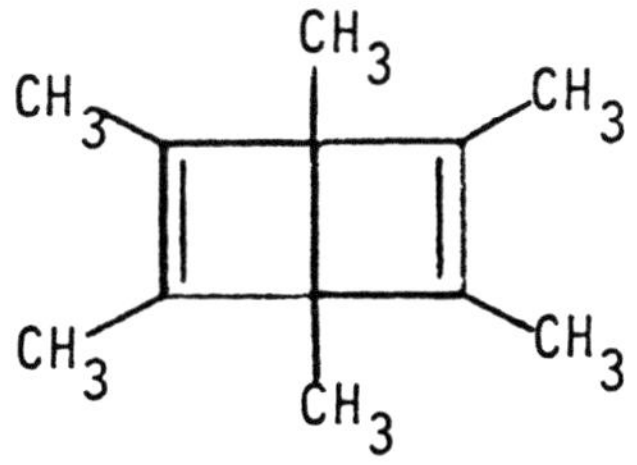

80. [structure]

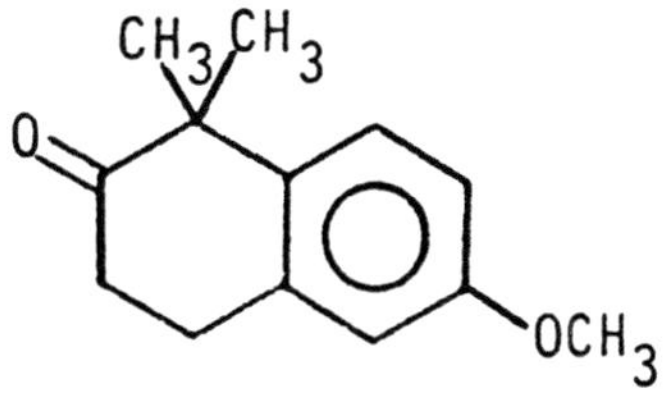

81. [structure]

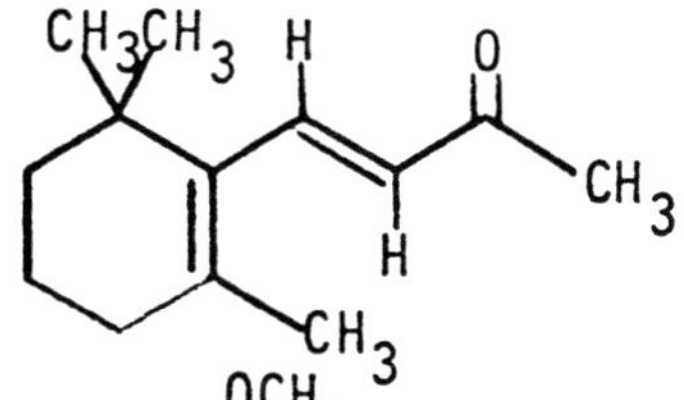

82A. [structure]

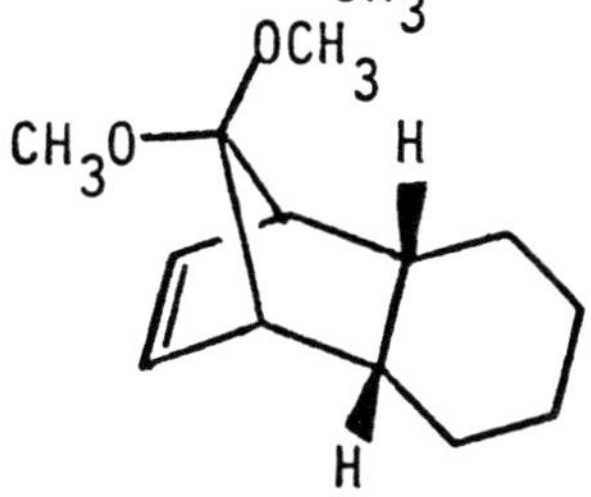

82B. [structure: methyladamantane acetic acid]

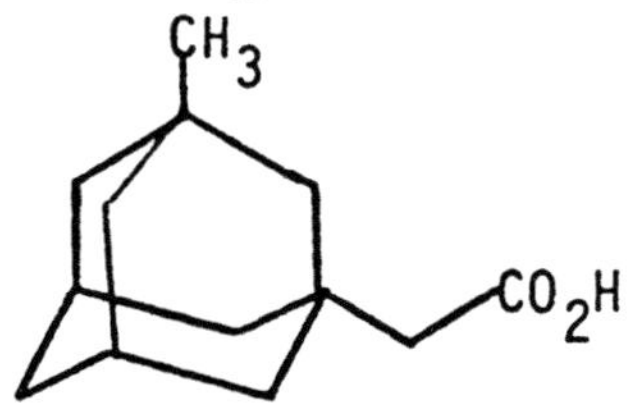

83. (CH₃)₂N, N(CH₃)₂ on naphthalene (1,8-positions)

84. 4,5-dimethoxy-2-acetylphenylacetic acid ethyl ester (CH₃O, CH₃O on ring; COCH₃ and CH₂CO₂C₂H₅ substituents)

85. 2,6-di-sec-butylphenol (OH, with two CH(CH₃)CH₂CH₃ groups ortho)

86. N-(3-phenylpropyl)phthalimide

87. *endo/exo* bicyclic: C₆H₅CH₂O–CH₂ group and CO₂H on norbornene

88. 2-(phthalimido)-4-phenylbutanoic acid (phthalimide-N-CH(CO₂H)-CH₂-C₆H₅)

89A. 1,3-diphenylbenzene (C₆H₅, C₆H₅)

89B. 1,2-diphenylbenzene (C₆H₅, C₆H₅)

89C. 1,1-diphenyl on cyclopentadiene (C₆H₅, C₆H₅ at exocyclic =C)

90. 1,1-bis(3,4-dimethylphenyl)ethane (CH·CH₃ linking two 3,4-dimethylphenyl rings)

91. (CH₃)₂N–C₆H₄–C(=CH₂)–C₆H₄–N(CH₃)₂

92. methoxy-substituted tricyclic with CH₃, CH₃, CO₂CH₃ (podocarpane-type: CH₃O on aromatic ring; angular CH₃, CH₃ and CO₂CH₃)

93. triptycene

94. N(C₇H₁₅)₃

95. pentacyclic alkaloid with CH₃N, NH, NH, N–CH₃ (sparteine/yohimbine-type structure)

ANSWERS for

CARBON-13 NMR BASED
ORGANIC SPECTRAL PROBLEMS

by

Philip L. Fuchs
Charles A. Bunnell

John Wiley & Sons, Inc.
New York · Chichester · Brisbane · Toronto

Reproduction or translation of any part of
this work beyond that permitted by Sections
107 or 108 of the 1976 United States Copyright
Act without the permission of the copyright
owner is unlawful. Requests for permission
or further information should be addressed to
the Permissions Department, John Wiley & Sons, Inc.

ISBN 0 471 05138 1
Printed in the United States of America

10 9 8 7 6 5 4 3 2 1

This material may be reproduced for testing or
instructional purposes by people using the text.

Answer Book to

"C-13 NMR-Based Organic

Spectral Problems"

P. L. Fuchs, C. A. Bunnell

1. 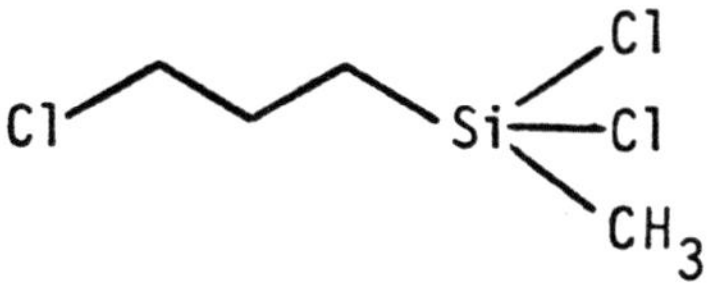 Cl-CH₂CH₂CH₂-Si(Cl)(Cl)-CH₃

2. 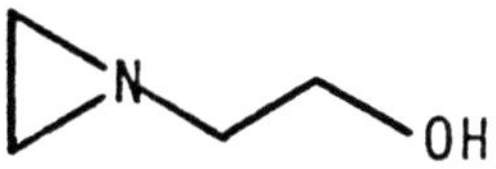 N-aziridinyl-CH₂CH₂-OH

3A. 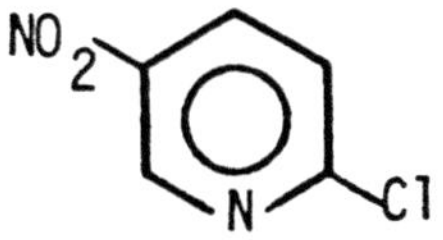 2-chloro-5-nitropyridine

3B. 2-chloro-3-nitropyridine

4. pyrrole-2-carboxaldehyde (N-H)

5. (E)-BrCH₂-CH=CH-CO₂CH₃

6. 3,4-dihydro-2H-pyran

7. δ-valerolactone (tetrahydro-2H-pyran-2-one)

8. 1-cyclopropylethanol

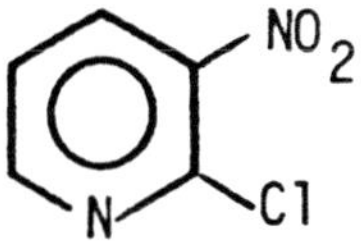

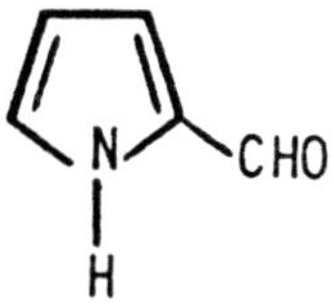

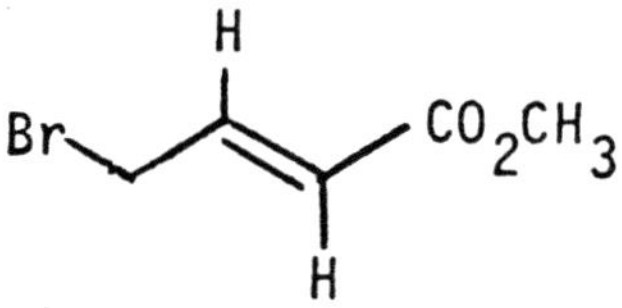

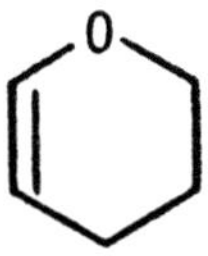

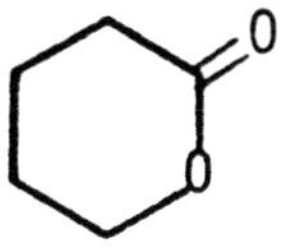

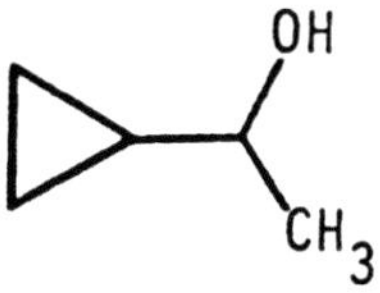

9. CH₃-CH(OCH₃)-CH₂-CH₂-OH

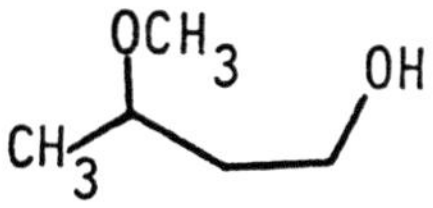

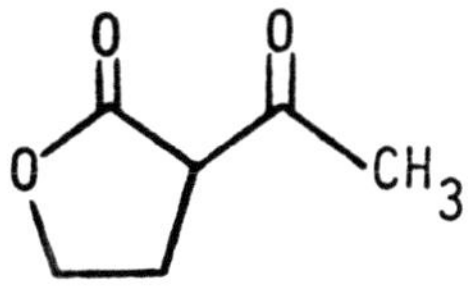

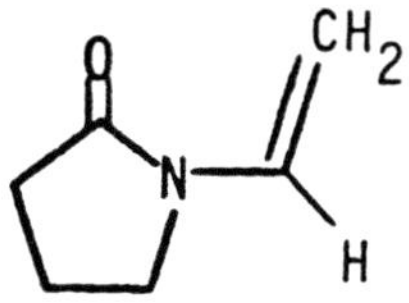

10. 2-acetyl-γ-butyrolactone

11. 1-vinyl-2-pyrrolidinone

12A. cyclohexanone

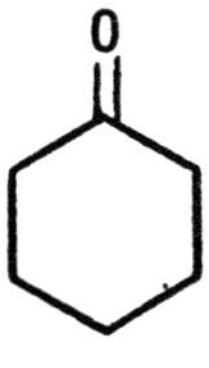

12B. HOCH₂CH₂C≡CCH₂CH₃

13A. 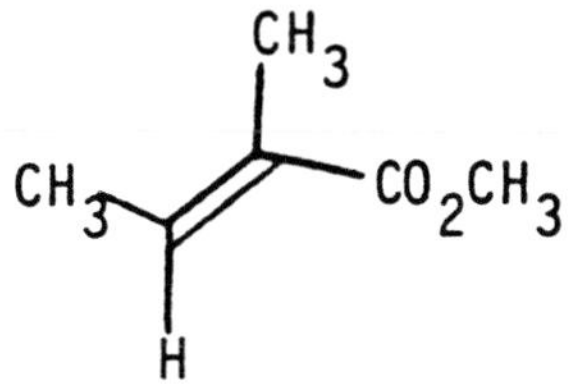 (E)-methyl 2-methyl-2-butenoate: CH₃-CH=C(CH₃)-CO₂CH₃ with H on the CH₃ side

13B. 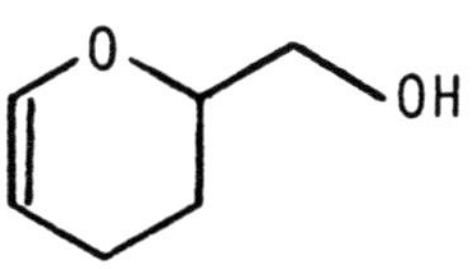 2-(hydroxymethyl)-3,4-dihydro-2H-pyran

14A.  Br-CH₂CH₂CH₂-CO₂C₂H₅

14B. 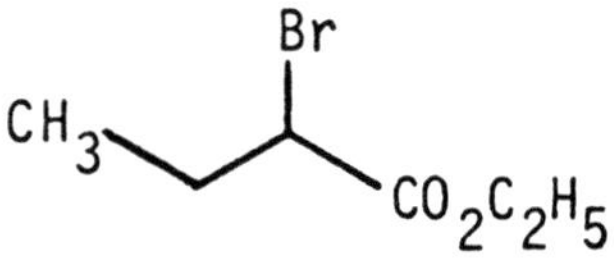 CH₃CH₂-CHBr-CO₂C₂H₅

15. 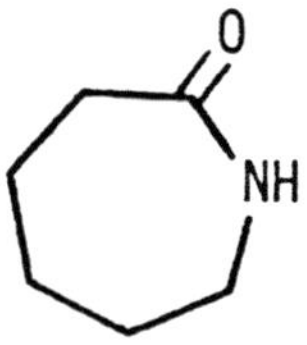 ε-caprolactam (7-membered ring lactam)

16. 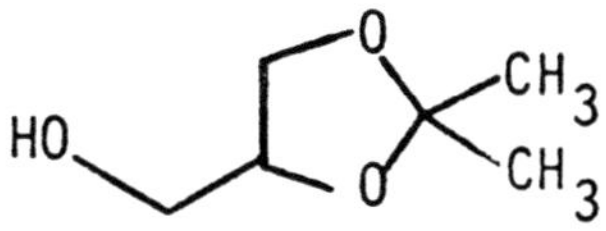 (2,2-dimethyl-1,3-dioxolan-4-yl)methanol: HOCH₂-CH(O-C(CH₃)₂-O-CH₂)

17. 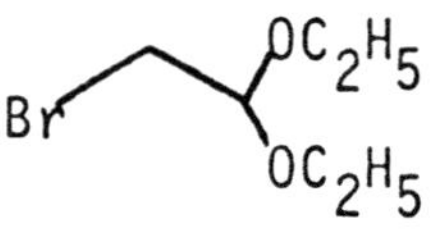 BrCH₂CH(OC₂H₅)₂

18. 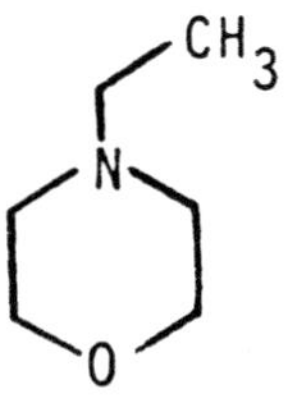 4-ethylmorpholine

19A. 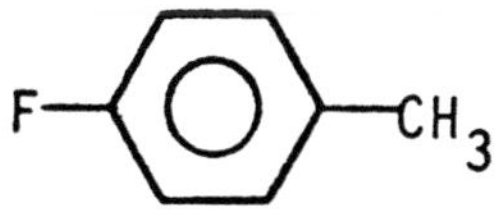 1-fluoro-4-methylbenzene (p-fluorotoluene)

19B. 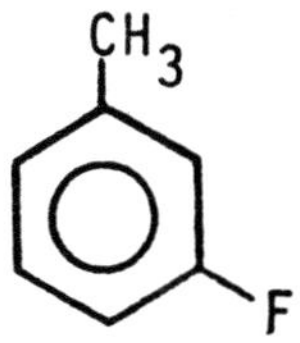 1-fluoro-3-methylbenzene (m-fluorotoluene)

19C. 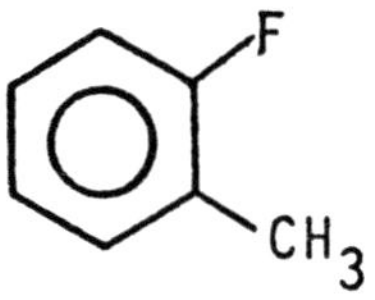 1-fluoro-2-methylbenzene (o-fluorotoluene)

20A. 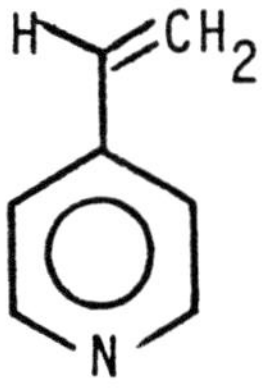 4-vinylpyridine

20B. 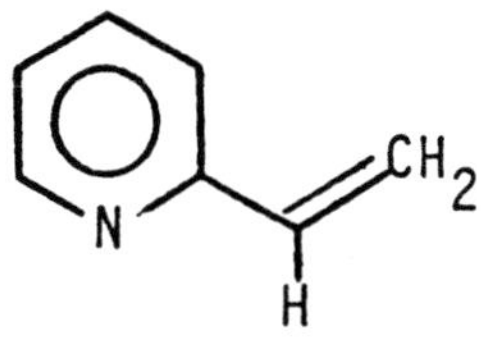 2-vinylpyridine

21. 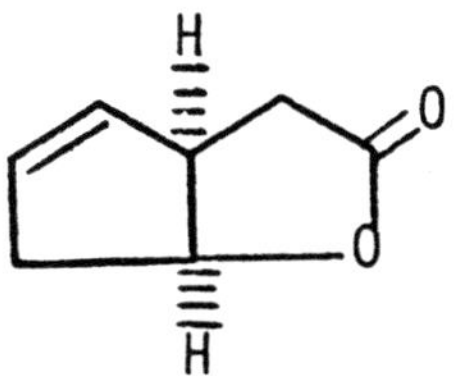 cis-bicyclic cyclopentene-fused γ-butyrolactone

22. 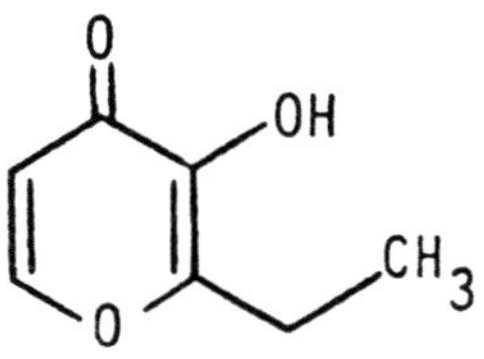 3-hydroxy-2-ethyl-4H-pyran-4-one

23. HOCH₂C≡CCH₂CH₂CN

24. norbornene (bicyclo[2.2.1]hept-2-ene)

25. 3-methoxycyclohex-2-enone

26A. 2-methylcyclohexanone

26B. 3-methylcyclohexanone

27. ethyl cyclobutanecarboxylate (cyclobutyl-CO$_2$C$_2$H$_5$)

28. CH$_3$CO-O-(CH$_2$)$_n$-Cl (6-chlorohexyl acetate)

29. 7a-methyl-tetrahydro-oxazolo[3,2-a]oxazole (bicyclic N,O,O-acetal with CH$_3$ at ring junction)

30. 2-(diethoxymethyl)oxirane — glycidaldehyde diethyl acetal; epoxide-CH(OC$_2$H$_5$)$_2$

31. CH$_3$O-CH(OCH$_3$)-CH$_2$-CH(OCH$_3$)-OCH$_3$ (1,1,3,3-tetramethoxybutane type — tetramethoxy compound)

32. C$_6$H$_5$-CH(H)- epoxide (styrene oxide)

33. piperonyl alcohol (1,3-benzodioxol-5-yl-CH$_2$OH)

34. C$_6$H$_5$-C(=O)-NHCH$_3$ (N-methylbenzamide)

35A. 3,4-dimethylaniline (NH$_2$ with CH$_3$, CH$_3$ at 3,4)

35B. 2,5-dimethylaniline (NH$_2$ with CH$_3$ ortho and CH$_3$ meta-para)

35C. 2,3-dimethylaniline

35D. 2,6-dimethylaniline

35E. 4-amino-3-methyl... 2-methyl-4-aminotoluene (2-methyl-1,4-...) — 1-amino-2-methyl-4-methylbenzene

35E. 4-methyl-2-methylaniline (2,4-dimethylaniline with NH₂)

35F. 3,5-dimethylaniline

36A. 1,3-cyclooctadiene

36B. 3-vinylcyclohexene

37. cyclooct-2-enone (cyclooctenone)

38. 4-isopropylidene-3,3-dimethyl-2-oxetanone

39. 1-(ethoxycarbonyl)-4-piperidinone

40. (CH₃)₂CHCH₂C(OH)(CH₃)C≡CH

41A. (Z)-CH₃CH=CHCH₂CO₂C₂H₅

41B. (E)-CH₃CH=CHCH₂CO₂C₂H₅

41C. bicyclic ether with OH

42. CH₃CH₂CH₂C(O)CH₂CO₂C₂H₅

43. CH₃C(O)CH(CH₃)P(O)(OC₂H₅)₂

44. (C₂H₅O)₂P(O)CH₂CO₂C₂H₅

45. CH₃(CH₂)₆CH₂OH

46. (CH₃)₂C(OCH₃)CH₂CH(OCH₃)₂ — with extra CH₃

47A. 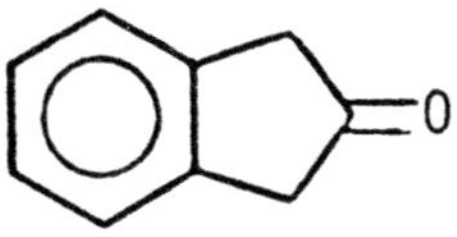 2-indanone

47B. 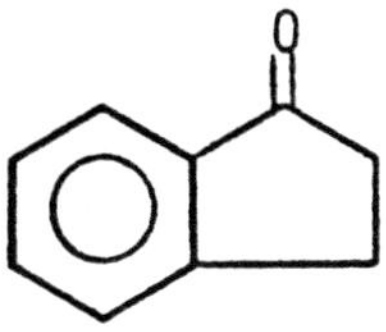 1-indanone

48. 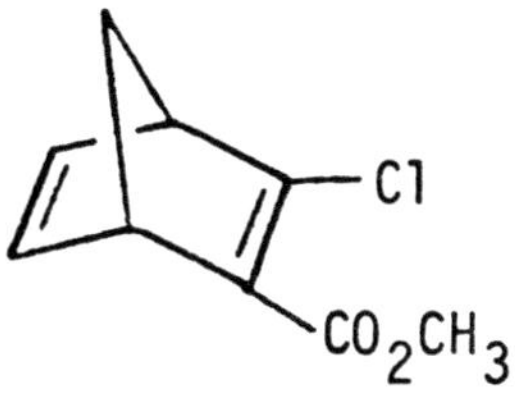 methyl 3-chlorobicyclo[2.2.1]hepta-2,5-diene-2-carboxylate

49. 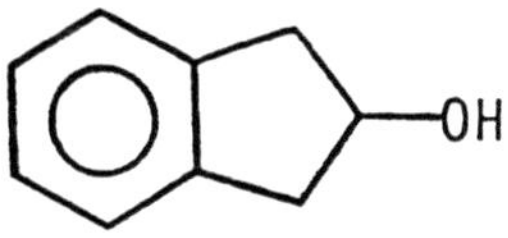 2-indanol

50A. 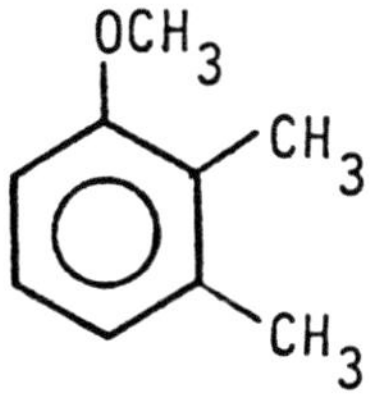 2,3-dimethylanisole

50B. 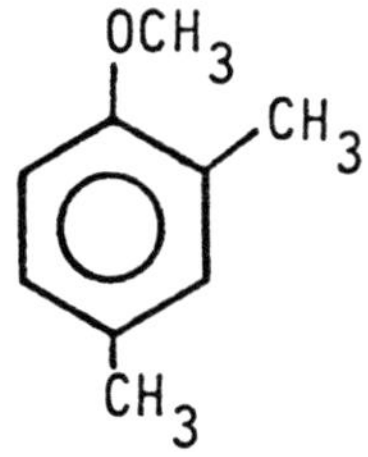 2,4-dimethylanisole

50C. 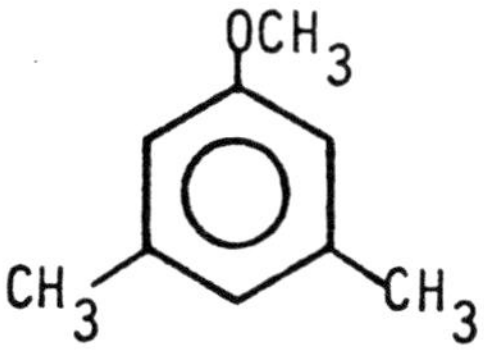 3,5-dimethylanisole

50D. 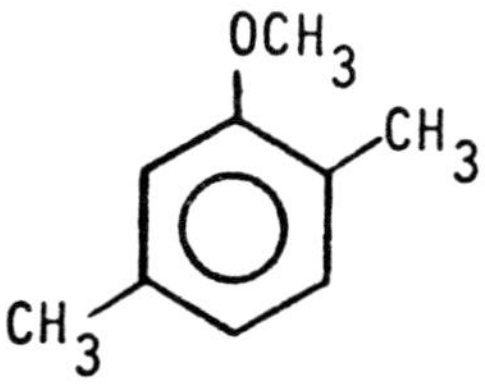 2,5-dimethylanisole

50E. 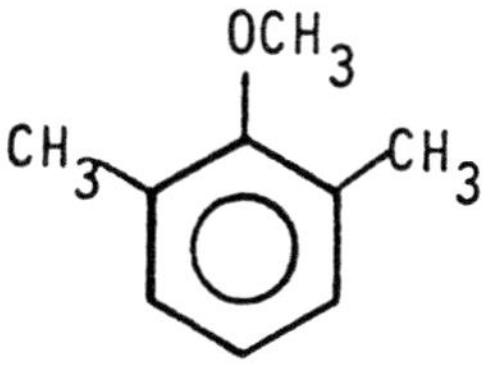 2,6-dimethylanisole

50F. 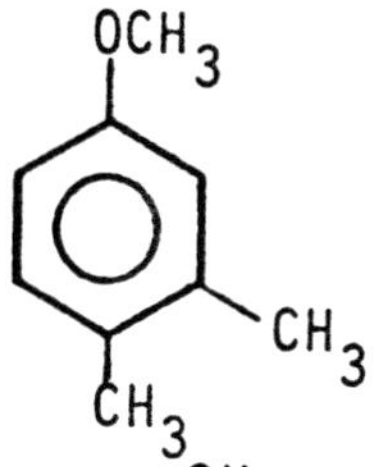 1,4-dimethyl-2,5... (2,5-dimethyl-1,4-dimethoxybenzene-like: 4-methoxy-2,3-dimethyl...)

51. 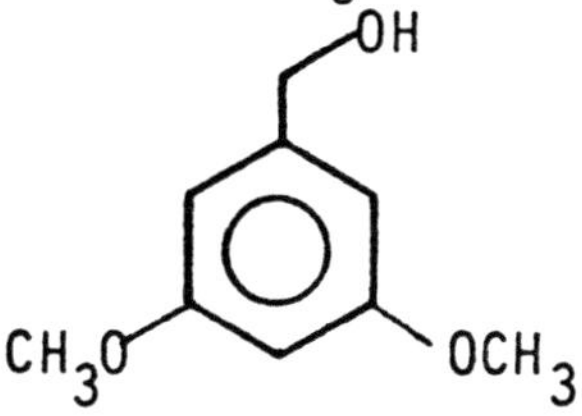 3,5-dimethoxybenzyl alcohol

52. 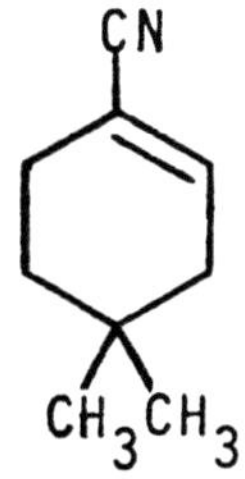 4,4-dimethylcyclohex-1-ene-1-carbonitrile

53. 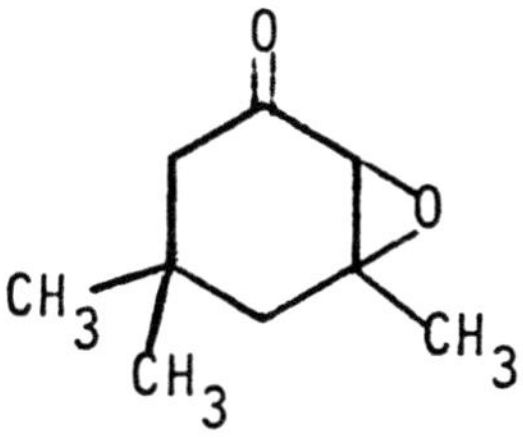 3,3,5-trimethyl-4,5-epoxycyclohexan-1-one

54. 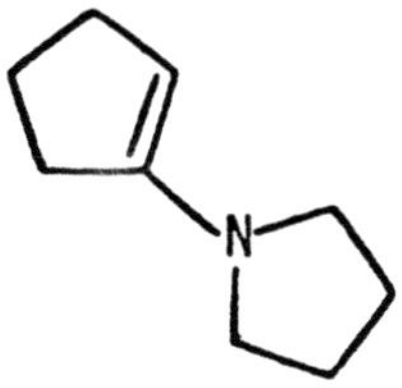 1-(cyclopent-1-en-1-yl)pyrrolidine

55. 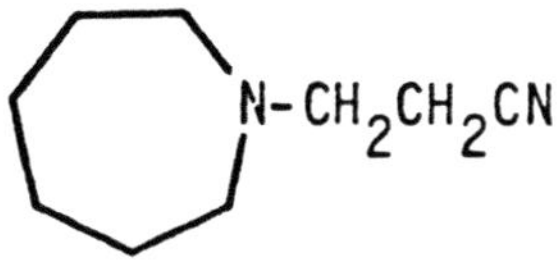 1-(2-cyanoethyl)hexahydroazepine

56. 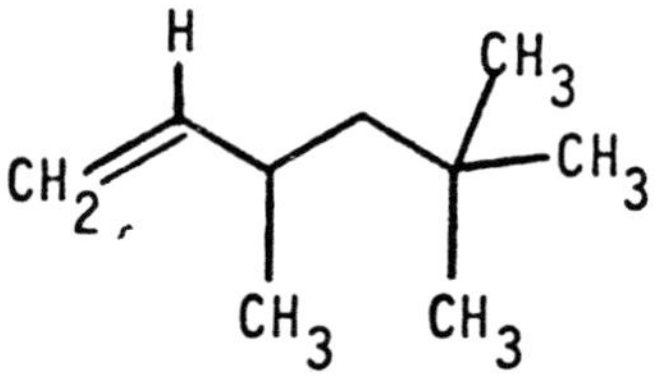 3,5,5-trimethyl-3-(vinyl)hexane... 3,5,5-trimethylhex-1-ene (approx)

-5-

68. 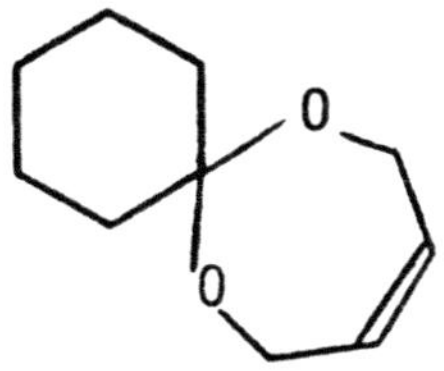 [structure]
69. 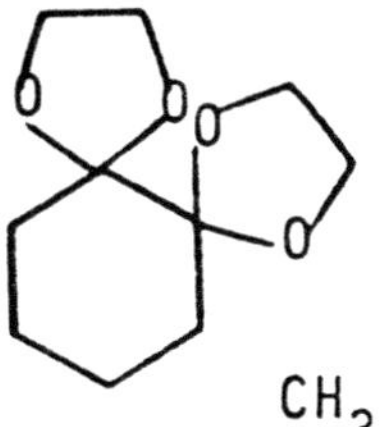 [structure]
70. 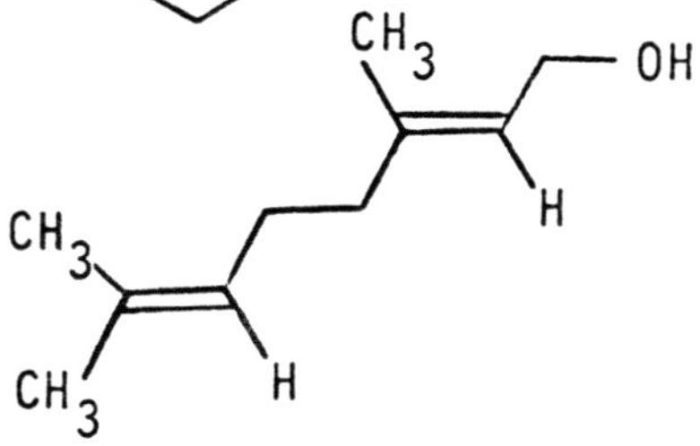 [structure]
71. 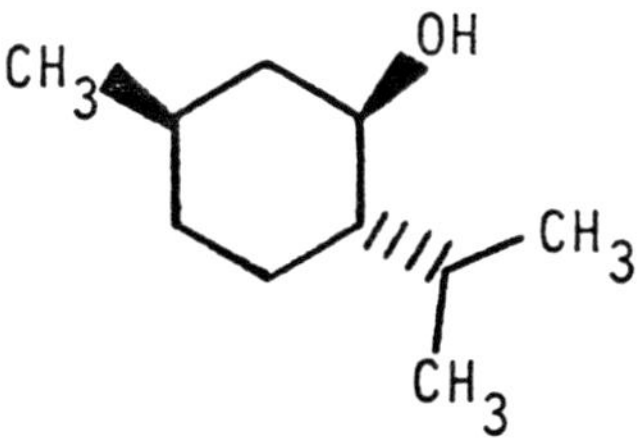 [structure]
72. 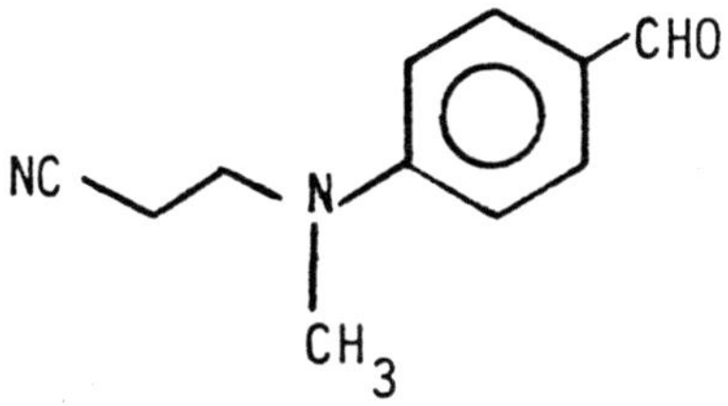 [structure]
73. 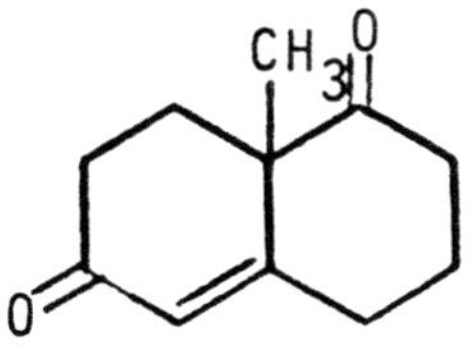 [structure]
74. 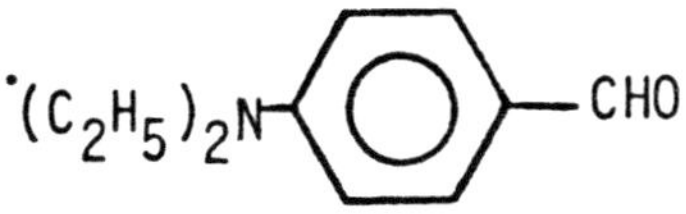 [structure]
75. 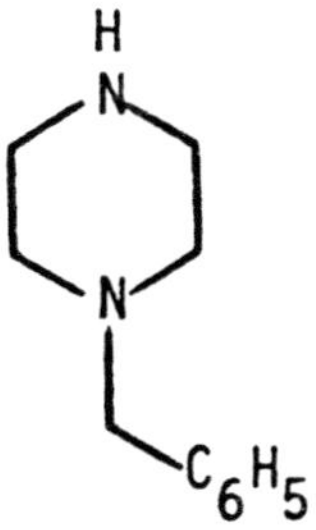 [structure]

76. 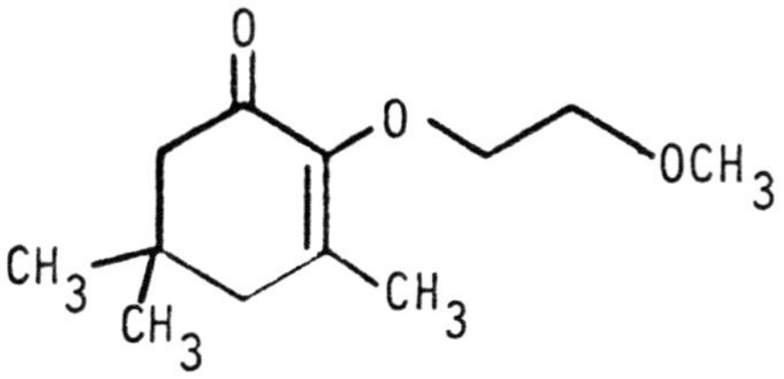 [structure]
77. 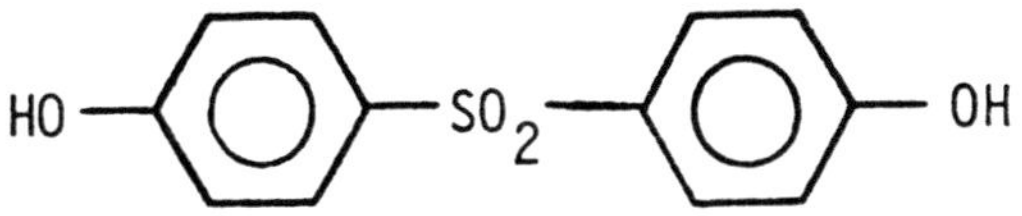 [structure]
78. 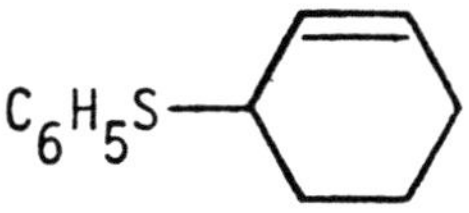 [structure]
79. 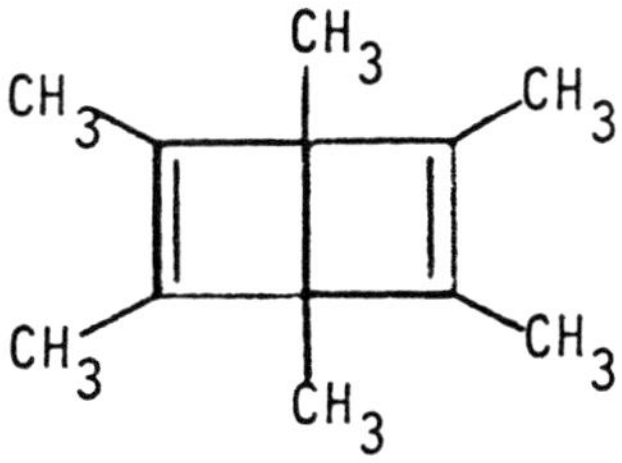 [structure]
80. 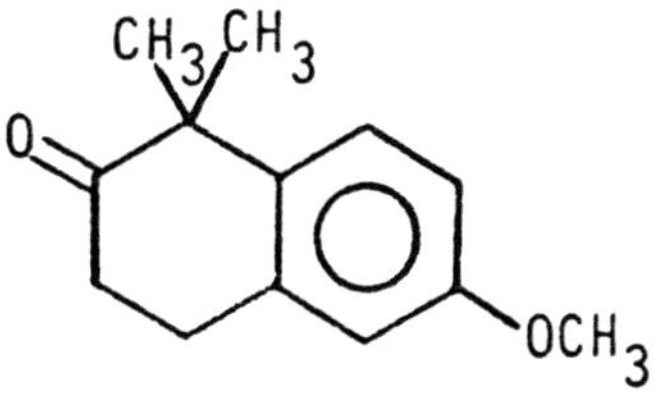 [structure]
81. 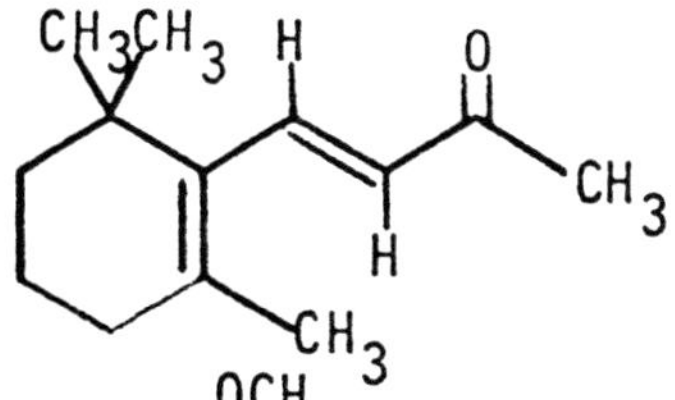 [structure]
82A. 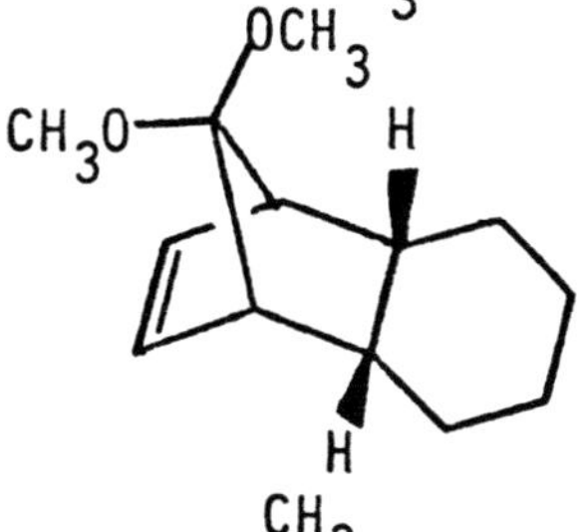 [structure]
82B. 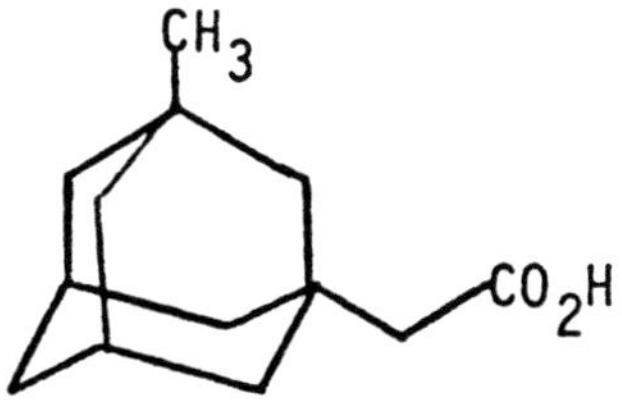 [structure]

83. 1,8-bis(dimethylamino)naphthalene

84. 1-(4,5-dimethoxy-2-(ethoxycarbonylmethyl)phenyl)ethanone (structure: dimethoxybenzene with acetyl group (COCH$_3$) and CH$_2$CO$_2$C$_2$H$_5$ group)

85. 2,6-di-sec-butylphenol

86. N-(3-phenylpropyl)phthalimide

87. bicyclic norbornene with C$_6$H$_5$CH$_2$O-CH$_2$- substituent and CO$_2$H substituent

88. 2-(phthalimido)-4-phenylbutanoic acid (N-phthalimide with CH(CO$_2$H)CH$_2$C$_6$H$_5$)

89A. 1,3-diphenylbenzene

89B. 1,2-diphenylbenzene

89C. 5-(diphenylmethylene)cyclopenta-1,3-diene (C$_6$H$_5$)$_2$C= on cyclopentadiene

90. 1,1-bis(3,4-dimethylphenyl)ethane (CH(CH$_3$) bridging two 3,4-dimethylphenyl groups)

91. (CH$_3$)$_2$N—C$_6$H$_4$—C(=CH$_2$)—C$_6$H$_4$—N(CH$_3$)$_2$

92. methoxy-substituted trans-decalin fused to aromatic ring with angular CH$_3$ groups and CO$_2$CH$_3$ substituent

93. triptycene

94. N(C$_7$H$_15$)$_3$

95. (stereochemical structure with CH$_3$N, NH, NH, NCH$_3$ groups — alkaloid-type cage structure)

Date: Mon, Jun 26, 1995 20:47 EDT
From: PaulNB
Subj: Fwd: Spectral Interp column/Sept & Oct
To: Lankindc, JohnC79051, 72727.1774@compuserve.com

Hi !

This is an FYI copy of a note Nancy sent to me.

Best Regards

Paul

- - - - - - - - - - - - -
Forwarded Message:

Date: Mon, Jun 26, 1995 14:57 EDT
From: Spectrscpy
Subj: Spectral Interp column/Sept & Oct
To: PaulNB

Hi, Paul:

I'm hoping to catch you before you begin typing up the bibliography for Part II of the Spectral Interpretation References series.

The following shows the house style for Spectroscopy references; if you are typing your references from scratch and can follow this style, you will save us A TON of time (and tedium) on our end. If you've already completed the inputting, then don't worry about it. If some of the inputting is done and some isn't, perhaps I could talk you into using this format for the ones you do from here on out. (In other words, don't go back and redo anything that's done already.)

Examples [Note that authors come first, with first initials before last name. No quote marks are used for book titles; they should be in italic, and significant words are capitalized]:

N.P.G. Roeges, A Guide to the Complete Interpretation of Infrared Spectra of Organic Structures (John Wiley & Sons, New York, 1994).

W.J. Criddle and G.P. Ellis, Spectral and Chemical Characterization of Organic Compounds: A Laboratory Handbook (John Wiley & Sons, New York, 1990).

H. Ishida, ed., Fourier Transform Infrared Characterization of Polymers (Plenum Press, New York, 1987).

R.A. Nyquist, Interpretation of Vapor Phase Infrared Spectra, Vol. 1: Group Frequency Data (Sadtler Research Laboratories, Philadelphia, PA, 1984).

Re: the September column -- I'll be sending the edited version to John within the next few days. Would you like to see the edited version as well? We give the regular columnist this option, but don't require you to do it. If you would like to see it, just let me know; otherwise, I'll send it to John only.

Thanks. --Nancy

ROBERT P. BORRIS

CARBON-13 NMR BASED

ORGANIC SPECTRAL PROBLEMS

PHILIP L. FUCHS
Purdue University
West Lafayette, Indiana

CHARLES A. BUNNELL
Tippecanoe Laboratories
Eli Lilly & Company
Lafayette, Indiana

JOHN WILEY & SONS, INC.

New York · Chichester · Brisbane · Toronto

PREFACE

The practice of organic spectroscopy is being continuously advanced. In addition to the more familiar tools of ultraviolet, infrared and low resolution mass spectroscopy, the organic chemist currently involved in research typically uses high resolution mass spectroscopy as well as "routine" 90-100 MHz proton nuclear magnetic resonance spectroscopy (with hetero- or homonuclear spin decoupling, when appropriate).

The advent of sensitive and reliable instrumentation capable of rapidly producing high-quality natural abundance ^{13}C nuclear magnetic resonance spectra has caused a revolution in the characterization of organic molecules. Carbon-13 NMR enables the analyst to directly observe the number of magnetically non-equivalent carbons in an organic structure; additionally, it defines the number of hydrogen atoms attached to each of these carbons. This outstanding tool has vastly simplified organic problem solving when used in conjunction with the previously existing methodology.

The teaching of modern organic structure determination is also being improved by including the subject of ^{13}C-NMR spectroscopy. Just as with the more familiar forms of spectroscopy, the problem-solving approach seems to be an especially attractive way to teach carbon-13 NMR. At the inception of our project five years ago, there were no problem books available which incorporated ^{13}C-NMR as an essential part of organic spectroscopy; therefore, we were forced by necessity to obtain the spectra (MS, IR, ^{1}H-NMR and ^{13}C-NMR) of hundreds of organic compounds, which were compiled into problem sets. These problem sets were "student-tested" (most of them over a five-year period), and, after further editing, evolved into the 125 problems presented in this book.

The work involved in obtaining the numerous spectra has been vastly aided by the conscientious efforts of numerous teaching assistants over the past few years. We wish to thank A. Borel, A. Erickson, M. Lusch, F. Palensky, V. Tisdale, P. Seemuth, T. Nylund, B. Kaufmann, T. Zarella, P. Marek, M. Fifolt, R. Kjonaas, R. Nelson, D. Snyder, L. Moore and D. Clark. Final manuscript preparation assistance was provided by Dr. J. Soderquist, Dr. O. Dailey, Dr. D. Barton, R. Donaldson, P. Conrad, R. Pariza, D. Hedstrand, and J. Saddler whom we wish to thank. Special thanks are due to J. J. Lu and A. Coddington for preparing the majority of the spectra included in this book. We also wish to thank our resident spectroscopists, Professors J. B. Grutzner (NMR) and R. G. Cooks (MS), for their valuable assistance and advice. Thanks are also due to Professor H. C. Brown for allowing us the use of his mass spectrum plotter. The National Science Foundation provided departmental grants for the purchase of the 90 MHz ^{1}H-NMR (NSF #8370) and the 20 MHz ^{13}C-NMR (NSF #7842) spectrometers. Finally, we wish to thank Pat Leburg for typing the manuscript and our wives for patience and proofreading.

Philip L. Fuchs, Ph.D.
Associate Professor of Chemistry
Purdue University
West Lafayette, Indiana 47907

Charles A. Bunnell, Ph.D.
Senior Organic Chemist
Tippecanoe Laboratories
Eli Lilly & Company
Lafayette, Indiana 47905

Copyright © 1979 by John Wiley & Sons, Inc.

All rights reserved. Published simultaneously in Canada.

Reproduction or translation of any part of
this work beyond that permitted by Sections
107 and 108 of the 1976 United States Copyright
Act without the permission of the copyright
owner is unlawful. Requests for permission
or further information should be addressed to
the Permissions Department, John Wiley & Sons.

Library of Congress Catalog Card Number: 78-20668

ISBN 0-471-04907-7

Printed in the United States of America

10 9 8 7 6 5 4 3 2 1

CONTENTS

INTRODUCTION   1

SPECTRAL PROBLEMS   3

APPENDICES

    APPENDIX I: A General Protocol for Solving Organic
    Spectral Problems   254

    APPENDIX II: Solved Problems

        Unknown 14B   265
        Unknown 30   268
        Unknowns 41A,B   271
        Unknown 53   275
        Unknown 65   285
        Unknown 73   288
        Unknown 83   296
        Unknown 84   299
        Unknown 90   304

    APPENDIX III: The Curphey-Morrison Additivity Constants
    for Proton NMR Chemical Shift Approximation   307

    APPENDIX IV: Selected Reference List   309

INTRODUCTION

This book is a collection of 125 spectral unknowns. The compounds included have many of the common functional groups encountered on a day-to-day basis by a typical organic graduate student. In addition to these "standard" unknowns, there are also less frequently encountered materials such as research samples, organometallics, and natural products.

The problems were selected so as to be workable by students who have had some introductory experience in solving simple spectroscopic problems. At Purdue these problems were used extensively during the final third of a one-semester course in organic spectroscopy taken by juniors, seniors, and beginning graduate students.

This book does not contain a compilation of the standard mass spectral, infrared, proton NMR, and carbon NMR spectral information, but rather was designed to be used in conjunction with several of the excellent references which provide that data. The specific companion texts used with these spectral unknowns at Purdue were: "Spectroscopic Identification of Organic Compounds," 3rd Edition, Silverstein, Bassler, and Morrill, (John Wiley, New York, 1974) and "Carbon-13 Nuclear Magnetic Resonance for Organic Chemists," Levy and Nelson, (John Wiley, New York, 1972). An additional list of useful reference books can be found in Appendix IV.

The problems in this book are arranged in order of increasing formula complexity $(C_w H_x N_y O_z)$ as they would be found in Chemical Abstracts. This expedient was employed in an attempt to organize the unknowns such that the "easier" problems appear earlier in the book. While it might seem obvious that a six-carbon unknown should be more easily solved than one with twelve carbon atoms, this is not always the case--especially if the C-12 unknown has high symmetry, while the C-6 compound lacks any symmetry element. All other attempts to organize the problems based upon "increasing difficulty" were highly dependent upon which particular problem-solver ranked the unknowns.

Where two or more of the unknowns have the same empirical formula, they are grouped together in "isomer families" by appending a letter to the number of the unknown of the isomer family. In several instances it will be a distinct advantage to work these families as a group, since the problem-solver can directly compare their spectra. This practice will often allow an unambiguous structural assignment to be made where some ambiguity would have otherwise existed if only one member of an isomer series had been examined.

A general protocol for solving organic spectroscopic problems is outlined in Appendix I. The method is illustrated in Appendix II by solving ten of the unknowns. It is strongly recommended that the group of unknowns from Appendix II should be the first problems attempted by the beginning problem-solver. It is possible in this way for the problem-solver to get some immediate feedback regarding the accuracy of the structural assignment. This exercise will hopefully also promote the development of a logical and systematic approach for spectral analysis. The answers to the remaining 114 problems are not given in the book since the authors have found through personal experience that such a practice tends to discourage extended effort. Course instructors can obtain an answer book from the publisher.

Each unknown is arranged such that all four spectra can be examined without having to turn the page. At the top of the left-facing page is a graphical representation of the mass spectrum of the unknown (obtained on a CEC-21-110 B High Resolution Mass Spectrometer) as well as a compound-information insert. The insert lists the unknown number, solvents in which the unknown was recorded, an exact mass of the highest m/e ion found in the mass spectrum, and the combustion analysis of all elements present in the unknown, except oxygen. The infrared spectrum (recorded on a Perkin-Elmer Infracord) is reproduced on the bottom of the left-facing page. The spectrum at the top of the right-facing page is a 90 MHz proton NMR, recorded on a Perkin-Elmer R-32 spectrometer. The proton NMR is either locked on TMS or contains TMS as an internal standard. Any special proton NMR experiments such as D_2O additions or homonuclear spin decoupling are included as labeled inserts on the parent spectra. The carbon-13 NMR was recorded on a 20 MHz Varian CFT-20 spectrometer and can be found at the bottom of the right-facing page. The bottom scan is the result of heteronuclear decoupling of all protons in the molecule, while the top scan is the off-resonance proton-decoupled spectra which allows assignment of the number of protons directly attached to each carbon type. Any carbonyl resonances further downfield than 200 p.p.m. are included as labeled inserts. If complex and overlapping patterns exist, a partial list of computer-generated chemical shifts *vs.* their peak intensity is also attached to the CMR spectrum.

SPECTRAL PROBLEMS

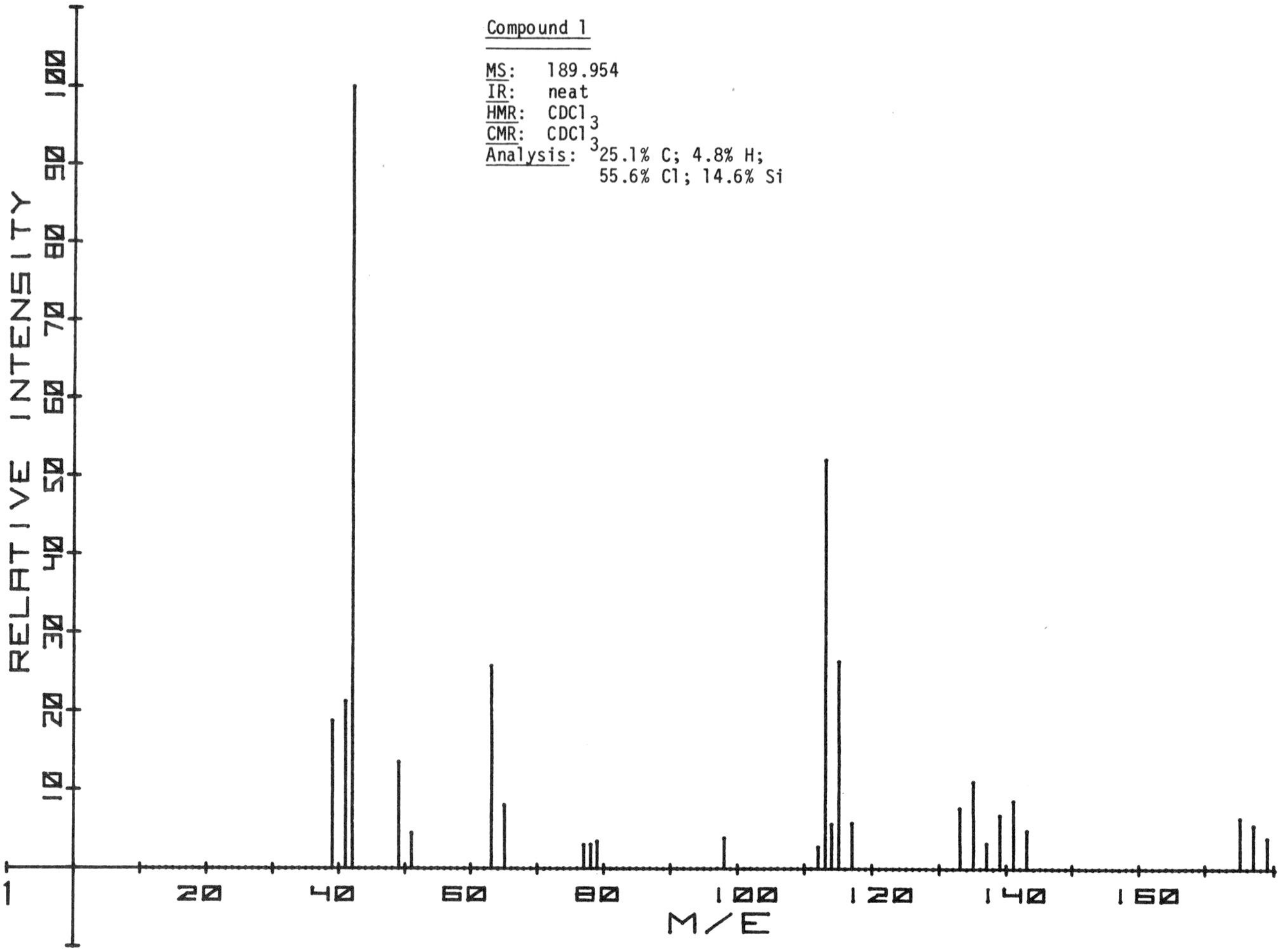

Compound 1
MS: 189.954
IR: neat
HMR: CDCl3
CMR: CDCl3
Analysis: 25.1% C; 4.8% H;
55.6% Cl; 14.6% Si
RELATIVE INTENSITY
M/E

CM-1
ABSORBANCE
WAVELENGTH (MICRONS)
1

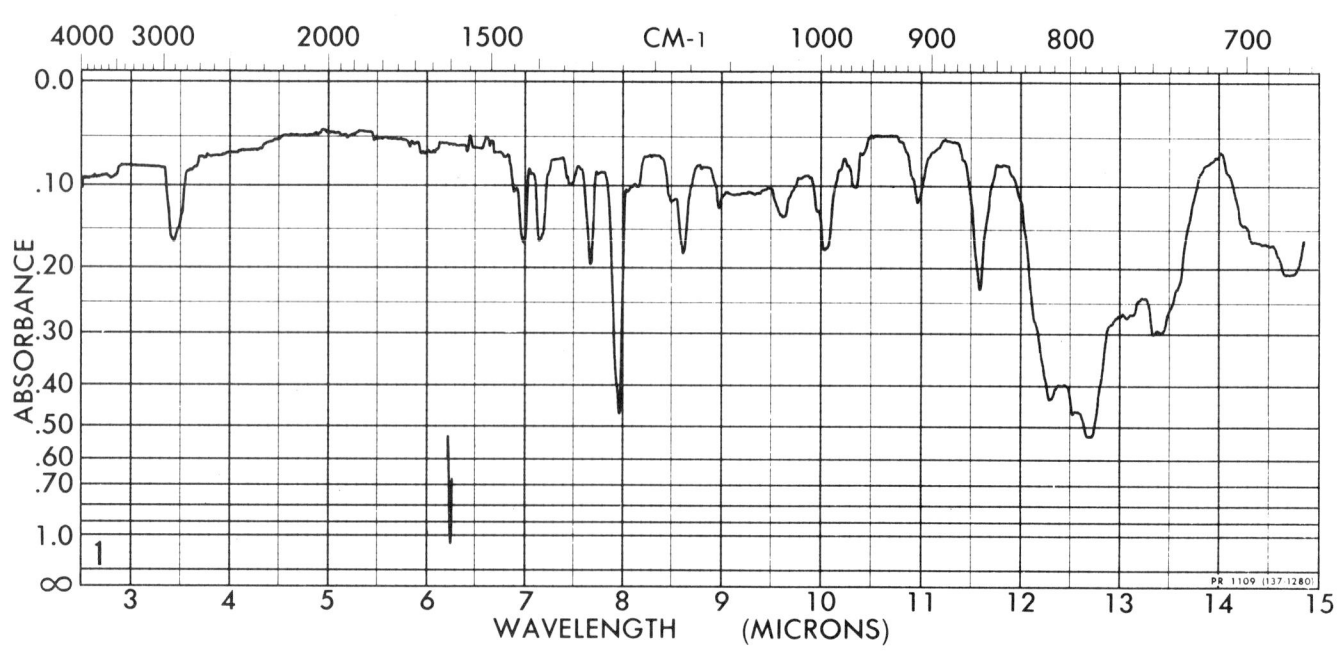

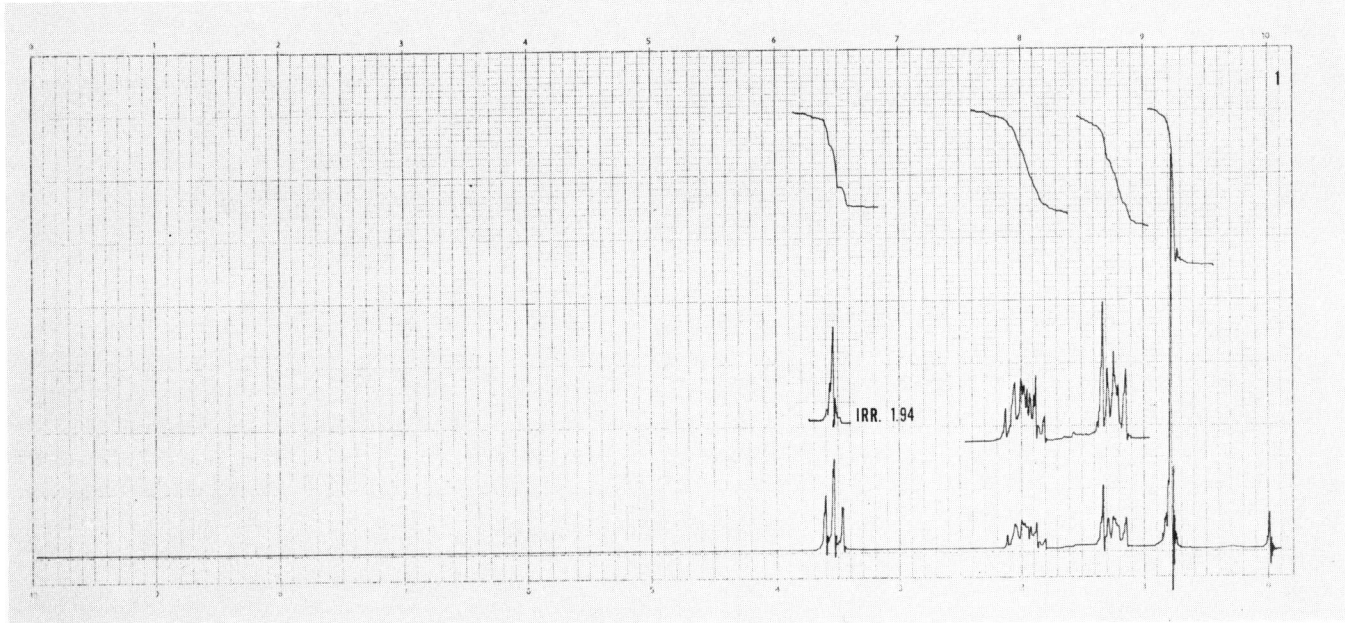

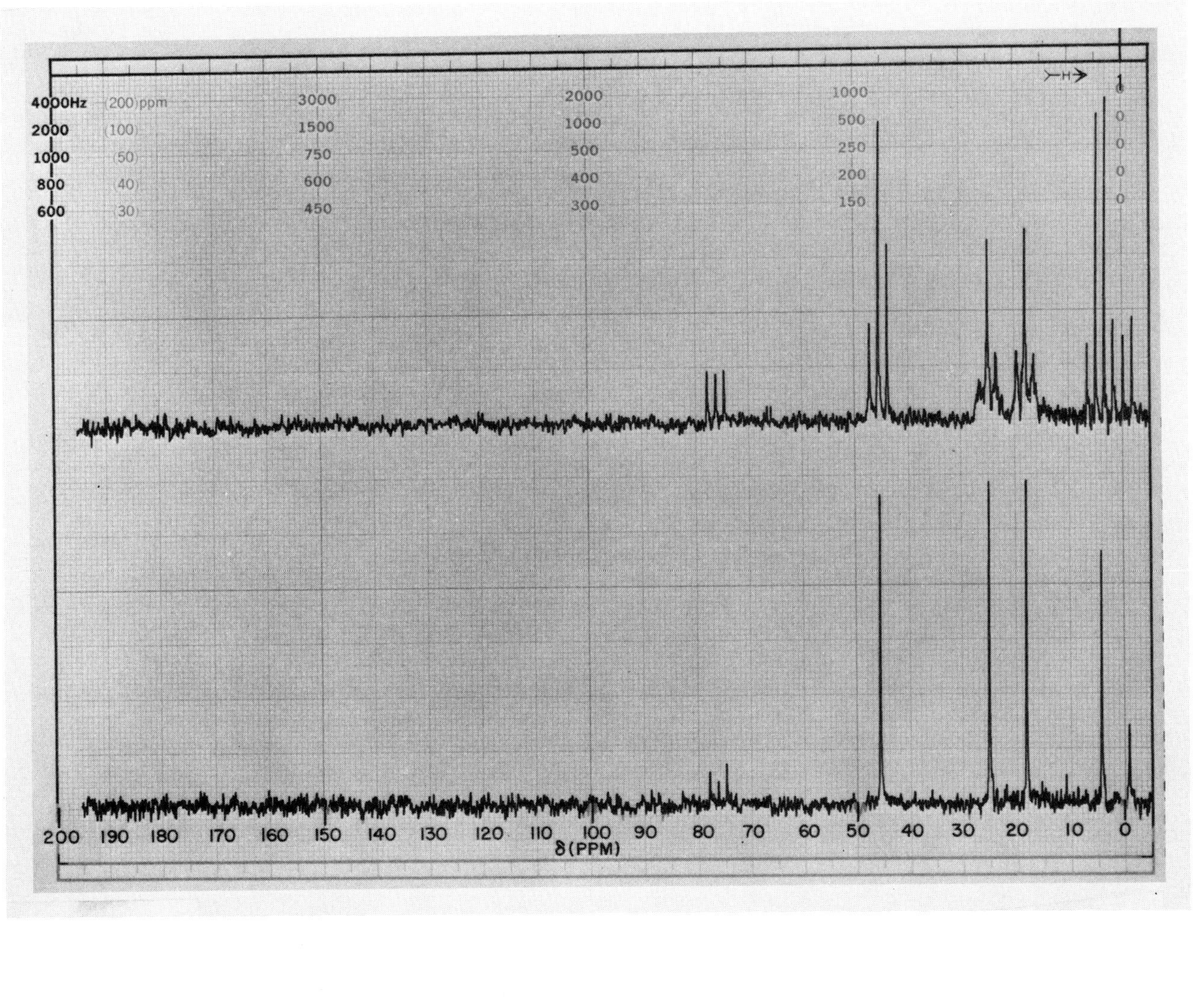

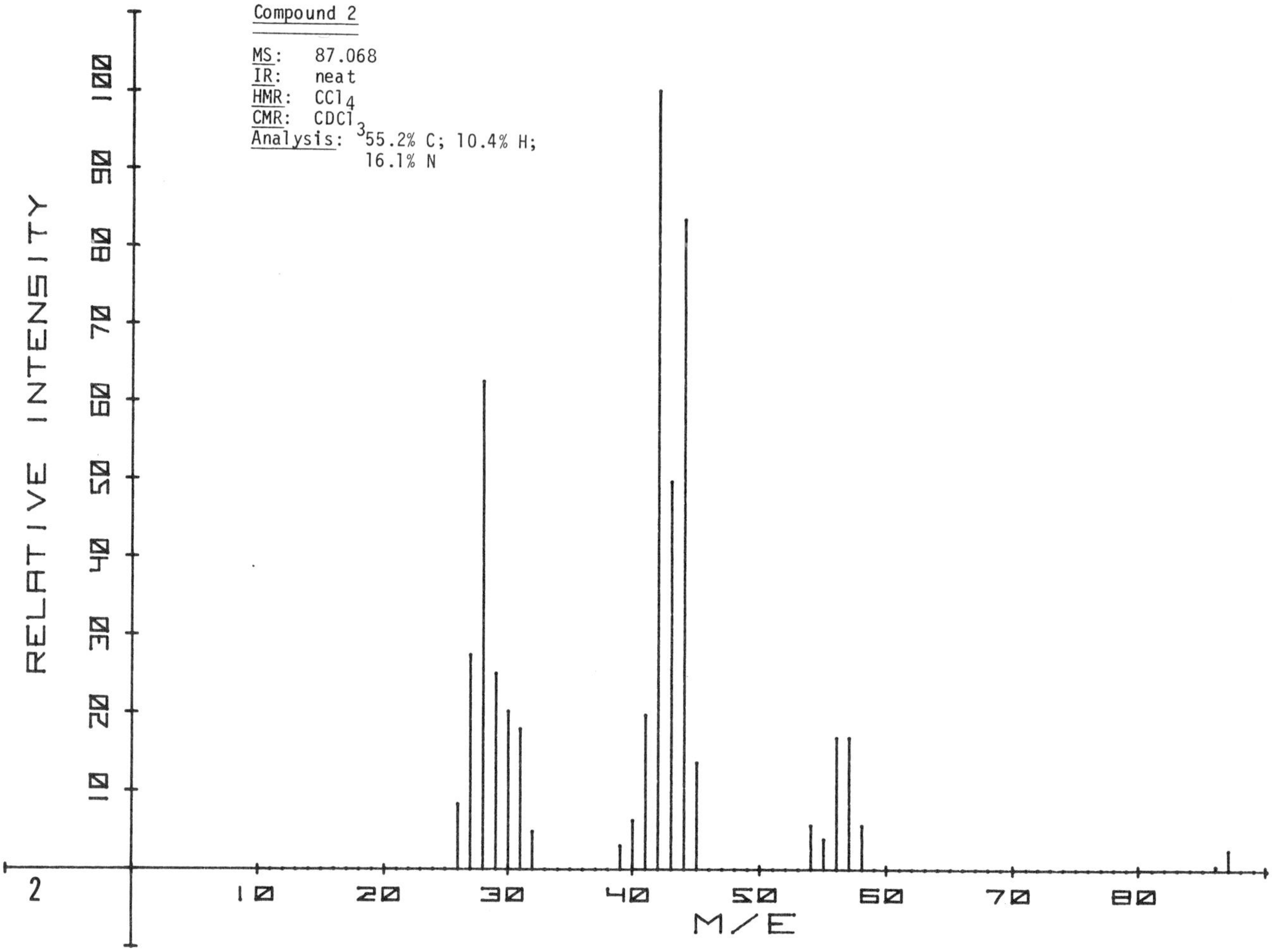

Compound 2
MS: 87.068
IR: neat
HMR: CCl_4
CMR: $CDCl_3$
Analysis: 55.2% C; 10.4% H; 16.1% N
RELATIVE INTENSITY
10 20 30 40 50 60 70 80 90 100
2
M/E

4000 3000 2000 1500 CM-1 1000 900 800 700
ABSORBANCE
0.0 .10 .20 .30 .40 .50 .60 .70 1.0 ∞
2
3 4 5 6 7 8 9 10 11 12 13 14 15
WAVELENGTH (MICRONS)

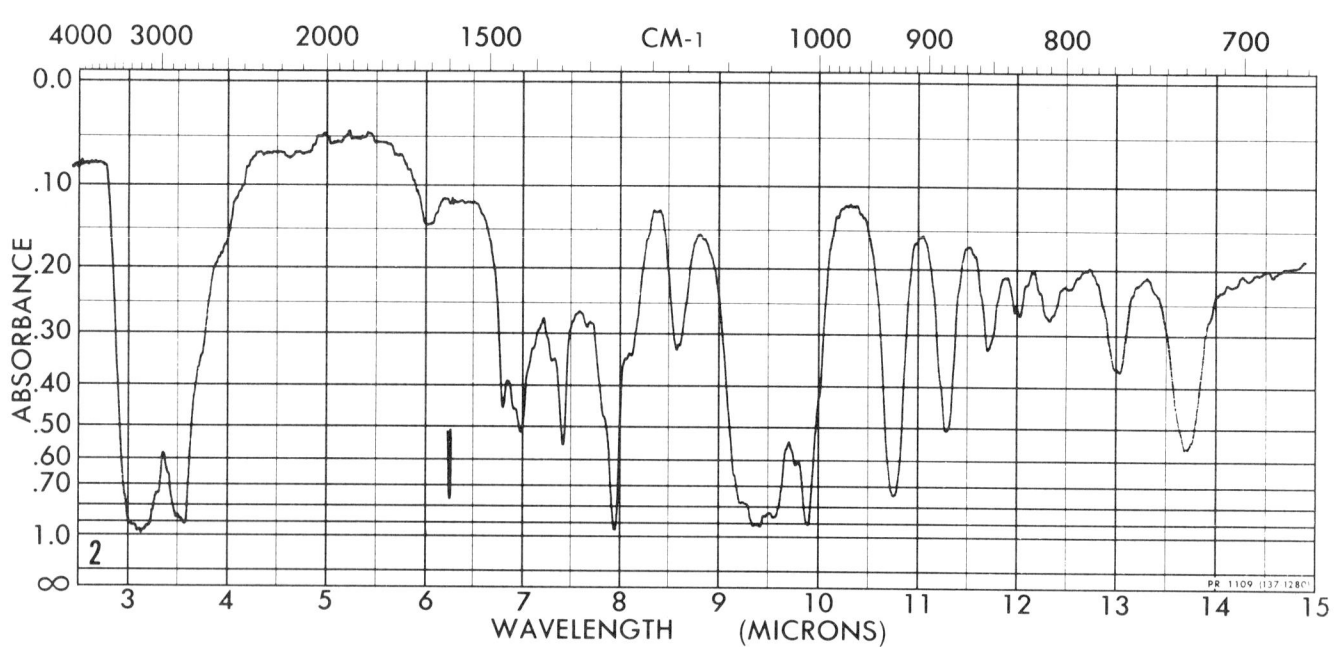

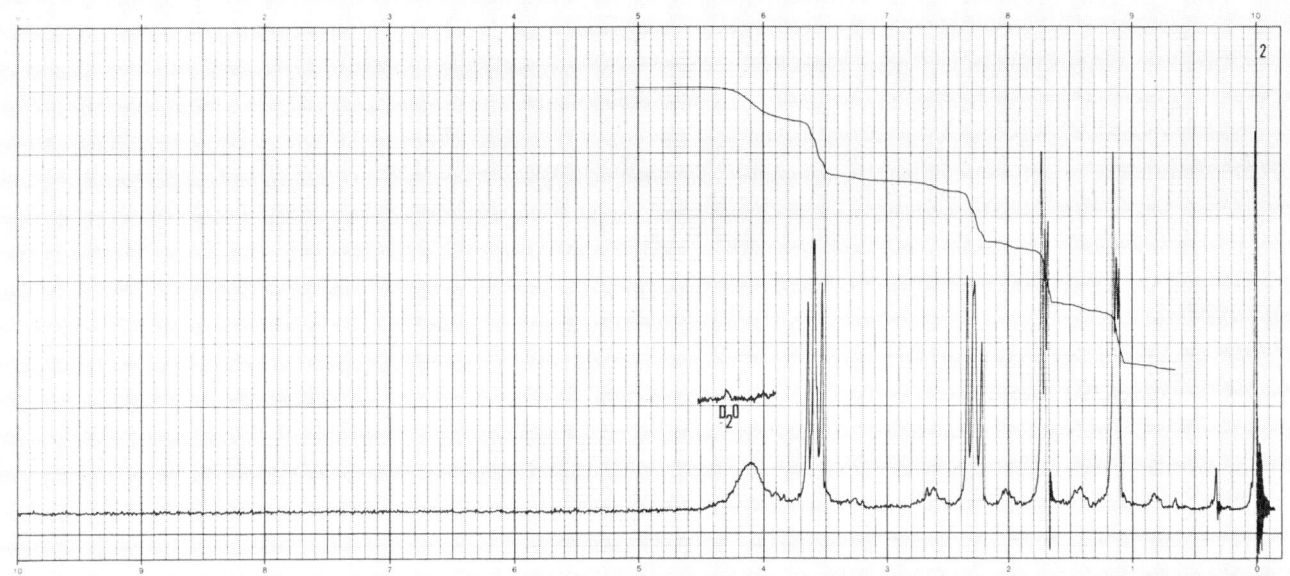

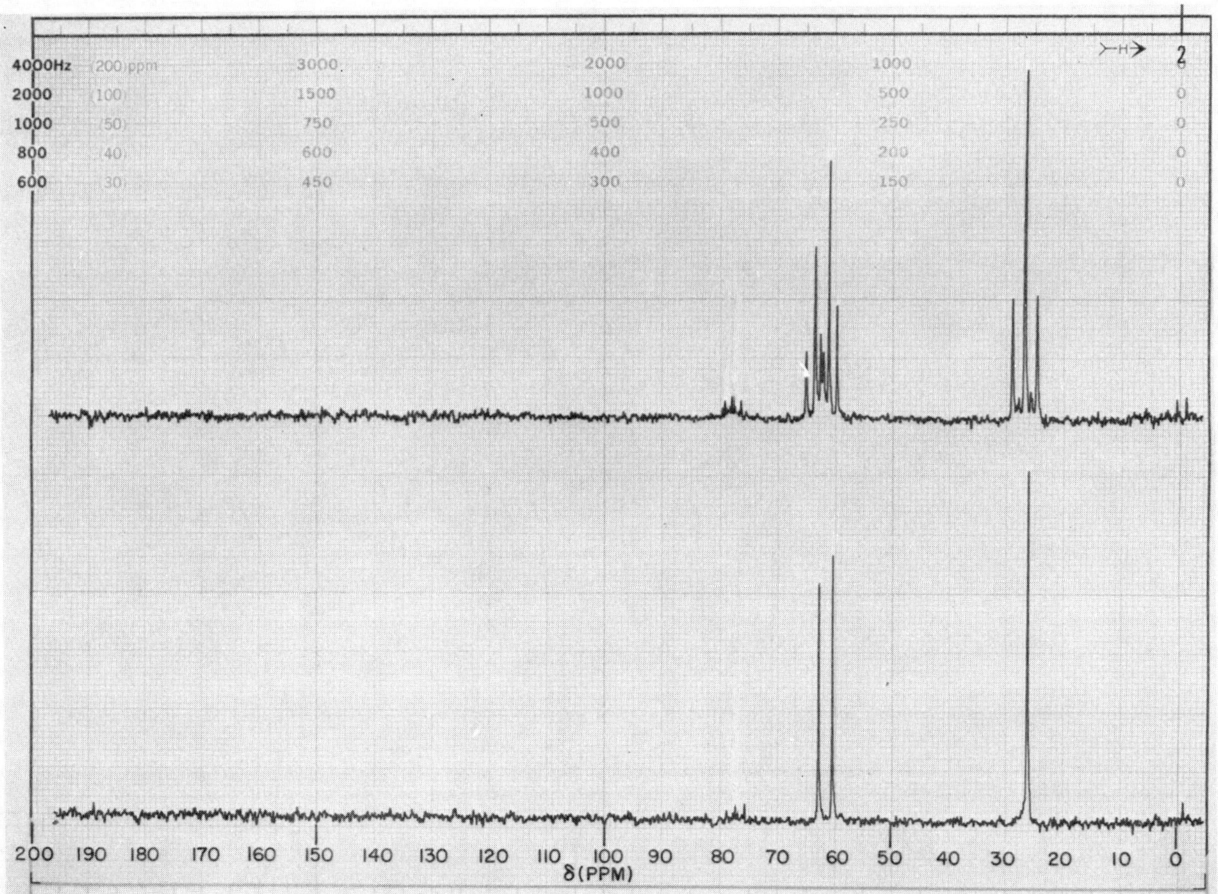

COMPOUND 3A

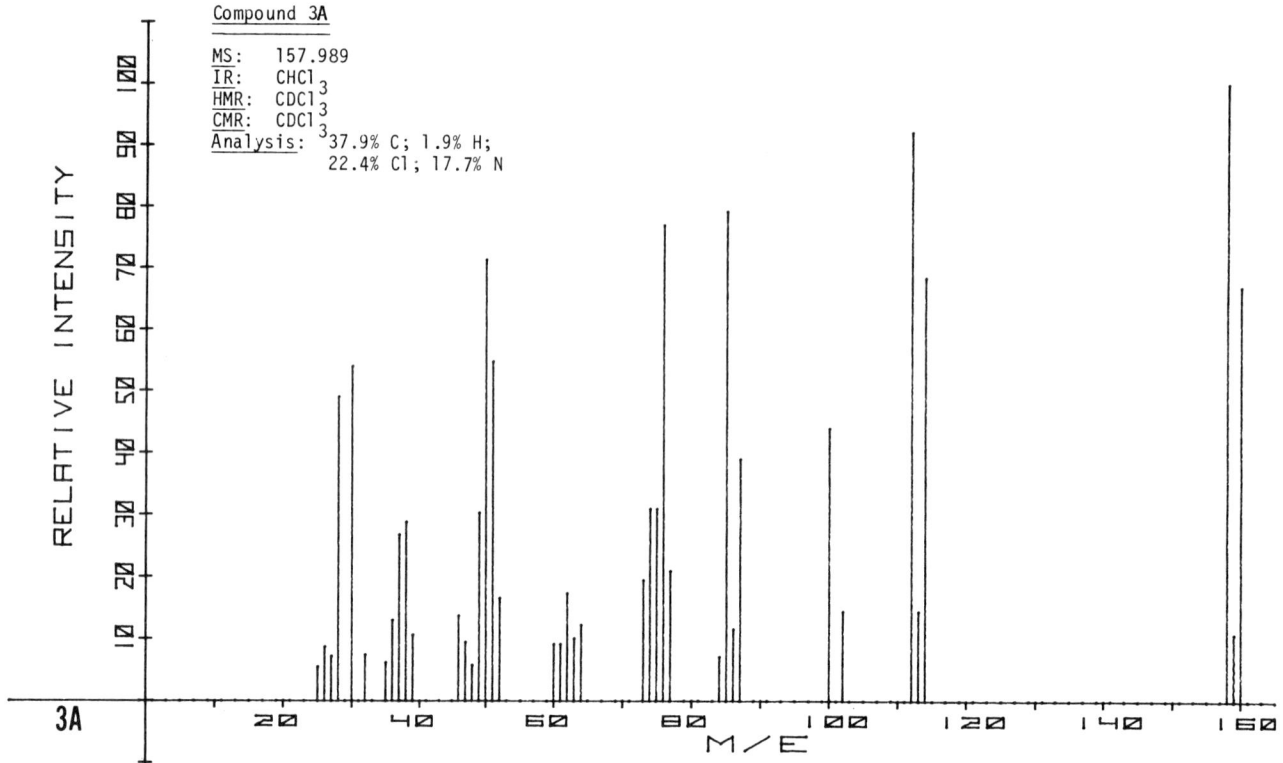

Compound 3A
MS: 157.989
IR: CHCl₃
HMR: CDCl₃
CMR: CDCl₃
Analysis: 37.9% C; 1.9% H; 22.4% Cl; 17.7% N

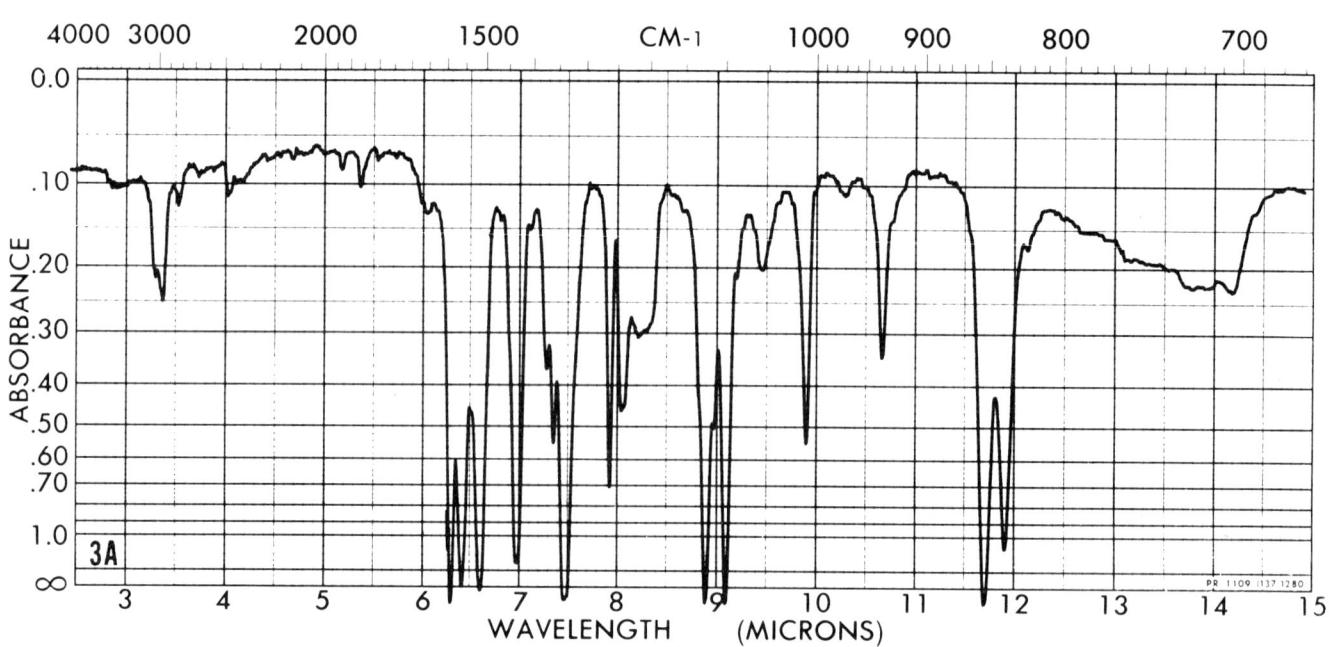

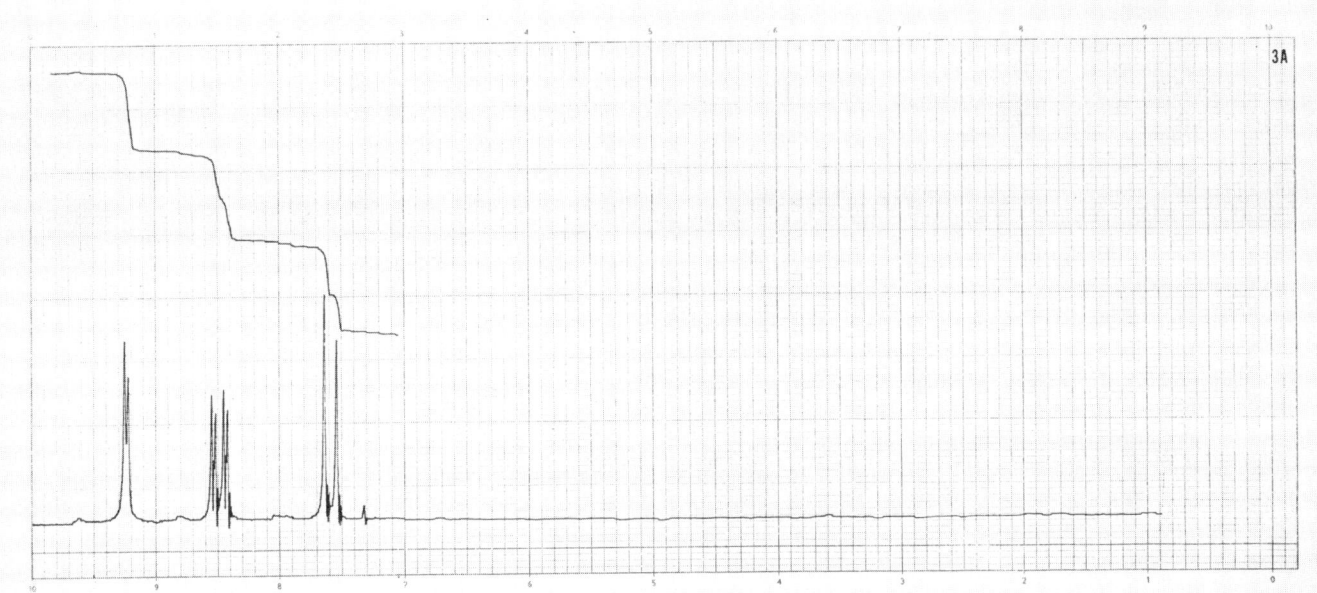

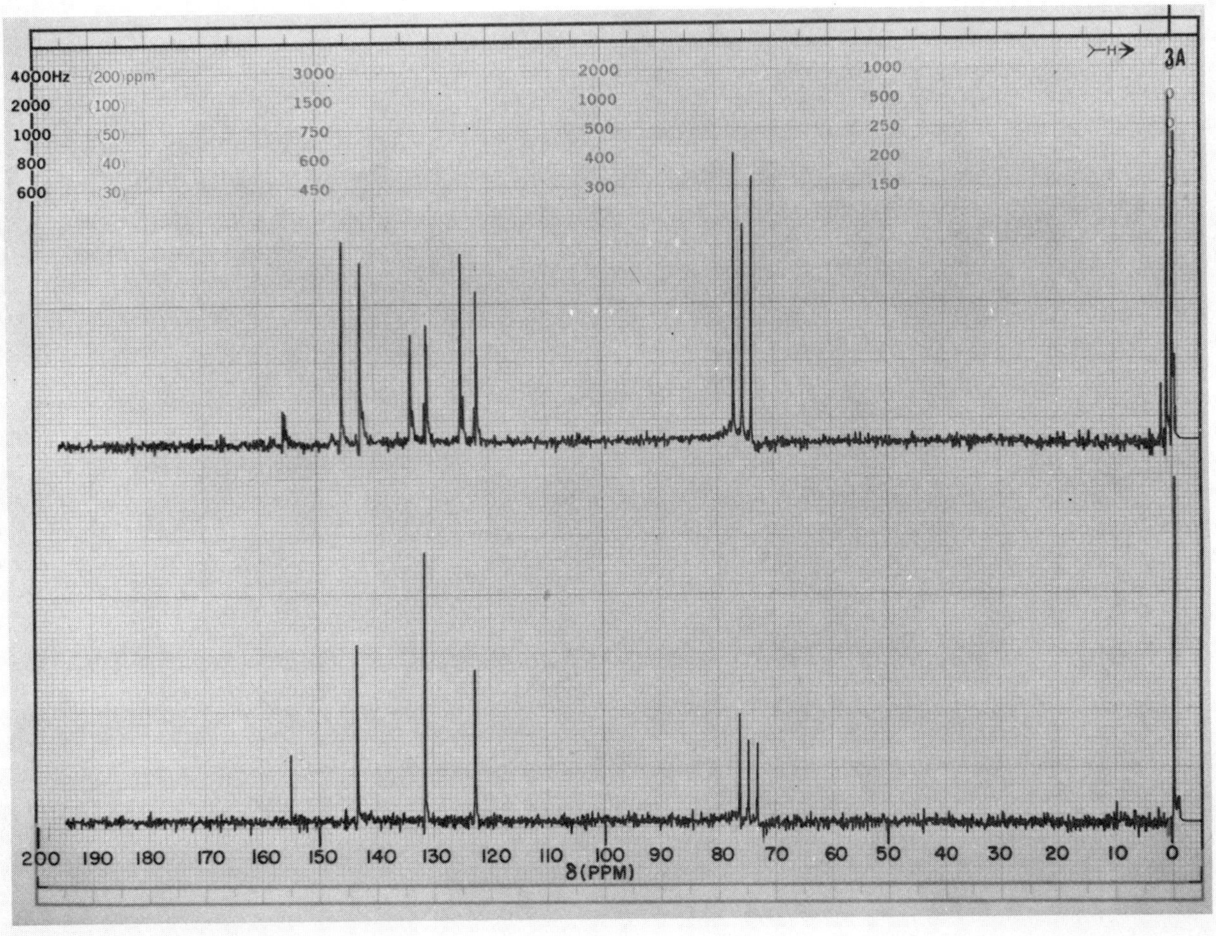

COMPOUND 3B

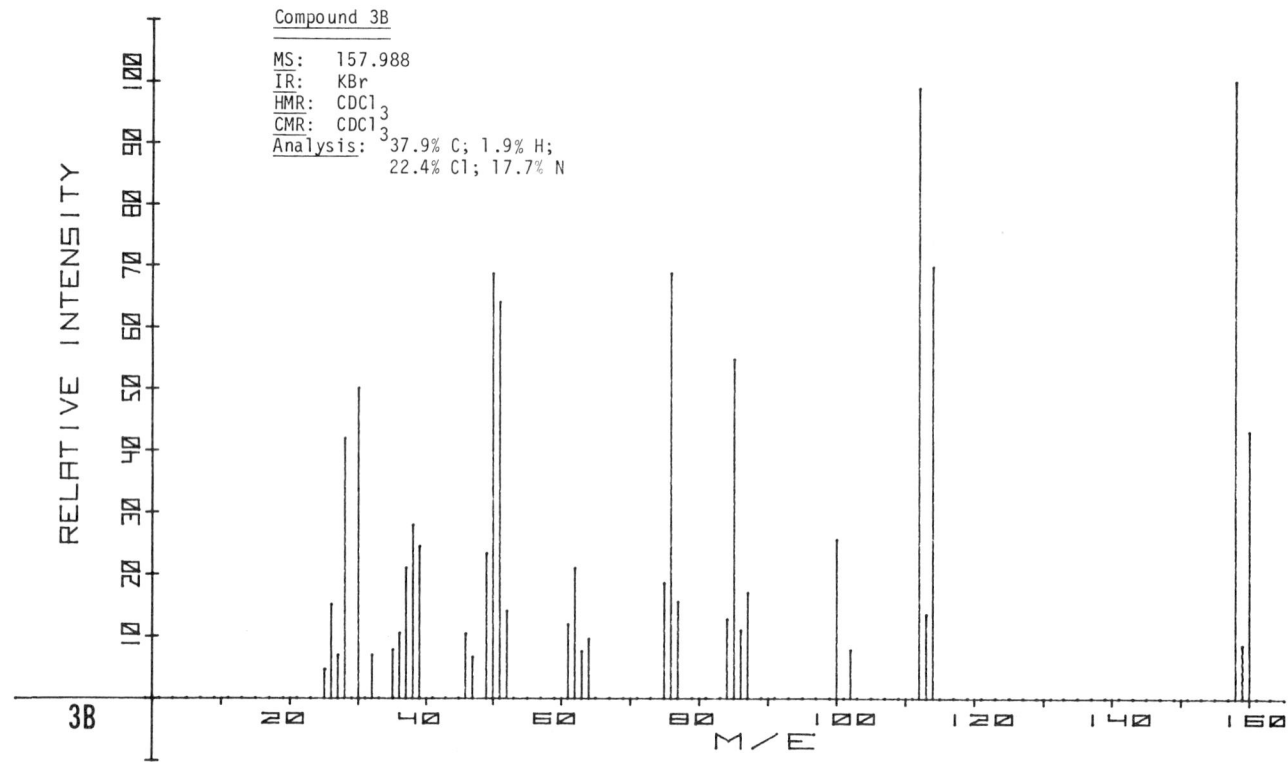

Compound 3B
MS: 157.988
IR: KBr
HMR: CDCl$_3$
CMR: CDCl$_3$
Analysis: 37.9% C; 1.9% H;
22.4% Cl; 17.7% N

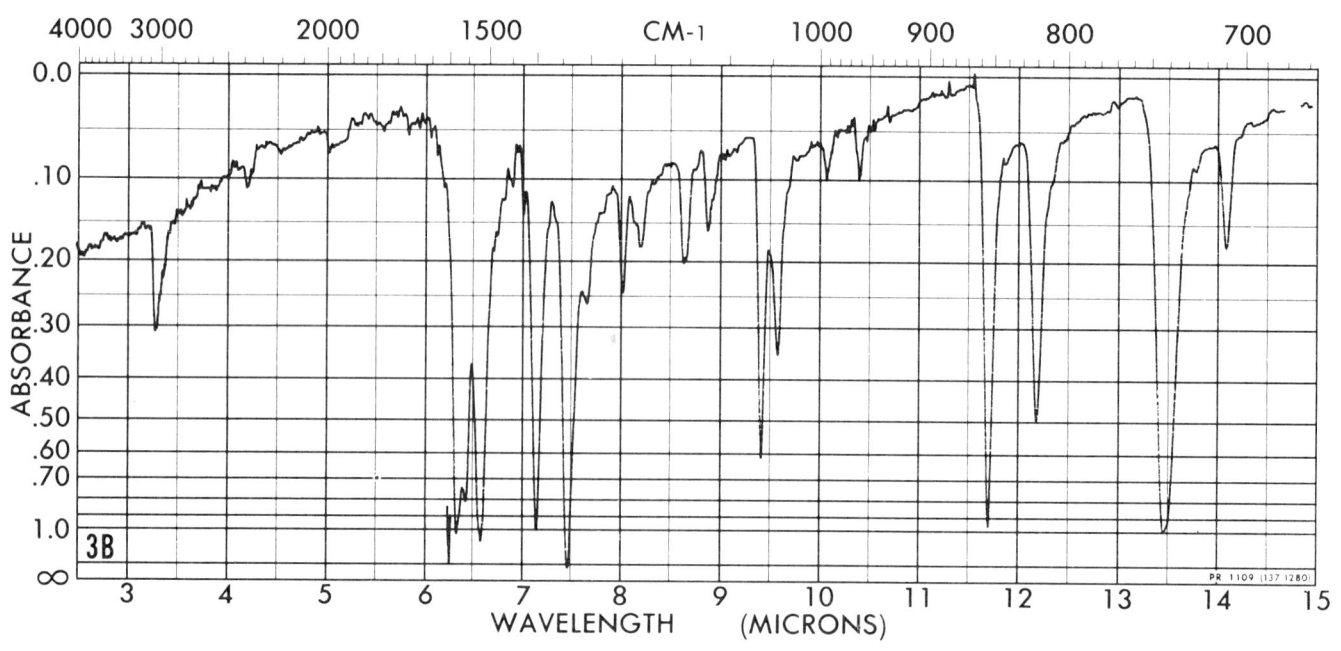

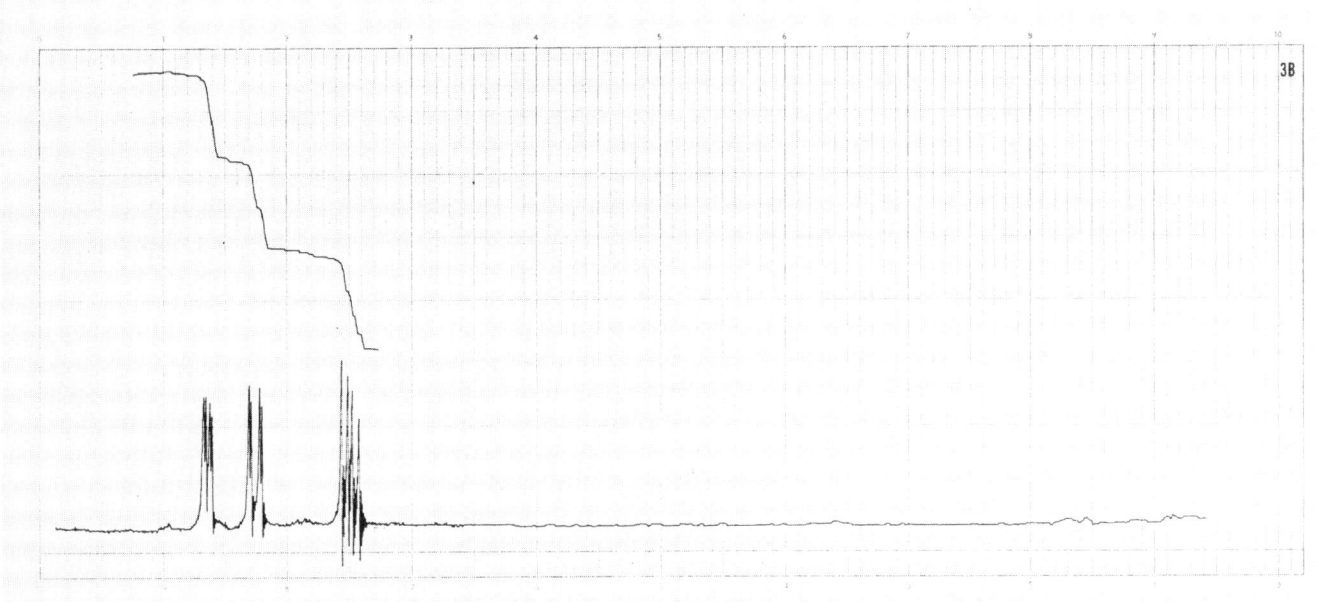

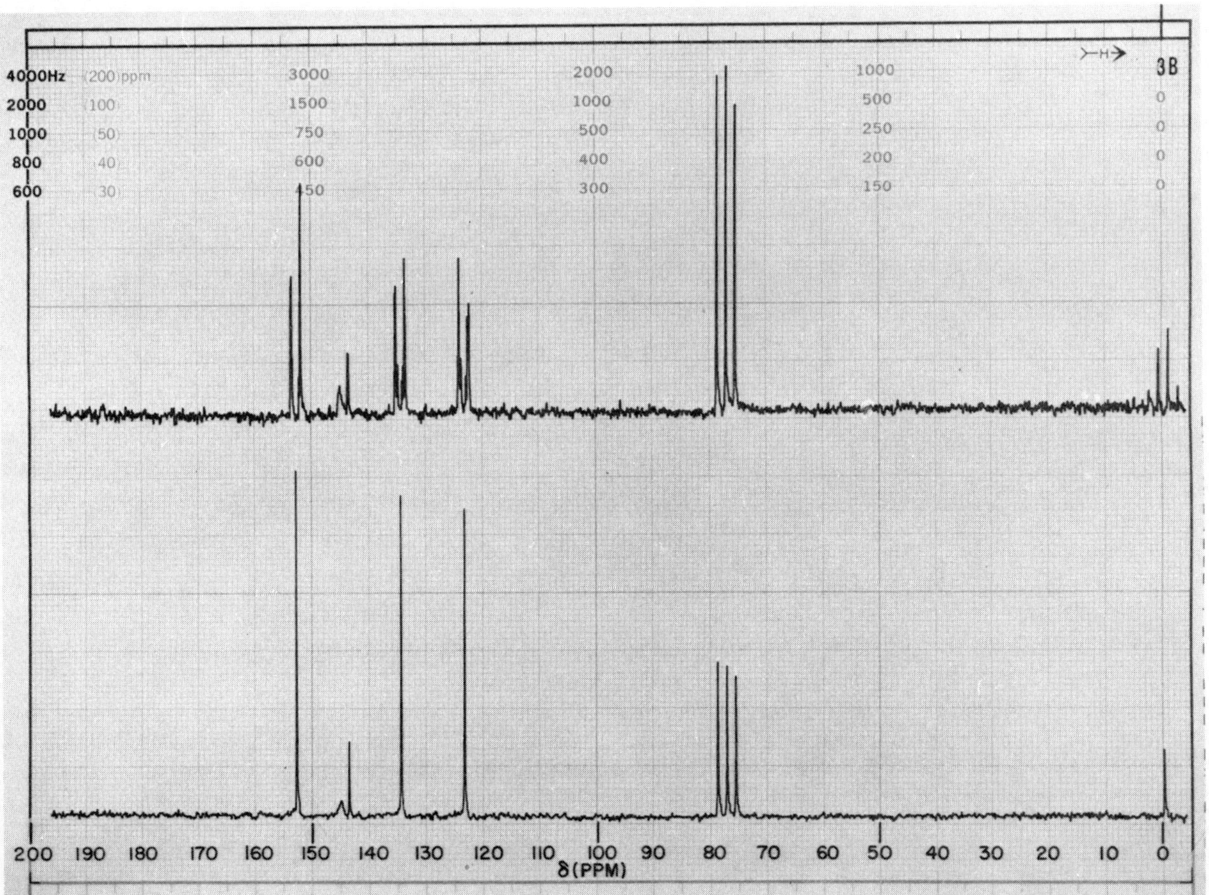

COMPOUND 4

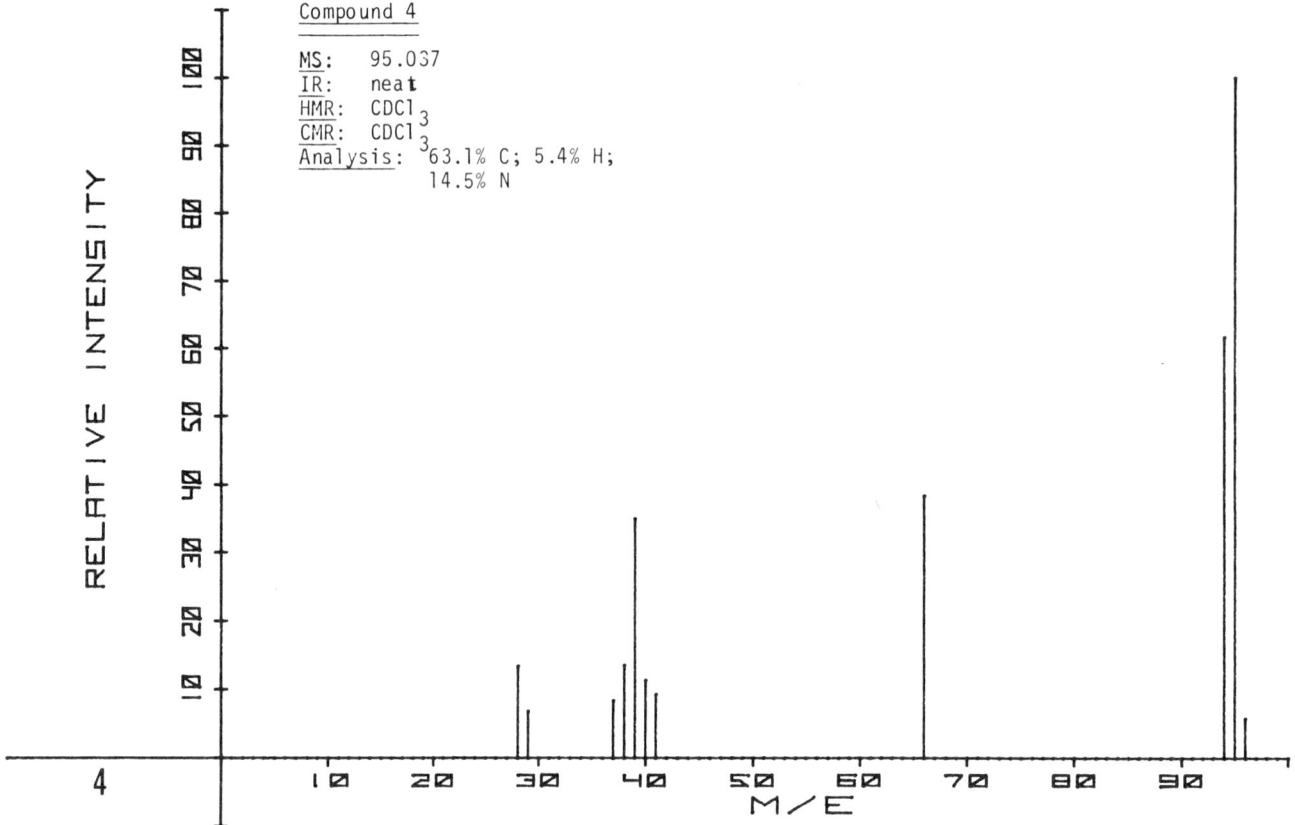

Compound 4

MS: 95.037
IR: neat
HMR: CDCl$_3$
CMR: CDCl$_3$
Analysis: 63.1% C; 5.4% H; 14.5% N

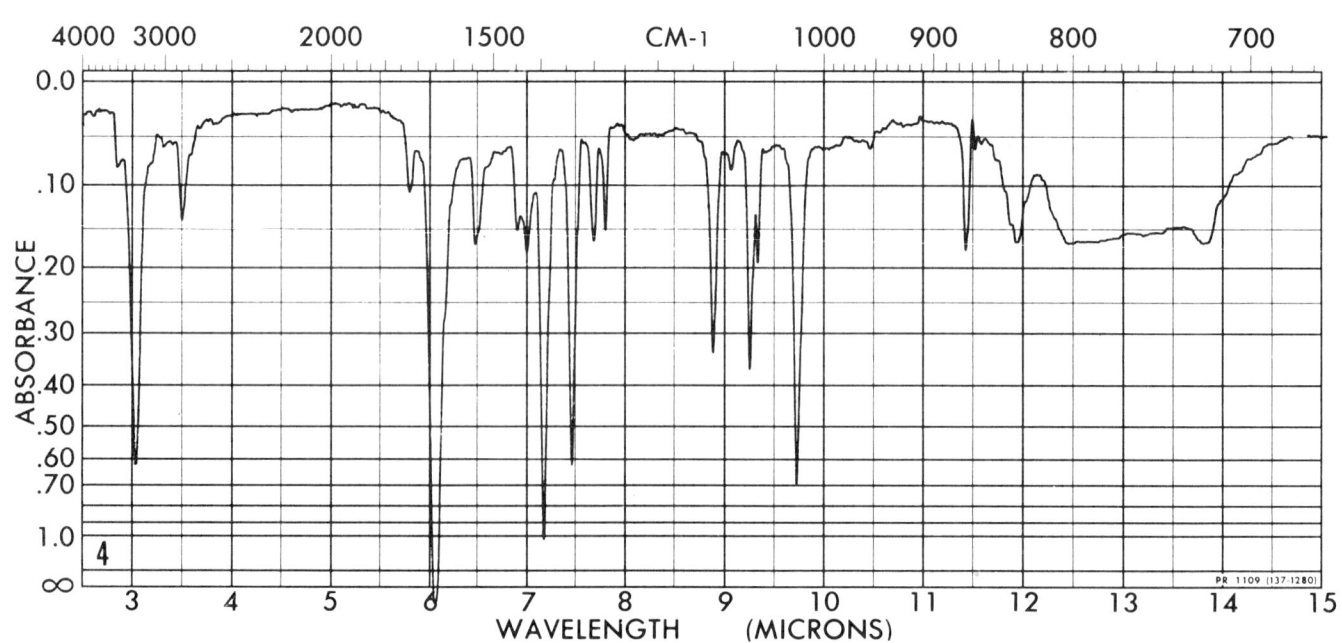

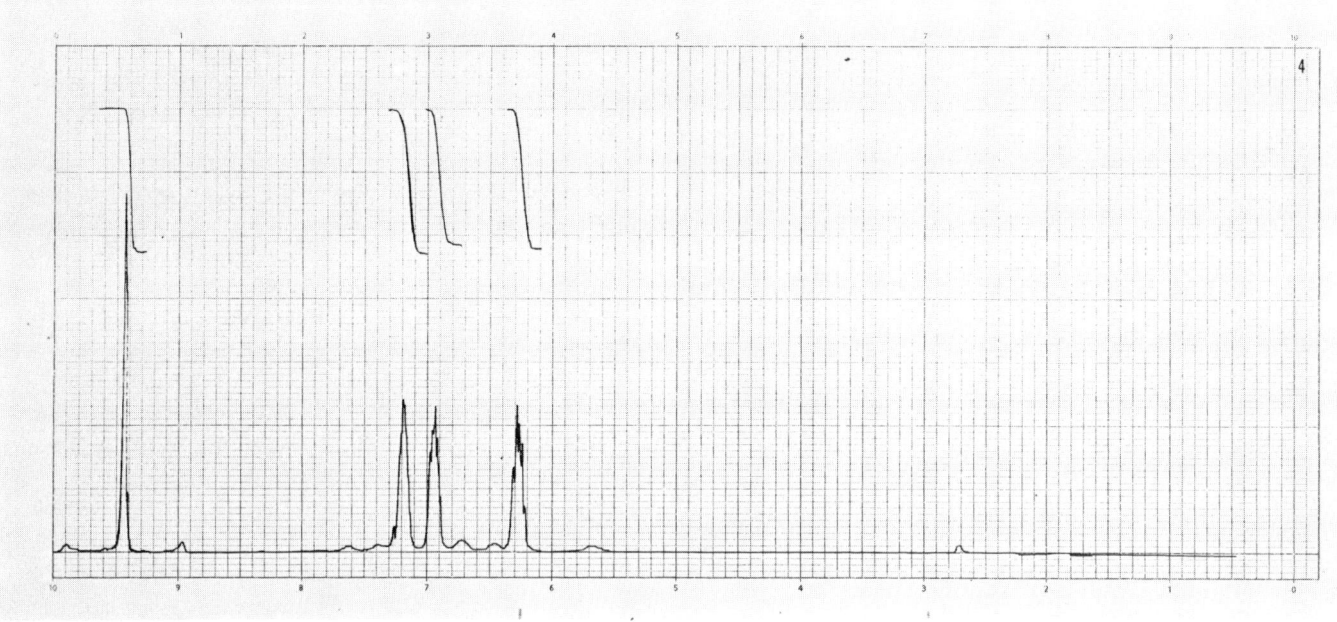

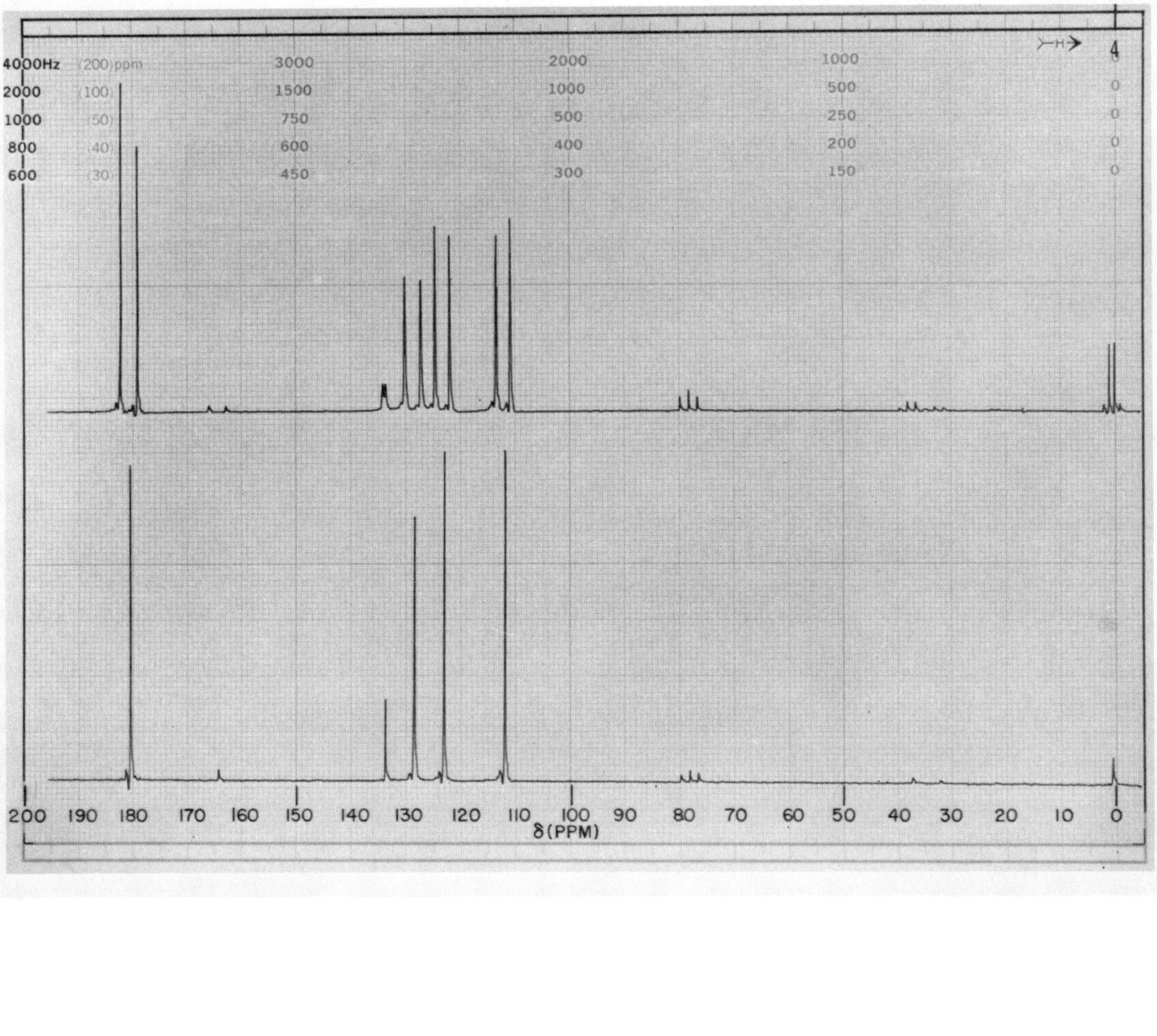

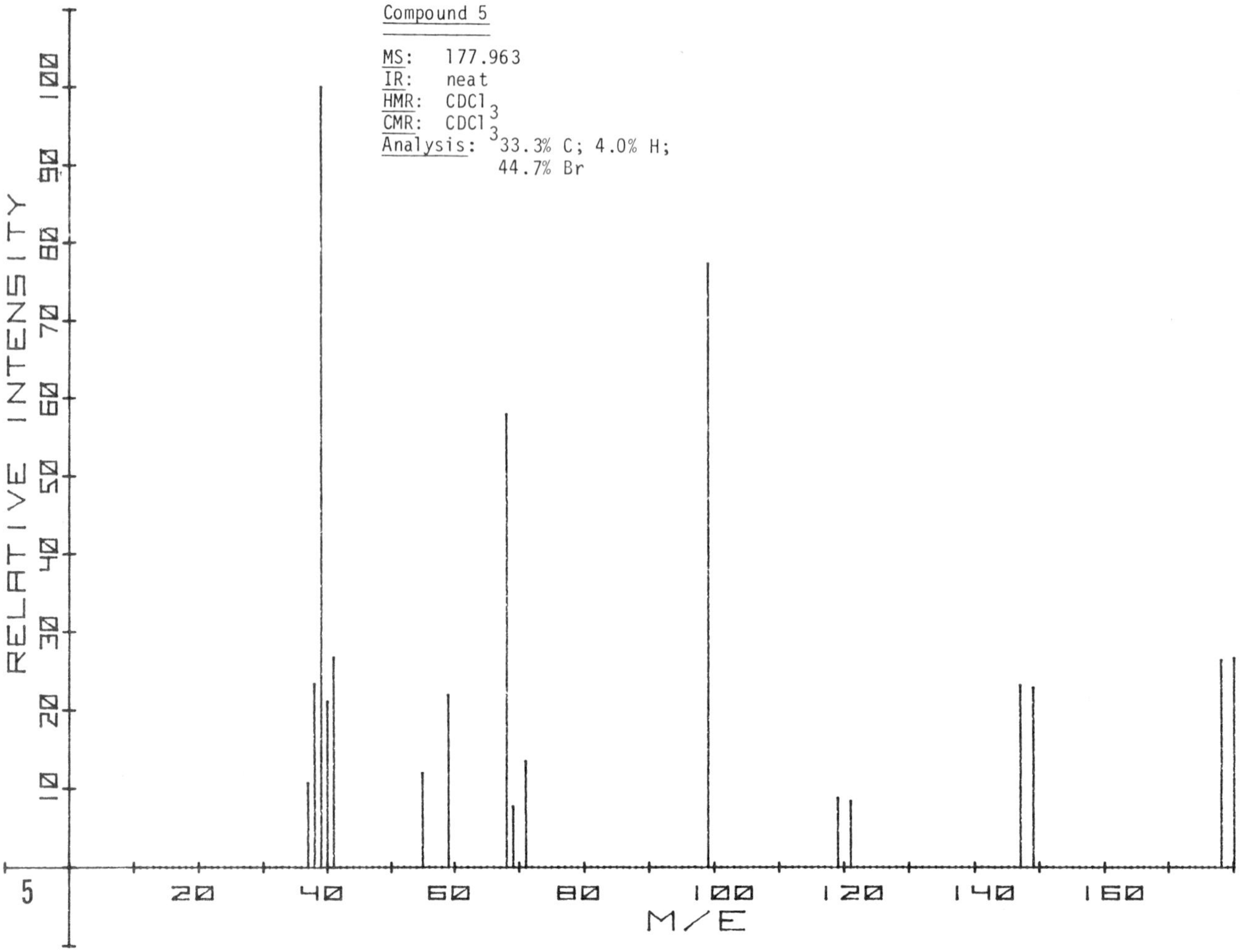

Compound 5
MS: 177.963
IR: neat
HMR: $CDCl_3$
CMR: $CDCl_3$
Analysis: 33.3% C; 4.0% H; 44.7% Br
RELATIVE INTENSITY
10 20 30 40 50 60 70 80 90 100
20 40 60 80 100 120 140 160
M/E
5

4000 3000 2000 1500 CM-1 1000 900 800 700
0.0 .10 .20 .30 .40 .50 .60 .70 1.0 ∞
ABSORBANCE
5
3 4 5 6 7 8 9 10 11 12 13 14 15
WAVELENGTH (MICRONS)
PR 1109 (137-1280)

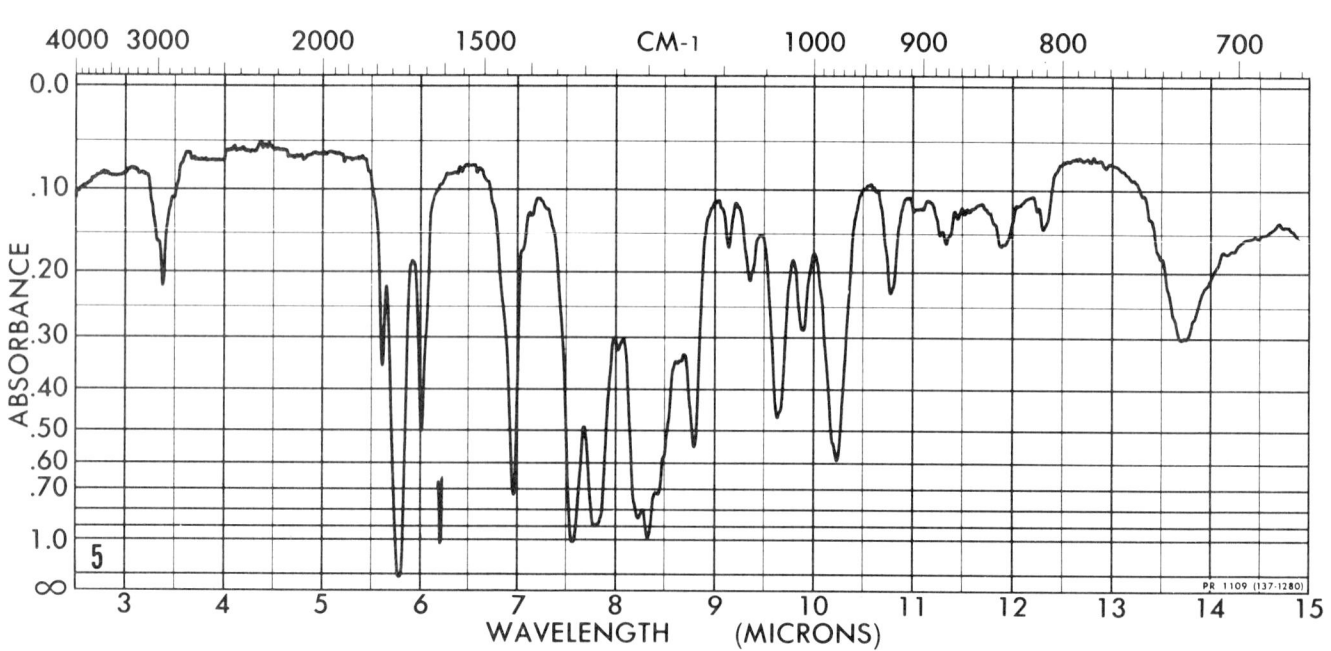

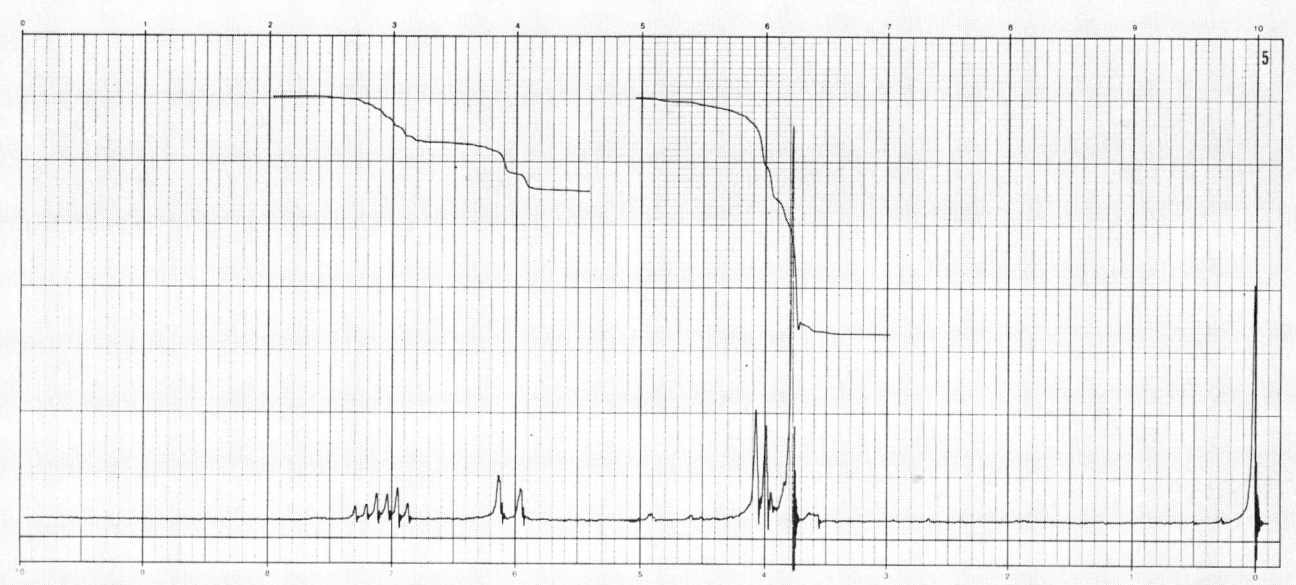

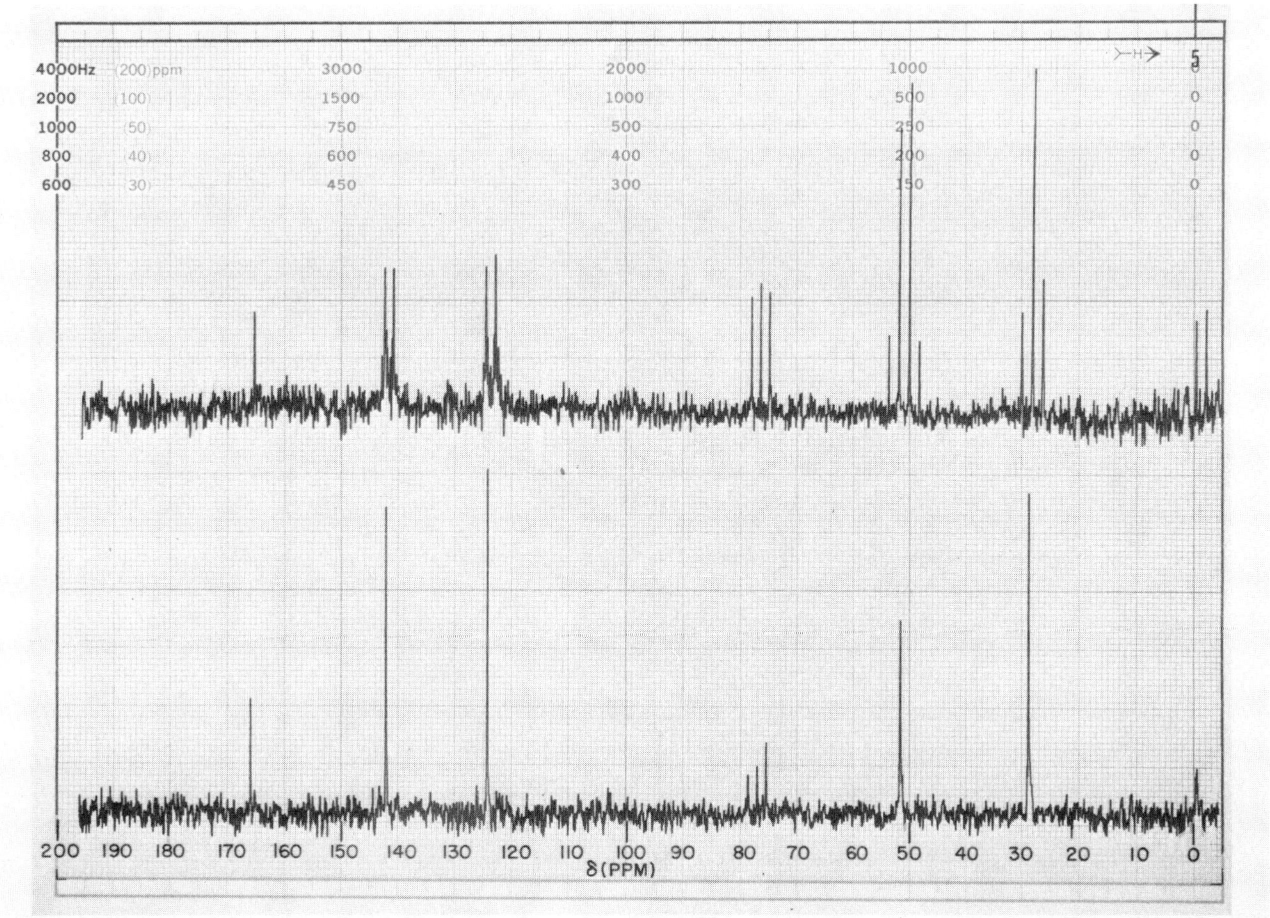

COMPOUND 6

Compound 6

MS: 84.058
IR: neat
HMR: CCl$_4$
CMR: neat
Analysis: 71.4% C; 9.7% H

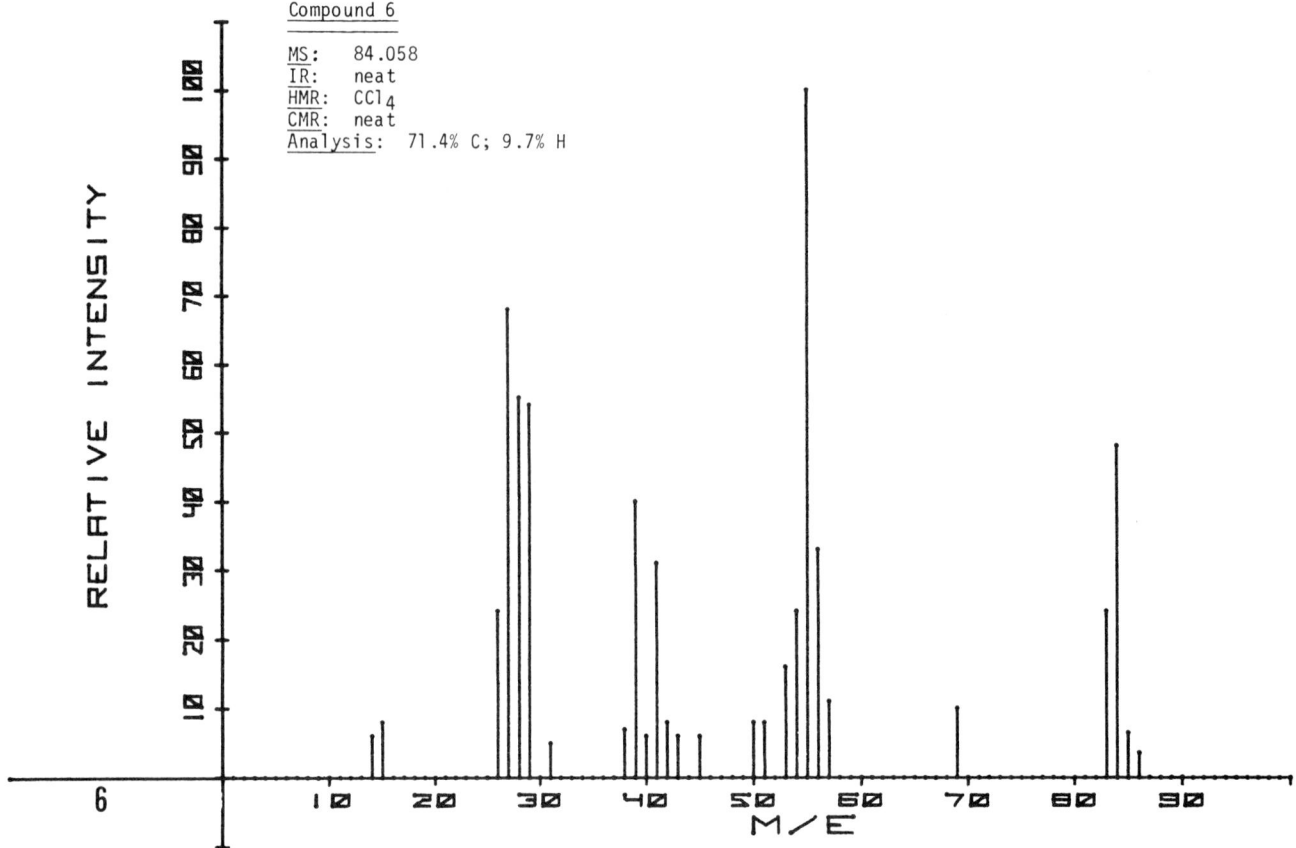

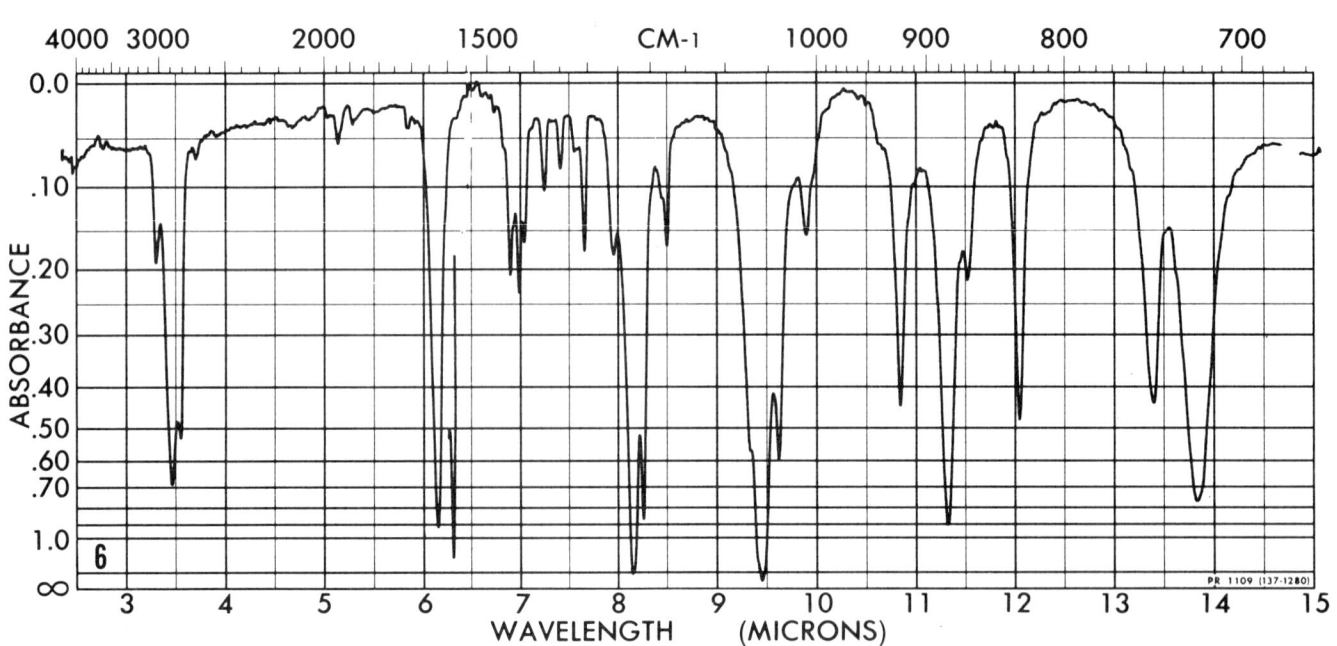

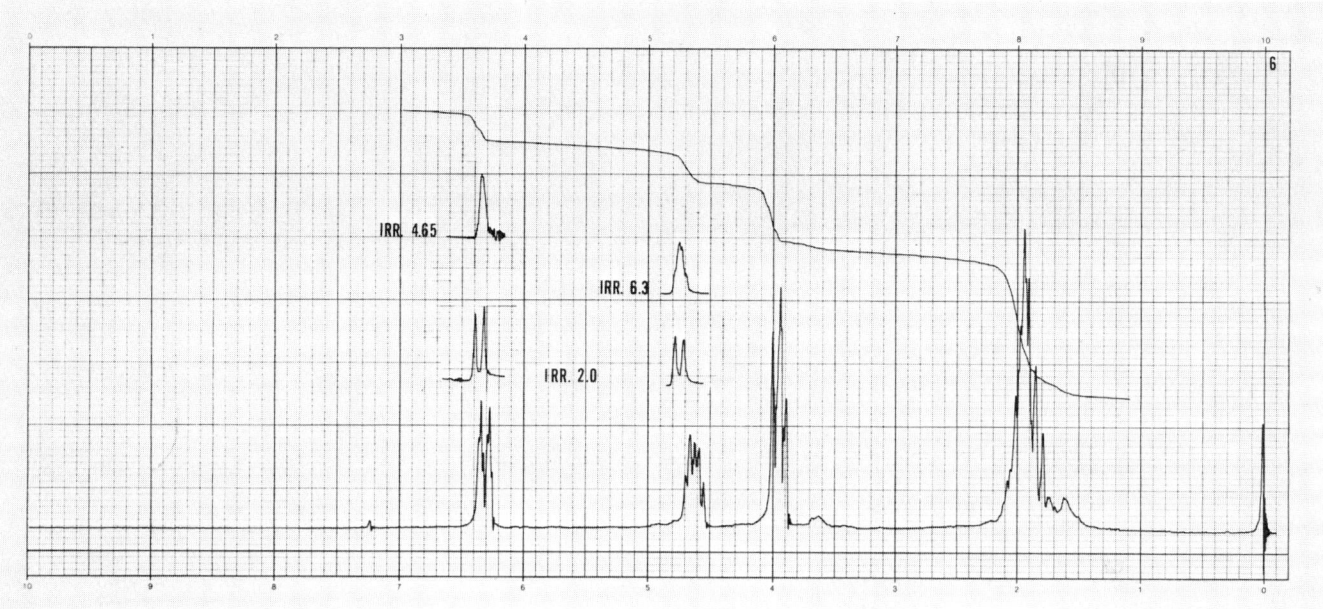

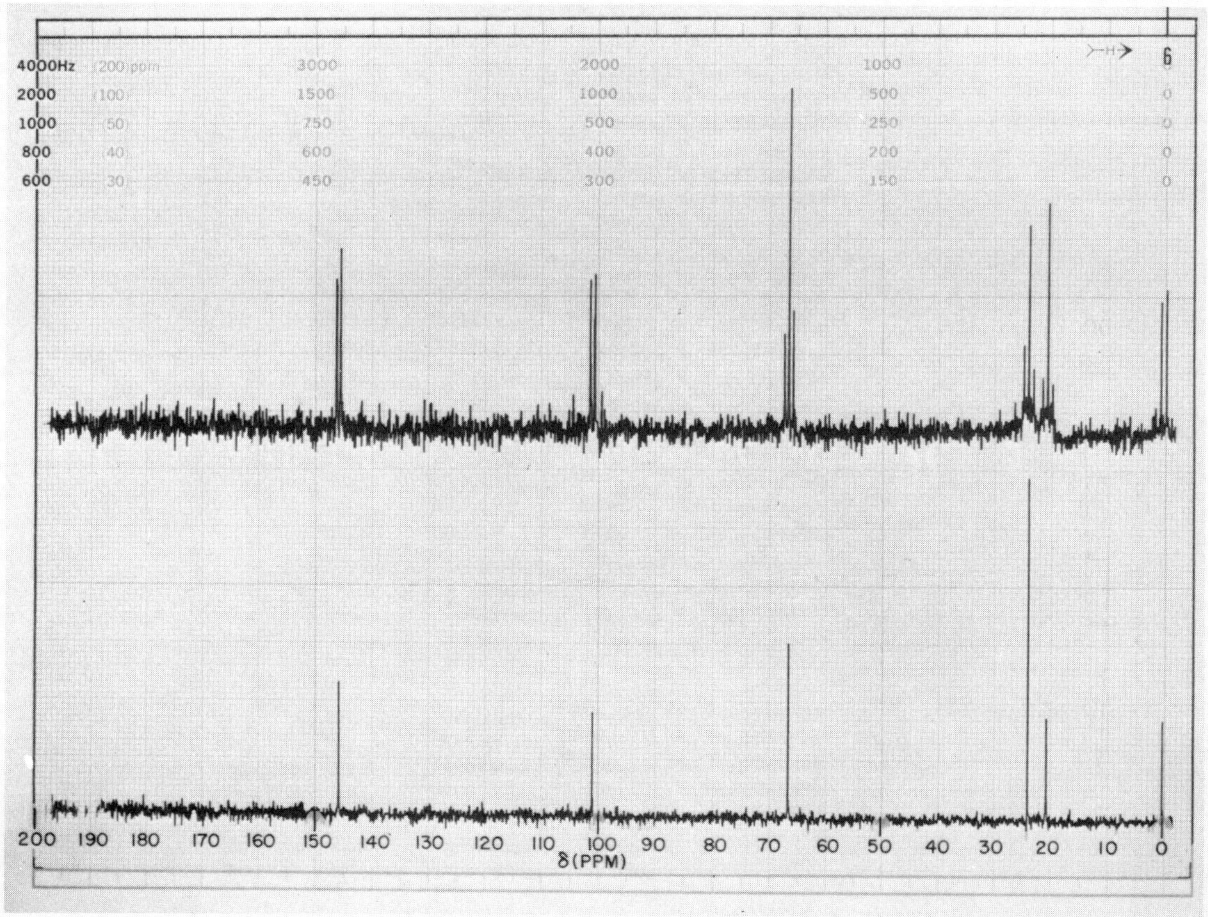

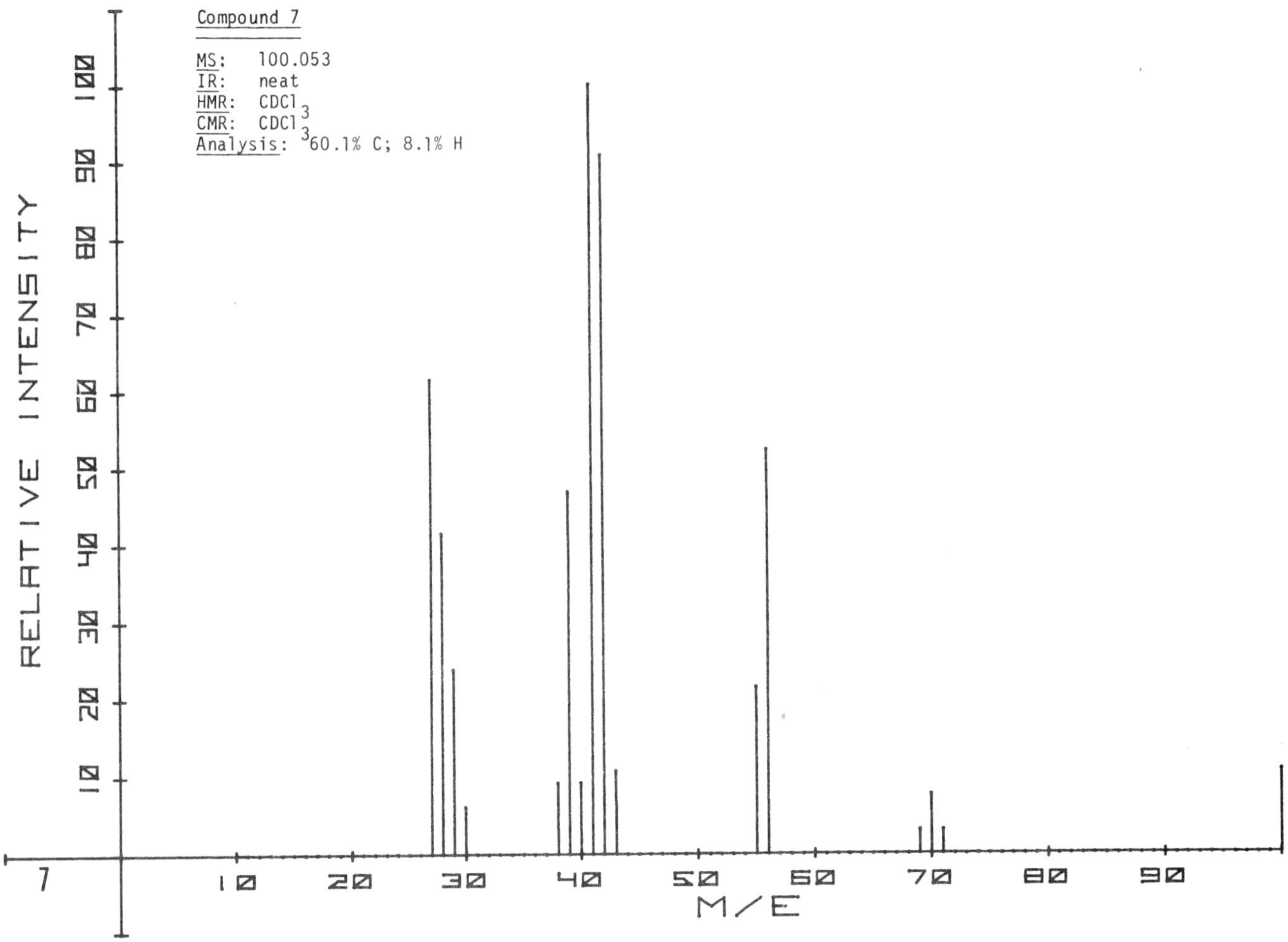

Compound 7
MS: 100.053
IR: neat
HMR: CDCl3
CMR: CDCl3
Analysis: 60.1% C; 8.1% H
RELATIVE INTENSITY
10 20 30 40 50 60 70 80 90 100
7
M/E

4000 3000 2000 1500 CM-1 1000 900 800 700
0.0 .10 .20 .30 .40 .50 .60 .70 1.0 ∞
ABSORBANCE
7
3 4 5 6 7 8 9 10 11 12 13 14 15
WAVELENGTH (MICRONS)

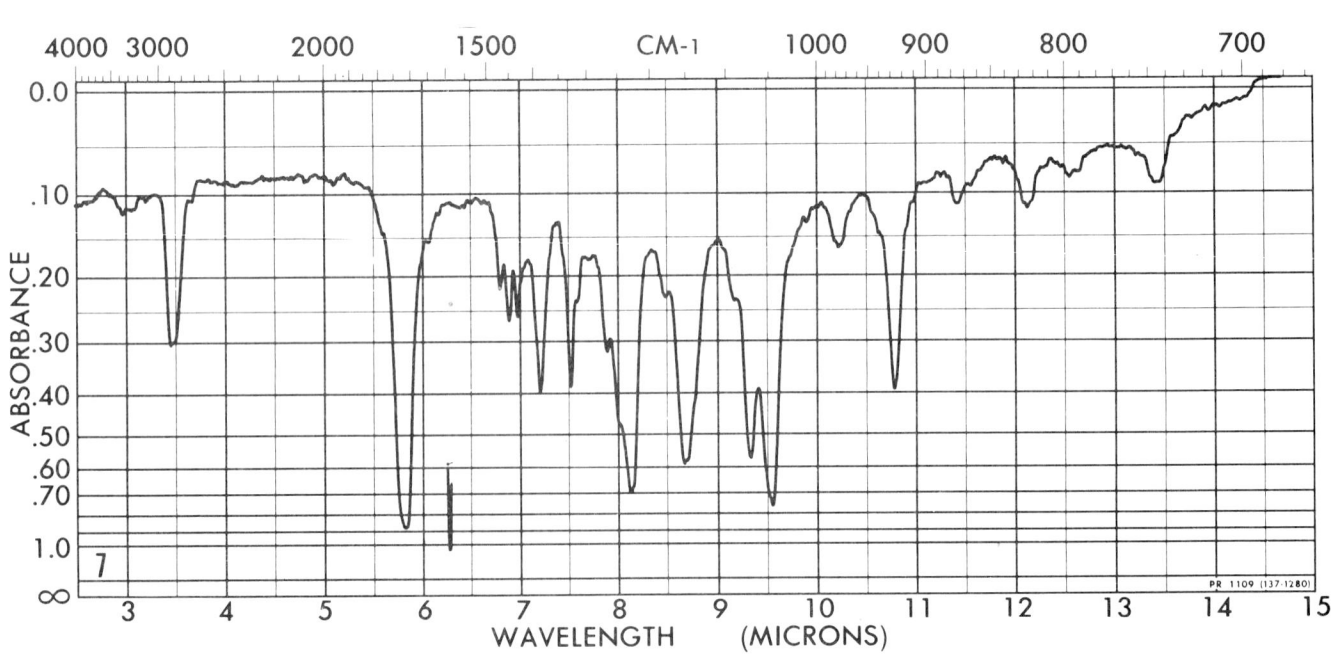

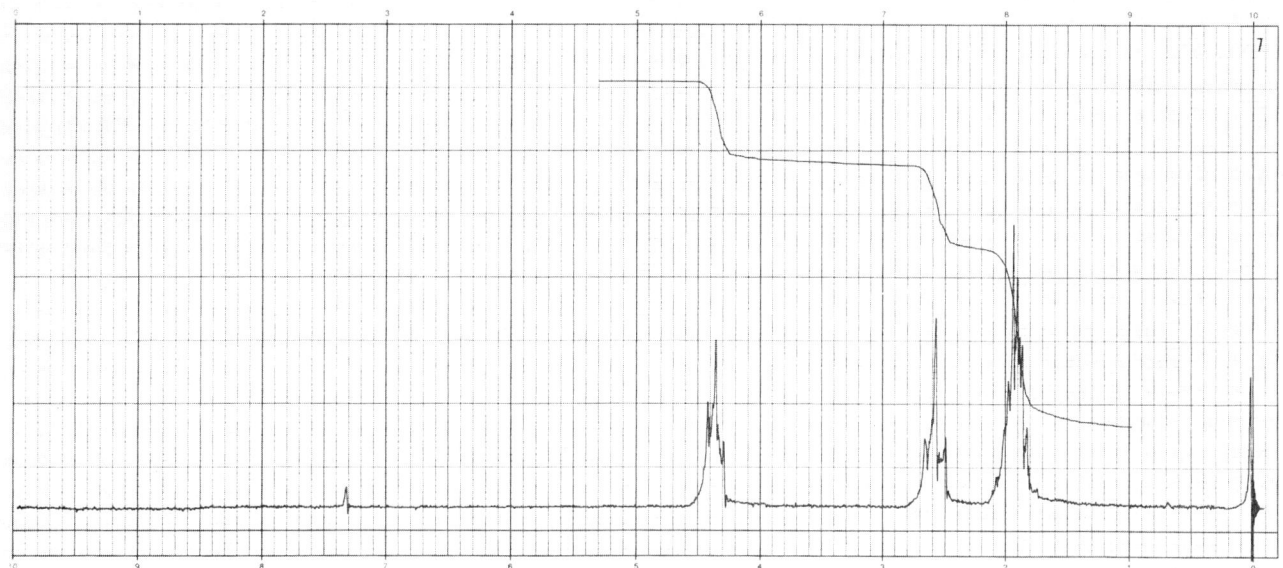

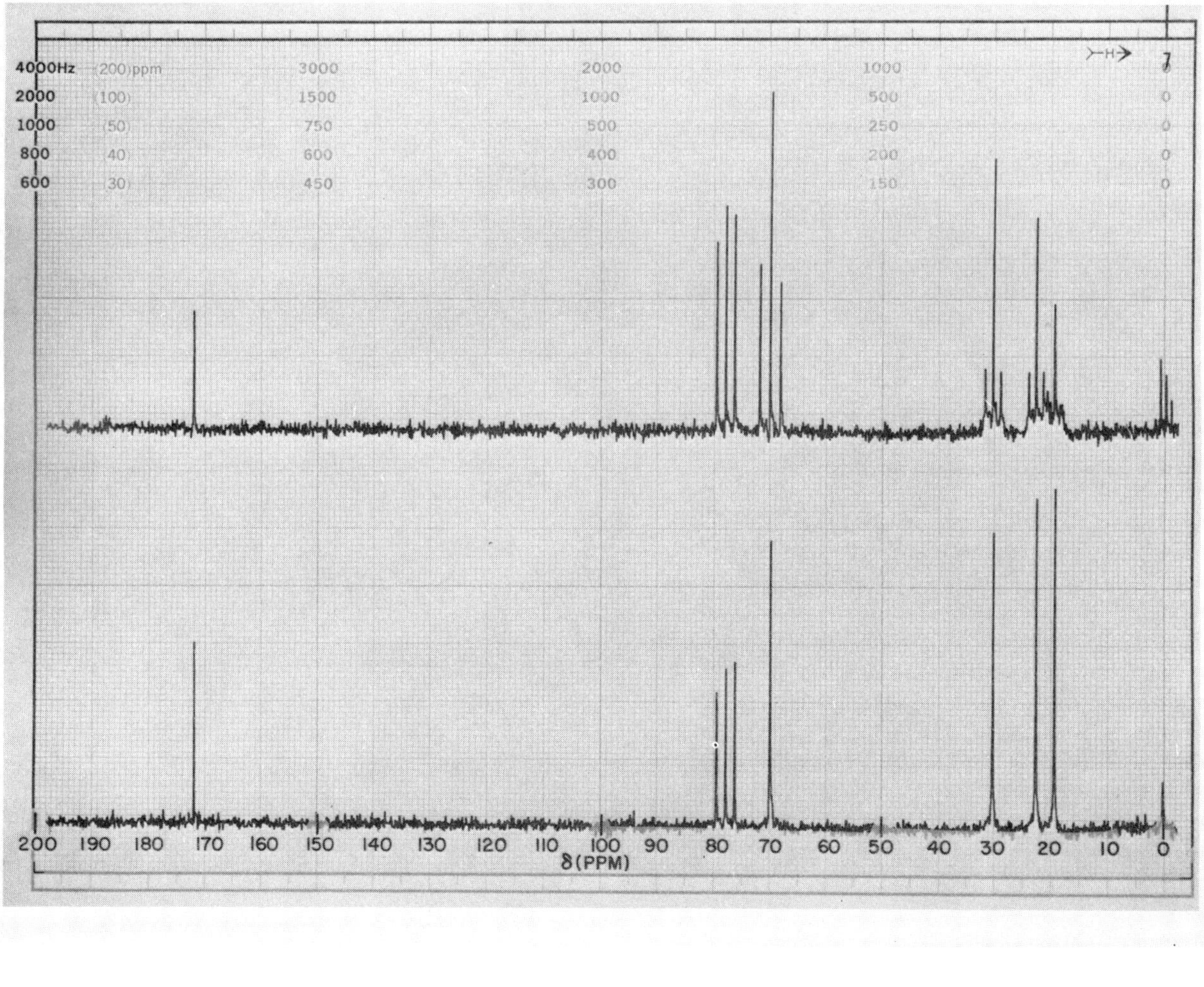

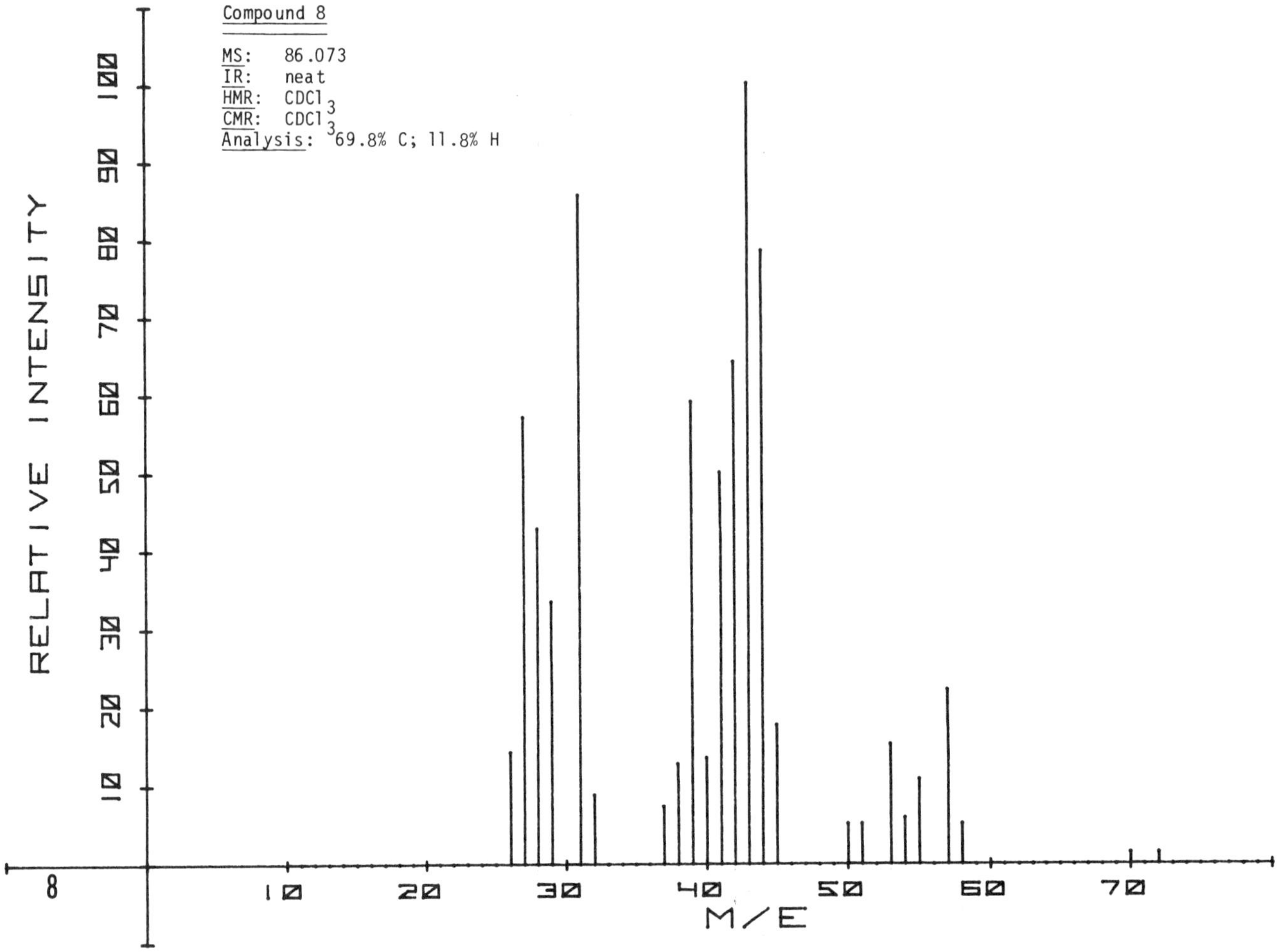

Compound 8
MS: 86.073
IR: neat
HMR: CDCl3
CMR: CDCl3
Analysis: 69.8% C; 11.8% H
RELATIVE INTENSITY
10 20 30 40 50 60 70 80 90 100
8
10 20 30 40 50 60 70
M/E

4000 3000 2000 1500 CM-1 1000 900 800 700
0.0 .10 .20 .30 .40 .50 .60 .70 1.0 ∞
ABSORBANCE
8
3 4 5 6 7 8 9 10 11 12 13 14 15
WAVELENGTH (MICRONS)

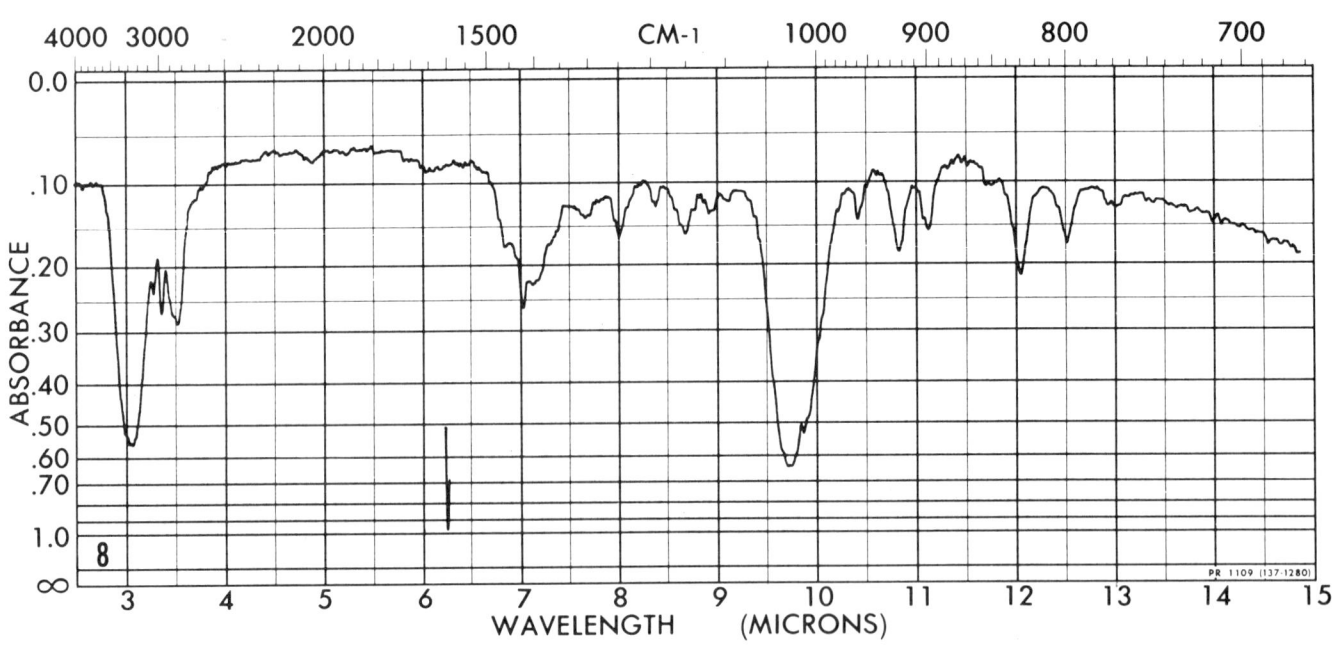

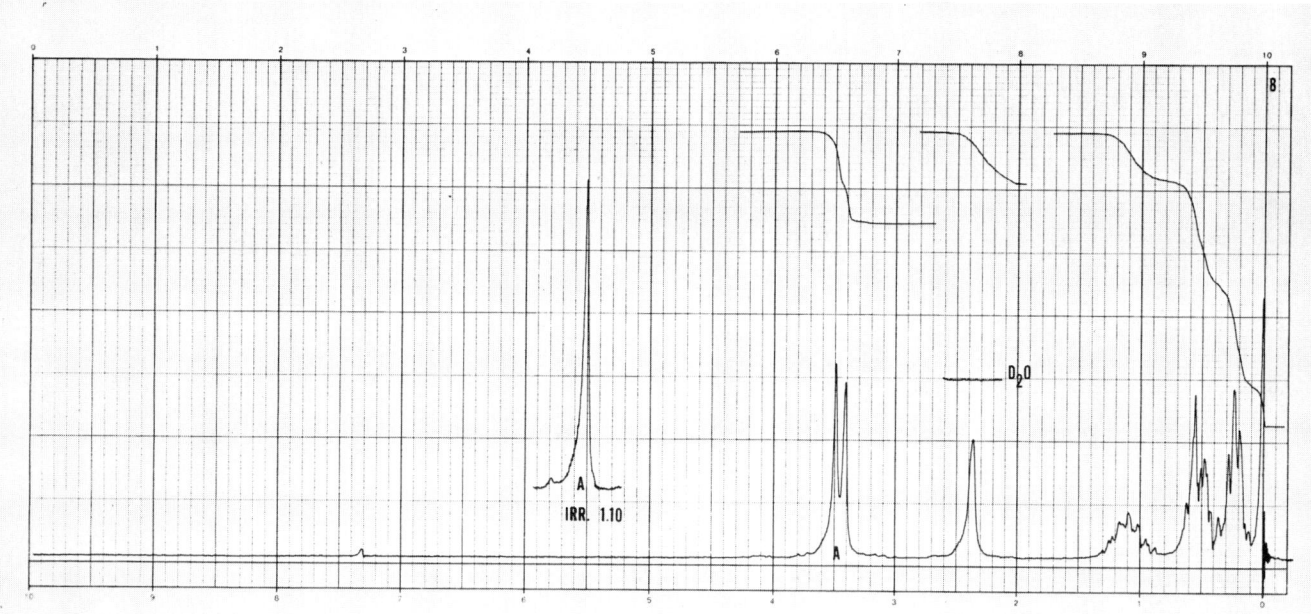

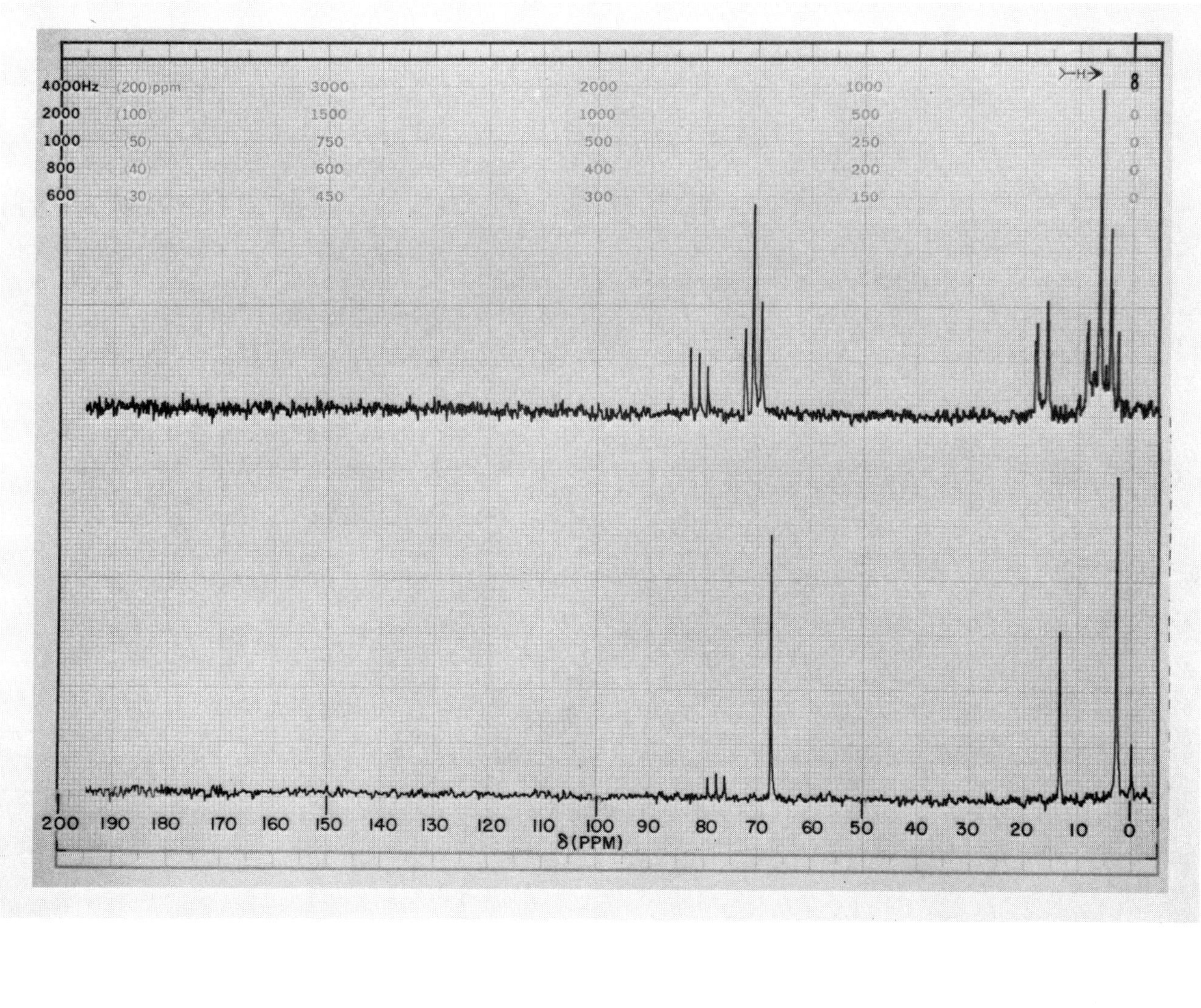

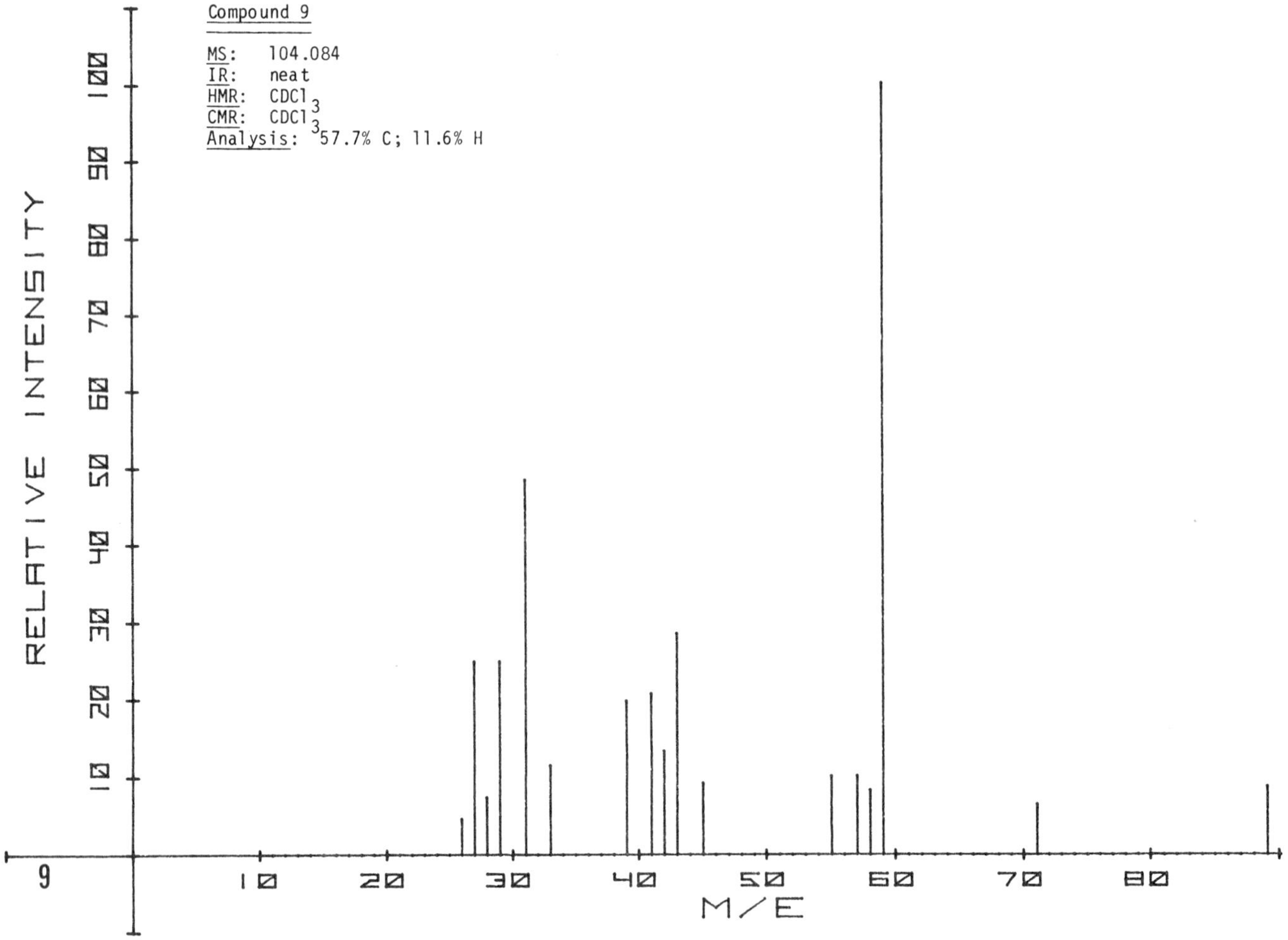

Compound 9
MS: 104.084
IR: neat
HMR: $CDCl_3$
CMR: $CDCl_3$
Analysis: 57.7% C; 11.6% H
RELATIVE INTENSITY
10 20 30 40 50 60 70 80 90 100
9
10 20 30 40 50 60 70 80
M/E

4000 3000 2000 1500 CM-1 1000 900 800 700
0.0 .10 .20 .30 .40 .50 .60 .70 1.0 ∞
ABSORBANCE
9
PR 1109 (137-1280)
3 4 5 6 7 8 9 10 11 12 13 14 15
WAVELENGTH (MICRONS)

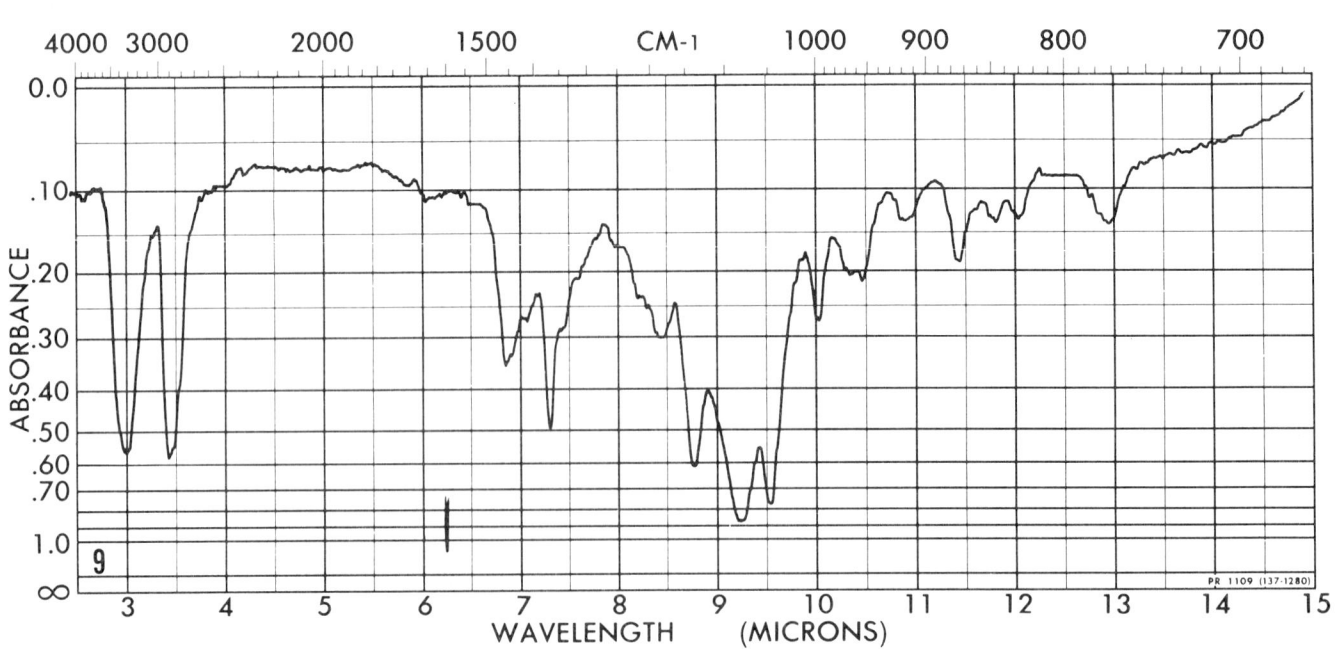

9
D2O

9
4000Hz 3000 2000 1000
2000 1500 1000 500 0
1000 750 500 250 0
800 600 400 200 0
600 450 300 150 0
PK. HT. PPM
19 61.4
46 59.9
21 58.4
12 58.2
42 56.7
41 55.2
10 53.7
PK. HT. PPM
52 75.6
60 59.9
33 55.9
56 39.1
39 19.0
200 190 180 170 160 150 140 130 120 110 100 90 80 70 60 50 40 30 20 10 0
δ(PPM)

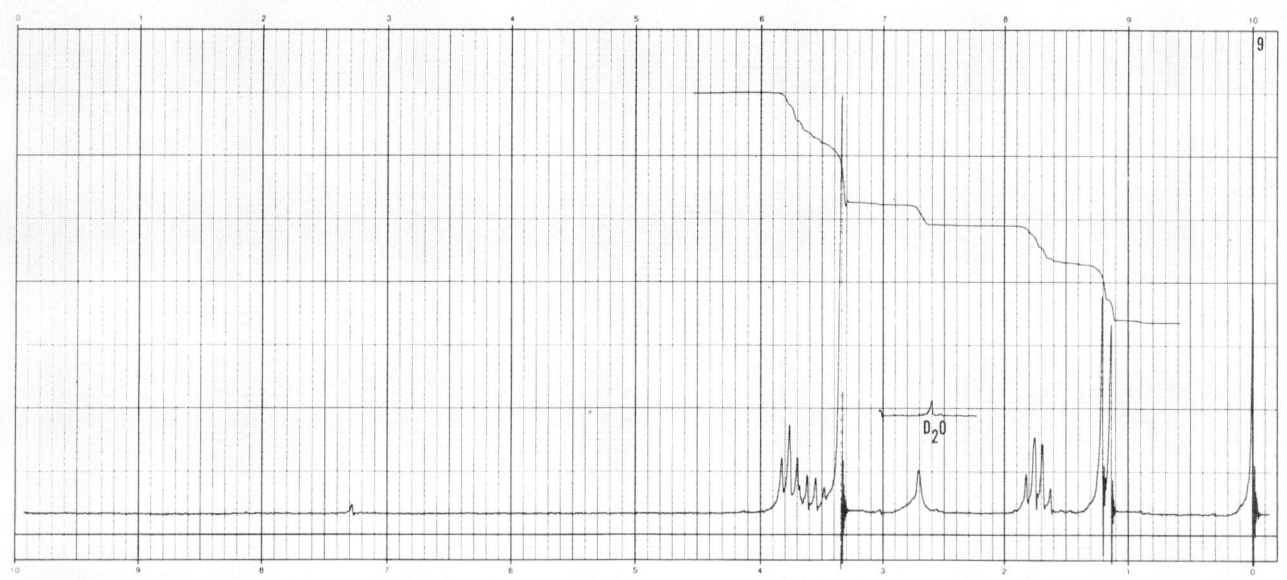

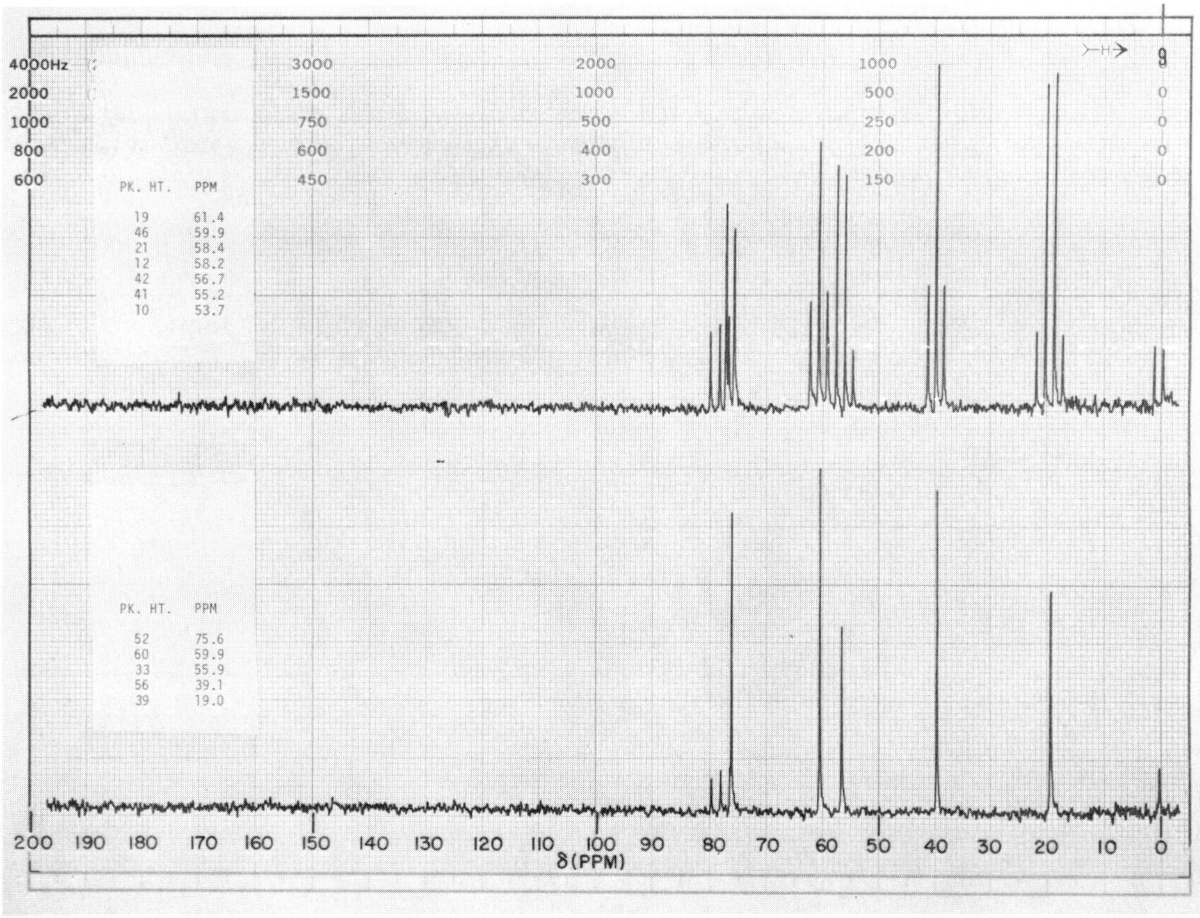

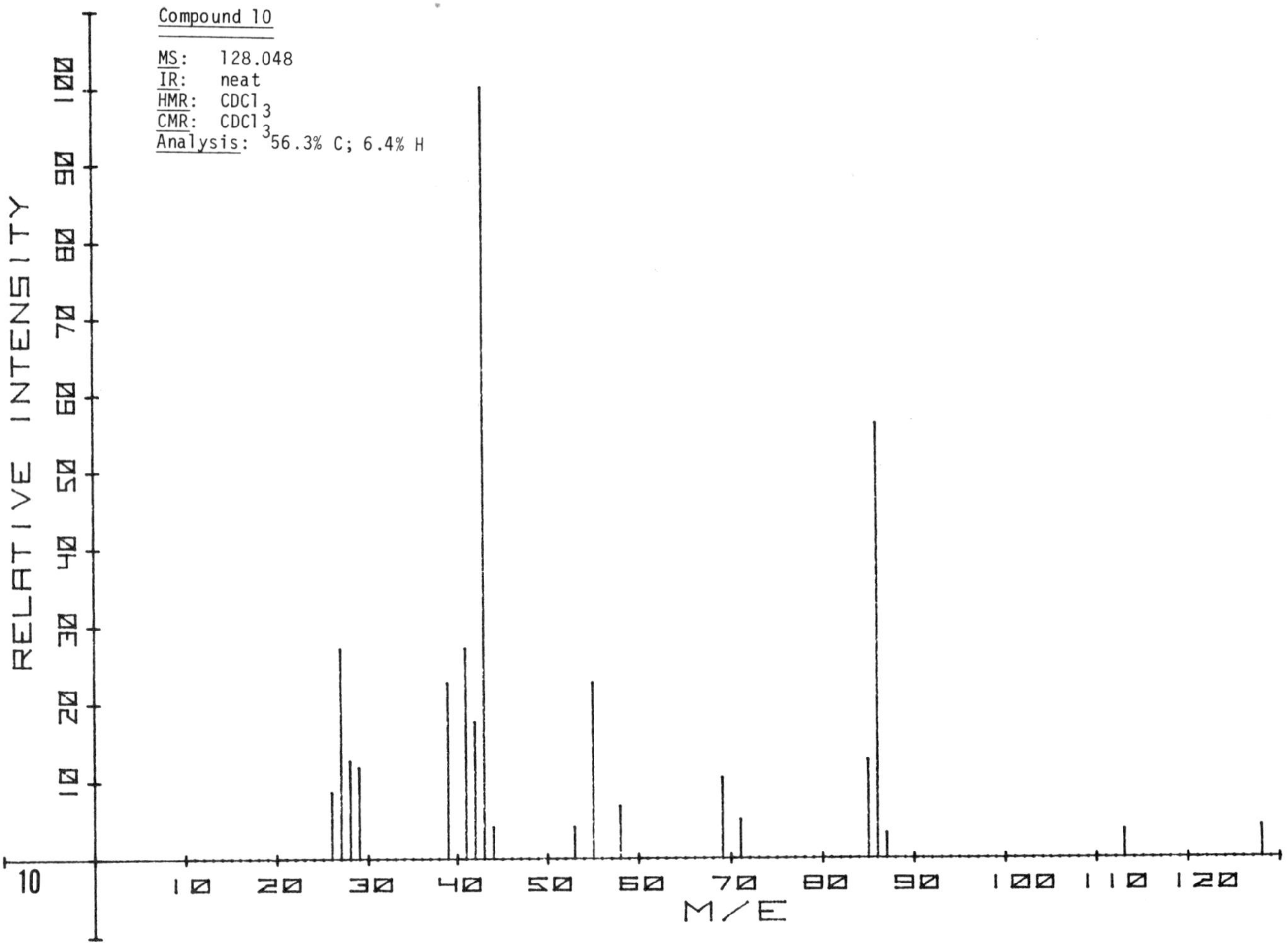

Compound 10
MS: 128.048
IR: neat
HMR: $CDCl_3$
CMR: $CDCl_3$
Analysis: 56.3% C; 6.4% H
RELATIVE INTENSITY
10
20
30
40
50
60
70
80
90
100
110
120
M/E

4000 3000 2000 1500 CM-1 1000 900 800 700
0.0
.10
.20
.30
.40
.50
.60
.70
1.0
∞
ABSORBANCE
10
3 4 5 6 7 8 9 10 11 12 13 14 15
WAVELENGTH (MICRONS)

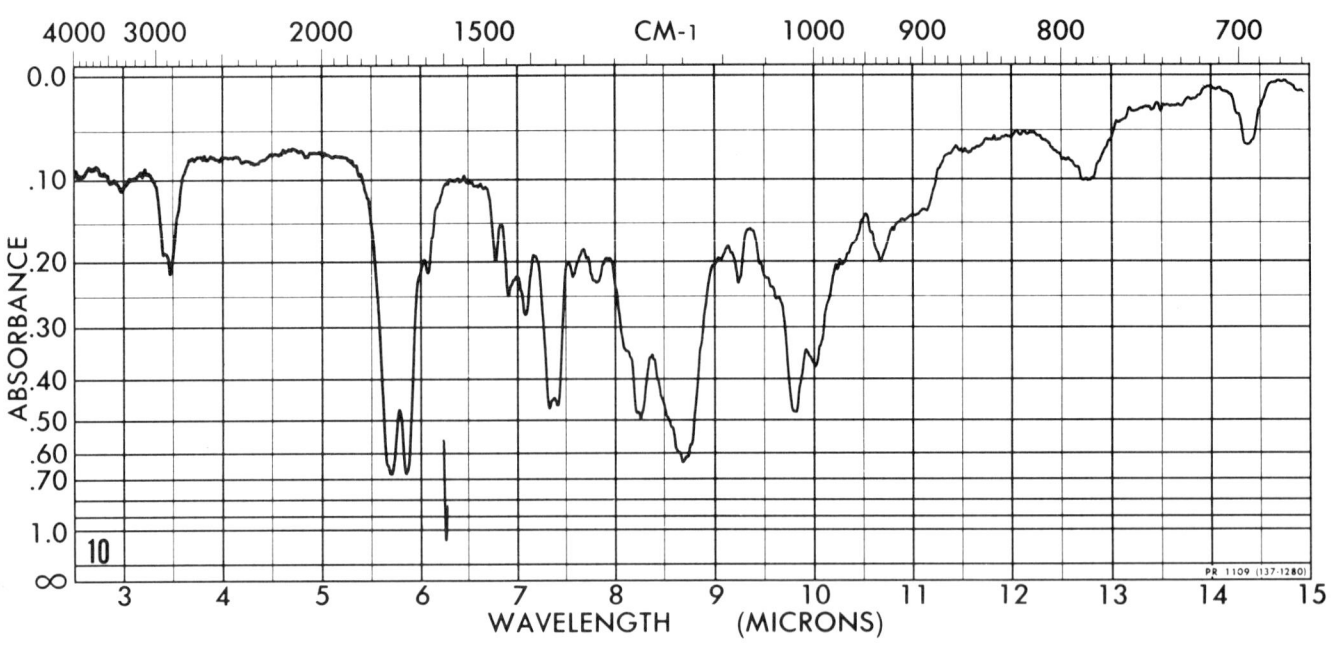

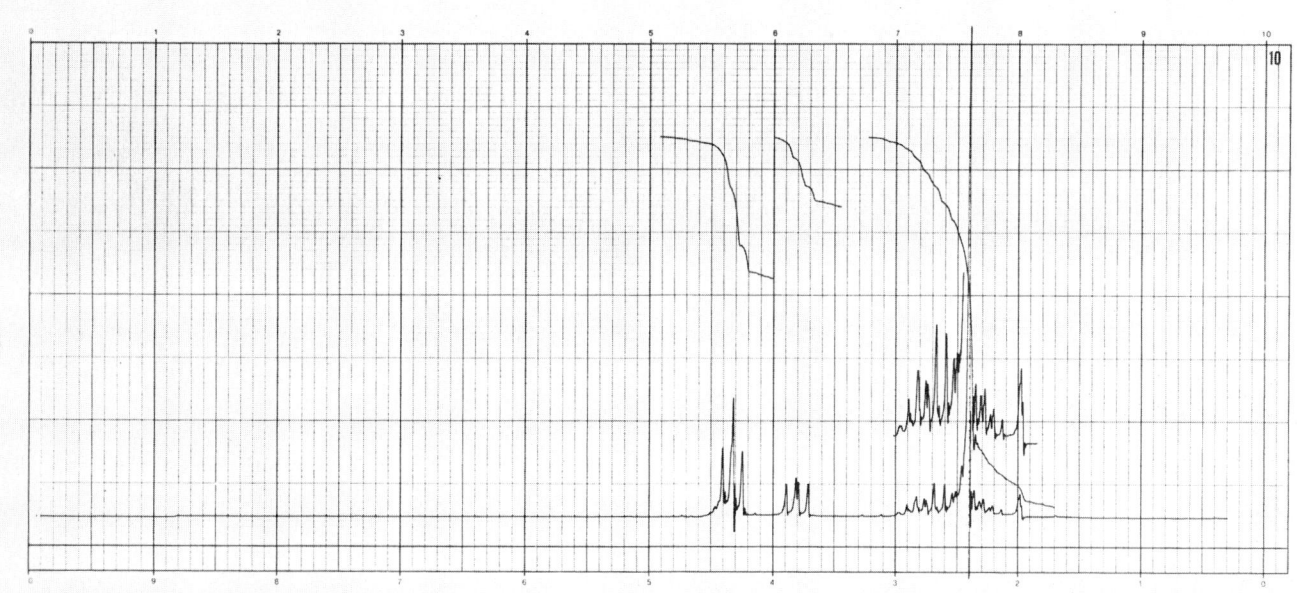

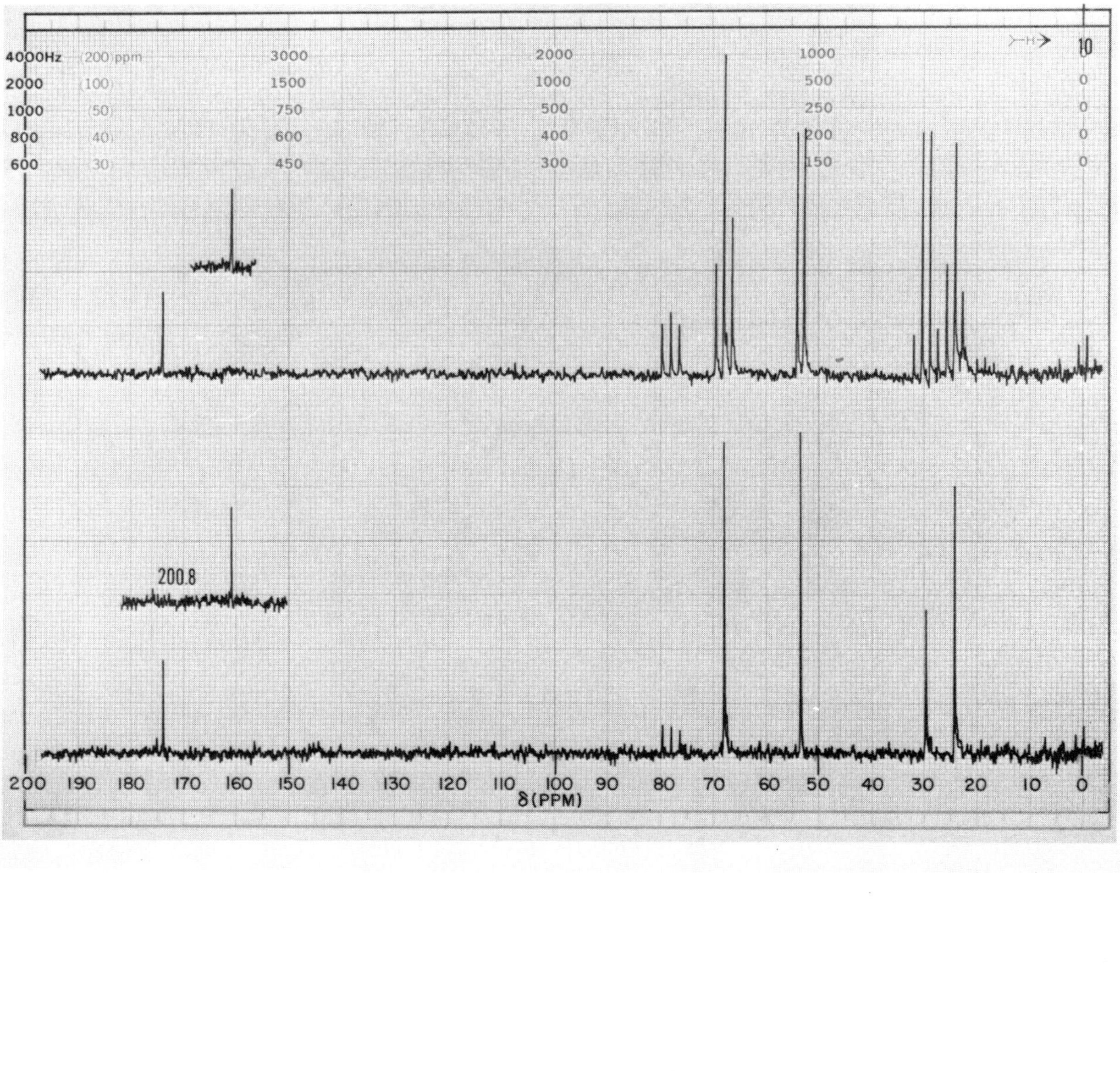

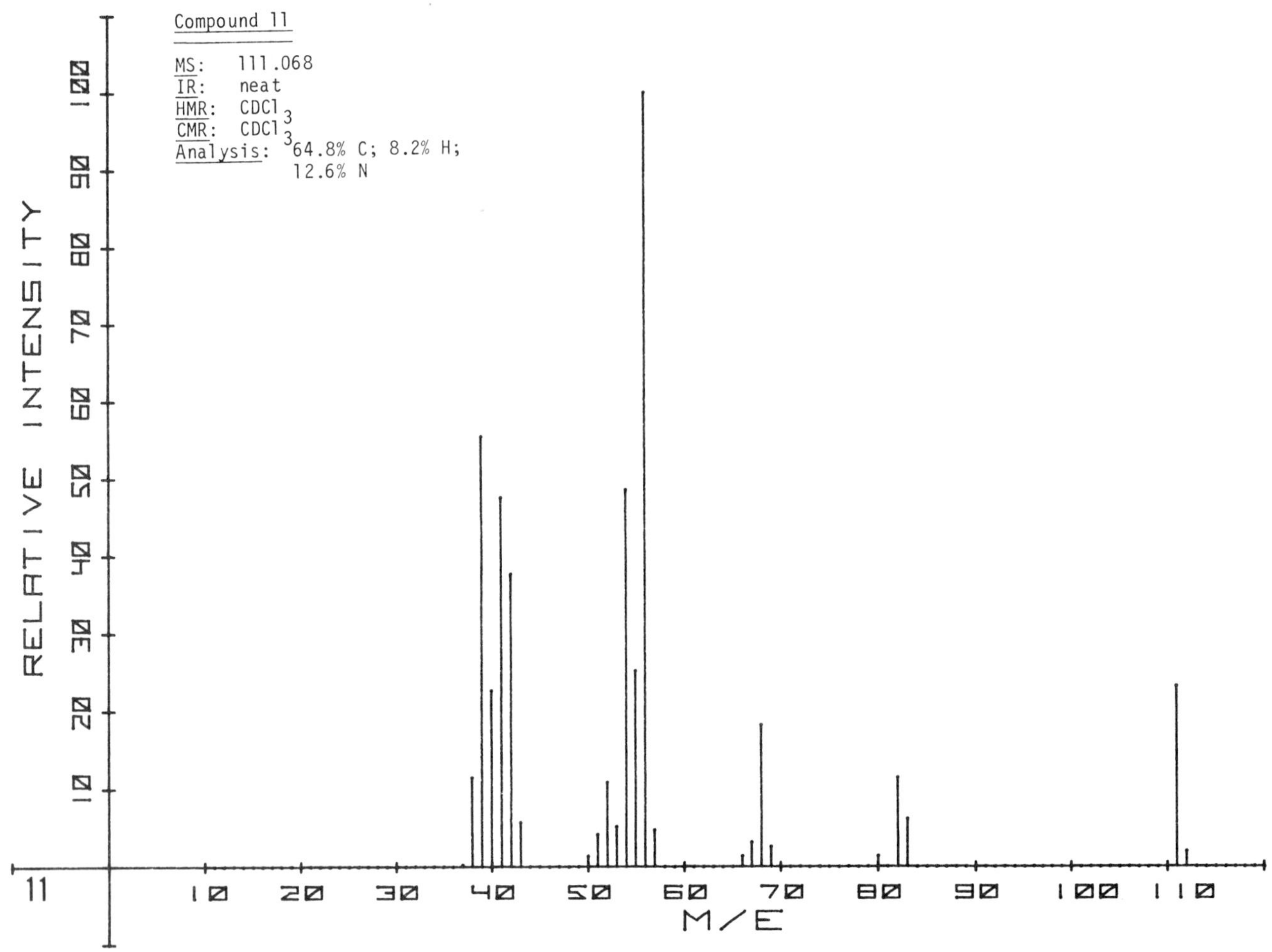

Compound 11
MS: 111.068
IR: neat
HMR: $CDCl_3$
CMR: $CDCl_3$
Analysis: 64.8% C; 8.2% H; 12.6% N
RELATIVE INTENSITY
M/E
11

CM-1
ABSORBANCE
WAVELENGTH (MICRONS)
11

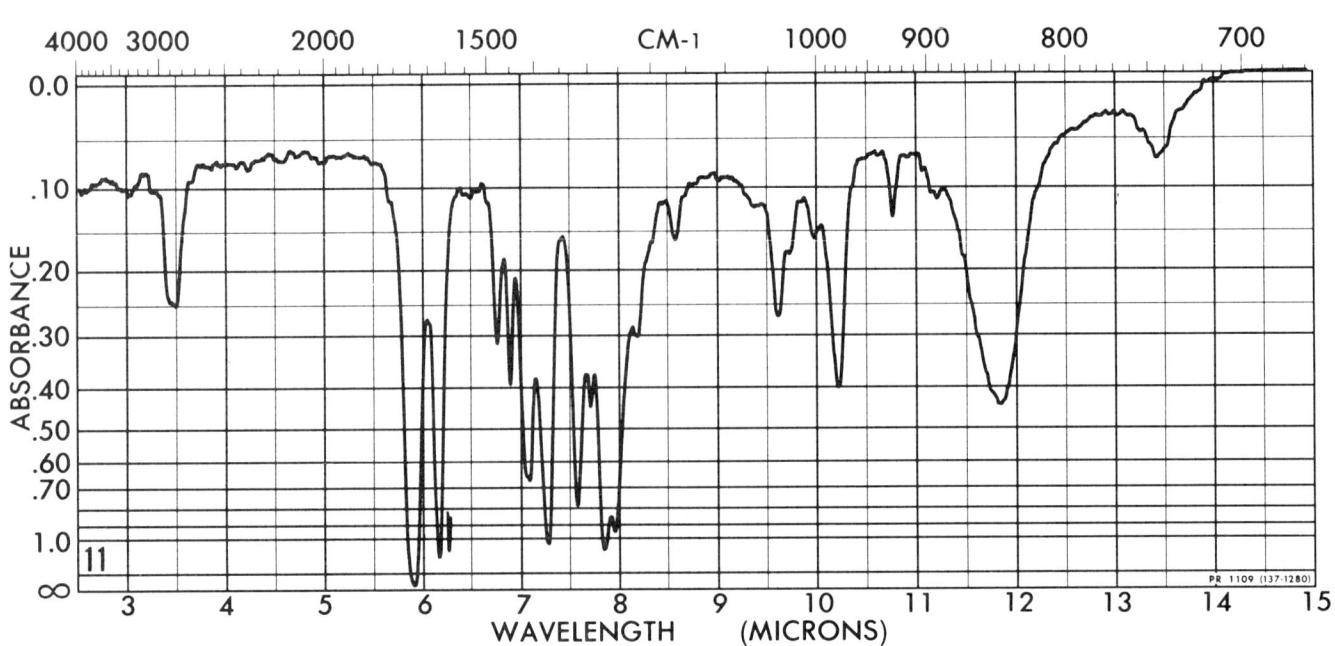

11
10 9 8 7 6 5 4 3 2 1 0

11
4000Hz (200)ppm 3000 2000 1000 0
2000 (100) 1500 1000 500 0
1000 (50) 750 500 250 0
800 (40) 600 400 200 0
600 (30) 450 300 150 0
PK. HT. PPM
16 173.0
39 130.2
50 128.4
60 93.9
45 44.4
33 31.2
34 17.3
PK. HT. PPM
16 173.0
60 129.3
57 93.9
52 44.4
51 31.2
50 17.3
200 190 180 170 160 150 140 130 120 110 100 90 80 70 60 50 40 30 20 10 0
δ(PPM)

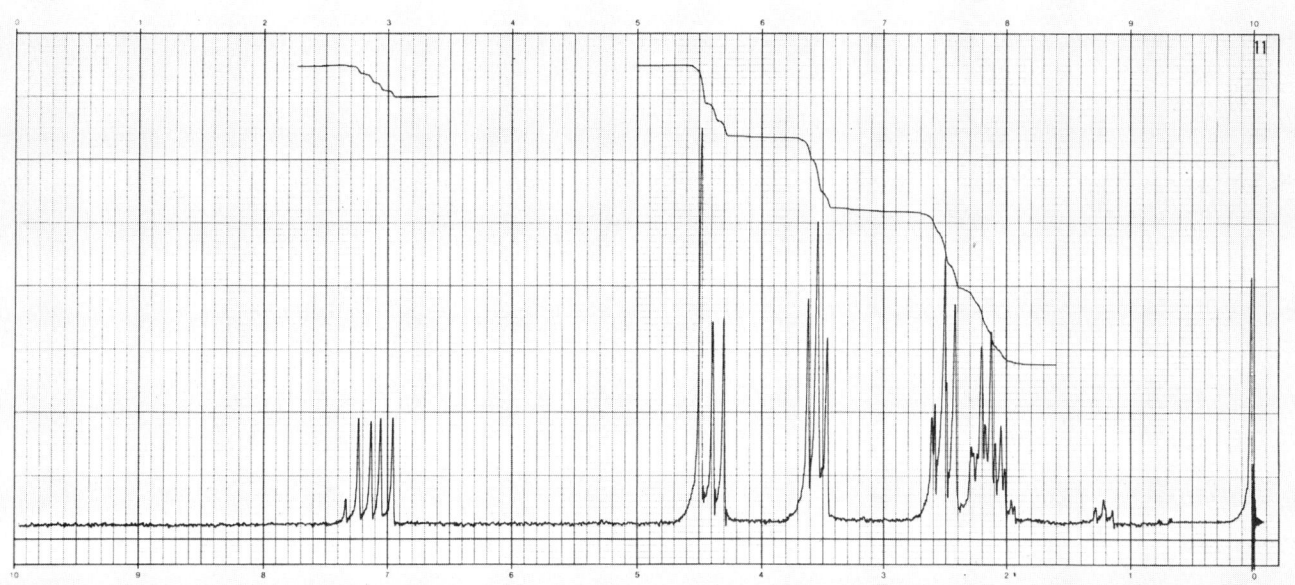

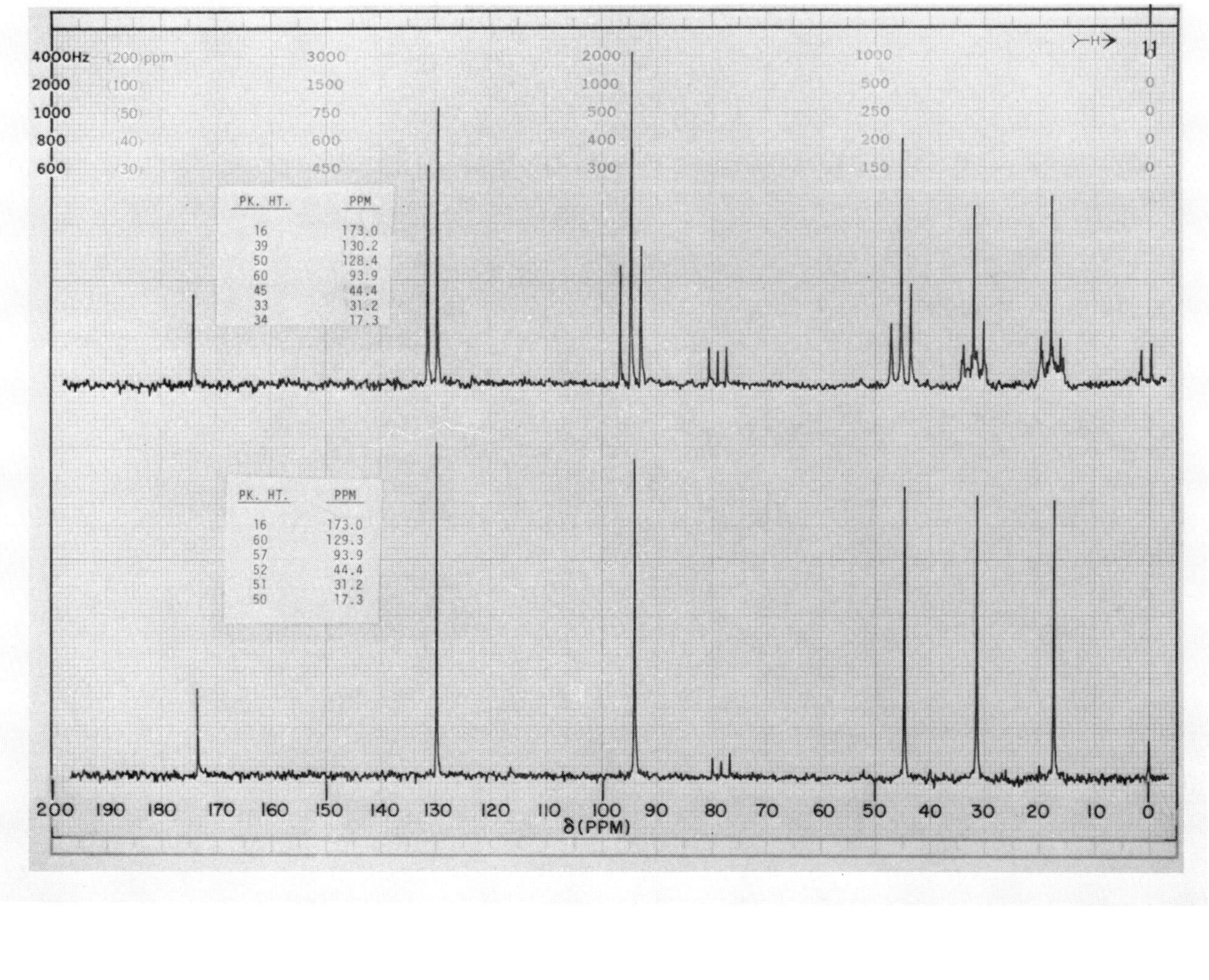

COMPOUND 12A

Compound 12A
MS: 98.074
IR: neat
HMR: CCl₄
CMR: CDCl₃
Analysis: 73.5% C; 10.3% H

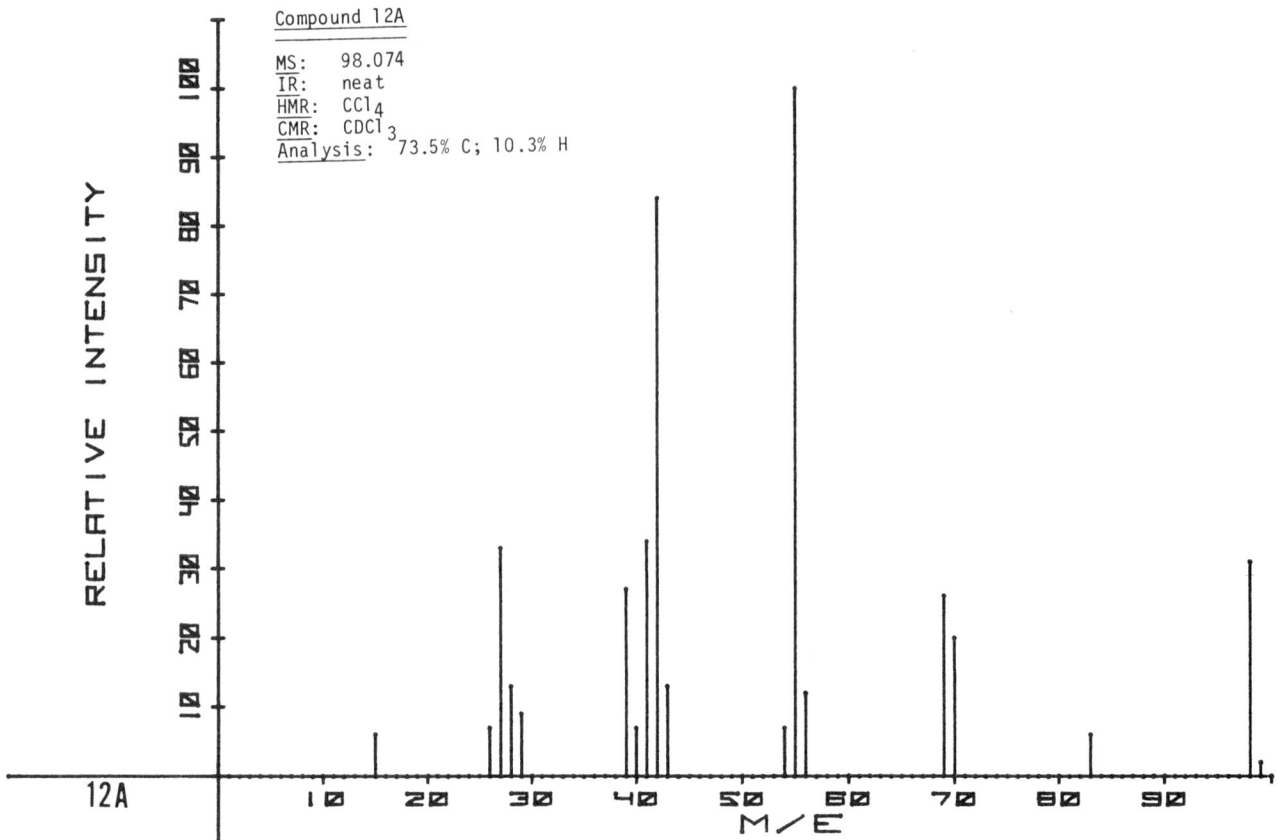

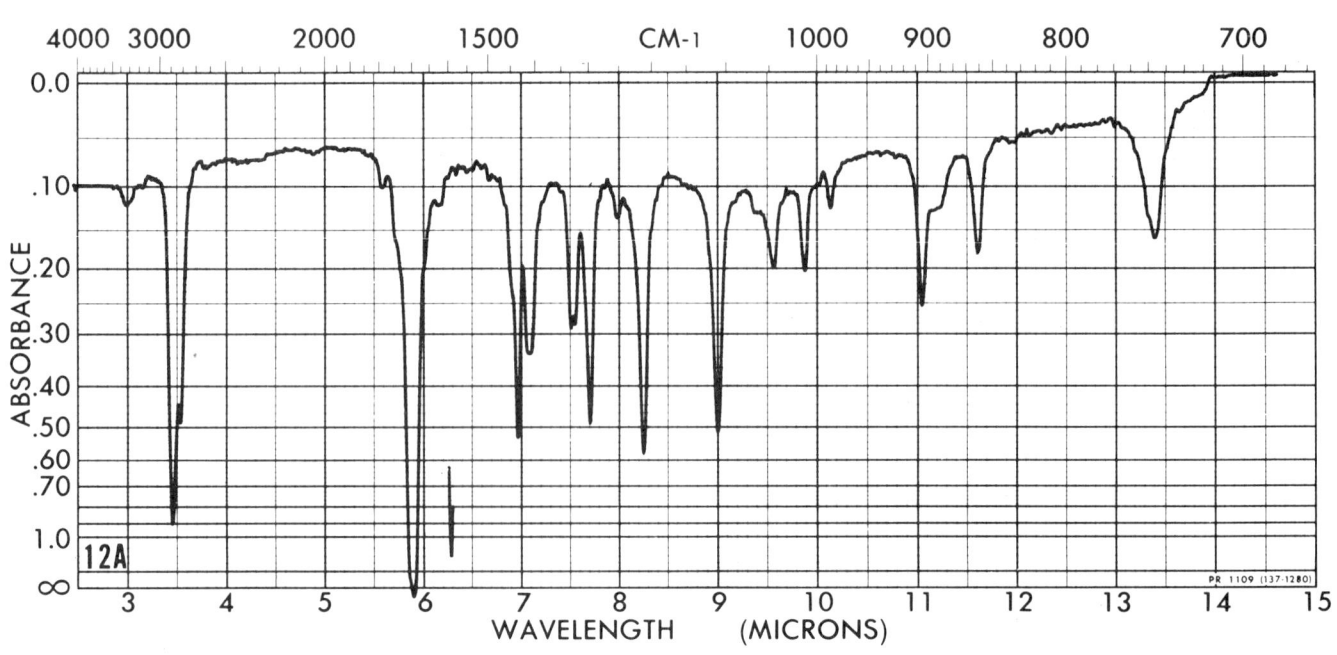

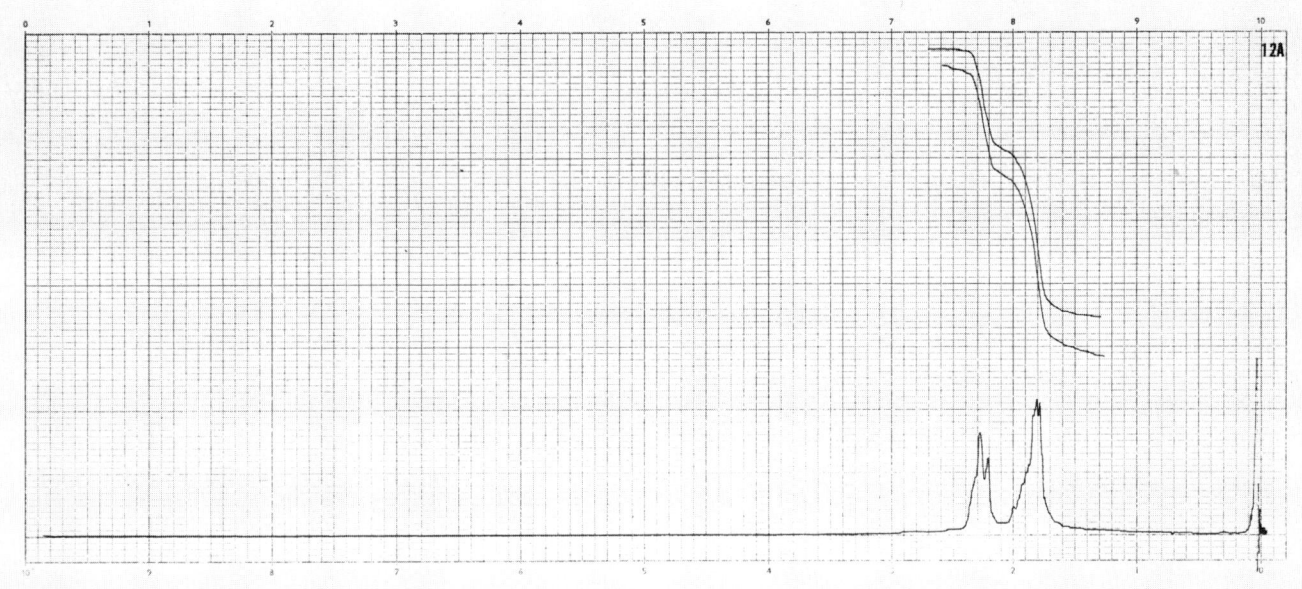

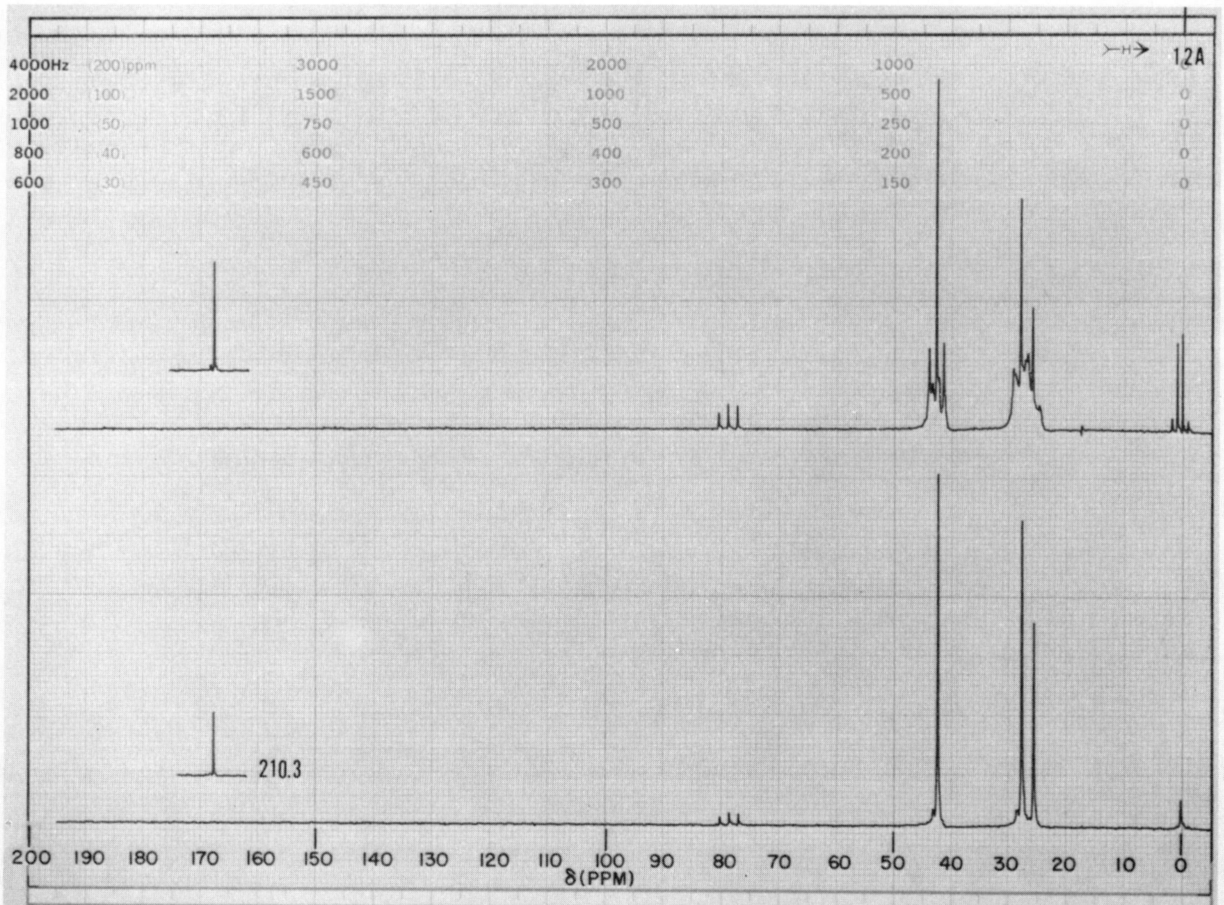

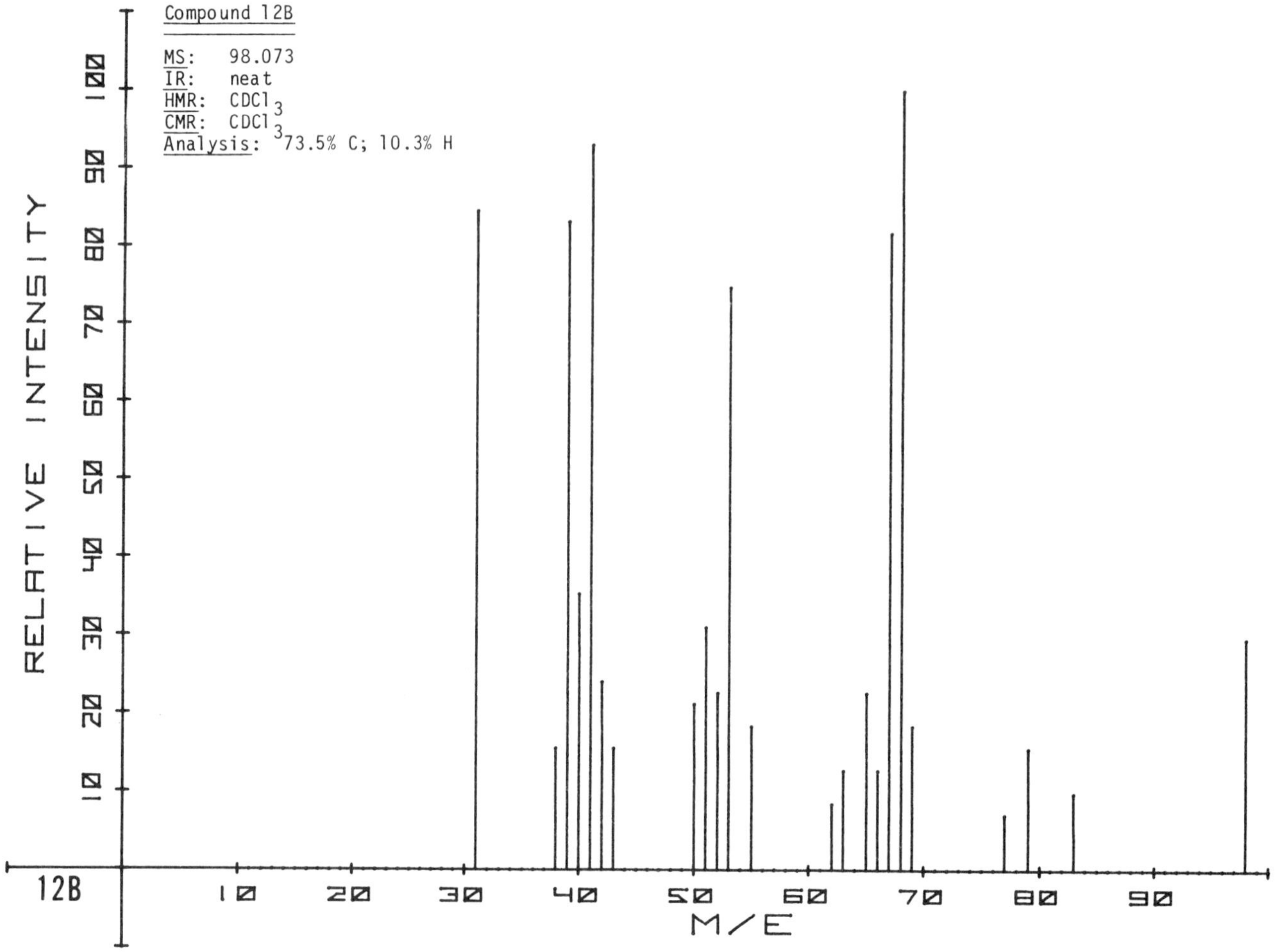

Compound 12B
MS: 98.073
IR: neat
HMR: CDCl3
CMR: CDCl3
Analysis: 73.5% C; 10.3% H
RELATIVE INTENSITY
10 20 30 40 50 60 70 80 90 100
12B
M/E

4000 3000 2000 1500 CM-1 1000 900 800 700
0.0 .10 .20 .30 .40 .50 .60 .70 1.0 ∞
ABSORBANCE
12B
3 4 5 6 7 8 9 10 11 12 13 14 15
WAVELENGTH (MICRONS)

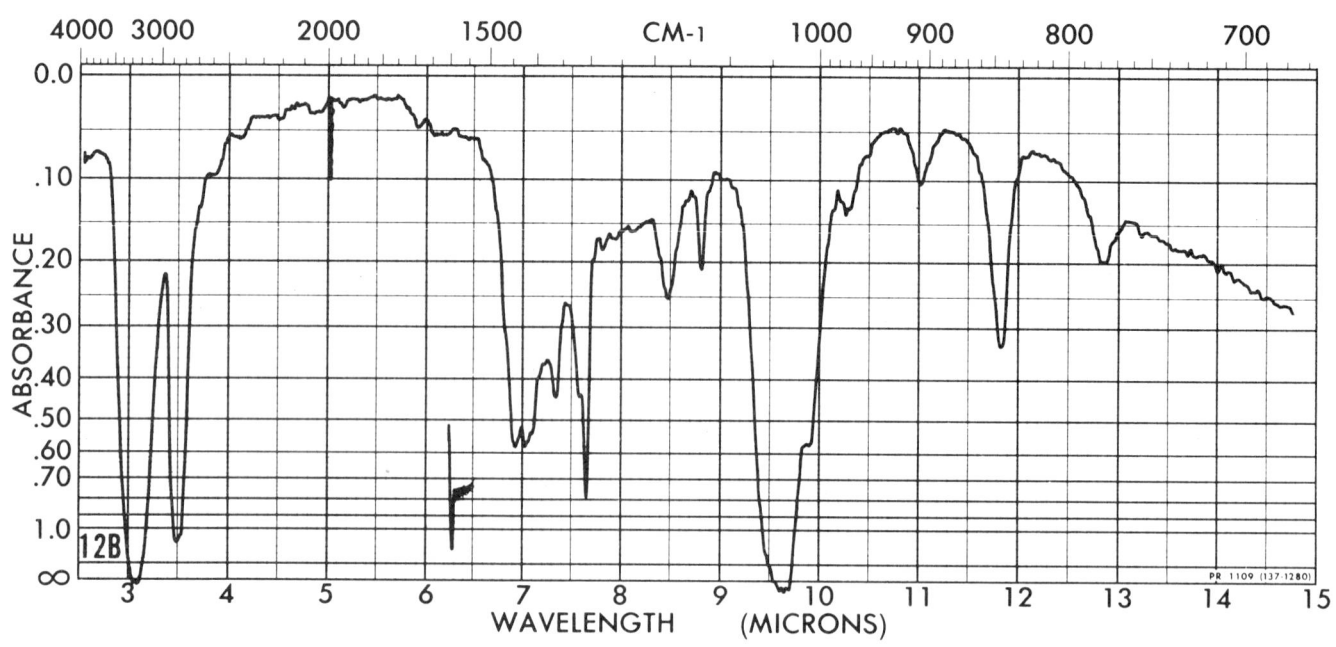

12B
D2O

12B
4000Hz (200) ppm
2000 (100)
1000 (50)
800 (40)
600 (30)
3000
1500
2000
1000
500
400
300
1000
500
250
200
150
PK. HT. PPM
6 16.7
25 15.1
11 14.1
26 13.5
20 12.5
8 11.9
17 10.9
PK. HT. PPM
17 83.5
16 76.1
60 61.5
60 23.1
38 14.3
40 12.5
200 190 180 170 160 150 140 130 120 110 100 90 80 70 60 50 40 30 20 10 0
δ(PPM)

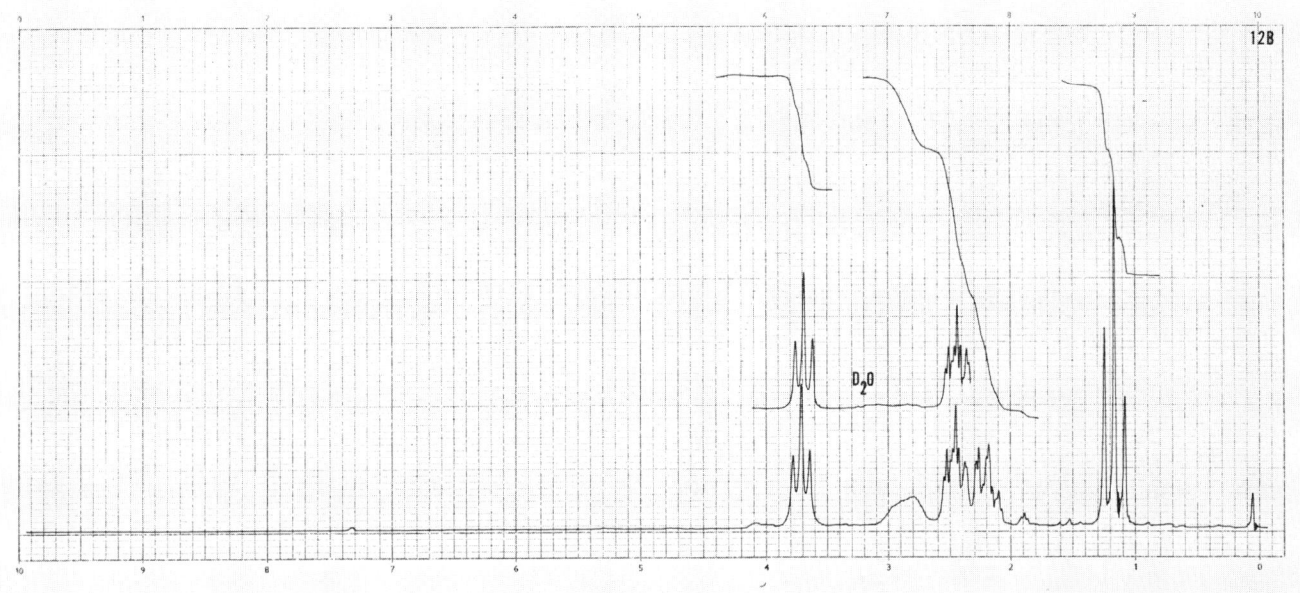

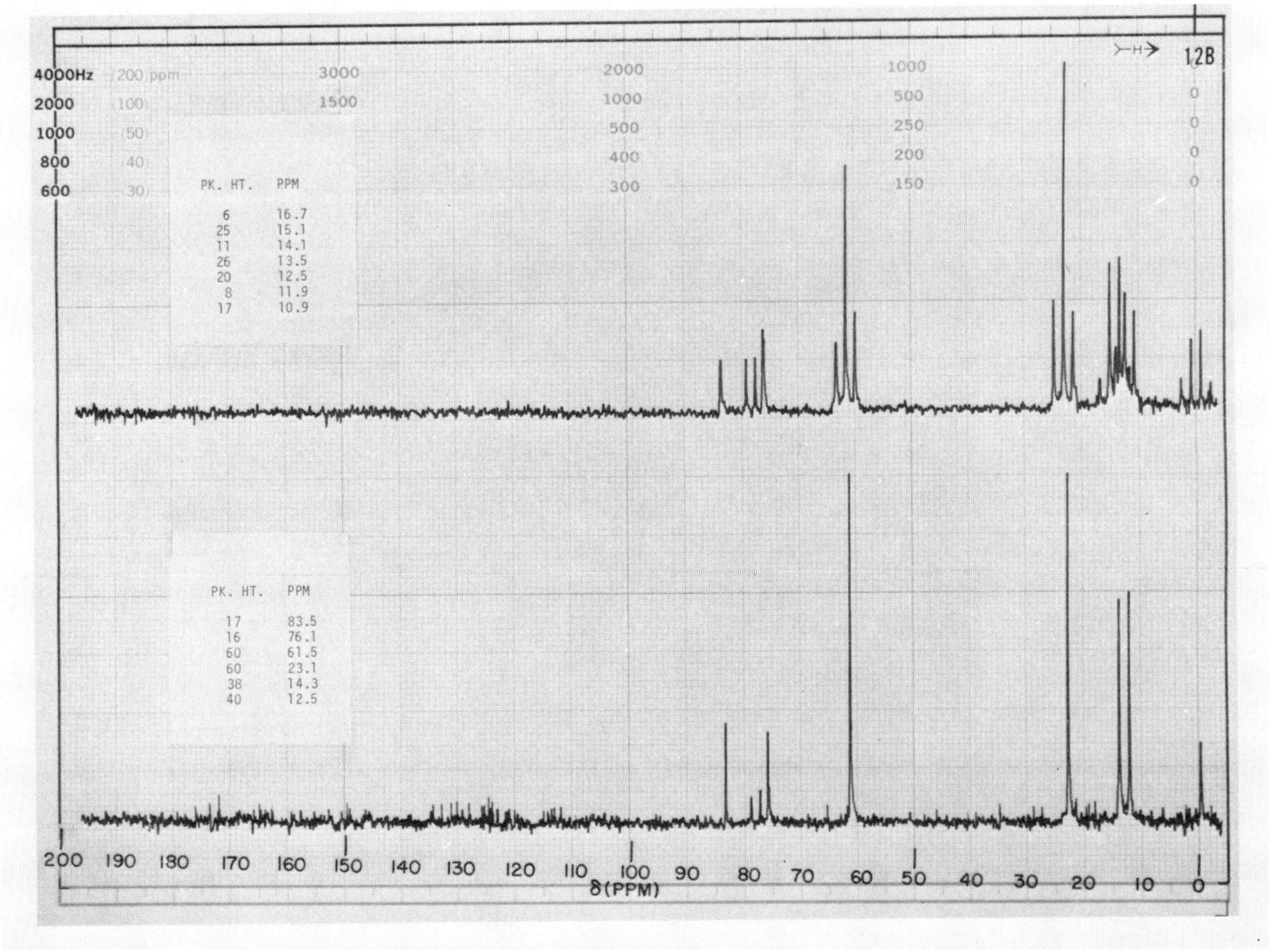

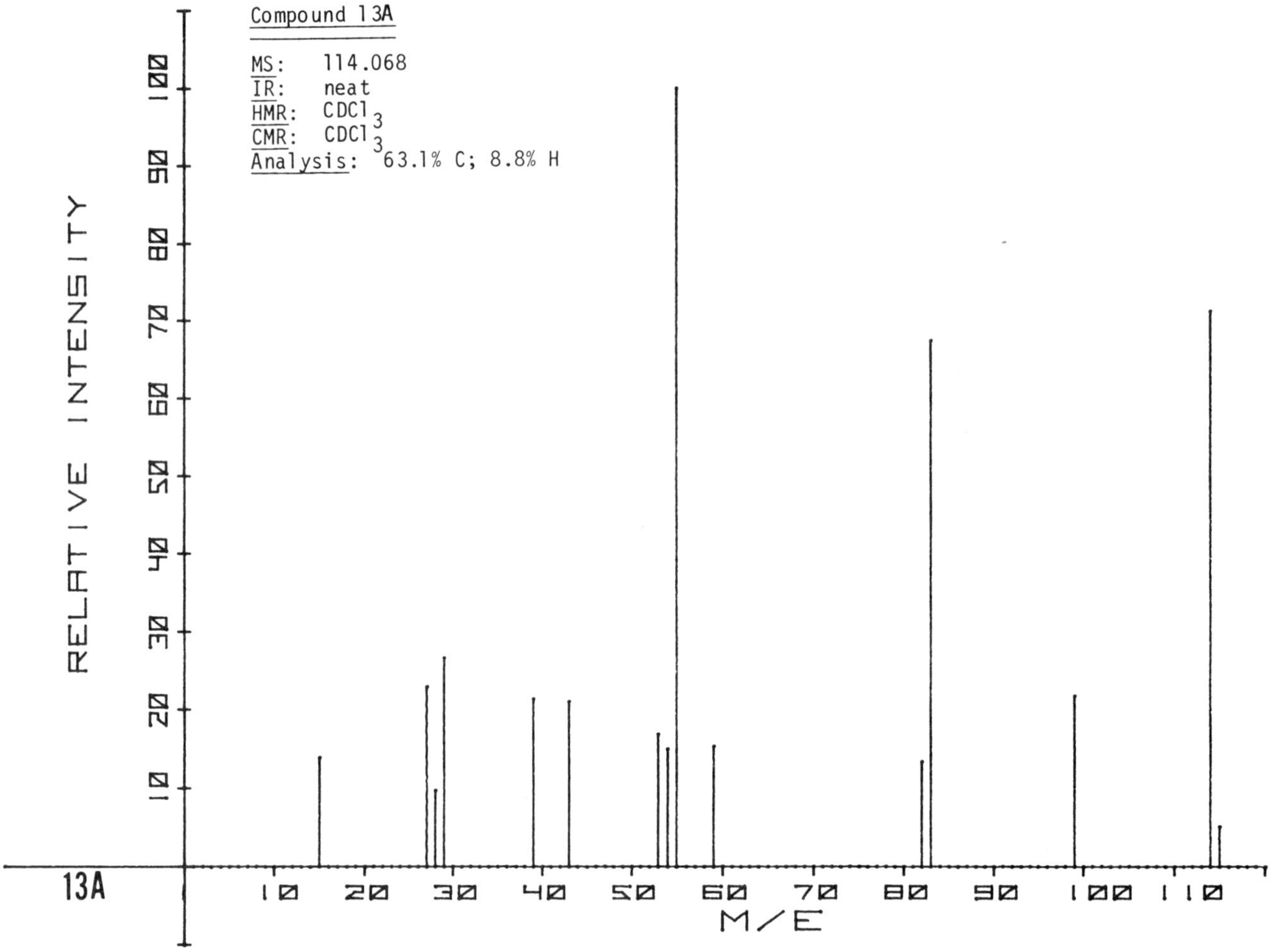

Compound 13A
MS: 114.068
IR: neat
HMR: CDCl3
CMR: CDCl3
Analysis: 63.1% C; 8.8% H
RELATIVE INTENSITY
M/E
13A

CM-1
ABSORBANCE
WAVELENGTH (MICRONS)
13A

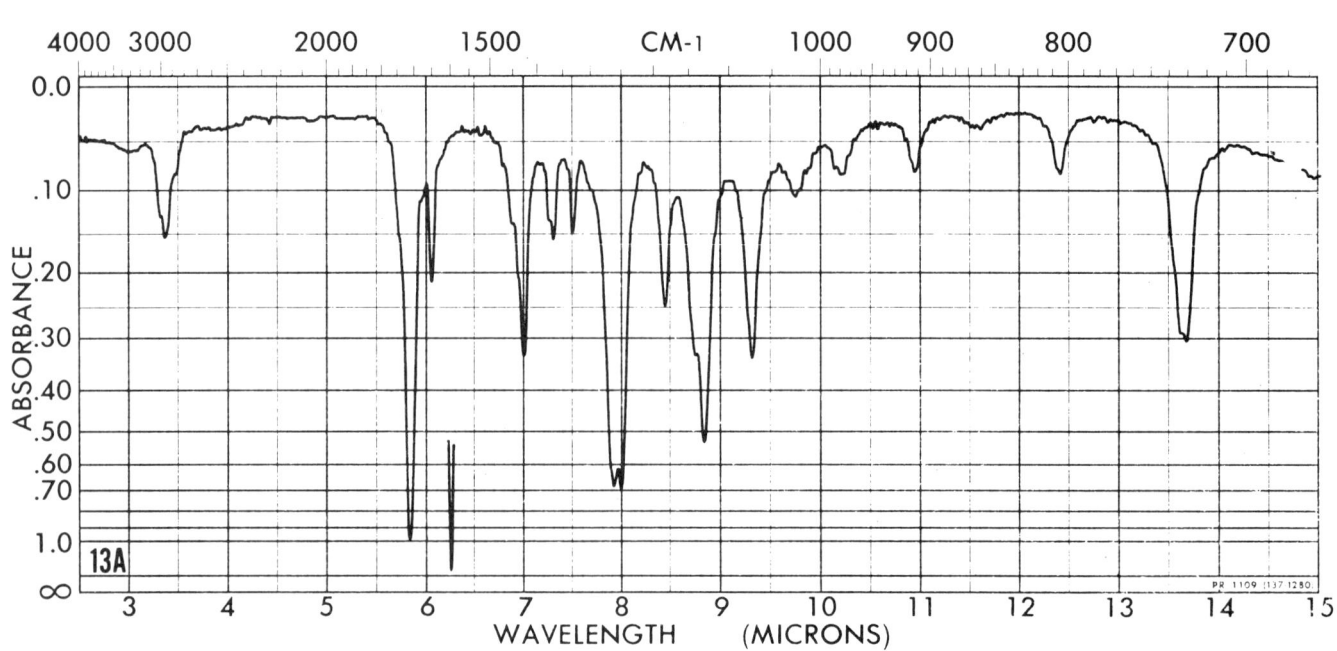

13A
10X

13A
4000Hz (200)ppm 3000 2000 1000
2000 (100) 1500 1000 500 0
1000 (50) 750 500 250 0
800 (40) 600 400 200 0
600 (30) 450 300 150 0
200 190 180 170 160 150 140 130 120 110 100 90 80 70 60 50 40 30 20 10 0
δ(PPM)

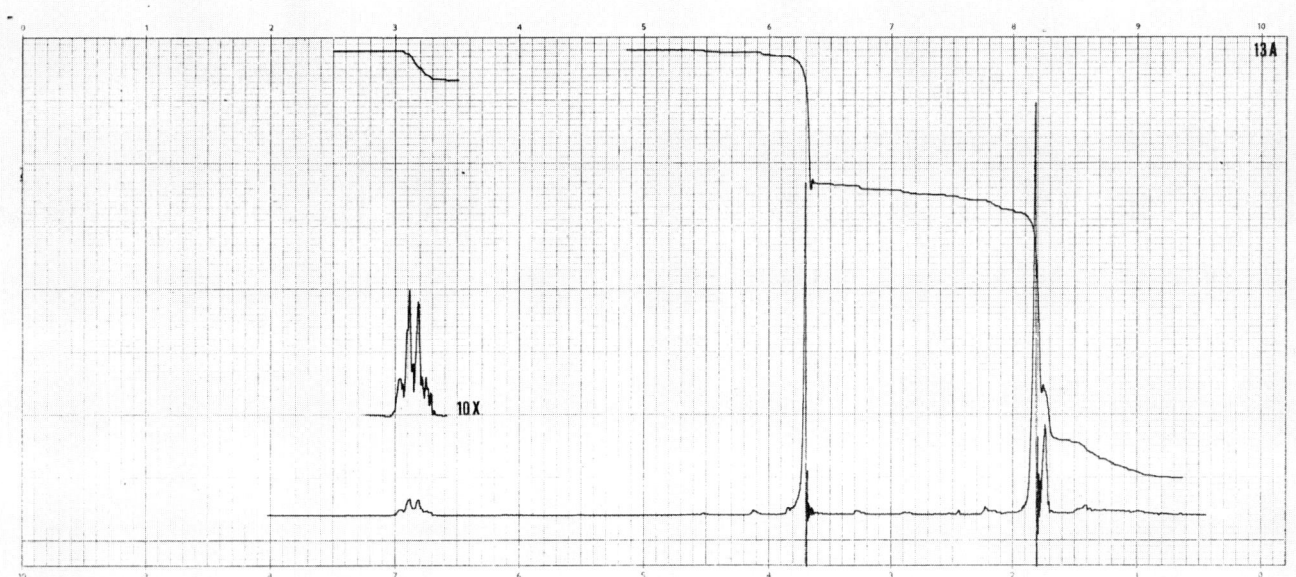

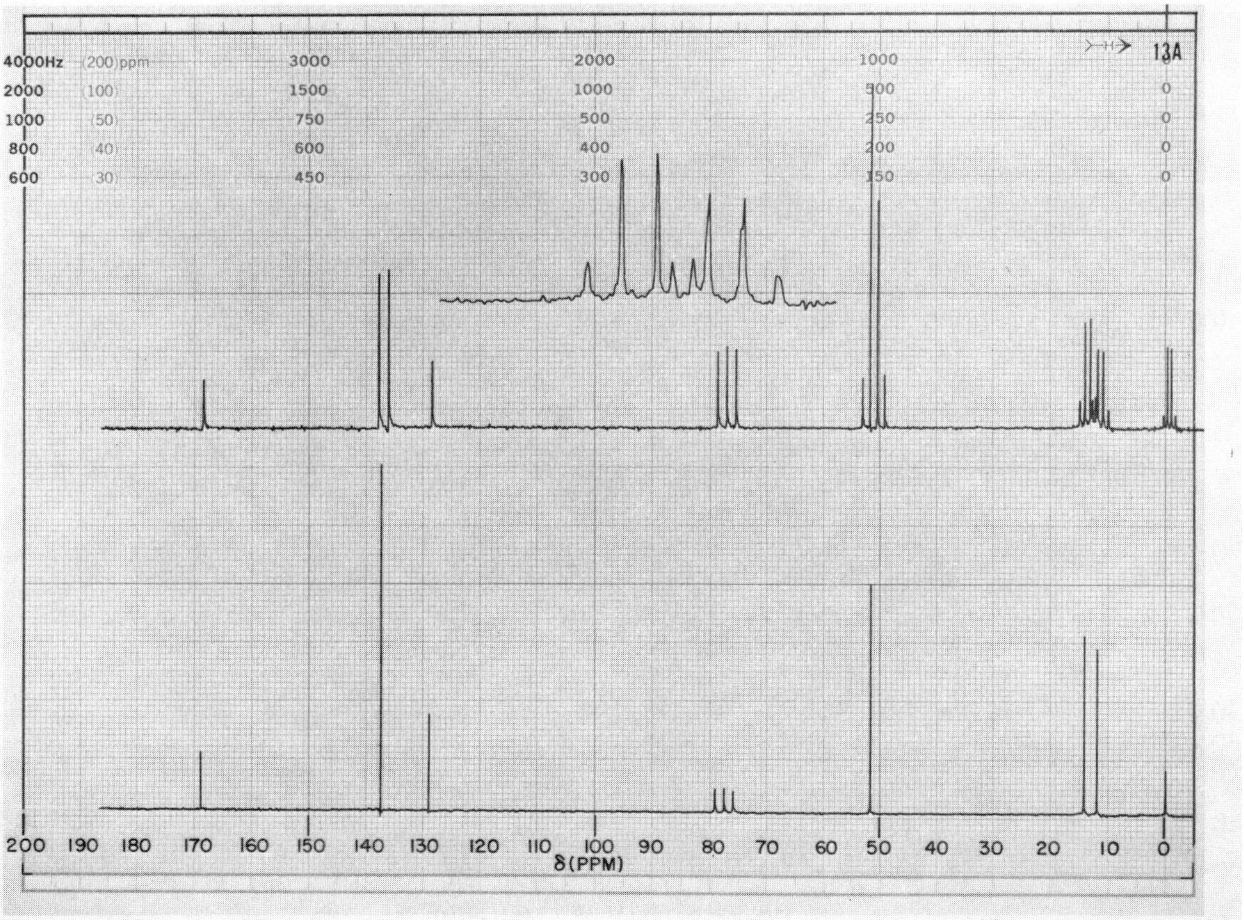

COMPOUND 13B

Compound 13B

MS: 114.070
IR: neat
HMR: CDCl$_3$
CMR: CDCl$_3$
Analysis: 63.3% C; 8.9% H

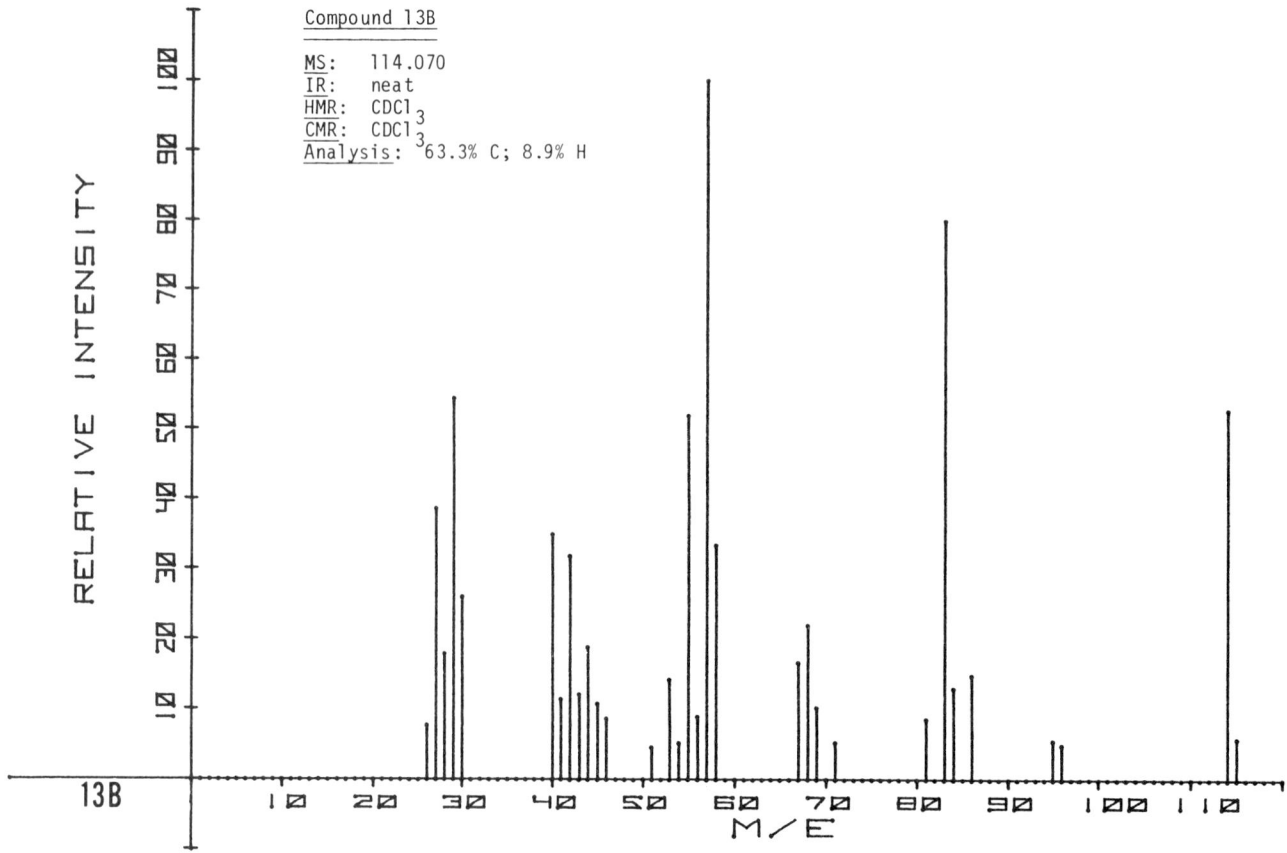

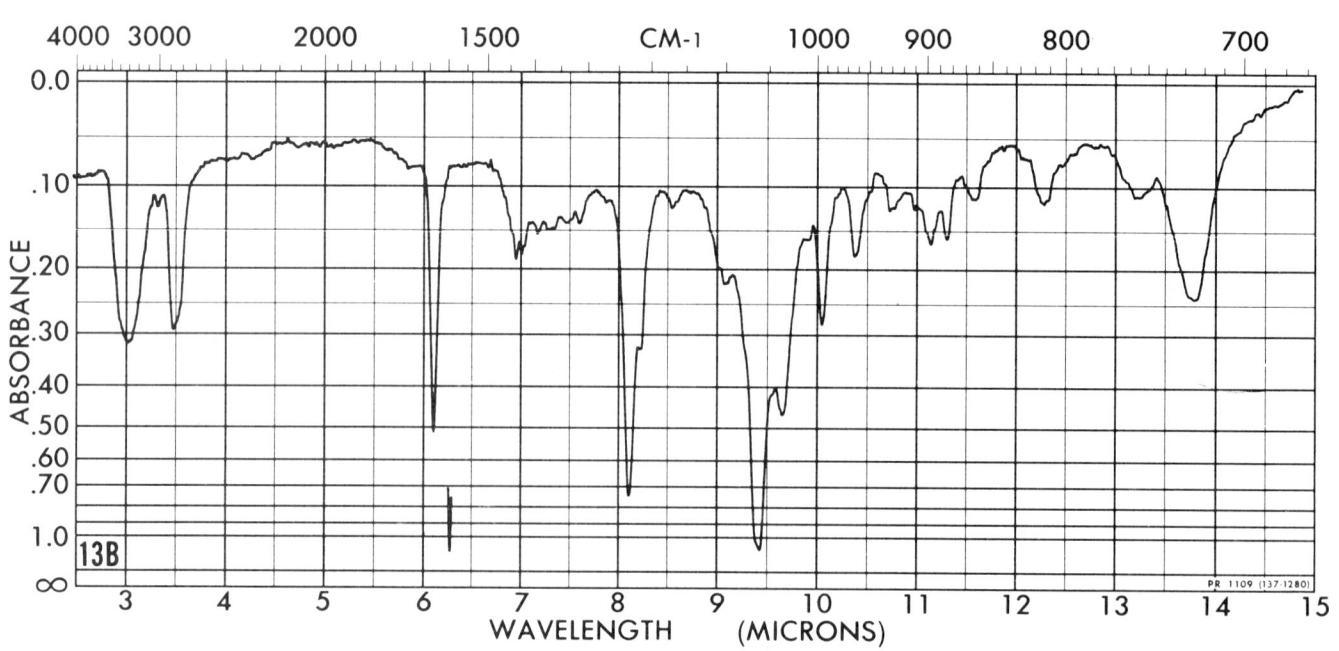

13B
D_2O

13B
4000Hz
2000
1000
800
600
3000
1500
750
600
450
2000
1000
500
400
300
1000
500
250
200
150
200 190 180 170 160 150 140 130 120 110 100 90 80 70 60 50 40 30 20 10 0
δ(PPM)

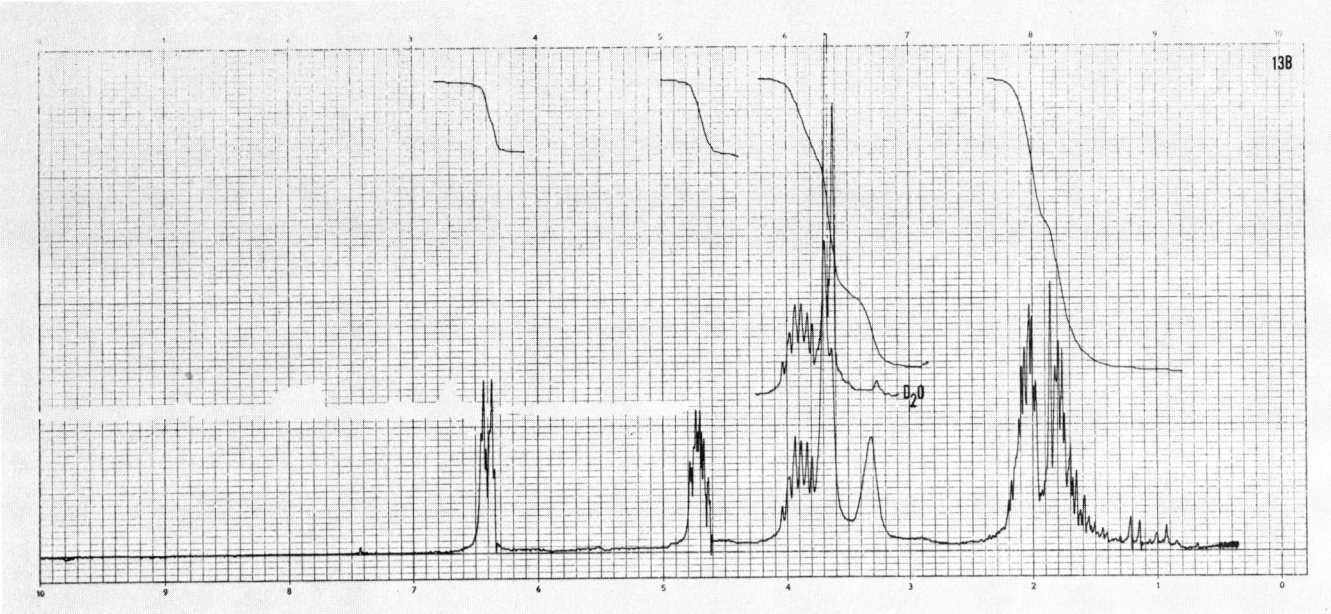

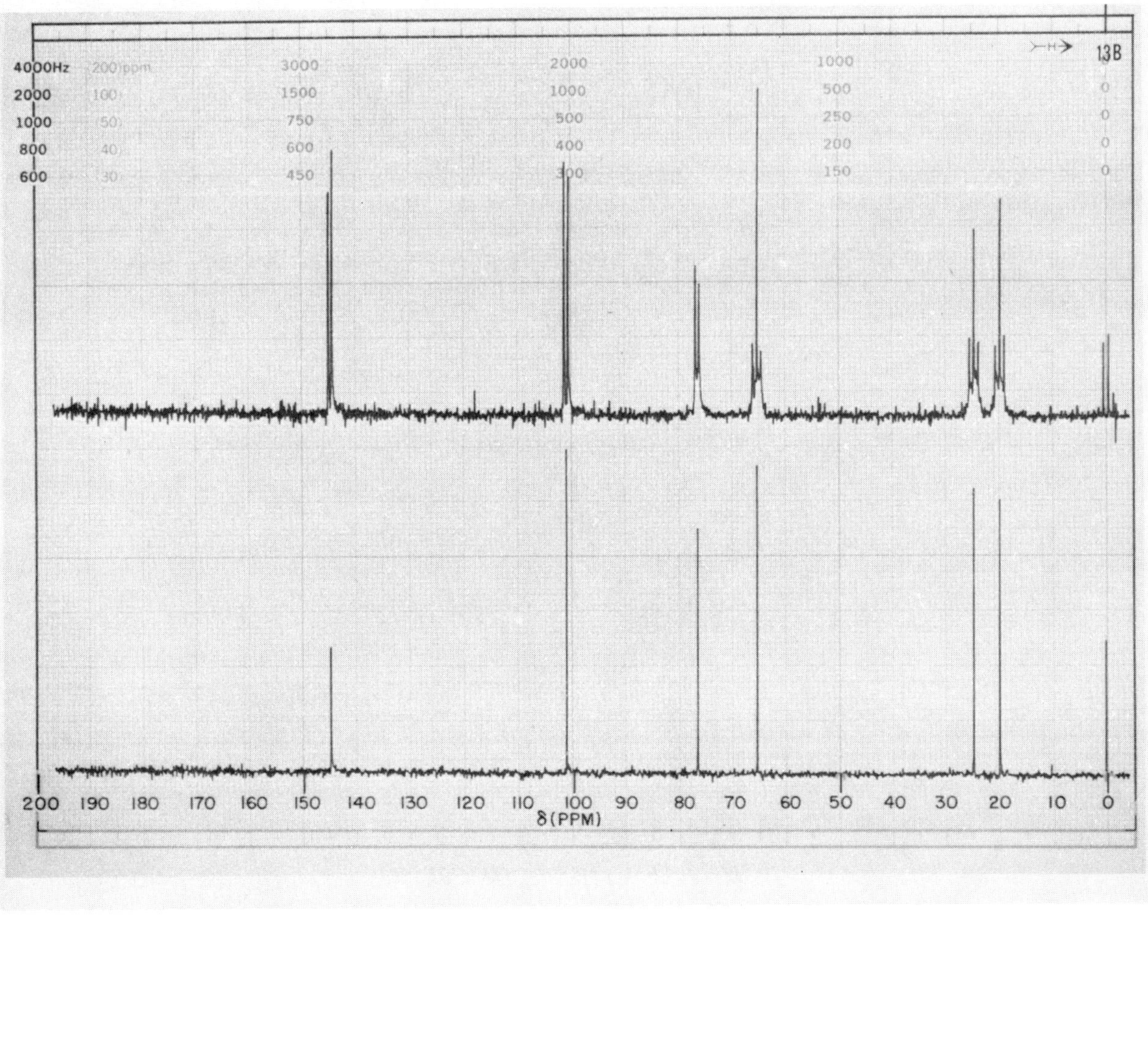

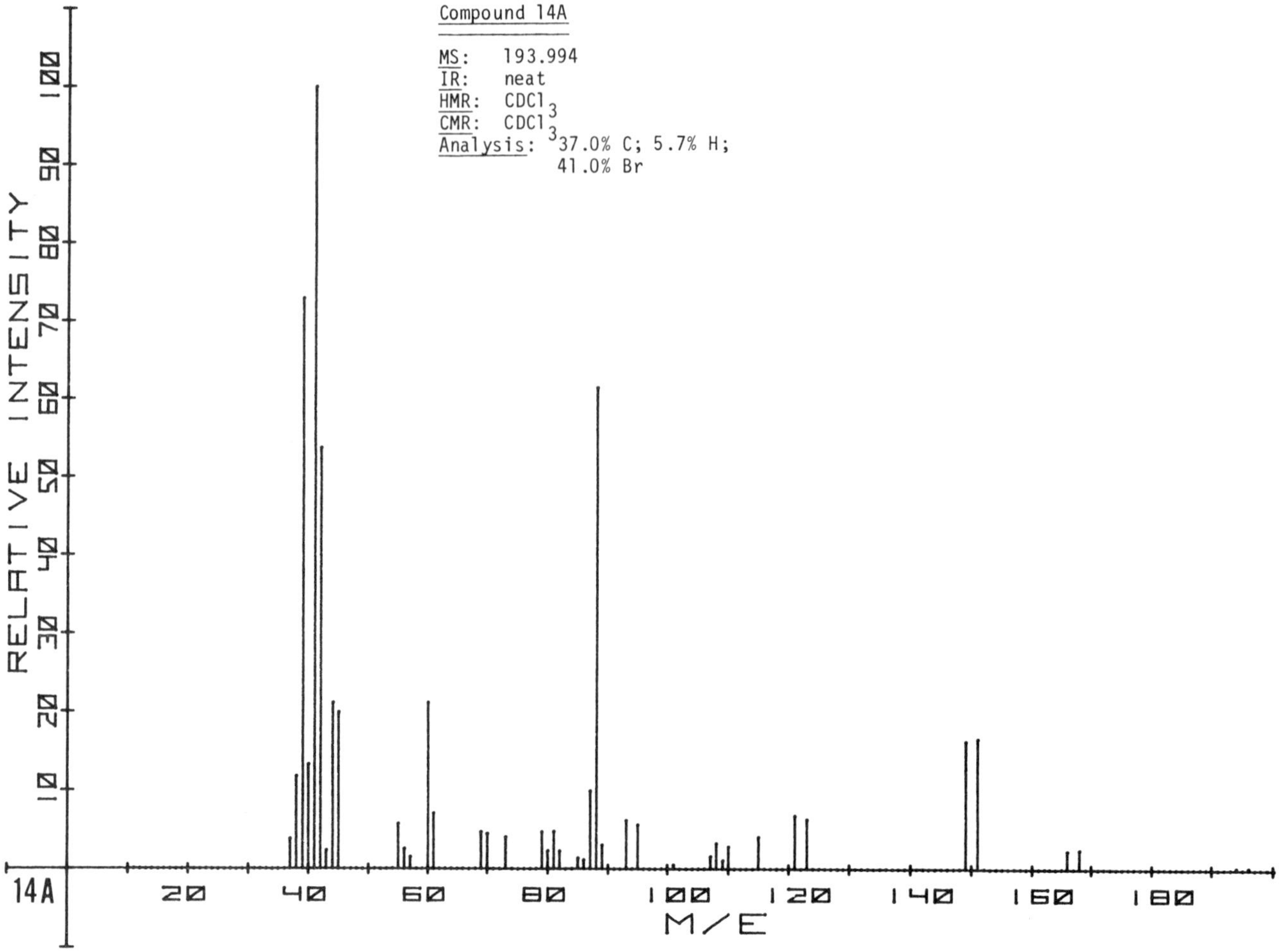

Compound 14A
MS: 193.994
IR: neat
HMR: $CDCl_3$
CMR: $CDCl_3$
Analysis: 37.0% C; 5.7% H; 41.0% Br
RELATIVE INTENSITY
10 20 30 40 50 60 70 80 90 100
14A
20 40 60 80 100 120 140 160 180
M/E

4000 3000 2000 1500 CM-1 1000 900 800 700
ABSORBANCE
0.0 .10 .20 .30 .40 .50 .60 .70 1.0 ∞
14A
3 4 5 6 7 8 9 10 11 12 13 14 15
WAVELENGTH (MICRONS)

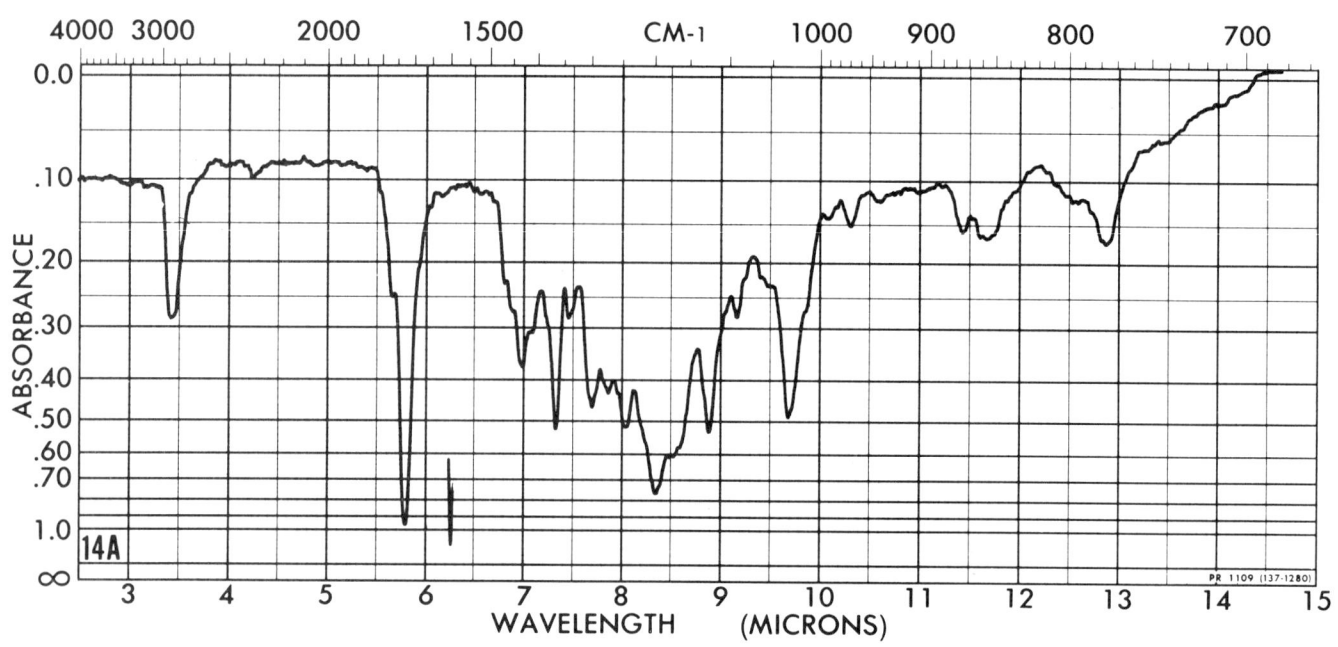

14A
IRR. 2.25
A
A

14A
4000Hz (200)ppm
2000 (100)
1000 (50)
800 (40)
600 (30)
PK. HT. PPM
13 34.5
12 34.2
43 32.8
40 32.5
24 31.2
11 30.9
12 29.7
32 28.0
10 26.3
PK. HT. PPM
20 172.3
48 60.5
53 32.8
60 32.5
49 28.0
39 14.3
δ(PPM)

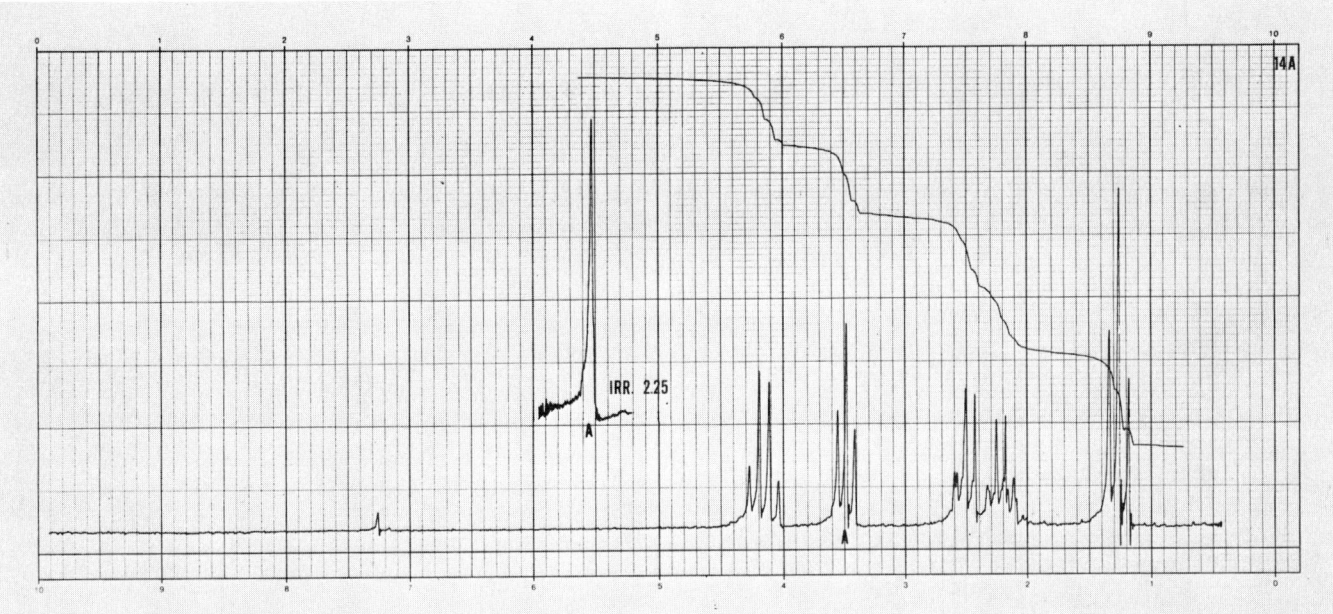

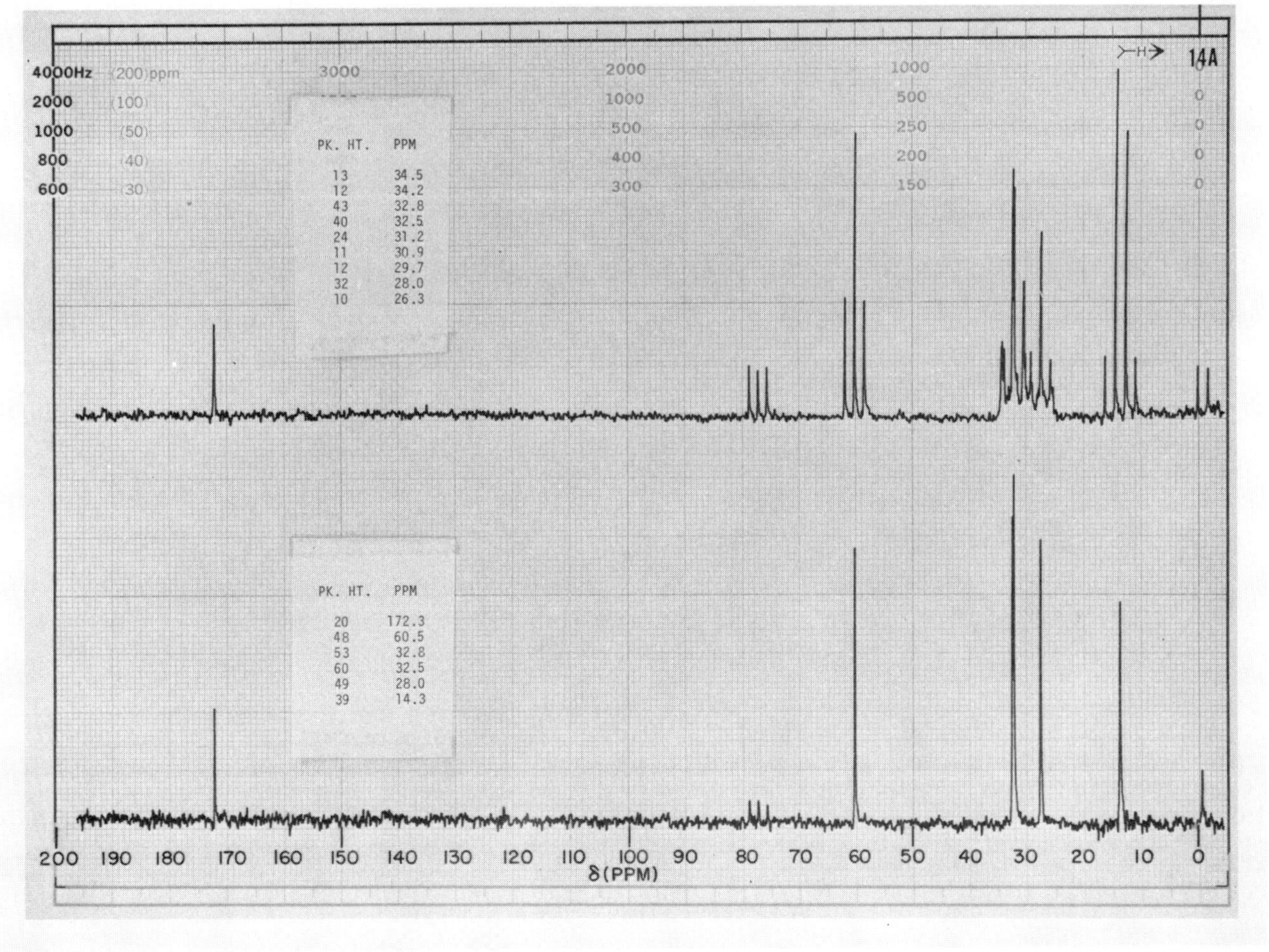

COMPOUND 14B

Compound 14B

MS: 193.994
IR: neat
HMR: CDCl$_3$
CMR: neat
Analysis: 36.9% C; 5.7% H; 41.0% Br

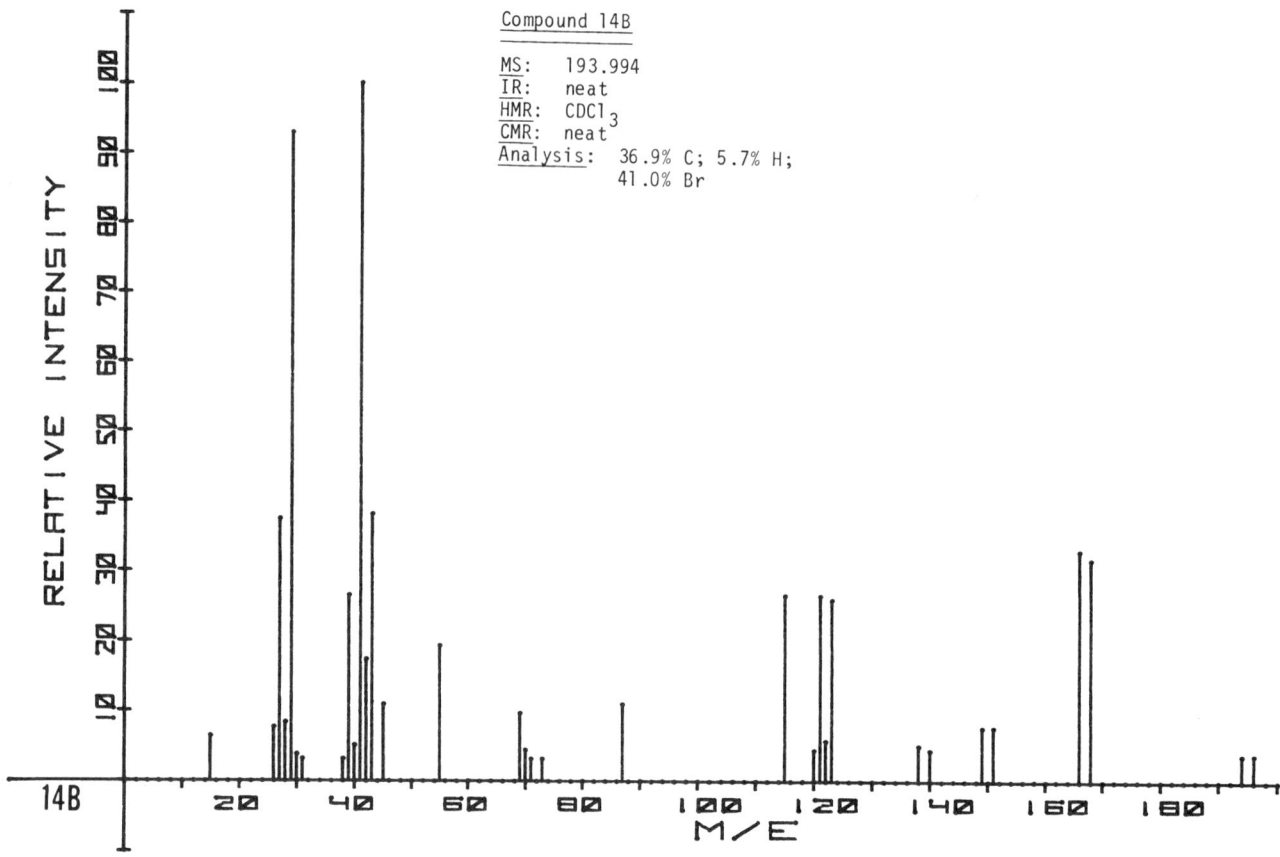

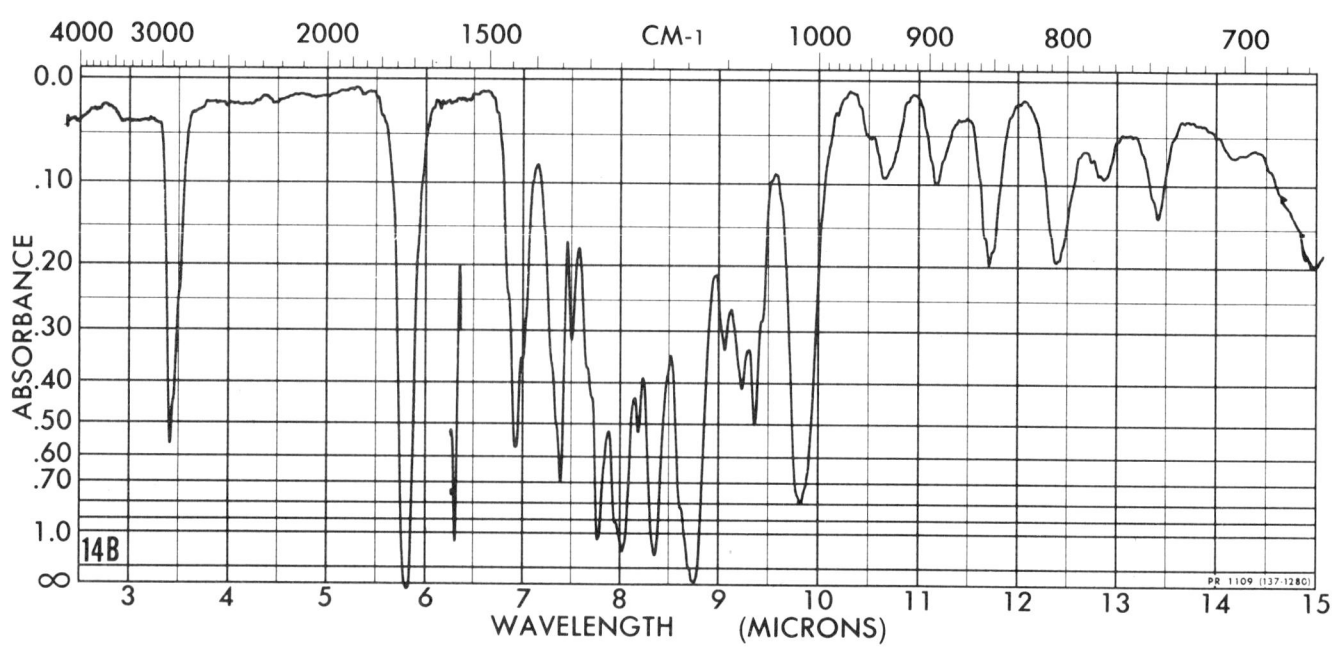

14B
IRR. 1.30
IRR. 2.05
C
C
A
B
A B
IRR. 2.05
10 9 8 7 6 5 4 3 2 1 0

14B
4000Hz (200)ppm 3000 2000 1000
2000 (100) 1500 1000 500
1000 (50) 500 250
800 (40) 400 200
600 (30) 300 150
PK. HT. PPM
11 16.4
49 15.6
57 14.8
7 14.2
14 14.1
34 13.4
32 12.6
11 11.8
PK. HT. PPM
10 170.3
46 62.8
49 49.1
59 29.7
34 15.2
60 13.0
200 190 180 170 160 150 140 130 120 110 100 90 80 70 60 50 40 30 20 10 0
δ(PPM)

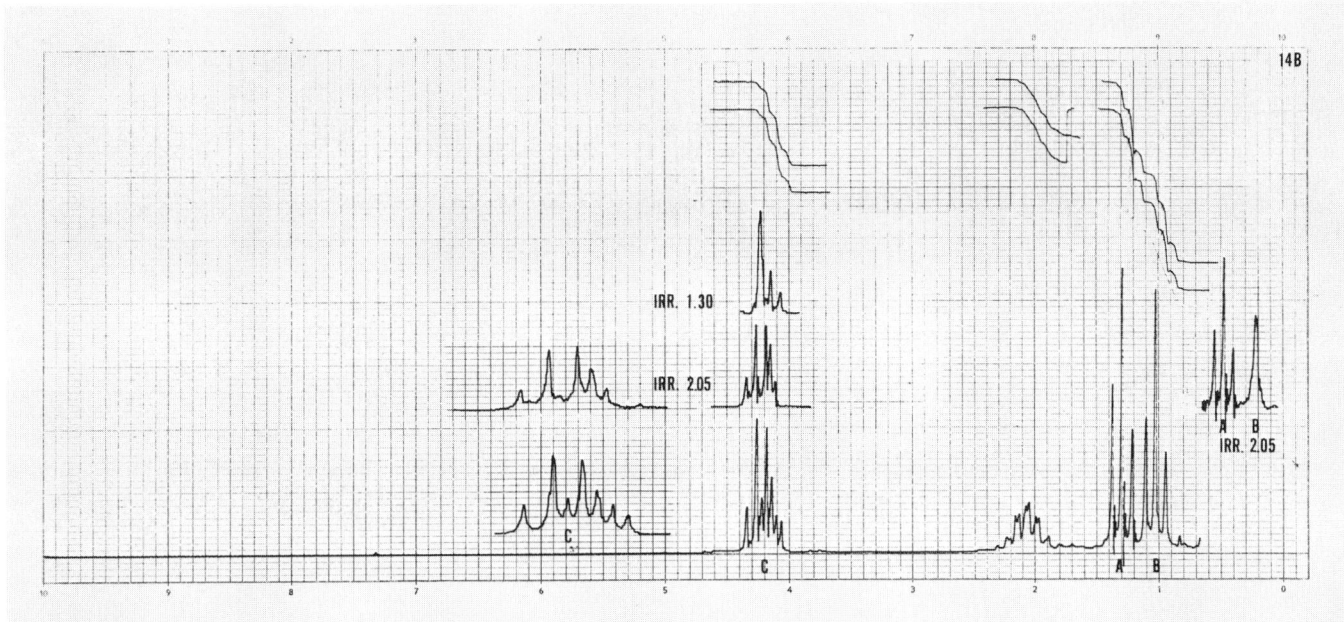

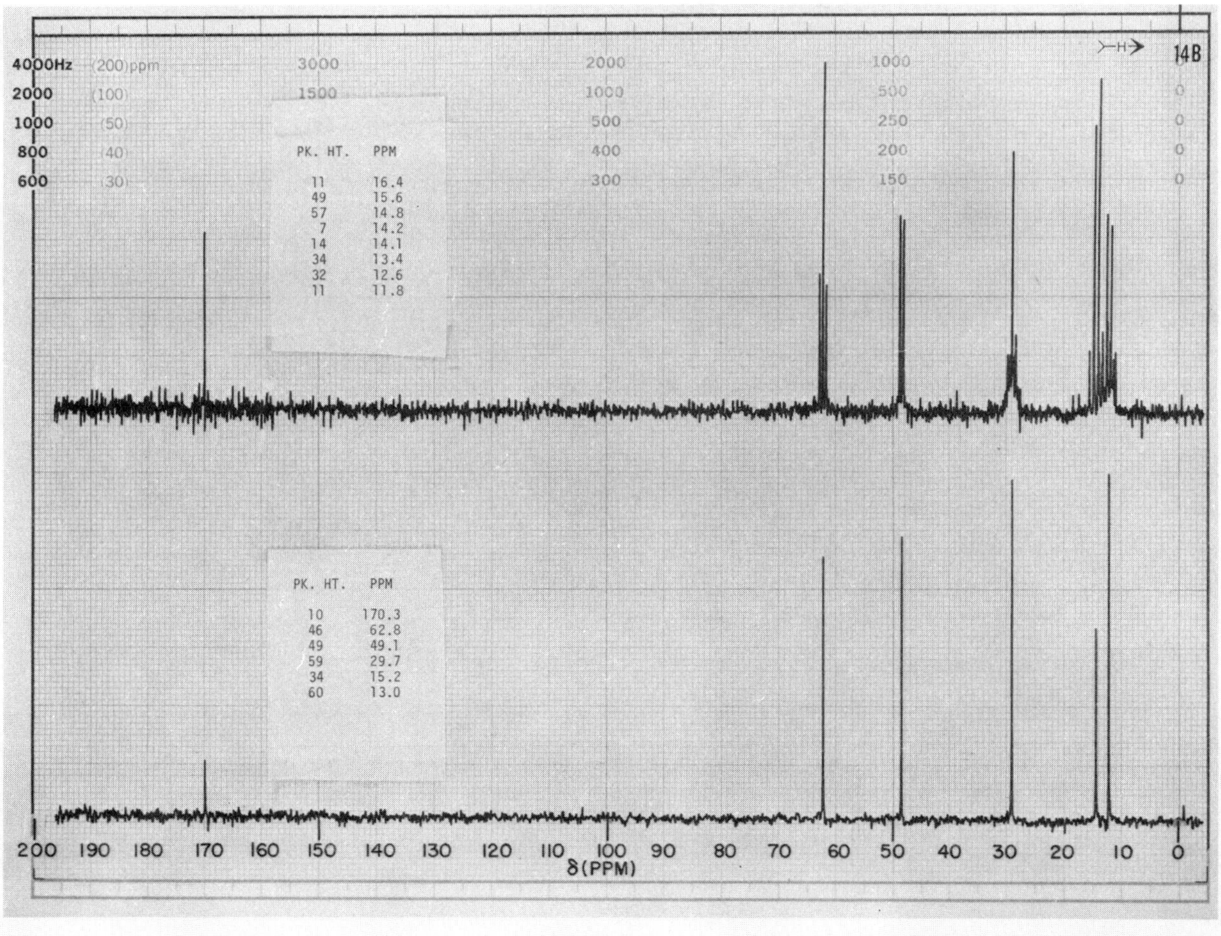

COMPOUND 15

Compound 15
MS: 113.085
IR: CHCl$_3$
HMR: CDCl$_3$
CMR: d$_6$ acetone
Analysis: 63.7% C; 9.8% H; 12.5% N

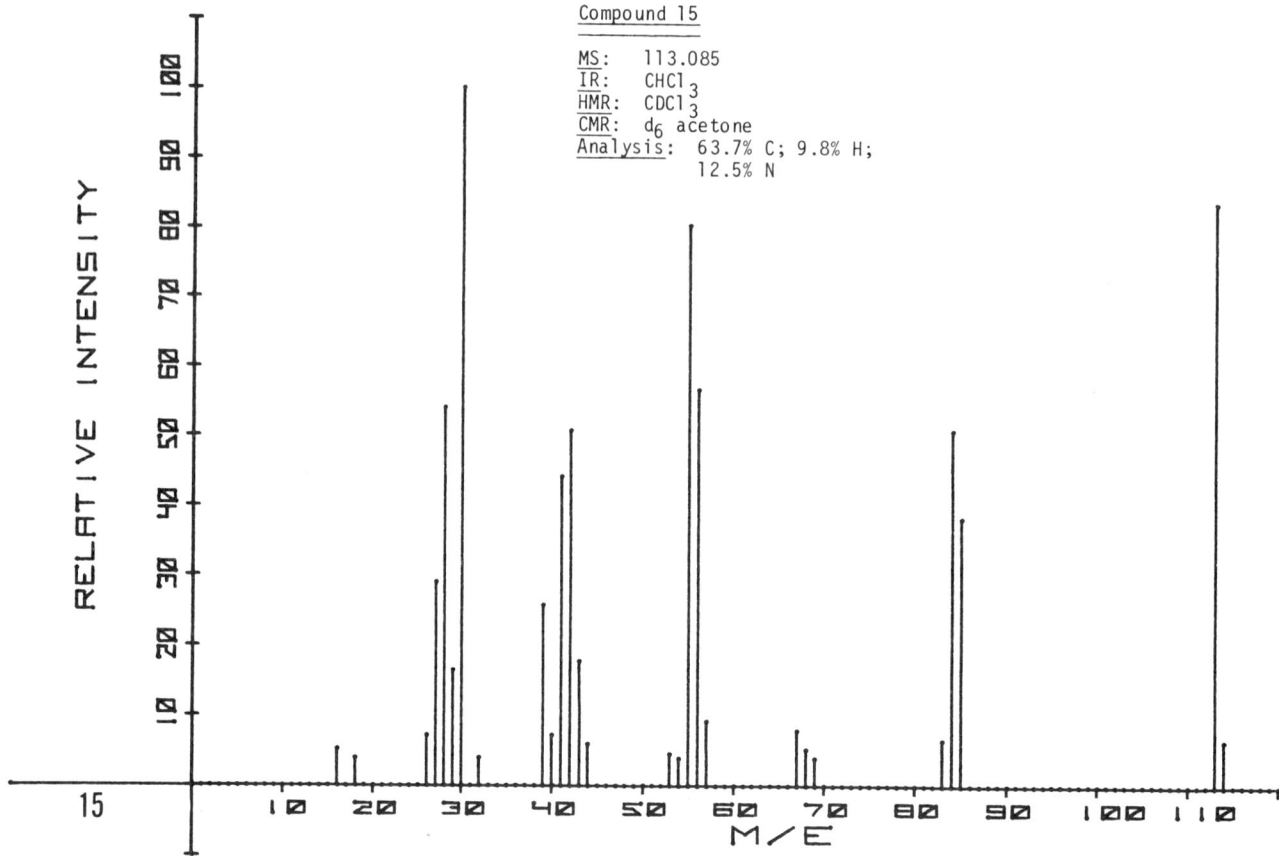

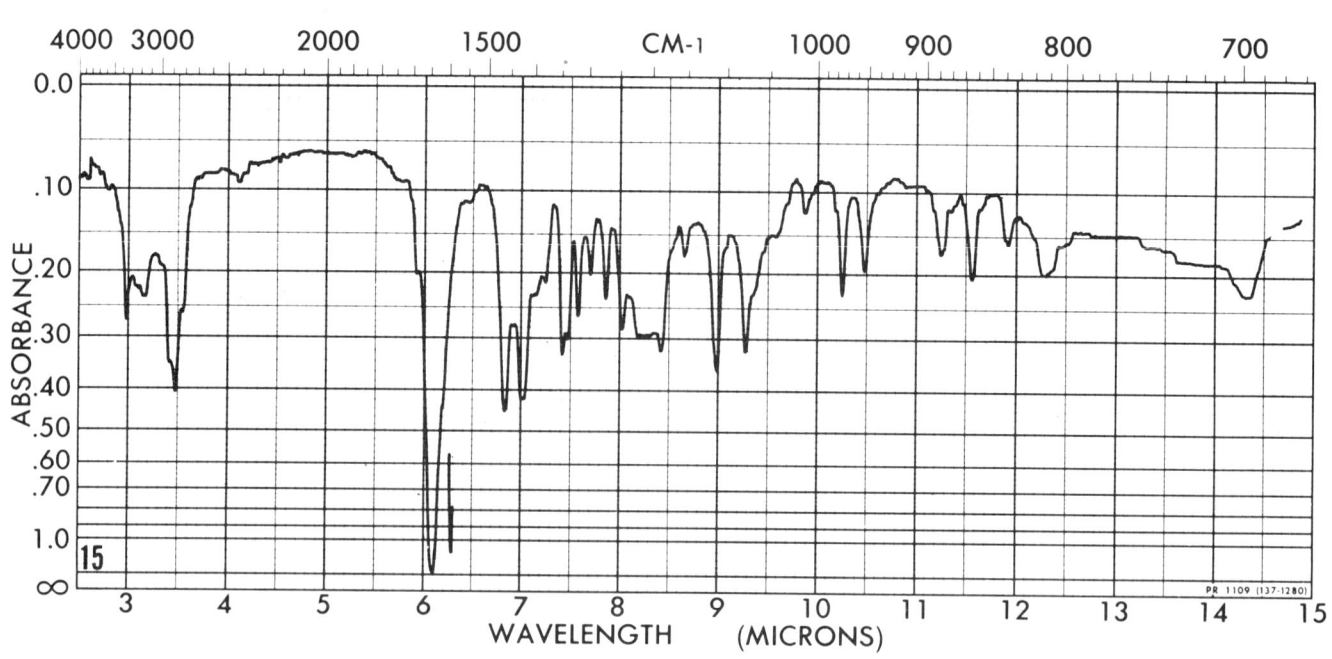

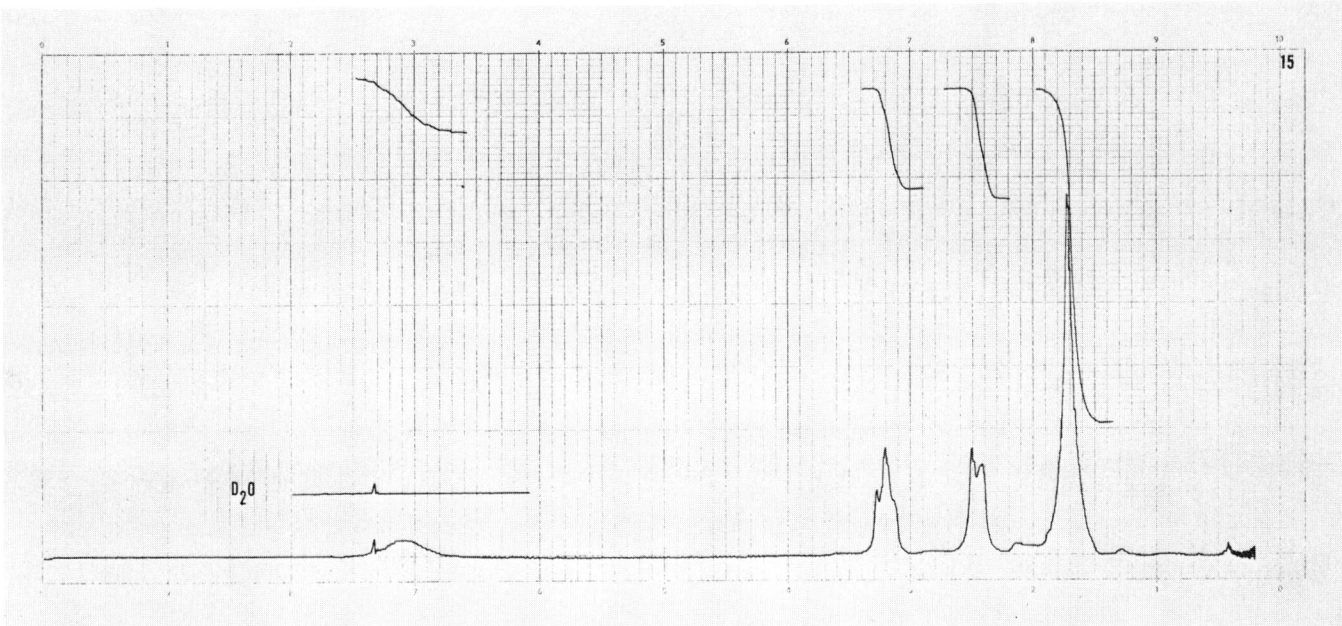

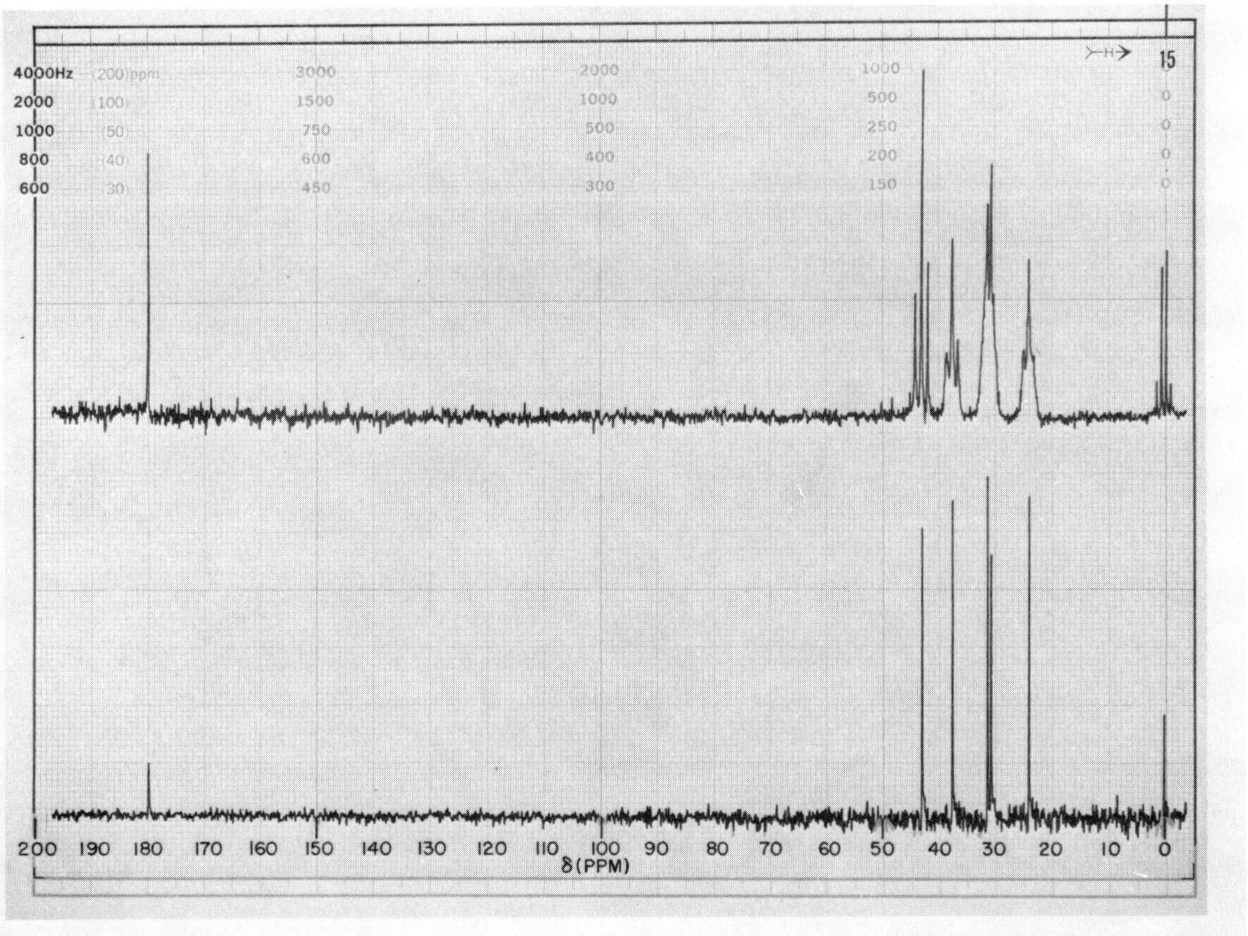

COMPOUND 16

Compound 16
MS: 117.055
IR: neat
HMR: CDCl$_3$
CMR: CDCl$_3$
Analysis: 54.5% C; 9.1% H

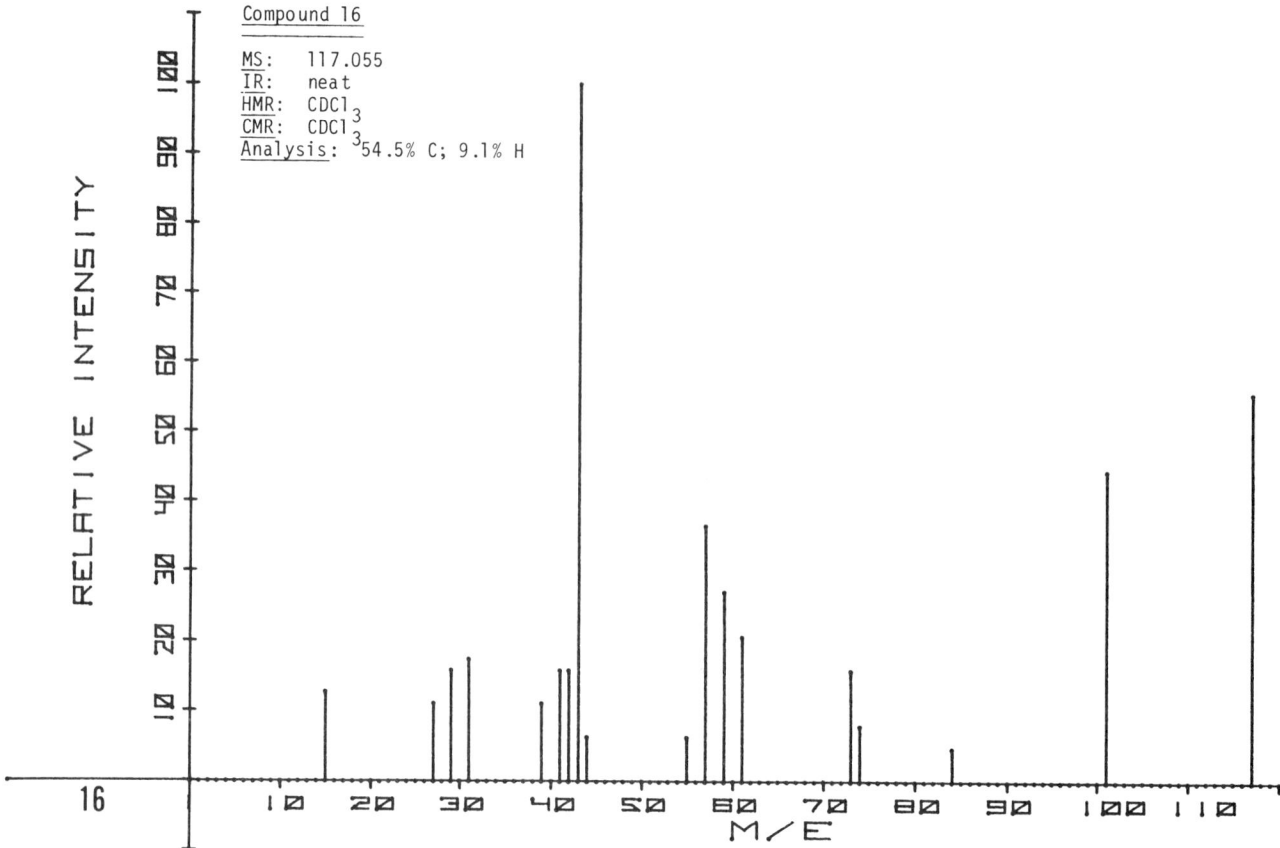

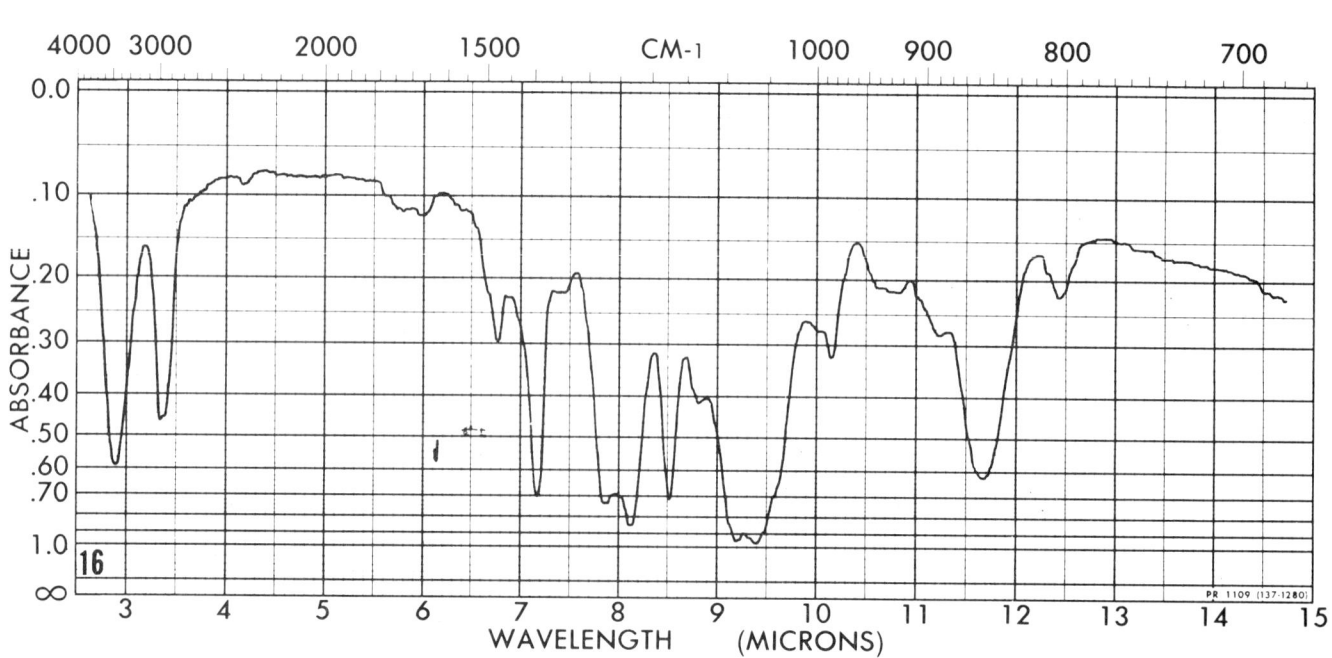

16
4
3
D2O

16
PK. HT. PPM
19 77.4
33 75.4
11 67.7
27 66.1
16 64.5
18 64.4
39 62.9
10 61.4
11 28.3
52 27.1
14 26.9
35 25.9
60 25.7
9 24.7
44 24.5
9 23.3
PK. HT. PPM
60 108.9
49 76.4
49 66.1
53 62.8
27 26.5
22 25.1
δ(PPM)

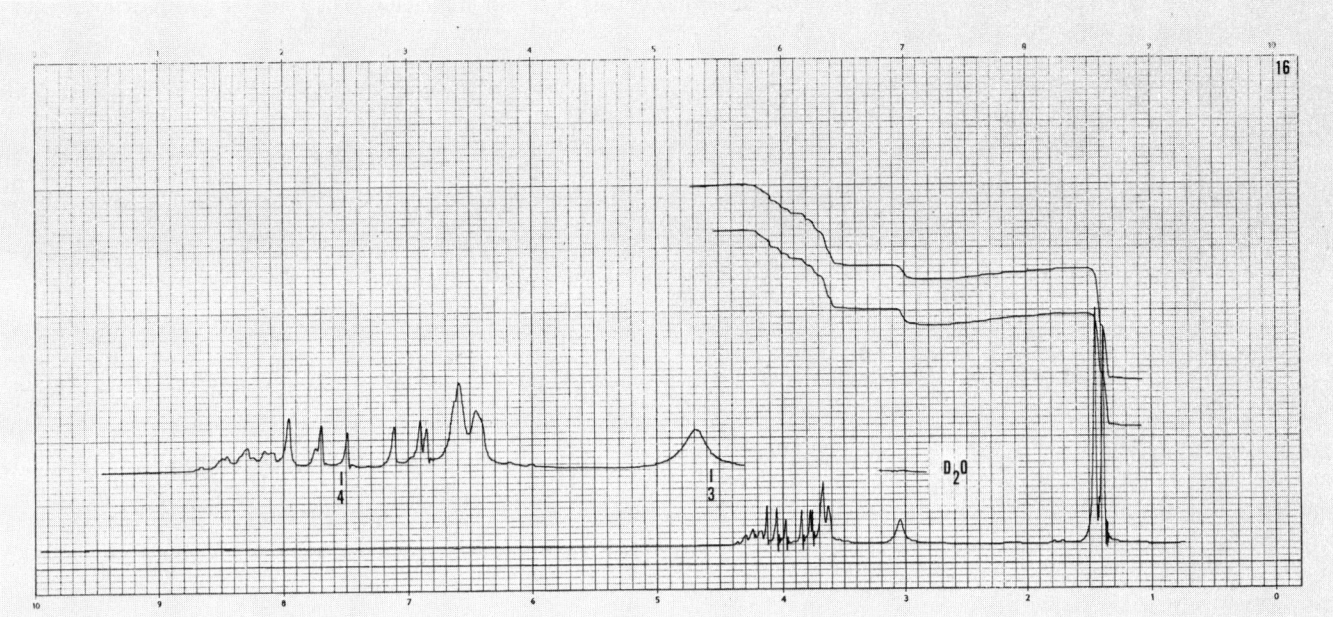

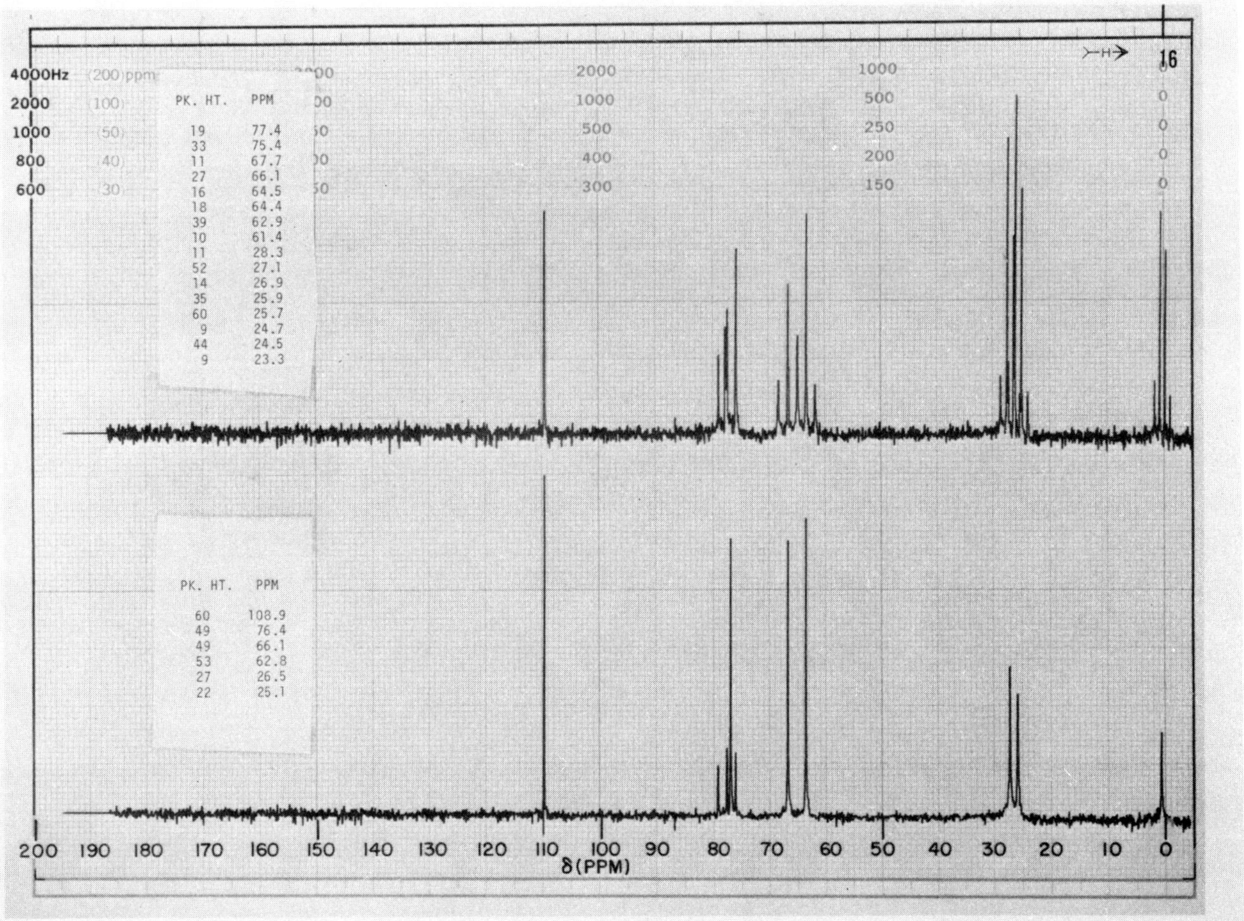

COMPOUND 17

Compound 17
MS: 196.009
IR: neat
HMR: CDCl$_3$
CMR: CDCl$_3$
Analysis: 36.6% C; 6.7% H; 40.6% Br

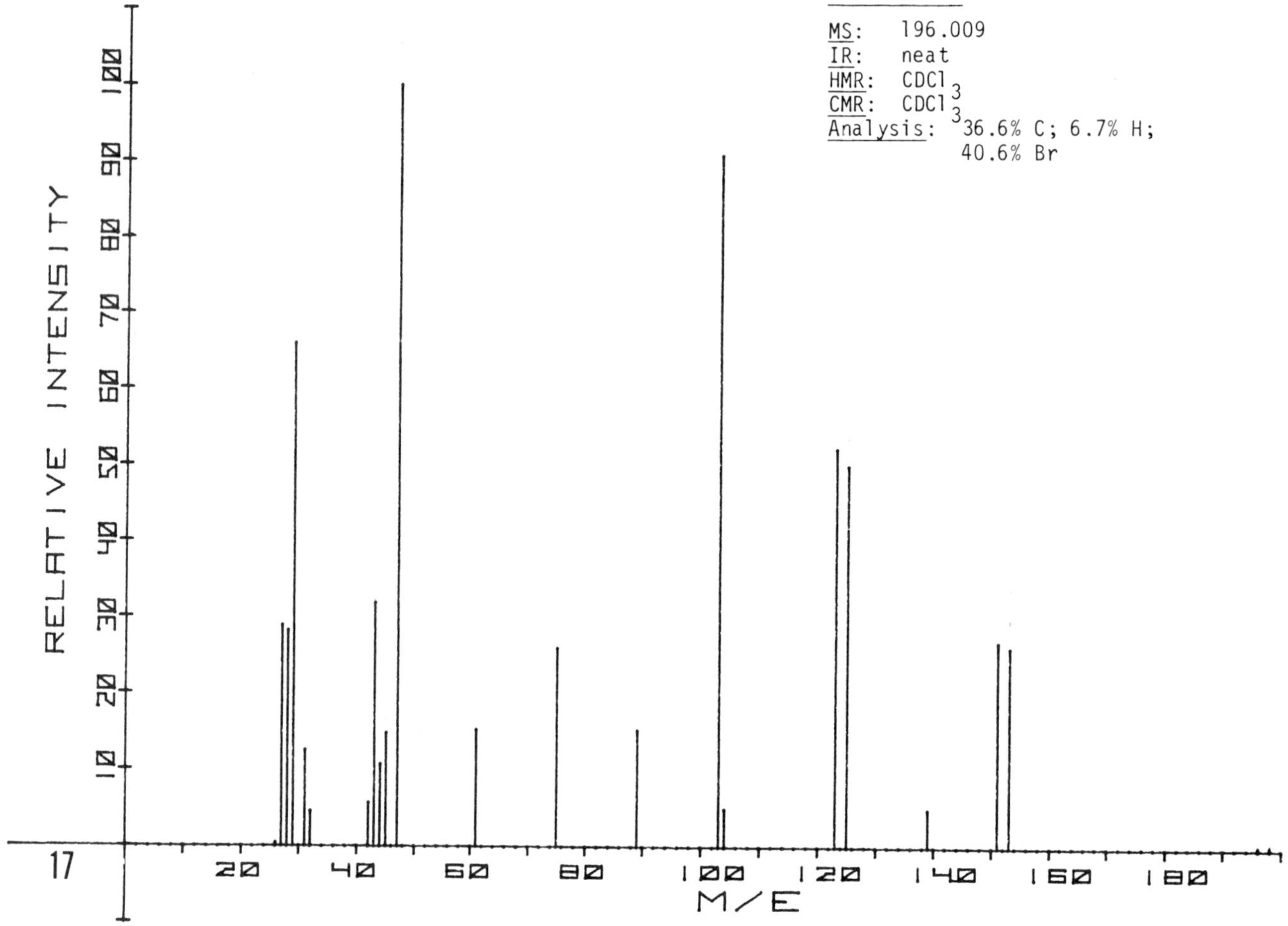

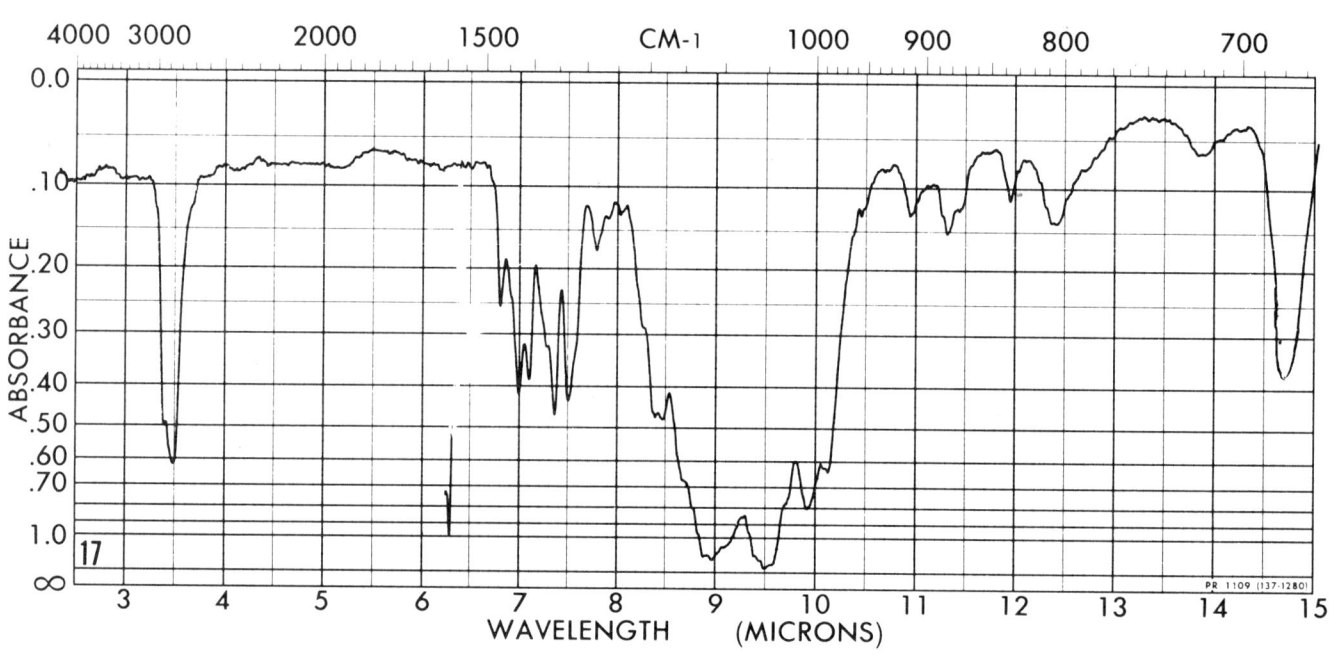

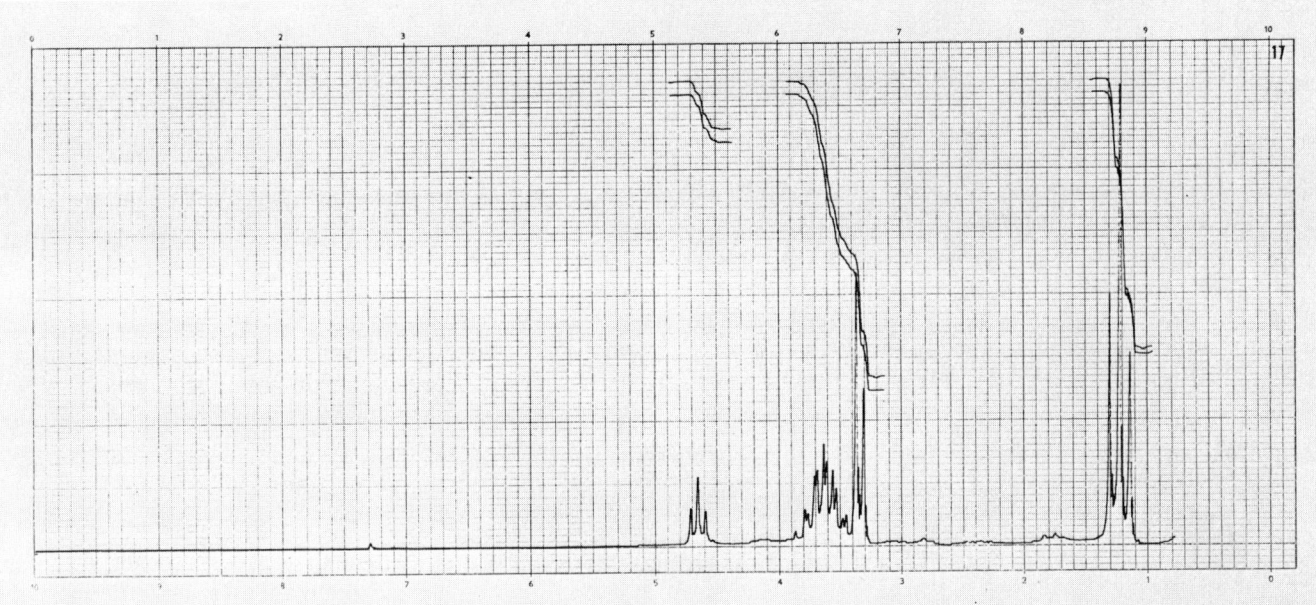

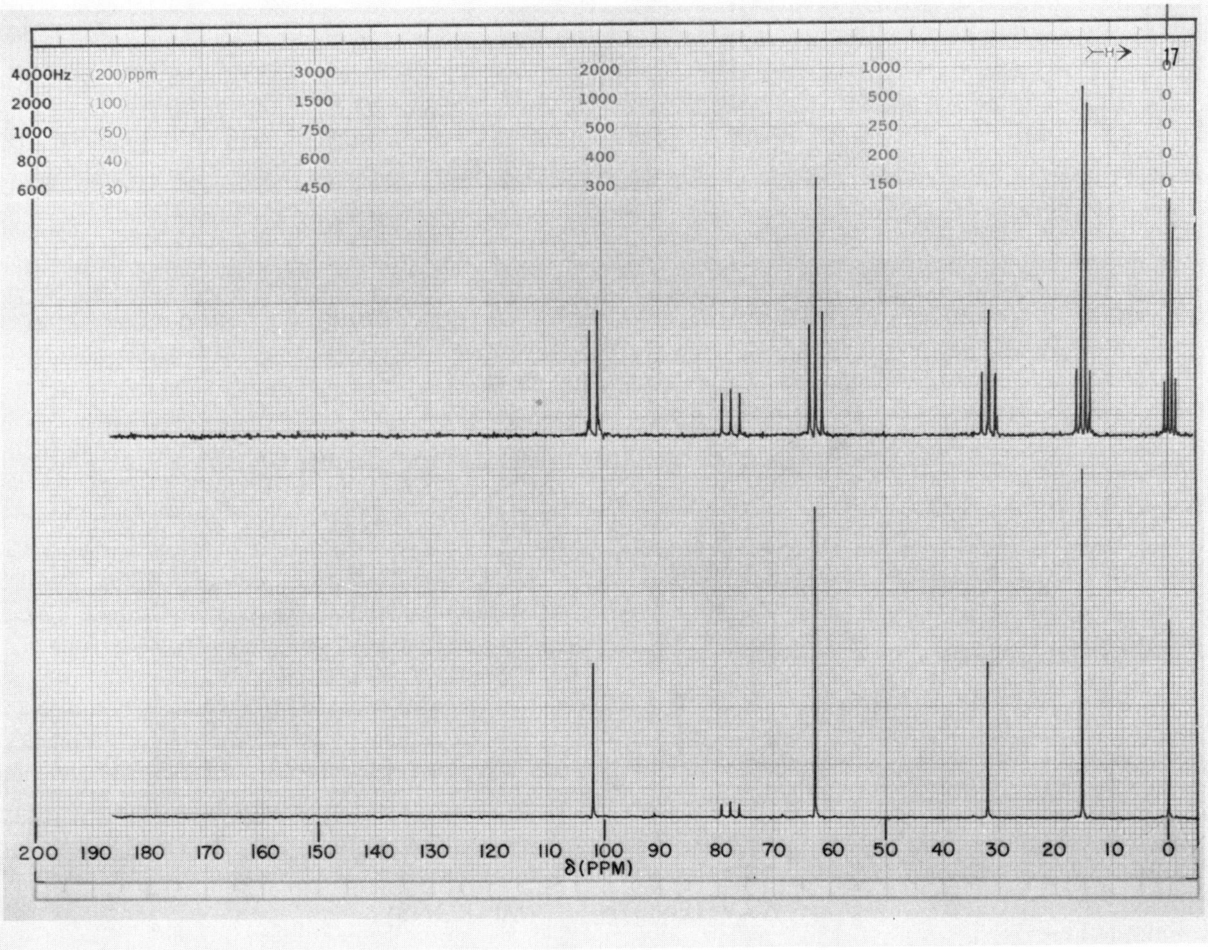

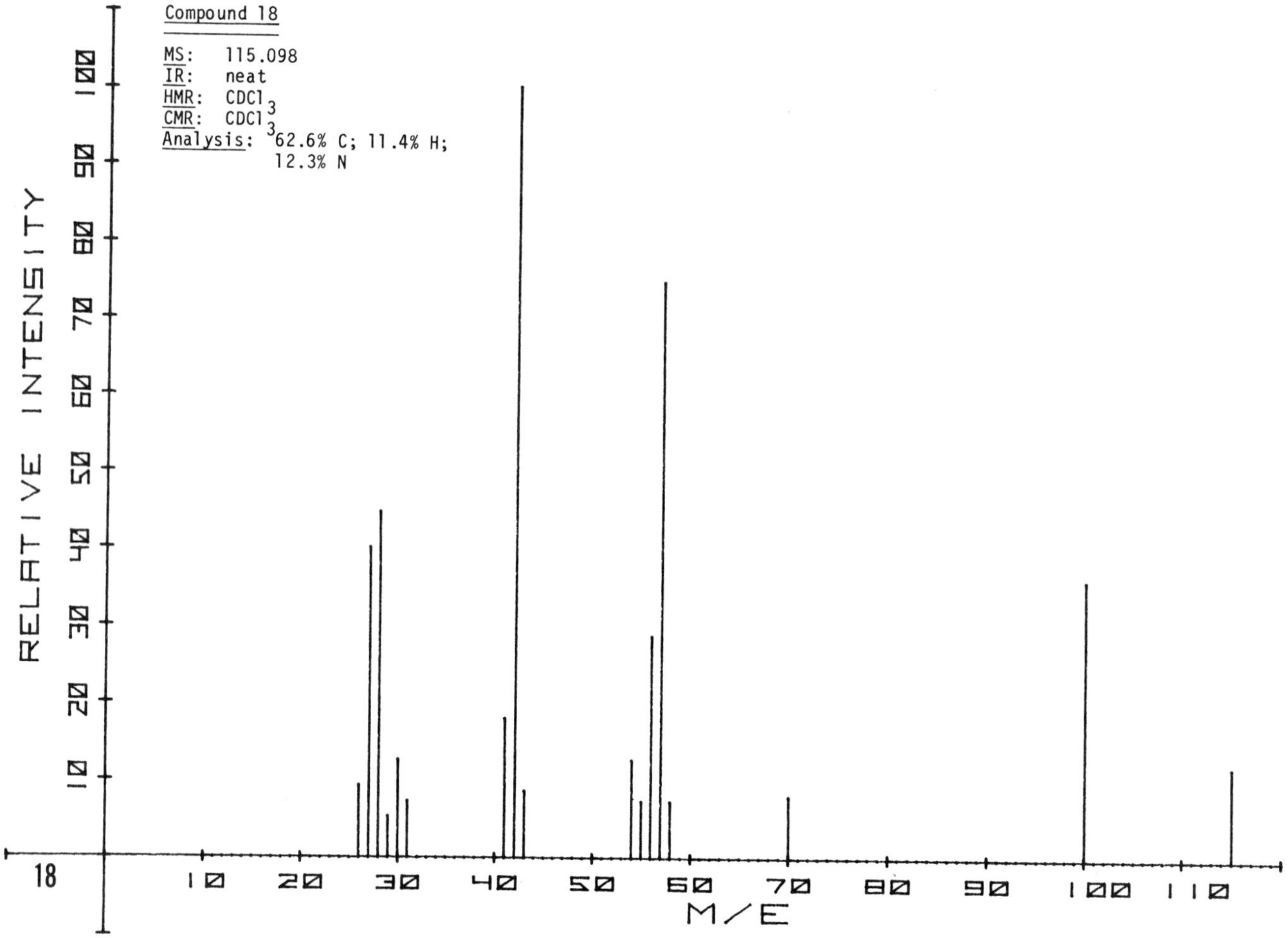

Compound 18
MS: 115.098
IR: neat
HMR: CDCl3
CMR: CDCl3
Analysis: 62.6% C; 11.4% H; 12.3% N
RELATIVE INTENSITY
M/E
18

CM-1
ABSORBANCE
WAVELENGTH (MICRONS)
18
PR 1109 (137-1280)

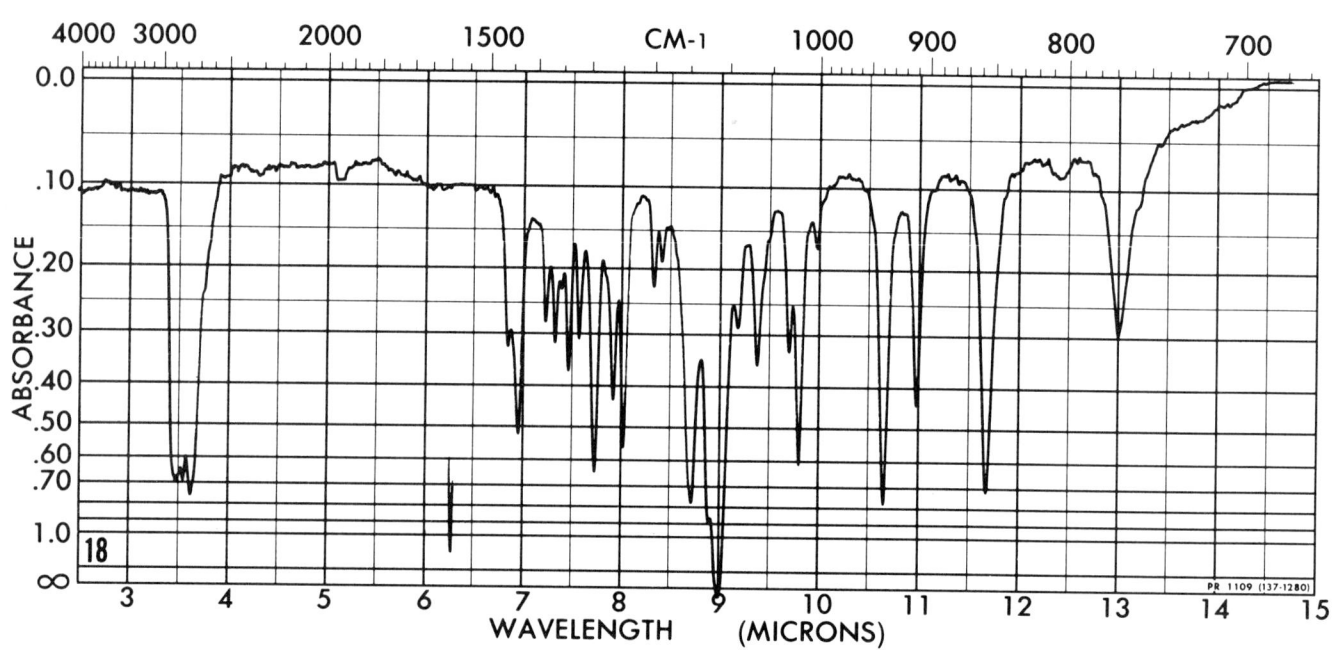

18
A
IRR. 2.45
B
IRR. 3.70
C
IRR. 2.40

18
4000Hz (200)ppm 3000 2000 1000 0
2000 (100) 1500 1000 500 0
1000 (50) 500 250 0
800 (40) 400 200 0
600 (30) 300 150 0
PK. HT. PPM
22 55.2
9 54.4
51 53.4
25 52.7
20 51.7
11 51.1
PK. HT. PPM
60 66.9
54 53.4
34 52.7
20 11.8
200 190 180 170 160 150 140 130 120 110 100 90 80 70 60 50 40 30 20 10 0
δ(PPM)

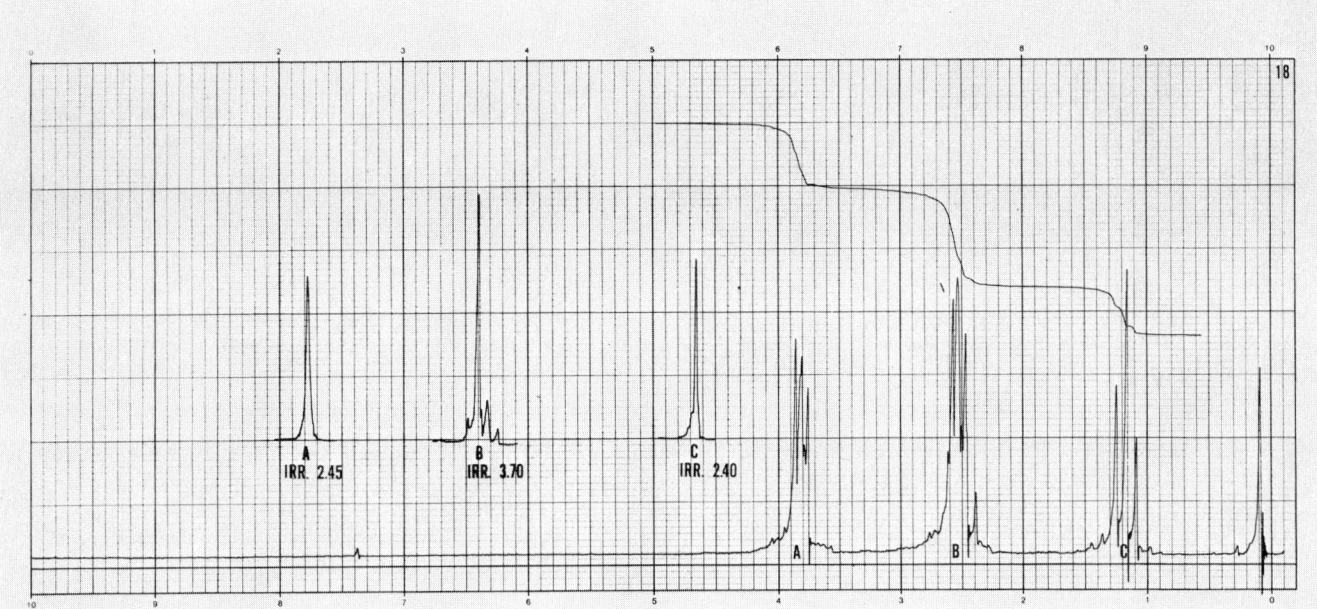

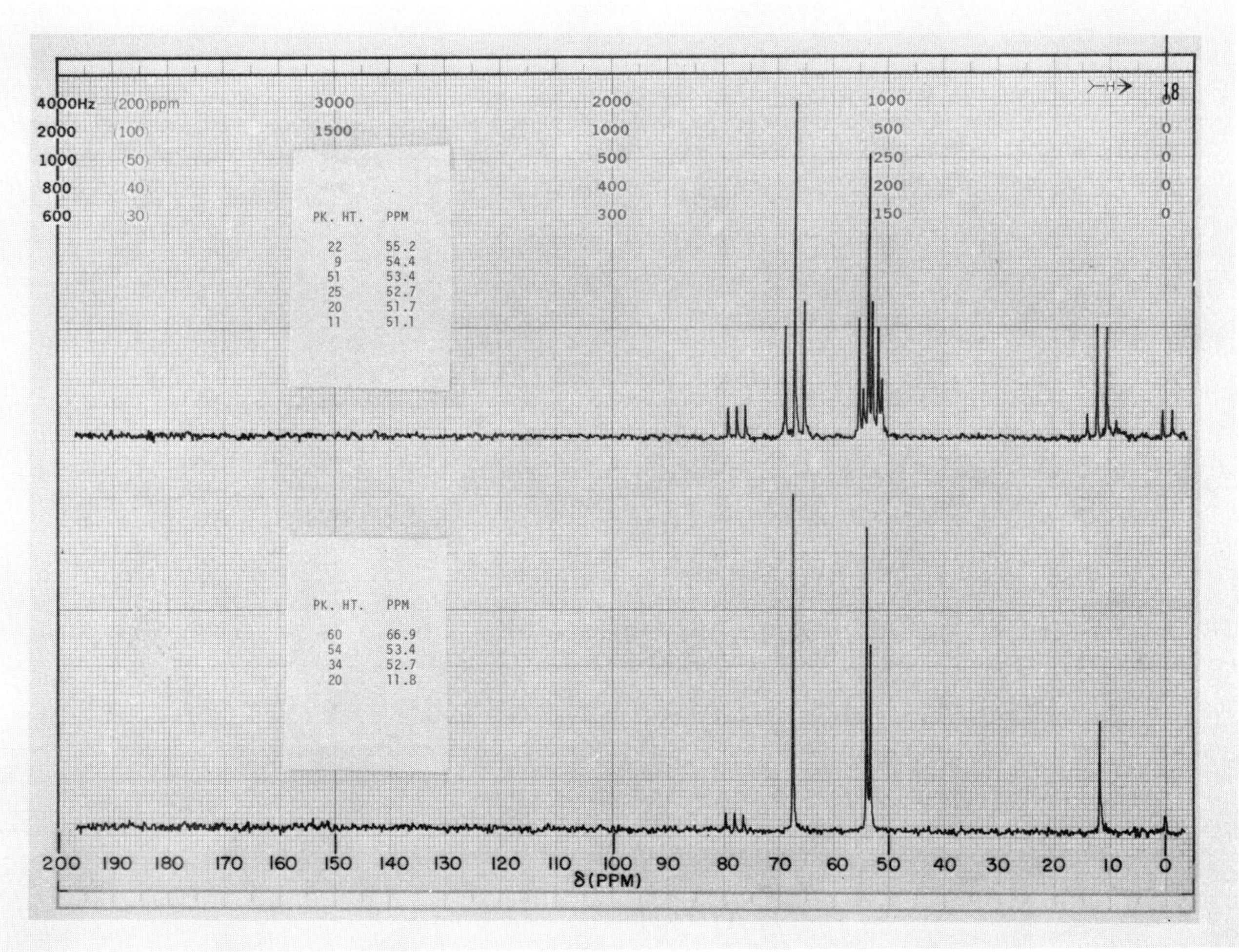

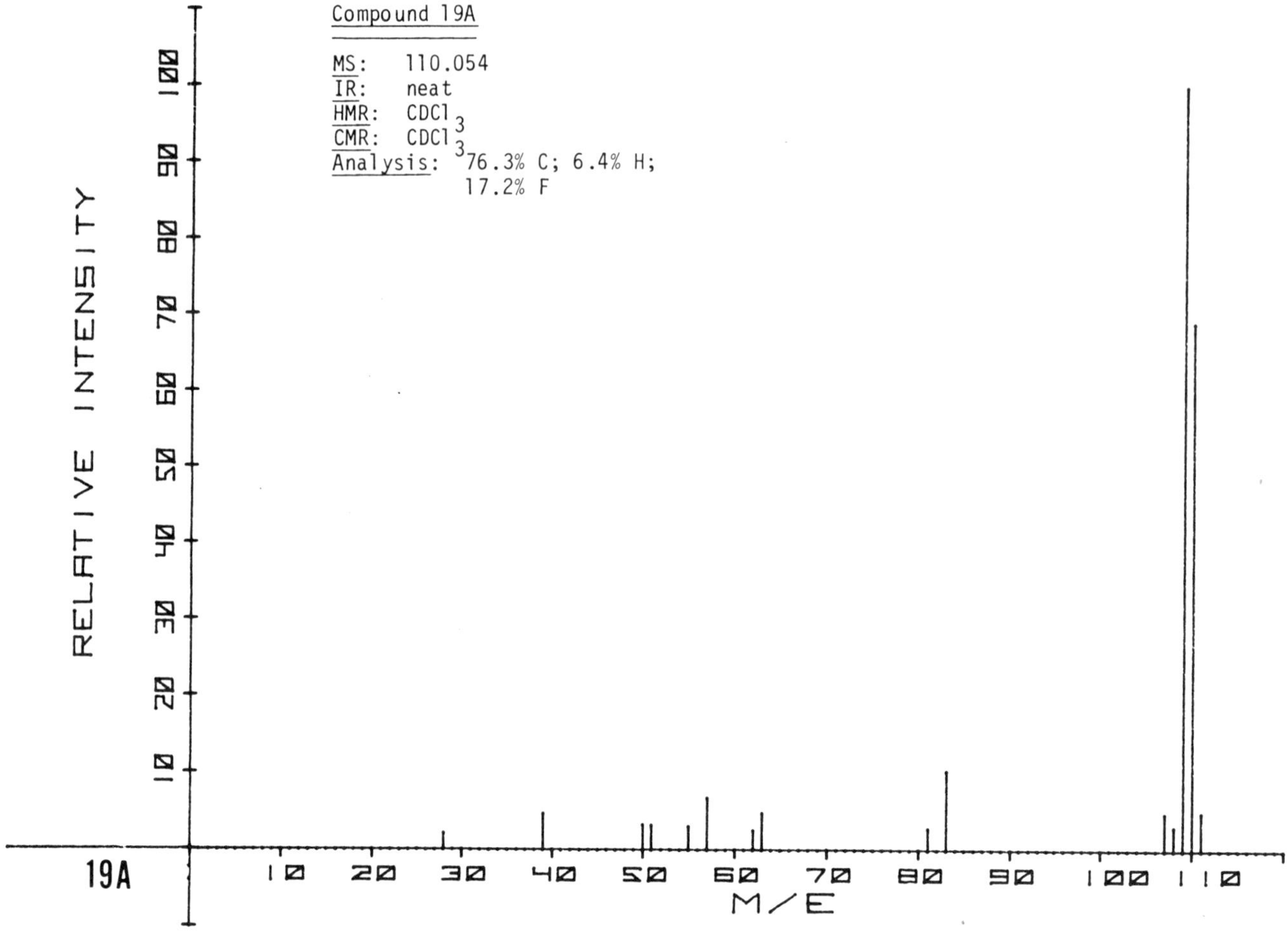

Compound 19A
MS: 110.054
IR: neat
HMR: CDCl3
CMR: CDCl3
Analysis: 76.3% C; 6.4% H; 17.2% F
RELATIVE INTENSITY
19A
M/E

CM-1
ABSORBANCE
WAVELENGTH (MICRONS)
19A

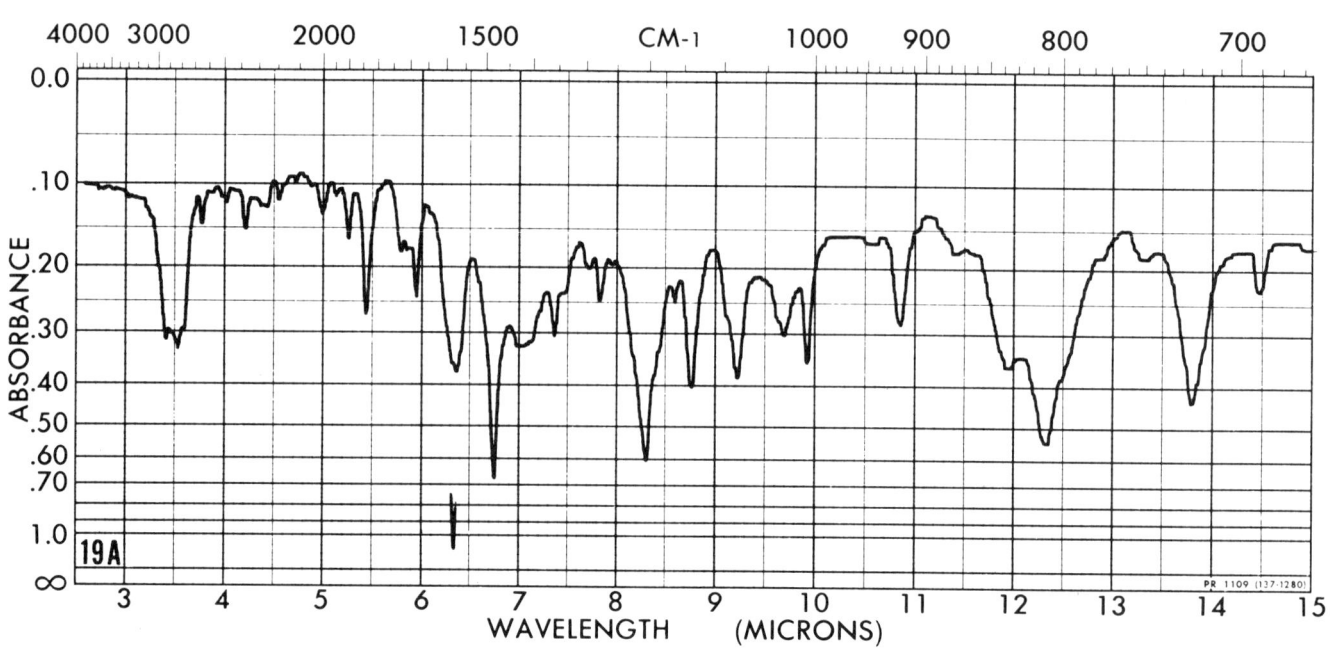

19A
19F-OBSERVE

19A
4000Hz (200)ppm
2000 (100)
1000 (50)
800 (40)
600 (30)
3000
2000 1000 500 400 300
1000 500 250 200 150
PK. HT. PPM
11 167.6
14 155.4
12 133.5
34 131.8
36 131.4
35 129.6
32 129.2
47 116.7
54 115.7
60 114.5
58 113.5
PK. HT. PPM
4 167.6
10 155.4
10 133.6
48 130.7
60 130.3
59 115.6
49 114.5
19 20.5
200 190 180 170 160 150 140 130 120 110 100 90 80 70 60 50 40 30 20 10 0
δ(PPM)

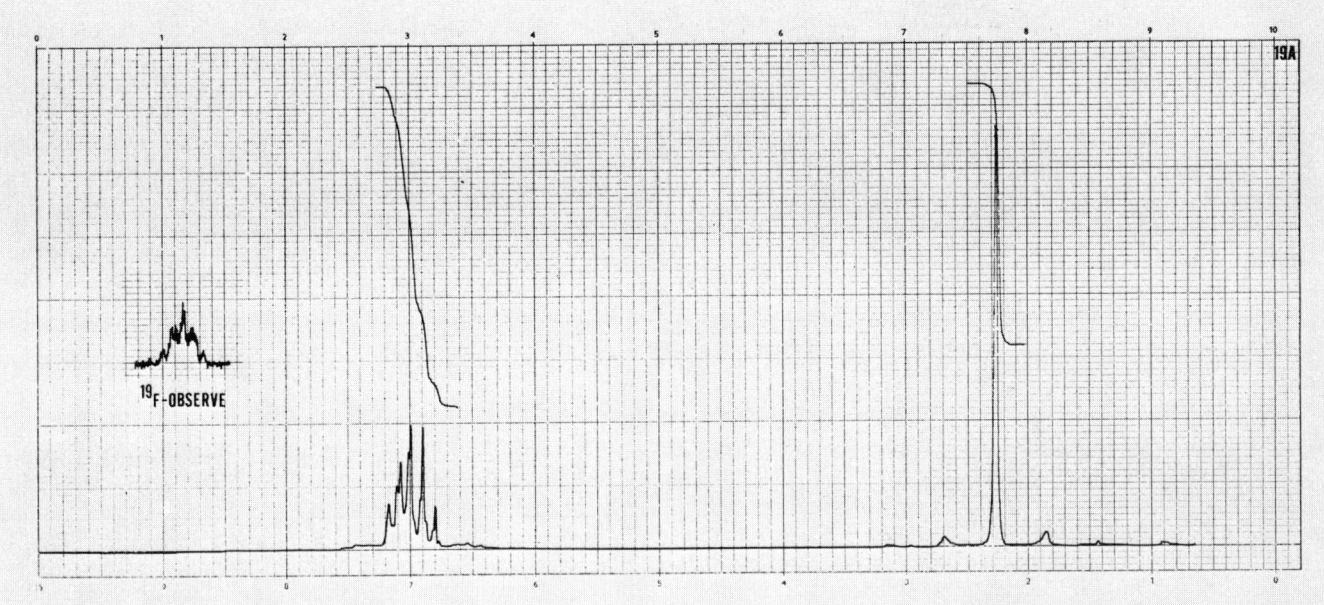

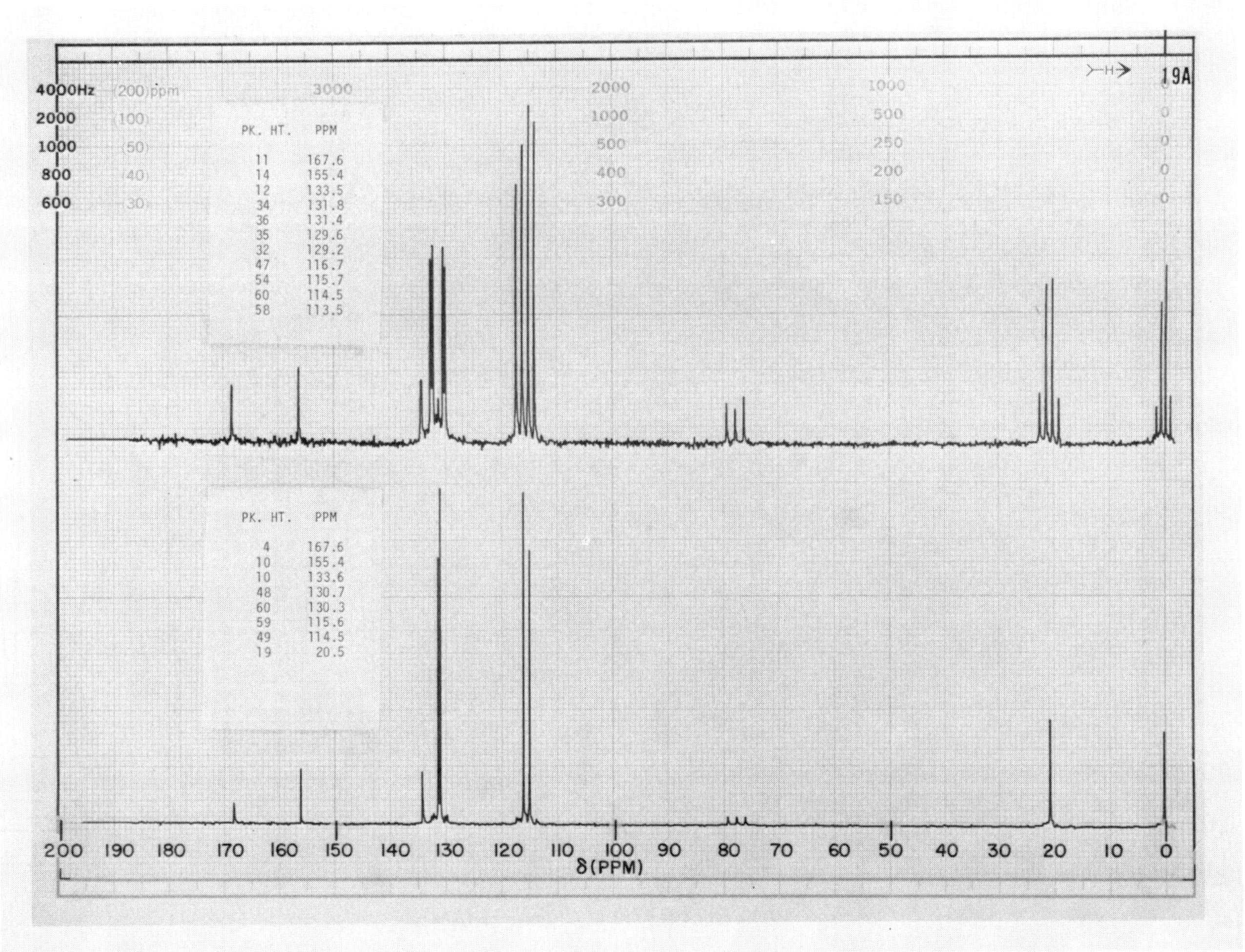

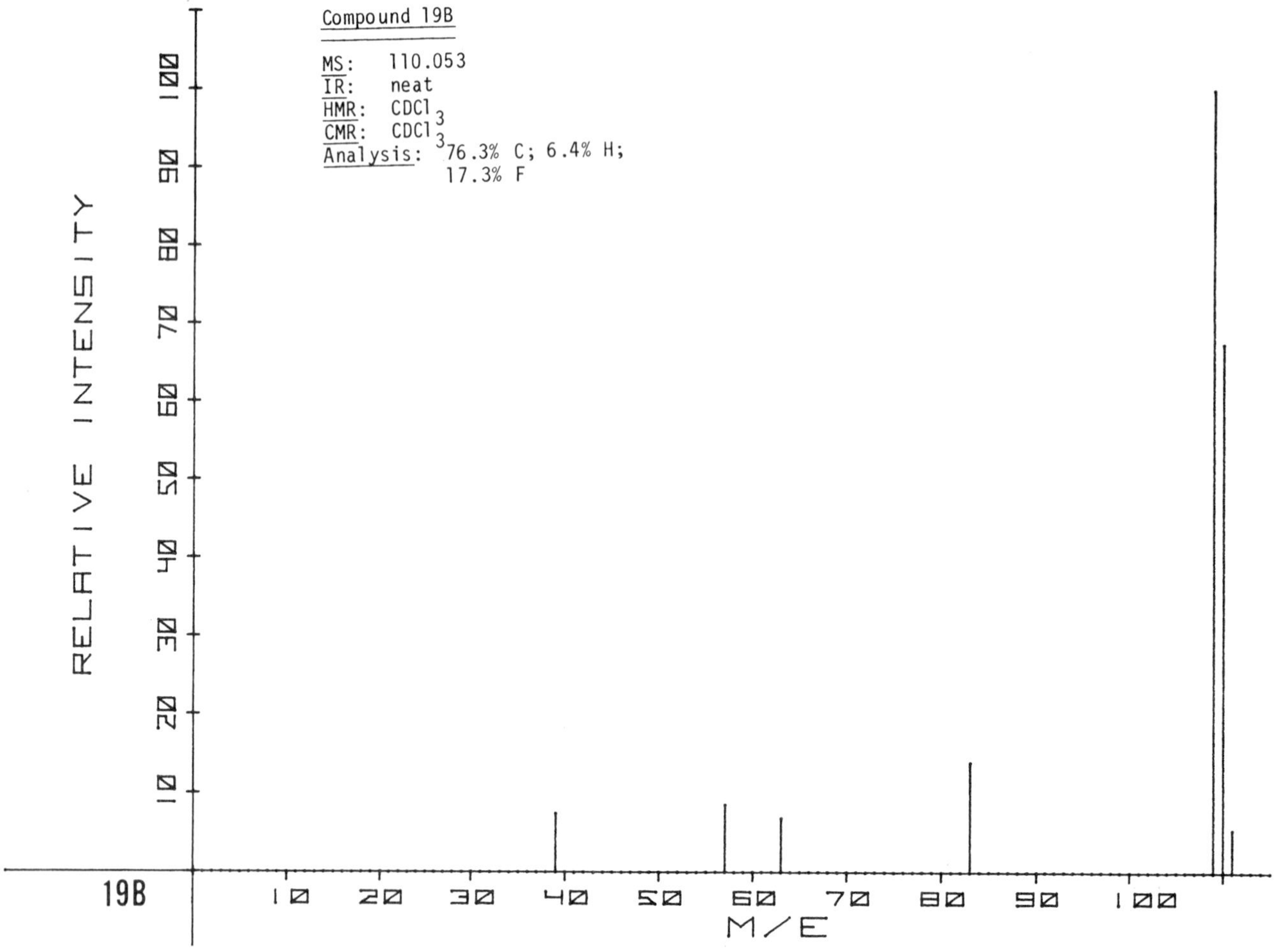

Compound 19B
MS: 110.053
IR: neat
HMR: CDCl3
CMR: CDCl3
Analysis: 76.3% C; 6.4% H; 17.3% F
RELATIVE INTENSITY
10 20 30 40 50 60 70 80 90 100
19B
M/E

4000 3000 2000 1500 CM-1 1000 900 800 700
ABSORBANCE
0.0 .10 .20 .30 .40 .50 .60 .70 1.0 ∞
19B
3 4 5 6 7 8 9 10 11 12 13 14 15
WAVELENGTH (MICRONS)
PR 1109 (137-1280)

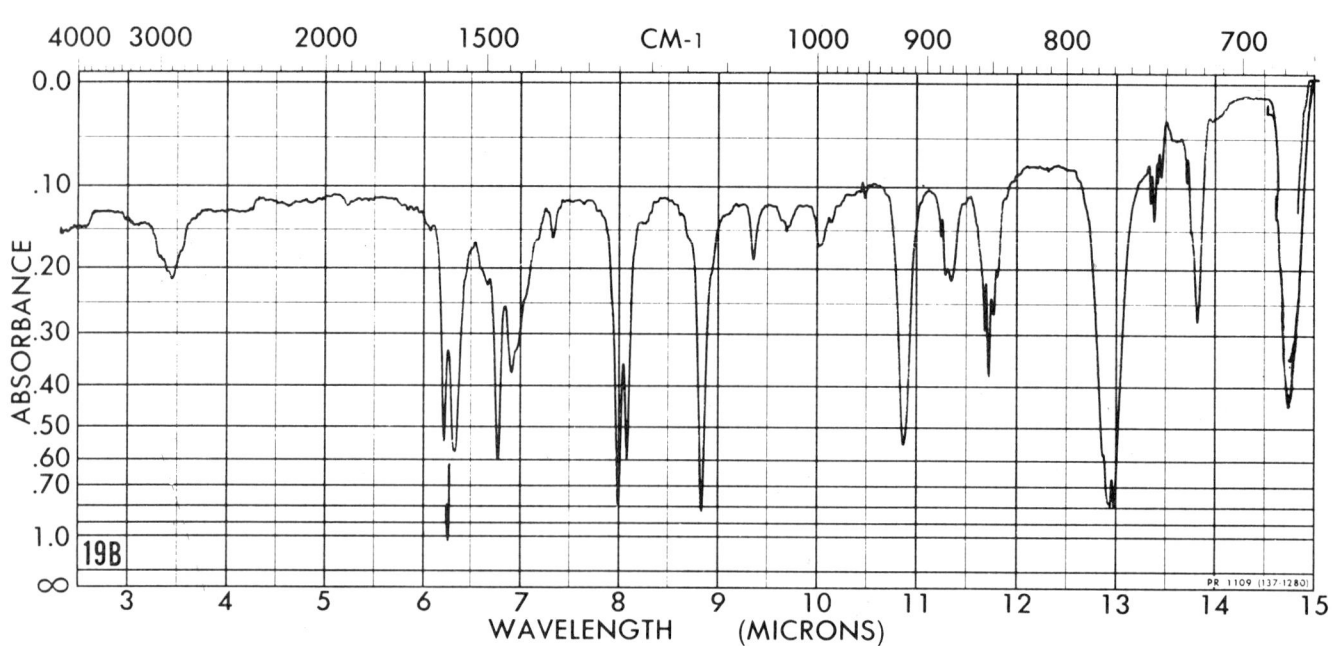

19B
19F-OBSERVE

19B
4000Hz (200) ppm
2000 (100)
1000 (50)
800 (40)
600 (30)
3000
1500
750
600
450
2000
1000
500
400
300
1000
500
250
200
150
PK. HT. PPM
12 165.8
9 153.6
7 153.5
8 137.2
10 136.8
14 127.5
12 127.1
13 127.0
12 125.2
12 125.1
14 124.7
11 124.7
22 122.4
19 120.2
17 114.0
16 113.0
14 111.8
13 110.8
19 110.4
18 109.3
19 108.1
5 107.8
21 107.0
PK. HT. PPM
32 168.6
30 156.4
33 140.0
29 139.6
42 129.1
35 128.7
43 124.1
56 124.0
45 115.7
49 114.7
52 112.0
55 111.0
200 190 180 170 160 150 140 130 120 110 100 90 80 70 60 50 40 30 20 10 0
δ(PPM)

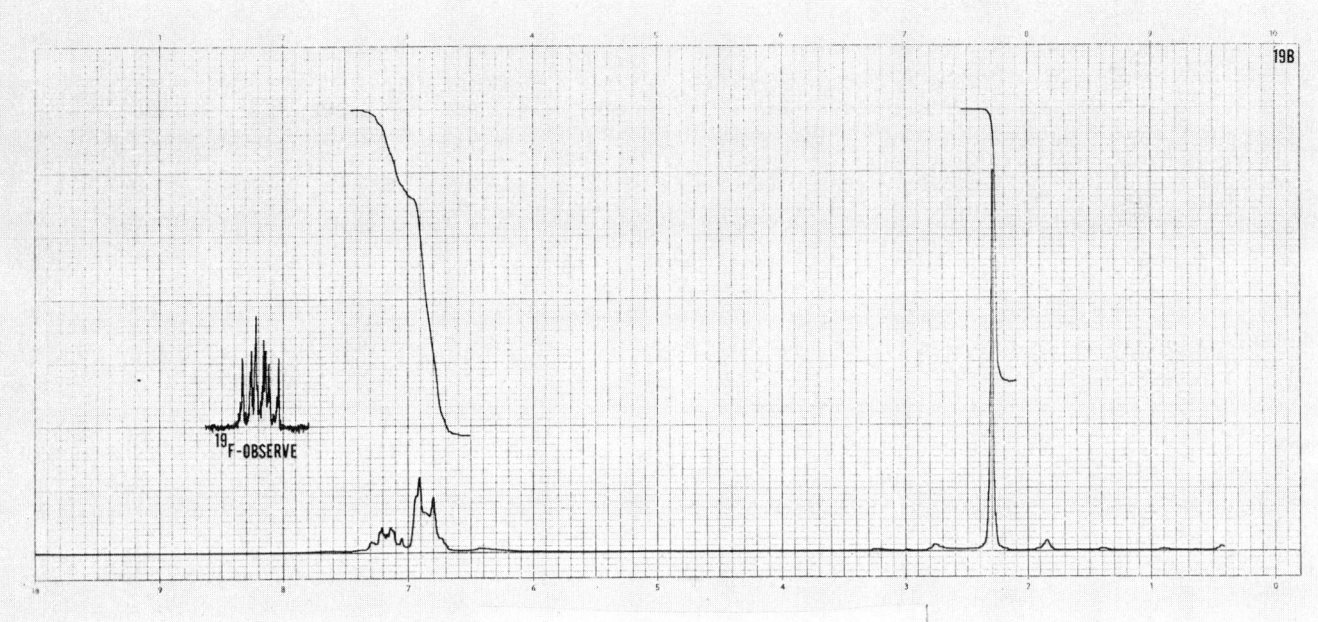

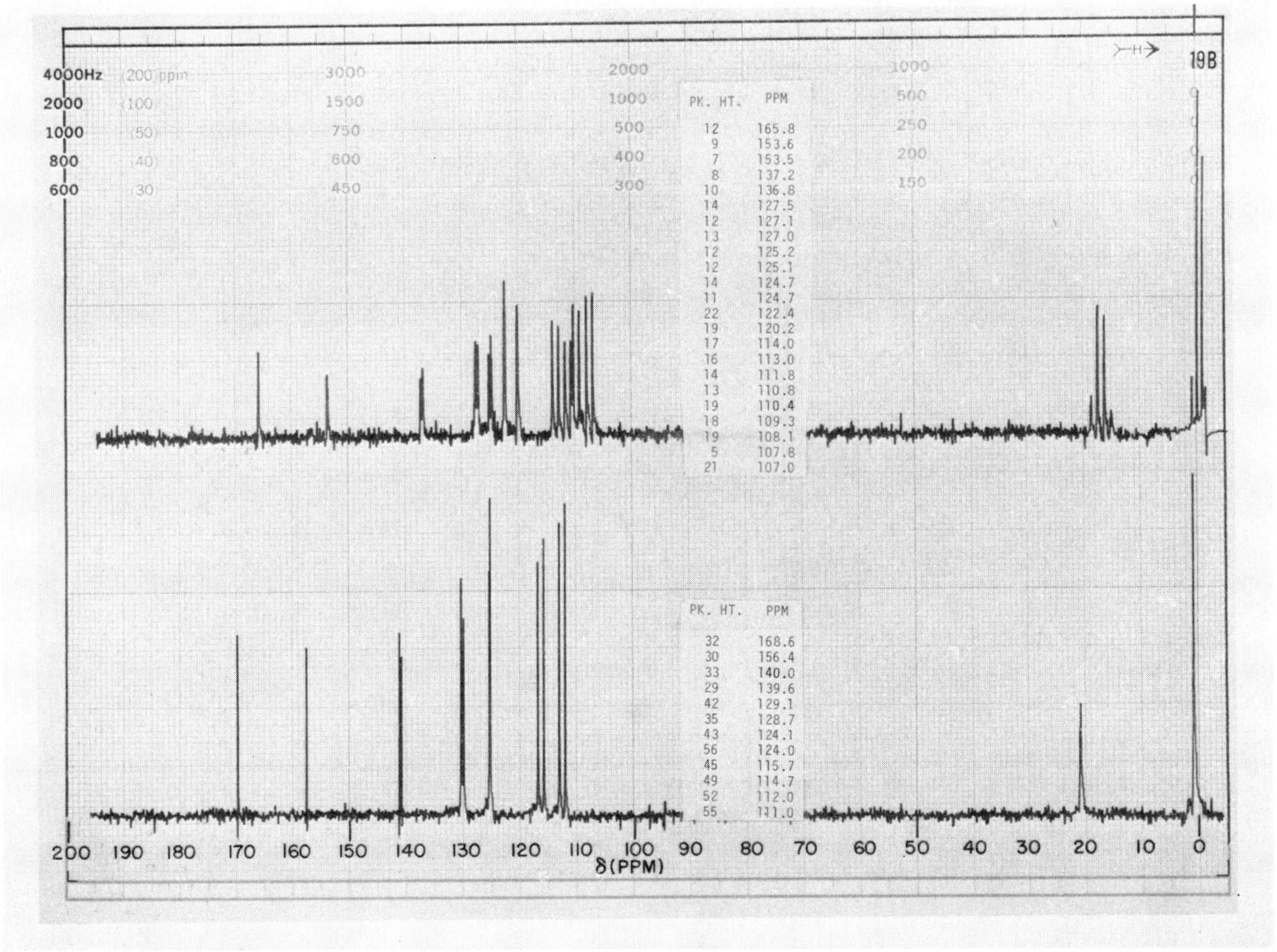

COMPOUND 19C

Compound 19C
MS: 110.053
IR: neat
HMR: CDCl₃
CMR: neat
Analysis: 76.2% C; 6.4% H; 17.3% F

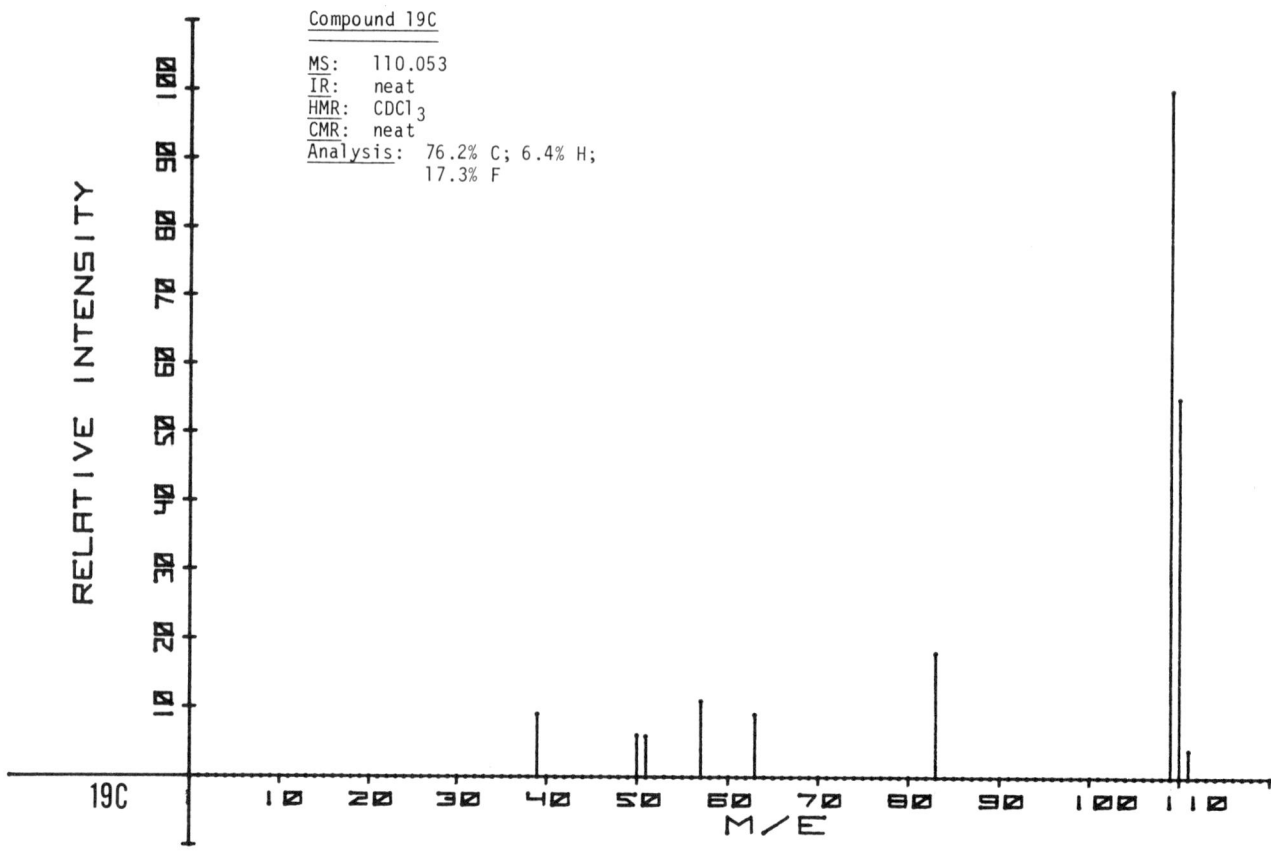

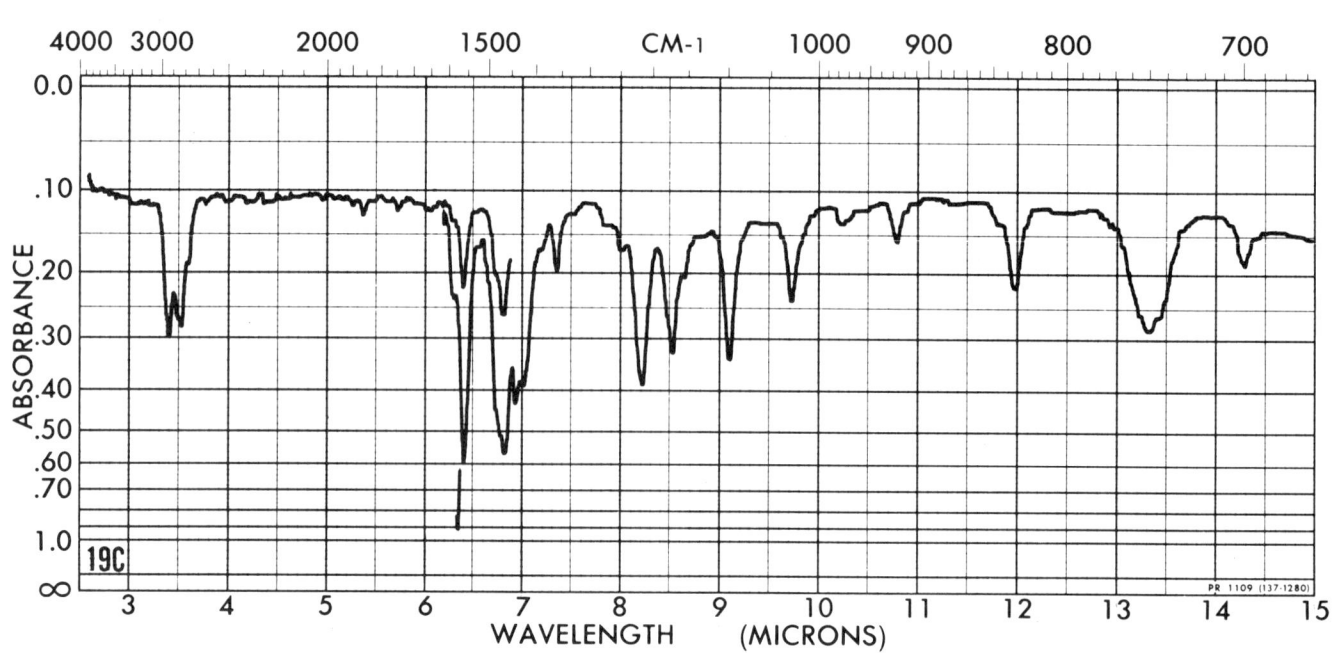

19C
19F-OBSERVE

19C
PK. HT. PPM
45 132.9
45 132.7
36 131.2
41 131.0
27 128.8
35 128.4
24 127.1
30 126.7
23 125.6
46 125.2
35 125.0
27 124.8
27 123.5
32 123.4
43 116.7
59 115.6
45 115.1
47 114.0
PK. HT. PPM
29 168.2
27 156.0
44 132.1
44 131.8
44 128.0
38 127.6
12 125.6
23 124.7
60 124.4
47. 124.2
38 115.9
53 114.8
41 14.4
39 14.2
δ(PPM)

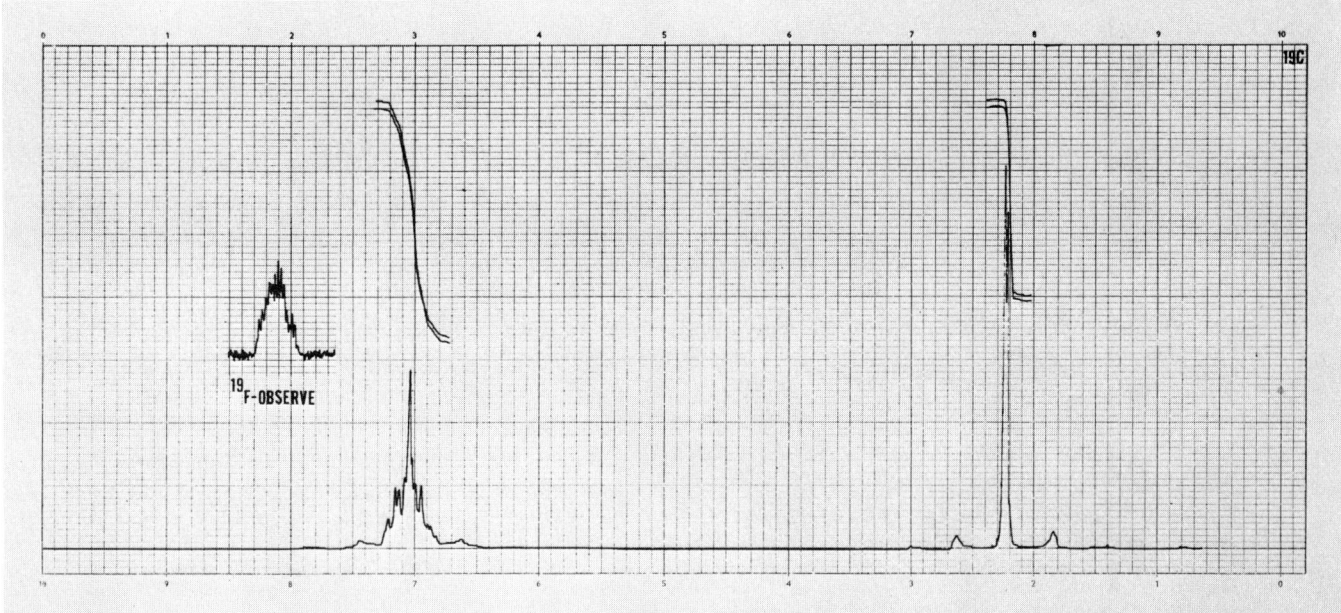

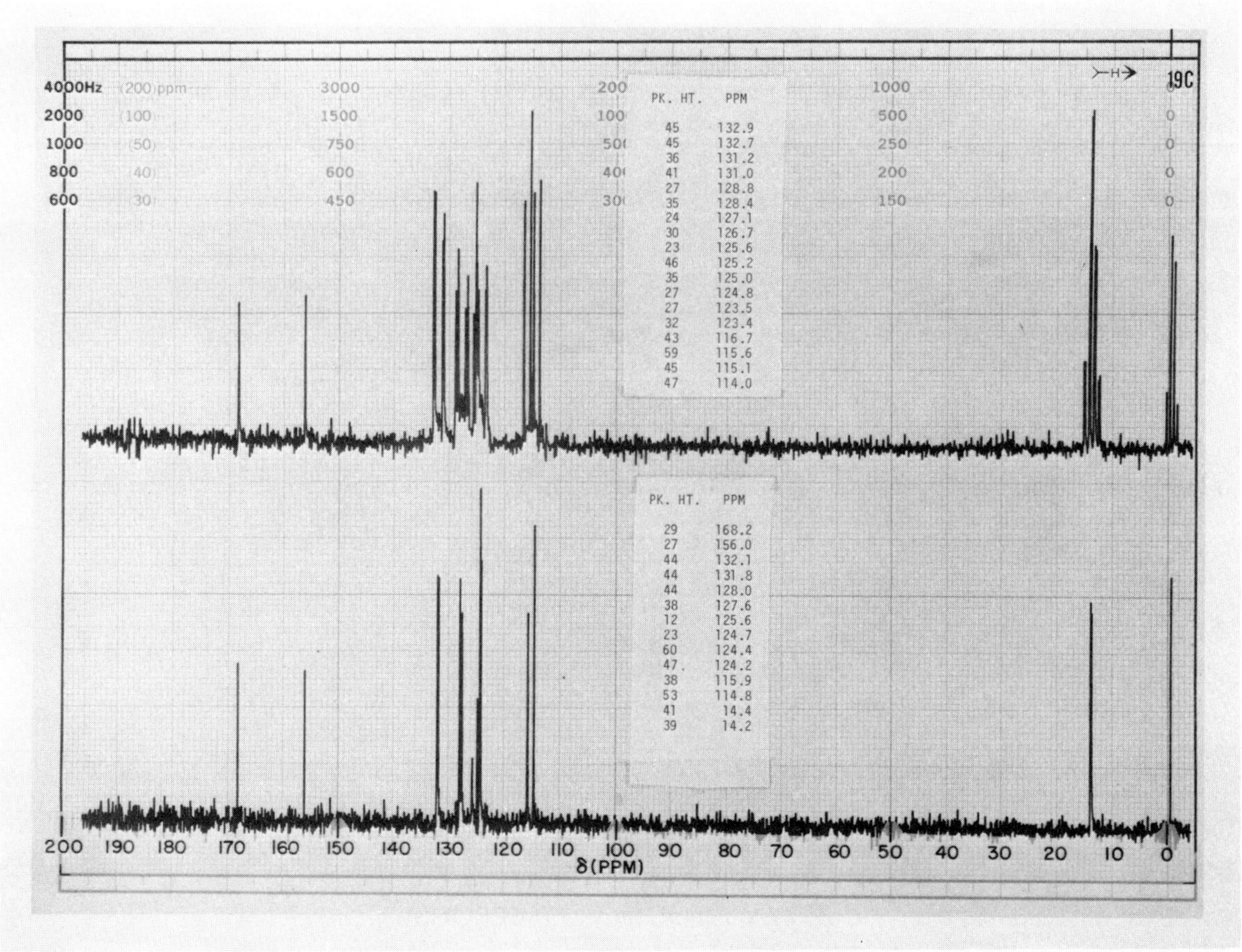

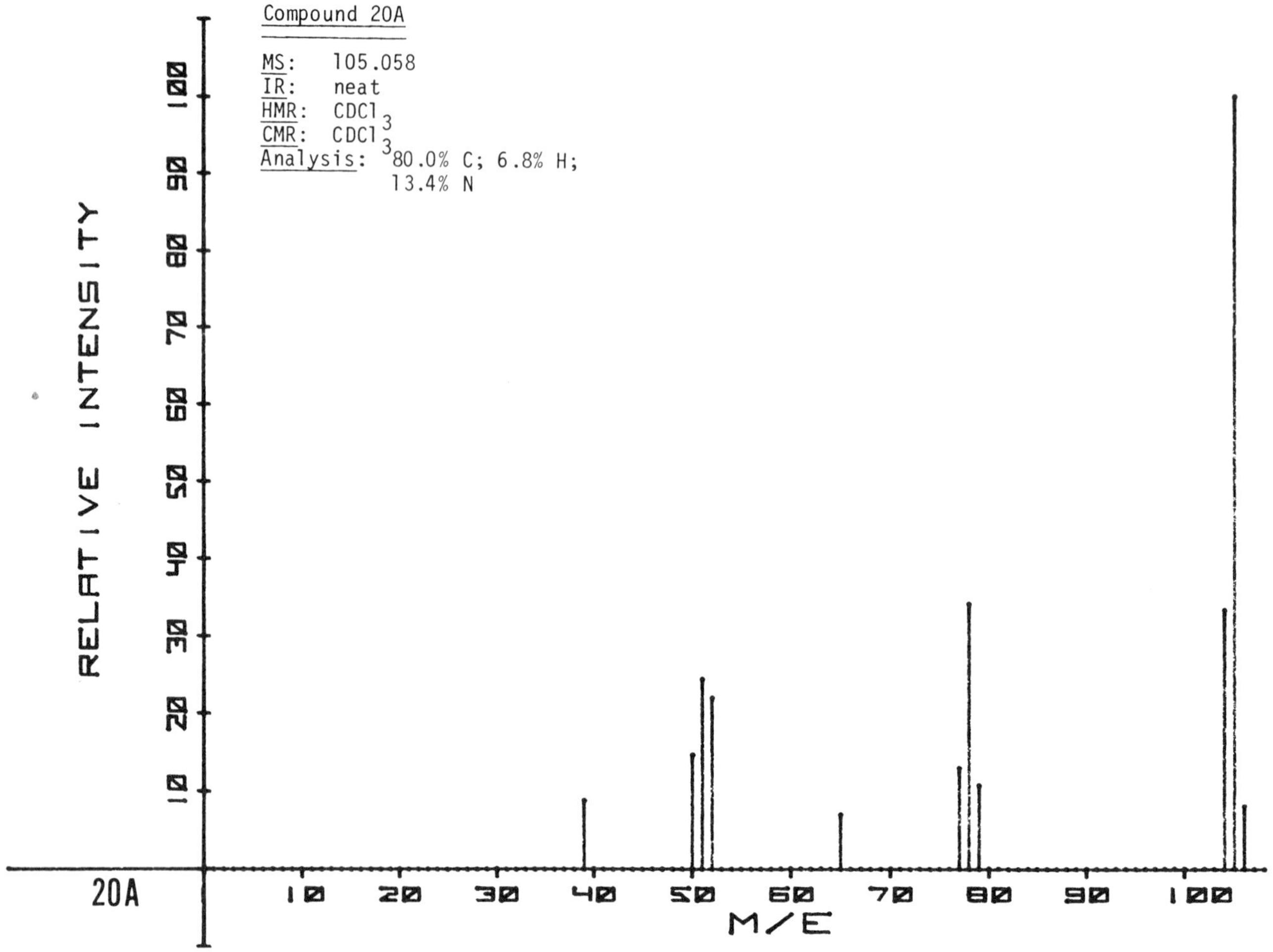

Compound 20A
MS: 105.058
IR: neat
HMR: CDCl3
CMR: CDCl3
Analysis: 80.0% C; 6.8% H; 13.4% N
RELATIVE INTENSITY
10 20 30 40 50 60 70 80 90 100
20A
M/E

4000 3000 2000 1500 CM-1 1000 900 800 700
ABSORBANCE
0.0 .10 .20 .30 .40 .50 .60 .70 1.0 ∞
20A
3 4 5 6 7 8 9 10 11 12 13 14 15
WAVELENGTH (MICRONS)

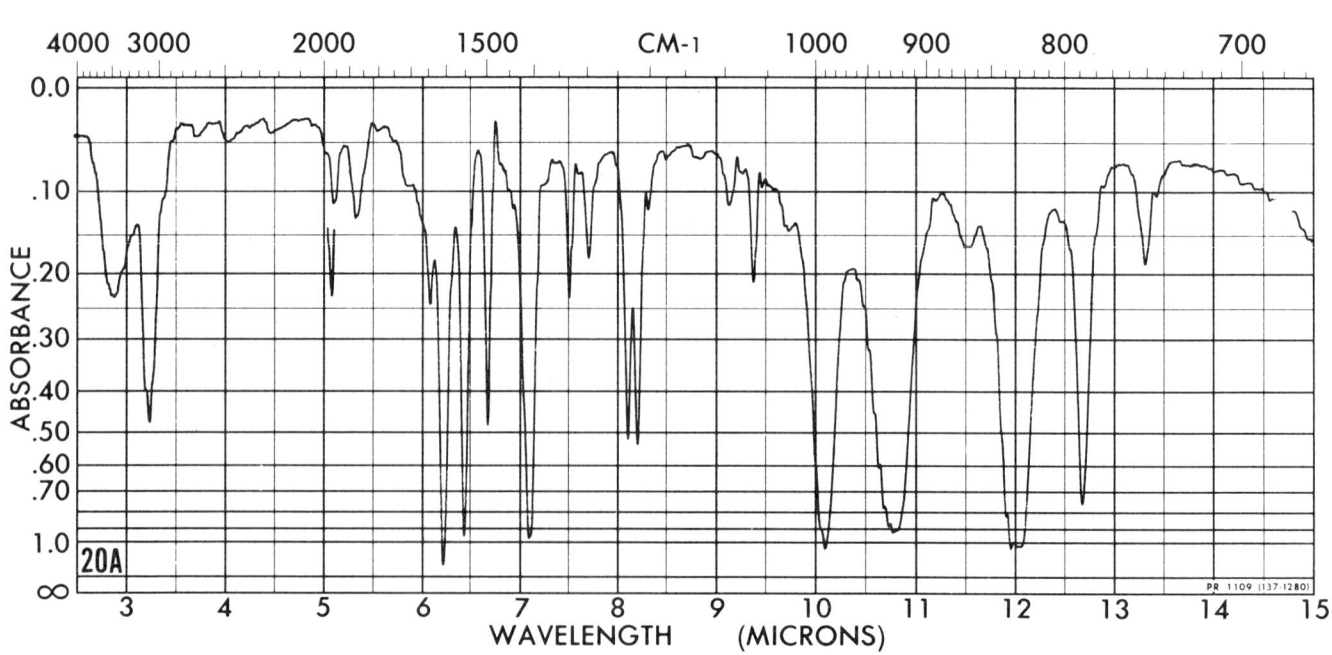

20A
IRR. 8.50
IRR. 5.88
IRR. 5.36
A
B
C
D
E
8
7
6
5
4
3
2
1
0

20A
4000Hz (200)ppm 3000 2000 1000 0
2000 (100) 1500 1000 500 0
1000 750 500 250 0
800 600 400 200 0
600 450 300 150 0
PK. HT. PPM
15 135.3
22 134.2
55 121.4
51 120.2
60 120.1
24 118.7
15 117.3
PK. HT. PPM
60 150.2
9 144.7
20 134.8
59 120.8
30 118.6
200 190 180 170 160 150 140 130 120 110 100 90 80 70 60 50 40 30 20 10 0
δ(PPM)

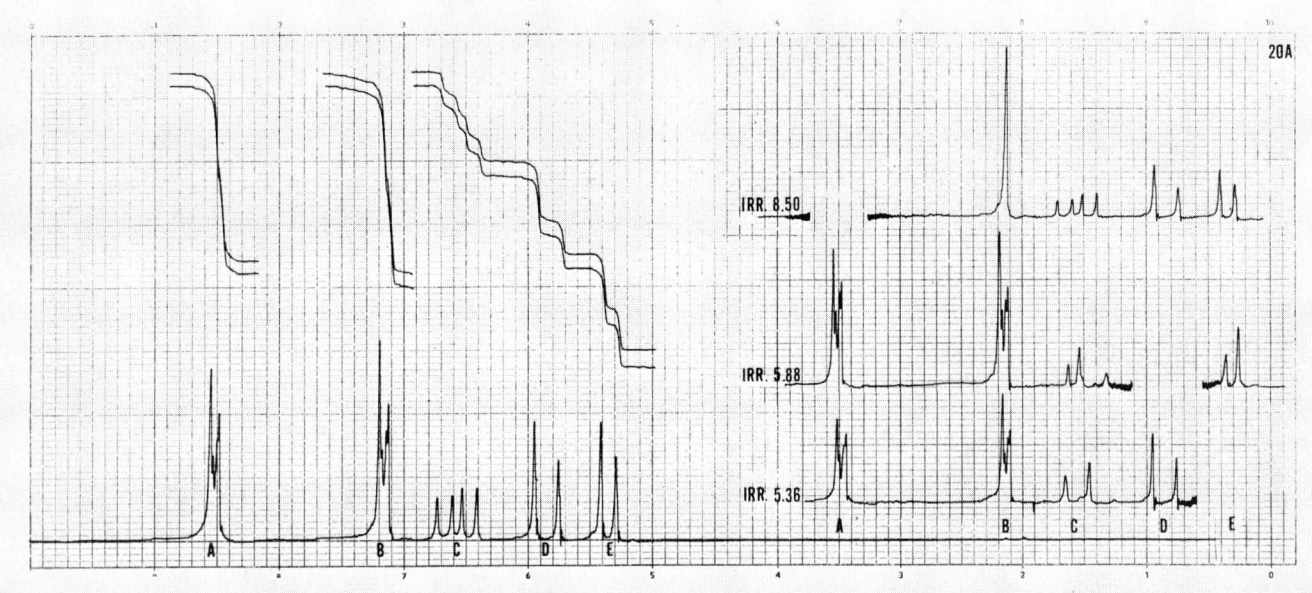

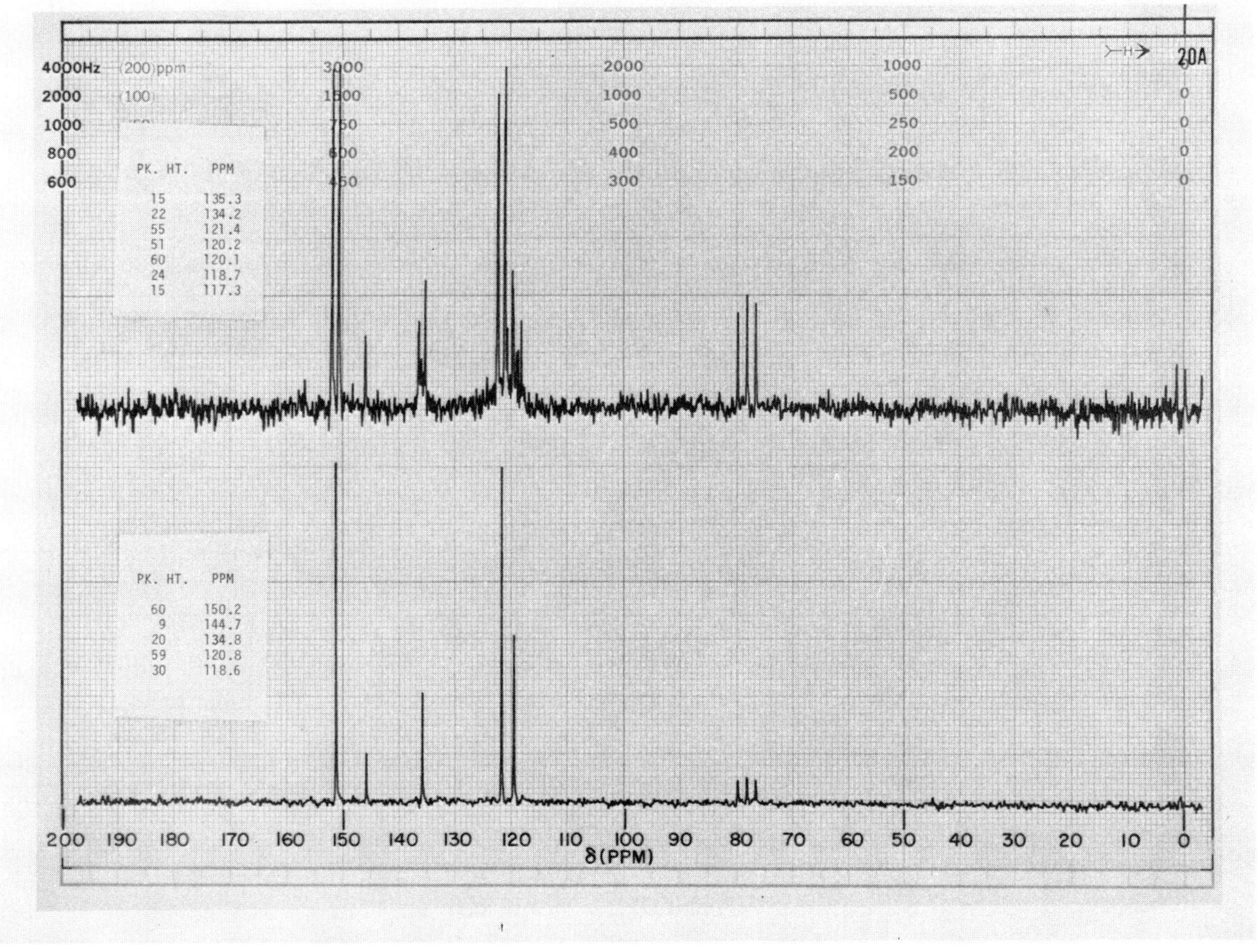

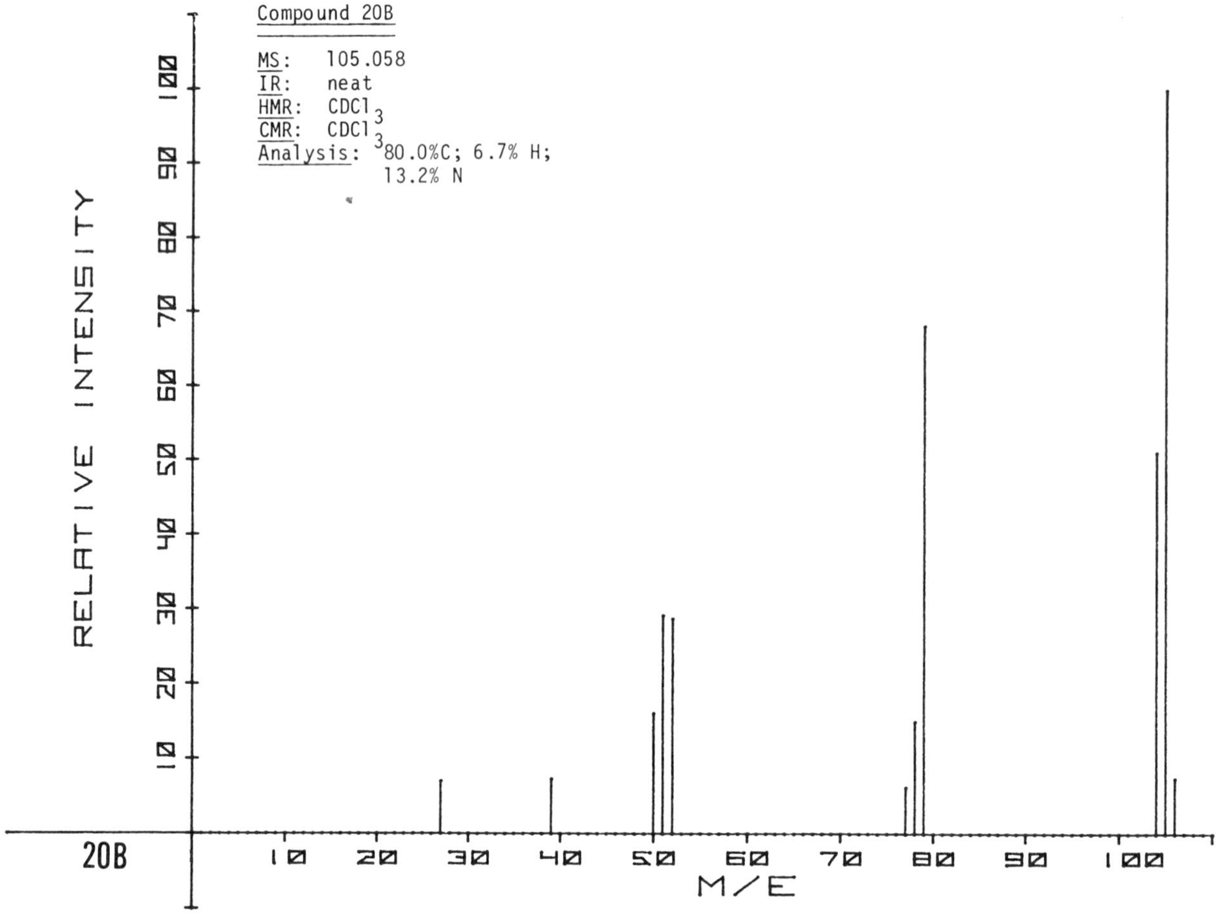

Compound 20B
MS: 105.058
IR: neat
HMR: CDCl3
CMR: CDCl3
Analysis: 80.0%C; 6.7% H; 13.2% N
RELATIVE INTENSITY
10 20 30 40 50 60 70 80 90 100
20B
M/E

4000 3000 2000 1500 CM-1 1000 900 800 700
ABSORBANCE
0.0 .10 .20 .30 .40 .50 .60 .70 1.0 ∞
20B
3 4 5 6 7 8 9 10 11 12 13 14 15
WAVELENGTH (MICRONS)

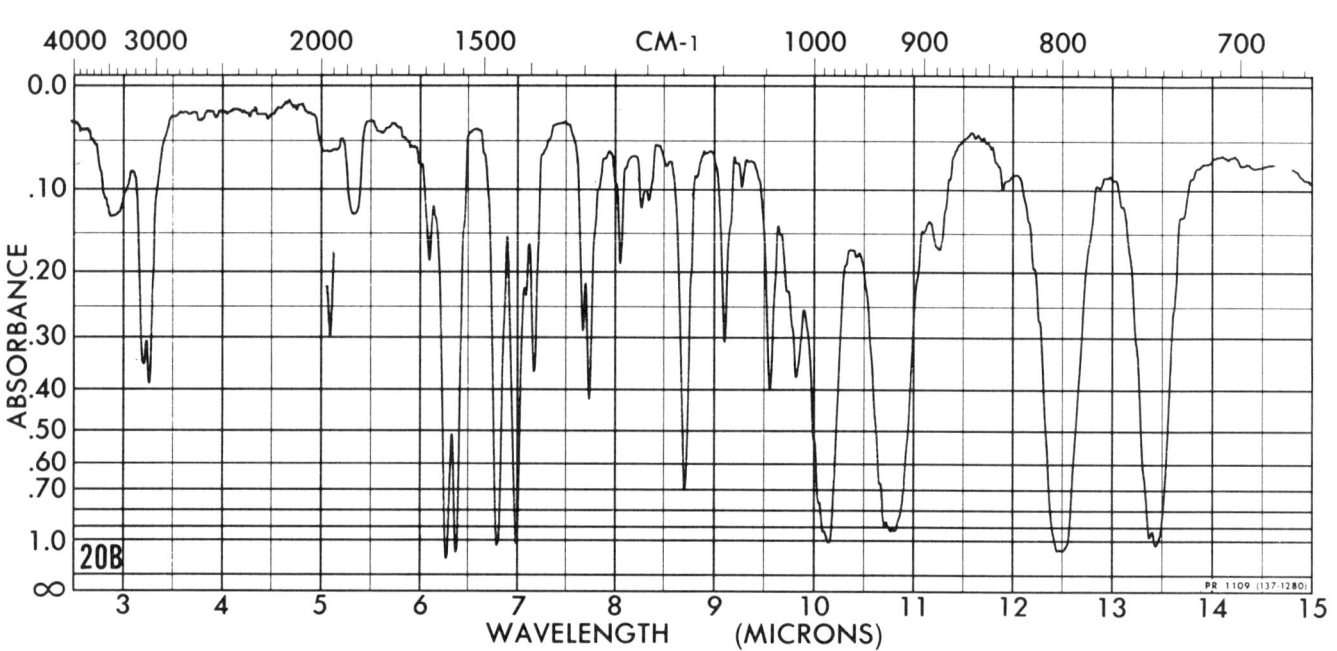

20B
OFFSET 5.3 PPM, SWEEP WIDTH 5 PPM
IRR. 5.43
IRR. 8.54

20B
PK. HT. PPM
22 139.8
33 139.0
21 138.1
28 137.0
39 125.1
50 124.0
43 123.1
48 122.0
28 121.3
60 119.7
22 117.9
PK. HT. PPM
12 157.5
33 151.3
37 139.0
56 138.0
60 124.1
43 123.0
47 119.6
δ(PPM)

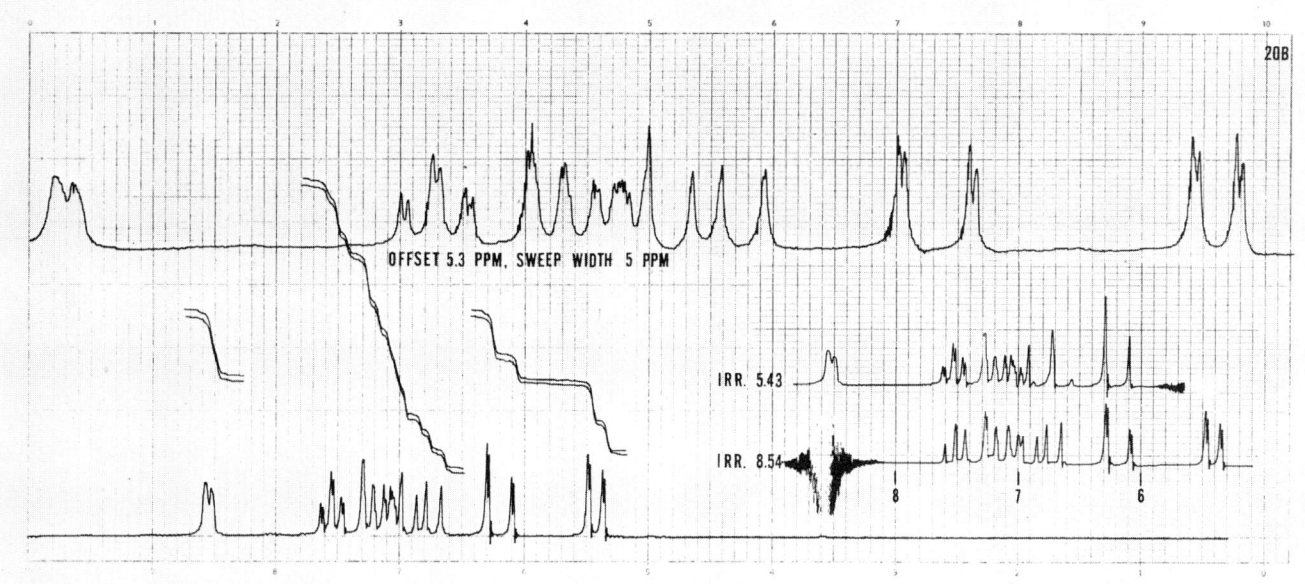

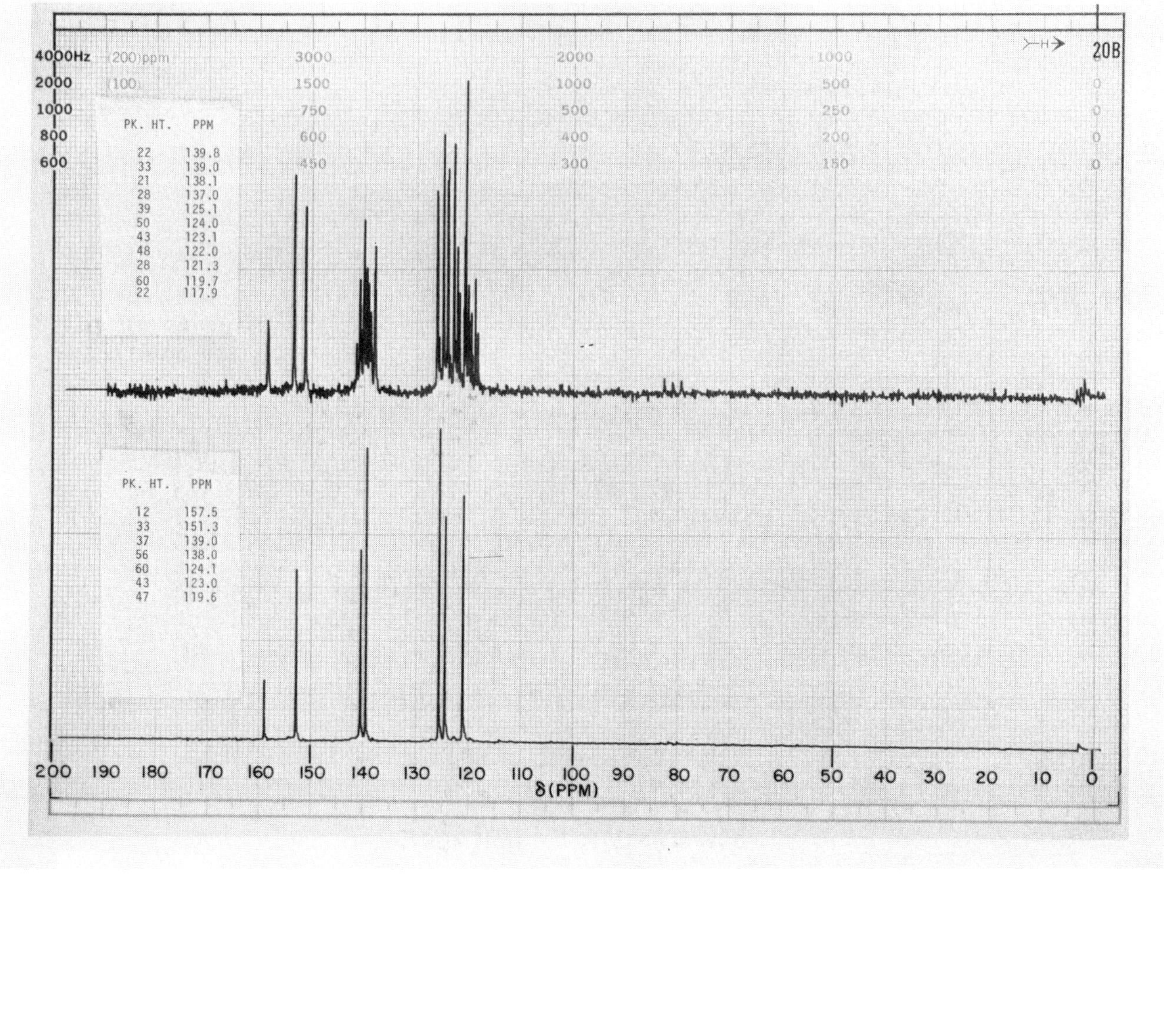

COMPOUND 21

Compound 21
MS: 124.052
IR: neat
HMR: CDCl₃
CMR: CDCl₃
Analysis: 67.8% C; 6.5% H

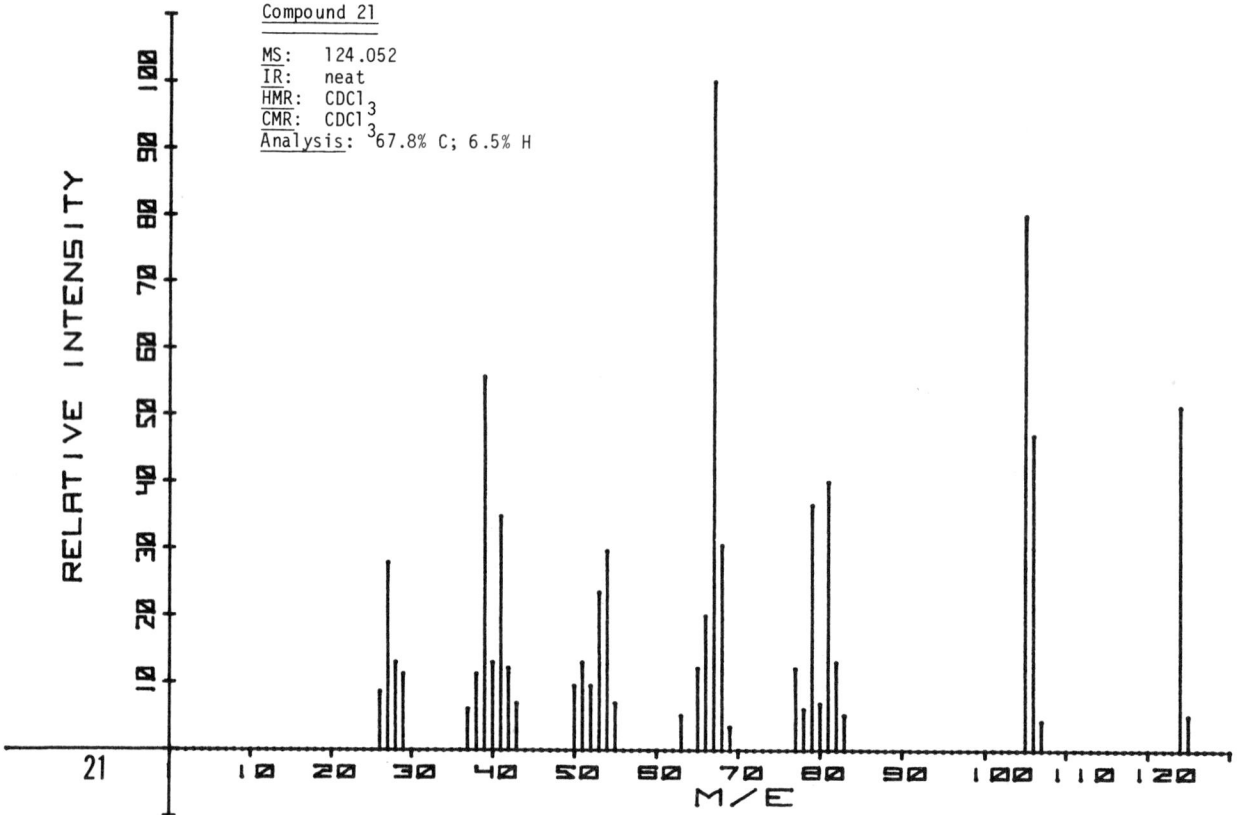

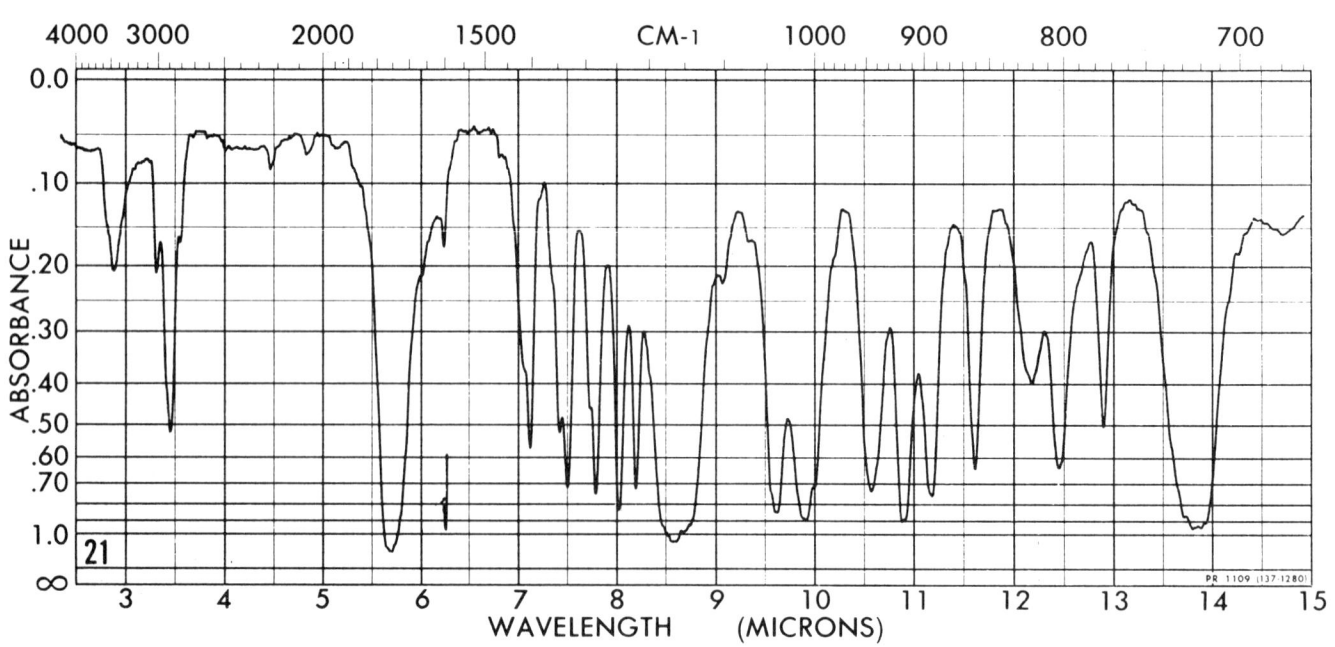

21
IRR. 5.80
IRR. 2.80
IRR. 5.60
IRR. 3.50
IRR. 5.17
A
B
A
B
B
10
9
8
7
6
5
4
3
2
1
0

21
4000Hz 200 ppm 3000 2000 1000
2000 100 1000 500 0
1000 50 500 250 0
800 40 400 200 0
600 30 300 150 0
Pk. Ht. PPM
14 176.7
28 132.7
39 130.9
35 130.1
39 128.4
33 84.3
38 81.8
27 46.7
28 44.5
26 41.7
60 39.5
25 37.3
32 35.5
49 33.2
31 31.0
Pk. Ht. PPM
11 176.7
51 131.4
55 129.7
47 83.1
52 45.6
60 39.6
59 33.3
200 190 180 170 160 150 140 130 120 110 100 90 80 70 60 50 40 30 20 10 0
δ(PPM)

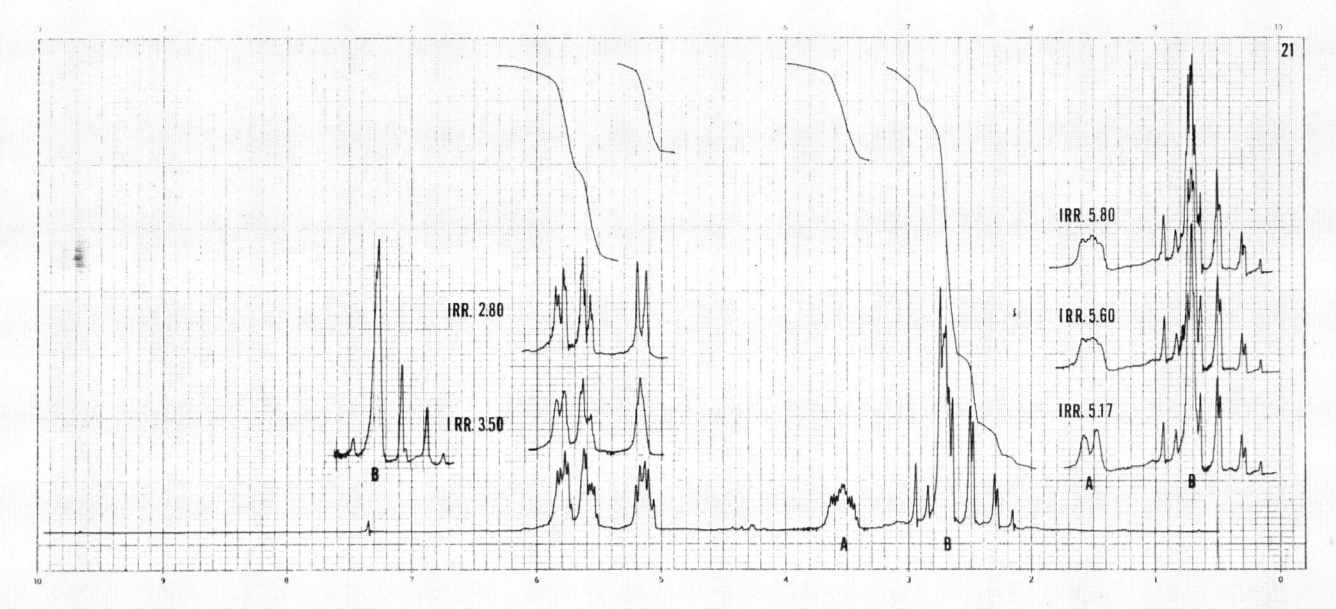

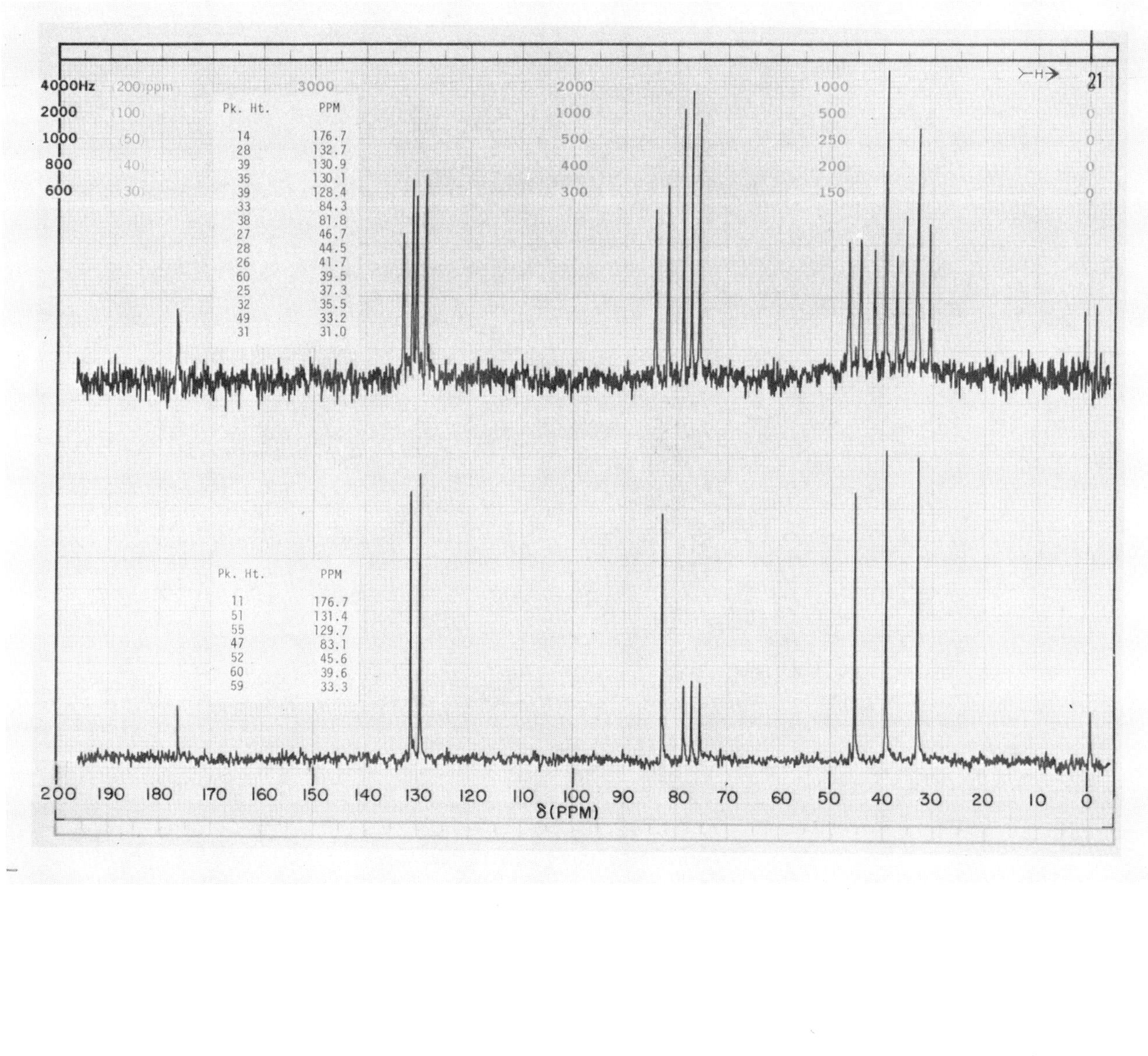

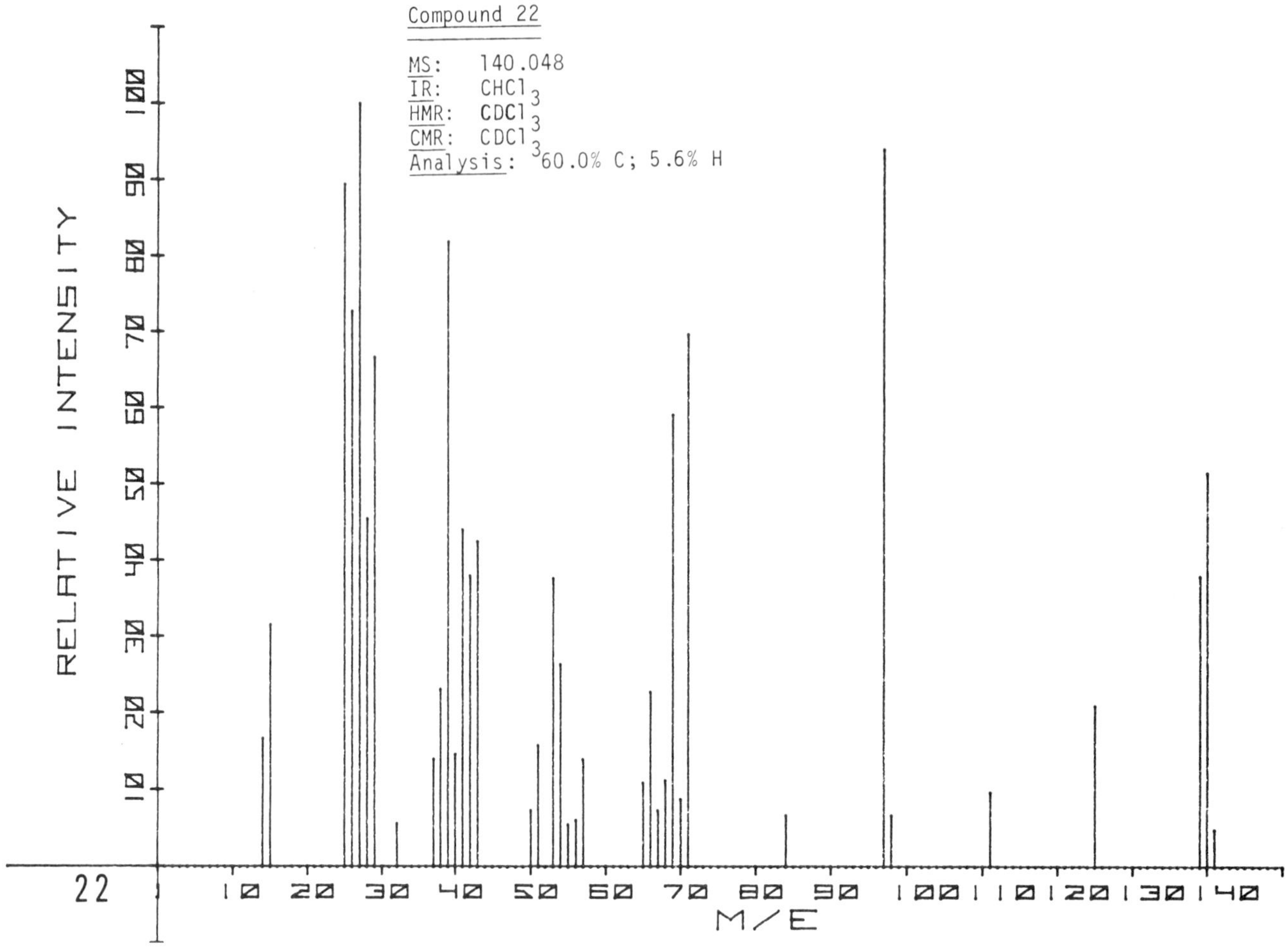

Compound 22
MS: 140.048
IR: $CHCl_3$
HMR: $CDCl_3$
CMR: $CDCl_3$
Analysis: 60.0% C; 5.6% H
RELATIVE INTENSITY
M/E
22

CM-1
ABSORBANCE
WAVELENGTH (MICRONS)
22
PR 1109 (137-1280)

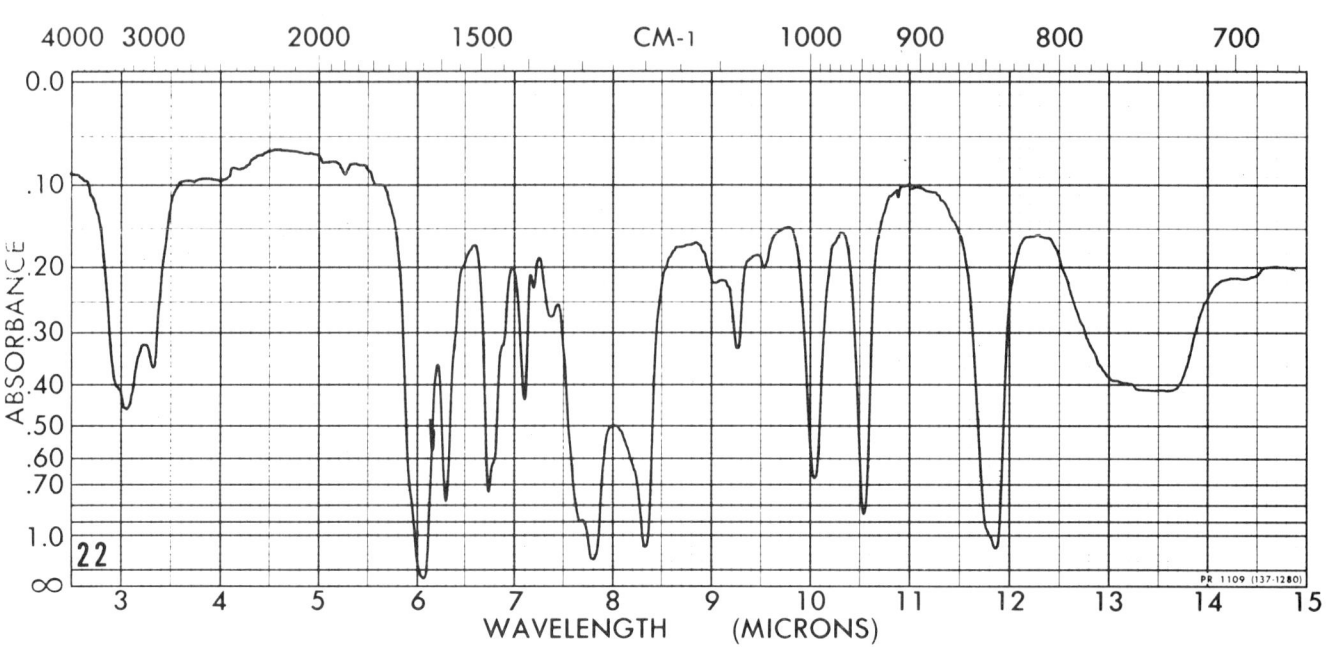

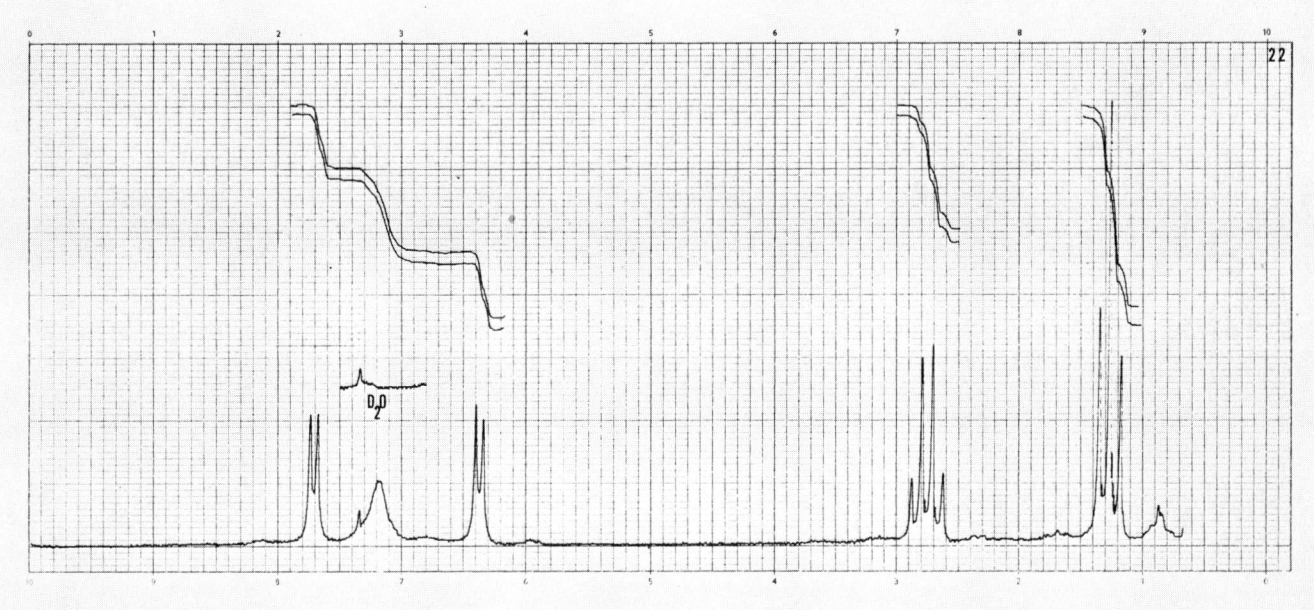

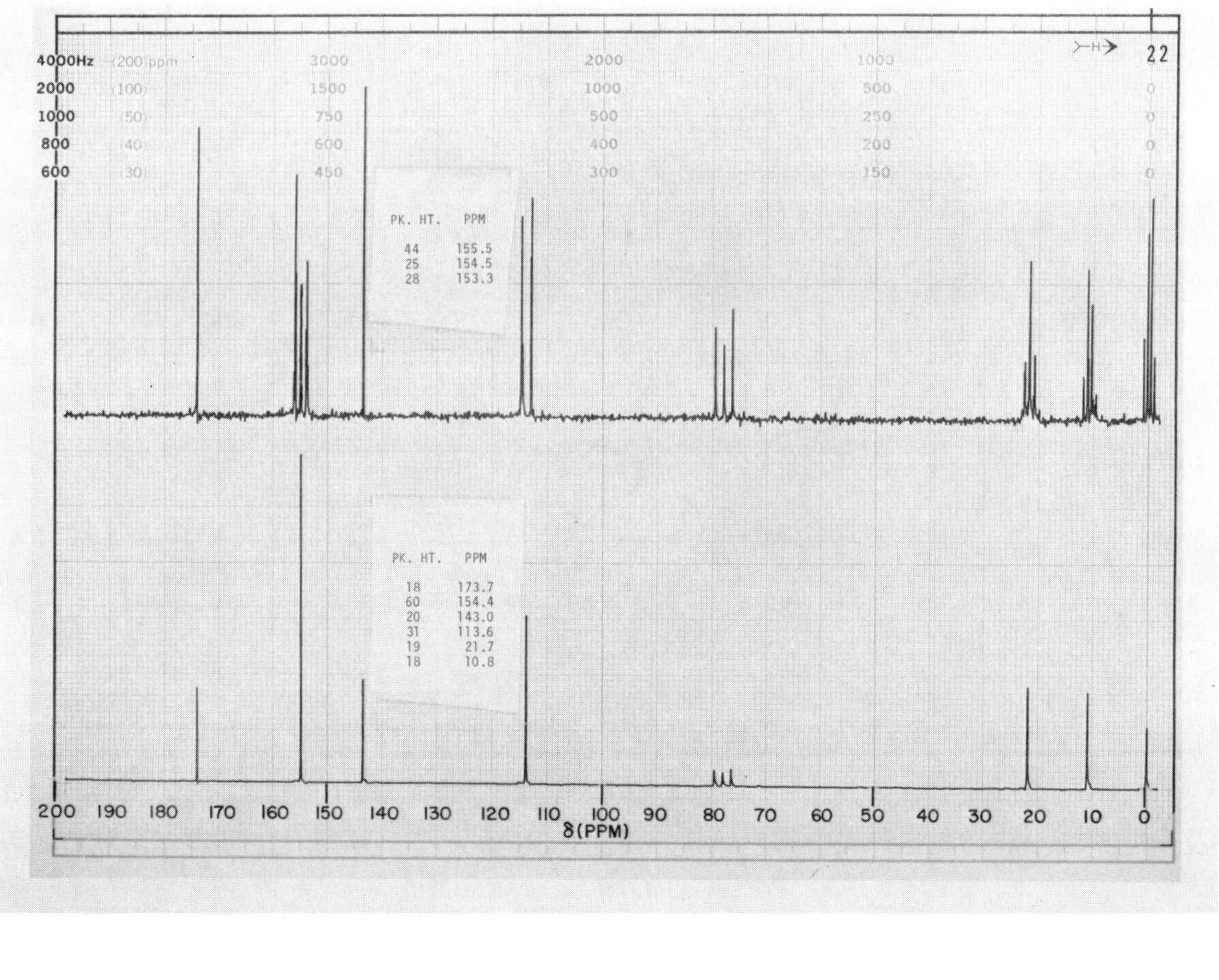

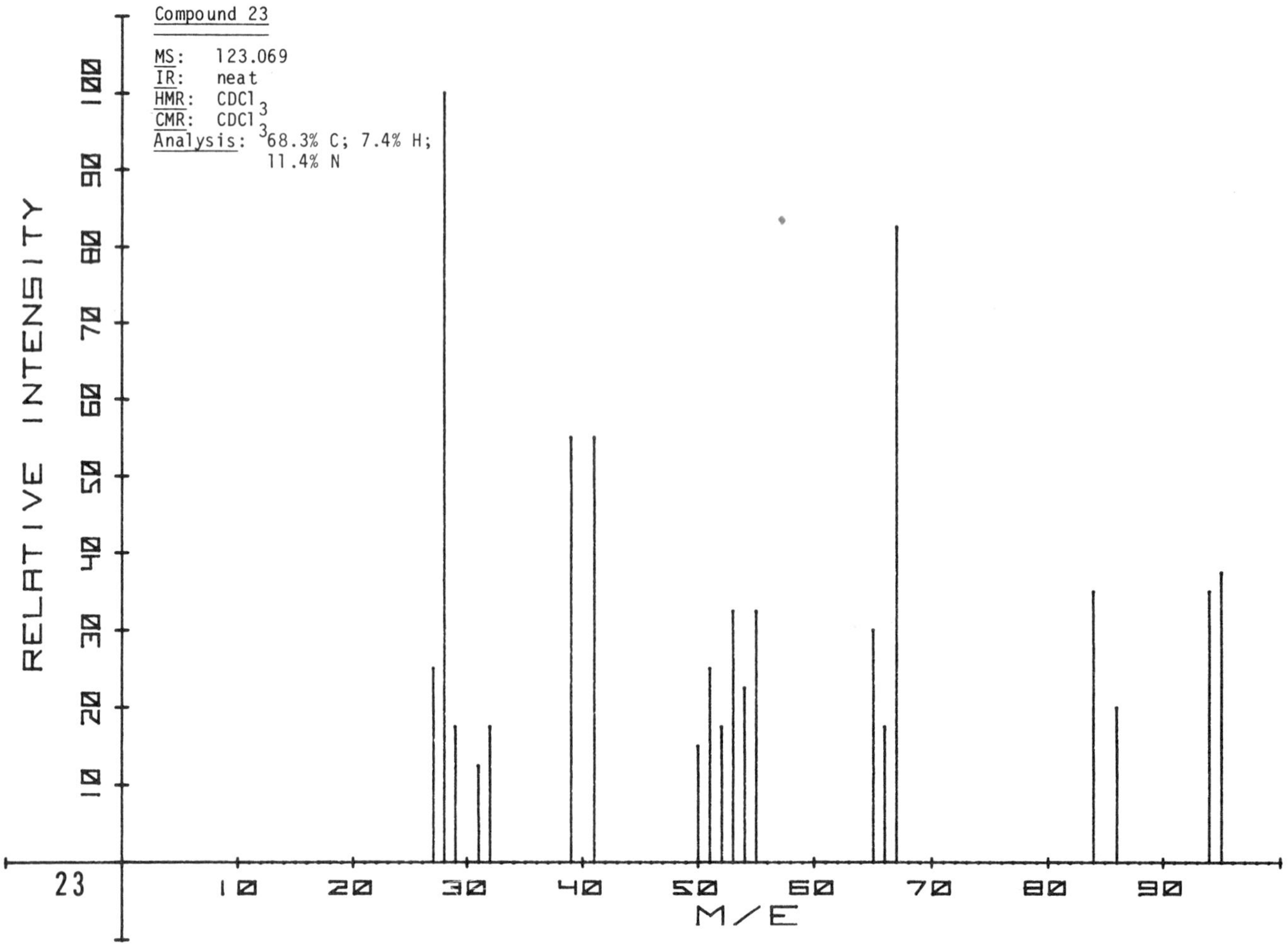

Compound 23
MS: 123.069
IR: neat
HMR: $CDCl_3$
CMR: $CDCl_3$
Analysis: 68.3% C; 7.4% H; 11.4% N
RELATIVE INTENSITY
M/E
23

CM-1
ABSORBANCE
WAVELENGTH (MICRONS)
23
PR 1109 (137-1280)

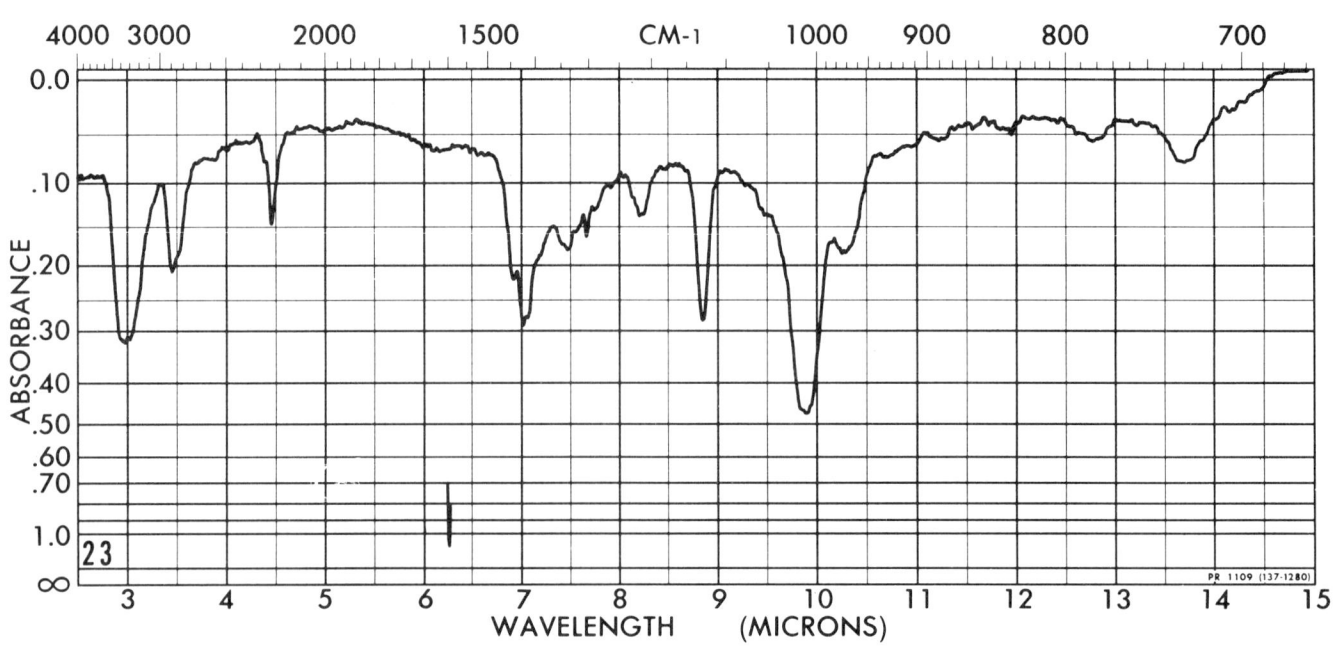

23
IRR. 2.45
D2O
A
A

23
4000Hz (200)ppm 3000 2000 1000 0
2000 (100) 1500 1000 500 0
1000 (50) 750 500 250 0
800 (40) 600 400 200 0
600 (30) 300 150 0
PK. HT. PPM
16 20.1
14 18.3
32 17.9
34 16.1
16 15.7
11 14.0
PK. HT. PPM
15 119.6
13 82.9
11 80.6
60 50.6
56 24.4
53 17.9
59 16.1
δ(PPM)

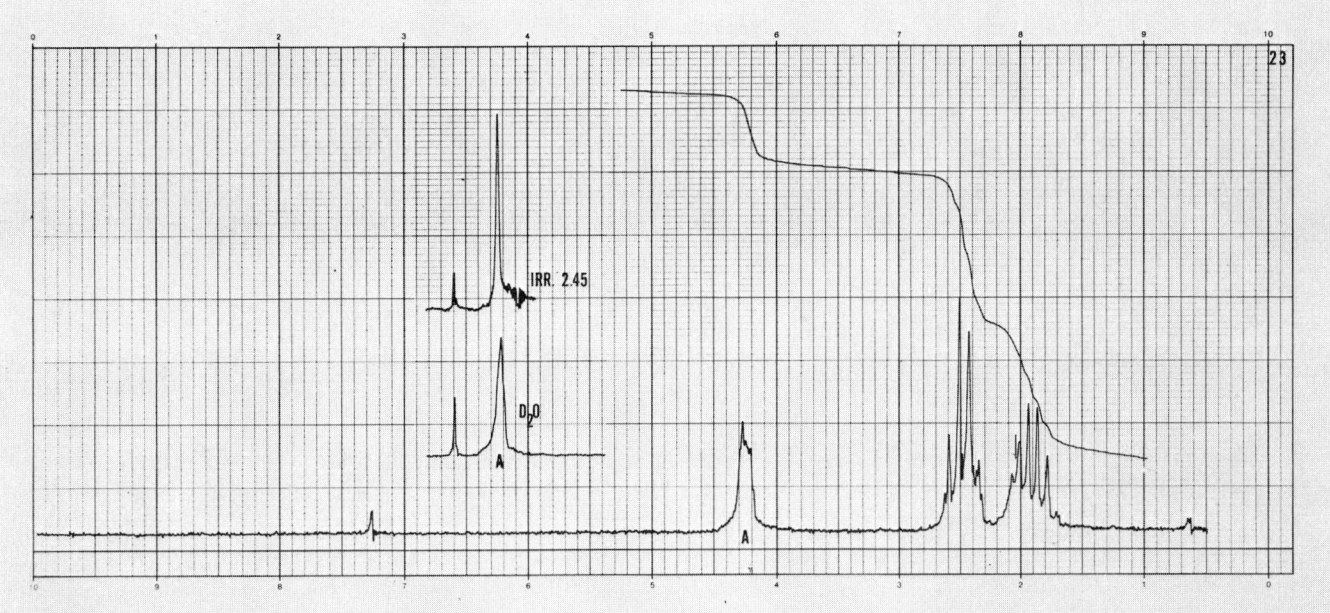

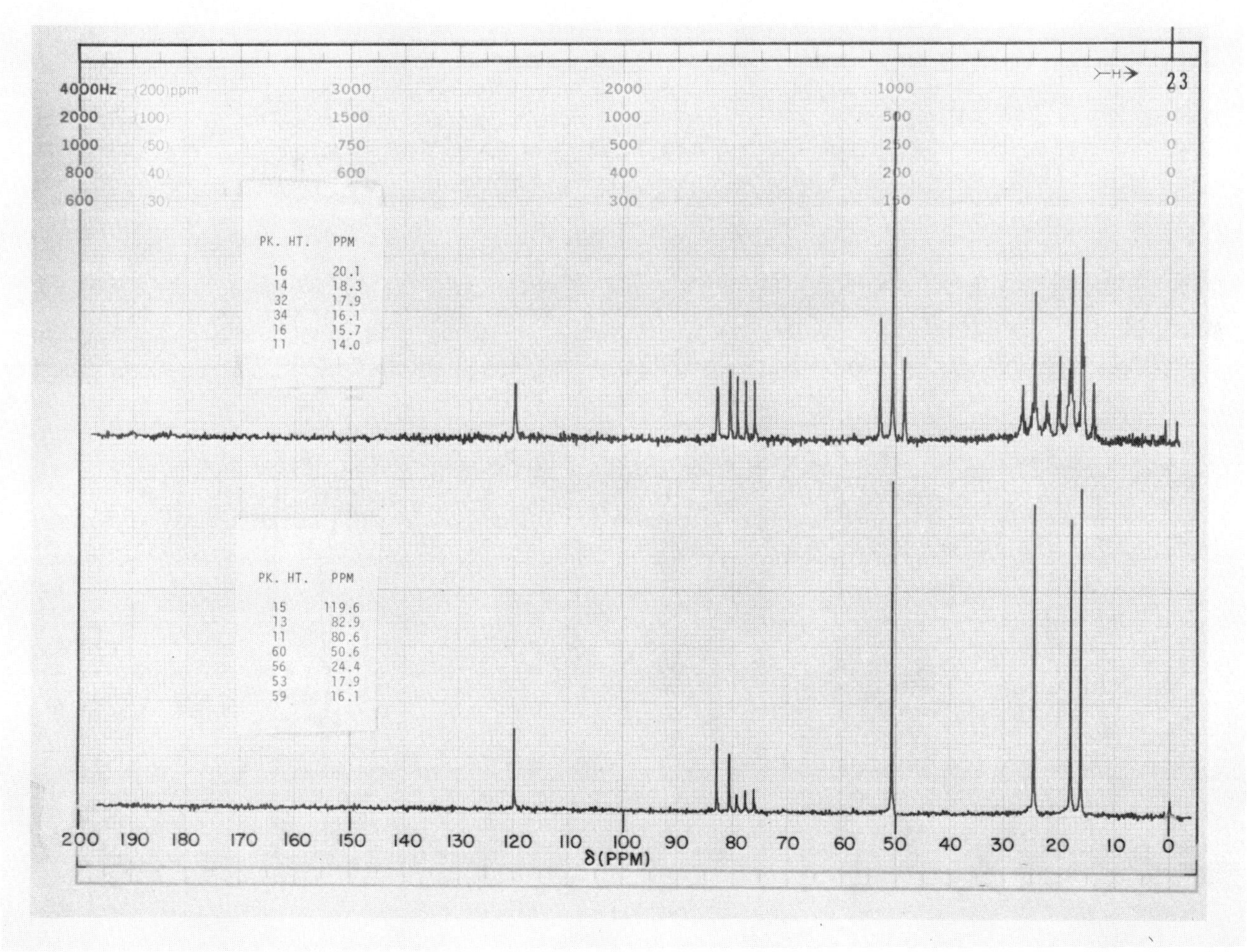

COMPOUND 24

Compound 24

MS: 94.079
IR: melt
HMR: CDCl_3
CMR: d_6 acetone
Analysis: 89.3% C; 10.7% H

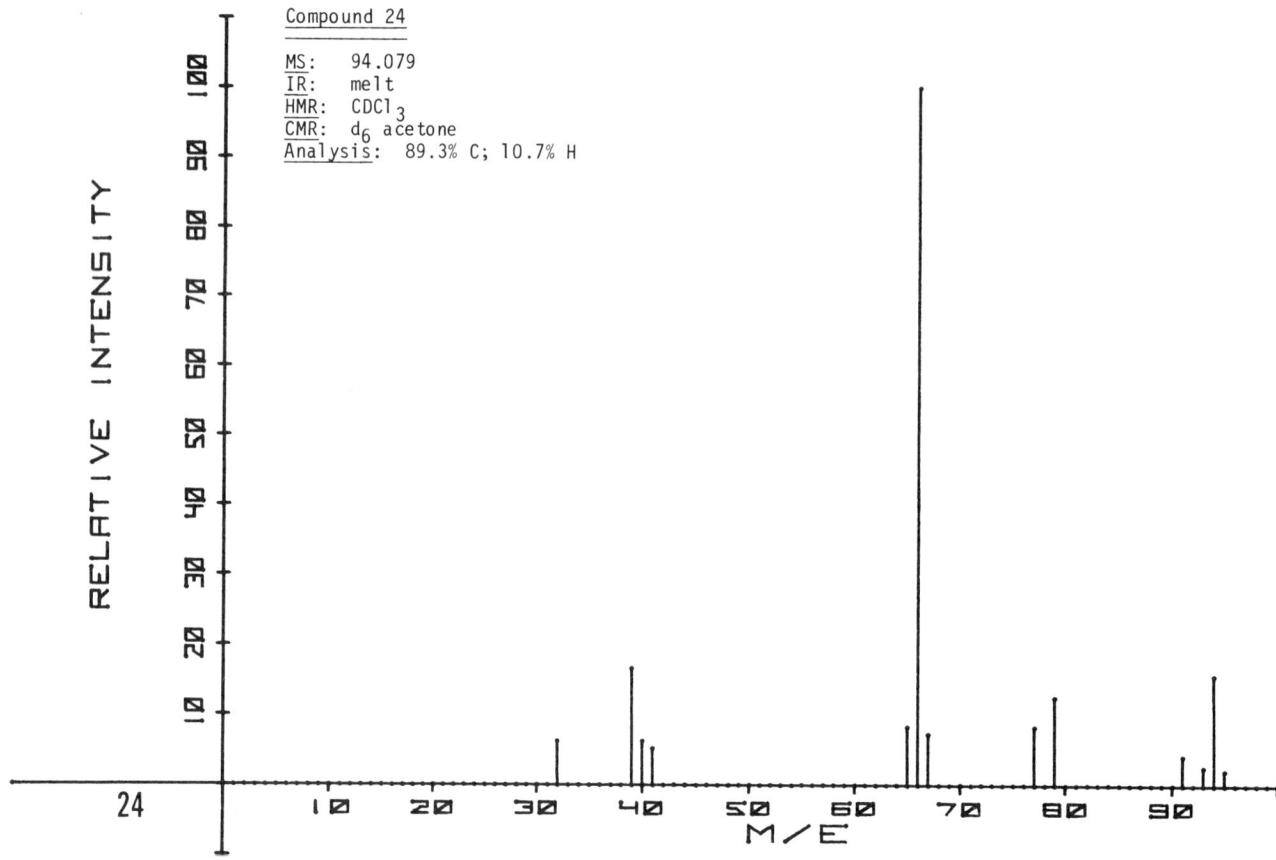

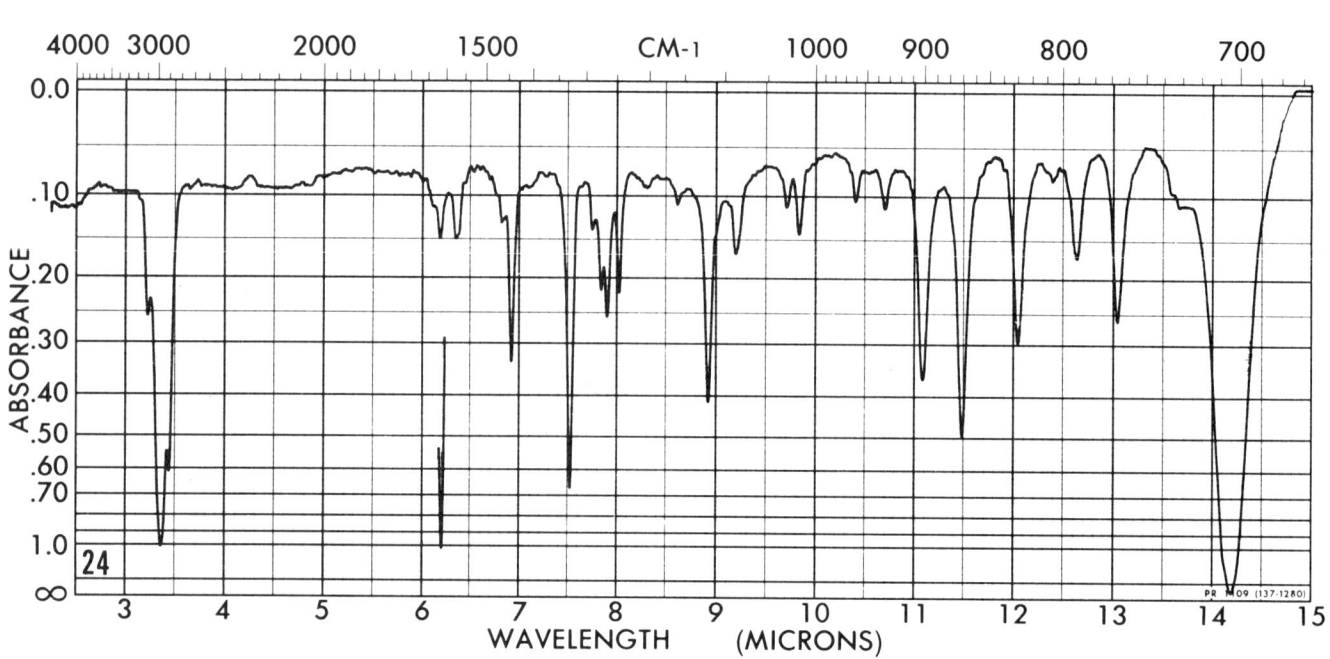

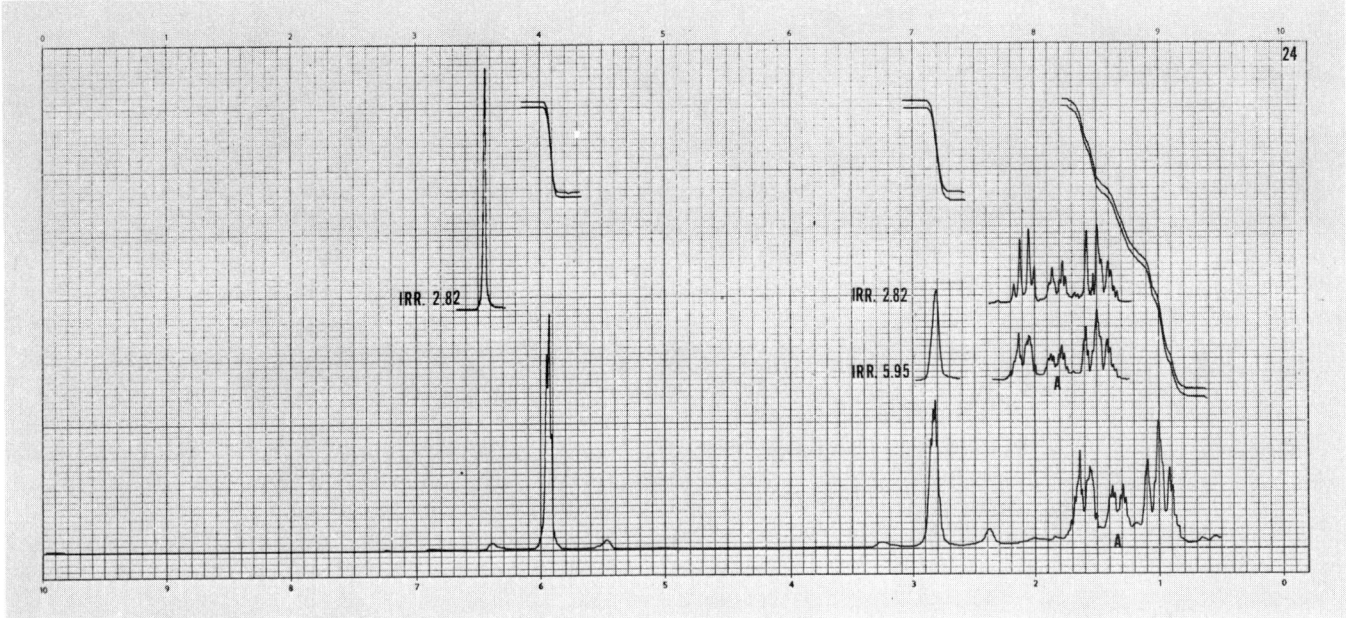

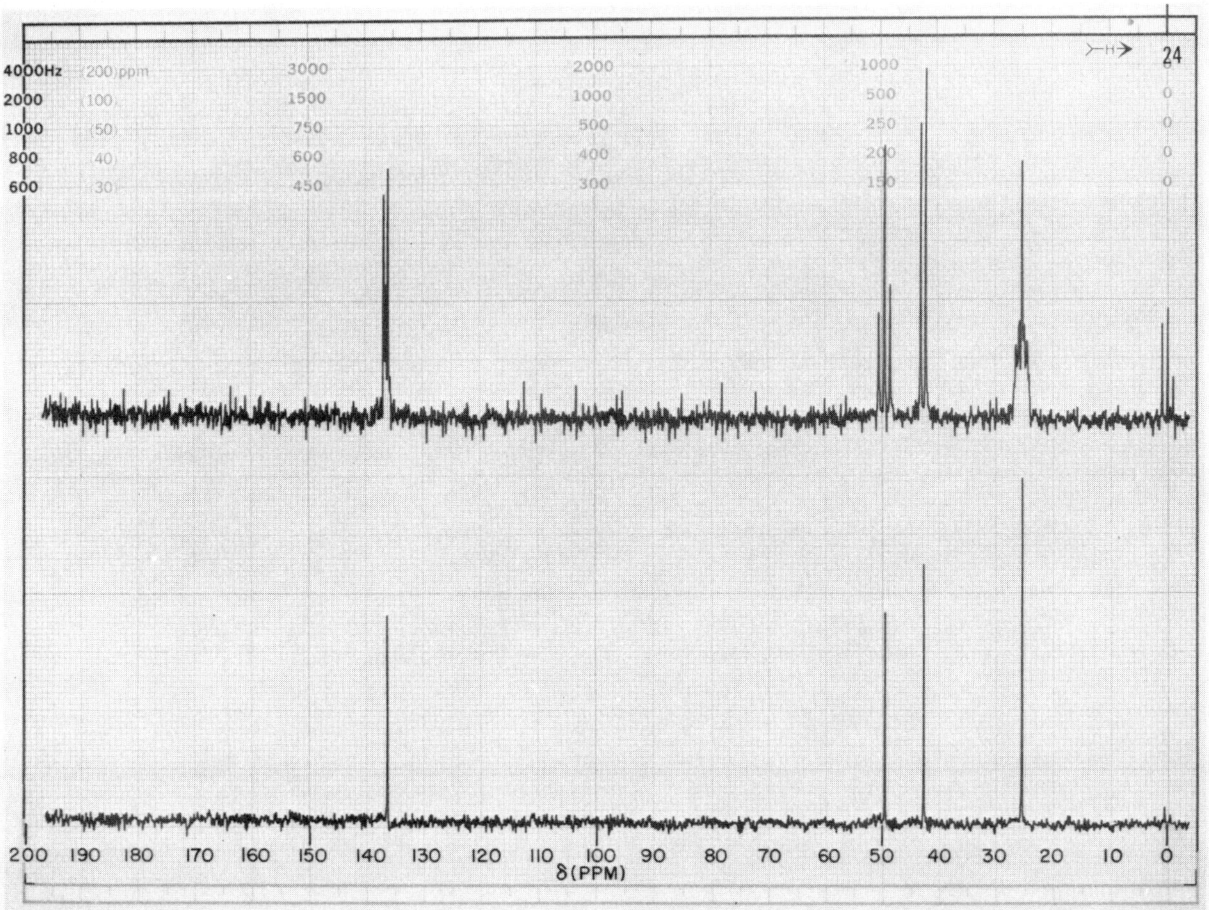

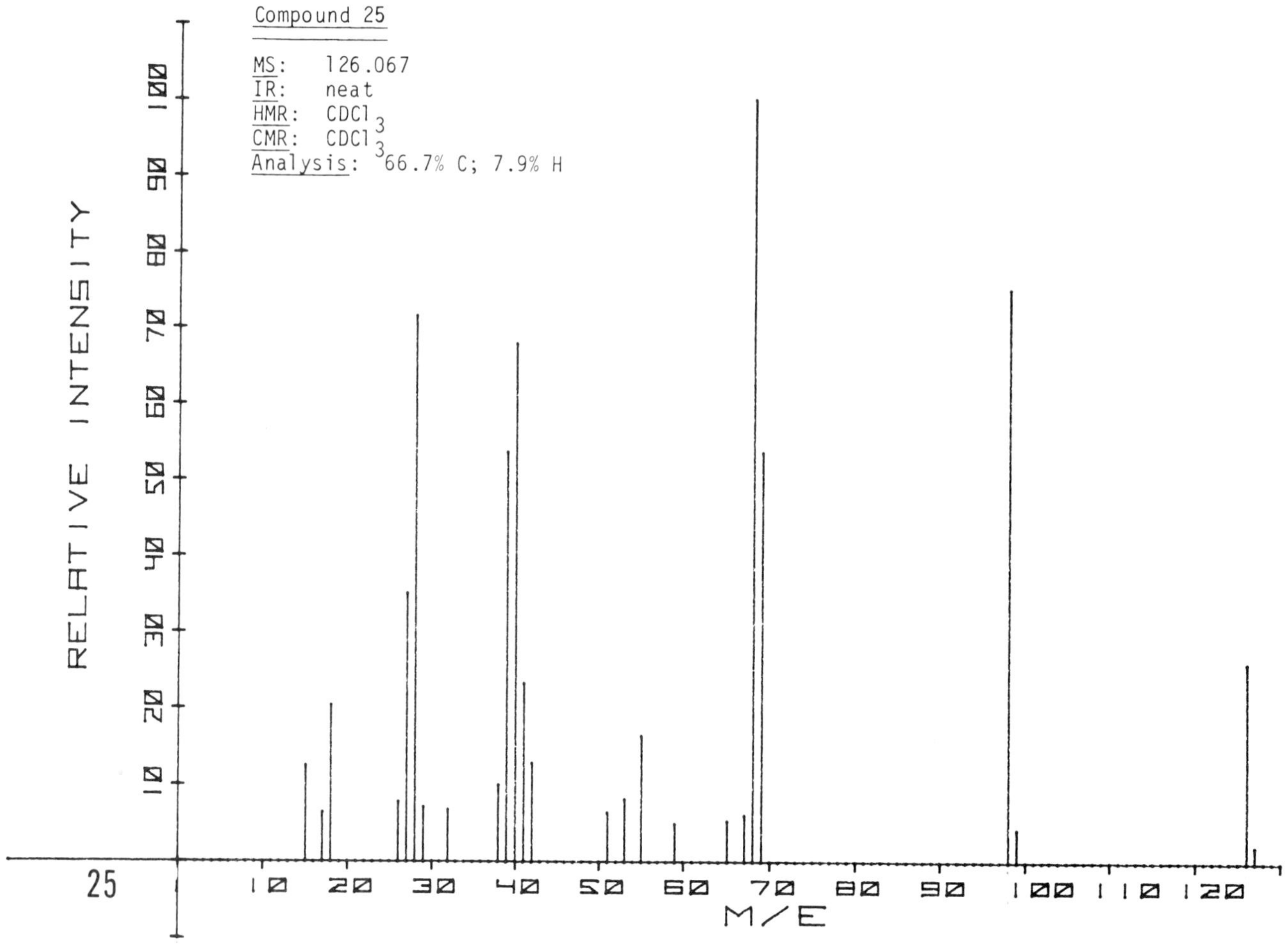

Compound 25
MS: 126.067
IR: neat
HMR: CDCl3
CMR: CDCl3
Analysis: 66.7% C; 7.9% H
RELATIVE INTENSITY
10 20 30 40 50 60 70 80 90 100
25
10 20 30 40 50 60 70 80 90 100 110 120
M/E

4000 3000 2000 1500 CM-1 1000 900 800 700
ABSORBANCE
0.0 .10 .20 .30 .40 .50 .60 .70 1.0 ∞
25
3 4 5 6 7 8 9 10 11 12 13 14 15
WAVELENGTH (MICRONS)

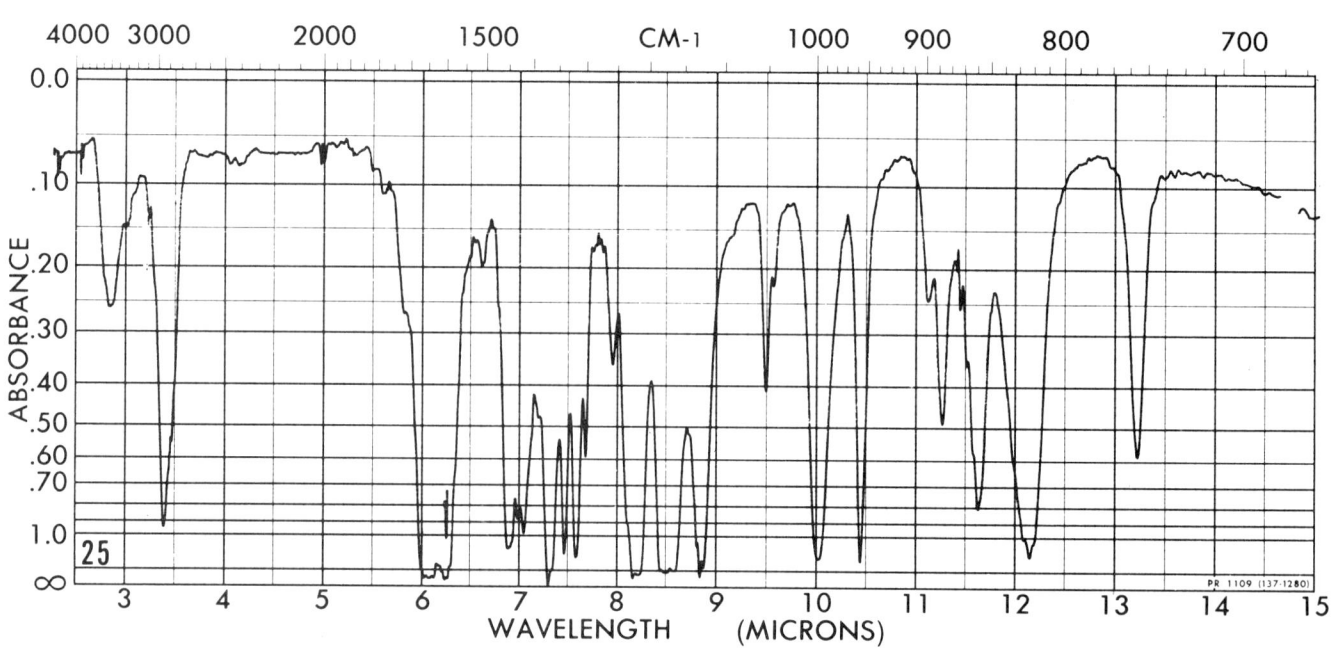

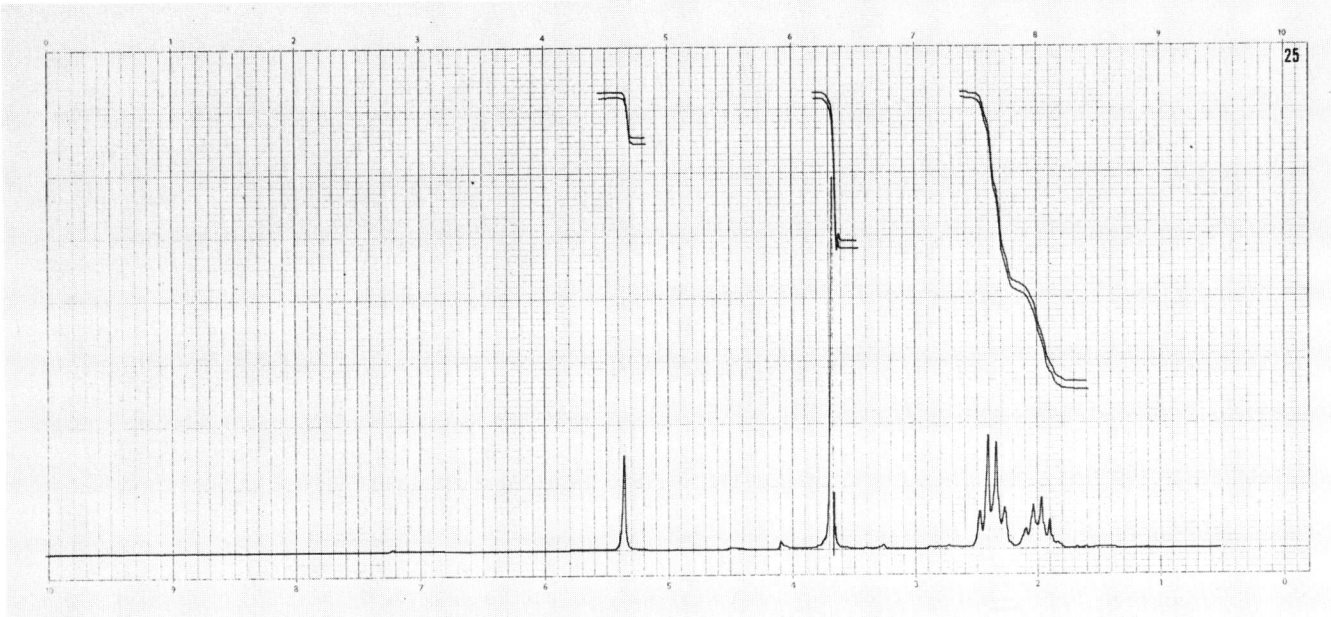

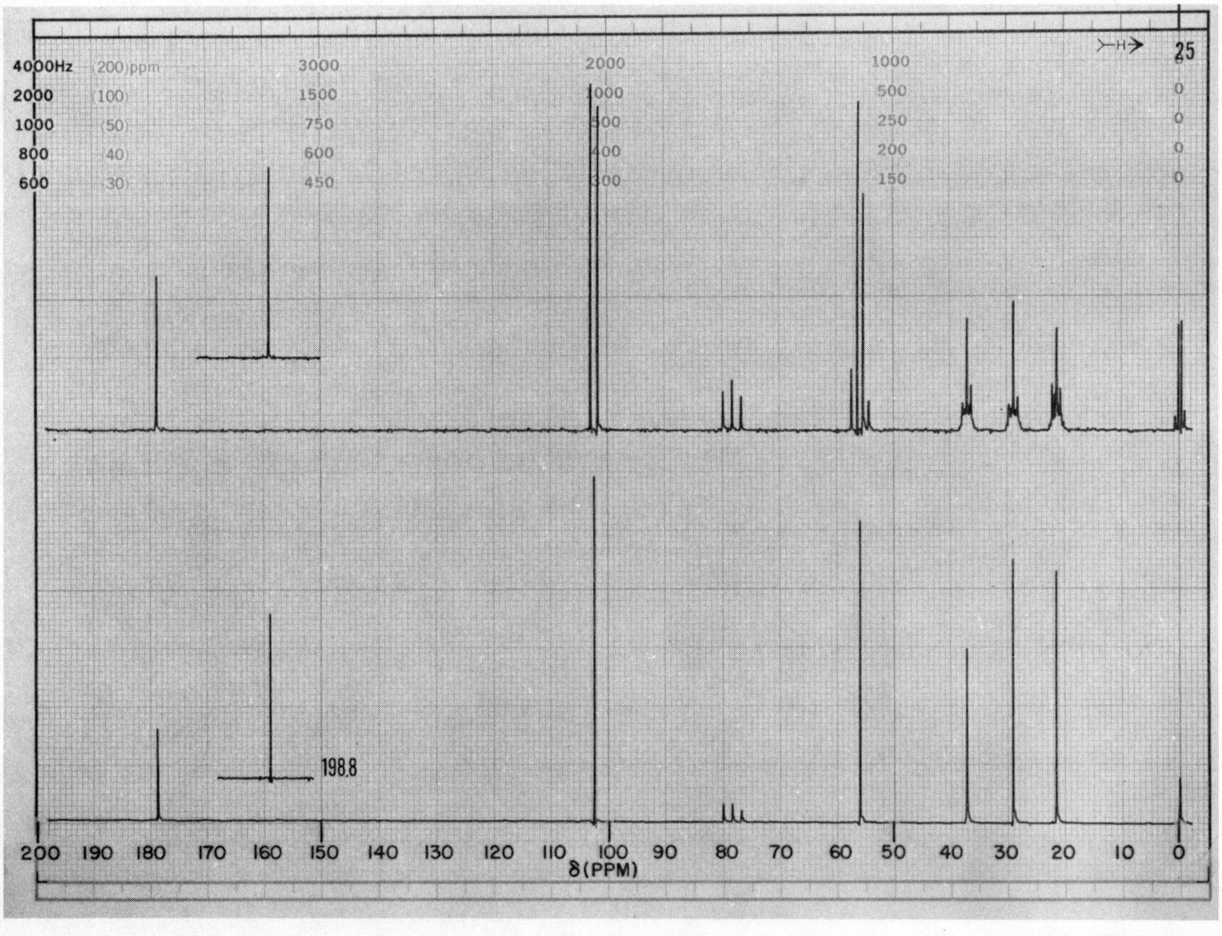

COMPOUND 26A

Compound 26A
MS: 112.088
IR: neat
HMR: CCl₄
CMR: neat
Analysis: 75.1% C; 10.8% H

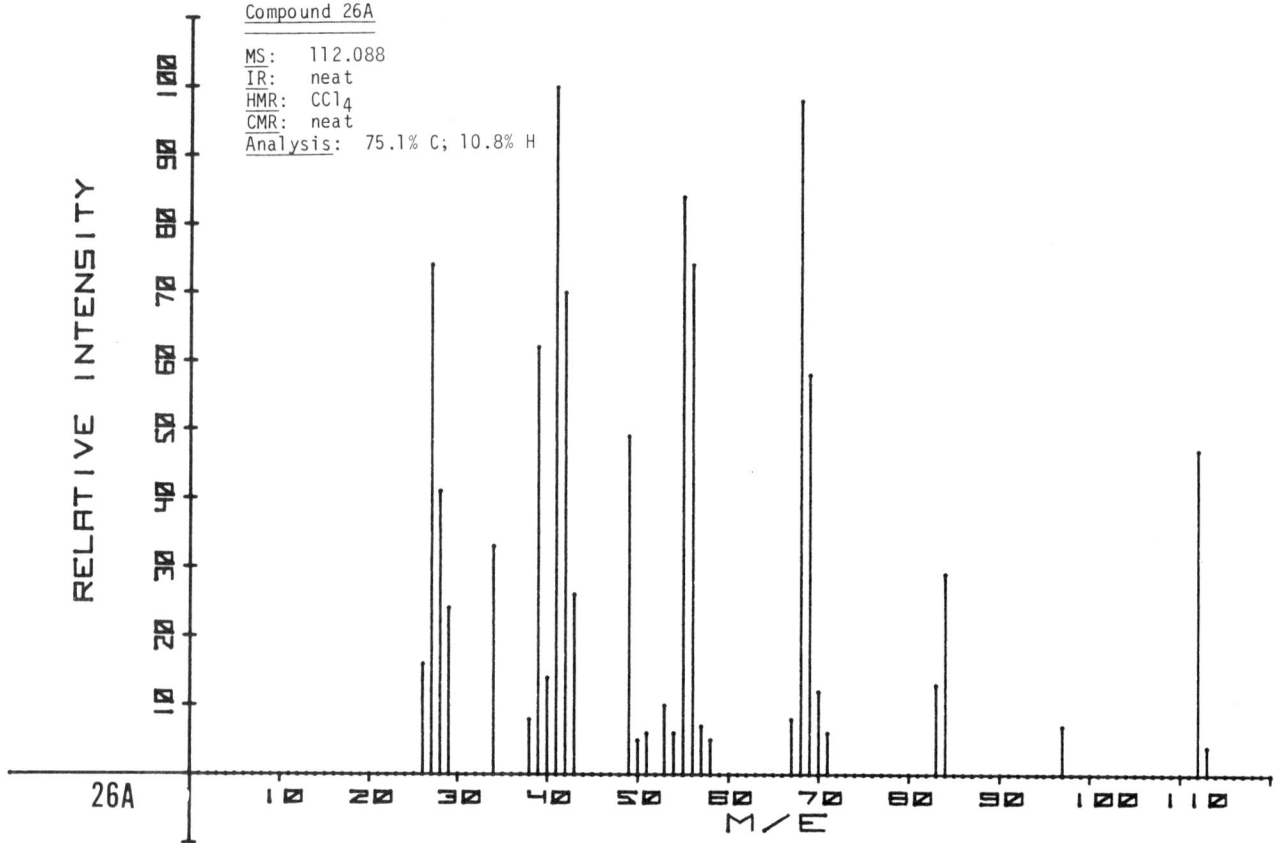

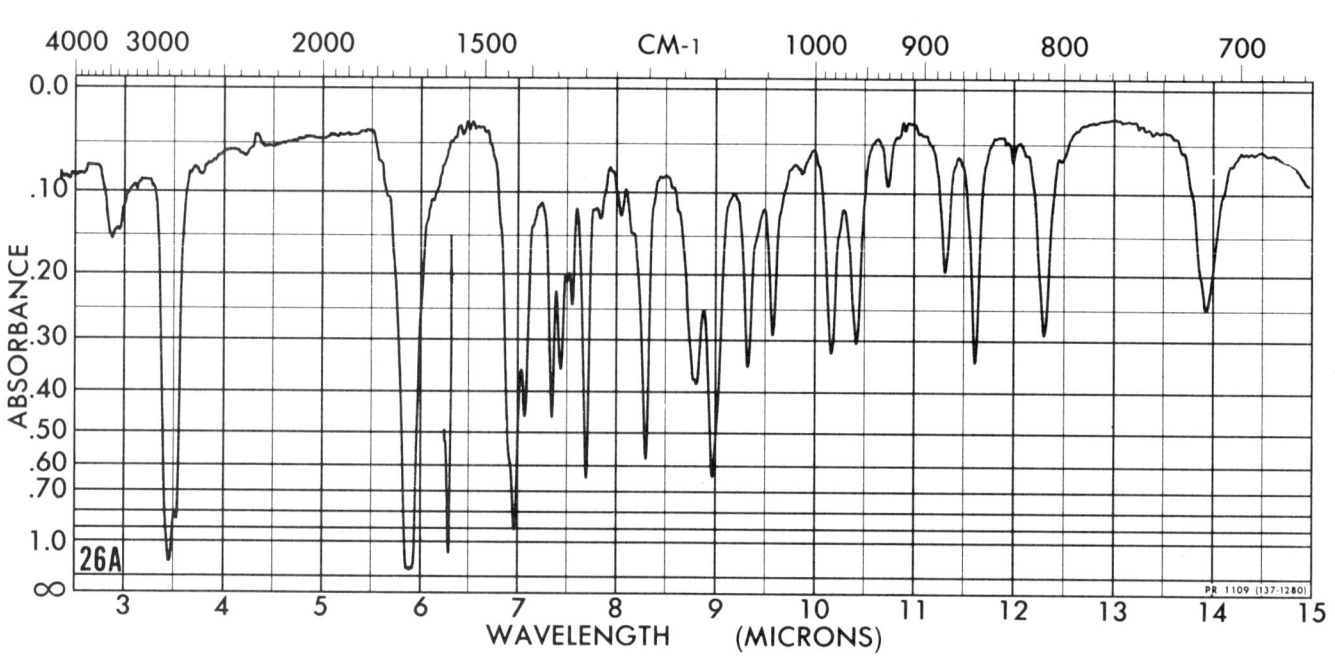

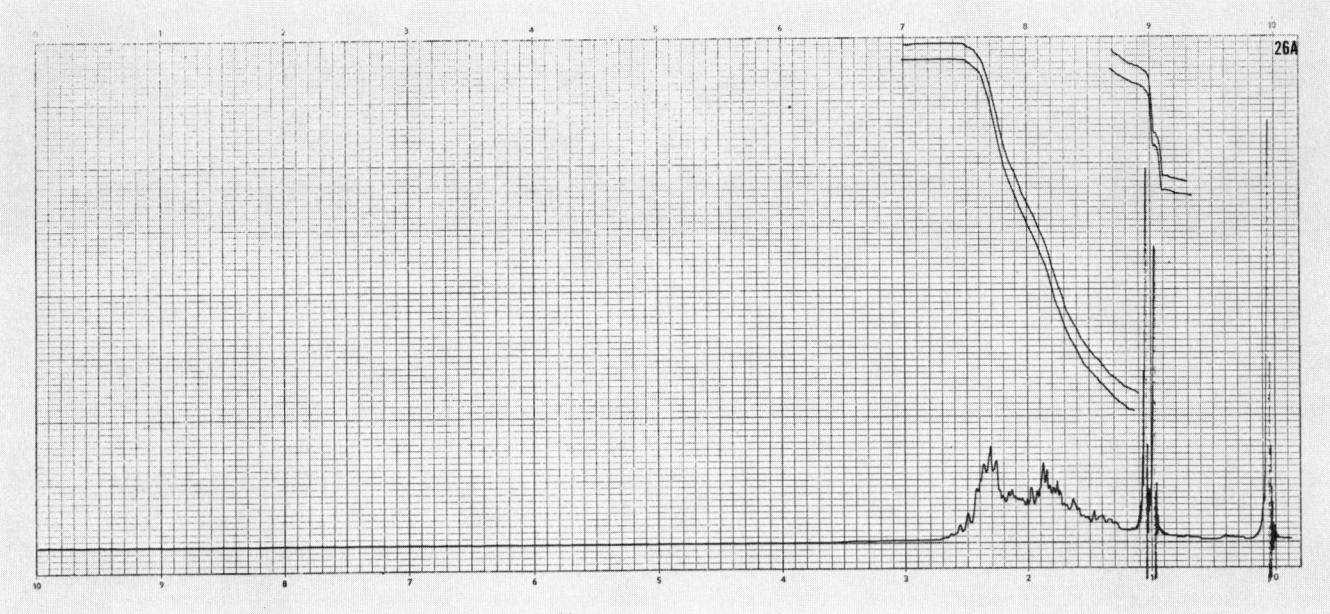

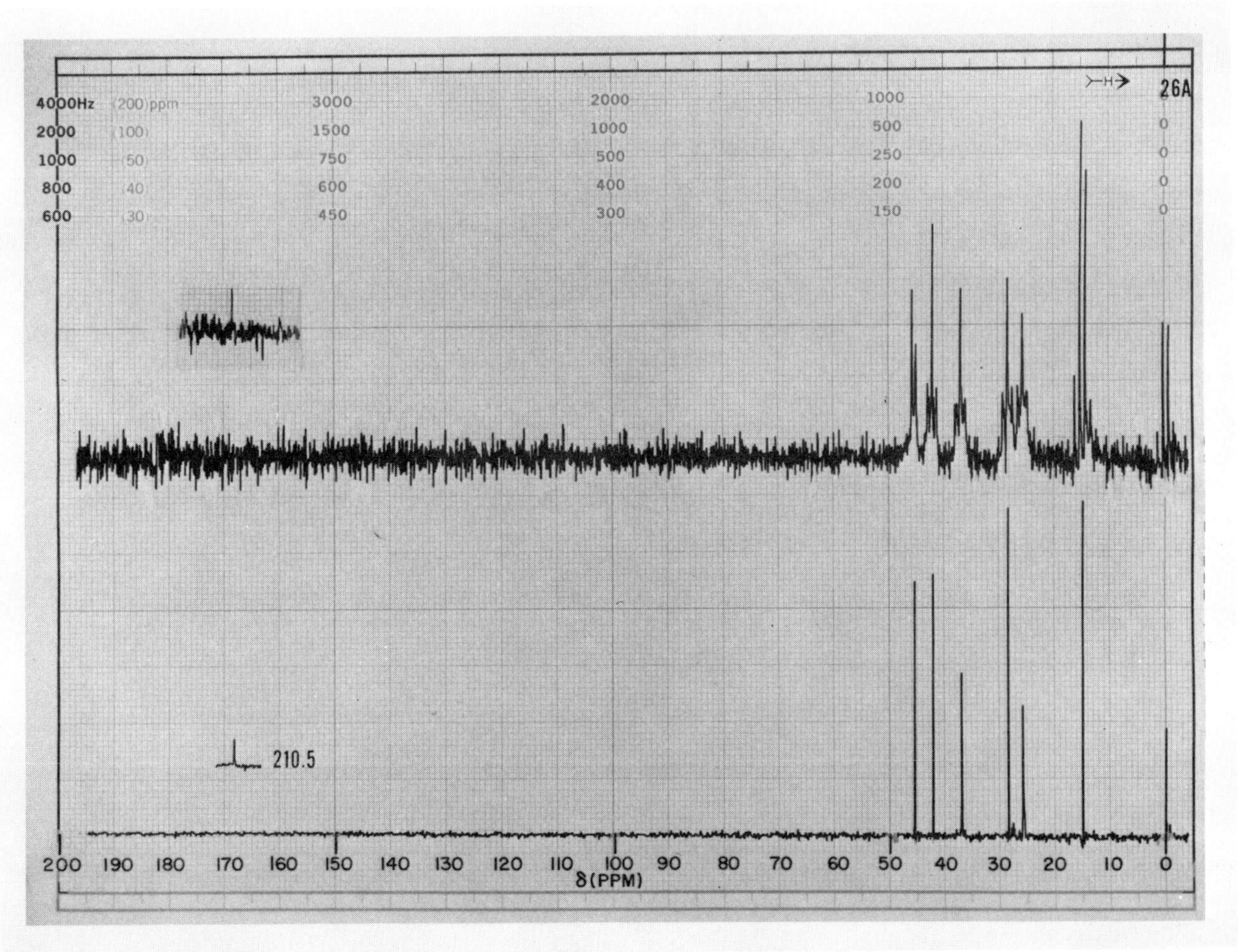

COMPOUND 26B

Compound 26B

MS: 112.088
IR: neat
HMR: CCl_4
CMR: CDCl_3
Analysis: 74.9% C; 10.7% H

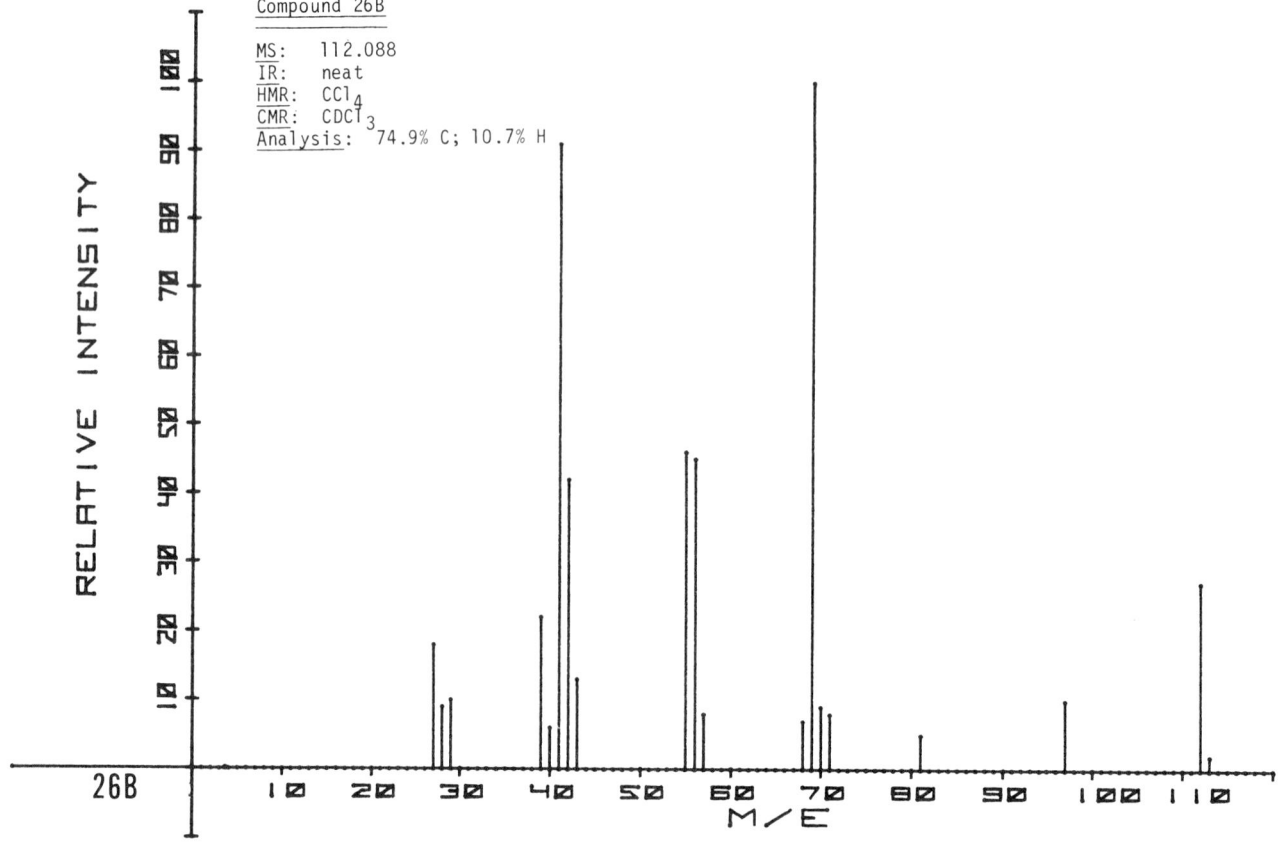

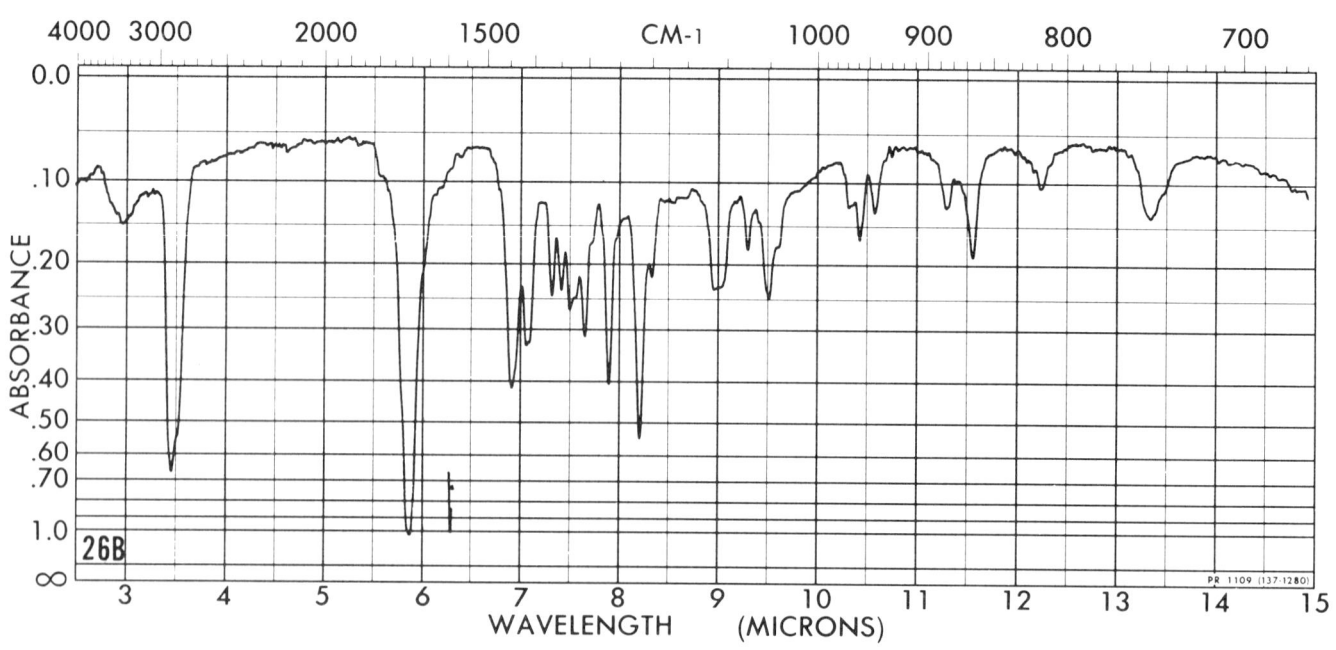

26B

26B
4000Hz (200)ppm 3000 2000 1000
2000 (100) 1500 500
1000 (50) 750 250
800 (40) 600 200
600 (30) 450 150
PK. HT. PPM
29 49.9
42 41.0
32 34.9
42 33.4
10 31.8
13 26.9
31 25.4
18 24.4
14 23.9
60 22.9
43 21.3
8 19.8
PK. HT. PPM
17 210.3
52 49.9
60 41.0
58 34.2
52 33.4
52 25.4
44 22.1
210.3
200 190 180 170 160 150 140 130 120 110 100 90 80 70 60 50 40 30 20 10 0
δ(PPM)

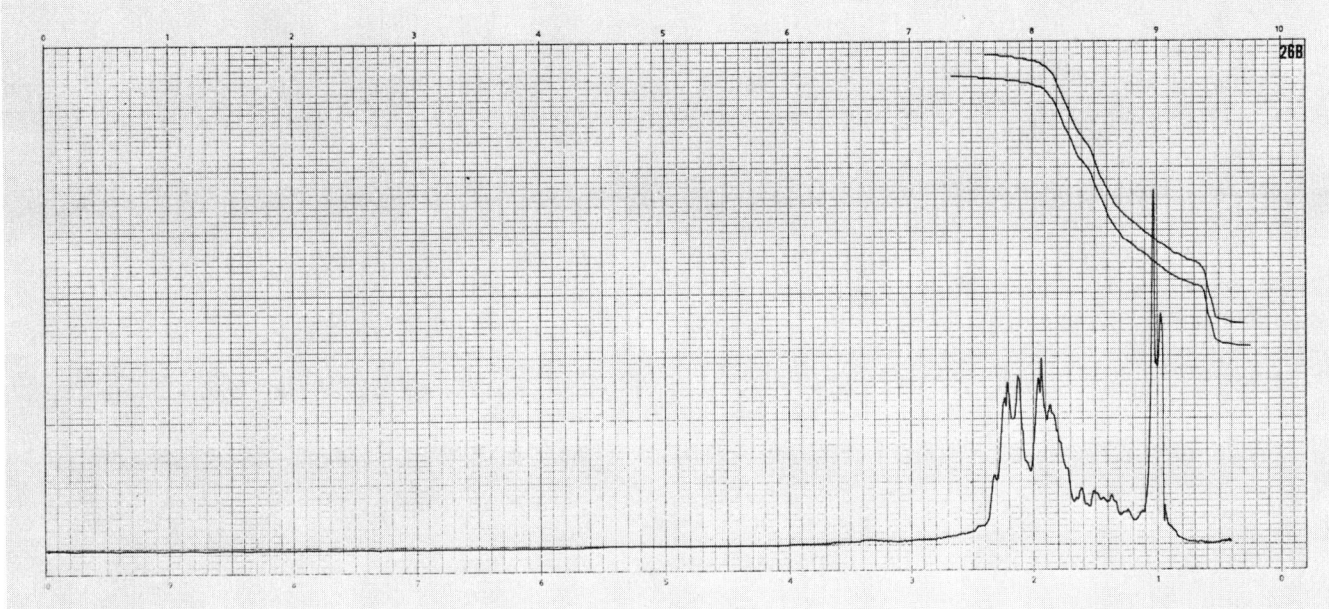

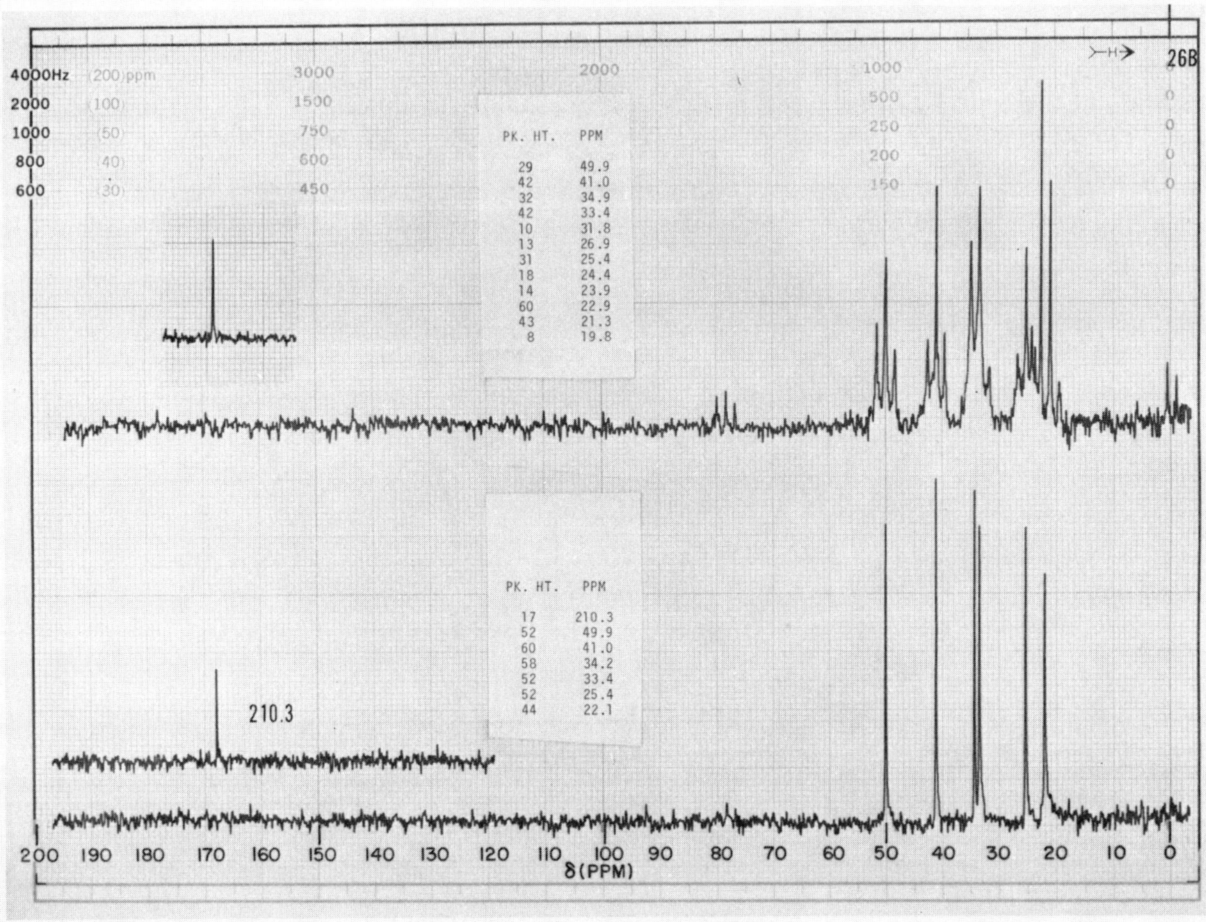

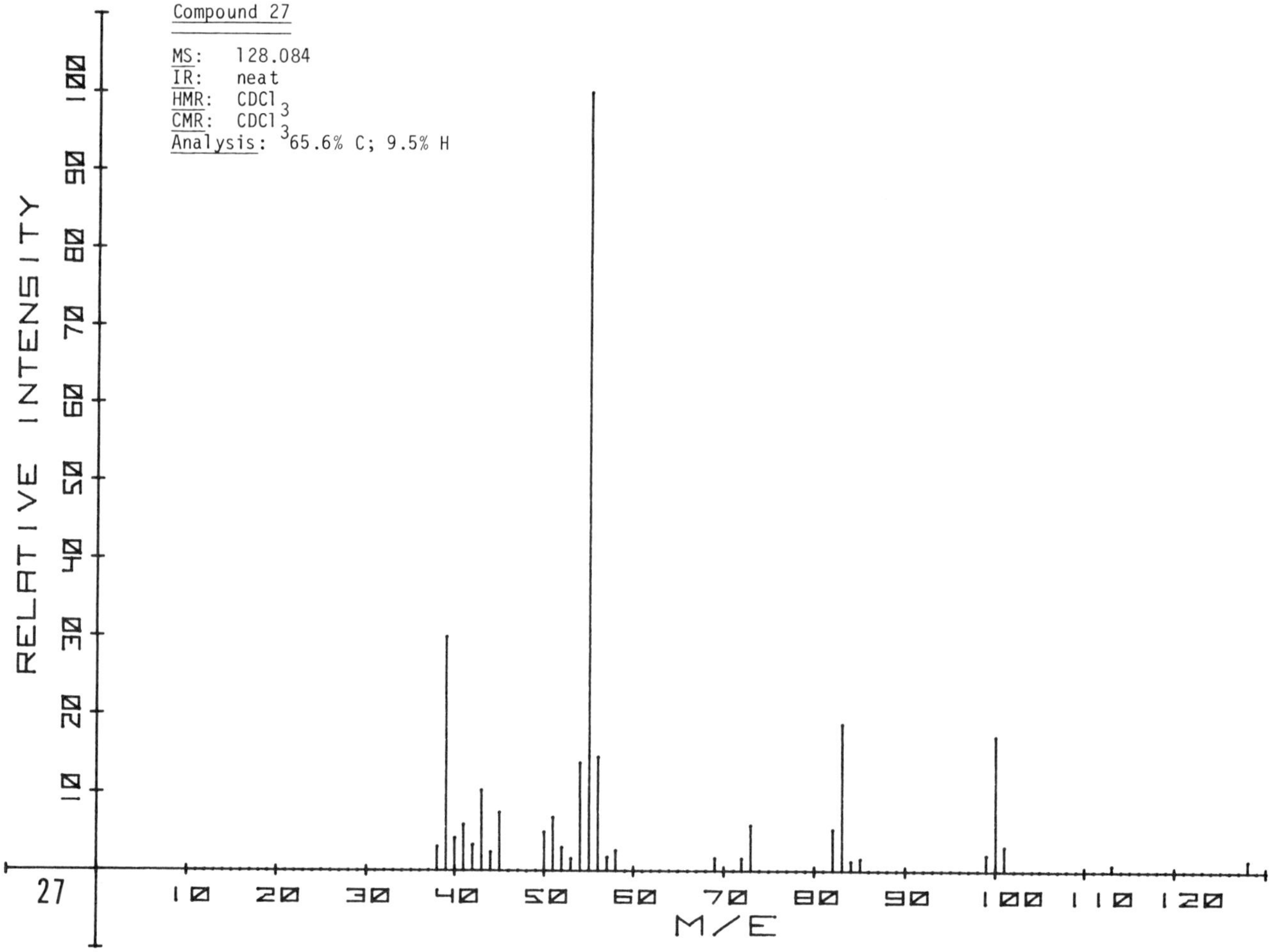

Compound 27
MS: 128.084
IR: neat
HMR: CDCl3
CMR: CDCl3
Analysis: 65.6% C; 9.5% H
RELATIVE INTENSITY
M/E
27

CM-1
ABSORBANCE
WAVELENGTH (MICRONS)
27
PR 1109 (137-1280)

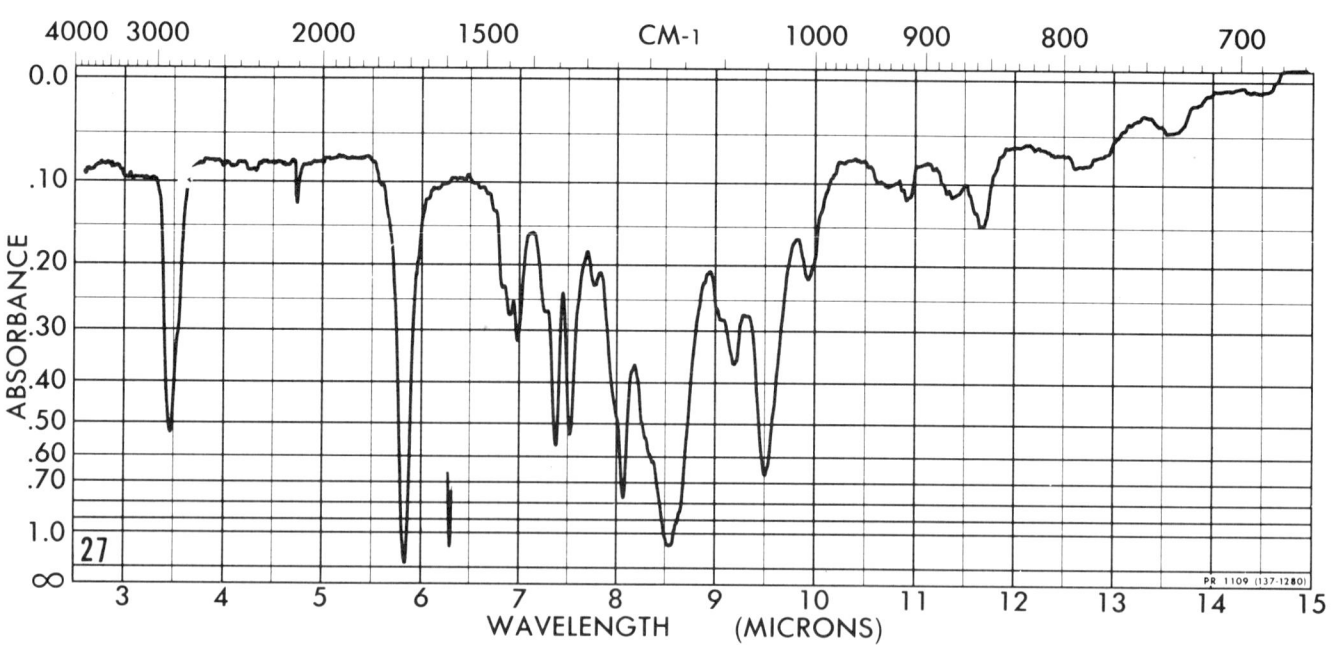

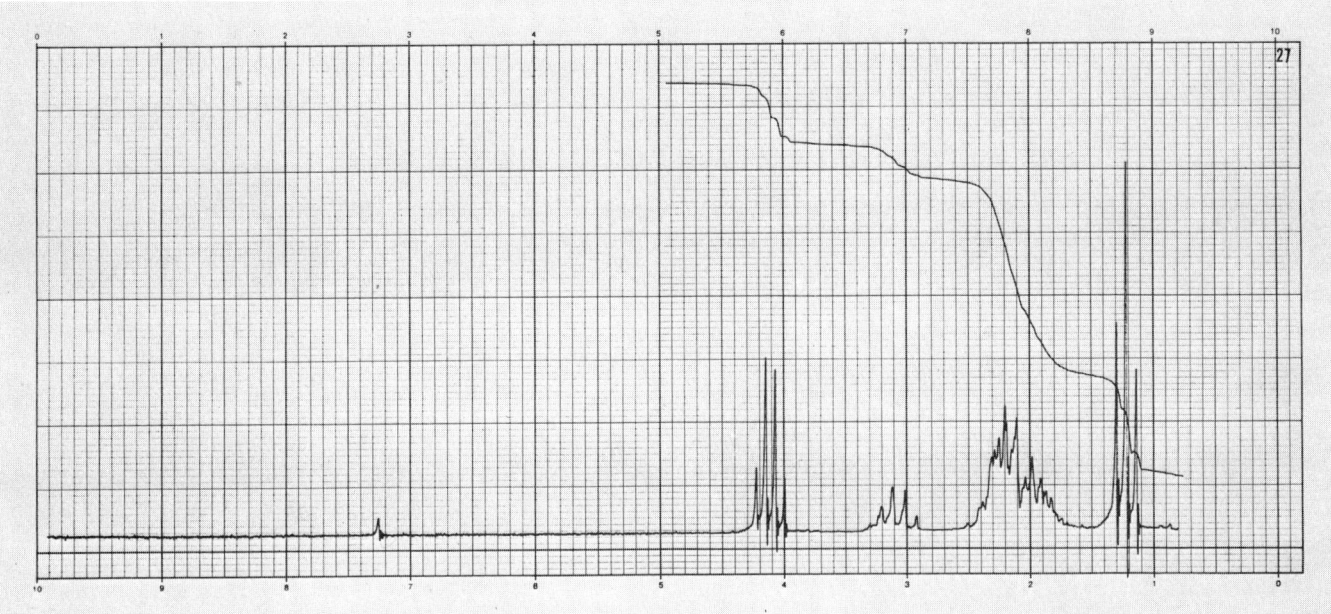

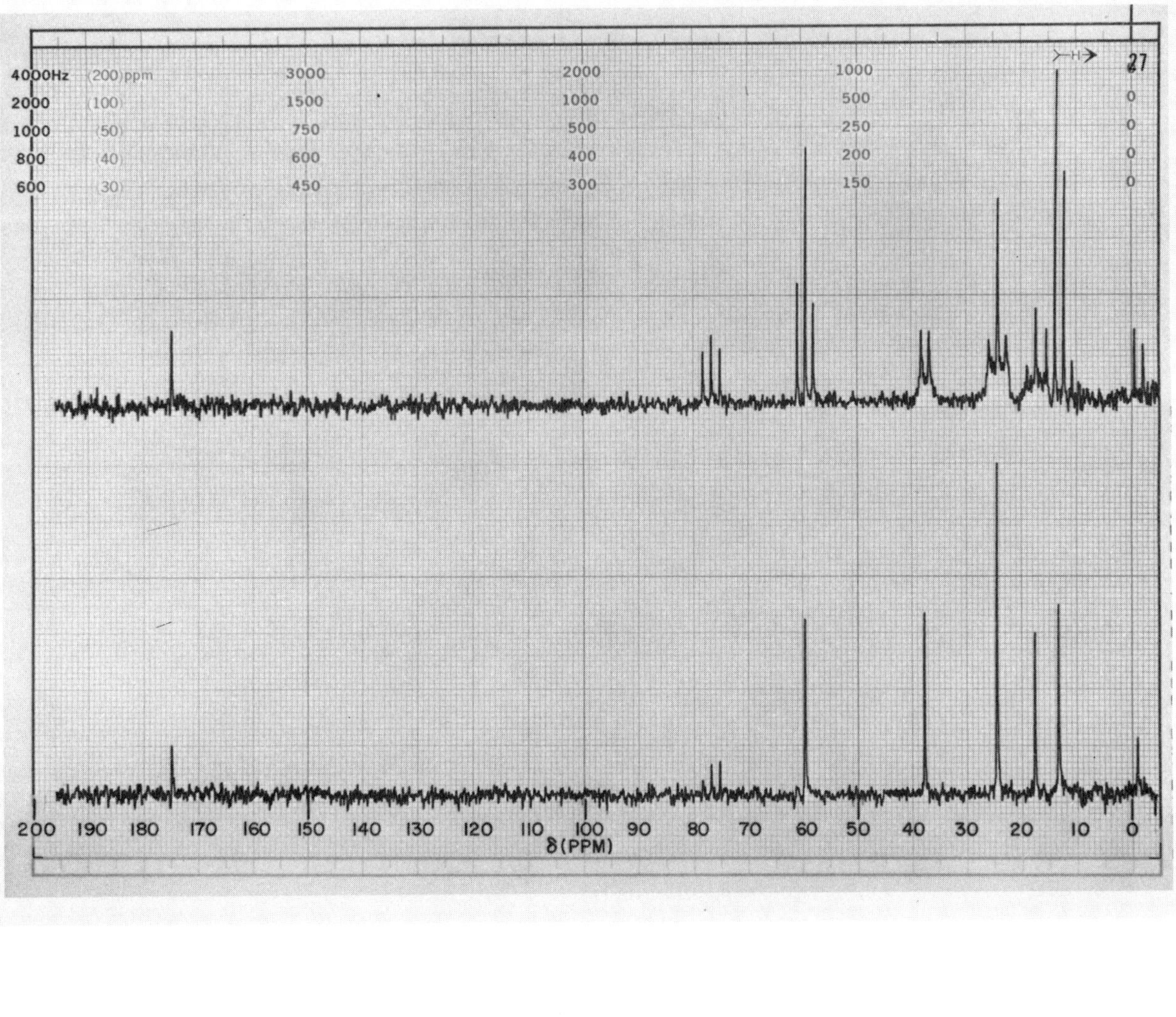

COMPOUND 28

Compound 28

MS: 164.060
IR: neat
HMR: CDCl$_3$
CMR: CDCl$_3$
Analysis: 51.1% C; 8.0% H; 21.6% Cl

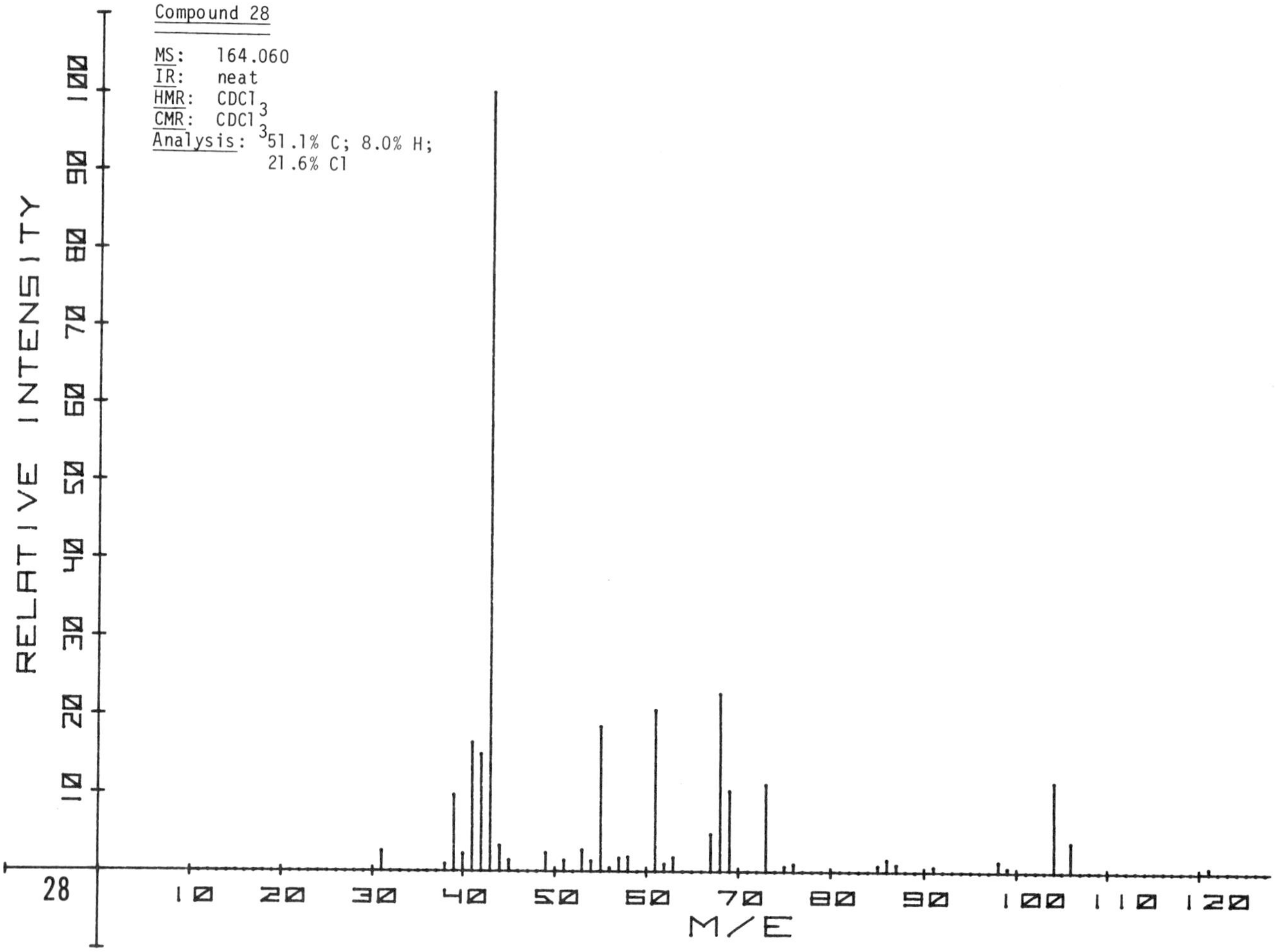

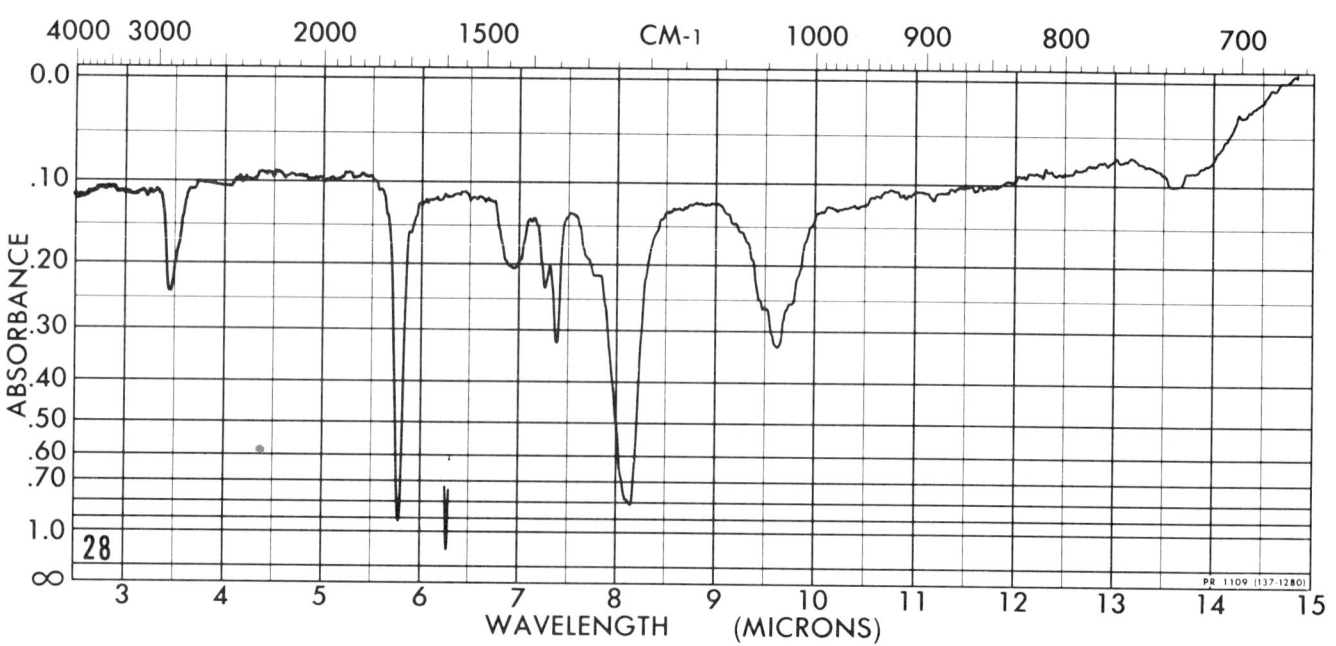

28
IRR. 1.65

4000Hz
2000
1000
800
600
200 ppm
100
50
40
30
2000
1000
500
400
300
1000
500
250
200
150
28
PK. HT. PPM
8 34.3
25 32.2
13 30.1
29 28.0
12 25.9
7 25.5
10 24.1
14 23.4
32 22.0
5 21.3
38 19.9
7 17.7
PK. HT. PPM
16 171.1
60 64.2
52 44.8
37 32.2
48 28.0
52 23.4
18 20.9
200 190 180 170 160 150 140 130 120 110 100 90 80 70 60 50 40 30 20 10 0
δ(PPM)

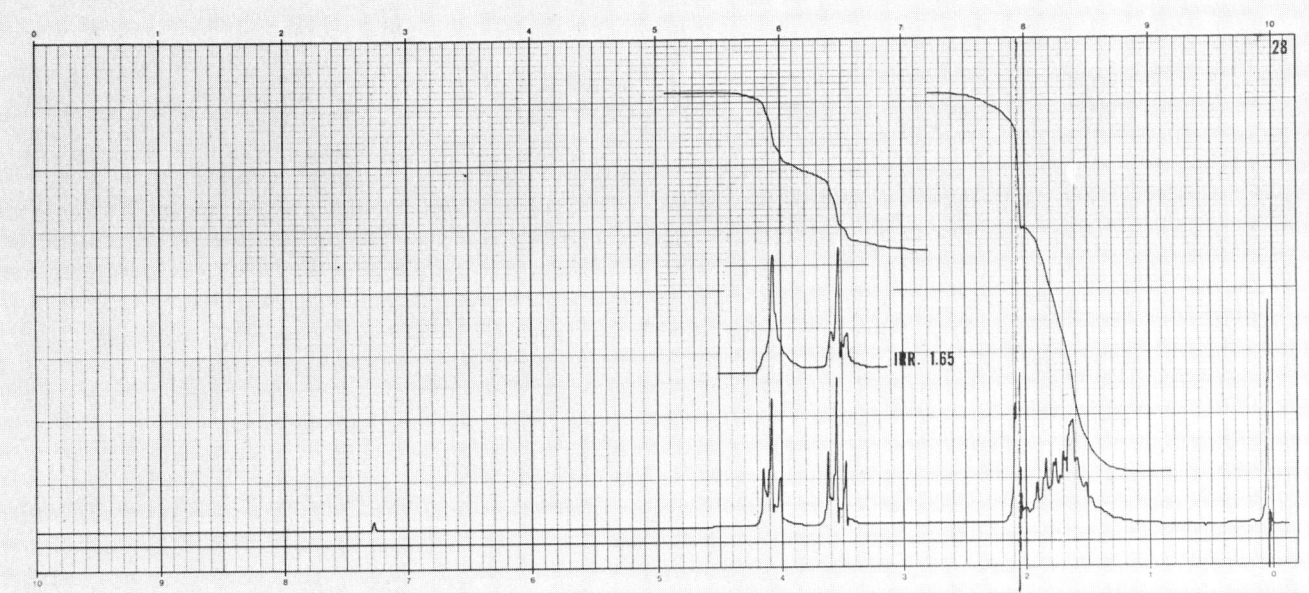

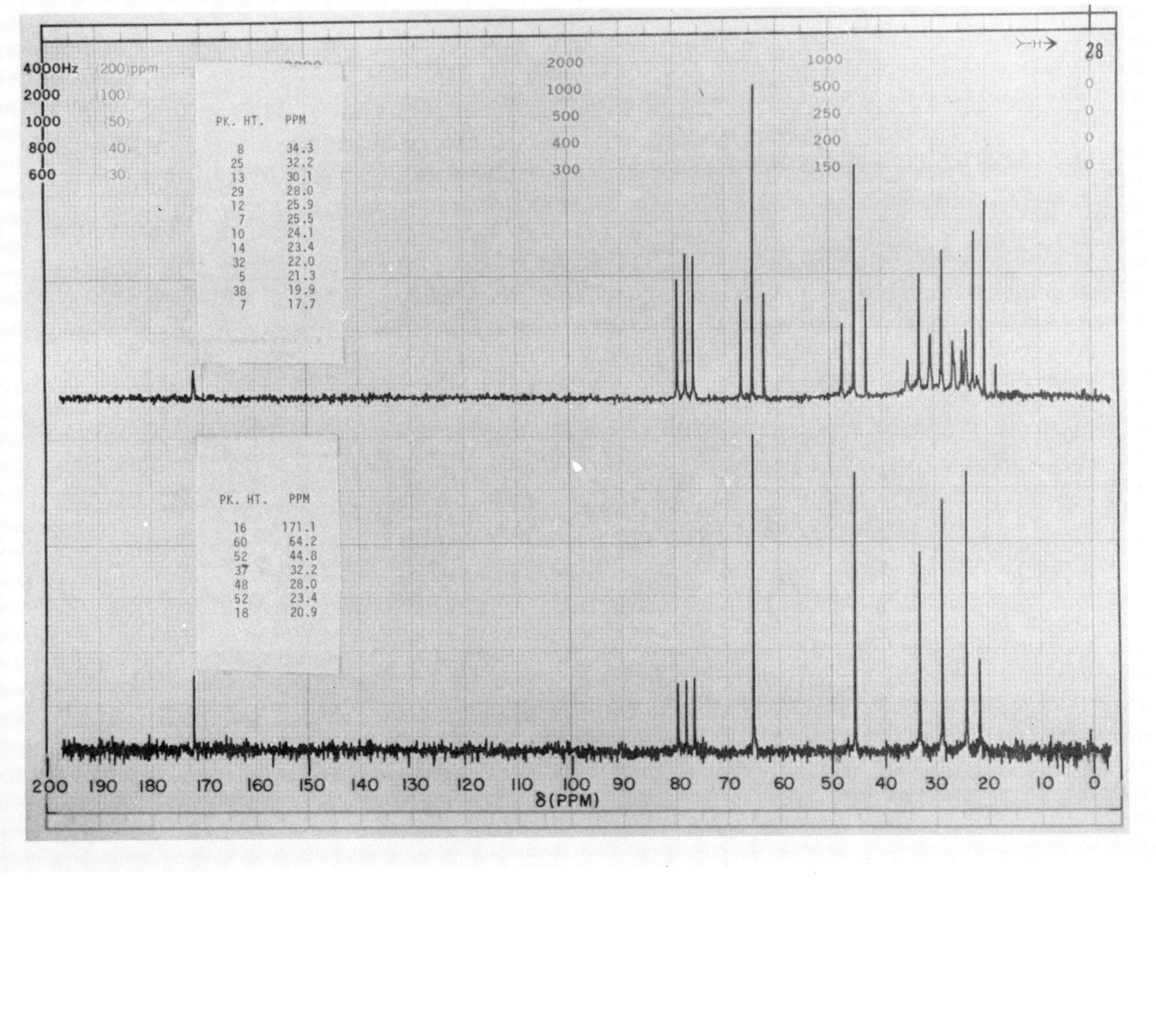

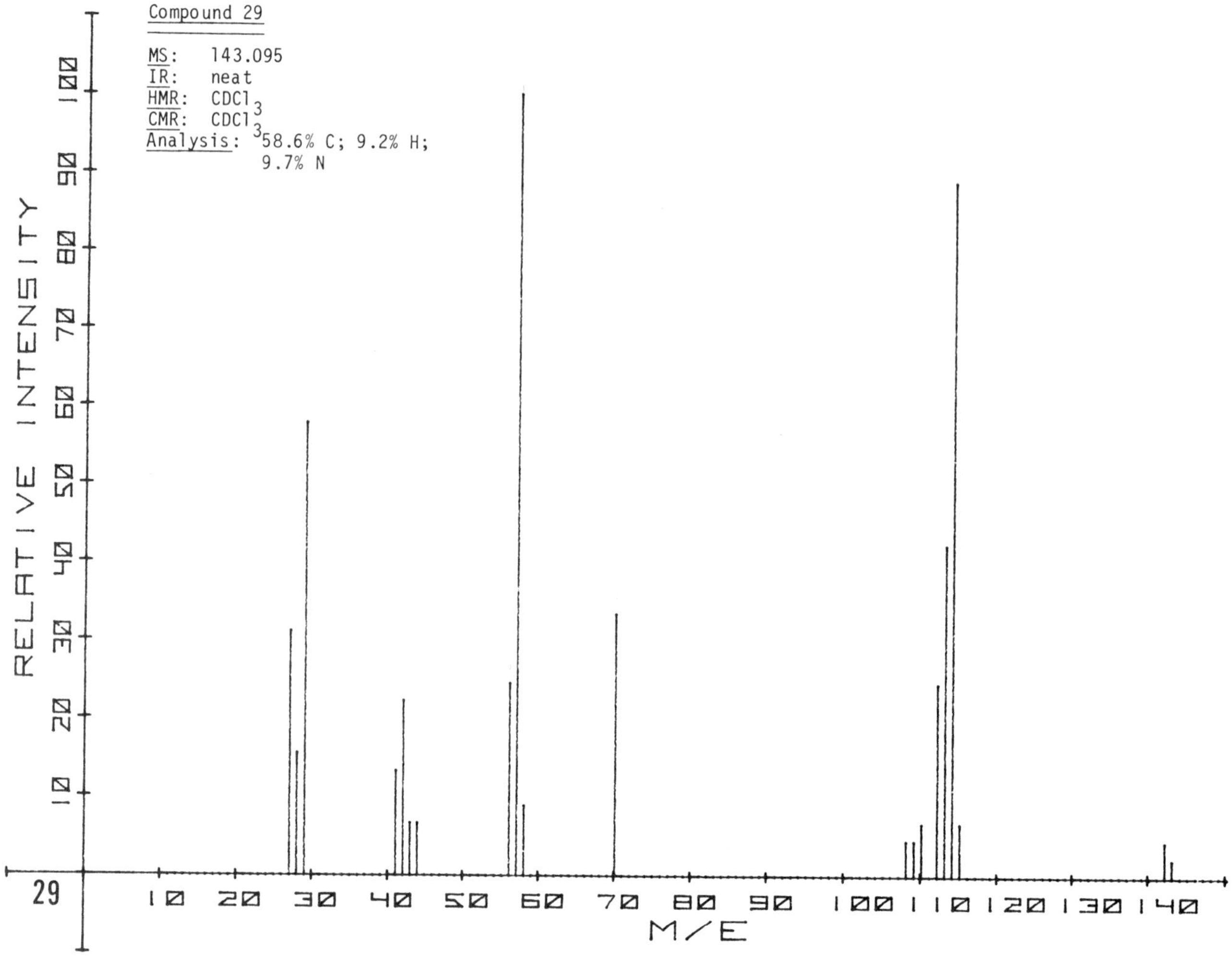

Compound 29
MS: 143.095
IR: neat
HMR: $CDCl_3$
CMR: $CDCl_3$
Analysis: 58.6% C; 9.2% H; 9.7% N
RELATIVE INTENSITY
10 20 30 40 50 60 70 80 90 100
29
10 20 30 40 50 60 70 80 90 100 110 120 130 140
M/E

4000 3000 2000 1500 CM-1 1000 900 800 700
0.0 .10 .20 .30 .40 .50 .60 .70 1.0 ∞
ABSORBANCE
29
3 4 5 6 7 8 9 10 11 12 13 14 15
WAVELENGTH (MICRONS)

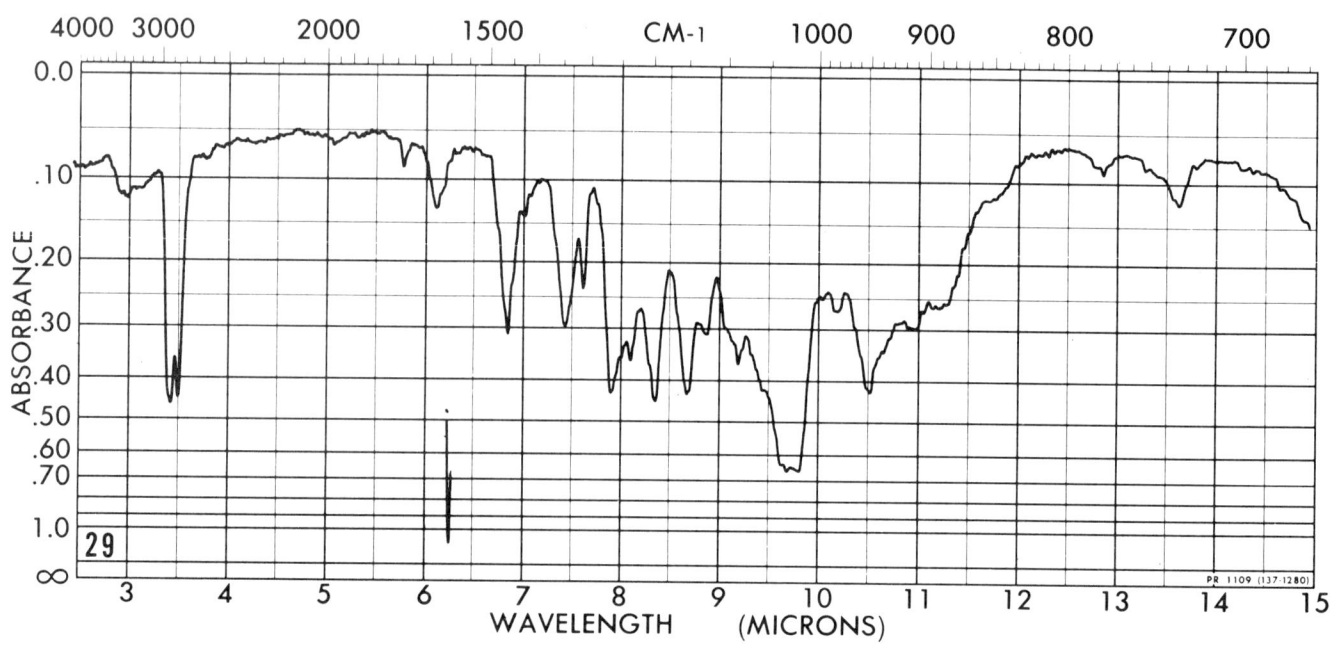

29
4
3

29
4000Hz (200)ppm 3000 2000 1000 0
2000 (100) 1500 1000 500 0
1000 (50) 750 500 250 0
800 (40) 600 400 200 0
600 (30) 450 300 150 0
200 190 180 170 160 150 140 130 120 110 100 90 80 70 60 50 40 30 20 10 0
δ(PPM)

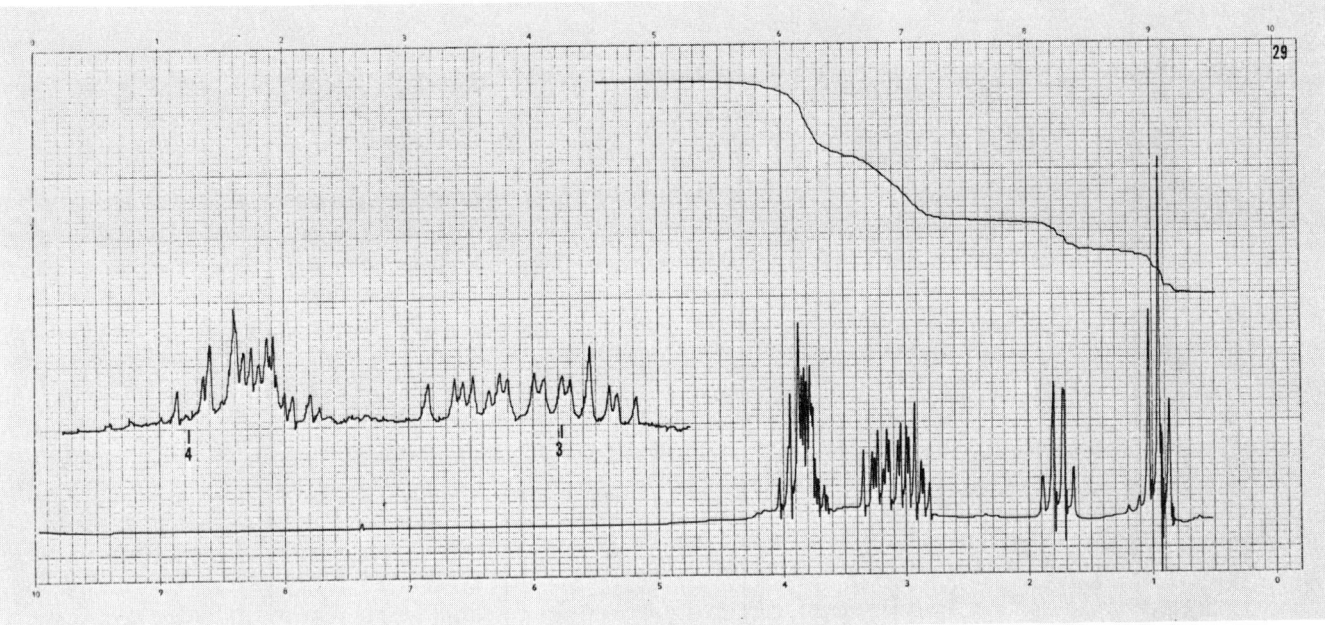

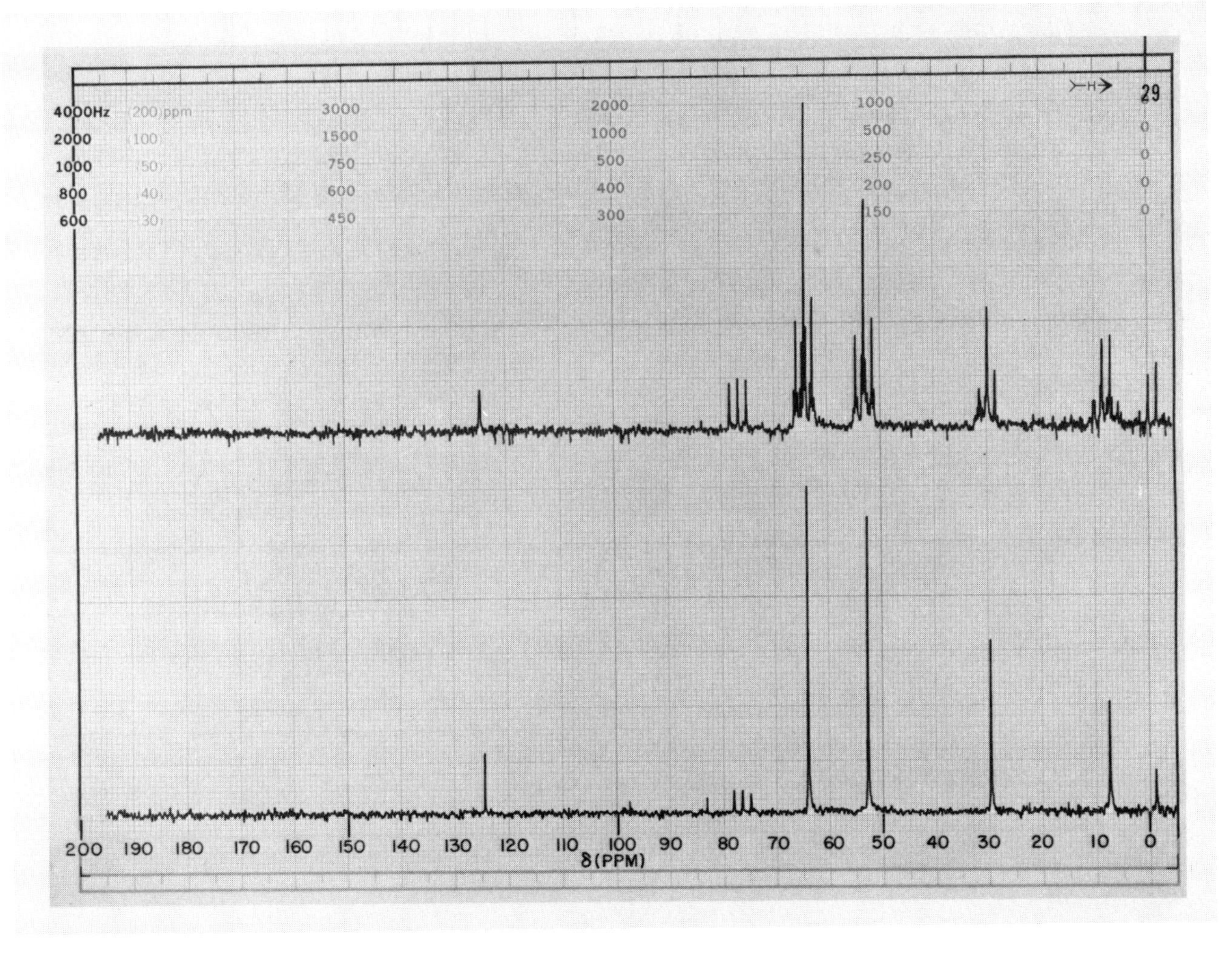

COMPOUND 30

Compound 30
MS: 103.076
IR: neat
HMR: CCl4
CMR: neat
Analysis: 57.6% C; 9.7% H

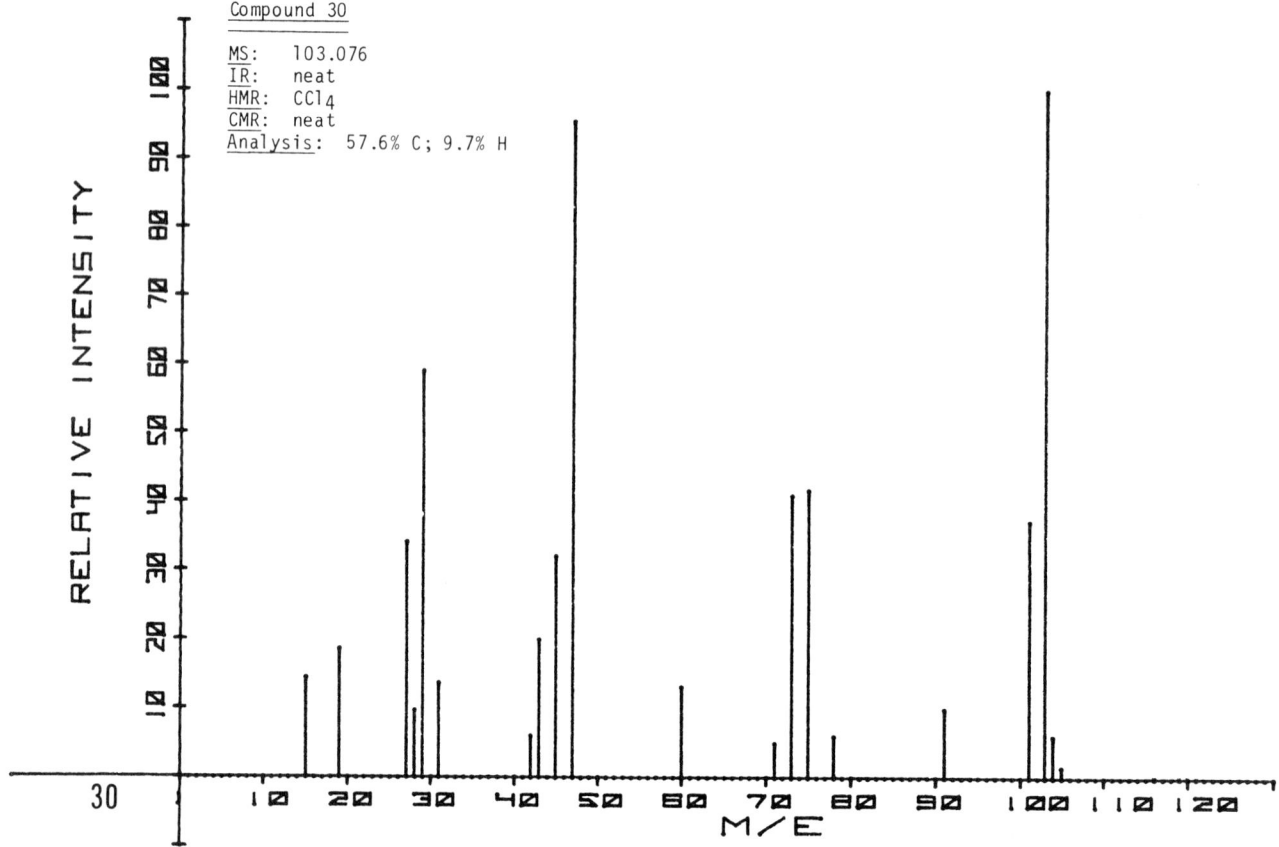

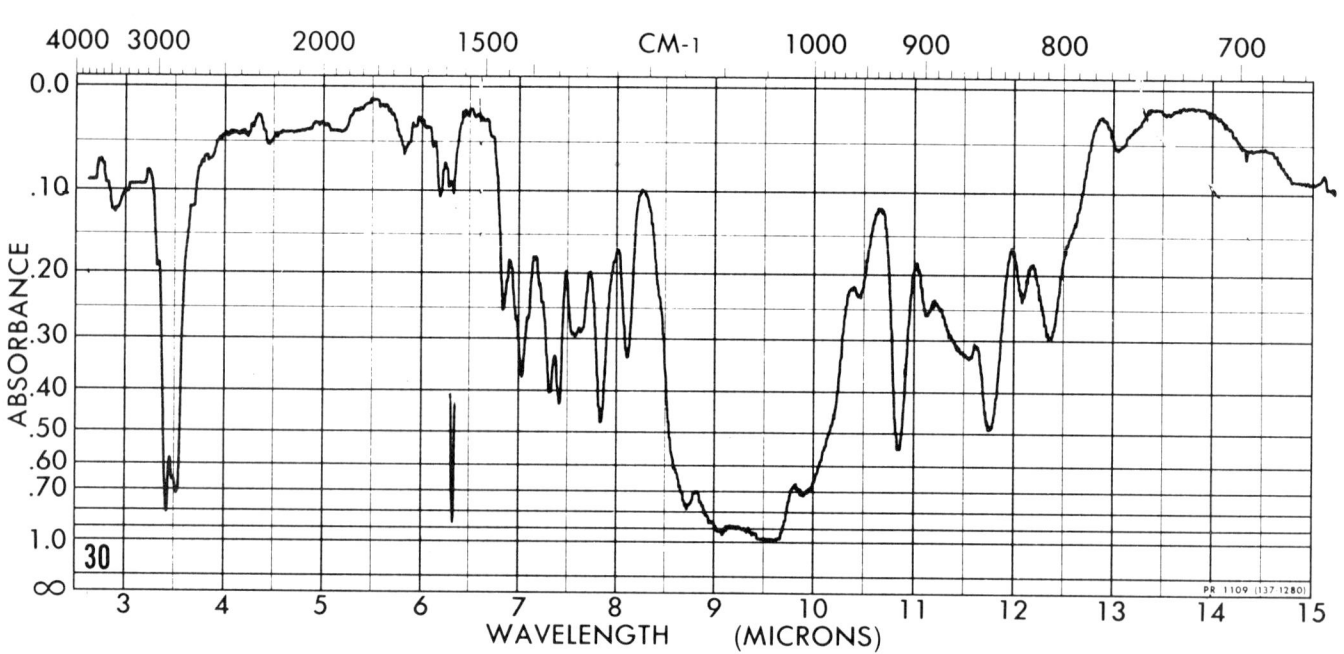

30
IRR. 1.20
IRR. 3.55
IRR. 4.16
IRR. 1.20
IRR. 3.55
A
B

30
4000Hz (200) ppm 3000 2000 1000
2000 (100) 1500 1000 500
1000 (50) 750 500 250
800 (40) 600 400 200
600 (30) 450 300 150
Pk. Ht. PPM
29 103.2
31 101.7
37 62.8
37 62.1
28 52.8
30 51.3
47 43.5
60 16.0
55 15.1
Pk. Ht. PPM
32 102.4
40 62.8
38 62.1
40 52.0
38 43.4
38 15.5
δ(PPM)

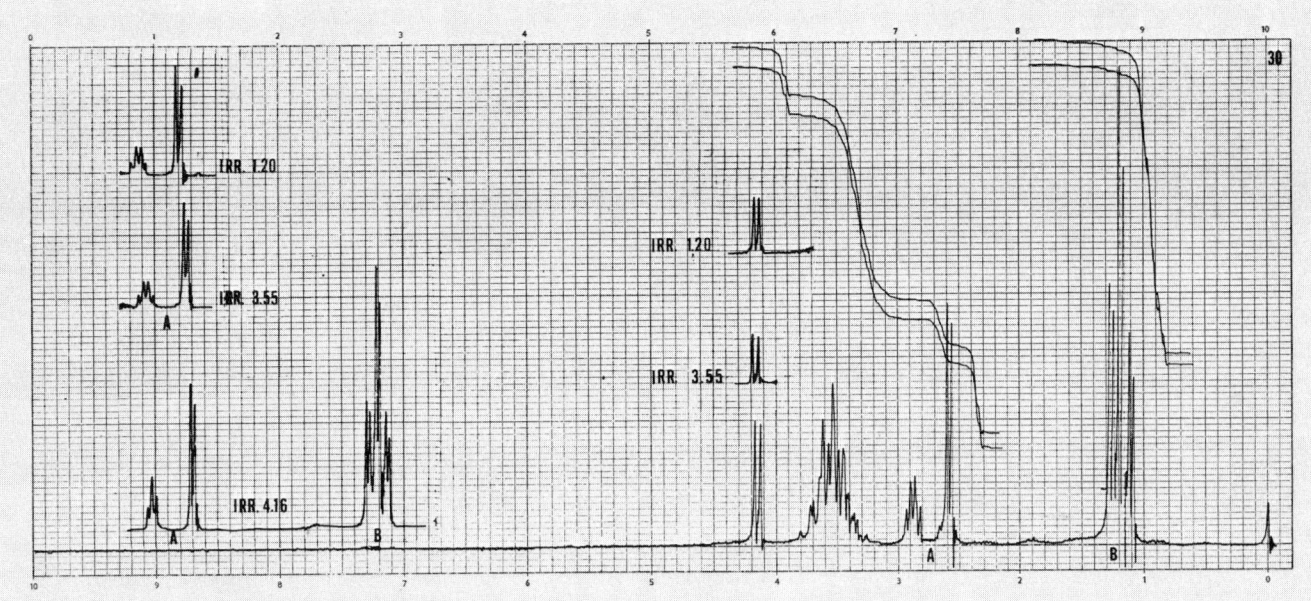

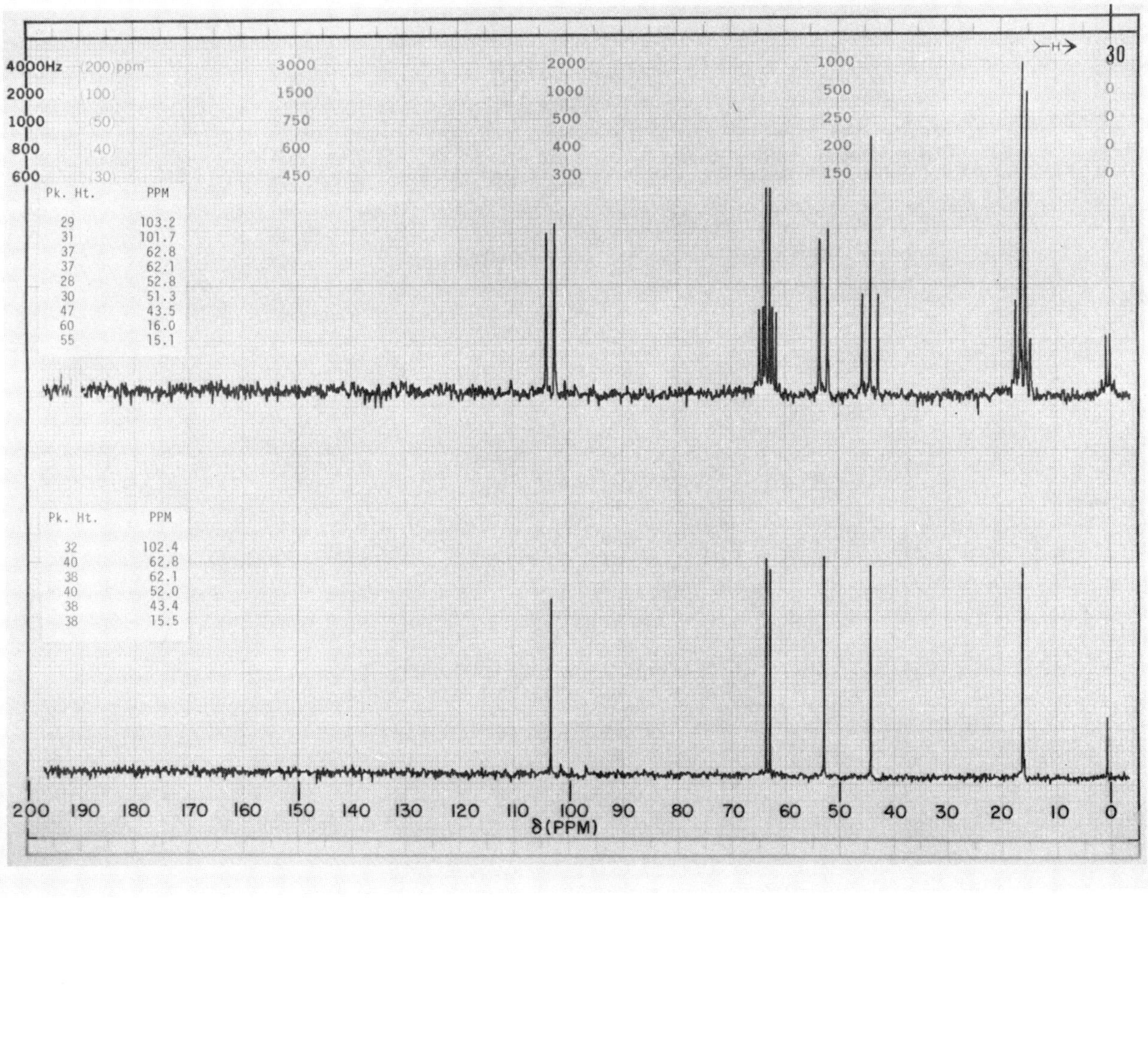

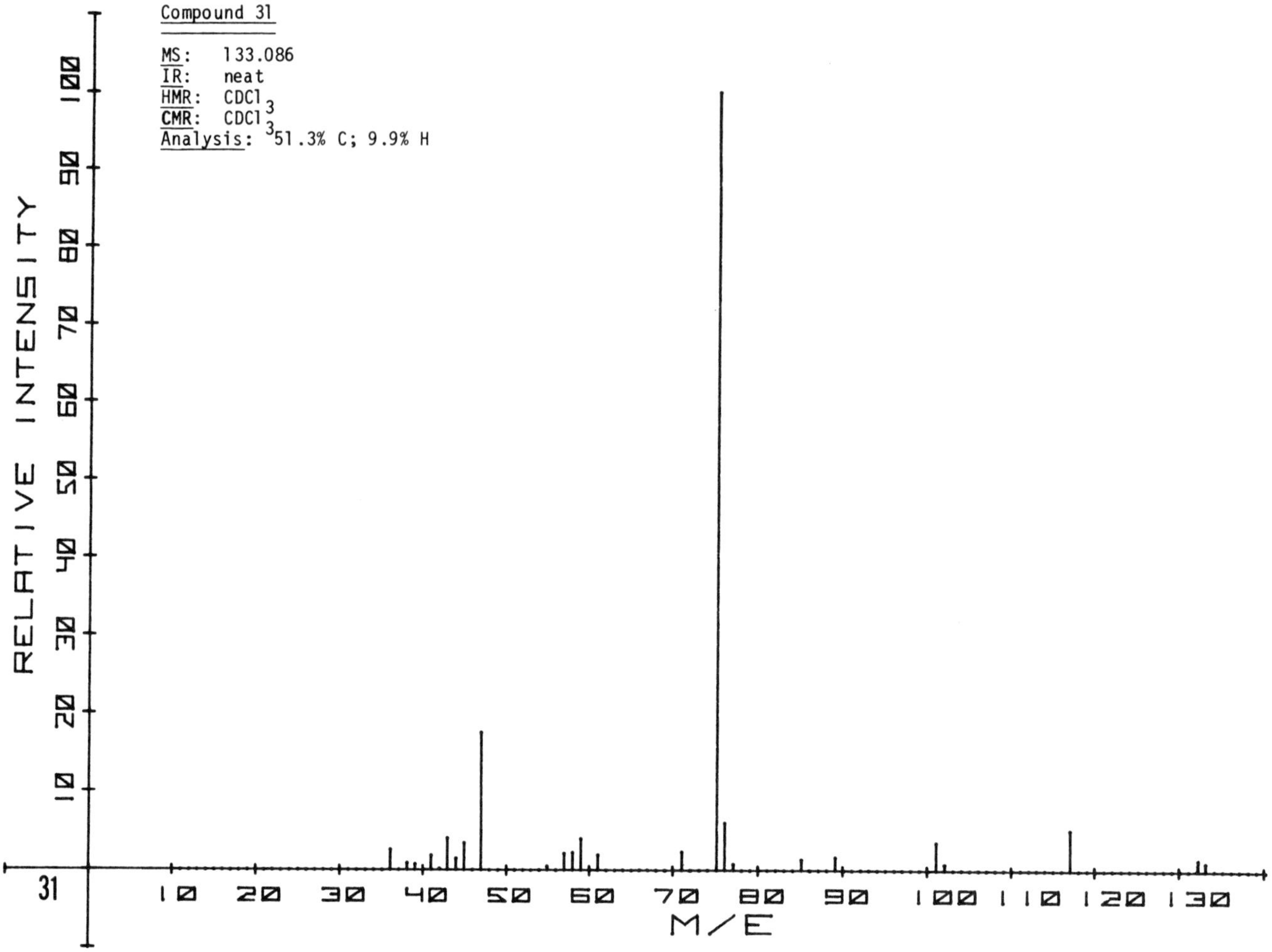

Compound 31
MS: 133.086
IR: neat
HMR: CDCl3
CMR: CDCl3
Analysis: 51.3% C; 9.9% H
RELATIVE INTENSITY
M/E
31

CM-1
ABSORBANCE
WAVELENGTH (MICRONS)
31

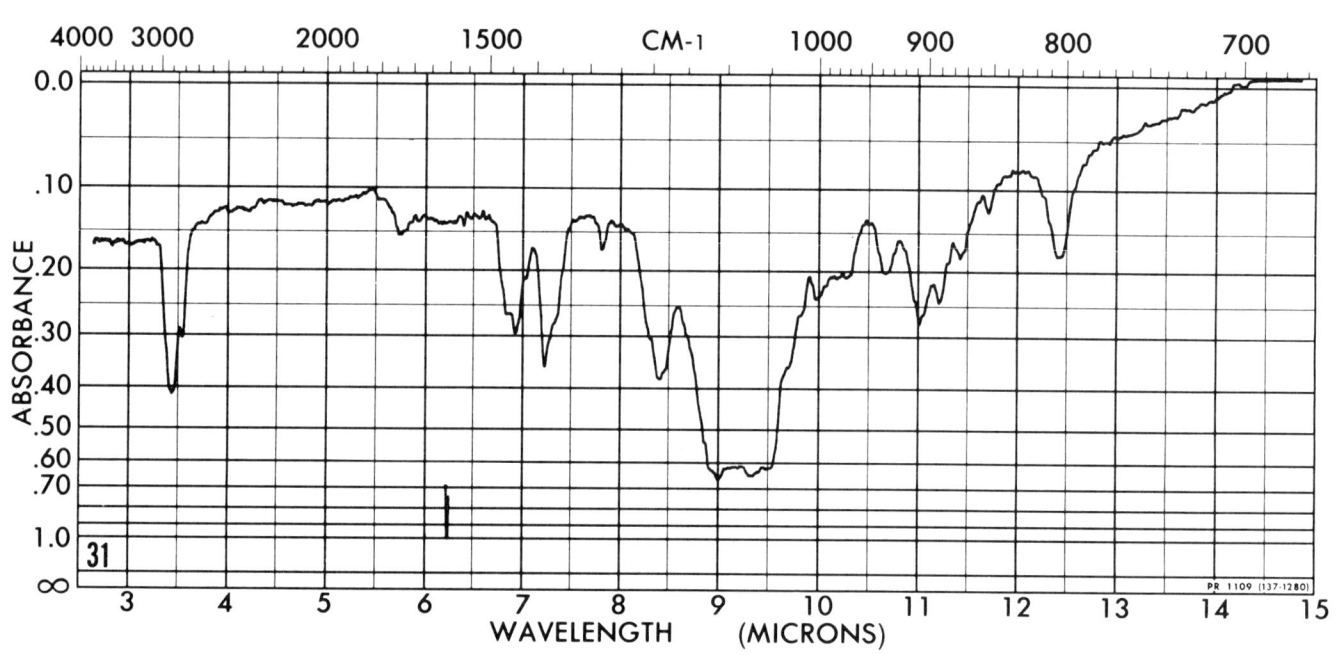

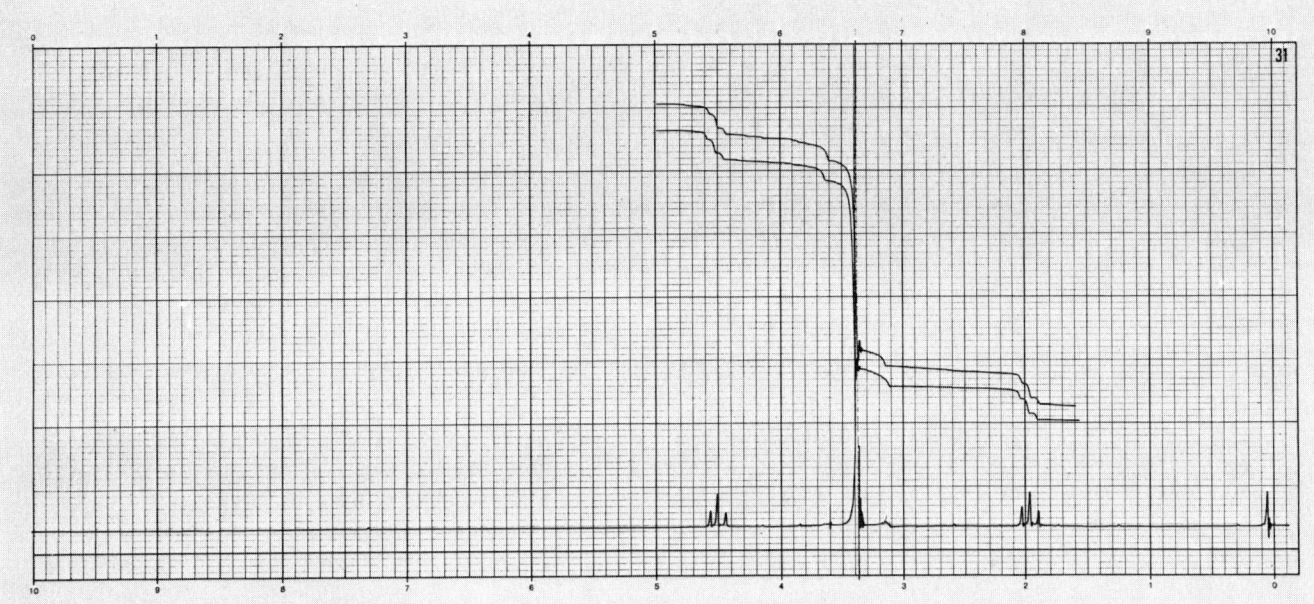

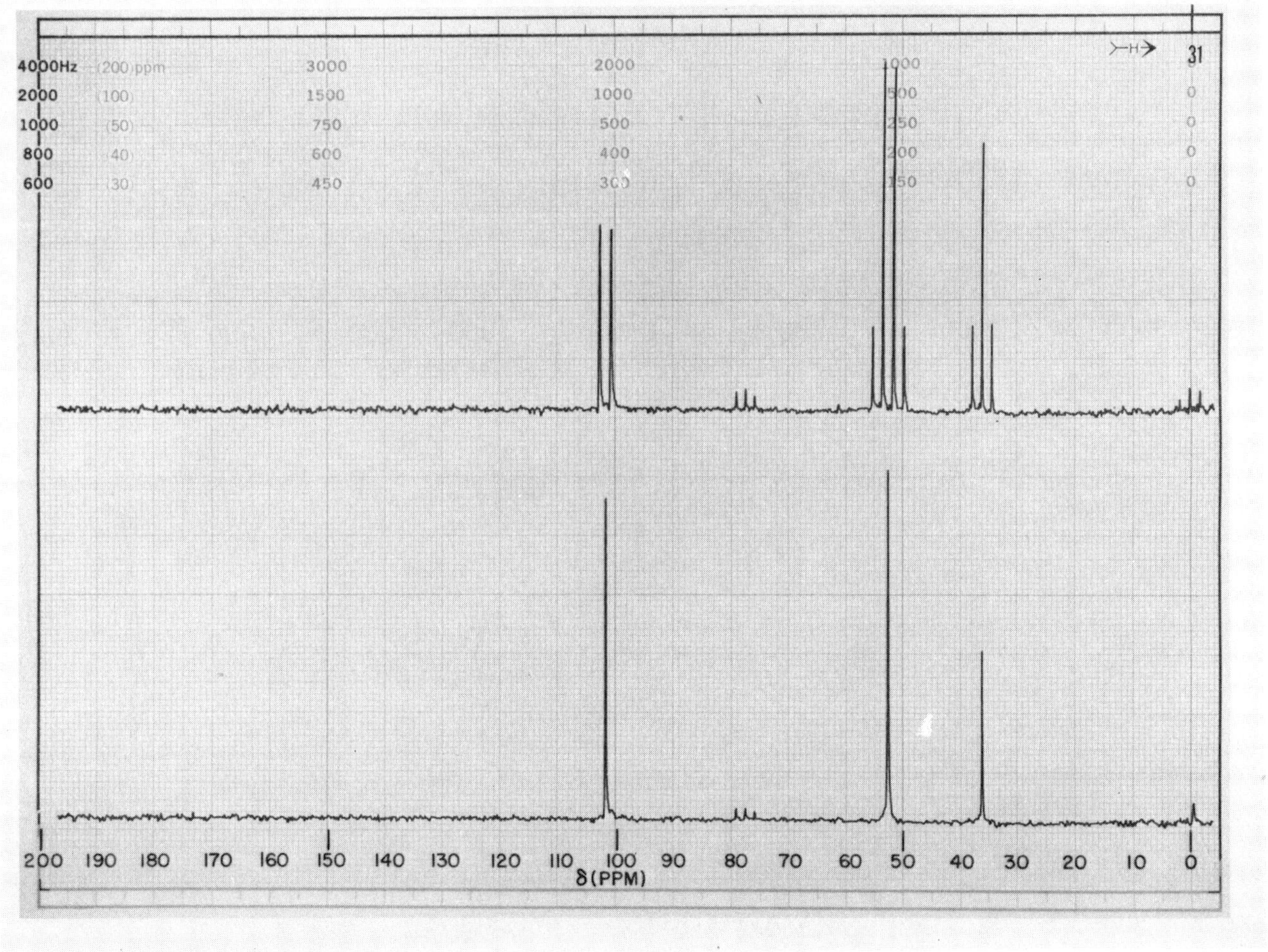

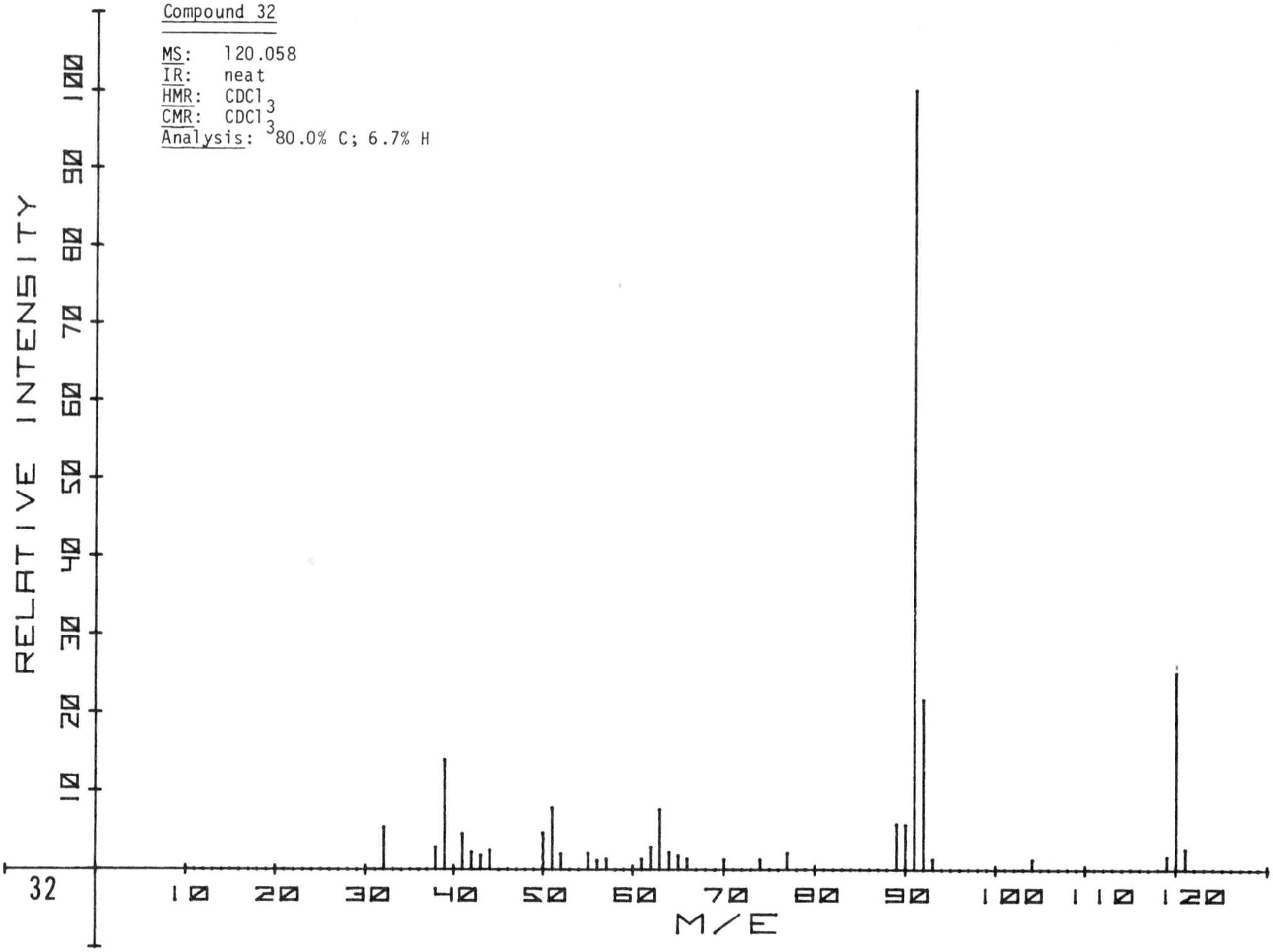

Compound 32
MS: 120.058
IR: neat
HMR: CDCl3
CMR: CDCl3
Analysis: 80.0% C; 6.7% H
RELATIVE INTENSITY
10 20 30 40 50 60 70 80 90 100
32
10 20 30 40 50 60 70 80 90 100 110 120
M/E

4000 3000 2000 1500 CM-1 1000 900 800 700
ABSORBANCE
0.0 .10 .20 .30 .40 .50 .60 .70 1.0 ∞
32
3 4 5 6 7 8 9 10 11 12 13 14 15
WAVELENGTH (MICRONS)

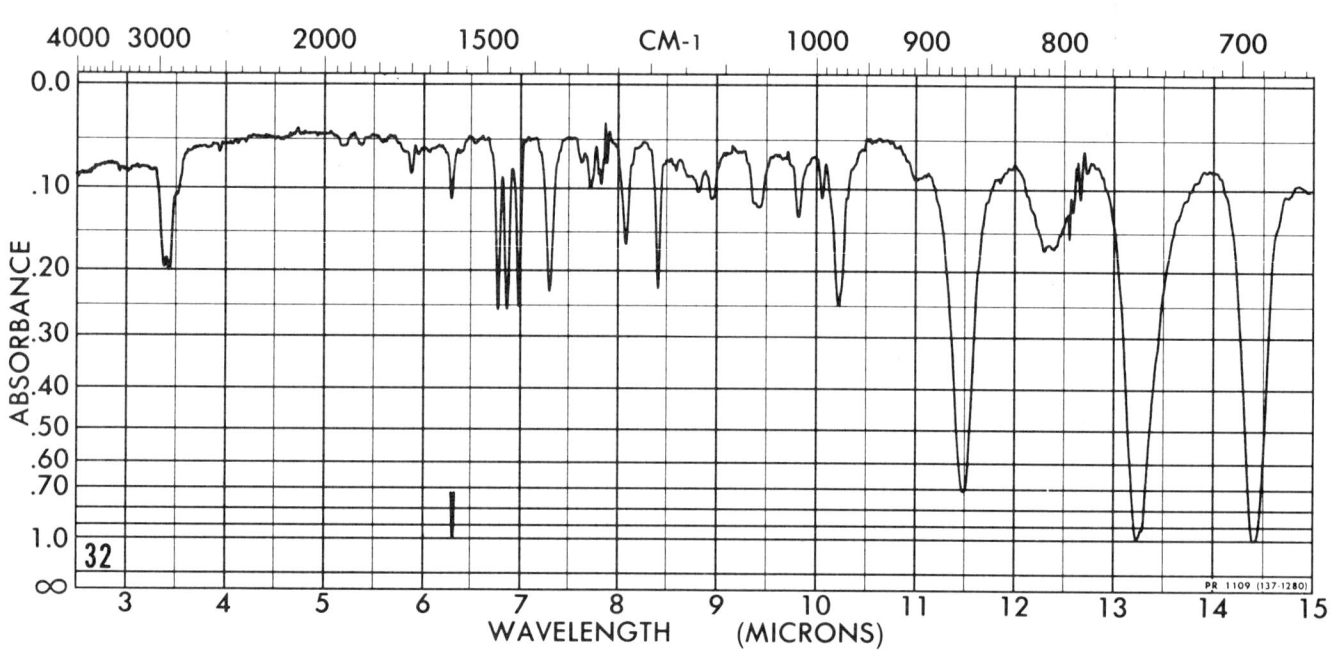

32
0 1 2 3 4 5 6 7 8 9 10
10 9 8 7 6 5 4 3 2 1 0

4000Hz (200)ppm 3000 2000 1000
2000 (100) 1000 500 0
1000 (50) 500 250 0
800 (40) 400 200 0
600 (30) 300 150 0
32
PK. HT. PPM
31 129.2
26 128.8
36 127.8
25 127.4
60 126.2
56 124.8
22 53.1
20 52.8
27 51.3
49 50.9
16 49.0
PK. HT. PPM
6 137.8
60 128.5
35 128.1
49 125.5
20 52.2
22 50.9
200 190 180 170 160 150 140 130 120 110 100 90 80 70 60 50 40 30 20 10 0
δ(PPM)

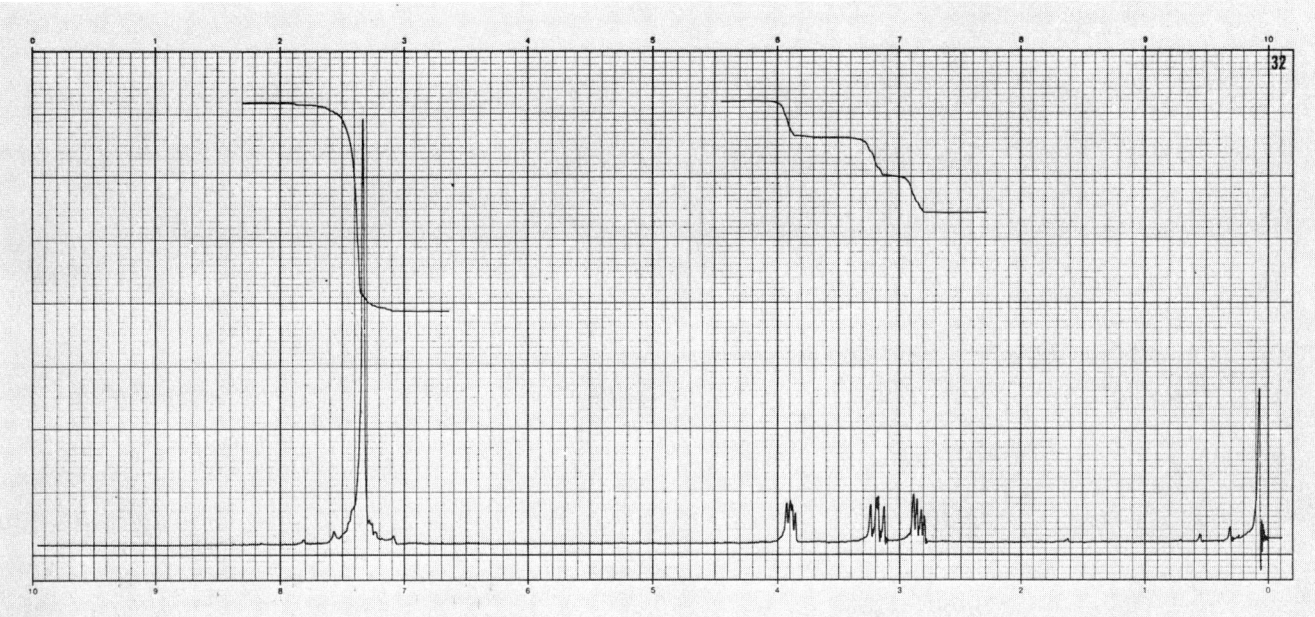

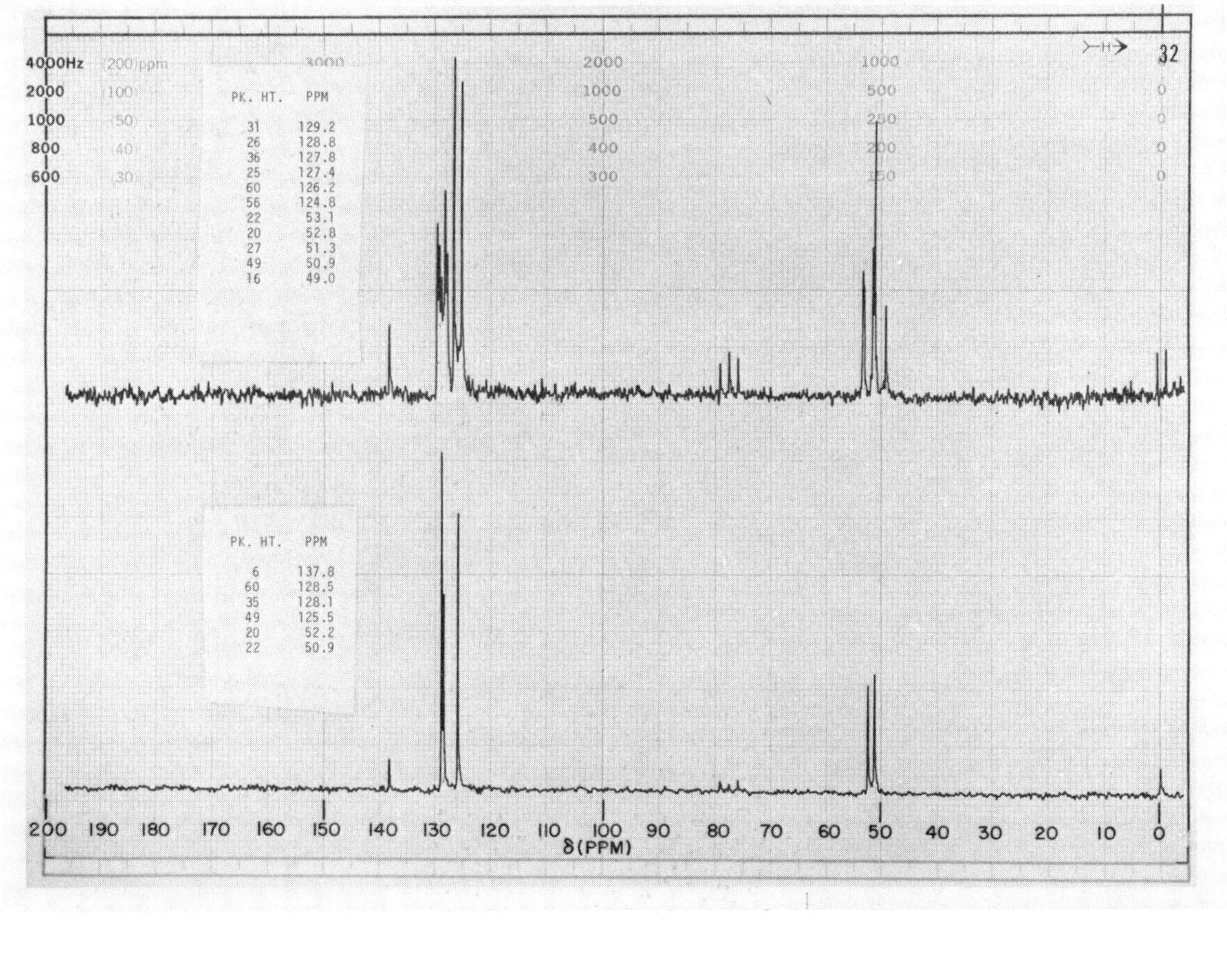

COMPOUND 33

Compound 33
MS: 152.049
IR: KBr
HMR: CDCl$_3$
CMR: CDCl$_3$
Analysis: 63.2% C; 5.3% H

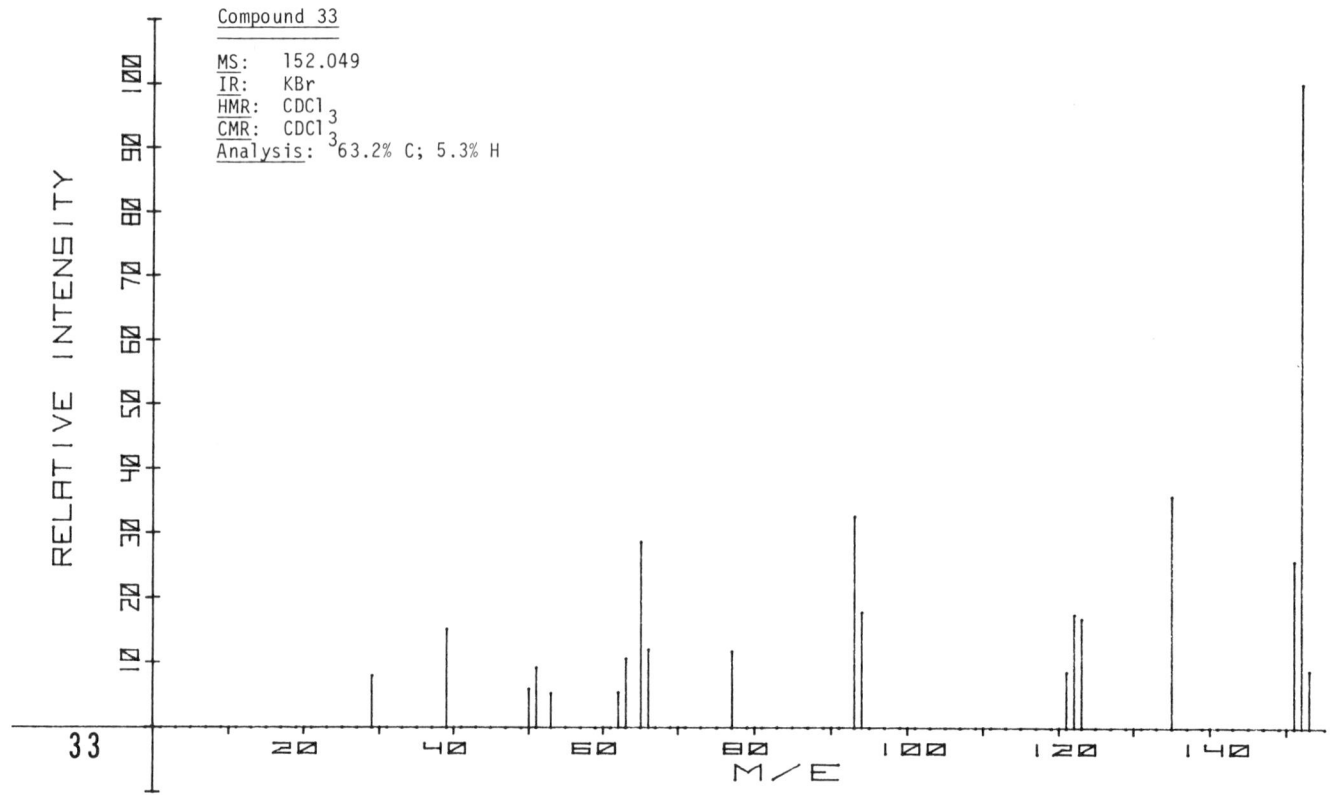

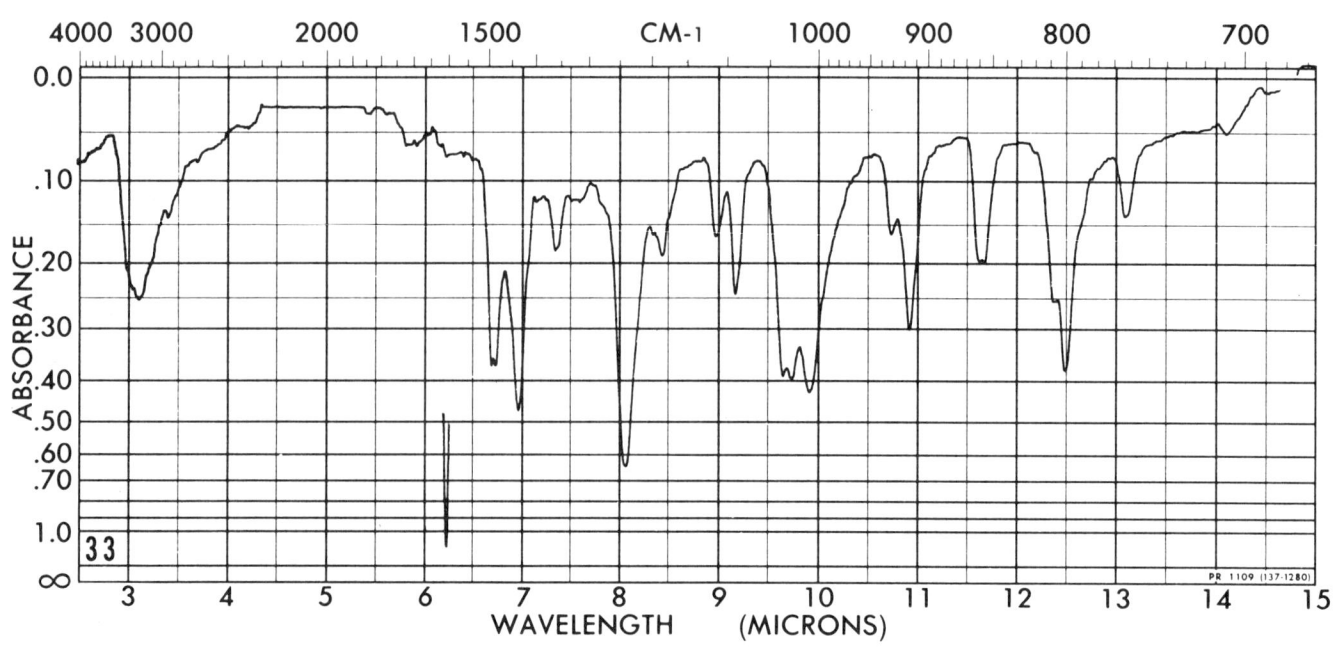

33
D_2O

4000Hz (200) ppm 3000 2000 1000
2000 (100) 1500 1000 500 0
1000 (50) 750 500 250 0
800 (40) 600 400 200 0
600 (30) 450 300 150 0
33
PK. HT. PPM
12 147.7
7 146.9
14 135.0
16 121.4
17 119.4
23 109.1
21 108.8
26 107.1
23 106.8
14 102.9
60 100.9
17 99.0
PK. HT. PPM
12 147.7
13 146.9
13 135.0
60 120.4
53 108.1
47 107.8
58 100.9
34 64.8
200 190 180 170 160 150 140 130 120 110 100 90 80 70 60 50 40 30 20 10 0
δ(PPM)

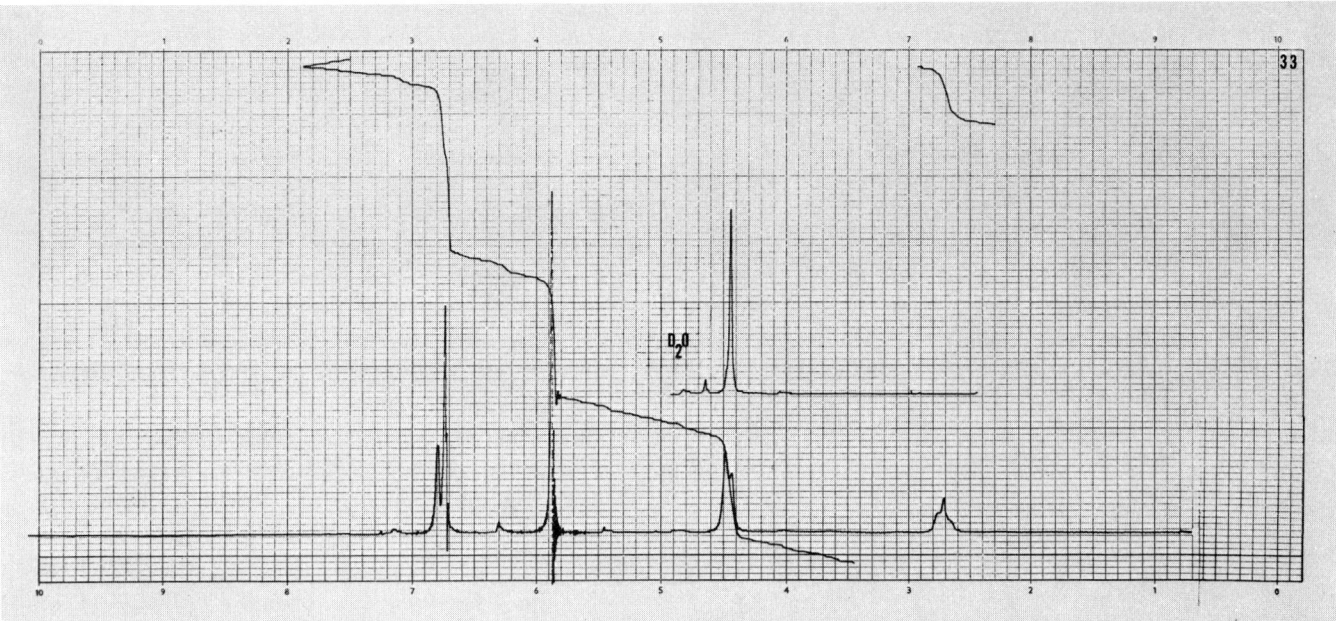

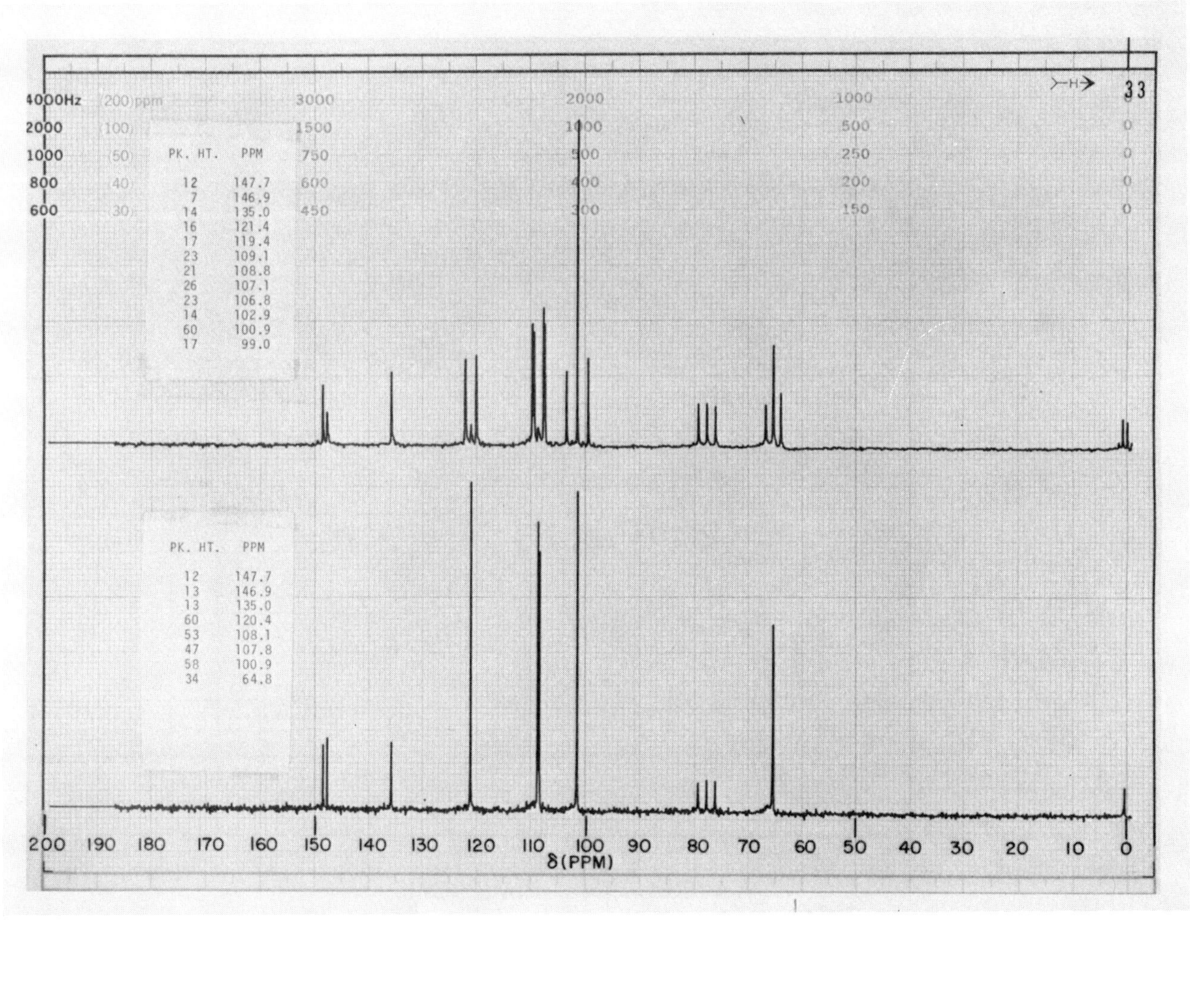

COMPOUND 34

Compound 34

MS: 135.069
IR: KBr
HMR: CDCl₃
CMR: CDCl₃
Analysis: 71.1% C; 6.7% H; 10.5% N

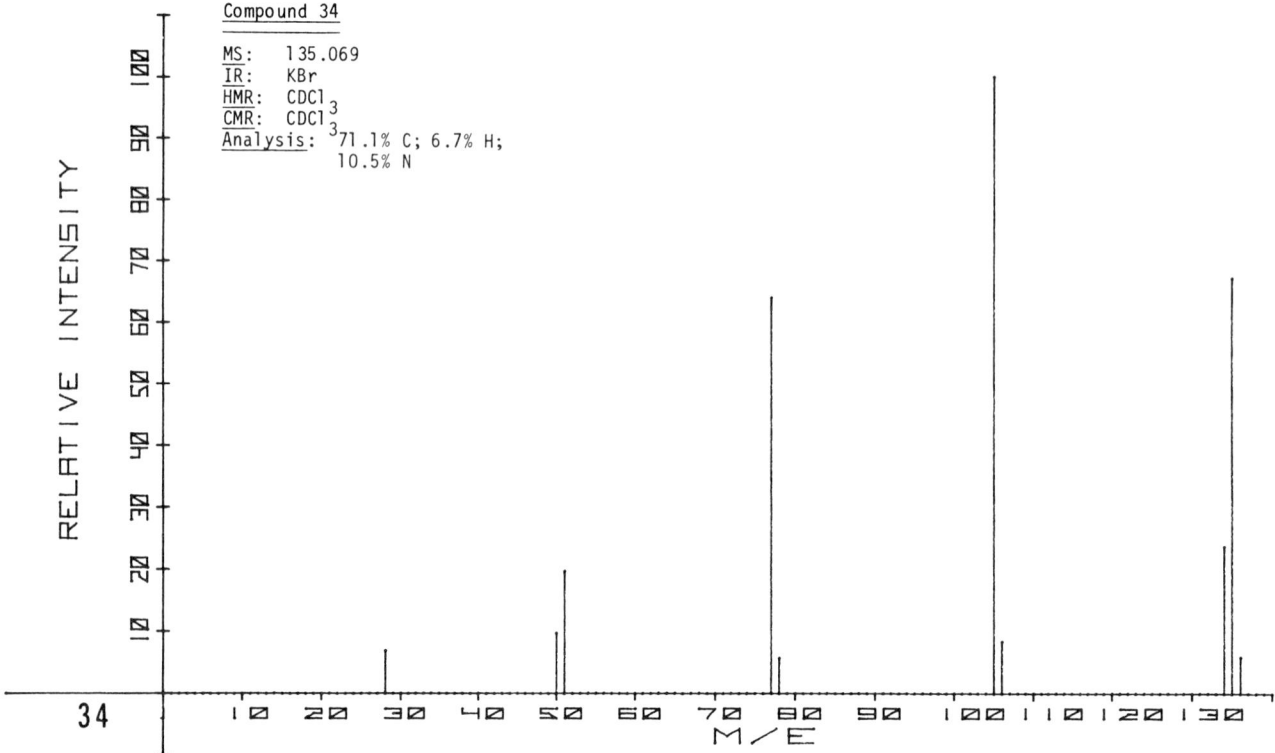

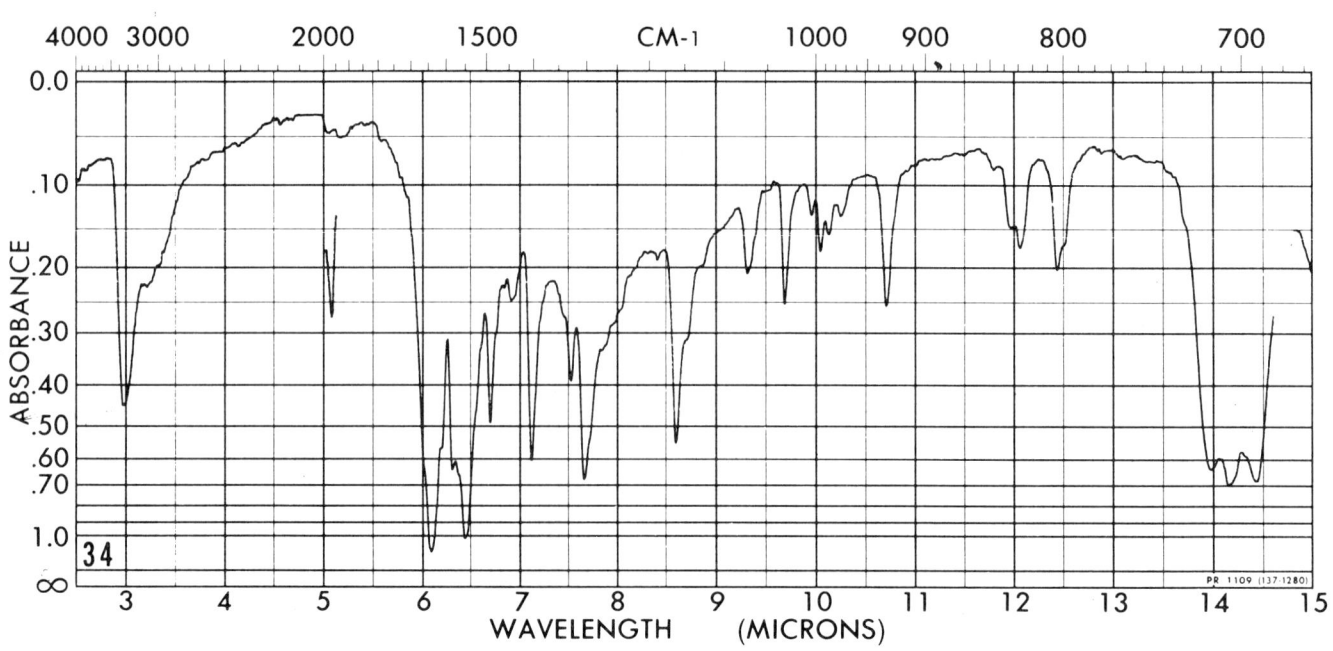

34

4000Hz (200) ppm 3000 2000 1000
2000 (100) 1000 500 0
1000 (50) 500 250 0
800 (40) 400 200 0
600 (30) 300 150 0
34
PK. HT. PPM
16 132.4
30 130.2
26 127.7
73 127.4
69 126.1
64 124.9
65 123.6
PK. HT. PPM
4 166.5
8 132.4
31 129.0
39 126.1
60 124.9
13 24.5
200 190 180 170 160 150 140 130 120 110 100 90 80 70 60 50 40 30 20 10 0
δ(PPM)

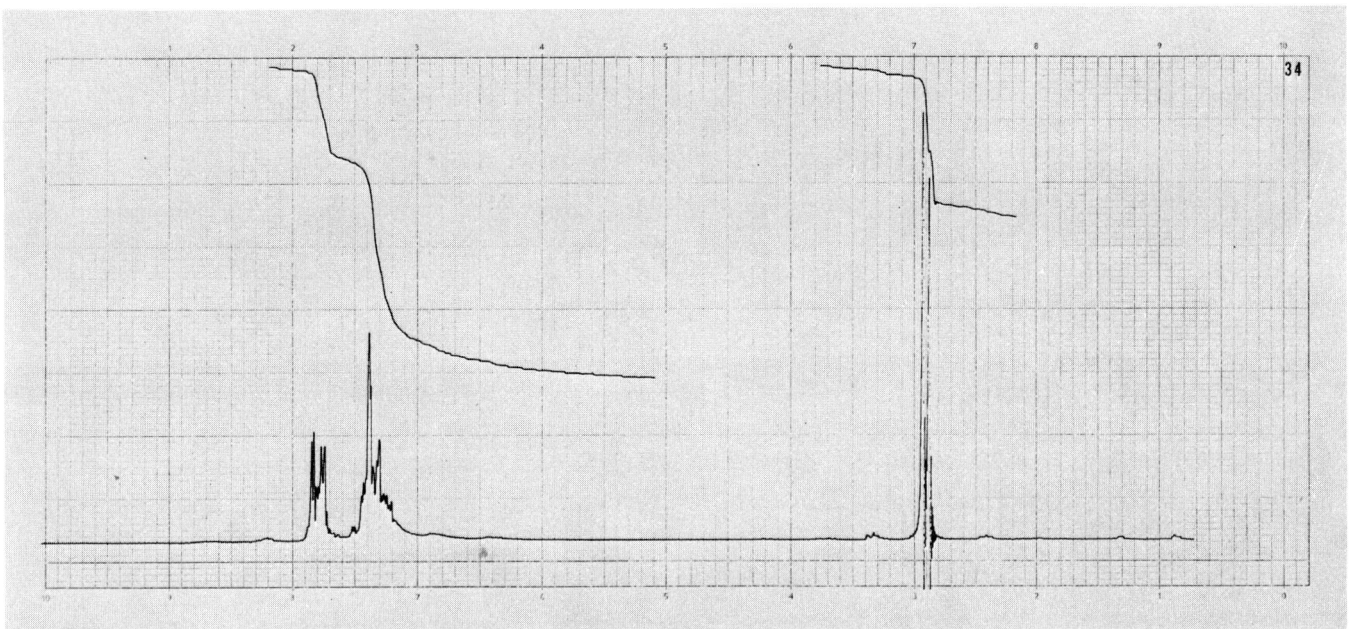

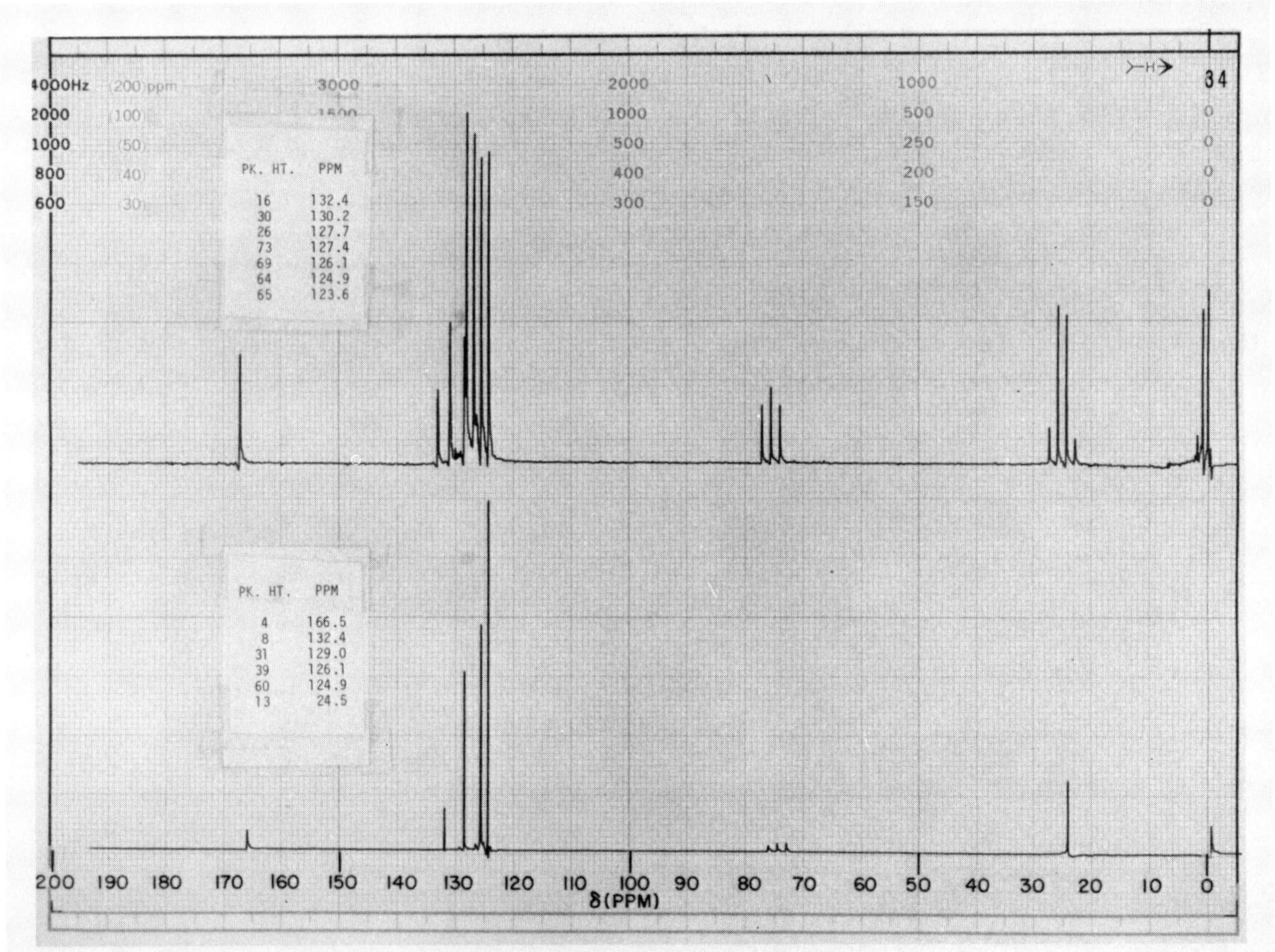

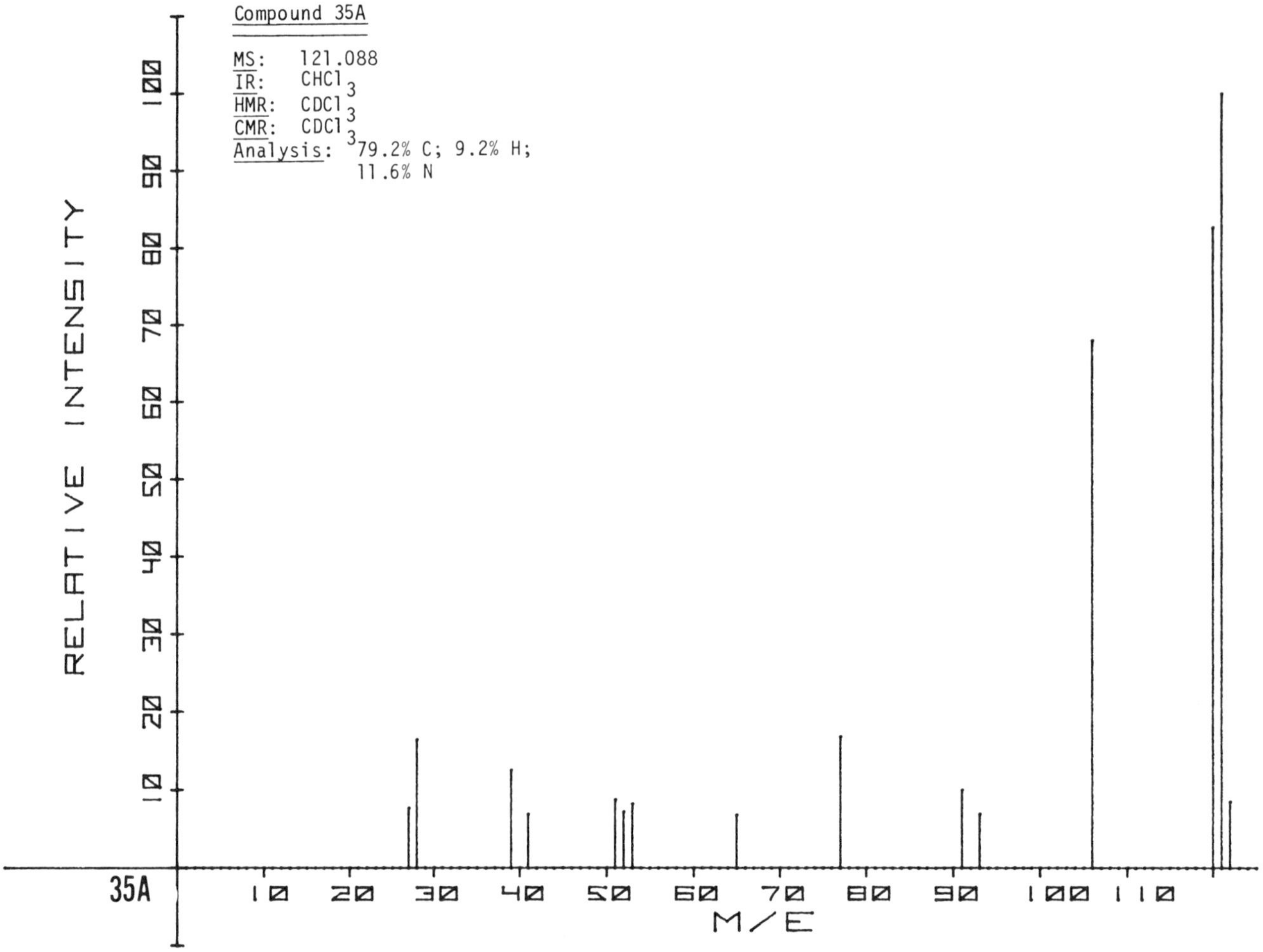

Compound 35A
MS: 121.088
IR: $CHCl_3$
HMR: $CDCl_3$
CMR: $CDCl_3$
Analysis: 79.2% C; 9.2% H; 11.6% N
RELATIVE INTENSITY
10 20 30 40 50 60 70 80 90 100
35A
10 20 30 40 50 60 70 80 90 100 110
M/E

4000 3000 2000 1500 CM-1 1000 900 800 700
0.0 .10 .20 .30 .40 .50 .60 .70 1.0 ∞
ABSORBANCE
35A
PR 1109 (137-1280)
3 4 5 6 7 8 9 10 11 12 13 14 15
WAVELENGTH (MICRONS)

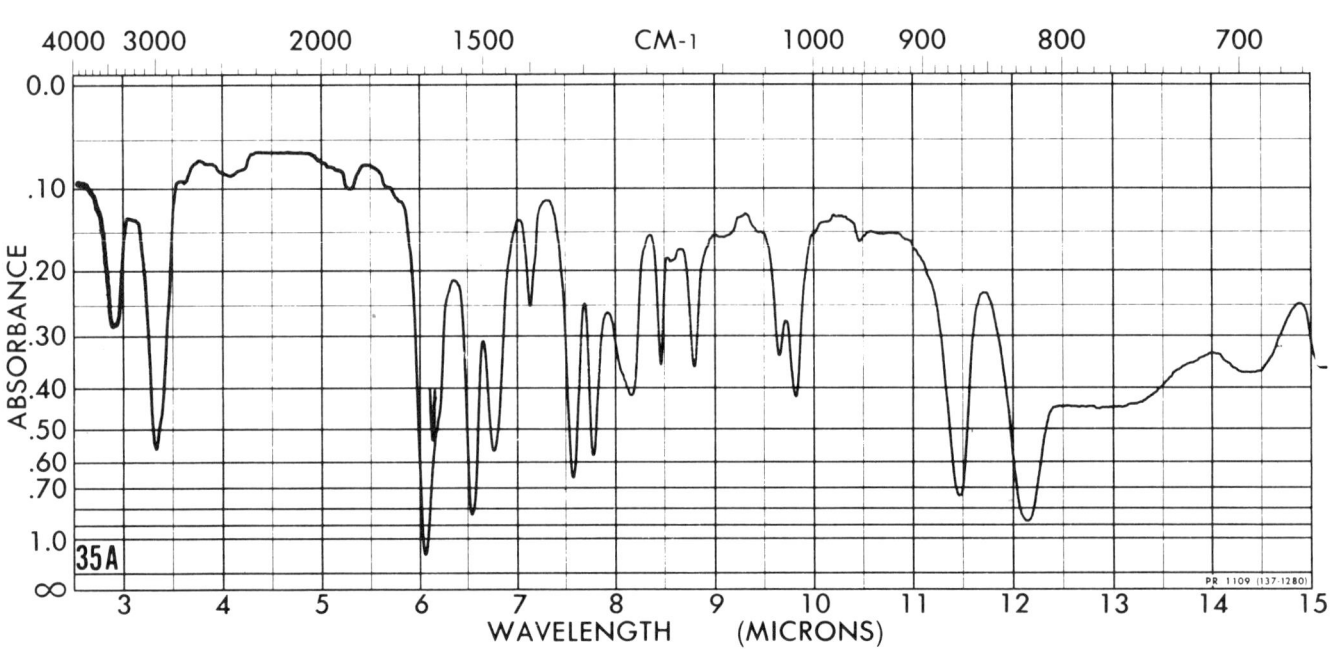

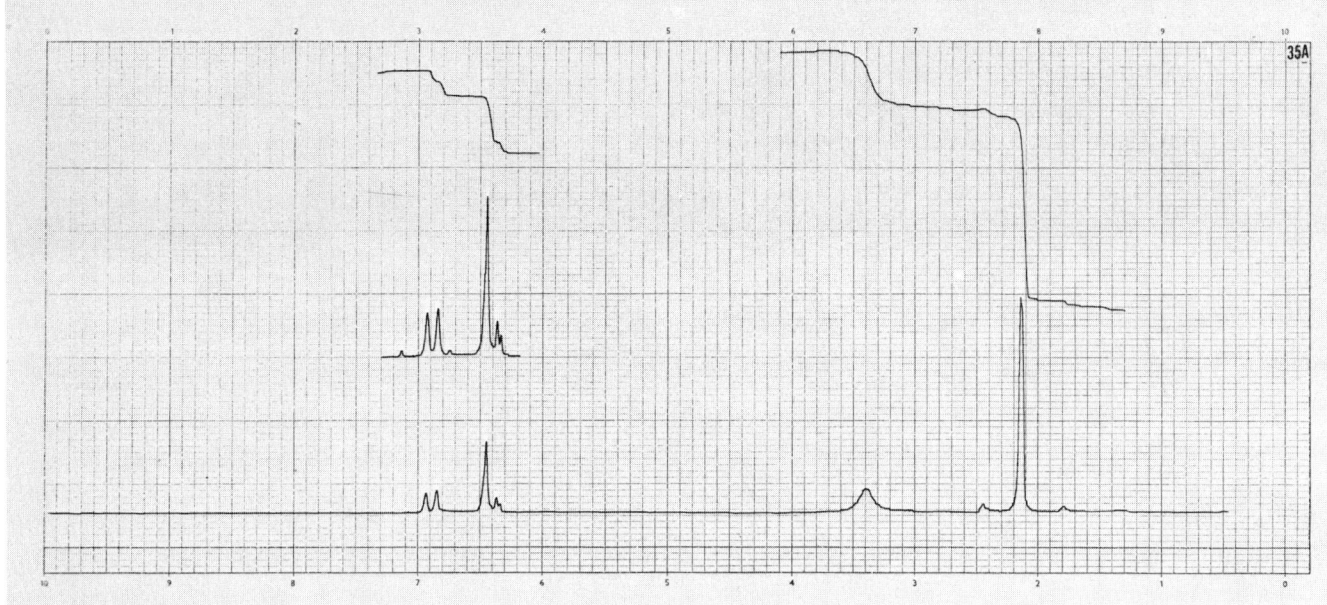

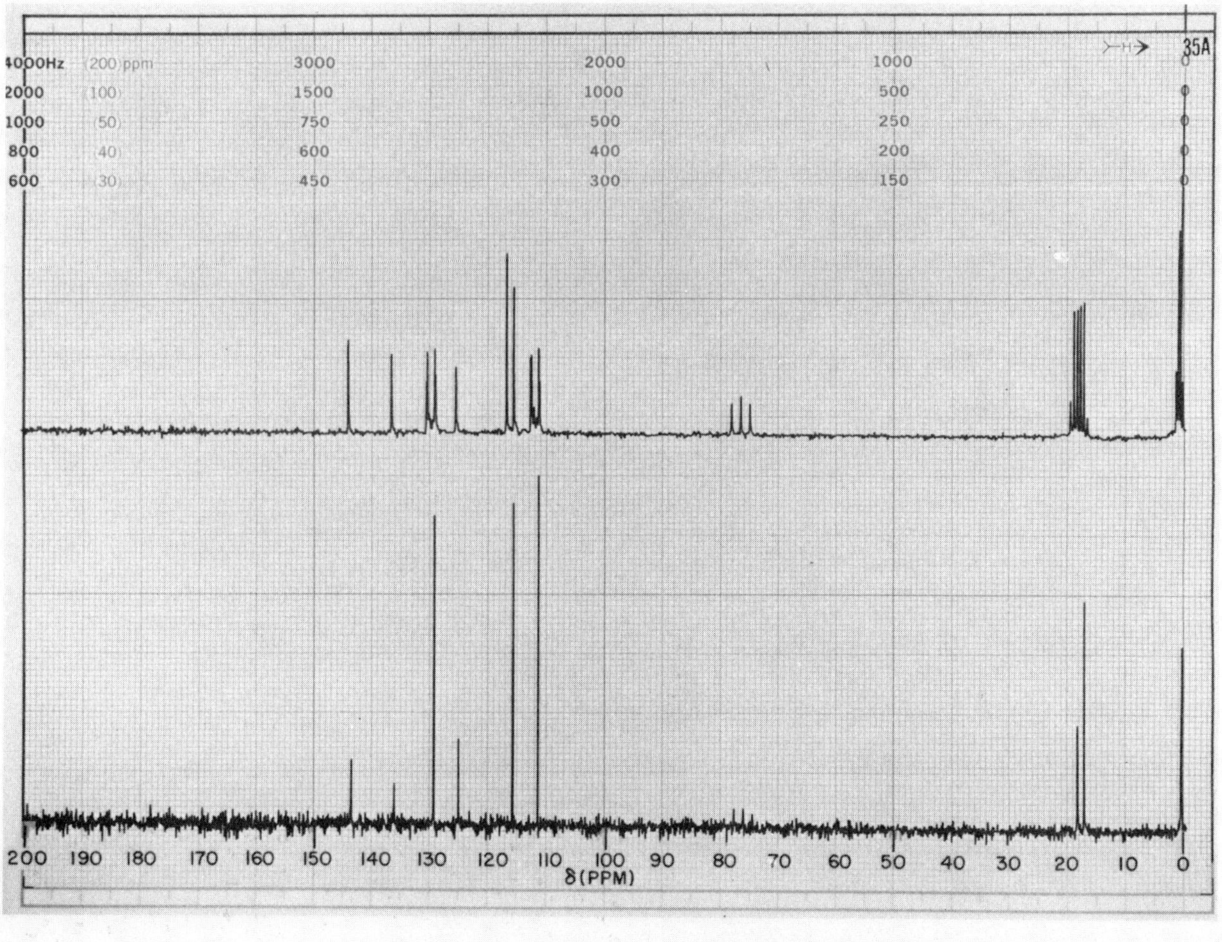

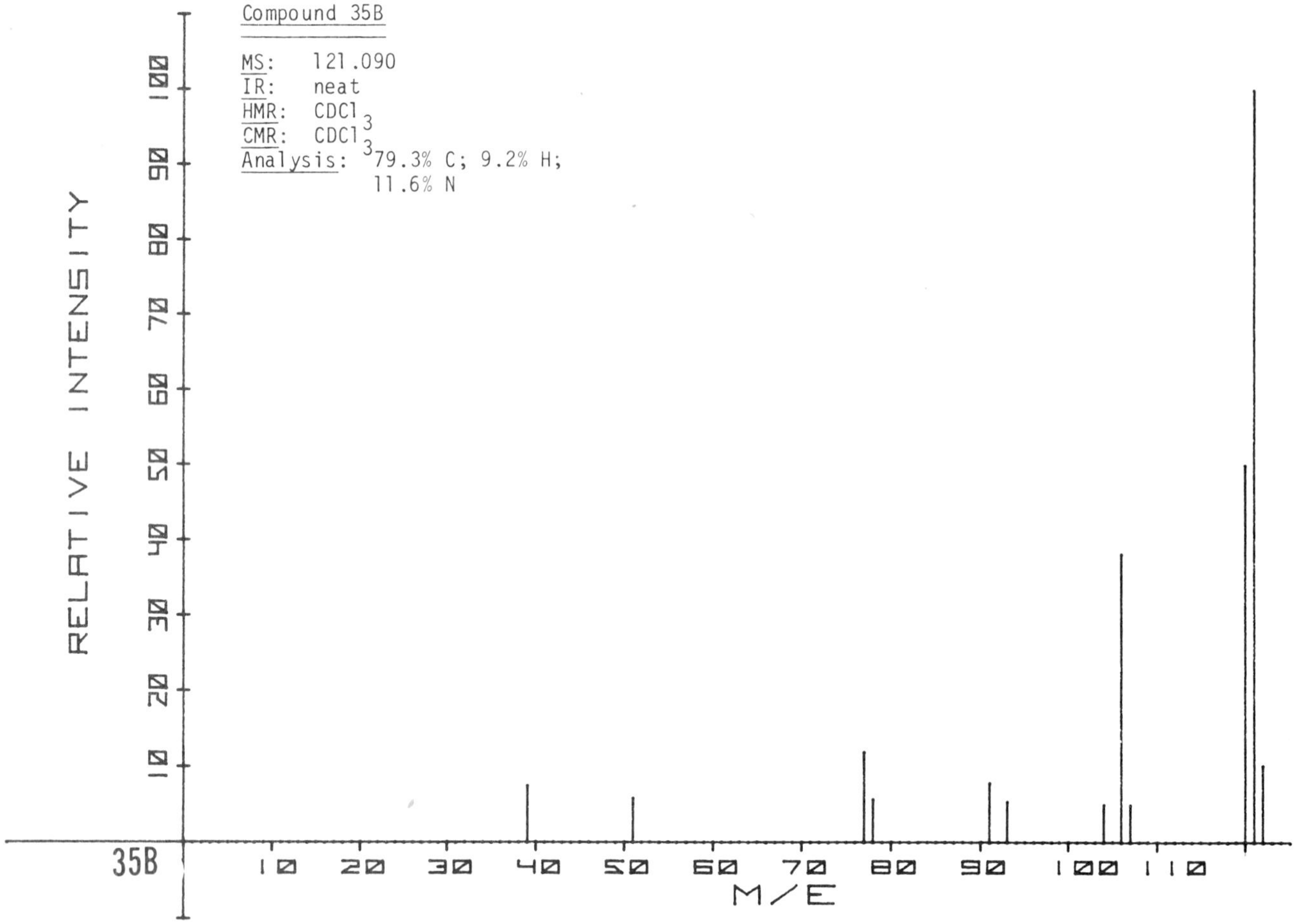

Compound 35B
MS: 121.090
IR: neat
HMR: CDCl3
CMR: CDCl3
Analysis: 79.3% C; 9.2% H; 11.6% N
RELATIVE INTENSITY
35B
M/E

CM-1
ABSORBANCE
35B
WAVELENGTH (MICRONS)

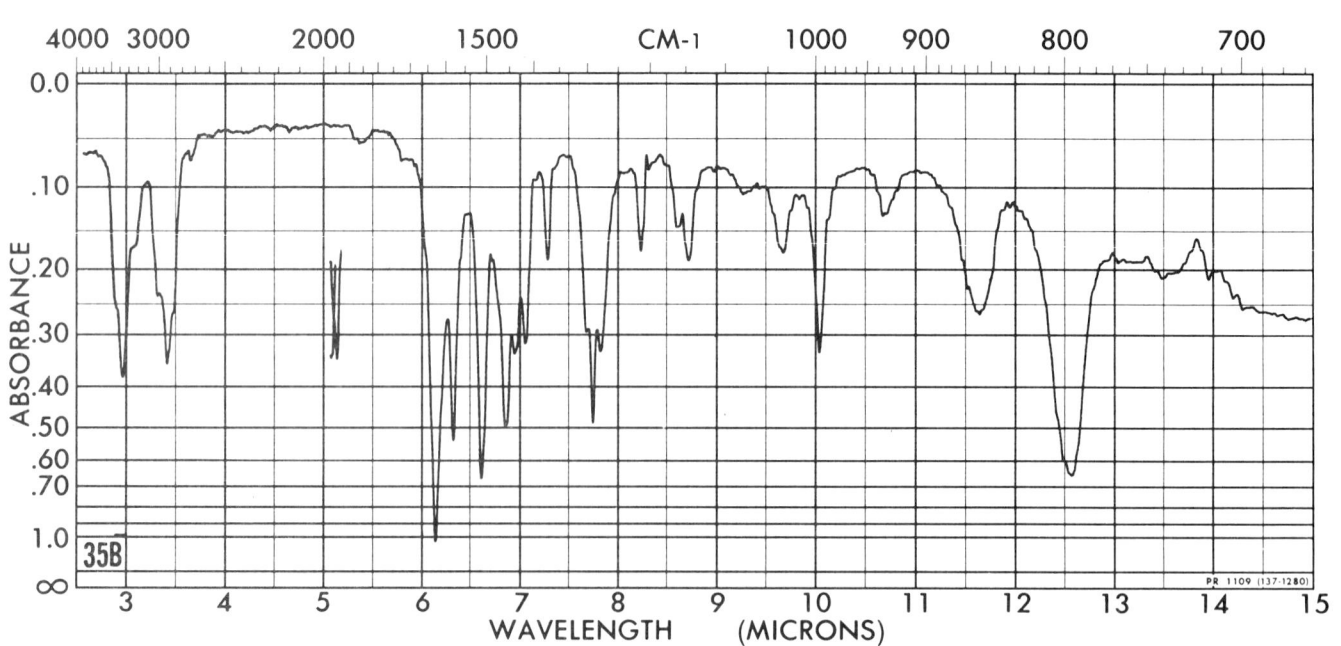

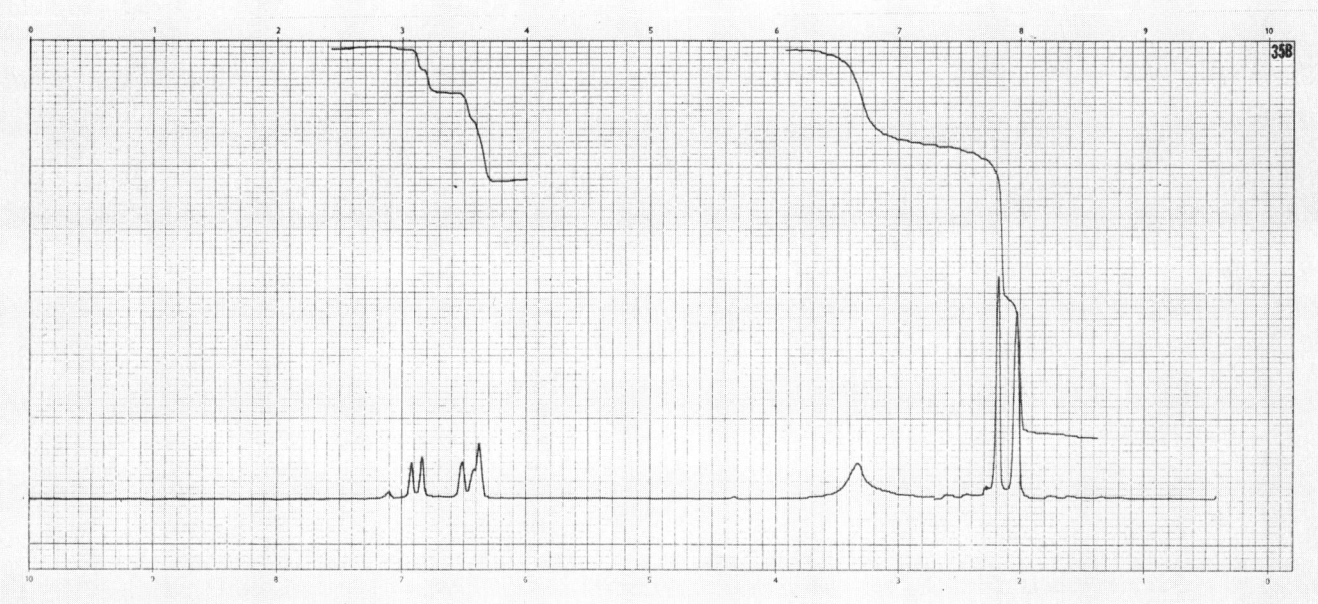

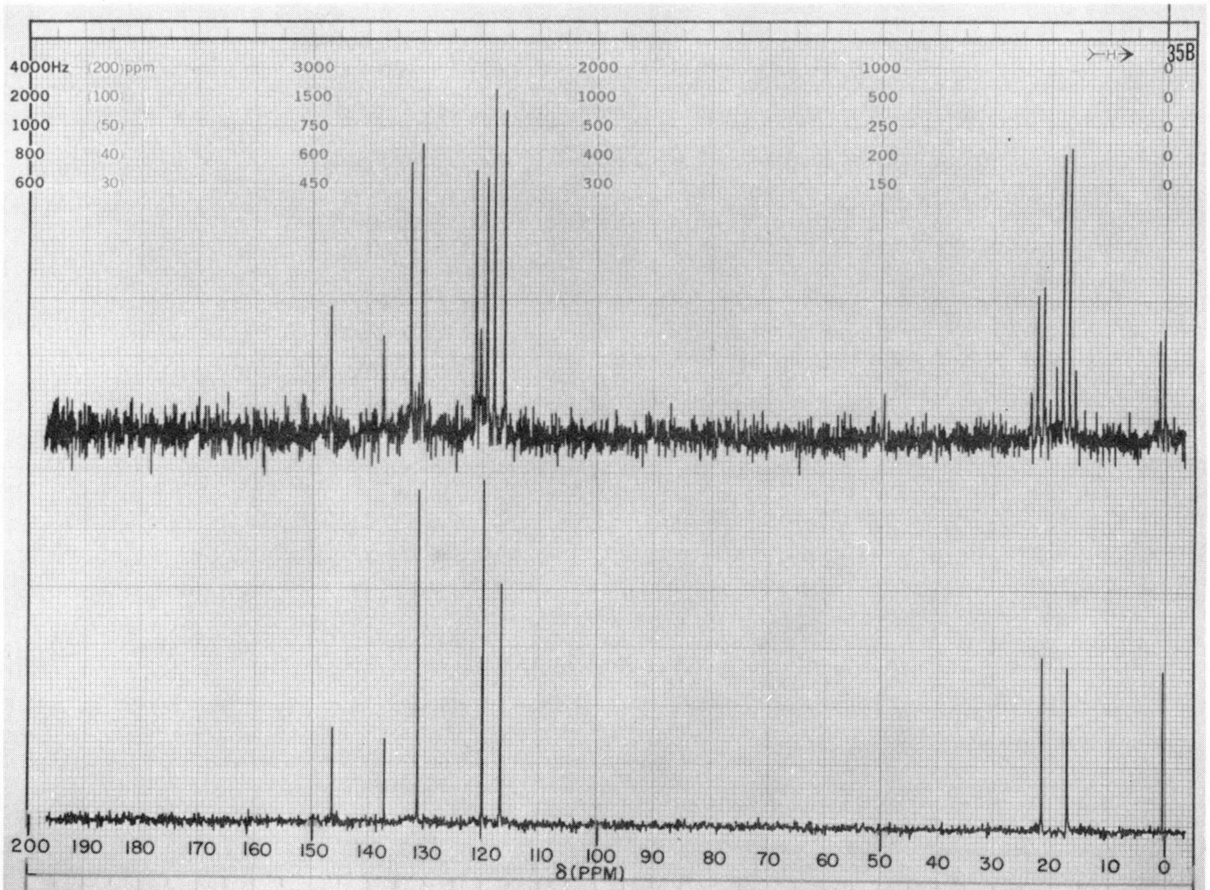

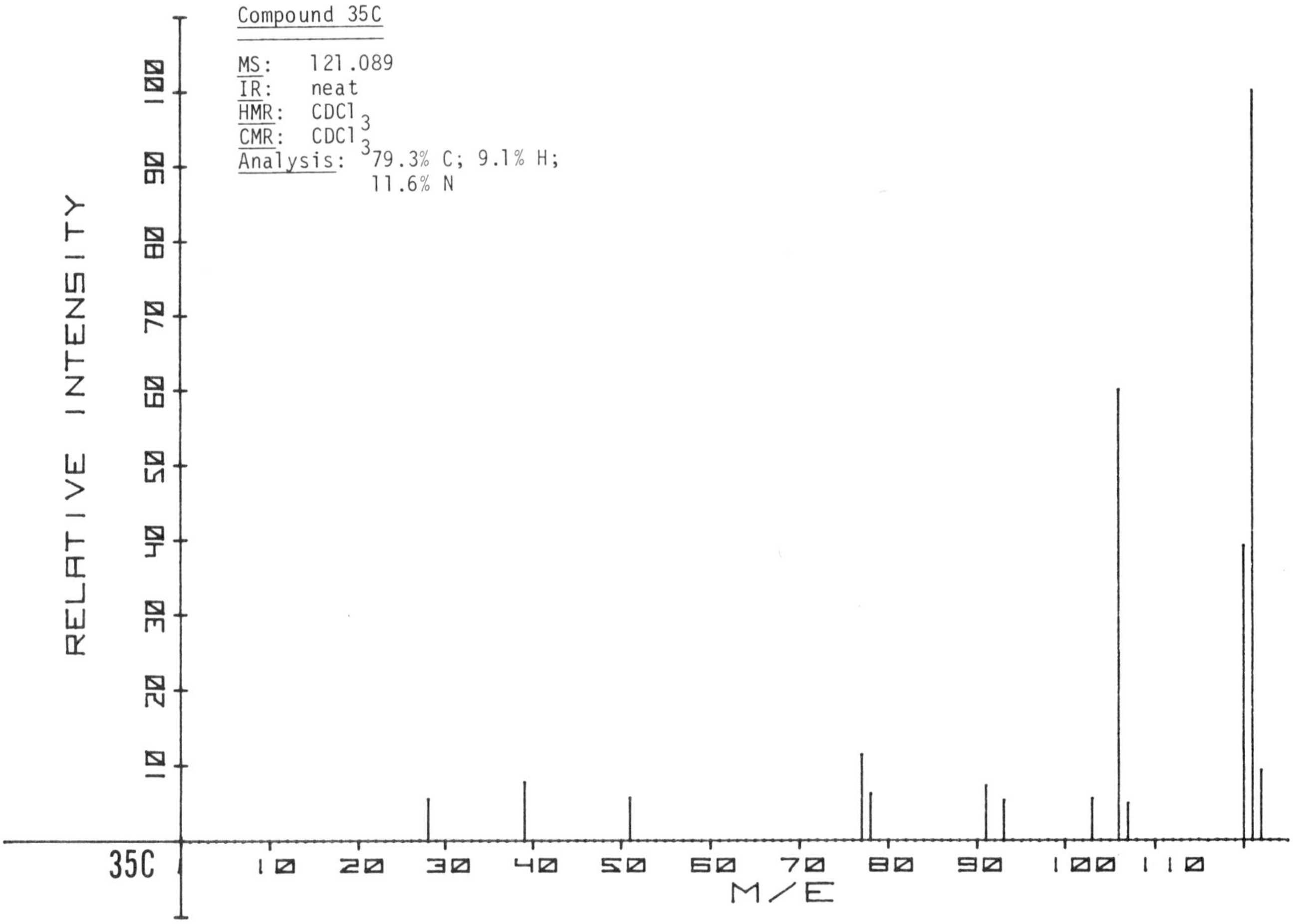

Compound 35C
MS: 121.089
IR: neat
HMR: CDCl3
CMR: CDCl3
Analysis: 79.3% C; 9.1% H; 11.6% N
RELATIVE INTENSITY
10 20 30 40 50 60 70 80 90 100
35C
10 20 30 40 50 60 70 80 90 100 110
M/E

4000 3000 2000 1500 CM-1 1000 900 800 700
0.0 .10 .20 .30 .40 .50 .60 .70 1.0 ∞
ABSORBANCE
35C
3 4 5 6 7 8 9 10 11 12 13 14 15
WAVELENGTH (MICRONS)

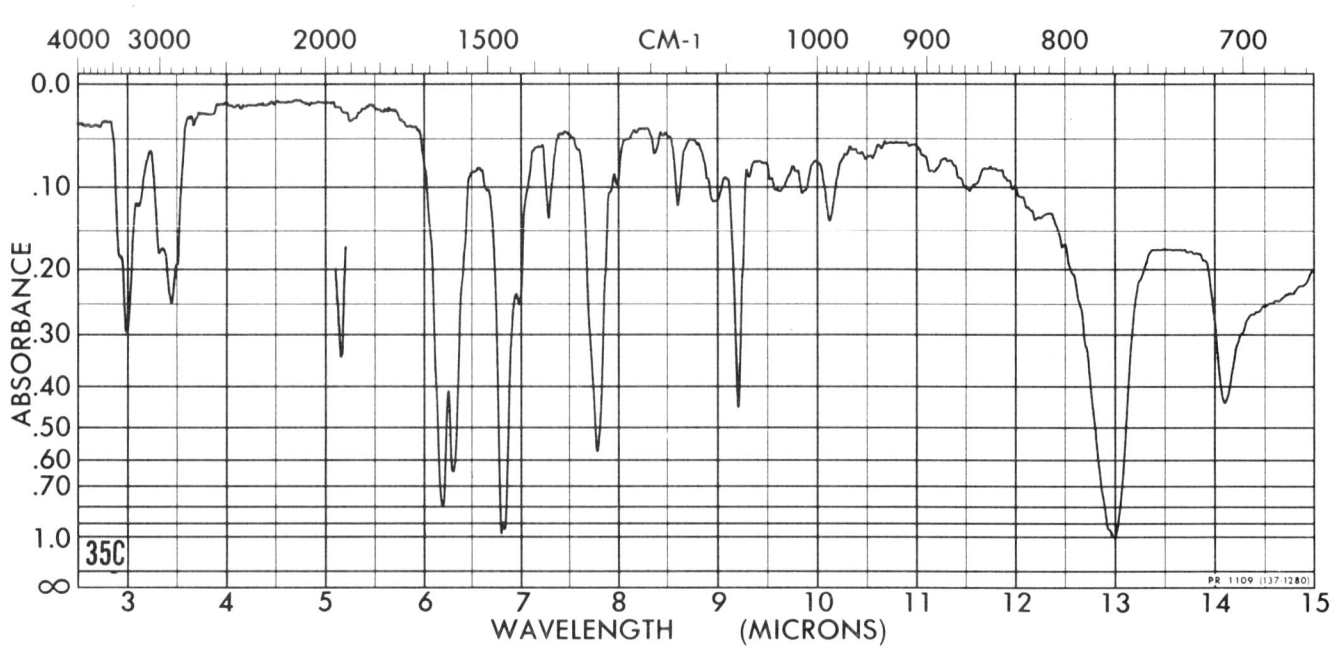

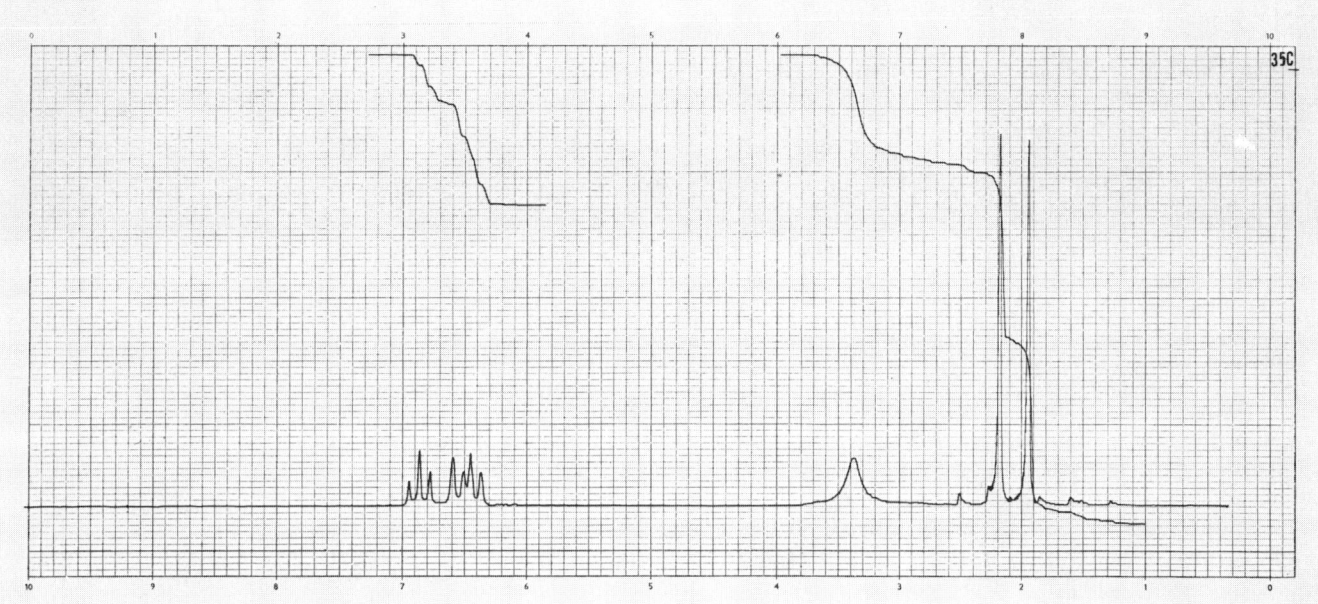

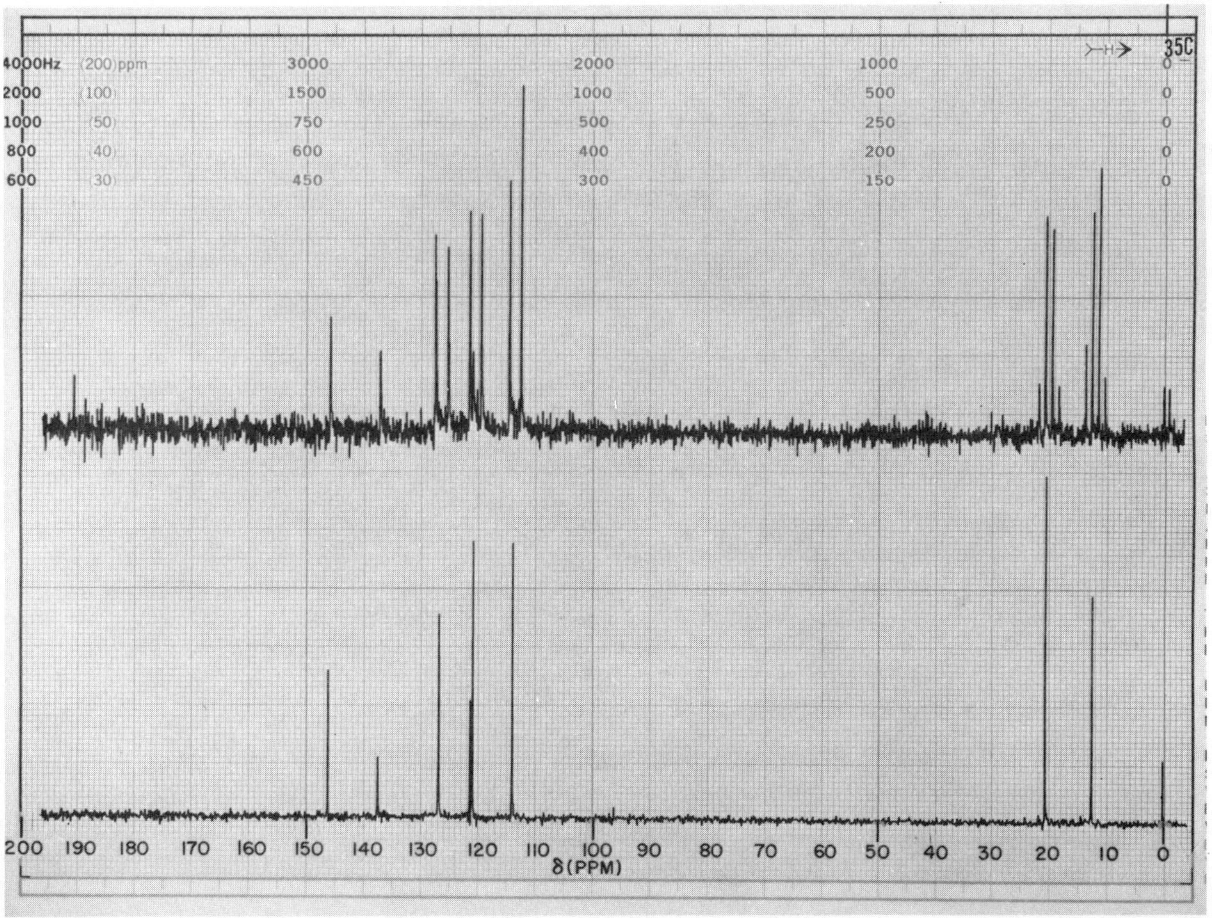

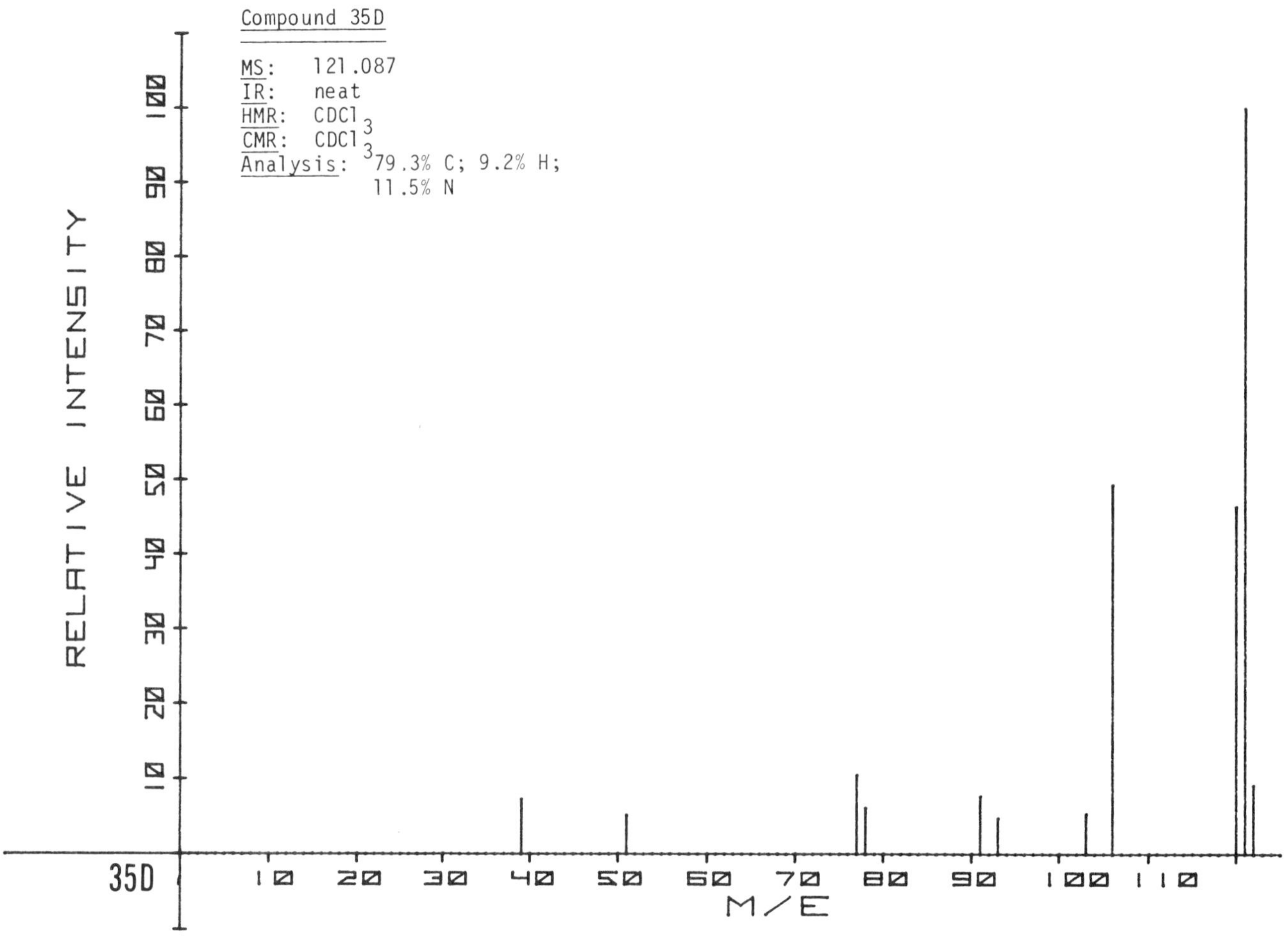

Compound 35D
MS: 121.087
IR: neat
HMR: CDCl3
CMR: CDCl3
Analysis: 79.3% C; 9.2% H; 11.5% N
RELATIVE INTENSITY
10 20 30 40 50 60 70 80 90 100
35D
10 20 30 40 50 60 70 80 90 100 110
M/E

4000 3000 2000 1500 CM-1 1000 900 800 700
ABSORBANCE
0.0 .10 .20 .30 .40 .50 .60 .70 1.0 ∞
35D
3 4 5 6 7 8 9 10 11 12 13 14 15
WAVELENGTH (MICRONS)
PR 1109 (137-1280)

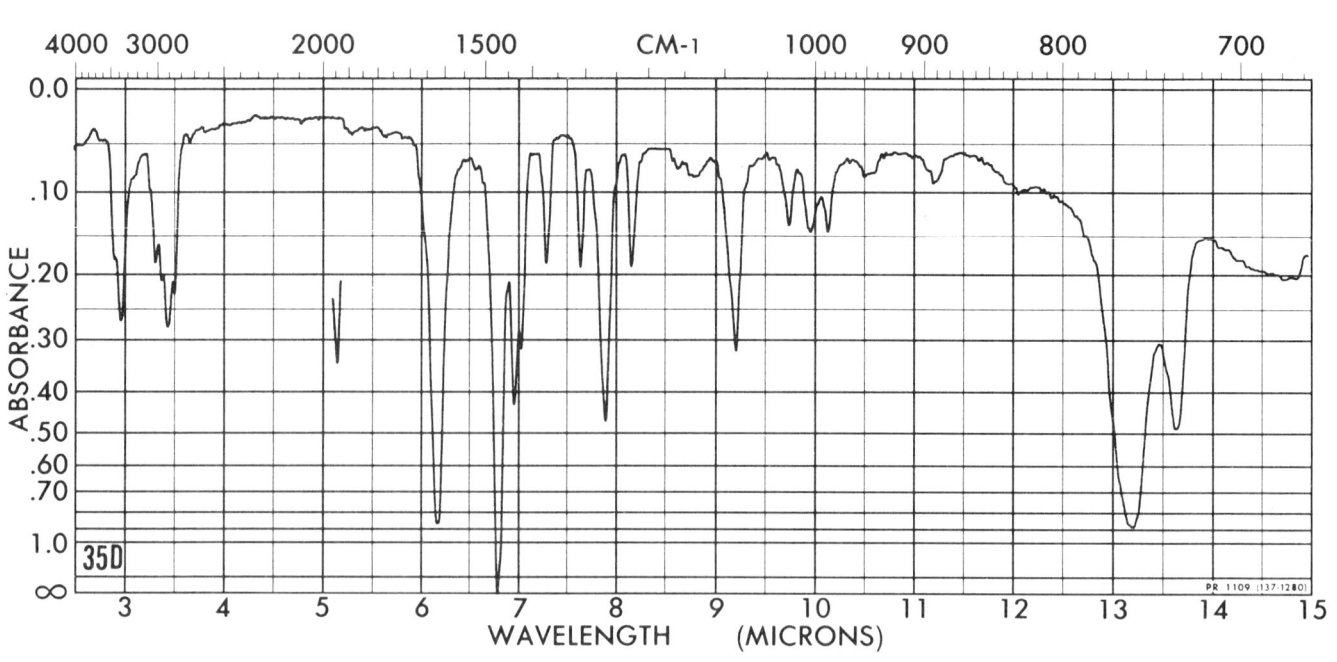

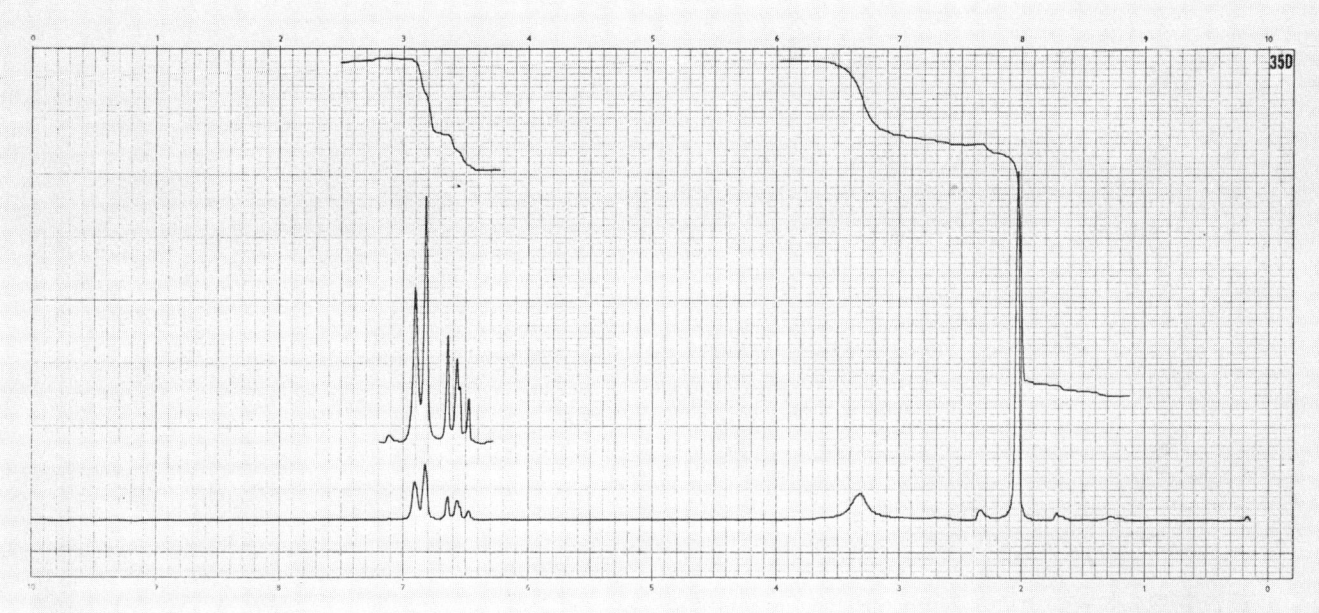

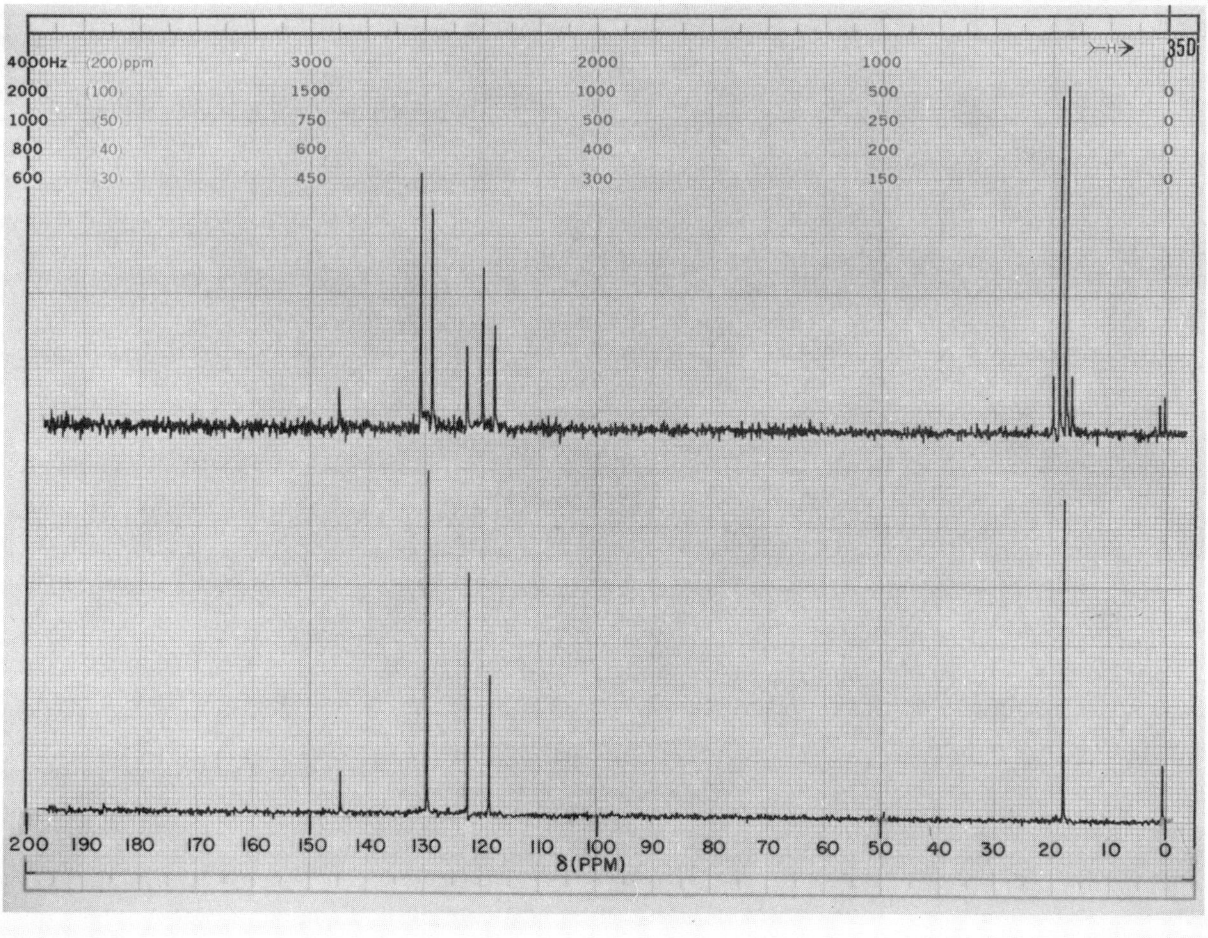

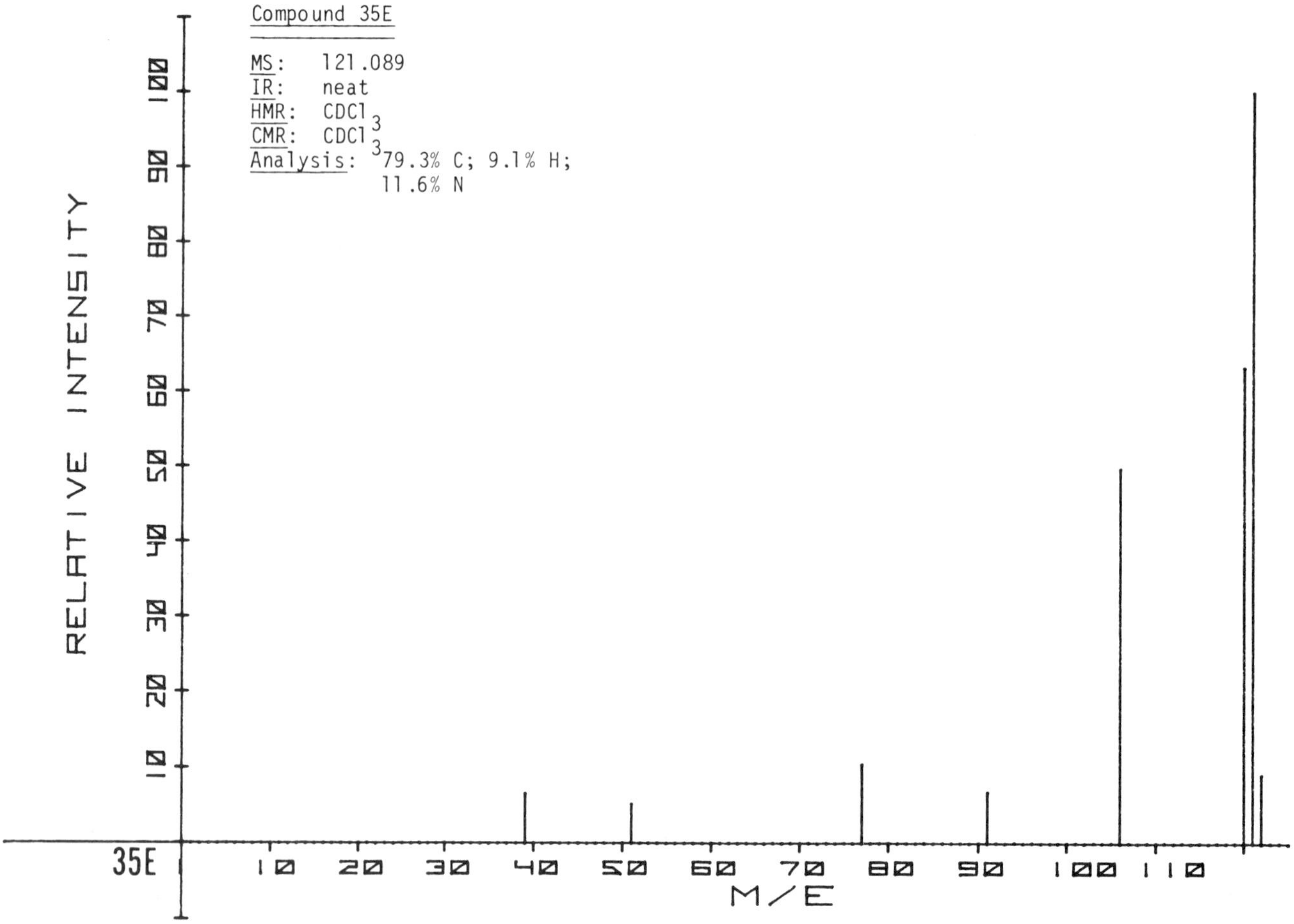

Compound 35E
MS: 121.089
IR: neat
HMR: $CDCl_3$
CMR: $CDCl_3$
Analysis: 79.3% C; 9.1% H; 11.6% N
RELATIVE INTENSITY
10 20 30 40 50 60 70 80 90 100
35E
M/E
10 20 30 40 50 60 70 80 90 100 110

4000 3000 2000 1500 CM-1 1000 900 800 700
ABSORBANCE
0.0 .10 .20 .30 .40 .50 .60 .70 1.0 ∞
35E
3 4 5 6 7 8 9 10 11 12 13 14 15
WAVELENGTH (MICRONS)

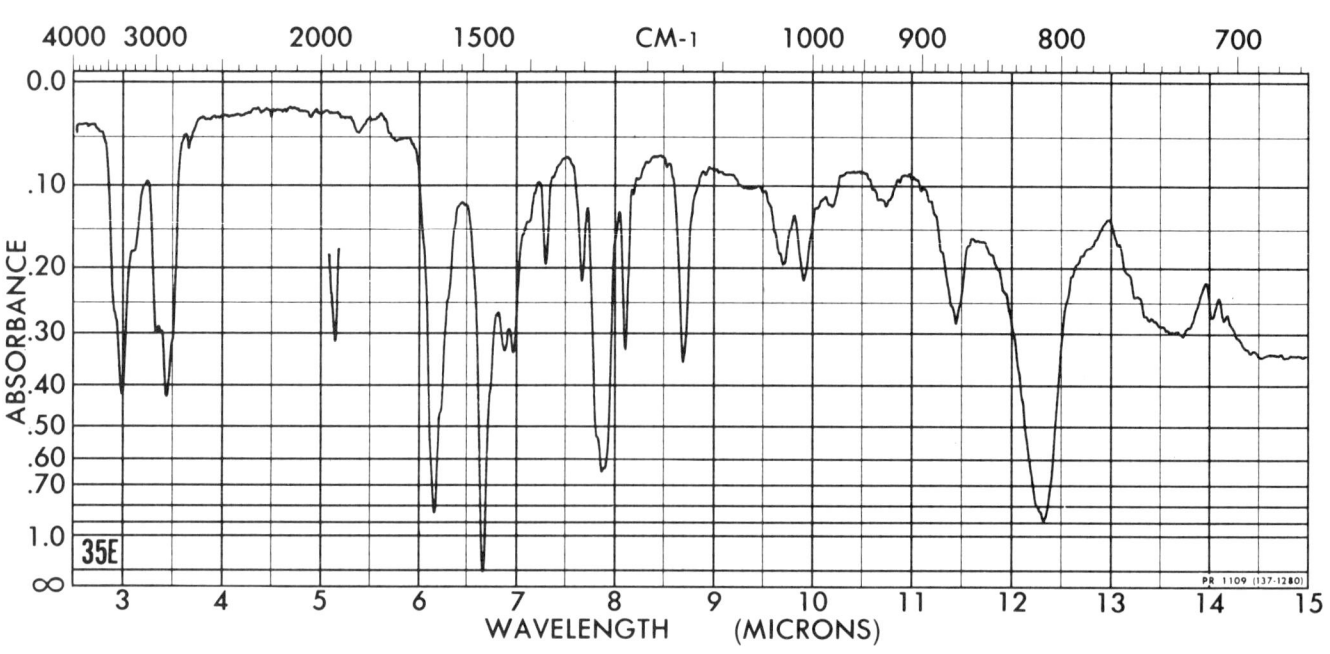

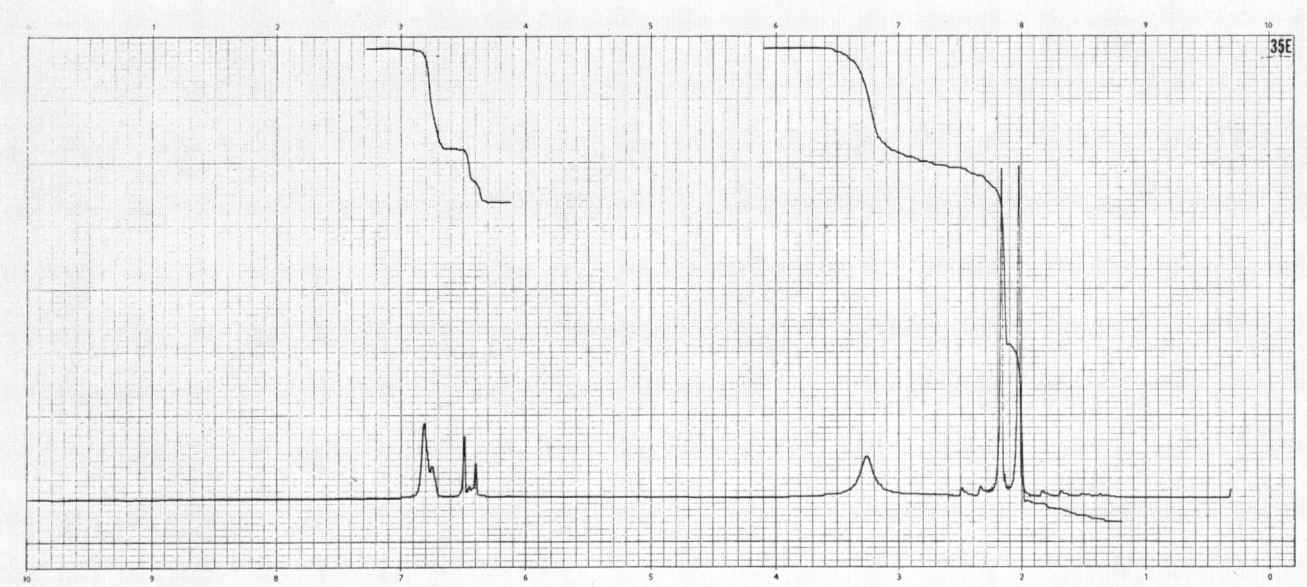

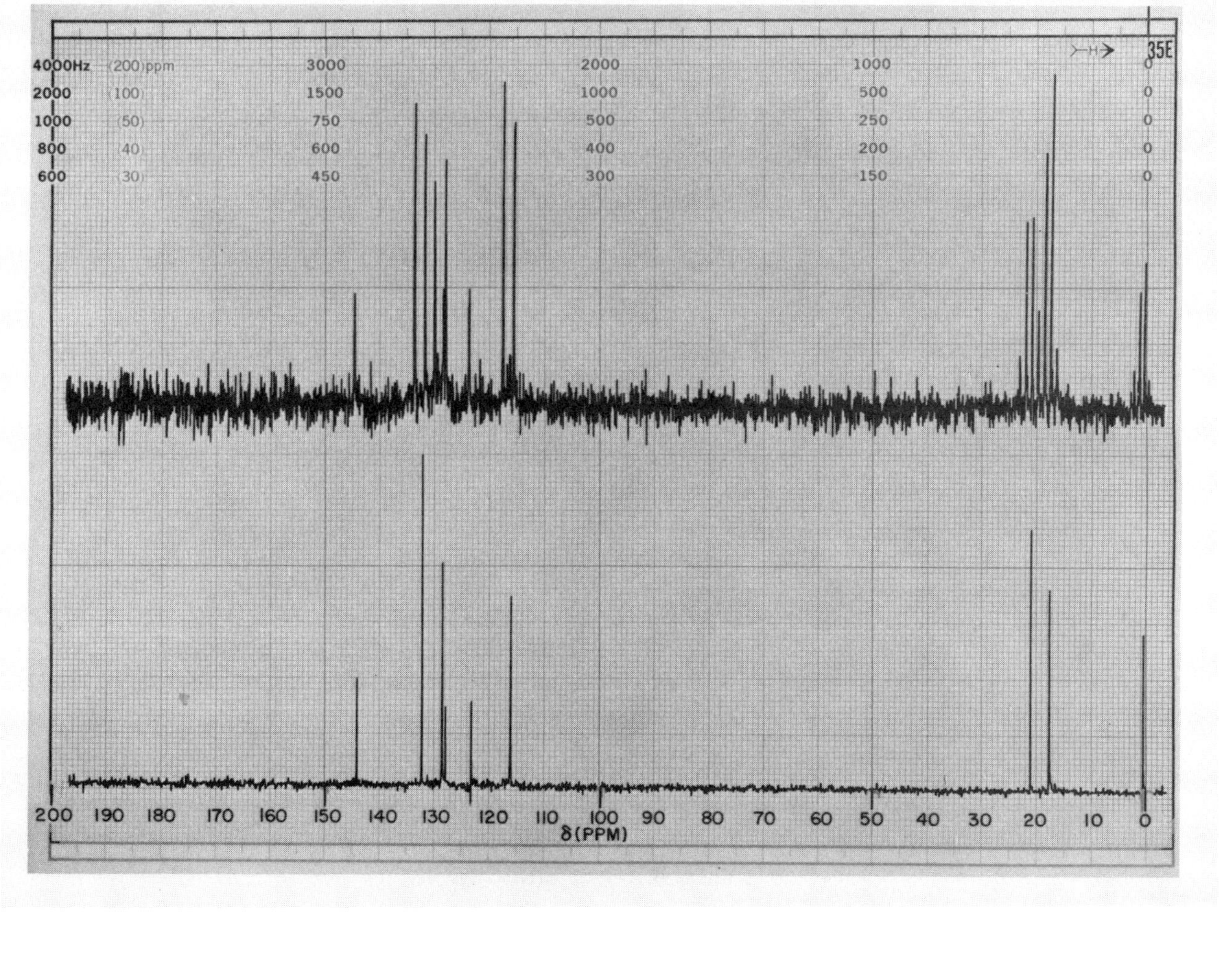

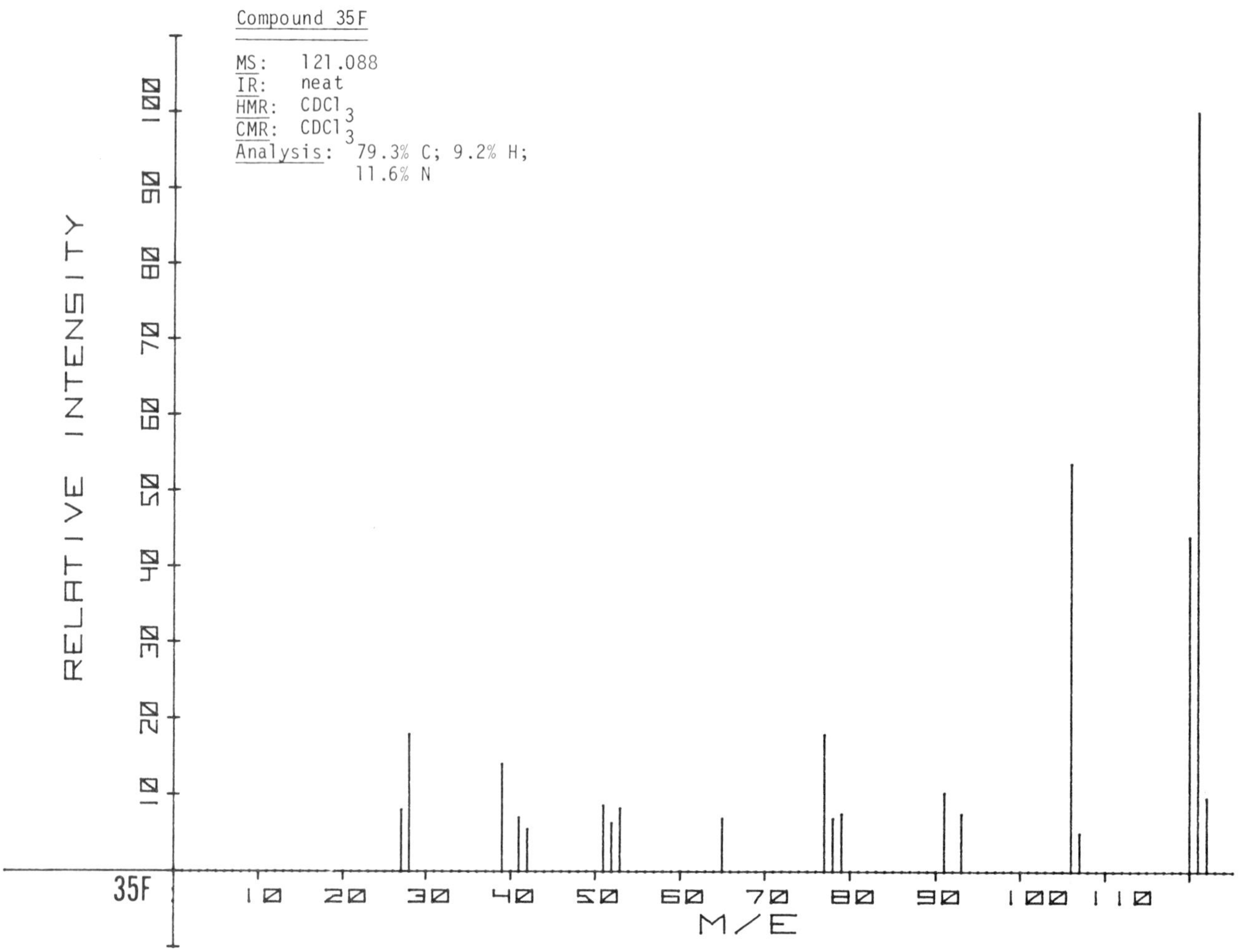

Compound 35F
MS: 121.088
IR: neat
HMR: $CDCl_3$
CMR: $CDCl_3$
Analysis: 79.3% C; 9.2% H; 11.6% N
RELATIVE INTENSITY
10 20 30 40 50 60 70 80 90 100
35F
M/E
10 20 30 40 50 60 70 80 90 100 110

4000 3000 2000 1500 CM-1 1000 900 800 700
ABSORBANCE
0.0 .10 .20 .30 .40 .50 .60 .70 1.0 ∞
35F
3 4 5 6 7 8 9 10 11 12 13 14 15
WAVELENGTH (MICRONS)

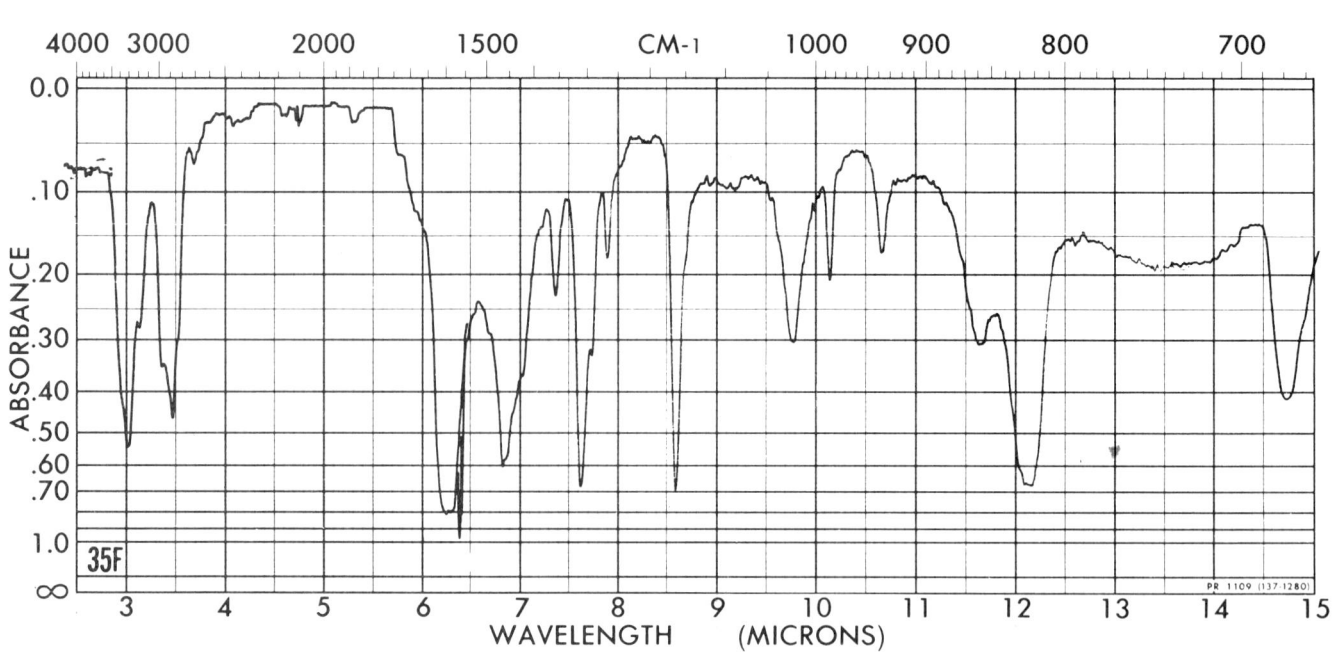

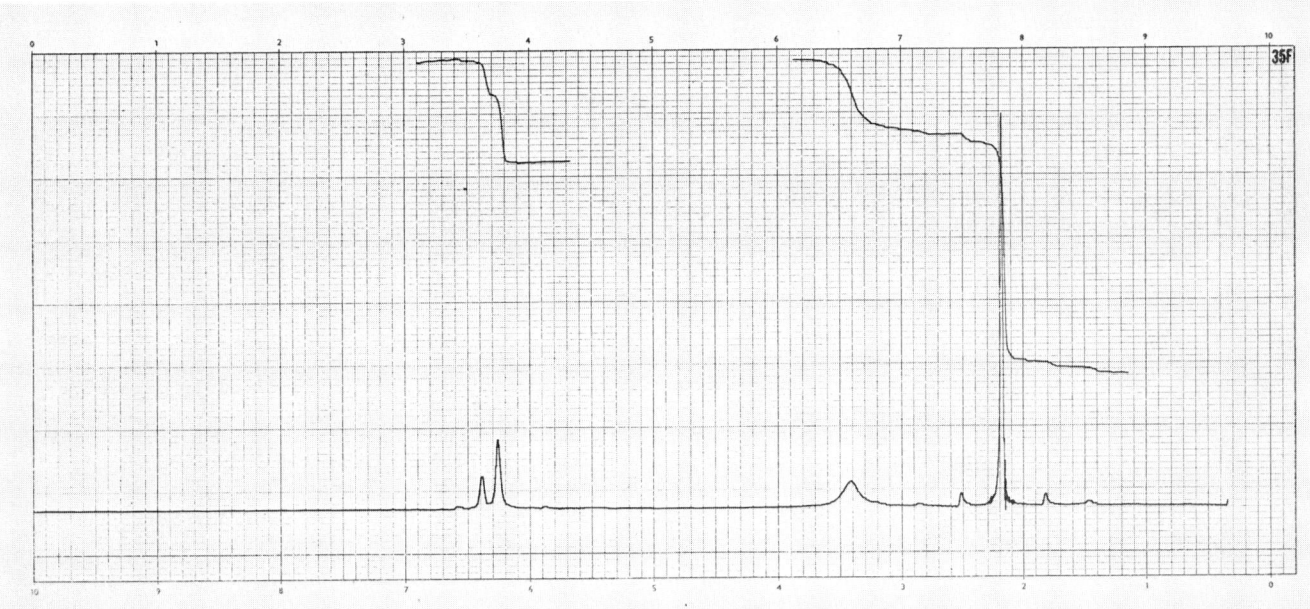

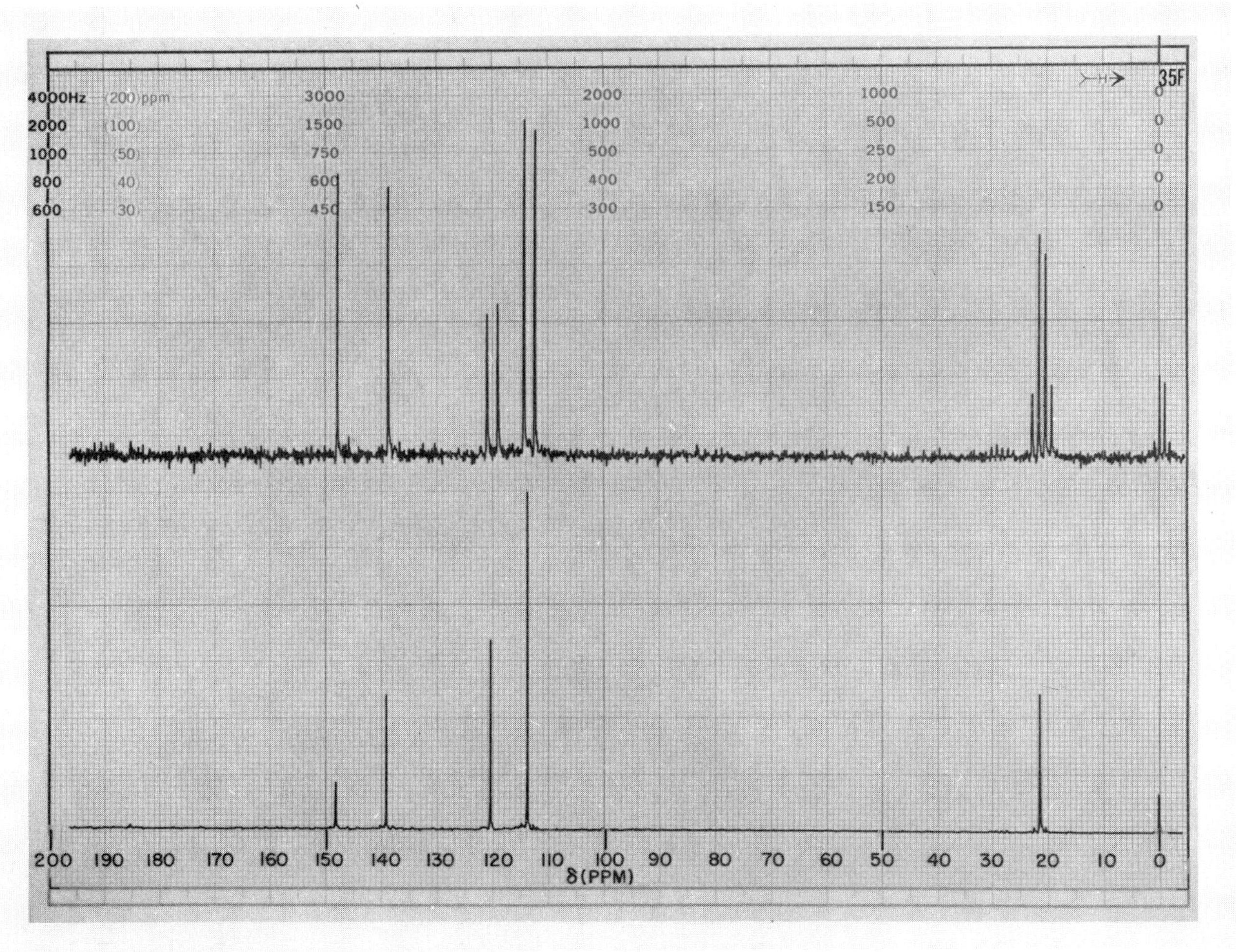

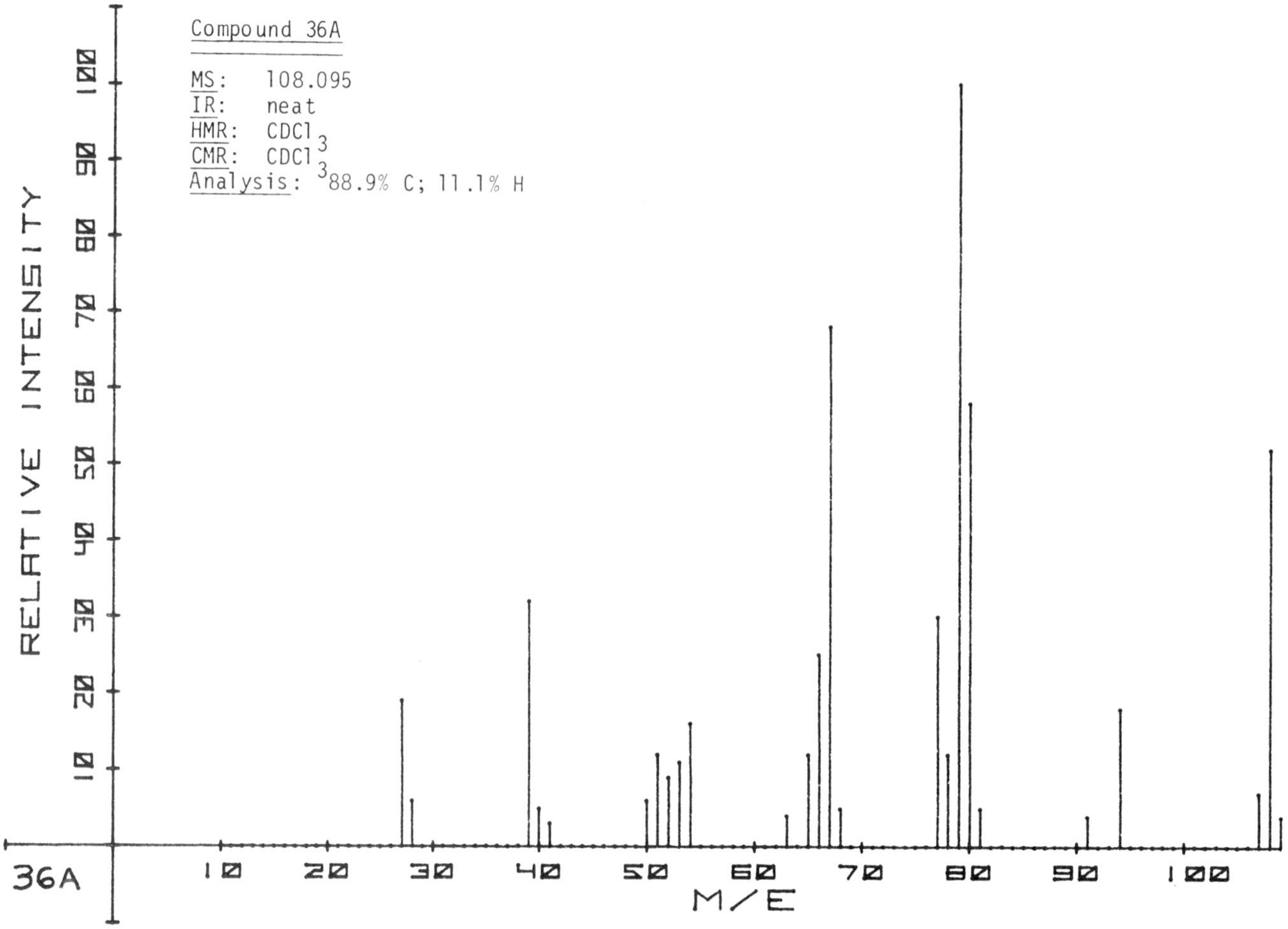

Compound 36A
MS: 108.095
IR: neat
HMR: CDCl3
CMR: CDCl3
Analysis: 88.9% C; 11.1% H
RELATIVE INTENSITY
10 20 30 40 50 60 70 80 90 100
36A
M/E

4000 3000 2000 1500 CM-1 1000 900 800 700
ABSORBANCE
0.0 .10 .20 .30 .40 .50 .60 .70 1.0 ∞
36A
3 4 5 6 7 8 9 10 11 12 13 14 15
WAVELENGTH (MICRONS)
PR 1109 (137-1280)

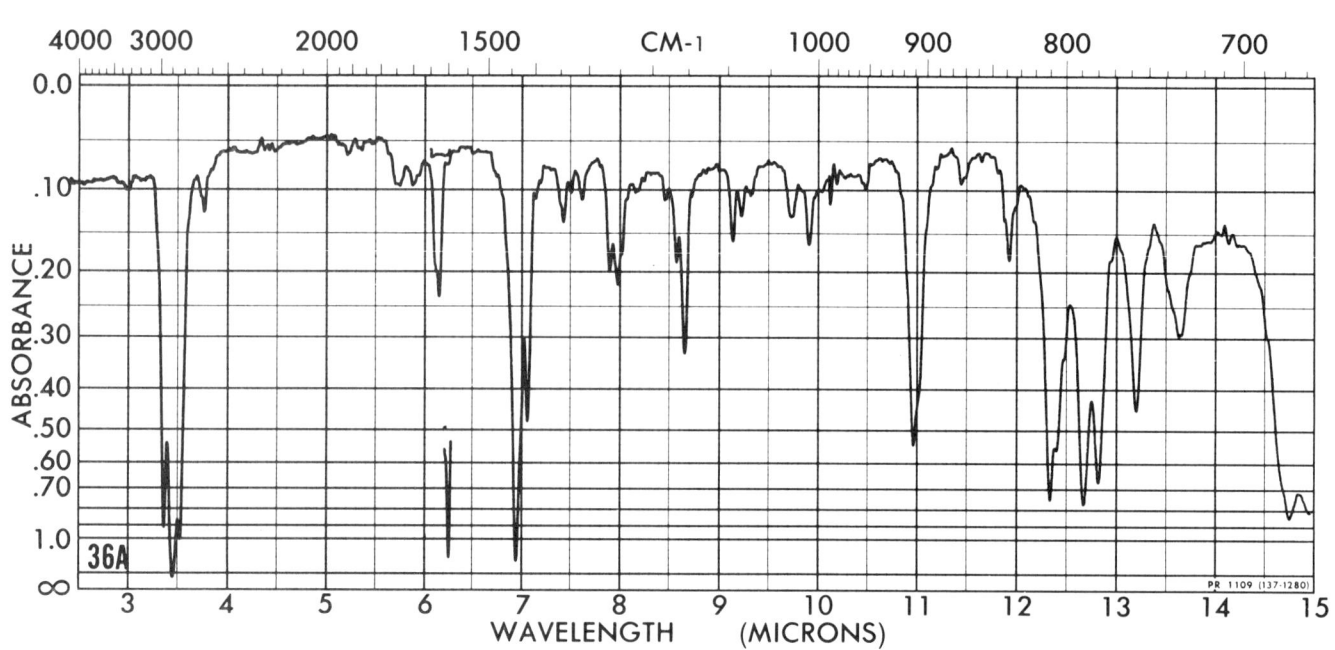

36A
IRR. 1.50
IRR. 2.16
IRR. 1.50
IRR. 2.16
A
B
A
B

36A
4000Hz (200)ppm 3000 2000 1000
2000 (100) 1500 1000 500
1000 (50) 750 500 250
800 (40) 600 400 200
600 450 300 150
Pk. Ht. PPM
45 132.5
45 130.7
38 127.6
36 125.7
48 28.7
42 23.9
Pk. Ht. PPM
45 131.2
60 126.3
52 28.3
38 23.5
200 190 180 170 160 150 140 130 120 110 100 90 80 70 60 50 40 30 20 10 0
δ(PPM)

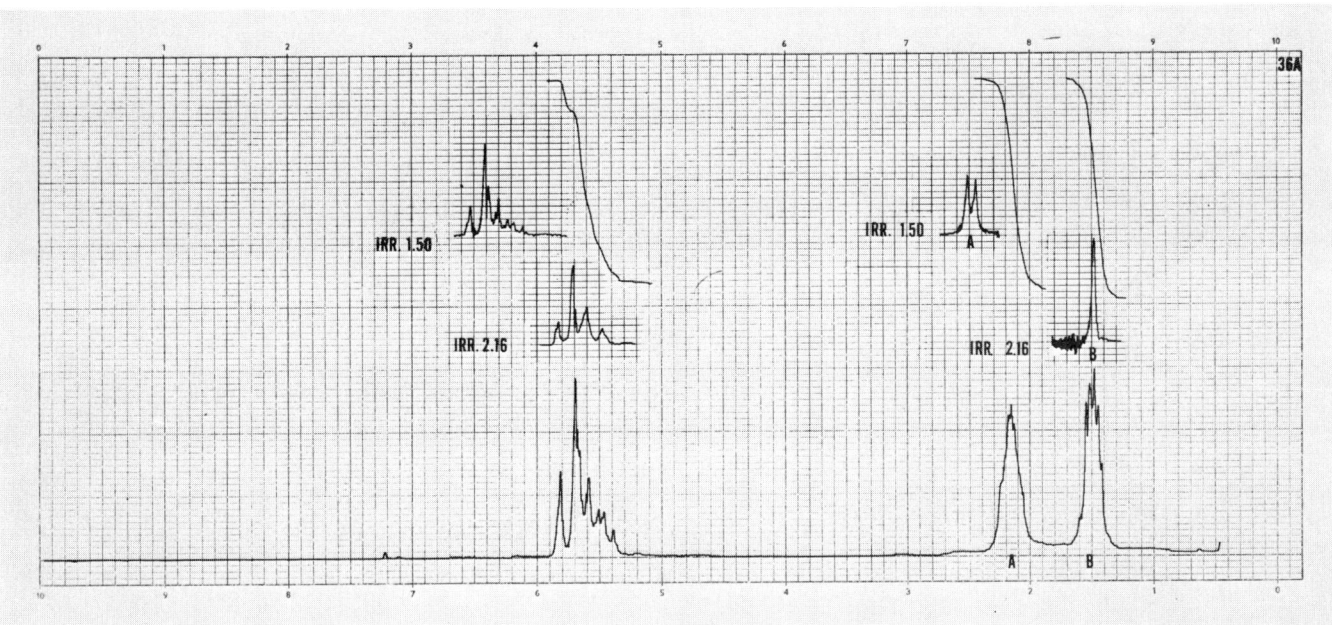

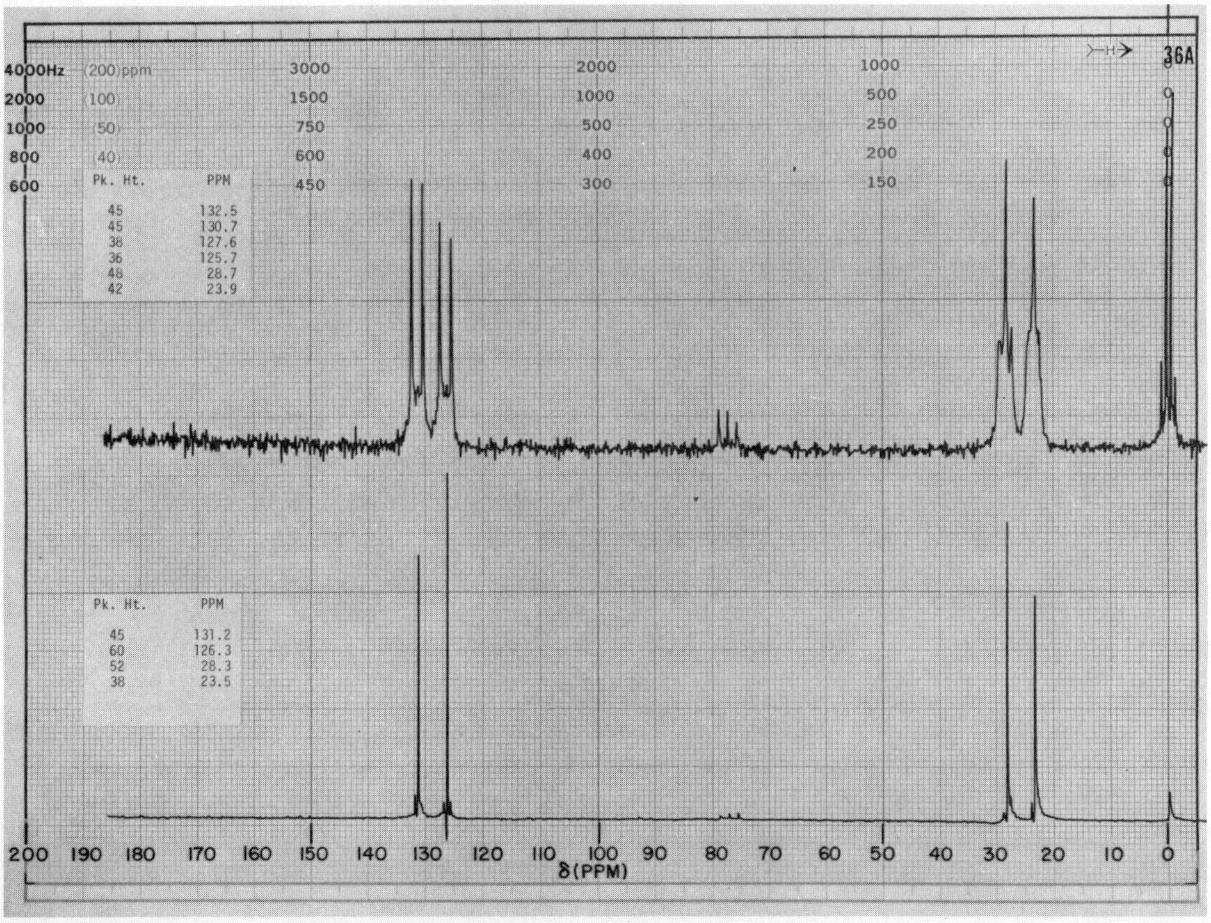

COMPOUND 36B

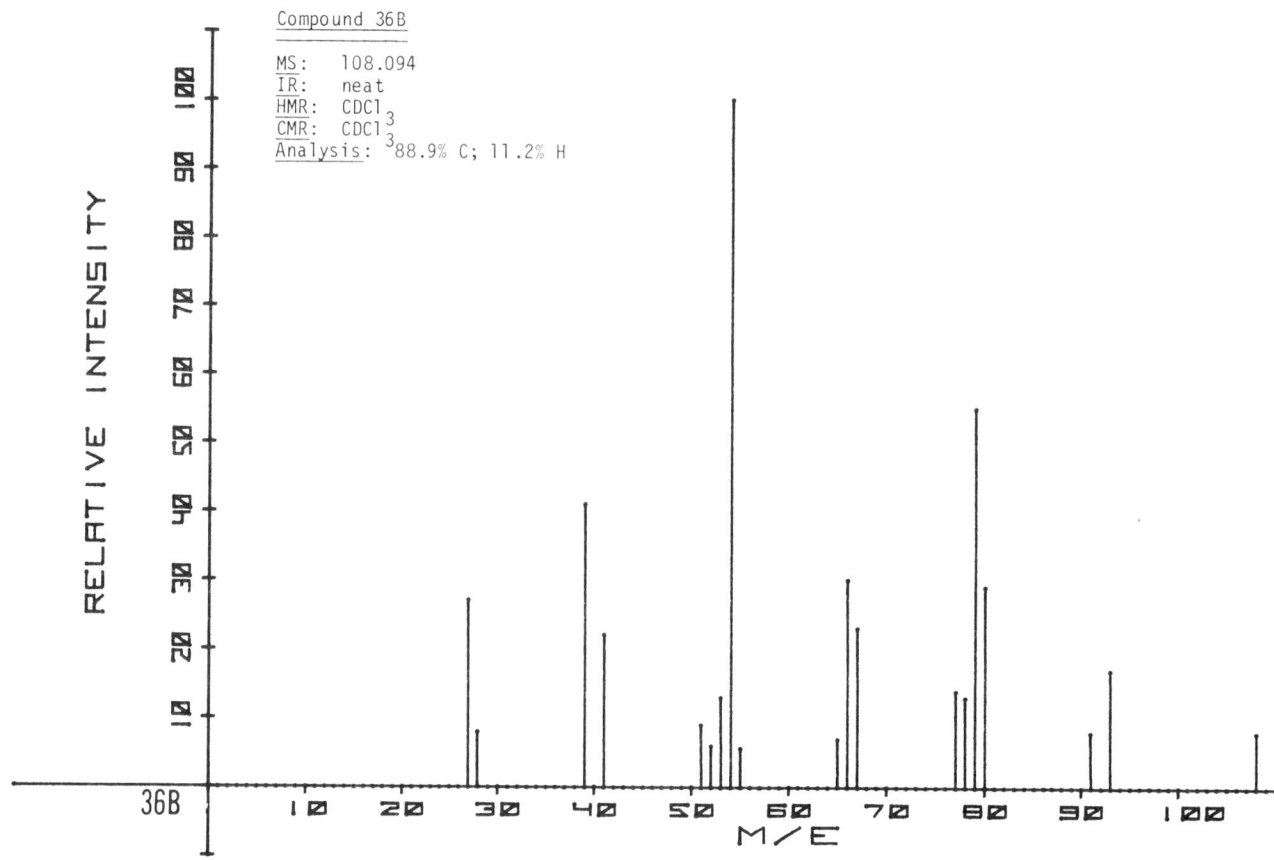

Compound 36B
MS: 108.094
IR: neat
HMR: CDCl$_3$
CMR: CDCl$_3$
Analysis: 88.9% C; 11.2% H

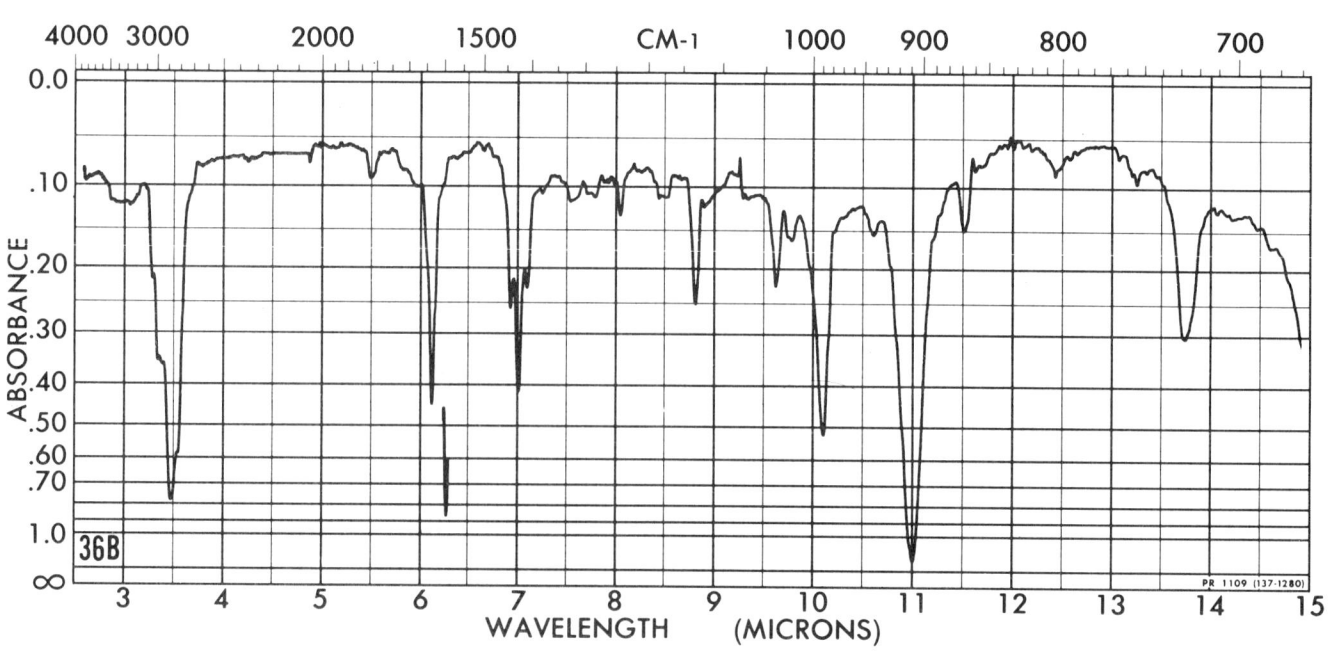

36B
10
9
8
7
6
5
4
3
2
1
0

36B
4000Hz (200)ppm
2000
1000
800
600
3000
1500
750
600
450
2000
1000
500
400
300
1000
500
250
200
150
PK. HT. PPM
48 127.5
52 126.8
58 126.3
44 125.4
21 32.5
52 31.2
30 29.8
57 28.6
17 27.1
19 26.3
57 25.0
22 23.6
PK. HT. PPM
36 143.7
60 126.9
47 126.1
52 112.4
38 37.8
42 31.2
48 28.6
51 25.0
200 190 180 170 160 150 140 130 120 110 100 90 80 70 60 50 40 30 20 10 0
δ(PPM)

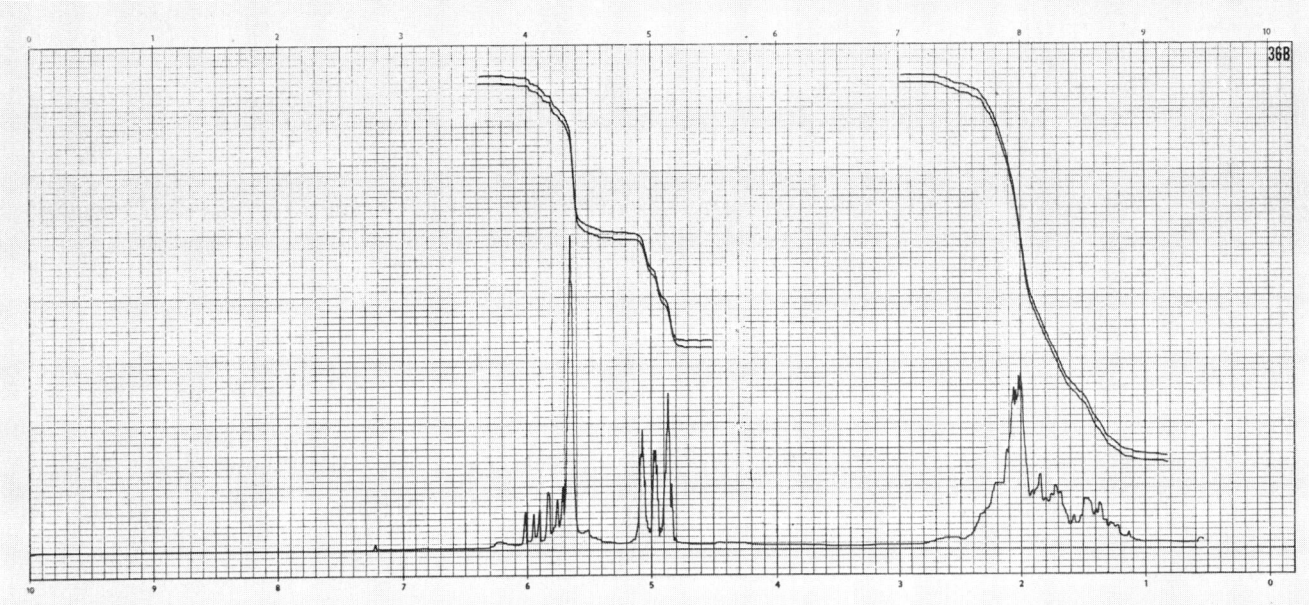

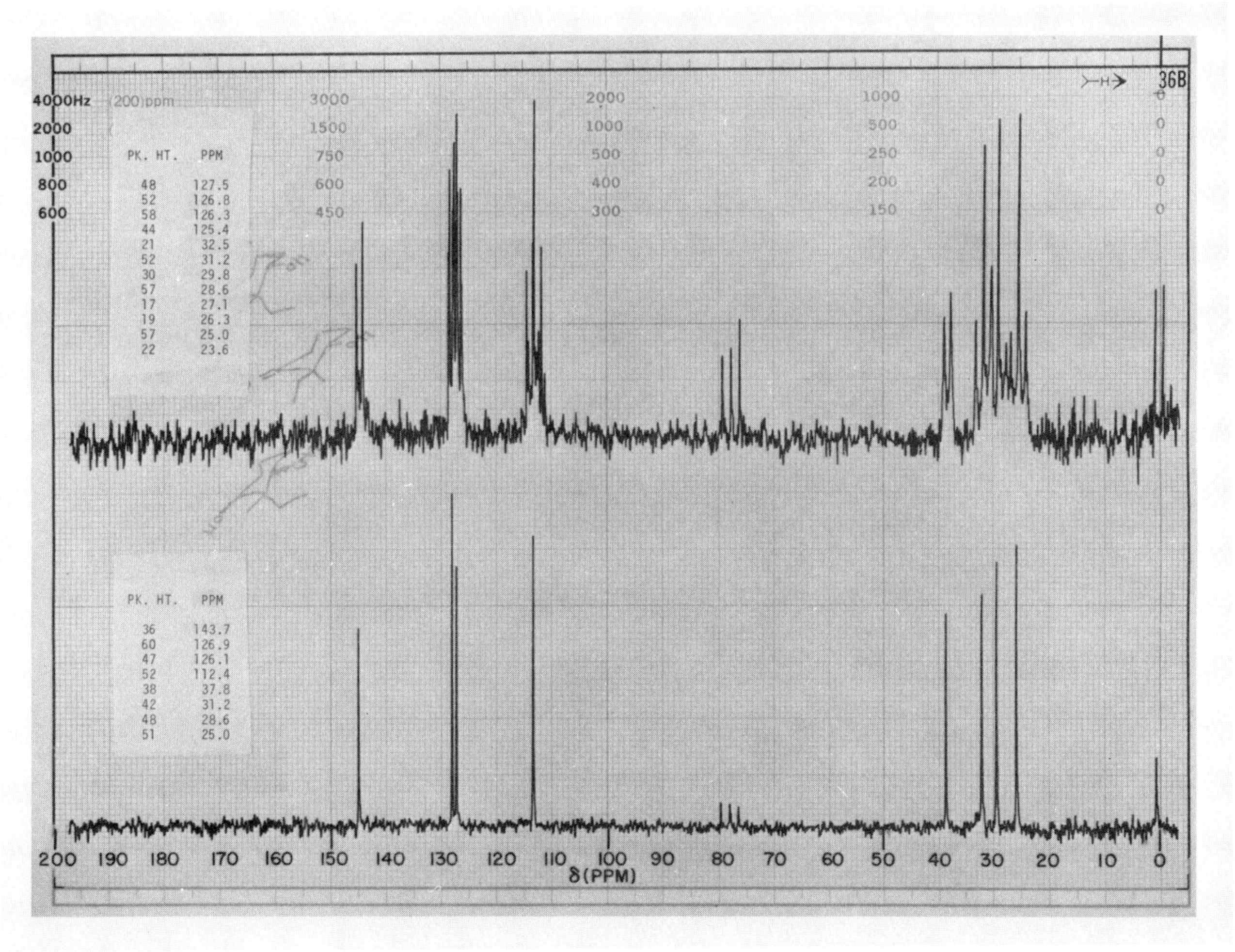

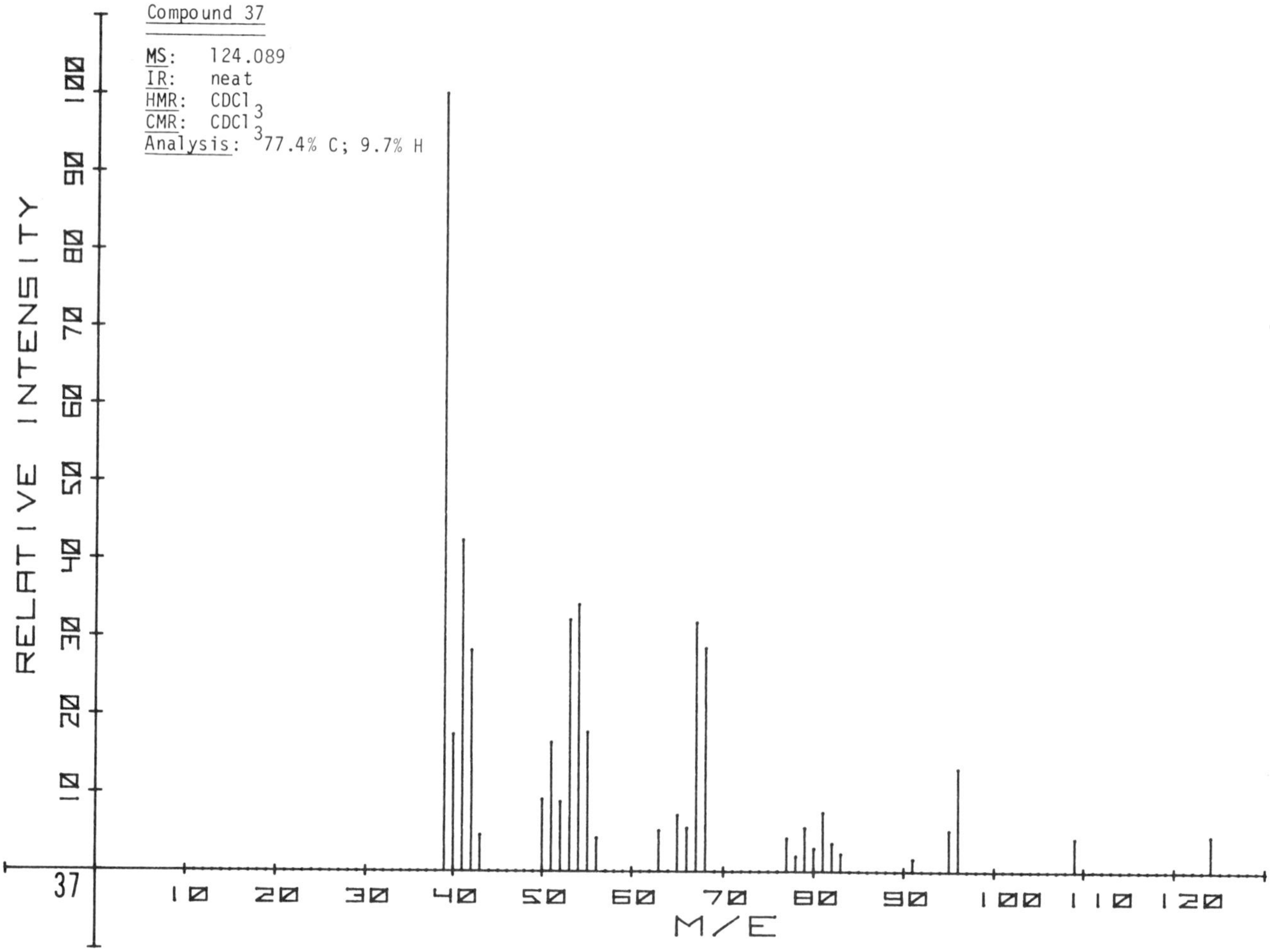

Compound 37
MS: 124.089
IR: neat
HMR: CDCl3
CMR: CDCl3
Analysis: 77.4% C; 9.7% H
RELATIVE INTENSITY
10 20 30 40 50 60 70 80 90 100
37
10 20 30 40 50 60 70 80 90 100 110 120
M/E

4000 3000 2000 1500 CM-1 1000 900 800 700
0.0 .10 .20 .30 .40 .50 .60 .70 1.0 ∞
ABSORBANCE
37
3 4 5 6 7 8 9 10 11 12 13 14 15
WAVELENGTH (MICRONS)

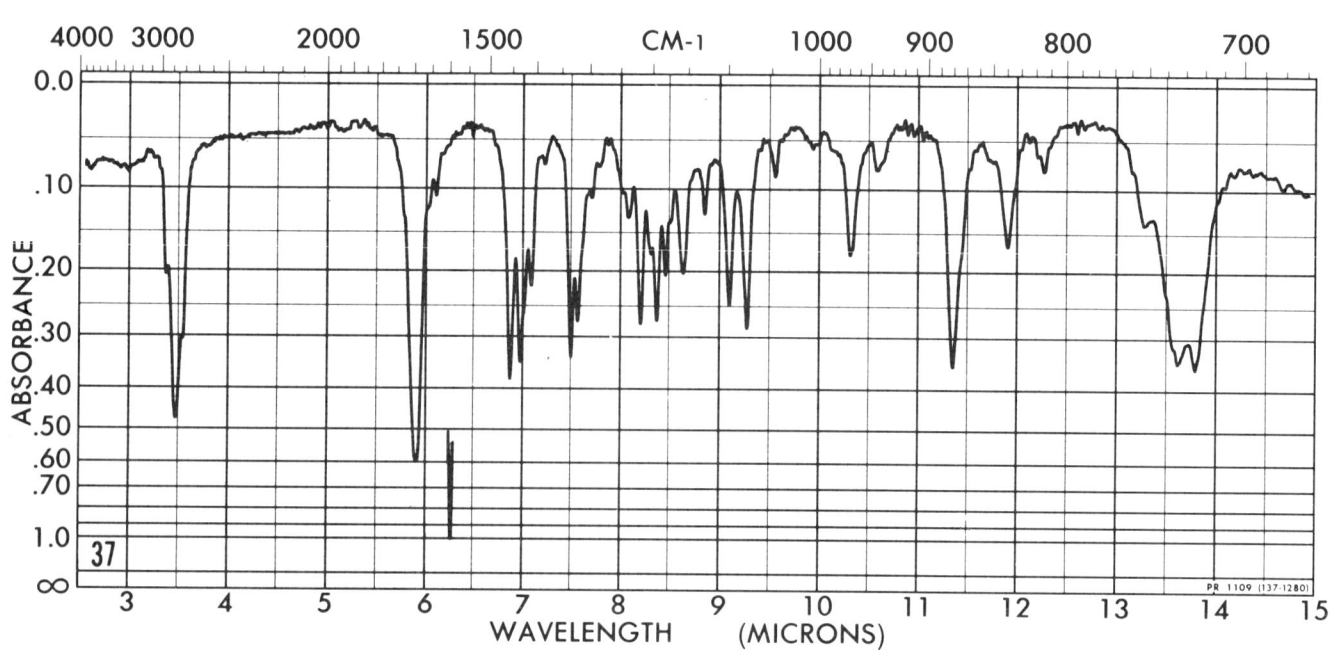

37

37
4000Hz (200)ppm 3000 2000 1000 0
2000 (100) 1500 1000 500 0
1000 (50) 750 500 250 0
800 (40) 600 400 200 0
600 (30) 450 300 150 0
PK. HT. PPM
9 214.6
10 131.4
12 130.9
12 130.2
10 129.7
19 47.3
14 40.4
11 26.3
10 23.9
12 21.9
PK. HT. PPM
11 214.6
48 130.8
49 130.3
51 47.3
60 40.4
58 26.4
56 23.9
54 21.9
214.6
200 190 180 170 160 150 140 130 120 110 100 90 80 70 60 50 40 30 20 10 0
δ(PPM)

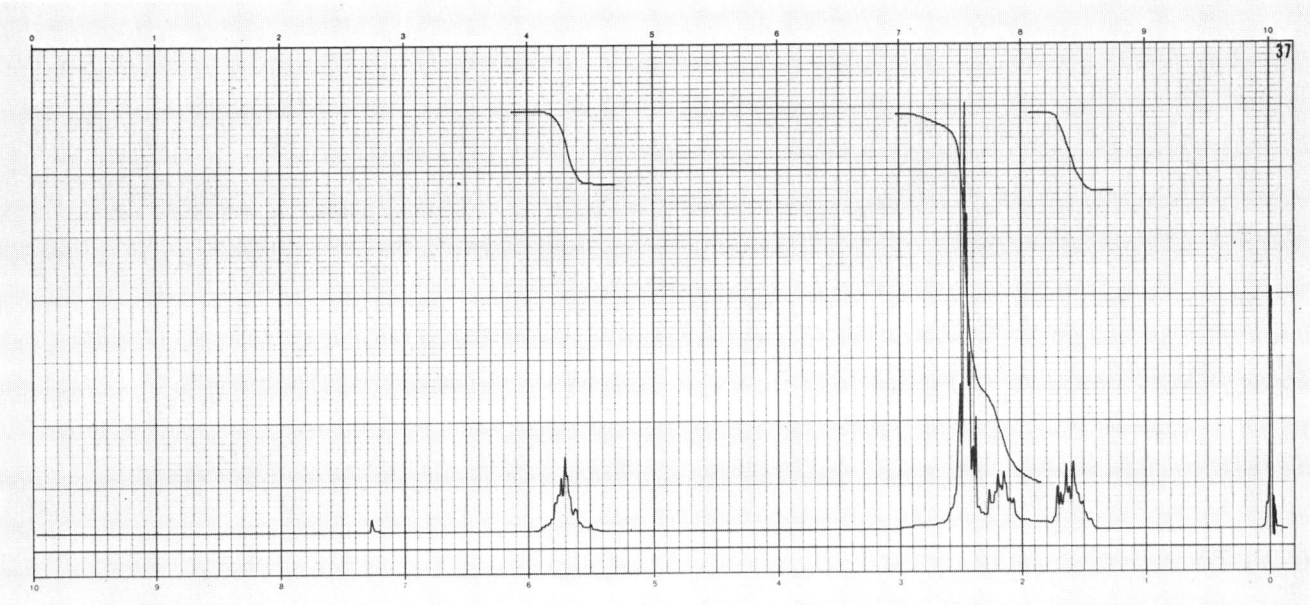

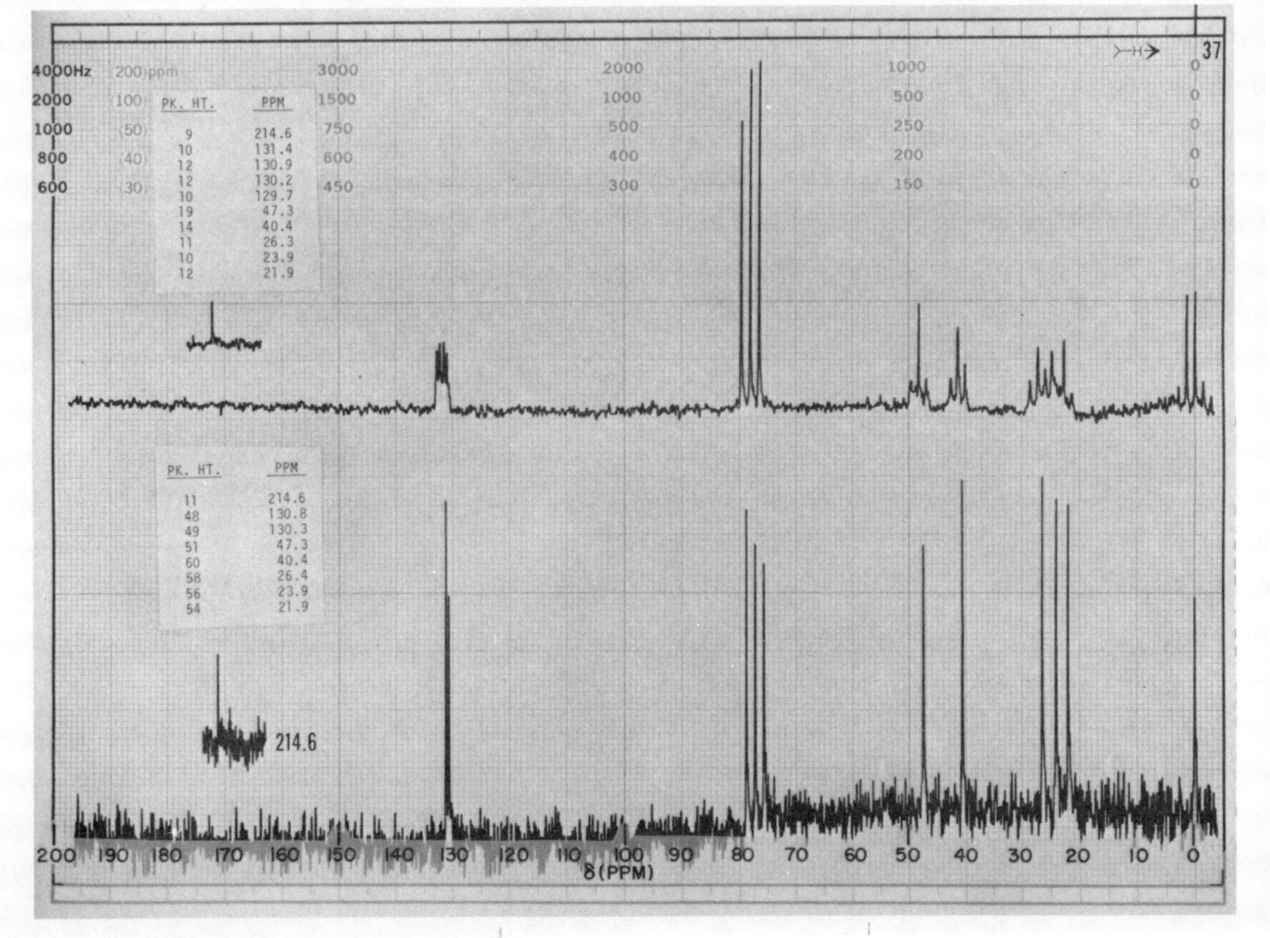

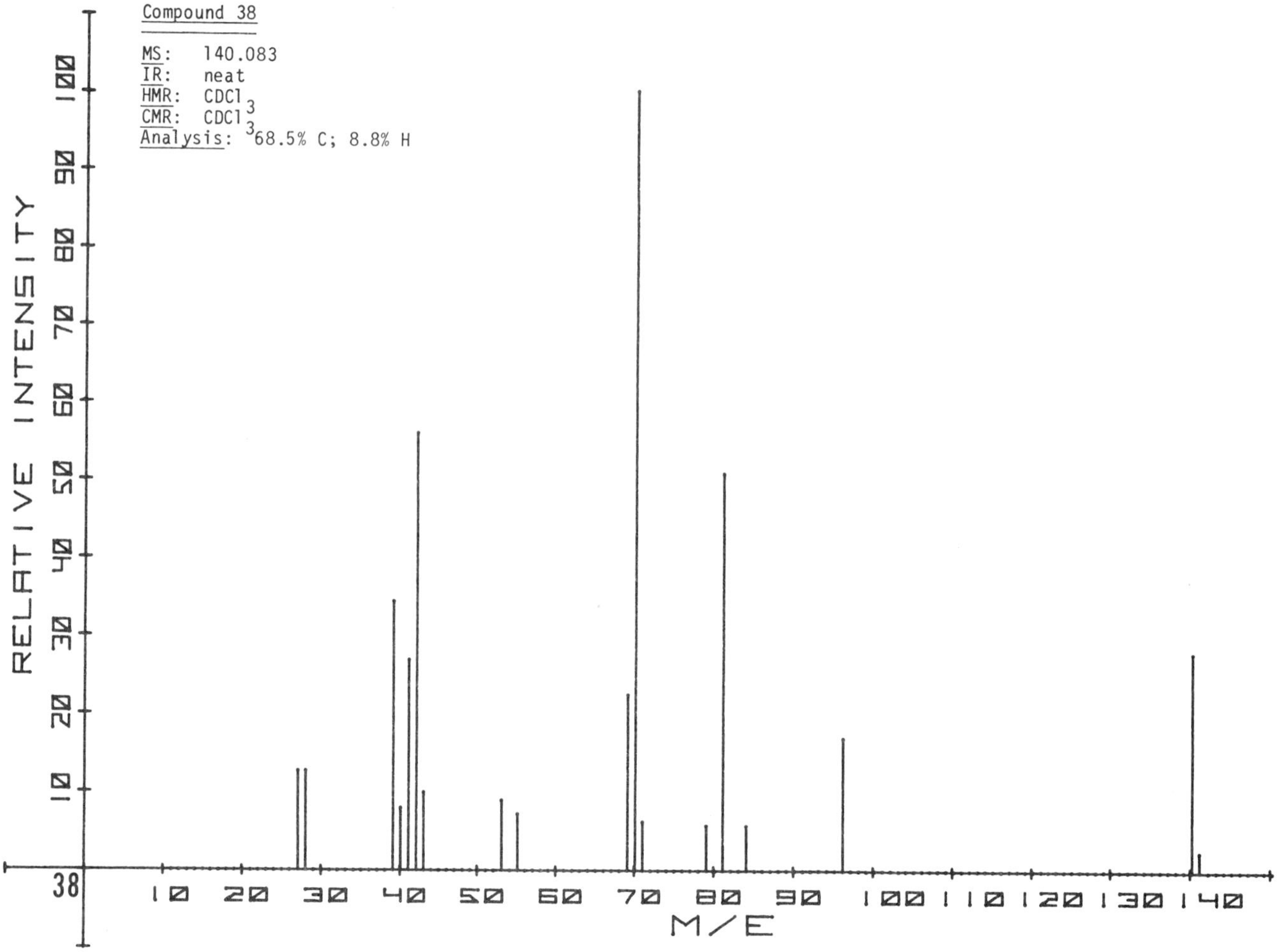

Compound 38
MS: 140.083
IR: neat
HMR: CDCl3
CMR: CDCl3
Analysis: 68.5% C; 8.8% H
RELATIVE INTENSITY
M/E
38

CM-1
ABSORBANCE
WAVELENGTH (MICRONS)
38
PR 1109 (137-1280)

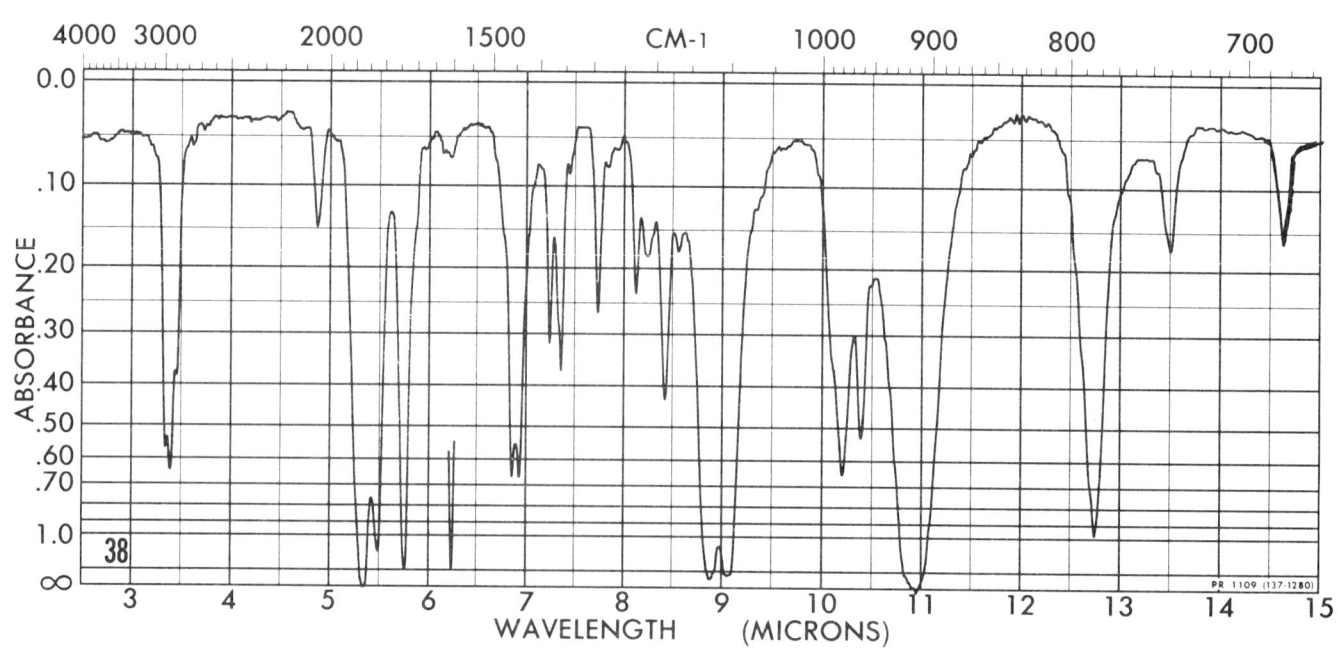

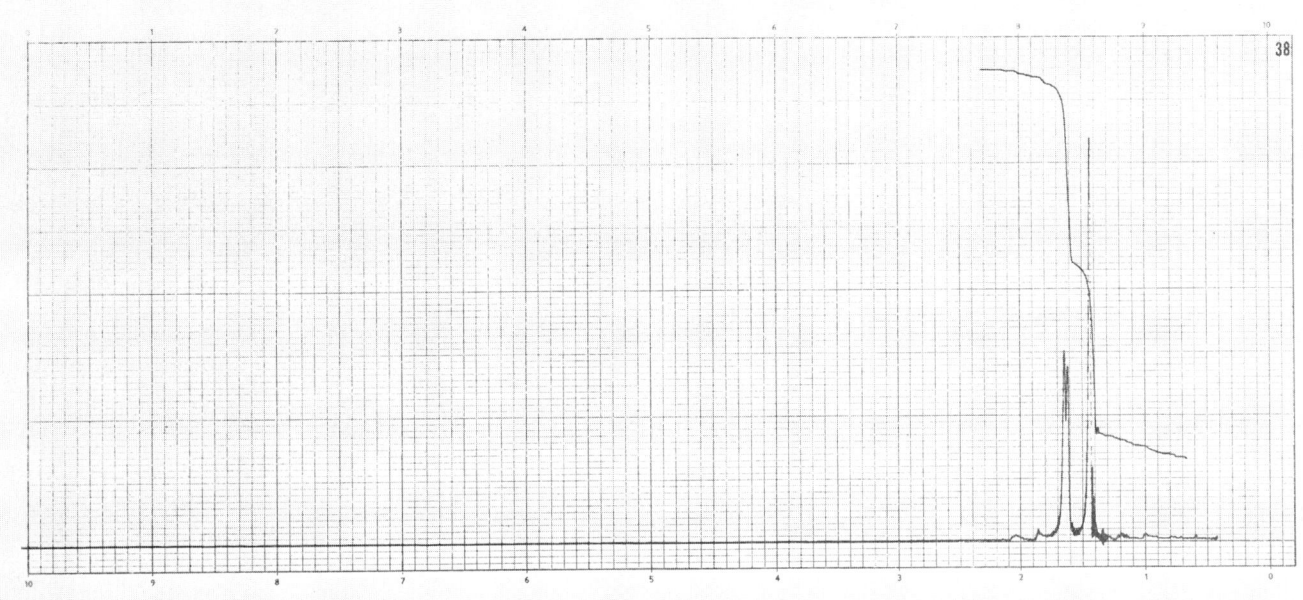

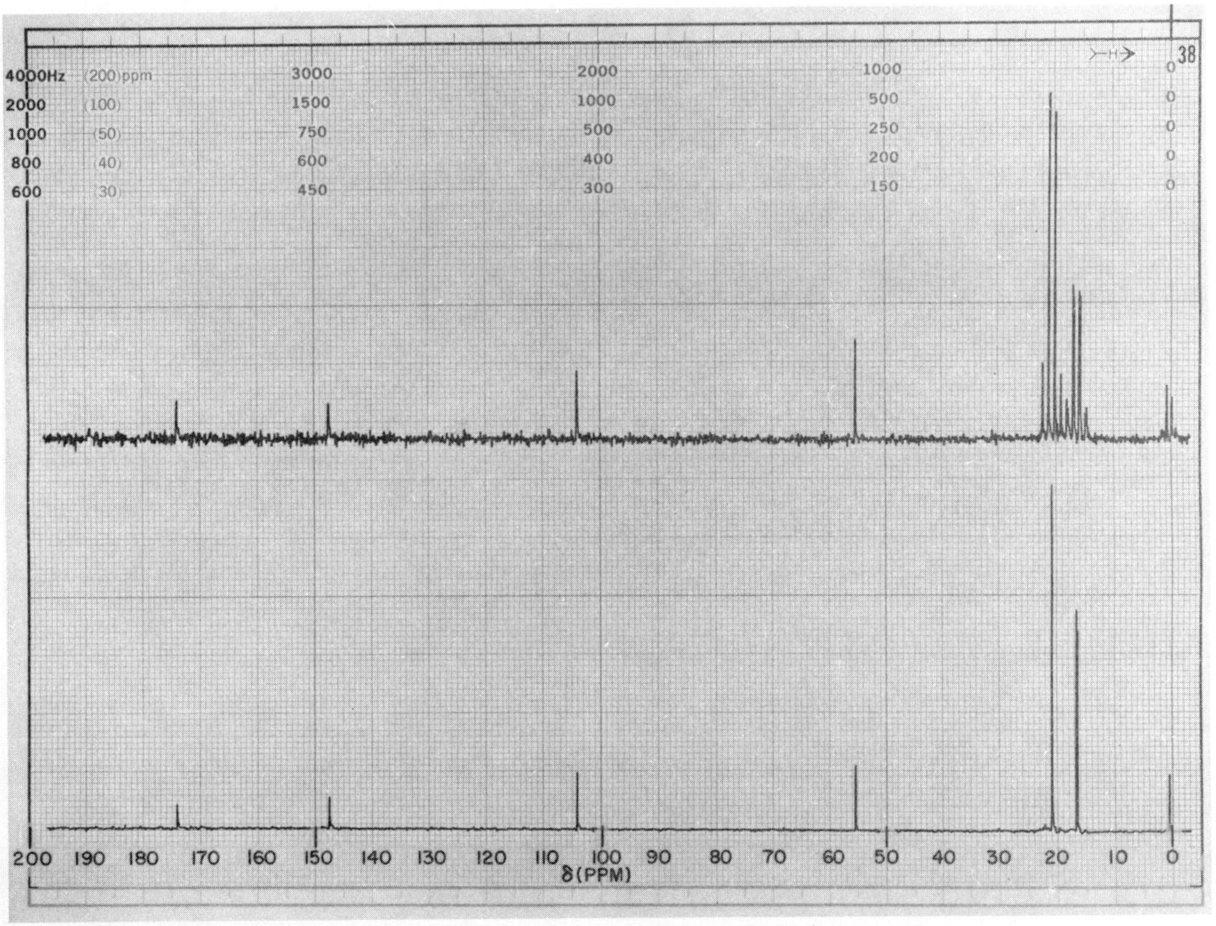

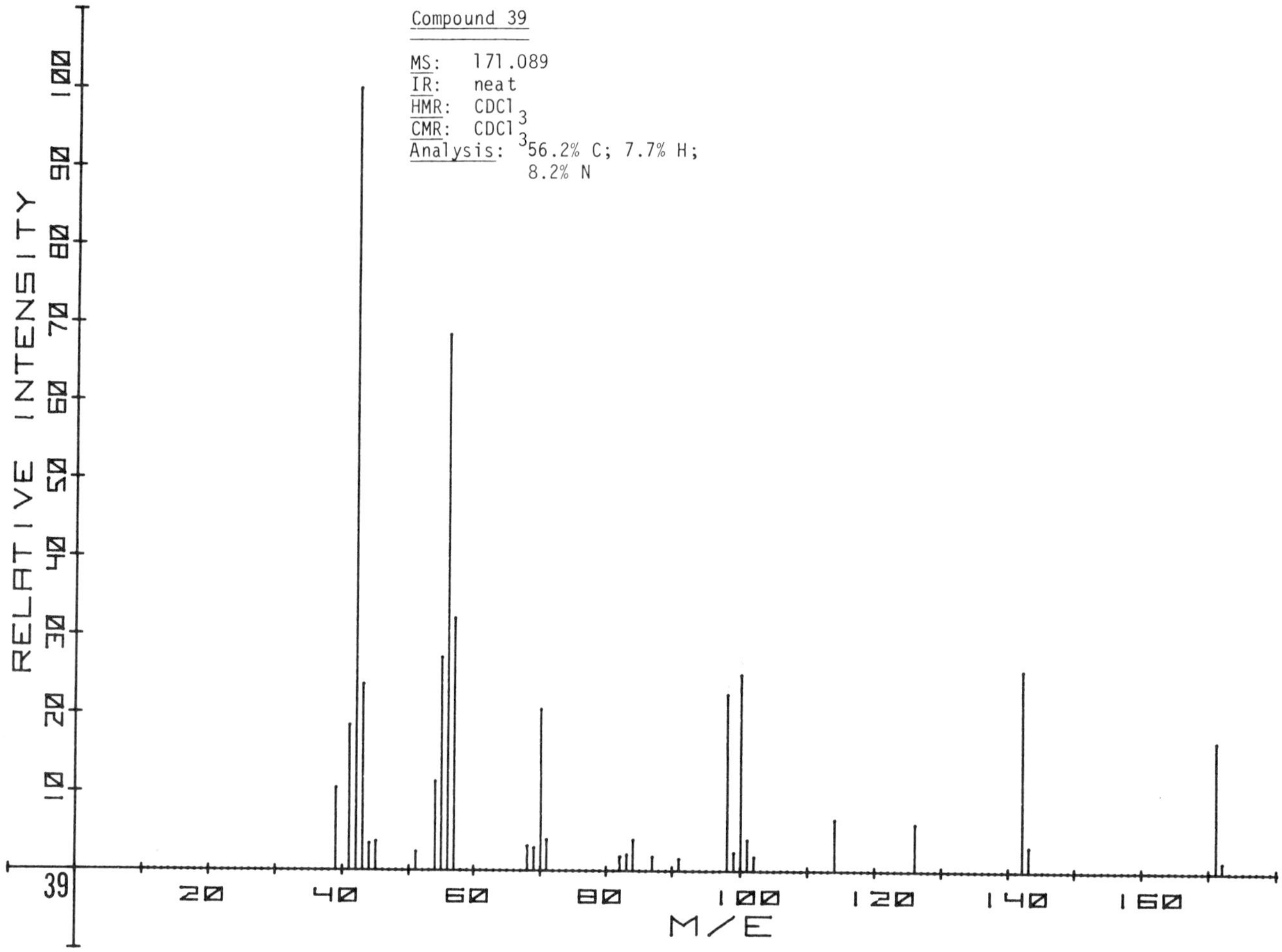

Compound 39
MS: 171.089
IR: neat
HMR: CDCl3
CMR: CDCl3
Analysis: 56.2% C; 7.7% H; 8.2% N
RELATIVE INTENSITY
10
20
30
40
50
60
70
80
90
100
39
20
40
60
80
100
120
140
160
M/E

4000 3000 2000 1500 CM-1 1000 900 800 700
0.0
.10
.20
.30
.40
.50
.60
.70
1.0
∞
ABSORBANCE
39
3 4 5 6 7 8 9 10 11 12 13 14 15
WAVELENGTH (MICRONS)
PR 1109 (137-1280)

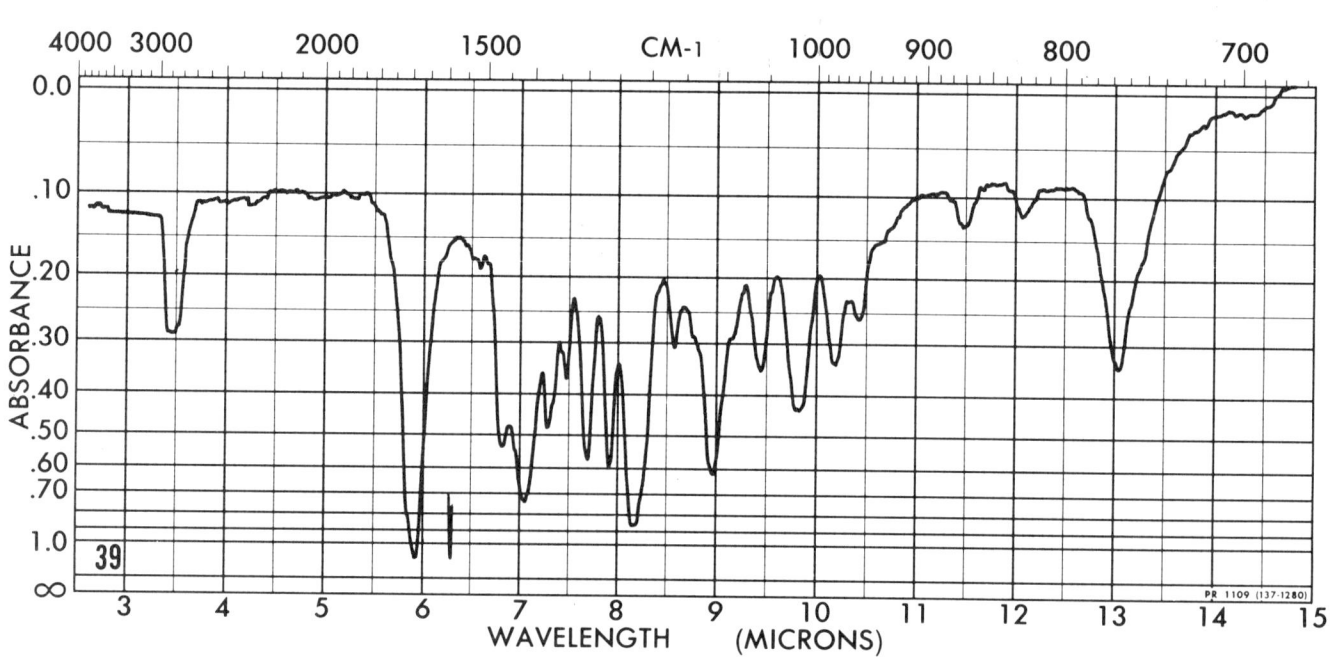

39

39
4000Hz (200) ppm
2000 (100)
1000 (50)
800 (40)
600 (30)
PK. HT. PPM
19 44.4
60 43.1
26 42.4
28 41.8
56 41.1
18 39.8
PK. HT. PPM
10 207.1
4 155.3
22 61.8
60 43.1
50 41.1
20 14.7
207.1
δ(PPM)

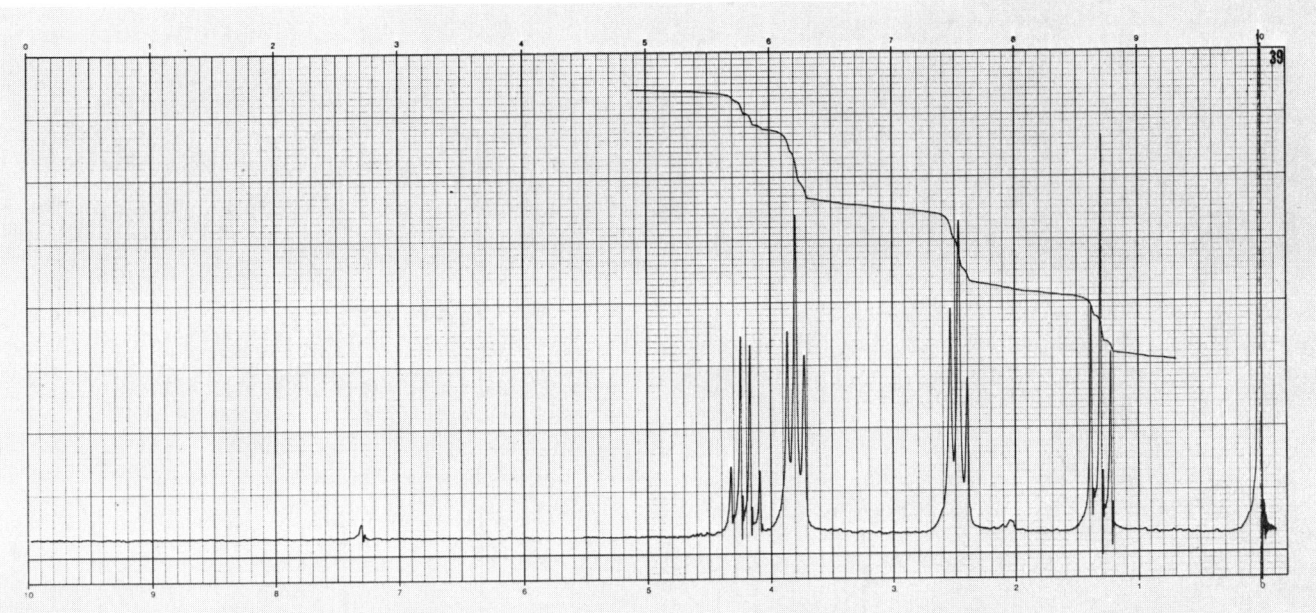

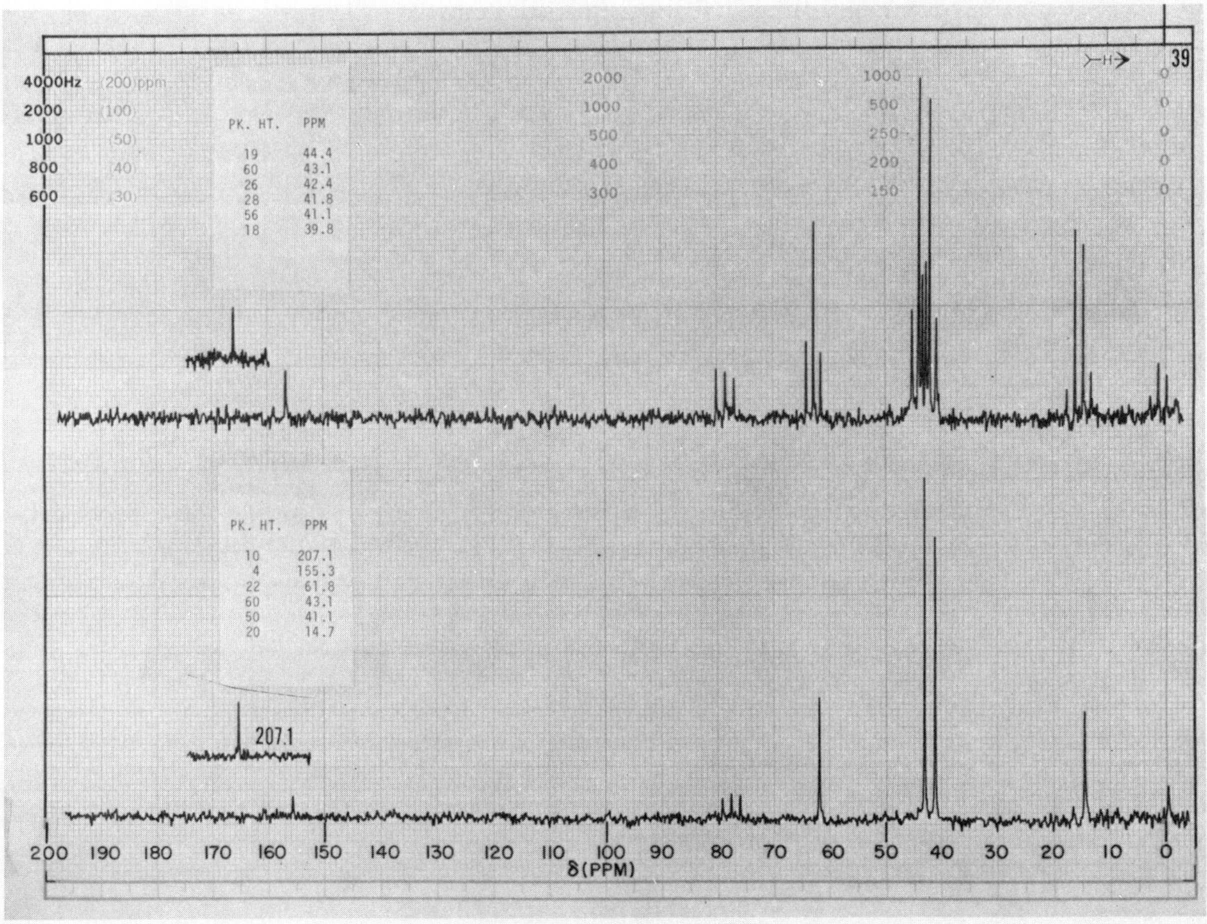

COMPOUND 40

Compound 40

MS: 111.081
IR: neat
HMR: CDCl$_3$
CMR: neat
Analysis: 76.2% C; 11.2% H

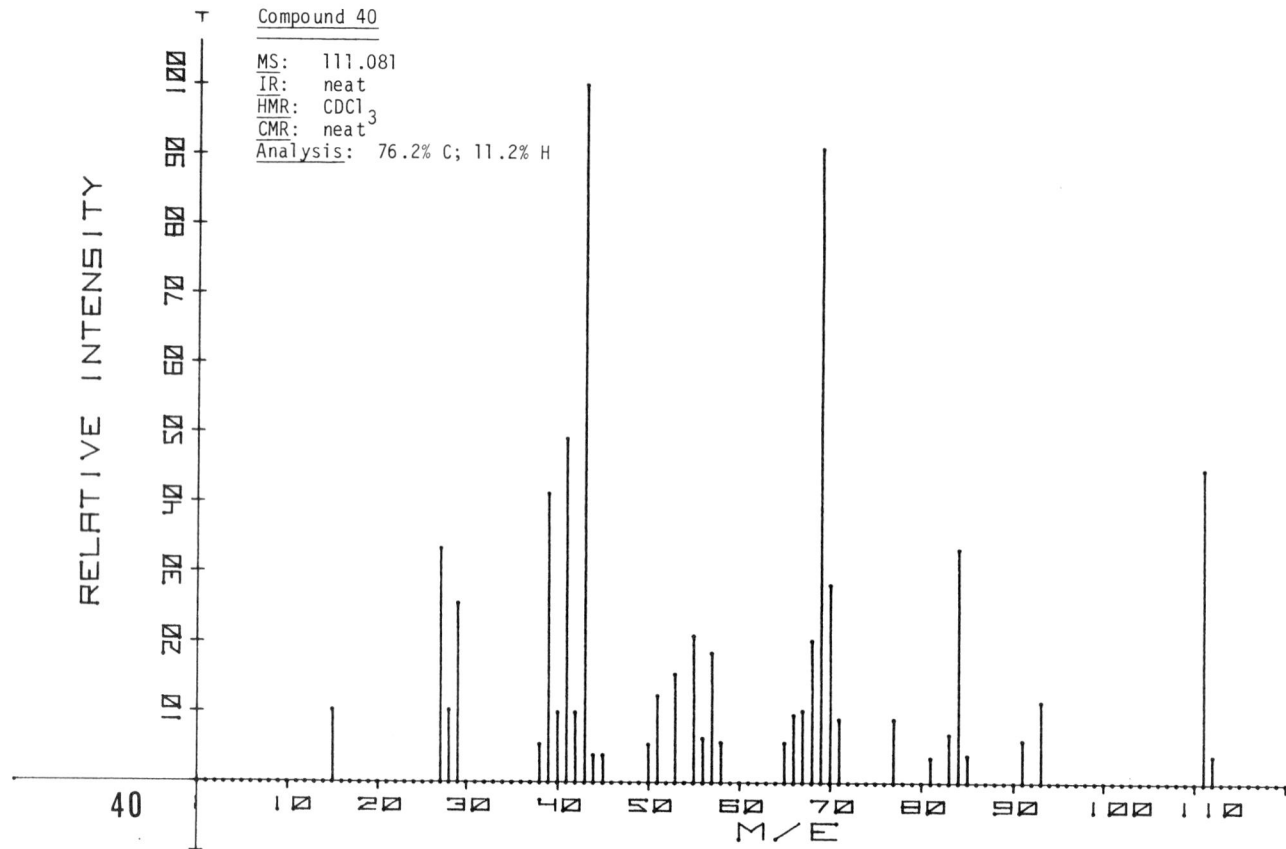

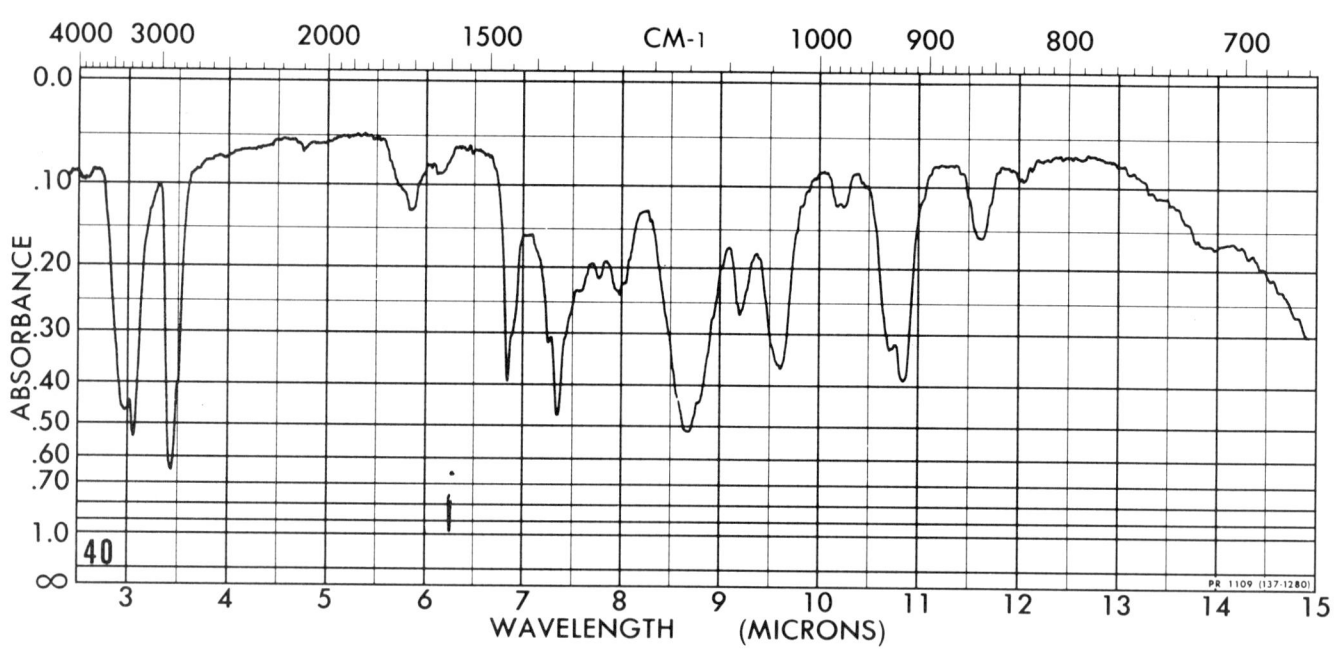

40
D_2O

40
4000Hz (200) ppm 3000 2000 1000 0
2000 (100) 1500 1000 500 0
1000 (50) 750 500 250 0
800 (40) 600 400 200 0
600 (30) 450 300 150 0
200 190 180 170 160 150 140 130 120 110 100 90 80 70 60 50 40 30 20 10 0
δ(PPM)

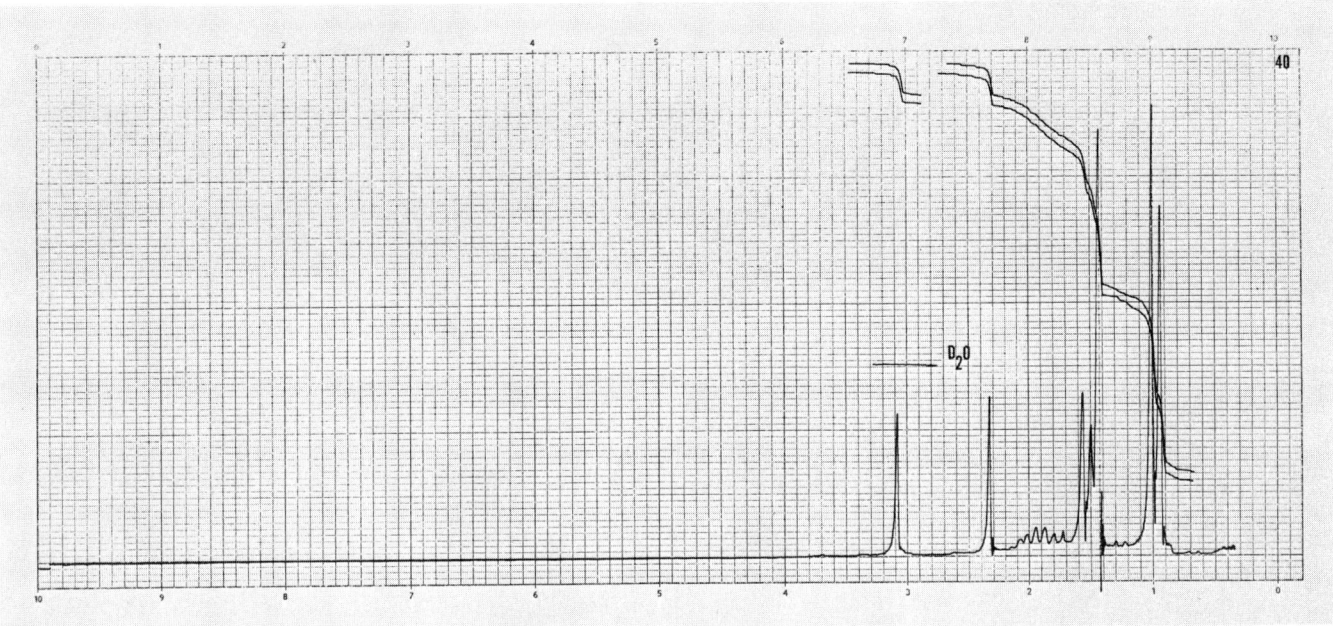

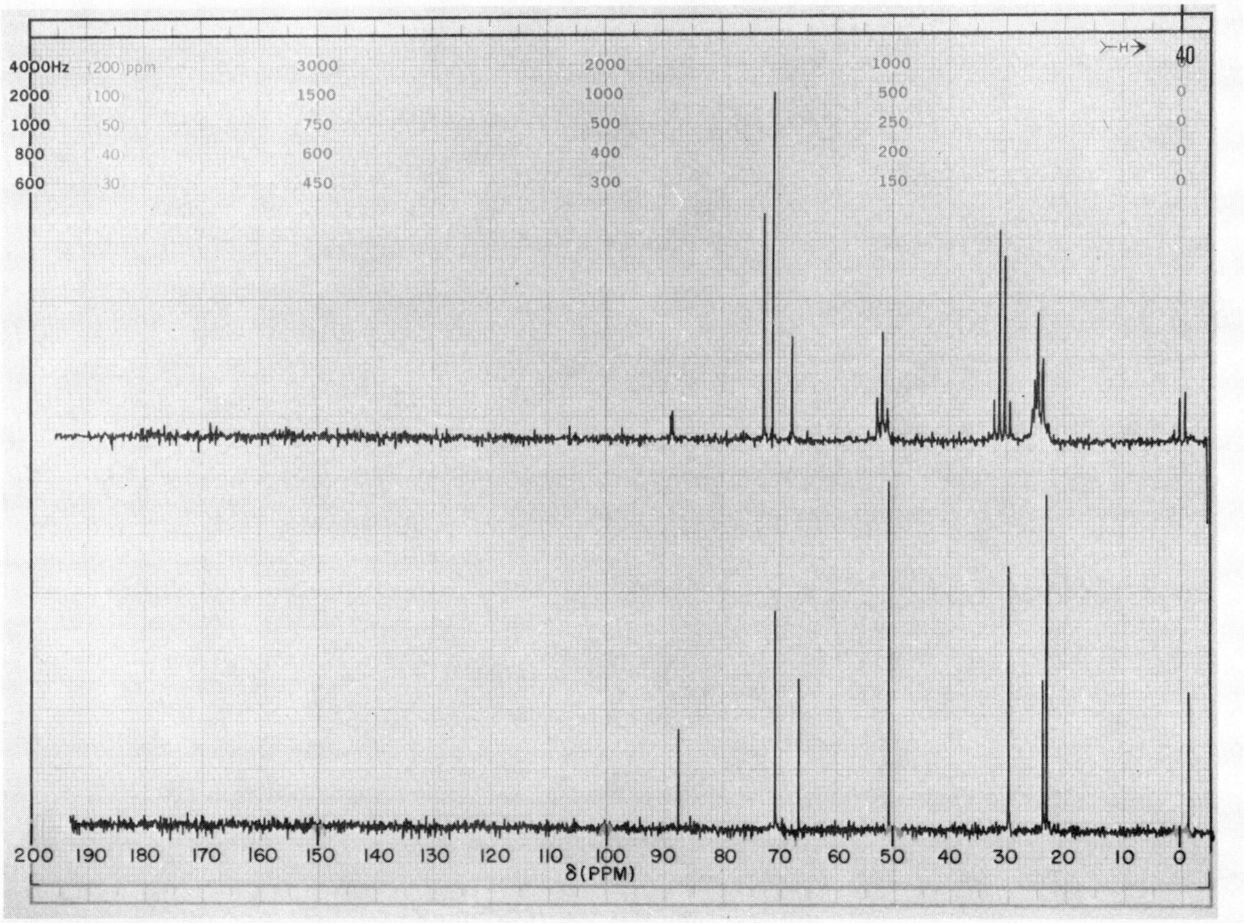

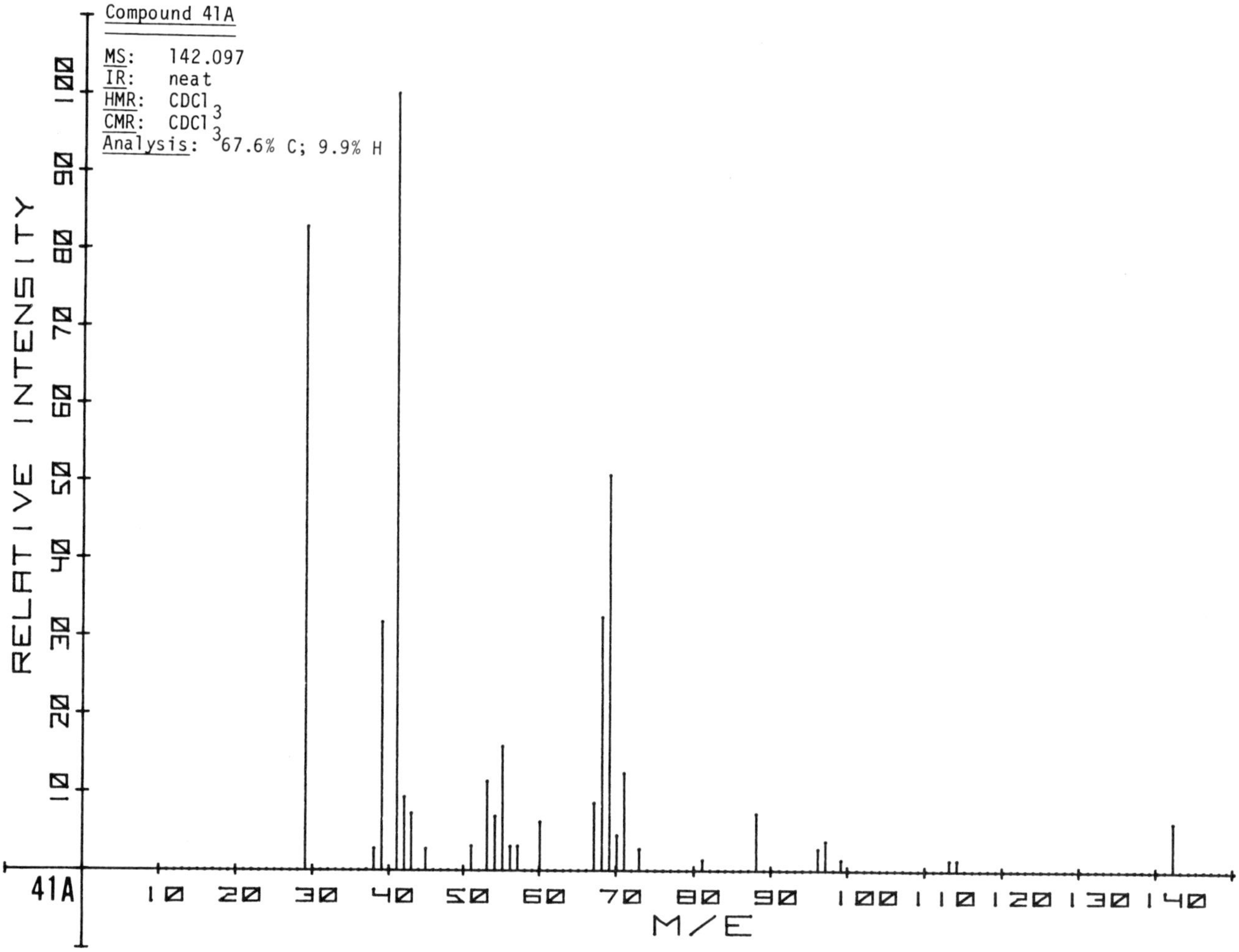

Compound 41A
MS: 142.097
IR: neat
HMR: CDCl3
CMR: CDCl3
Analysis: 67.6% C; 9.9% H
RELATIVE INTENSITY
41A
M/E

CM-1
ABSORBANCE
41A
WAVELENGTH (MICRONS)

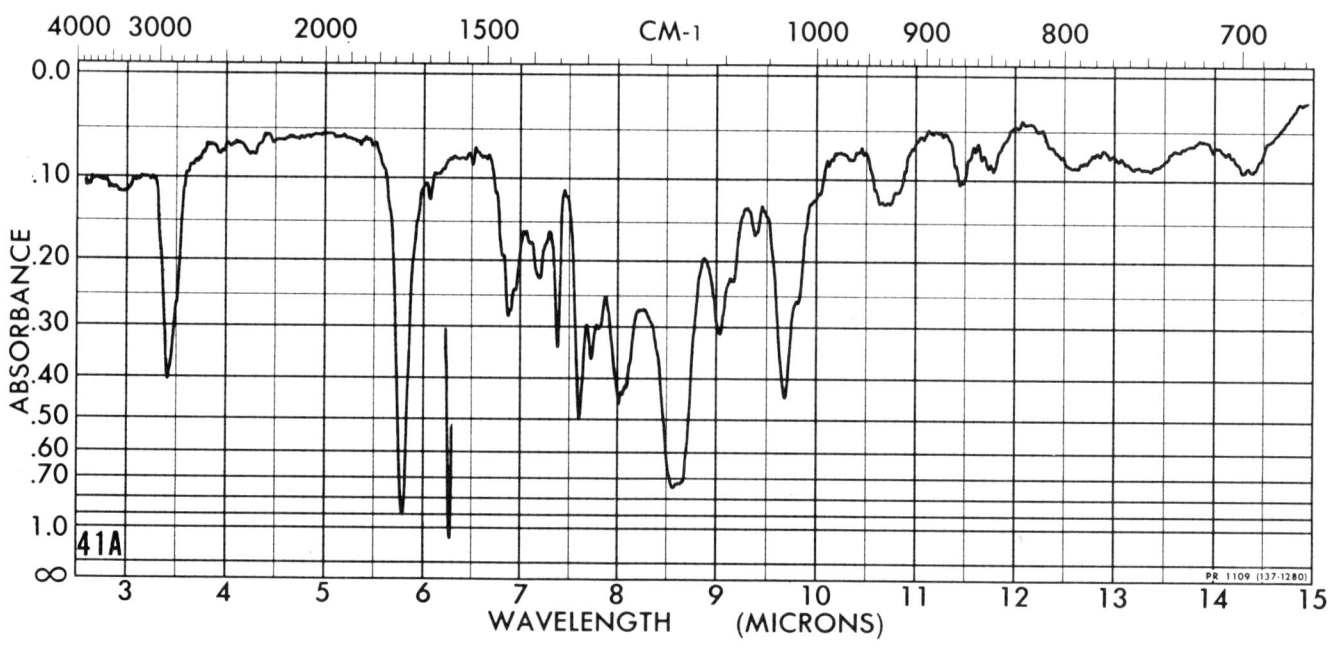

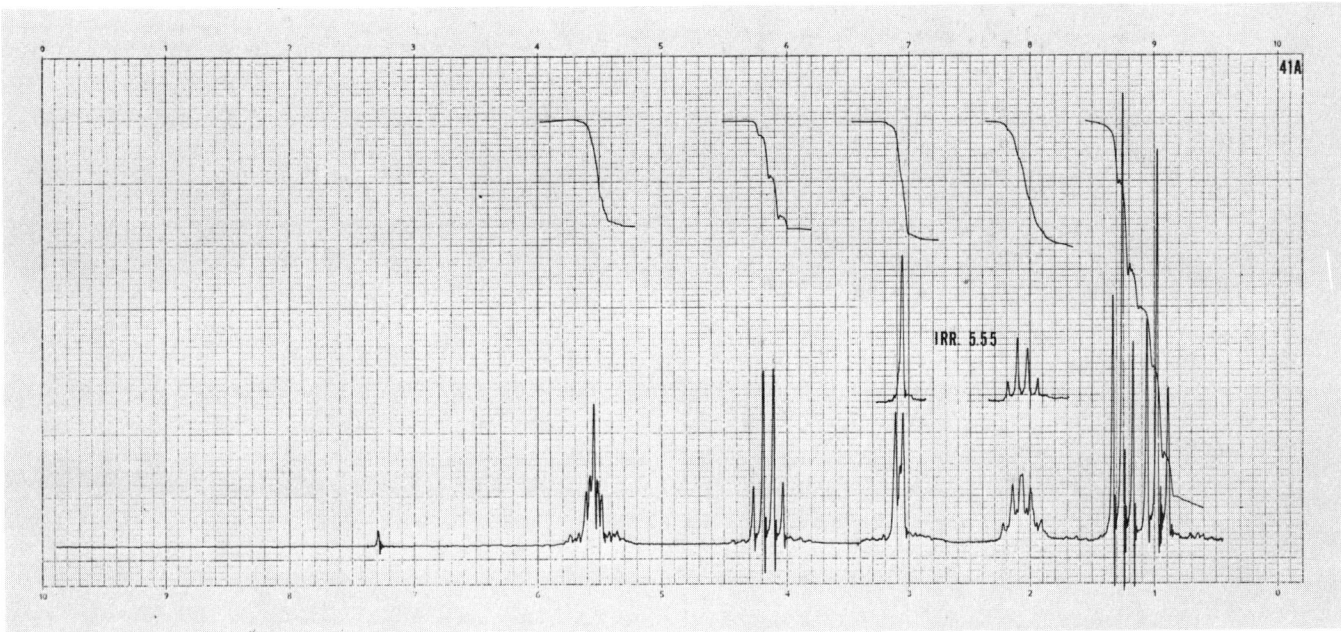

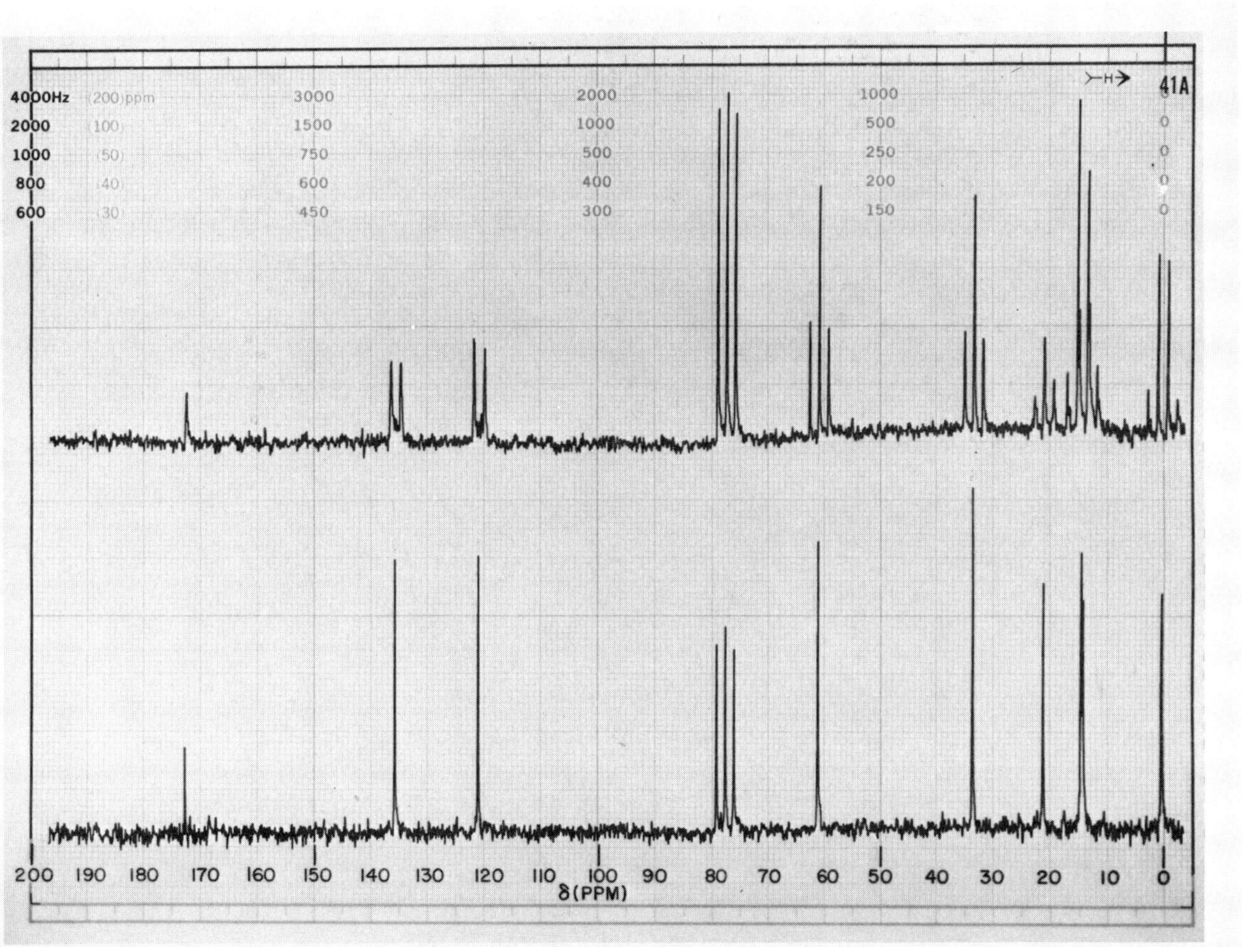

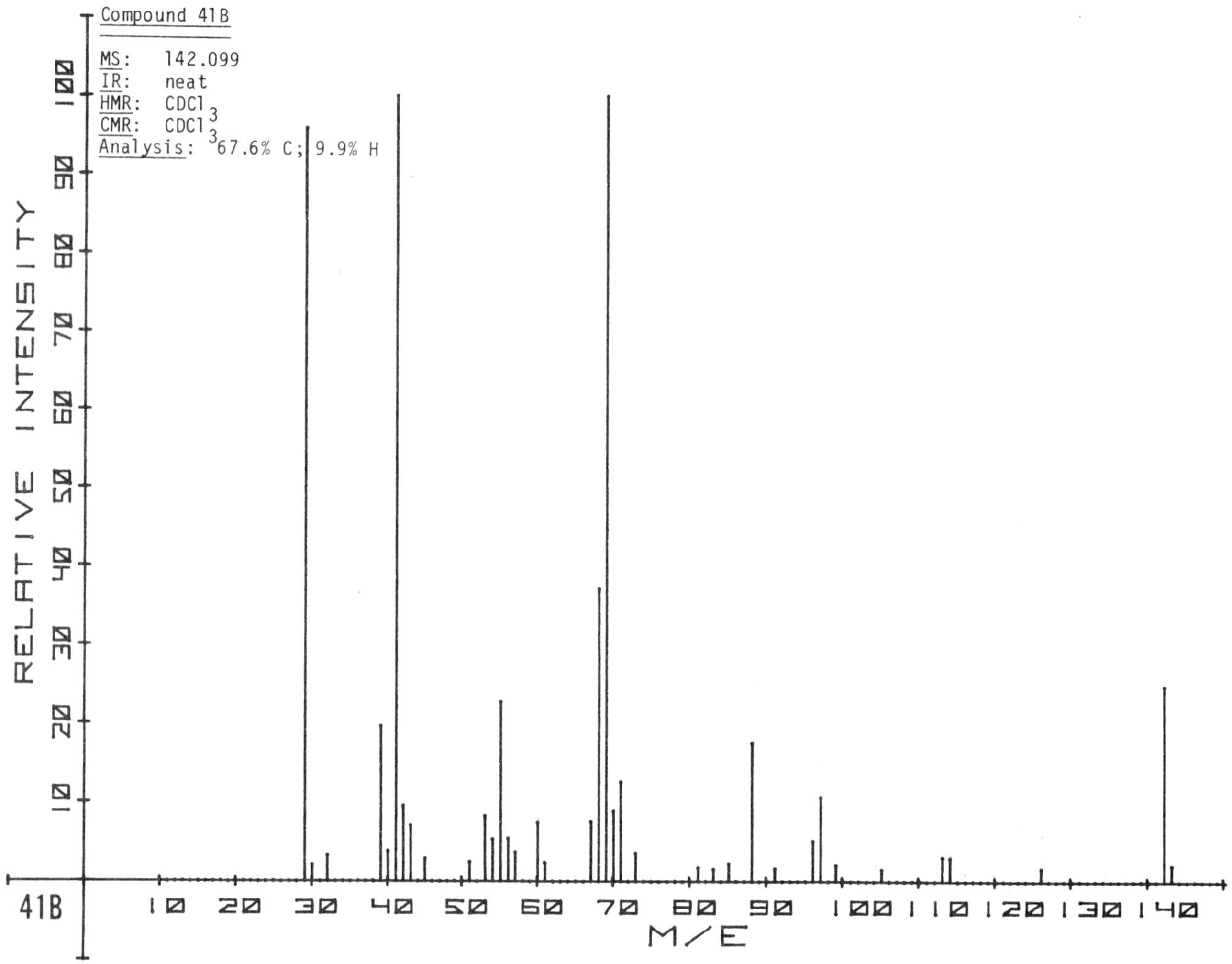

Compound 41B
MS: 142.099
IR: neat
HMR: $CDCl_3$
CMR: $CDCl_3$
Analysis: 67.6% C; 9.9% H
RELATIVE INTENSITY
M/E
41B

CM-1
ABSORBANCE
WAVELENGTH (MICRONS)
41B

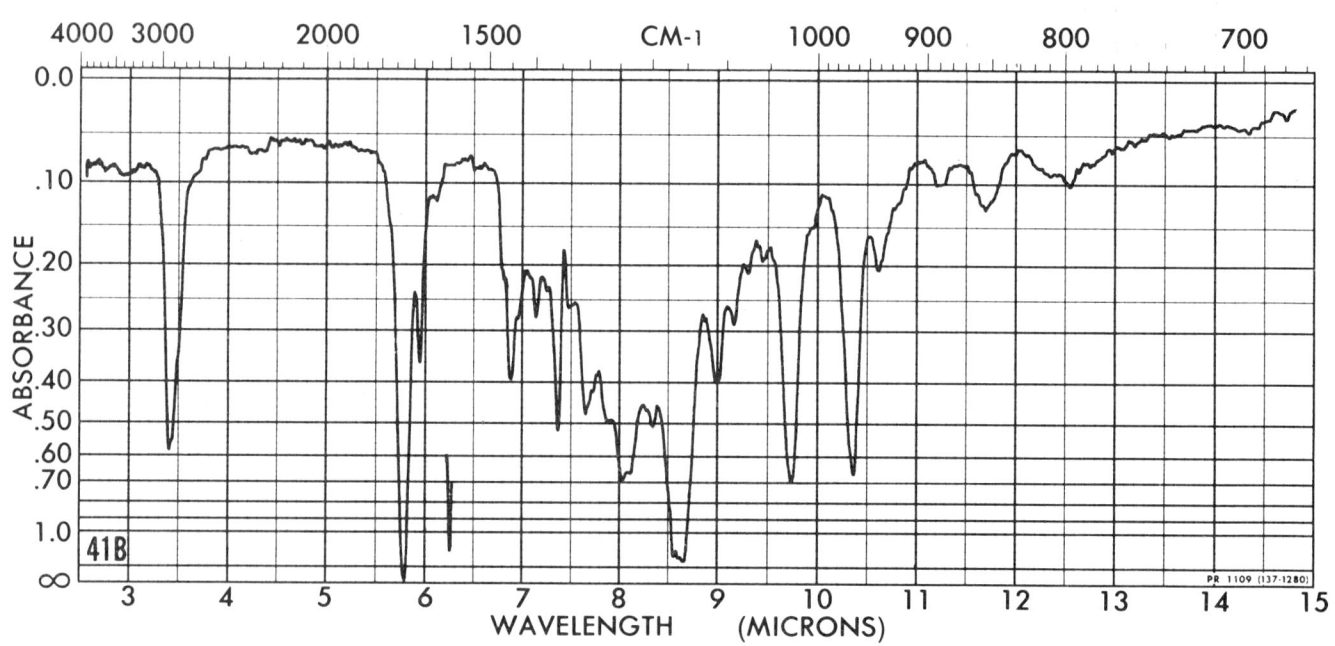

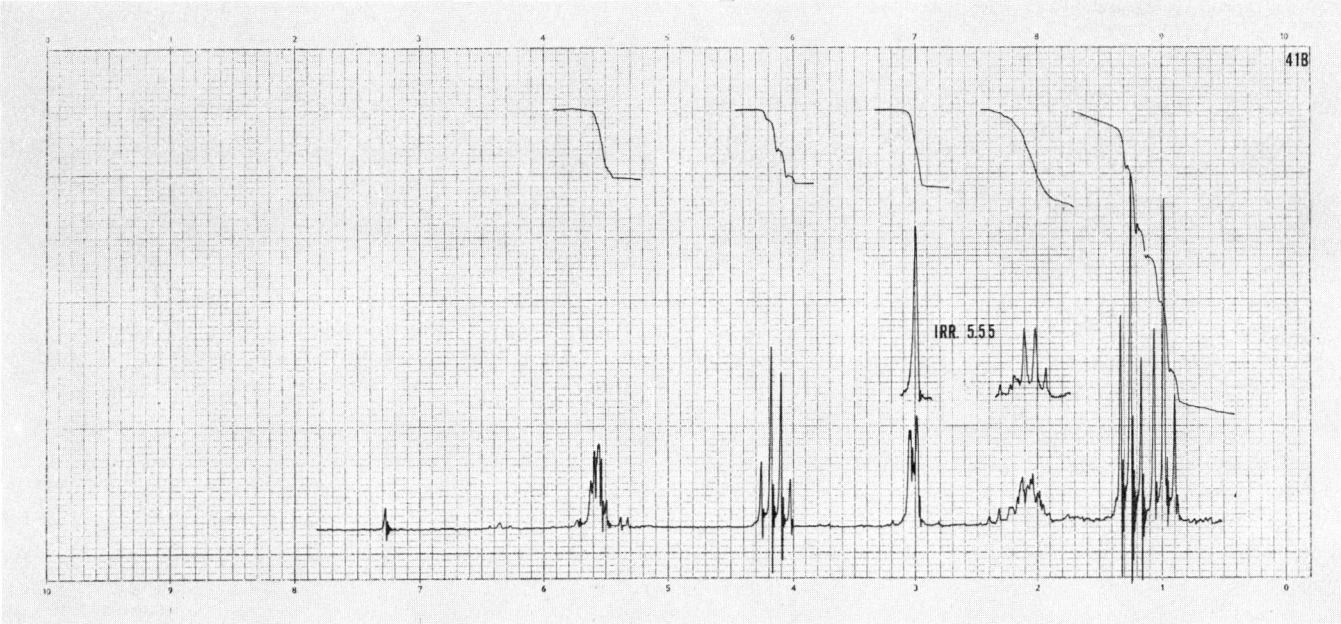

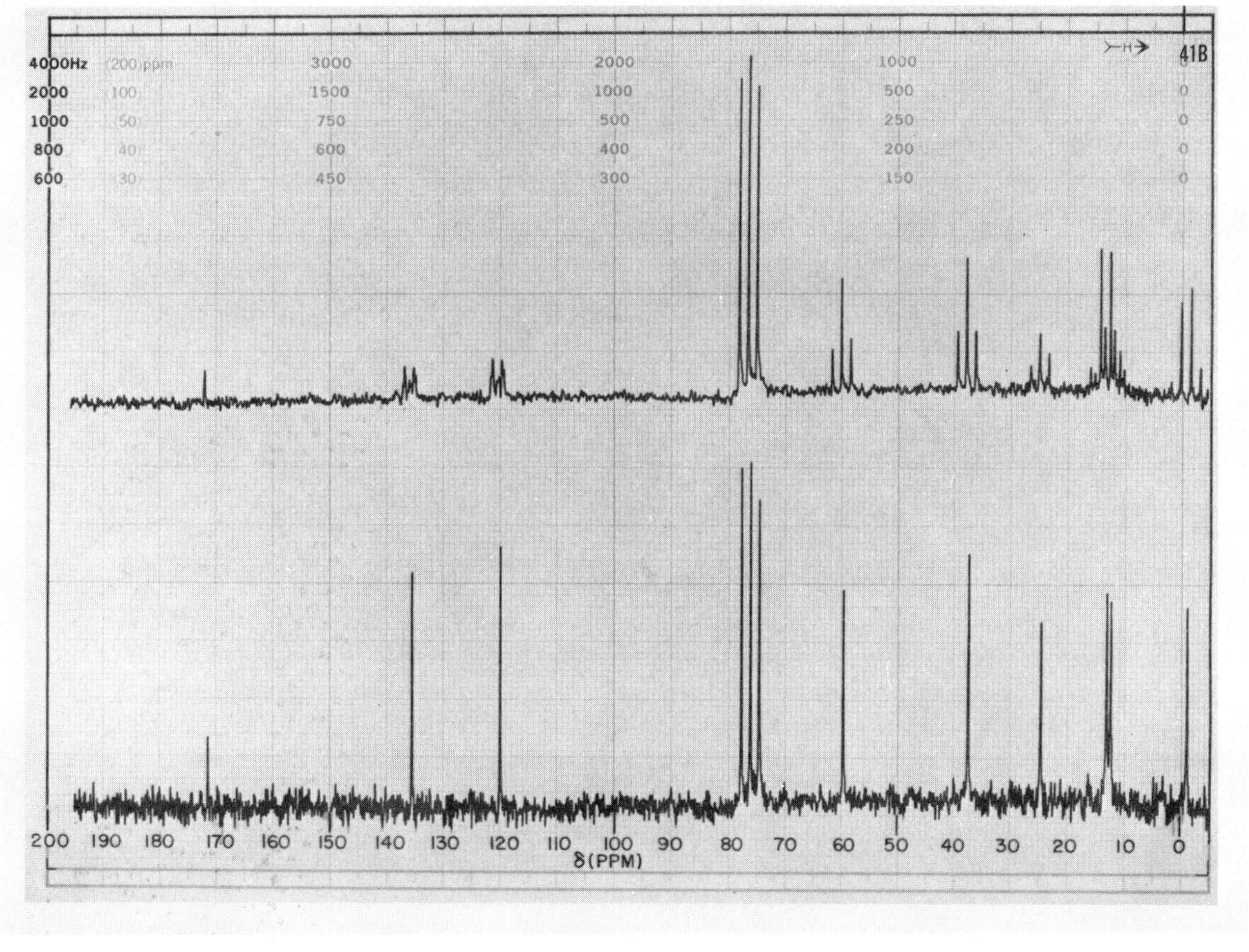

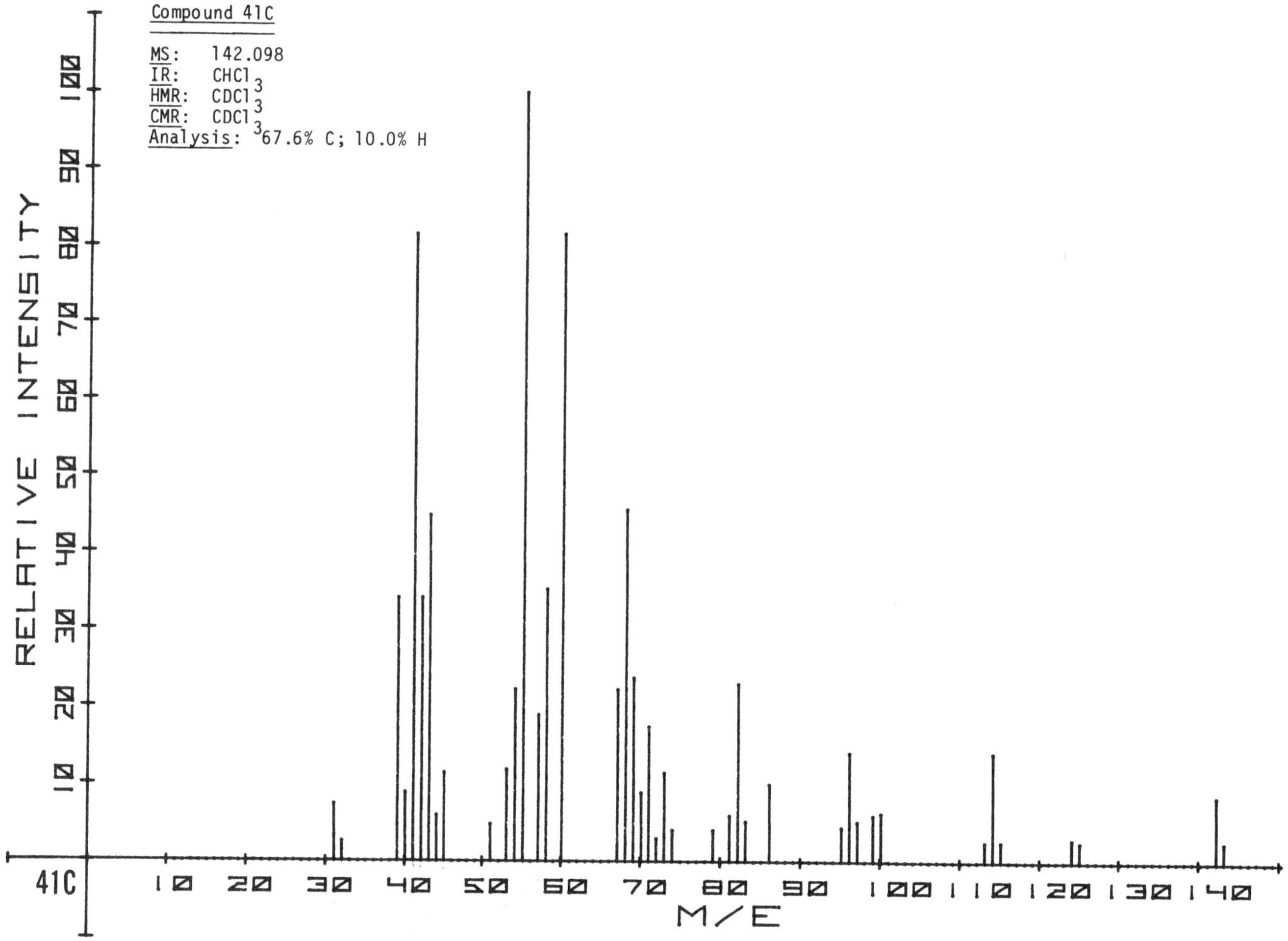

Compound 41C
MS: 142.098
IR: CHCl3
HMR: CDCl3
CMR: CDCl3
Analysis: 67.6% C; 10.0% H
RELATIVE INTENSITY
41C
M/E

CM-1
ABSORBANCE
41C
WAVELENGTH (MICRONS)

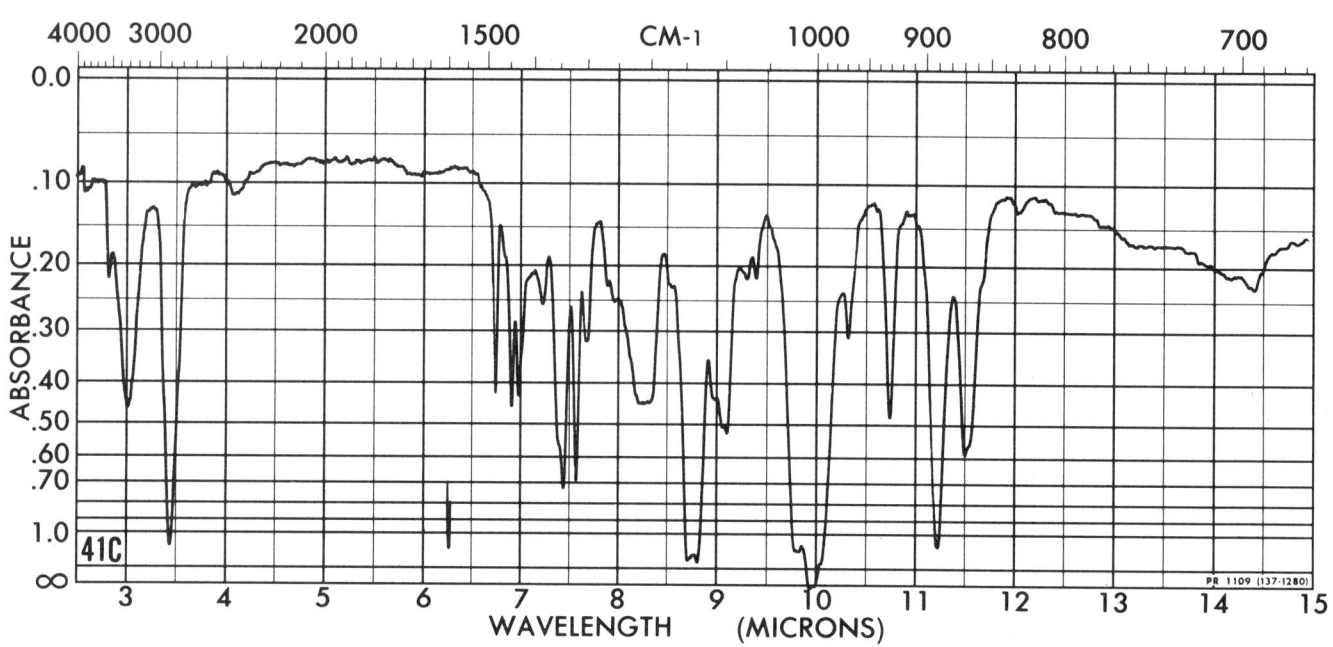

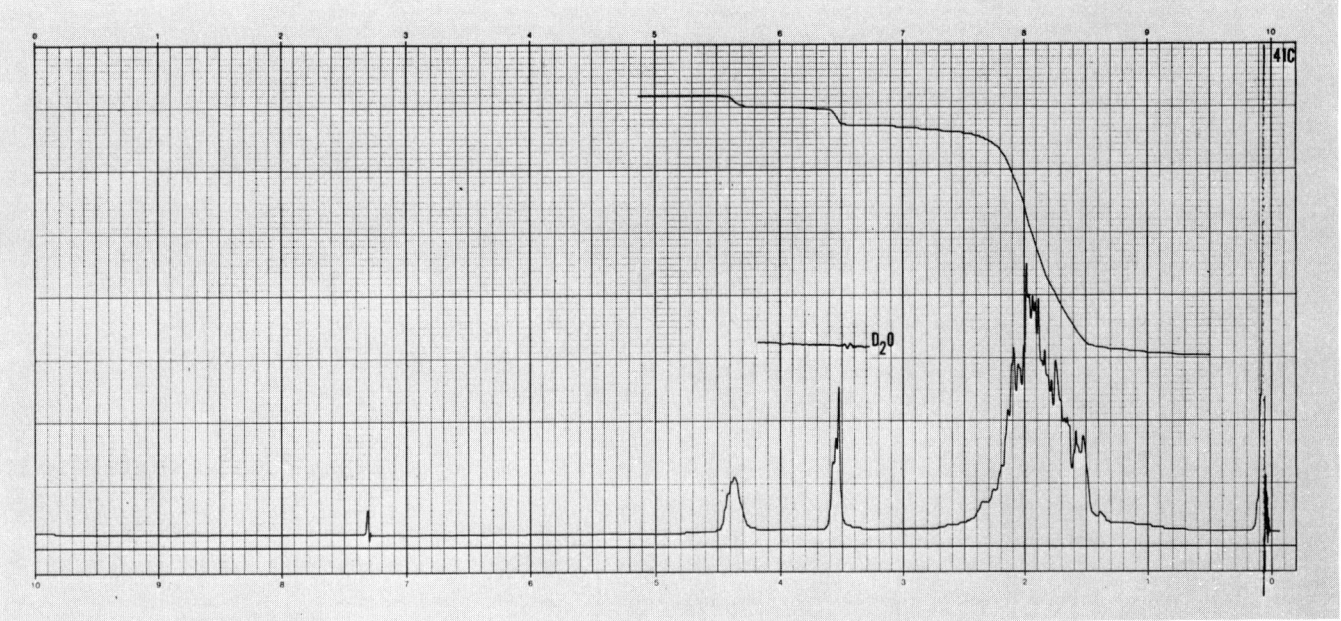

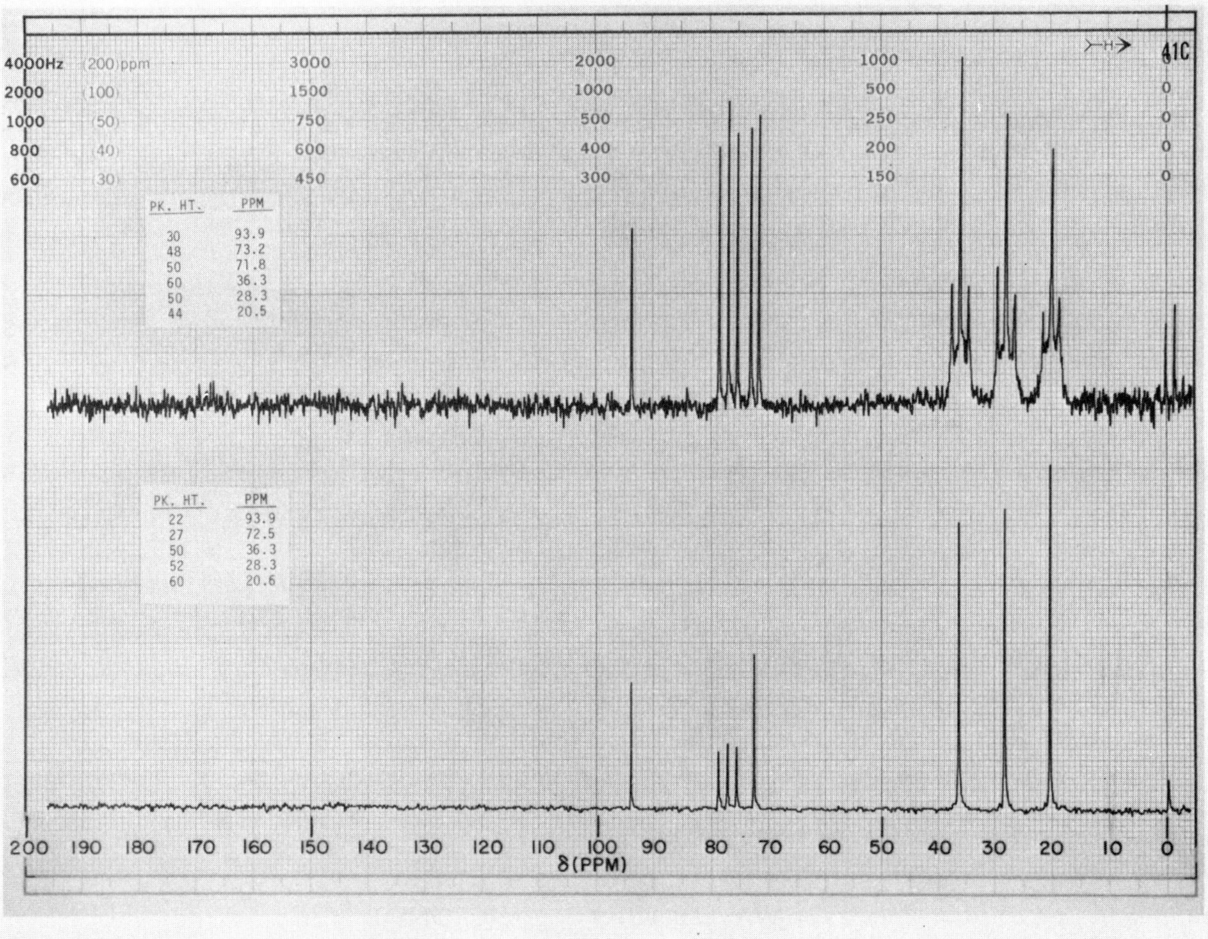

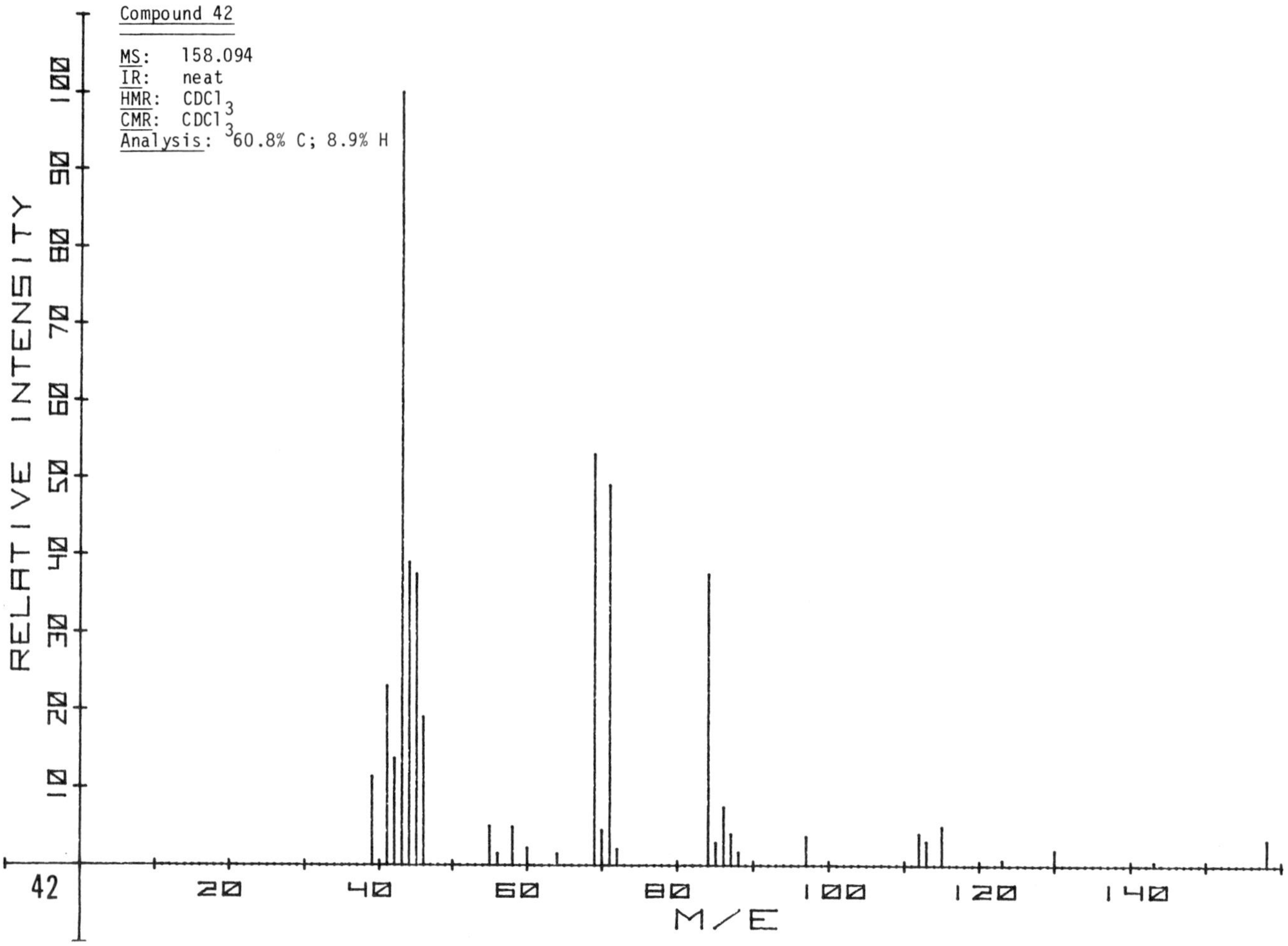

Compound 42
MS: 158.094
IR: neat
HMR: CDCl3
CMR: CDCl3
Analysis: 60.8% C; 8.9% H
RELATIVE INTENSITY
10 20 30 40 50 60 70 80 90 100
42
20 40 60 80 100 120 140
M/E

4000 3000 2000 1500 CM-1 1000 900 800 700
0.0 .10 .20 .30 .40 .50 .60 .70 1.0 ∞
ABSORBANCE
42
PR 1109 (137-1280)
3 4 5 6 7 8 9 10 11 12 13 14 15
WAVELENGTH (MICRONS)

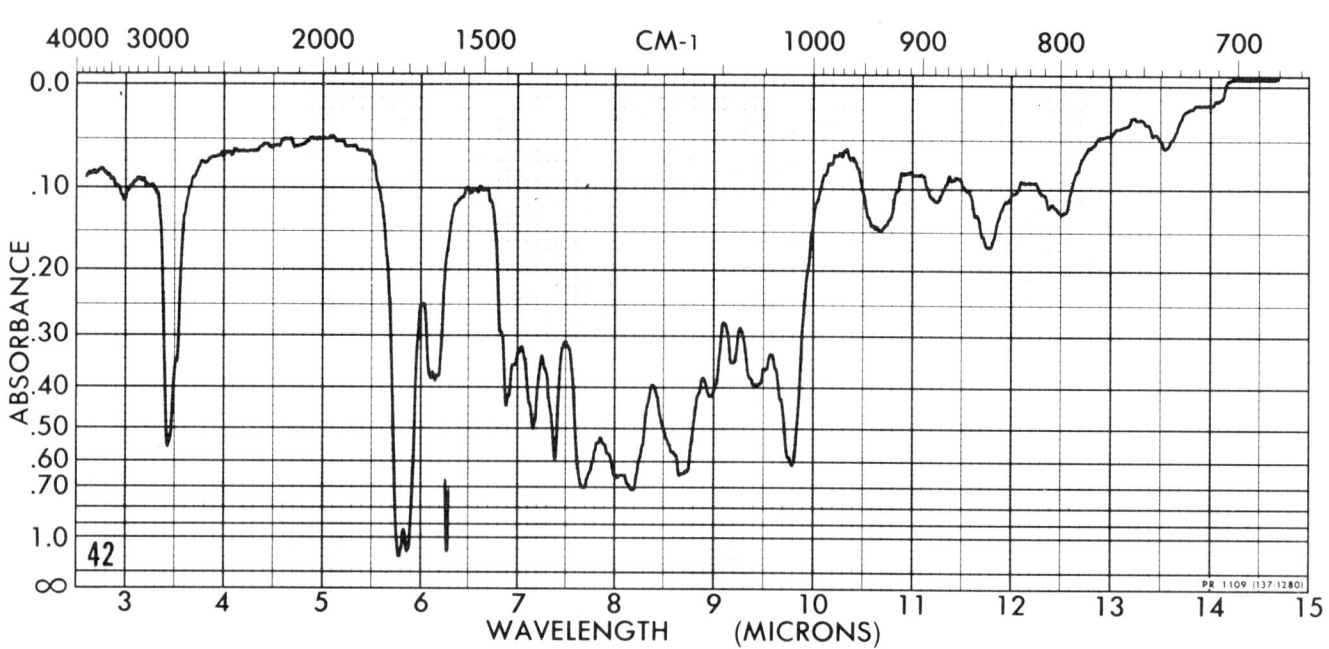

42

42
4000Hz (200)ppm 3000 1000 0
2000 (100) 1500 500 0
1000 (50) 750 250 0
800 (40) 600 200 0
600 (30) 450 150 0
PK. HT. PPM
4 18.5
15 17.1
10 16.4
8 15.7
27 14.9
12 14.3
37 13.5
13 12.9
8 12.1
4 11.5
A
A
PK. HT. PPM
27 202.7
27 167.4
42 61.2
60 49.4
52 44.9
51 17.1
43 14.2
44 13.6
202.7
200 190 180 170 160 150 140 130 120 110 100 90 80 70 60 50 40 30 20 10 0
δ(PPM)

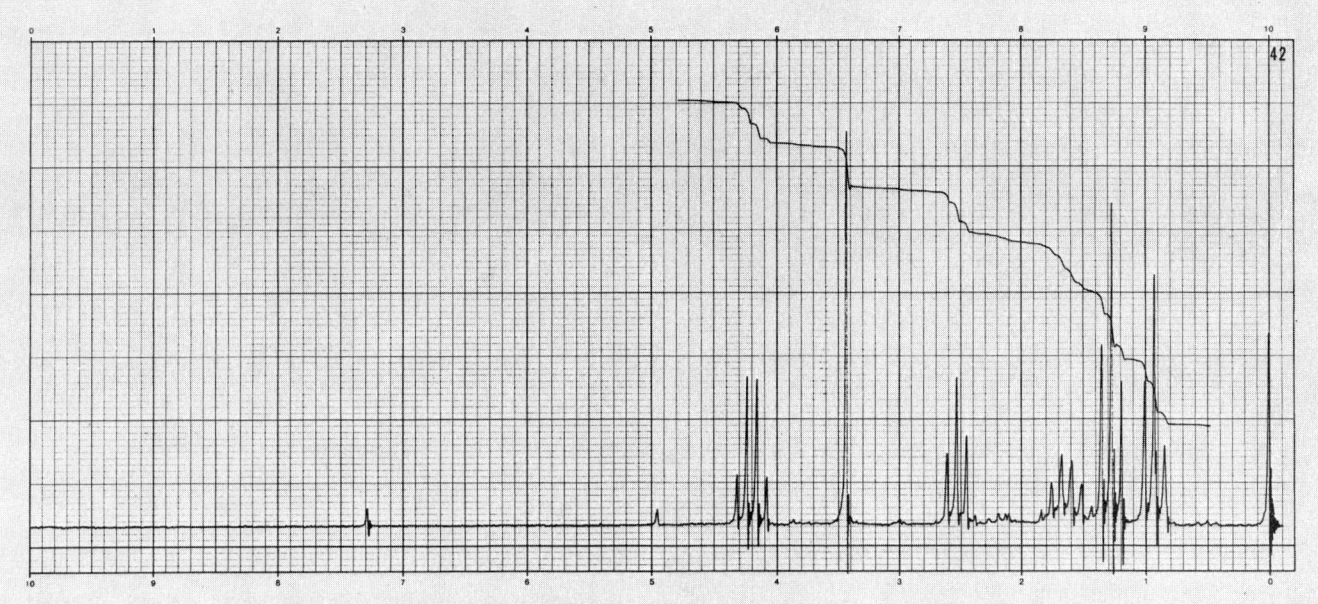

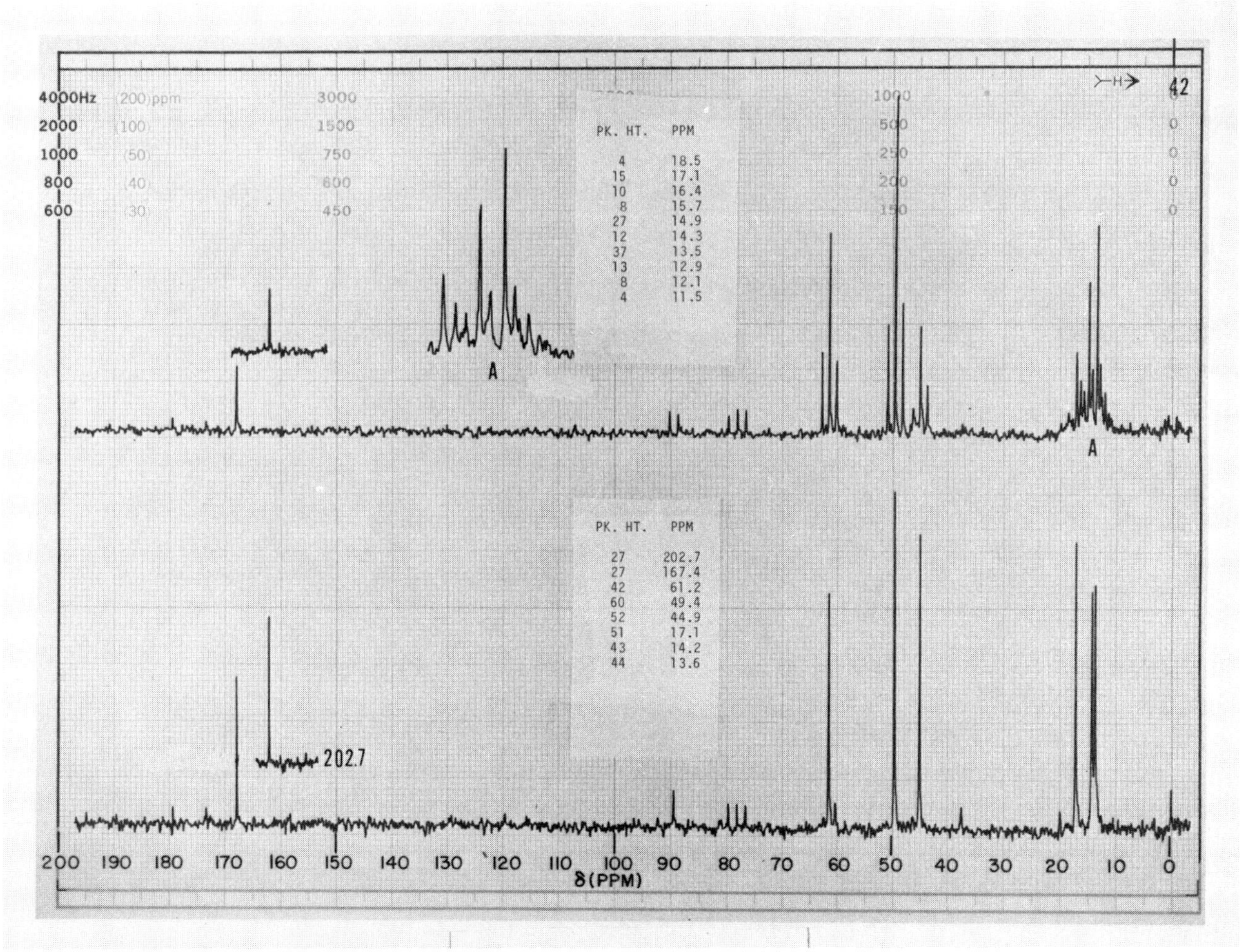

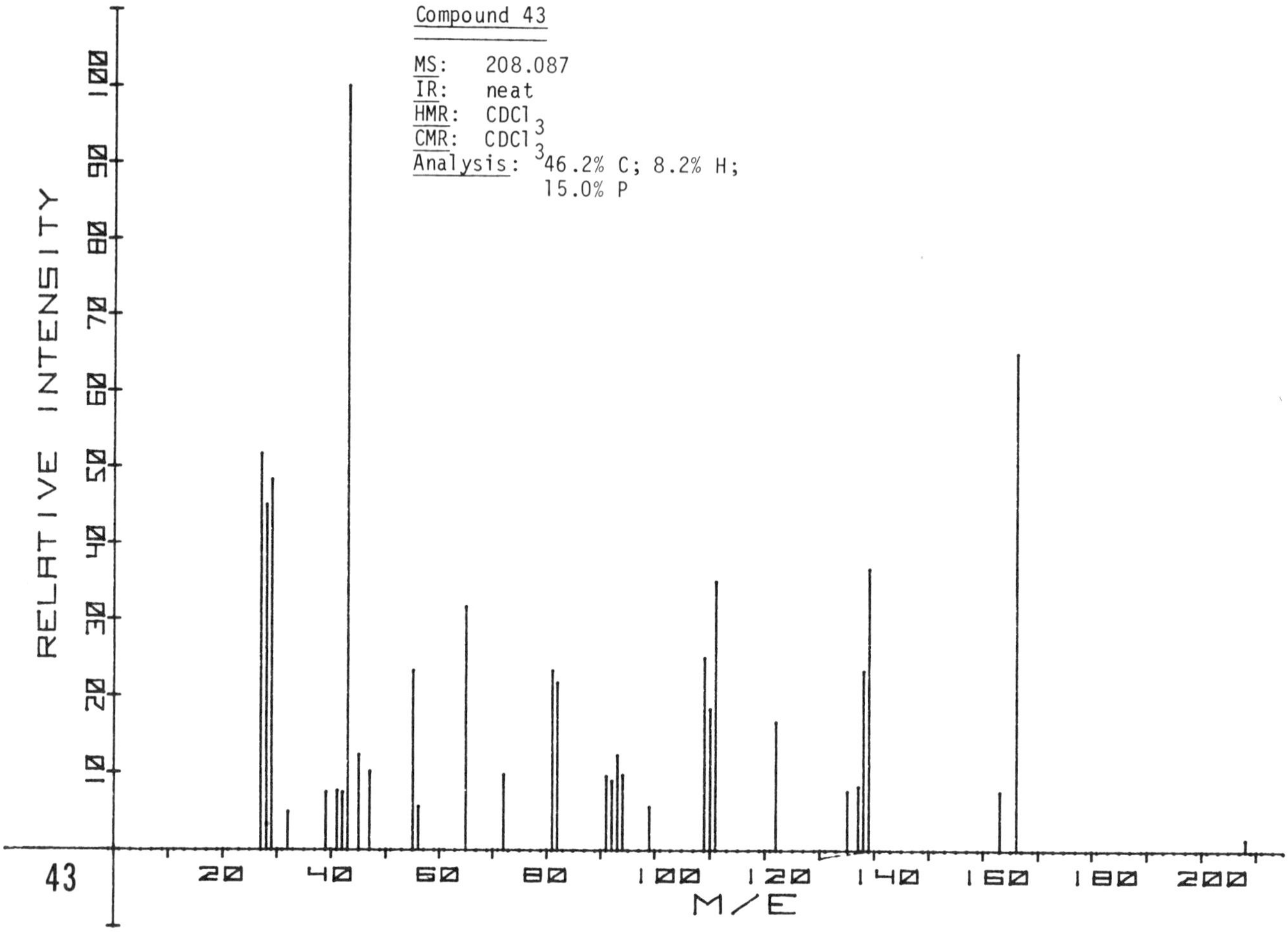

Compound 43
MS: 208.087
IR: neat
HMR: CDCl3
CMR: CDCl3
Analysis: 46.2% C; 8.2% H; 15.0% P
RELATIVE INTENSITY
10 20 30 40 50 60 70 80 90 100
20 40 60 80 100 120 140 160 180 200
M/E
43

4000 3000 2000 1500 CM-1 1000 900 800 700
0.0 .10 .20 .30 .40 .50 .60 .70 1.0 ∞
ABSORBANCE
43
3 4 5 6 7 8 9 10 11 12 13 14 15
WAVELENGTH (MICRONS)

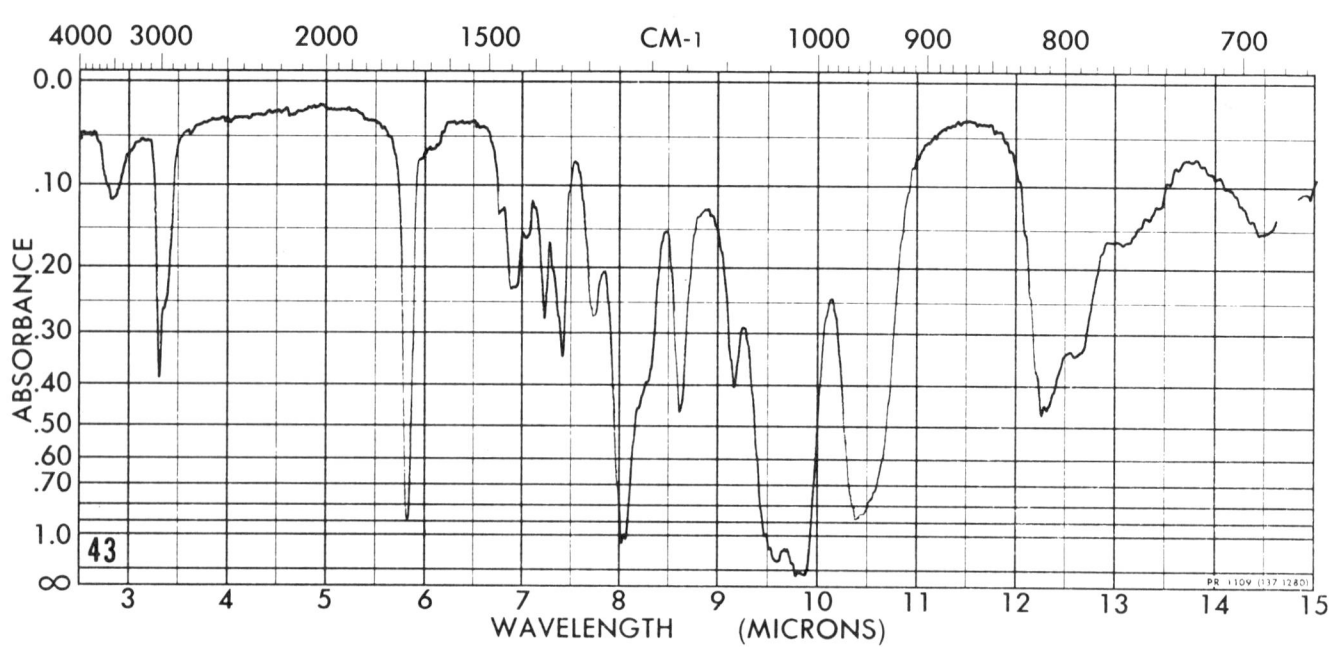

43
10 X

43
A
B
C
PK. HT. PPM
15 203.6
18 203.4
35 62.8
40 62.7
34 62.5
25 62.4
28 50.5
19 44.2
60 30.3
45 16.5
40 16.2
34 11.0
37 10.7
203.6
203.4
δ(PPM)

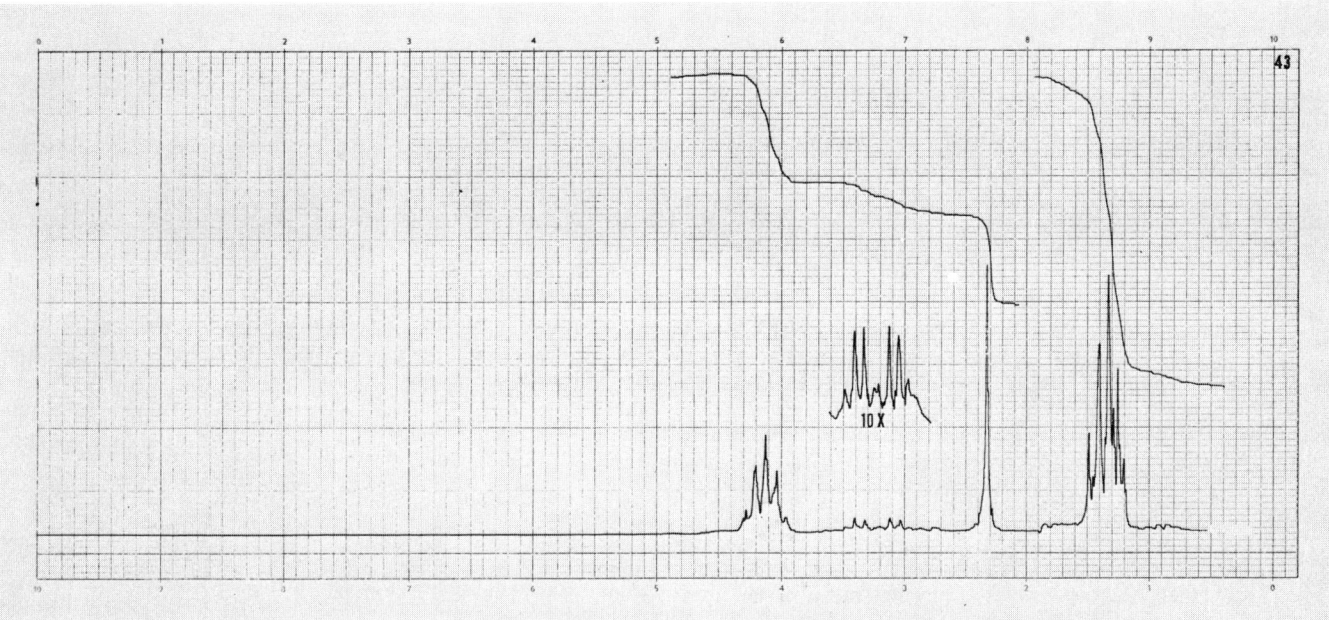

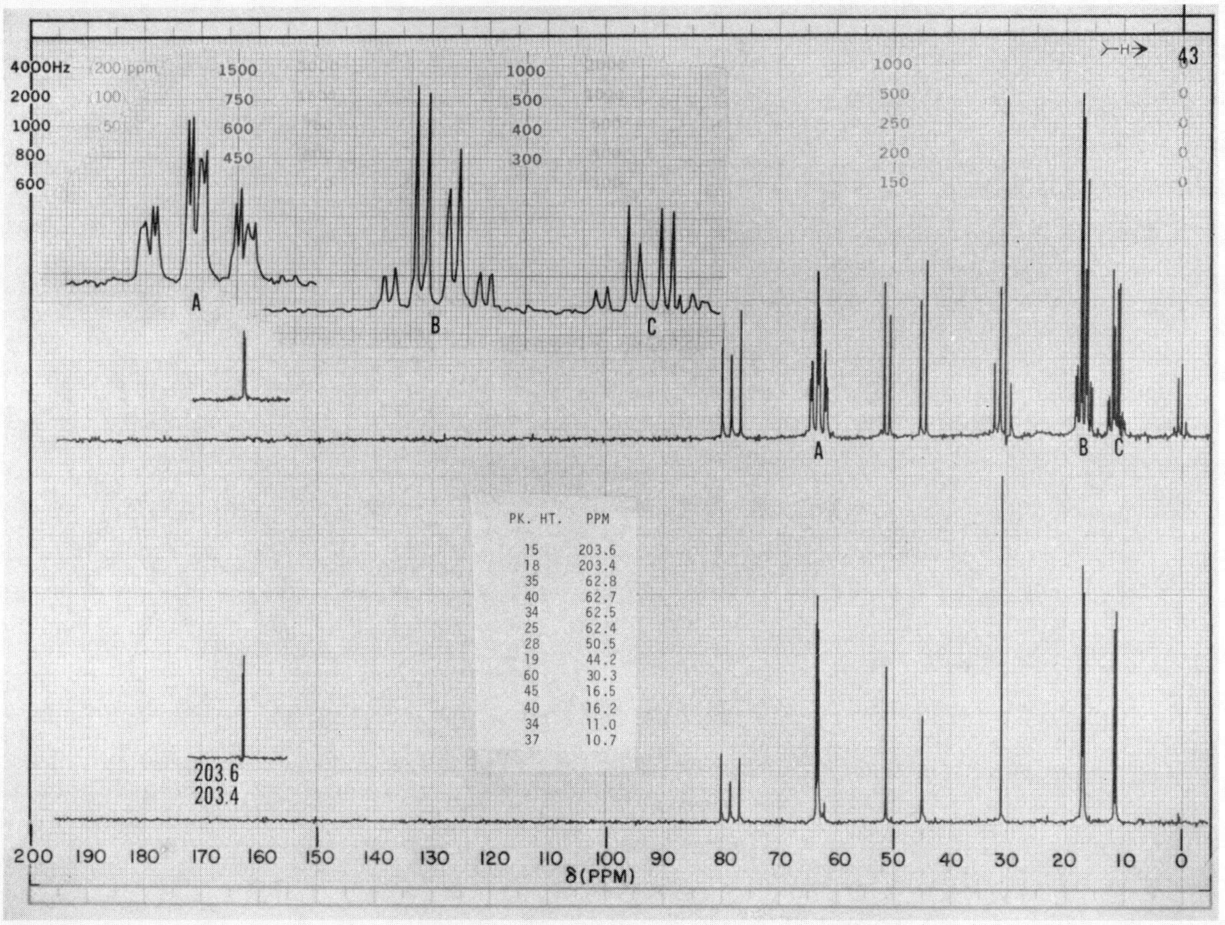

COMPOUND 44

Compound 44
MS: 224.081
IR: neat
HMR: CCl4
CMR: neat
Analysis: 42.9% C; 7.6% H; 13.9% P

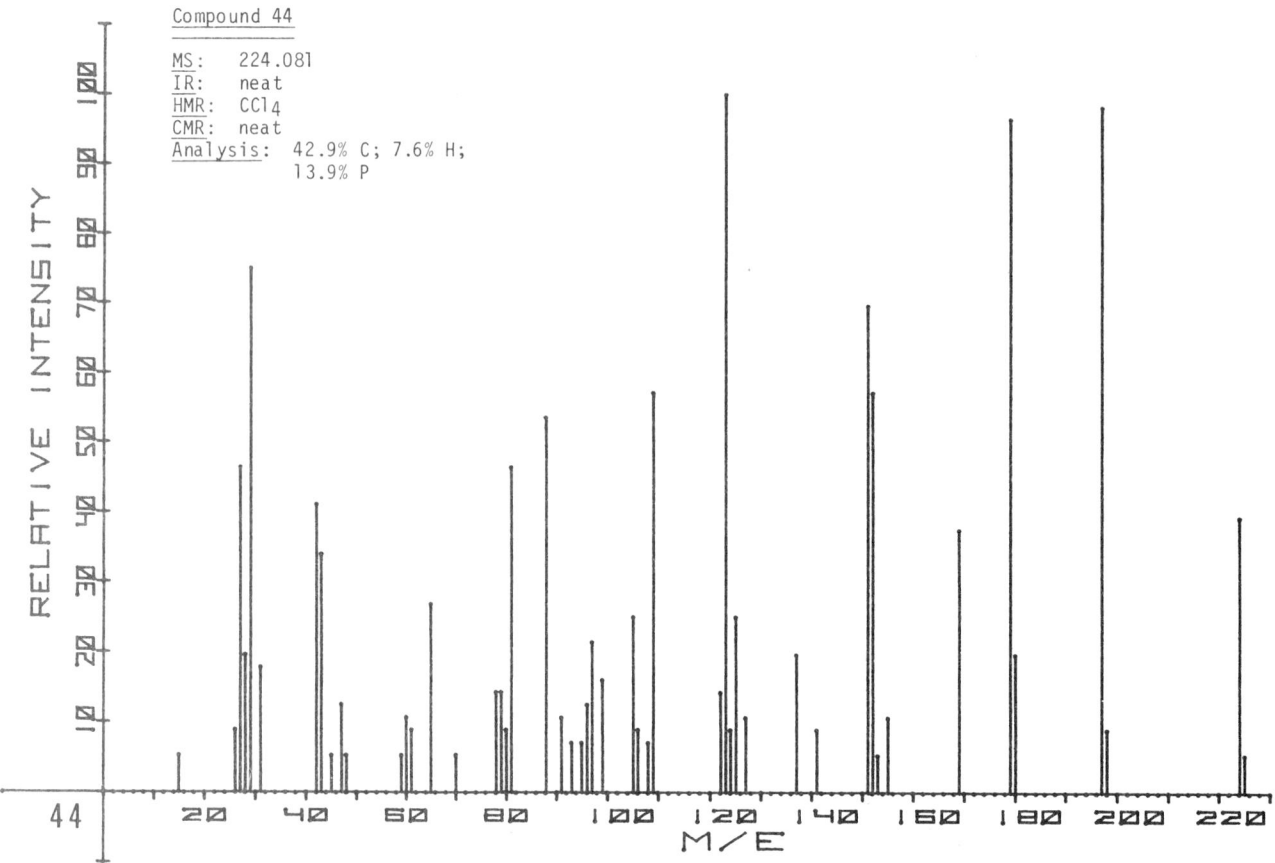

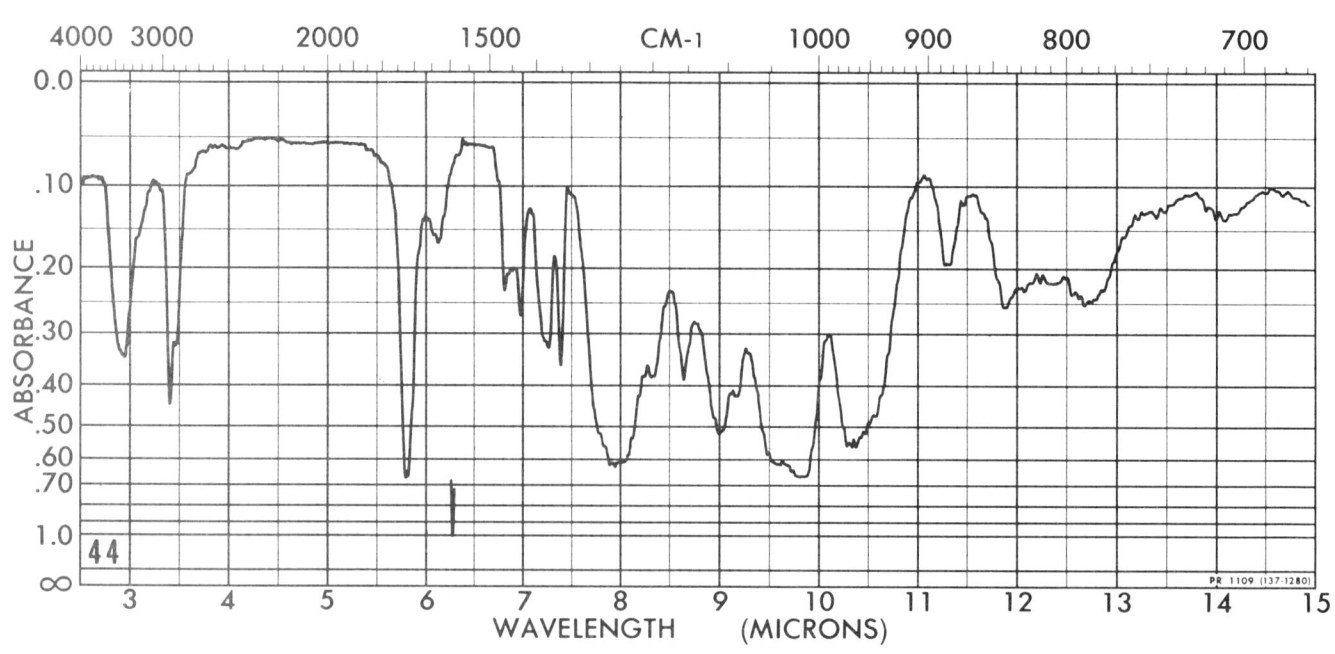

44

44
4000Hz (200) ppm
2000 (100)
1000 (50)
800 (40)
600 (30)
1000
500
250
200
150
PK. HT. PPM
18 63.9
18 63.6
41 62.7
34 62.4
47 61.4
22 61.1
22 60.1
9 18.0
8 17.7
44 17.1
43 16.8
29 16.3
25 16.0
10 15.6
7 15.4
6 15.1
37 14.7
45 13.9
5 13.0
17.0
14.3
PK. HT. PPM
13 166.4
14 166.1
60 62.7
53 62.4
52 61.4
58 37.8
40 31.2
44 16.7
37 16.4
38 14.3
200 190 180 170 160 150 140 130 120 110 100 90 80 70 60 50 40 30 20 10 0
δ(PPM)

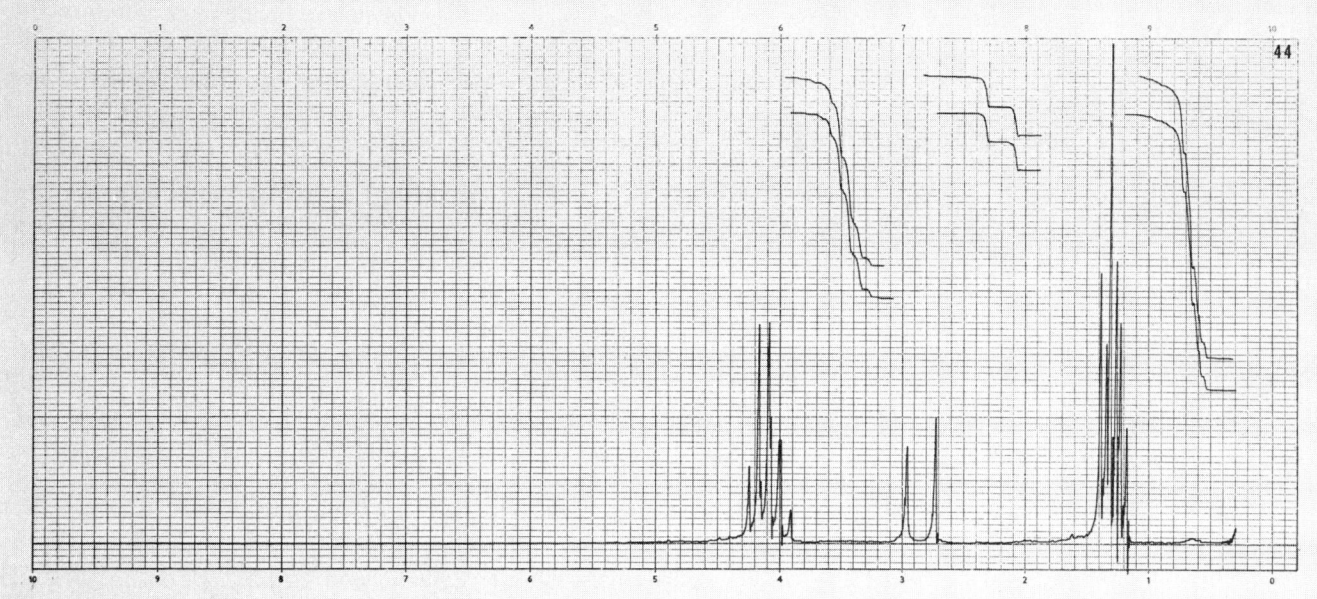

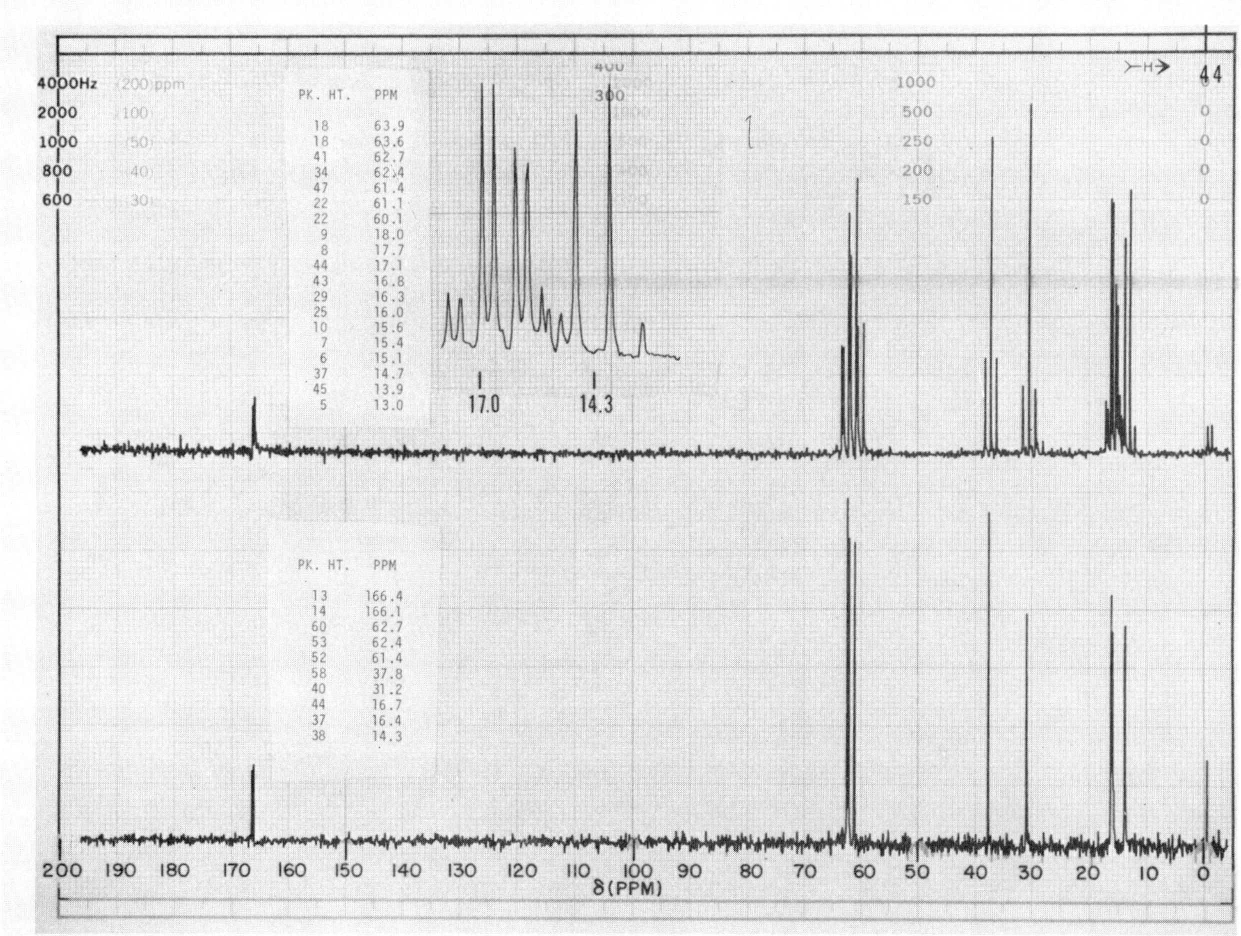

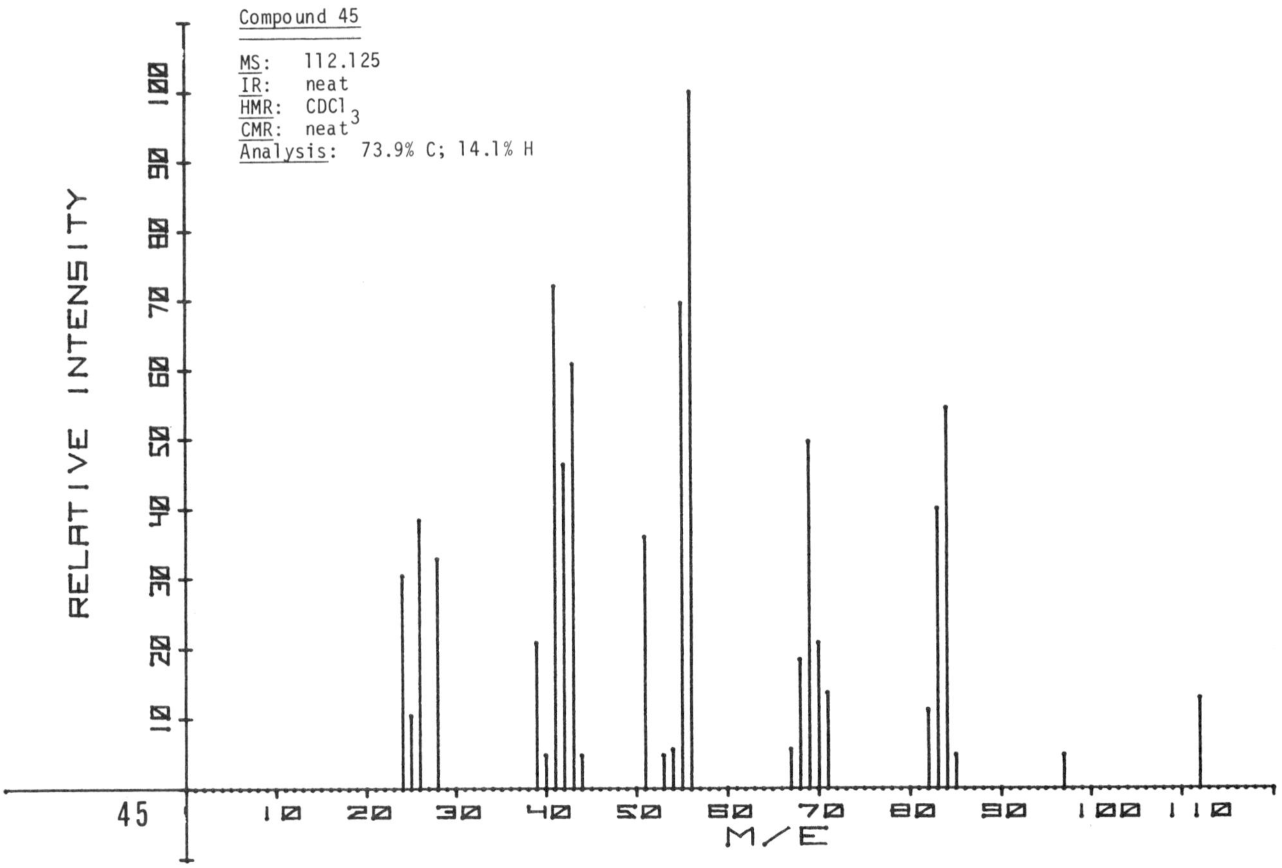

Compound 45
MS: 112.125
IR: neat
HMR: $CDCl_3$
CMR: neat
Analysis: 73.9% C; 14.1% H
RELATIVE INTENSITY
10 20 30 40 50 60 70 80 90 100
45
10 20 30 40 50 60 70 80 90 100 110
M/E

4000 3000 2000 1500 CM-1 1000 900 800 700
0.0 .10 .20 .30 .40 .50 .60 .70 1.0 ∞
ABSORBANCE
45
PR 1109 (137-1280)
3 4 5 6 7 8 9 10 11 12 13 14 15
WAVELENGTH (MICRONS)

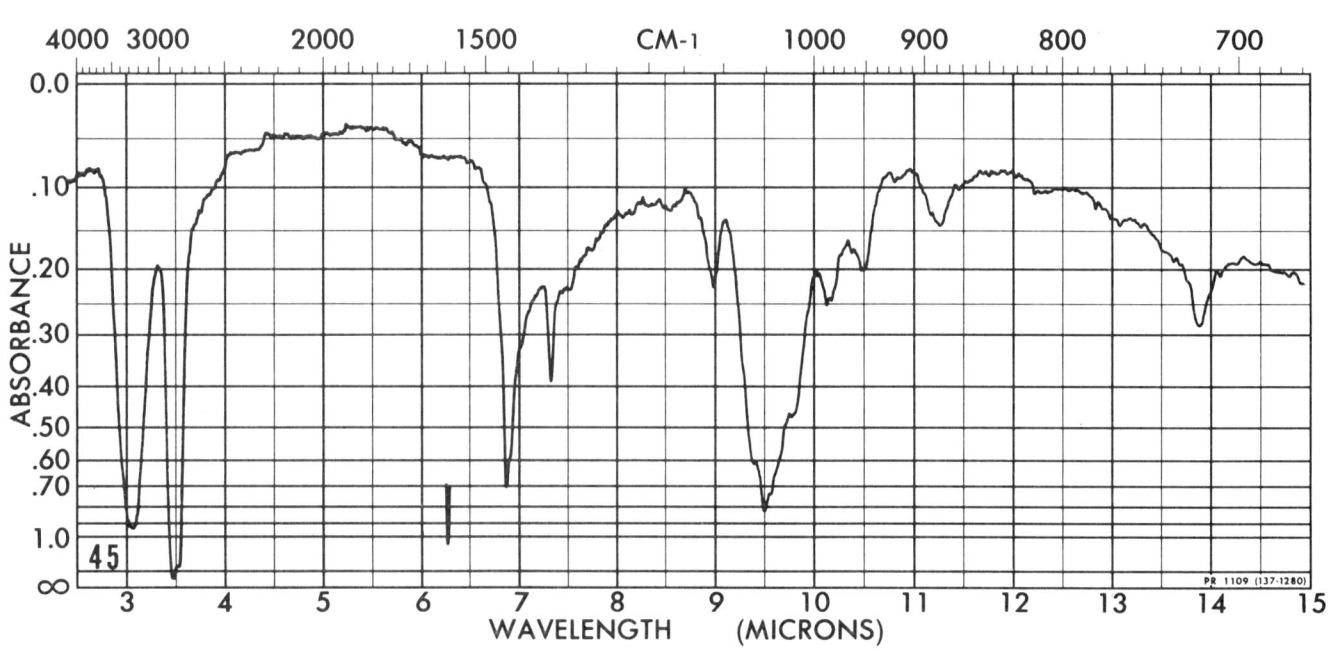

45
D2O

45
4000Hz (200)ppm
2000 (100)
1000 (50)
800 (40)
600 (30)
2000 1000
1000 500
500 250
400 200
300 150
PK. HT. PPM
15 62.7
60 62.2
18 61.7
11 33.7
43 33.1
34 32.4
11 32.0
42 30.0
42 29.9
30 26.4
29 23.1
4 15.3
14 14.6
13 13.9
4 13.3
PK. HT. PPM
54 62.2
51 33.1
43 32.4
55 30.1
60 29.9
53 26.4
45 23.1
43 14.3
200 190 180 170 160 150 140 130 120 110 100 90 80 70 60 50 40 30 20 10 0
δ(PPM)

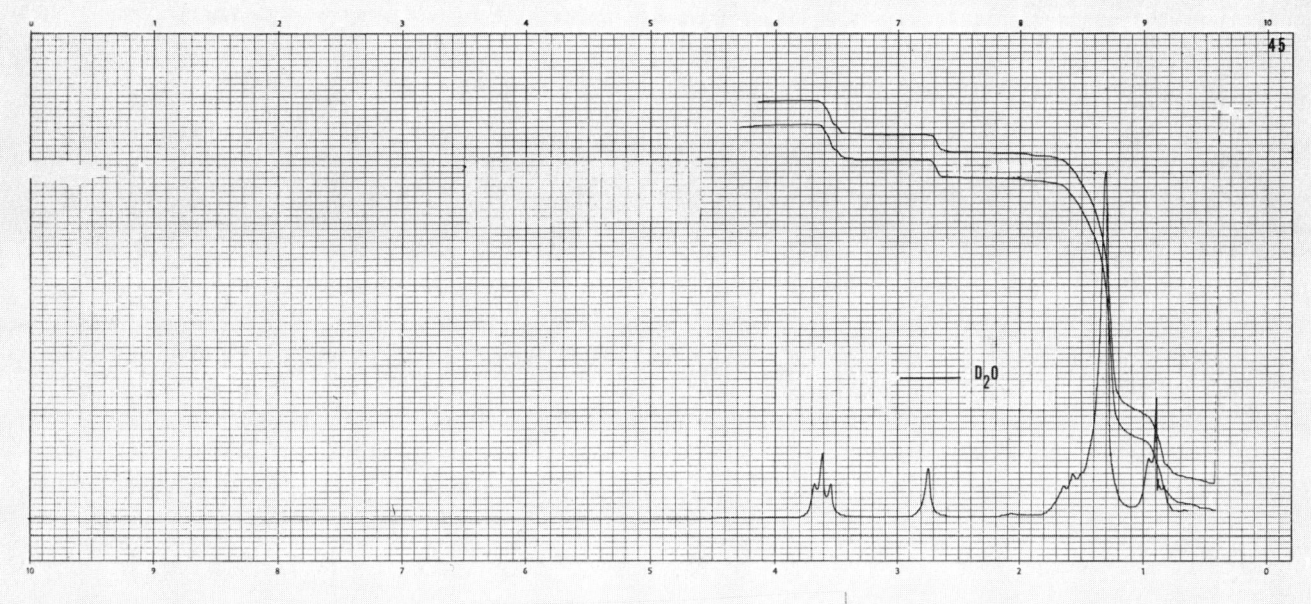

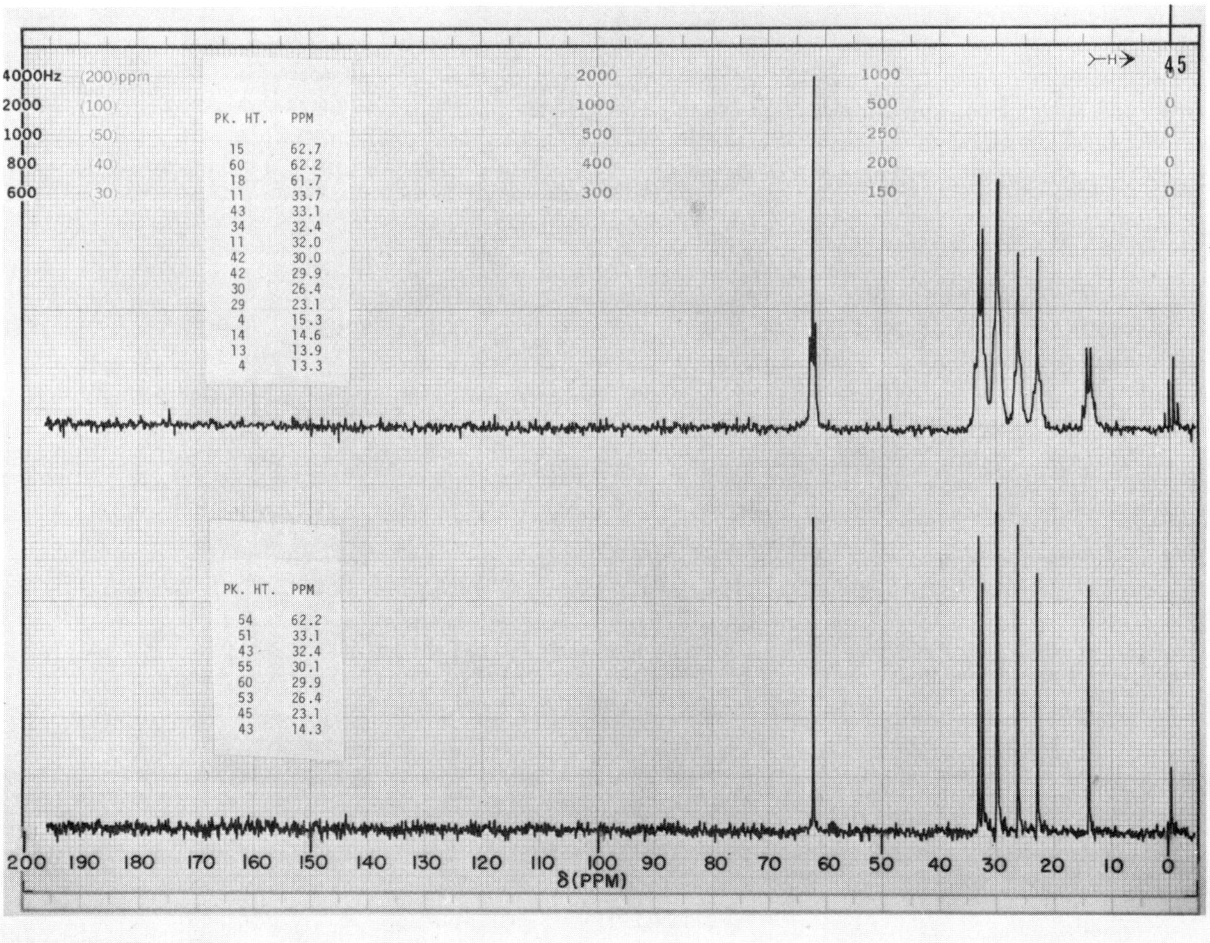

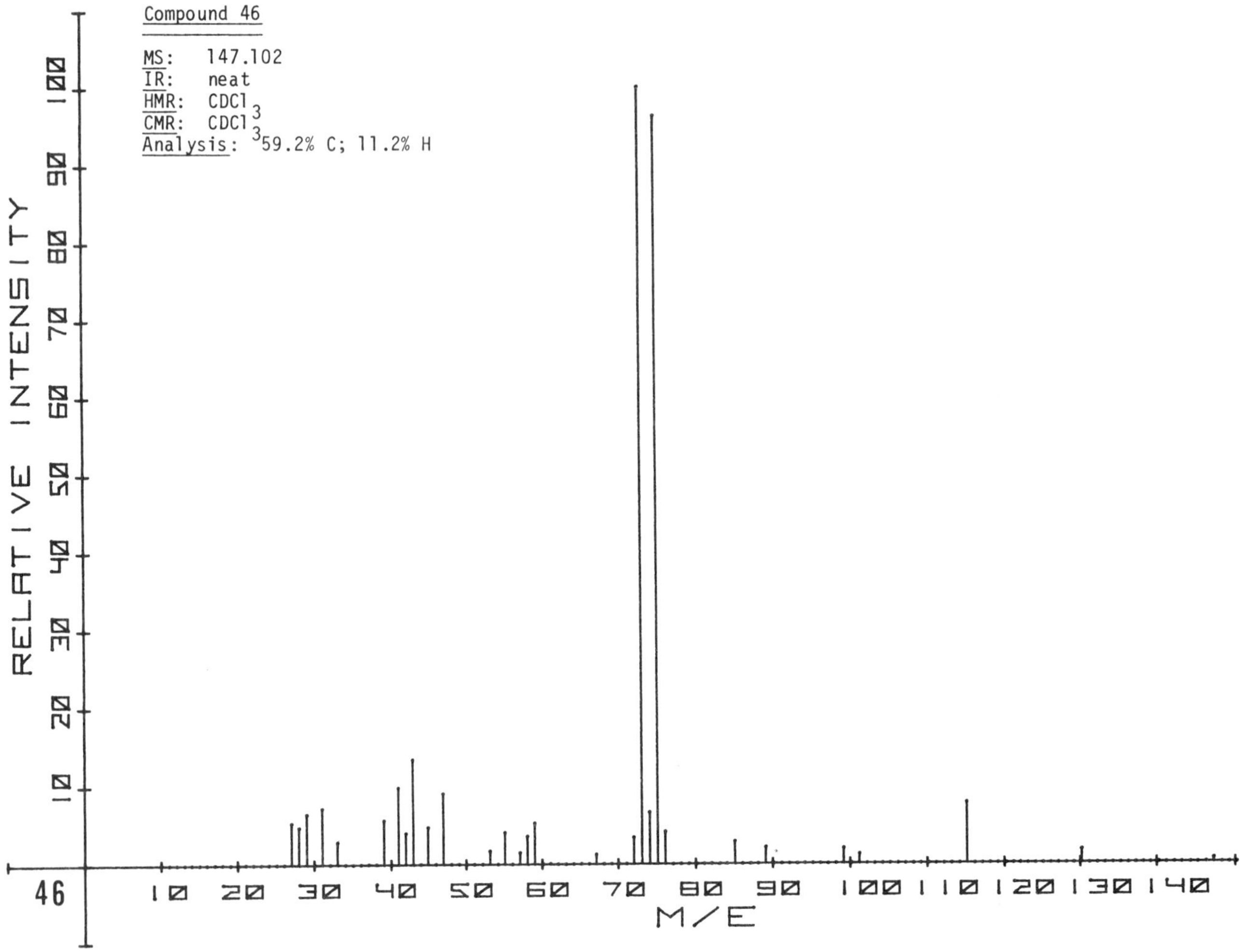

Compound 46
MS: 147.102
IR: neat
HMR: $CDCl_3$
CMR: $CDCl_3$
Analysis: 59.2% C; 11.2% H
RELATIVE INTENSITY
M/E
46

CM-1
ABSORBANCE
WAVELENGTH (MICRONS)
46
PR 1109 (137-1280)

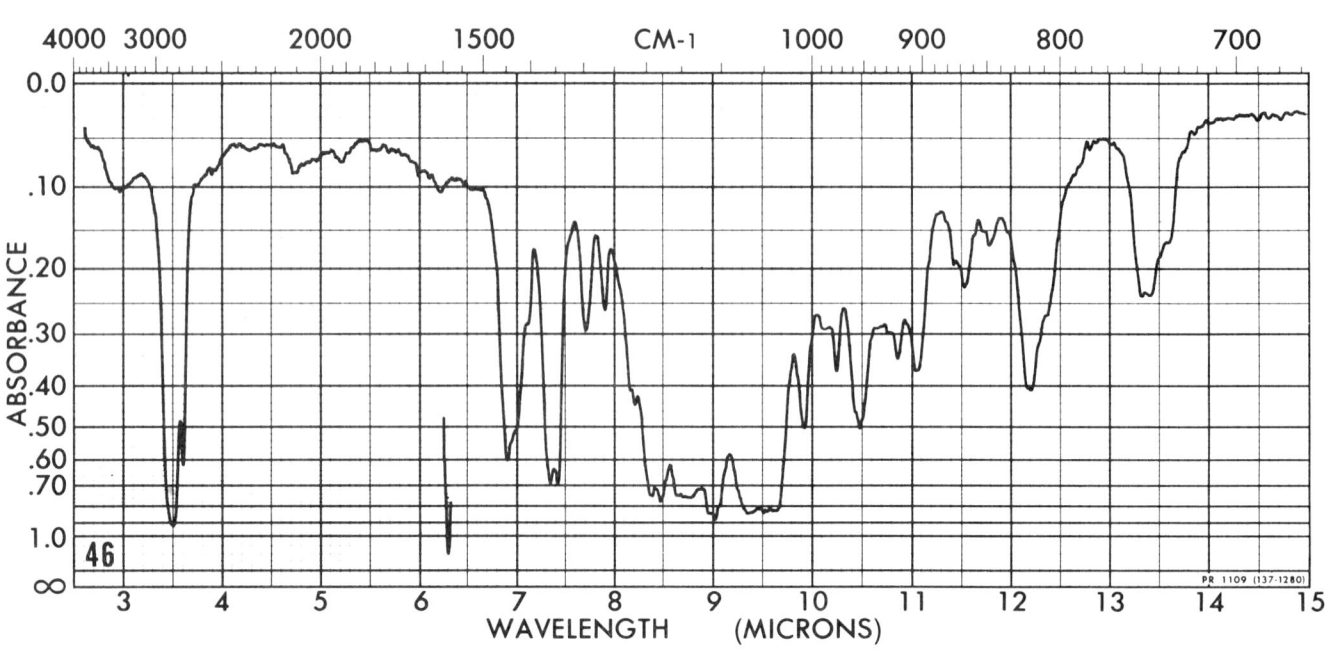

46

46
4000Hz (200)ppm 3000 2000 1000
2000 (100) 1500 1000 500
1000 (50) 750 500 250
800 (40) 400 200
600 (30) 300 150
PK. HT. PPM
10 55.1
41 53.2
7 51.8
50 51.4
26 50.0
16 49.5
31 48.1
8 46.2
PK. HT. PPM
39 102.1
23 73.3
40 52.3
19 49.0
32 42.9
60 25.4
200 190 180 170 160 150 140 130 120 110 100 90 80 70 60 50 40 30 20 10 0
δ(PPM)

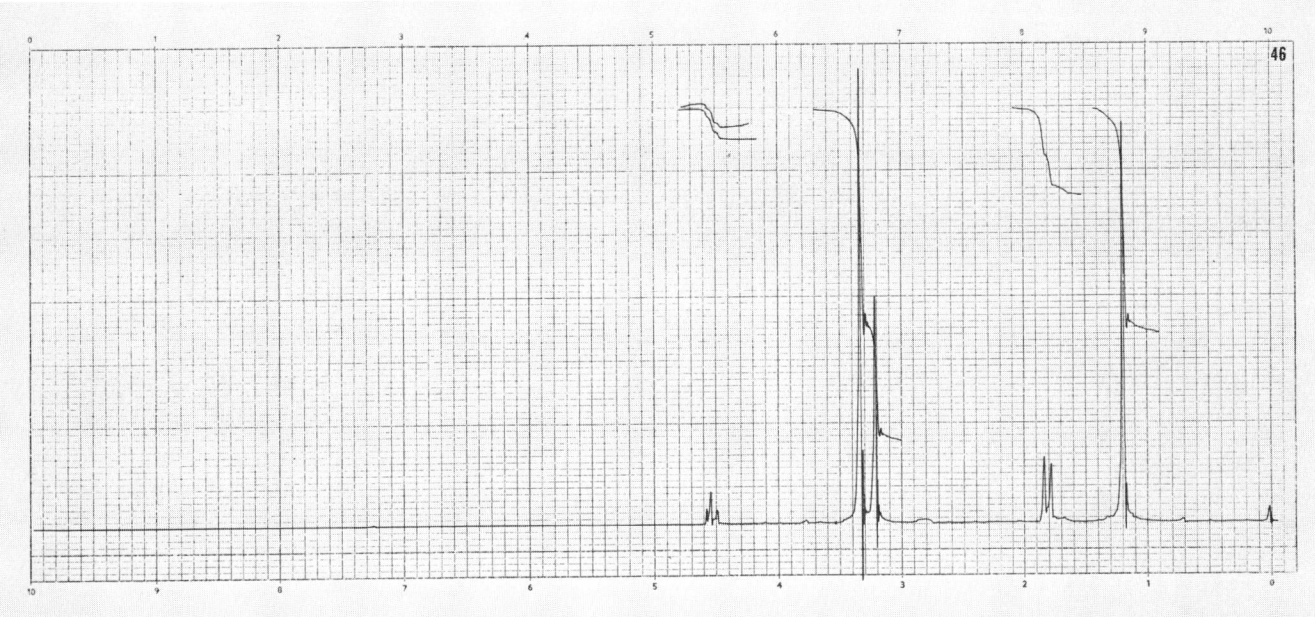

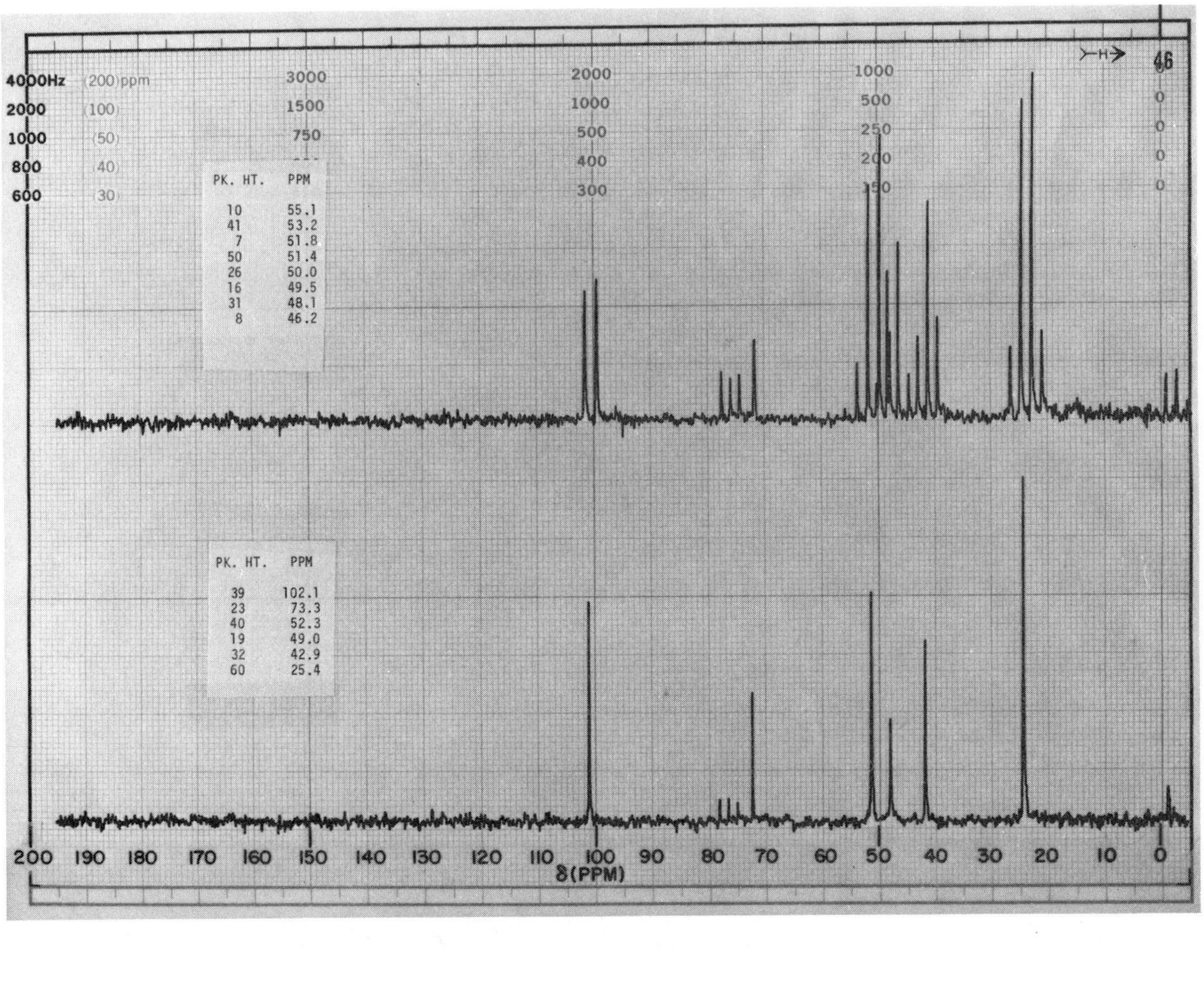

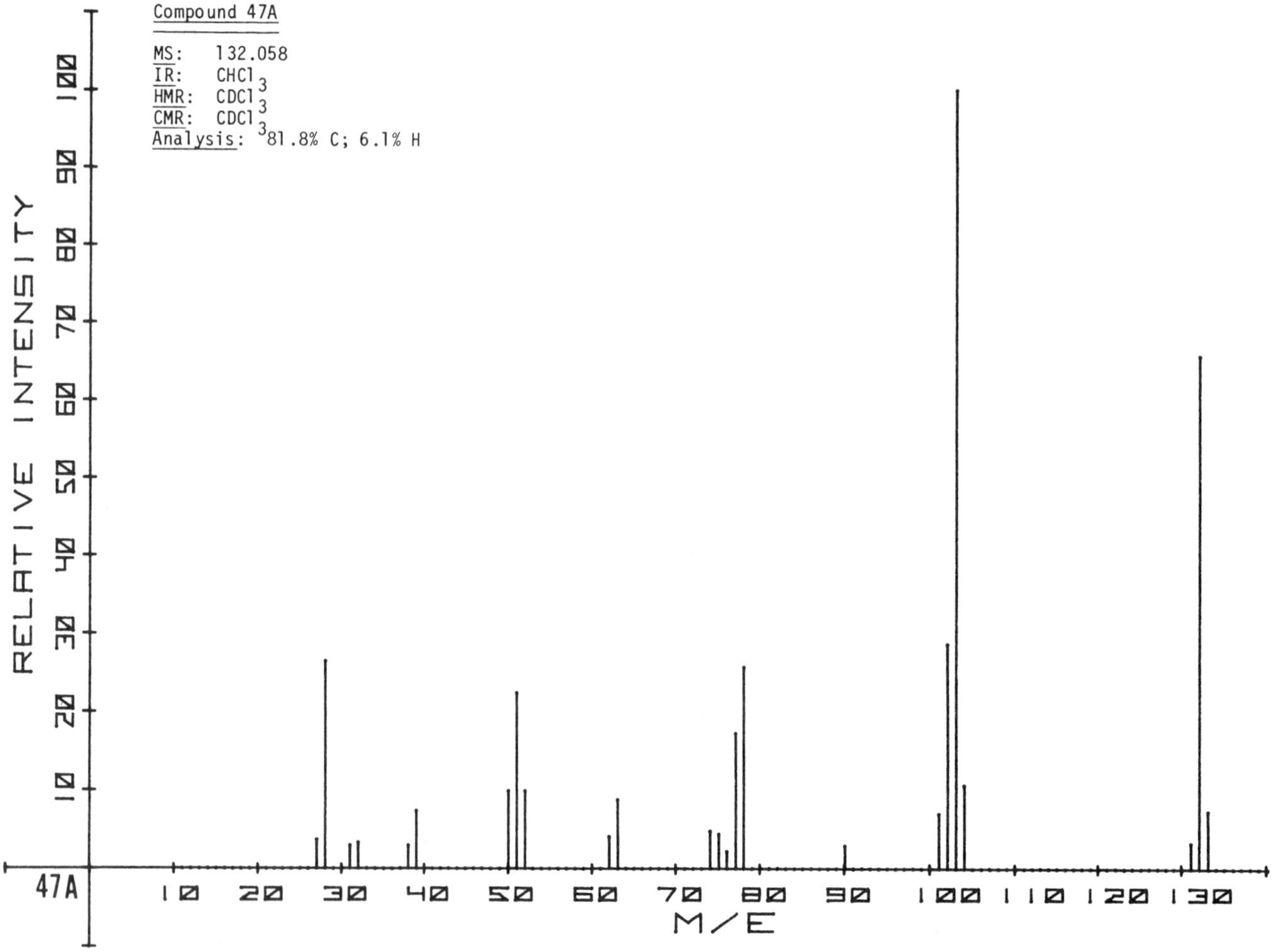

Compound 47A
MS: 132.058
IR: CHCl3
HMR: CDCl3
CMR: CDCl3
Analysis: 81.8% C; 6.1% H
RELATIVE INTENSITY
10 20 30 40 50 60 70 80 90 100
47A
10 20 30 40 50 60 70 80 90 100 110 120 130
M/E

4000 3000 2000 1500 CM-1 1000 900 800 700
0.0 .10 .20 .30 .40 .50 .60 .70 1.0 ∞
ABSORBANCE
47A
3 4 5 6 7 8 9 10 11 12 13 14 15
WAVELENGTH (MICRONS)
PR 1109 (137-1280)

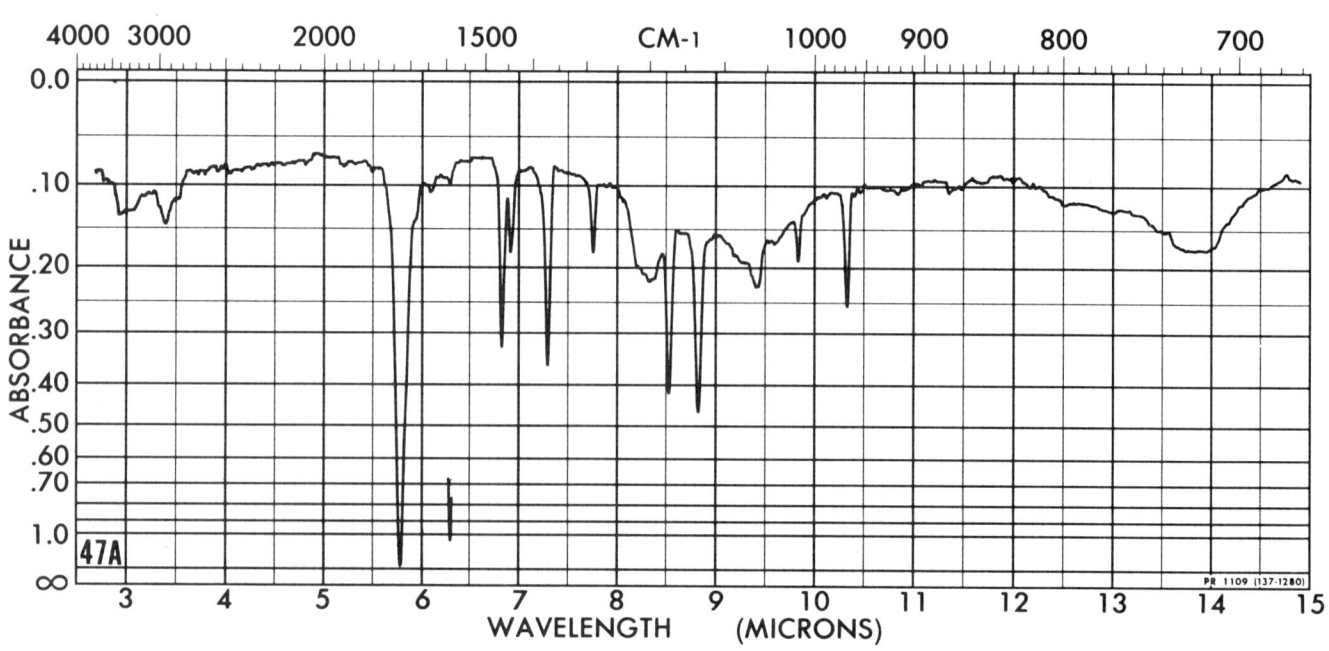

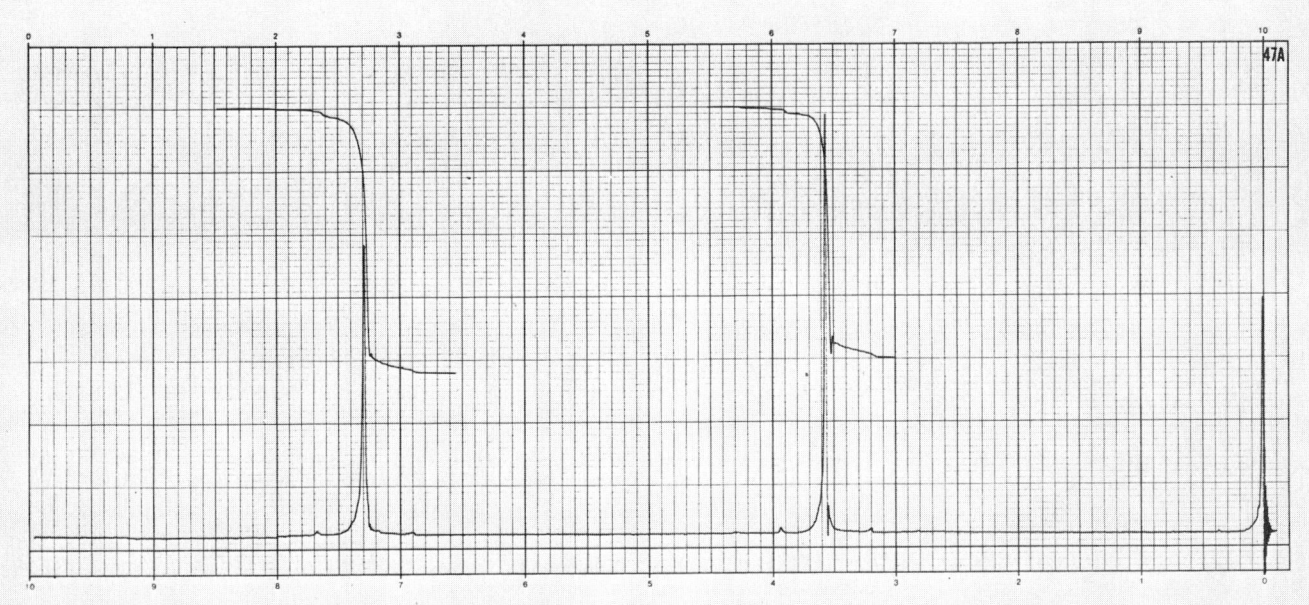

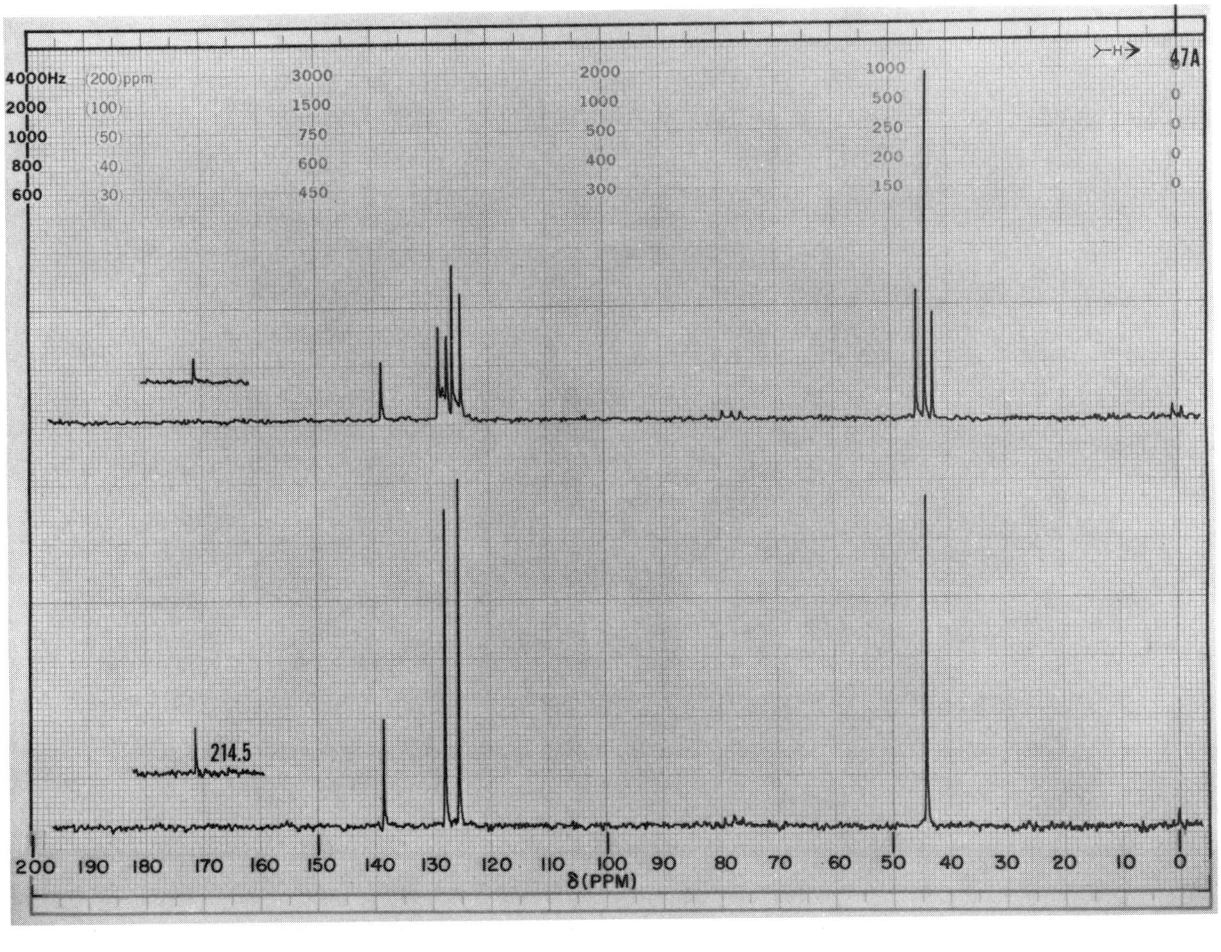

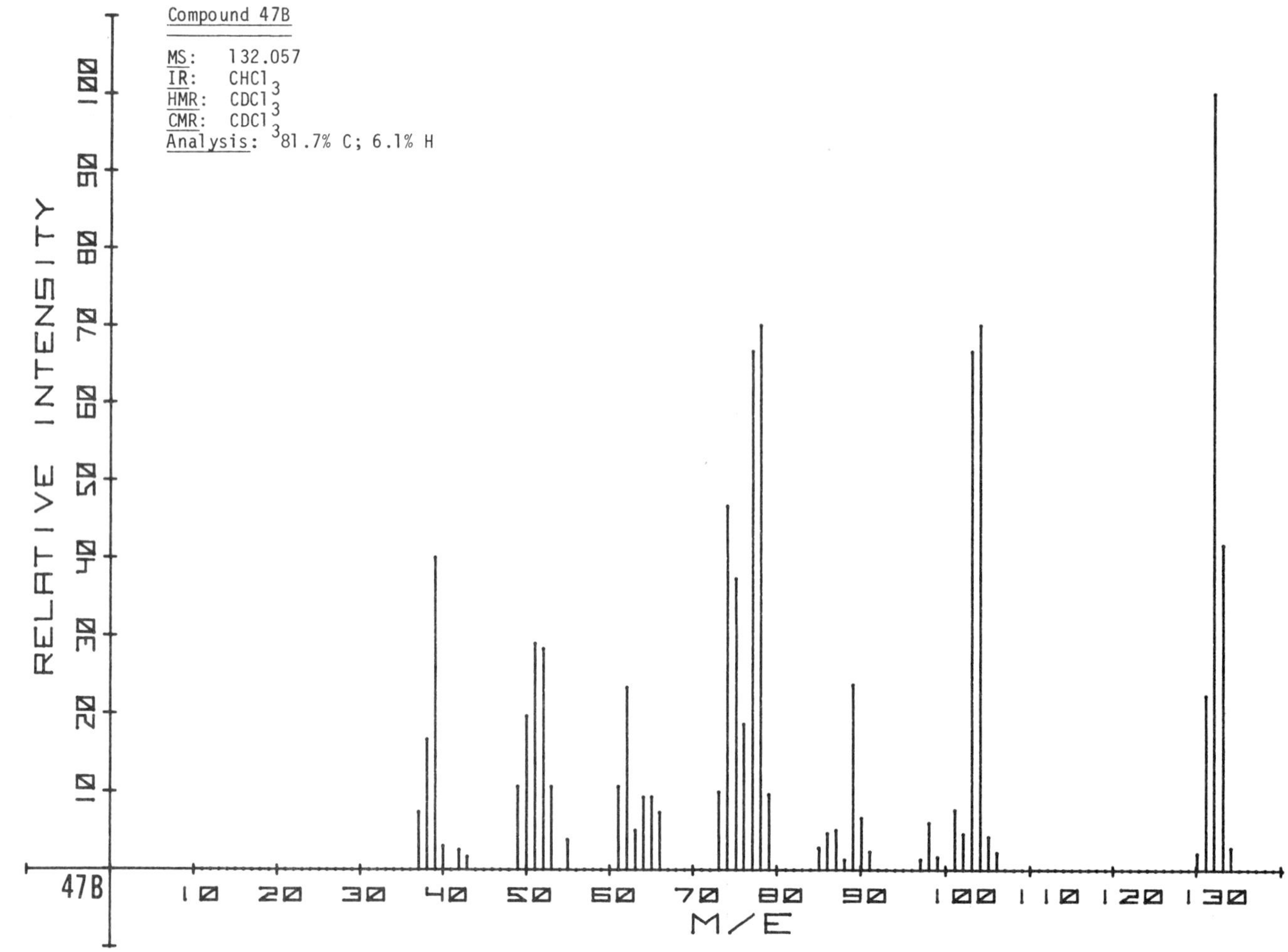

Compound 47B
MS: 132.057
IR: CHCl3
HMR: CDCl3
CMR: CDCl3
Analysis: 81.7% C; 6.1% H
RELATIVE INTENSITY
47B
M/E

CM-1
ABSORBANCE
47B
WAVELENGTH (MICRONS)
PR 1109 (137-1280)

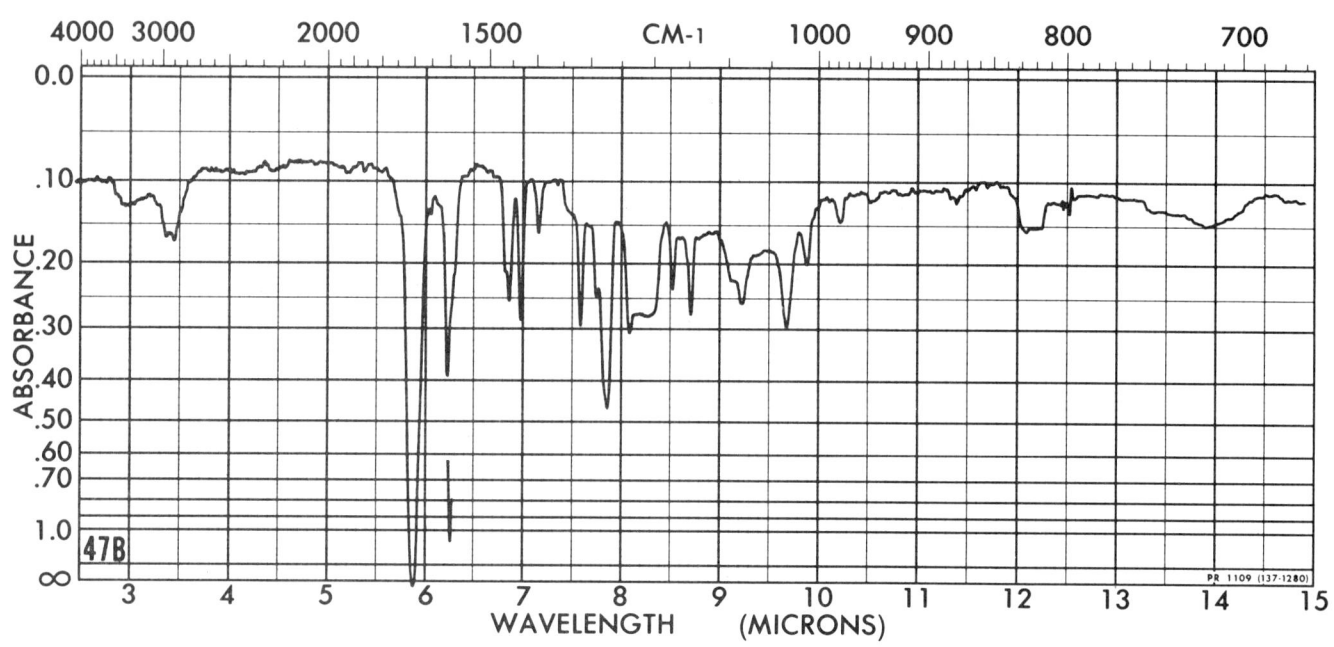

47B

47B
4000Hz (200)ppm 3000 2000 1000
2000 (100) 1500 1000 500
1000 (50) 750 500 250
800 (40) 600 200
600 (30) 450 150
PK. HT. PPM
16 137.0
38 135.3
32 133.6
40 128.0
58 127.5
34 126.3
60 125.8
48 124.3
59 122.6
PK. HT. PPM
8 206.5
11 155.0
14 137.0
60 134.4
41 127.1
58 126.7
42 123.4
35 36.1
37 25.7
206.5
200 190 180 170 160 150 140 130 120 110 100 90 80 70 60 50 40 30 20 10 0
δ(PPM)

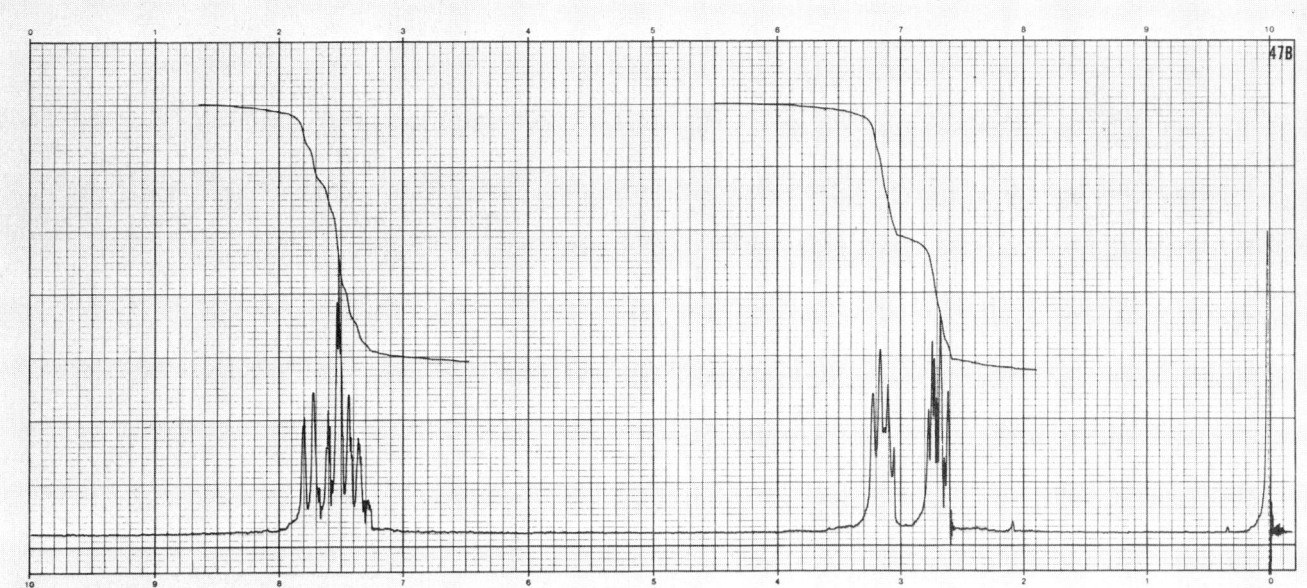

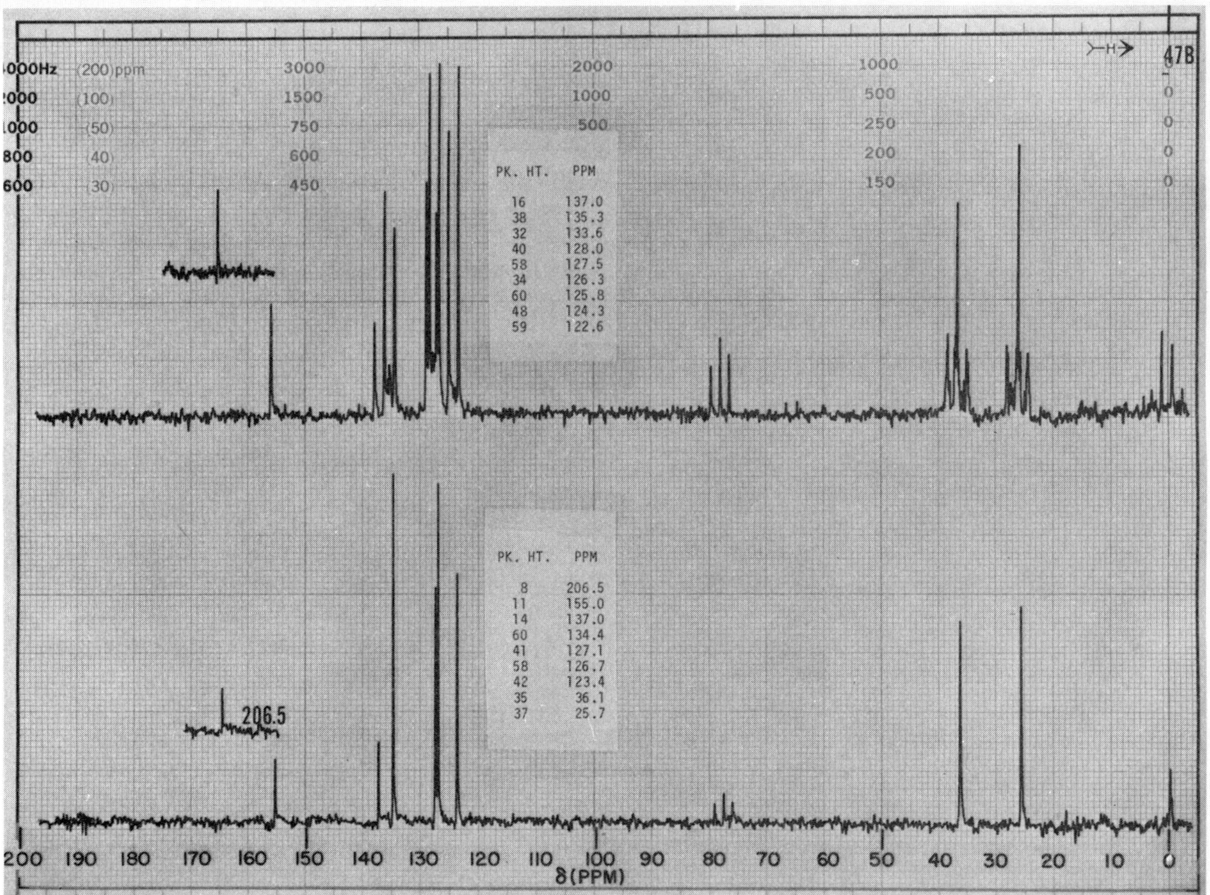

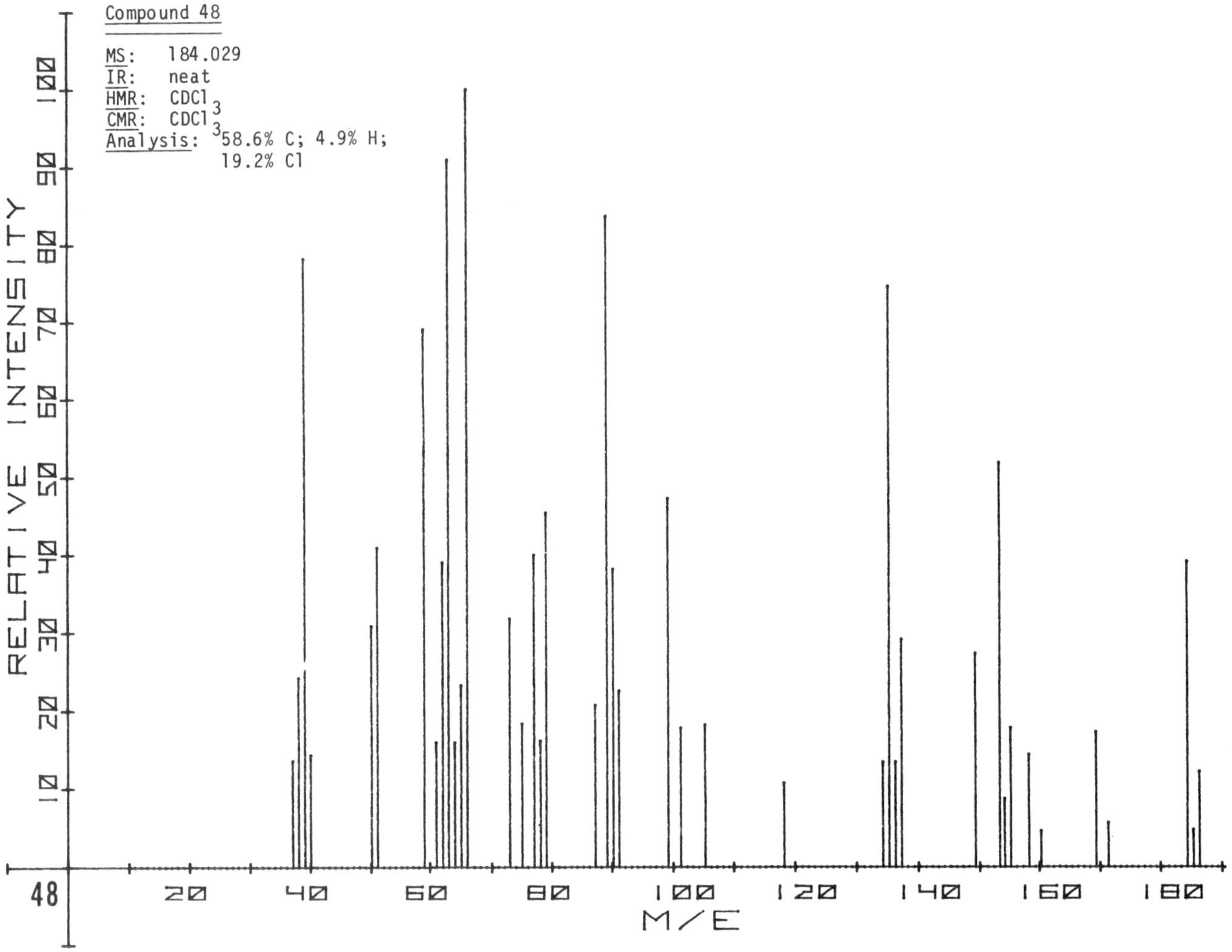

Compound 48
MS: 184.029
IR: neat
HMR: CDCl3
CMR: CDCl3
Analysis: 58.6% C; 4.9% H; 19.2% Cl
RELATIVE INTENSITY
10 20 30 40 50 60 70 80 90 100
48
20 40 60 80 100 120 140 160 180
M/E

4000 3000 2000 1500 CM-1 1000 900 800 700
0.0 .10 .20 .30 .40 .50 .60 .70 1.0 ∞
ABSORBANCE
48
PR 1109 (137-1280)
3 4 5 6 7 8 9 10 11 12 13 14 15
WAVELENGTH (MICRONS)

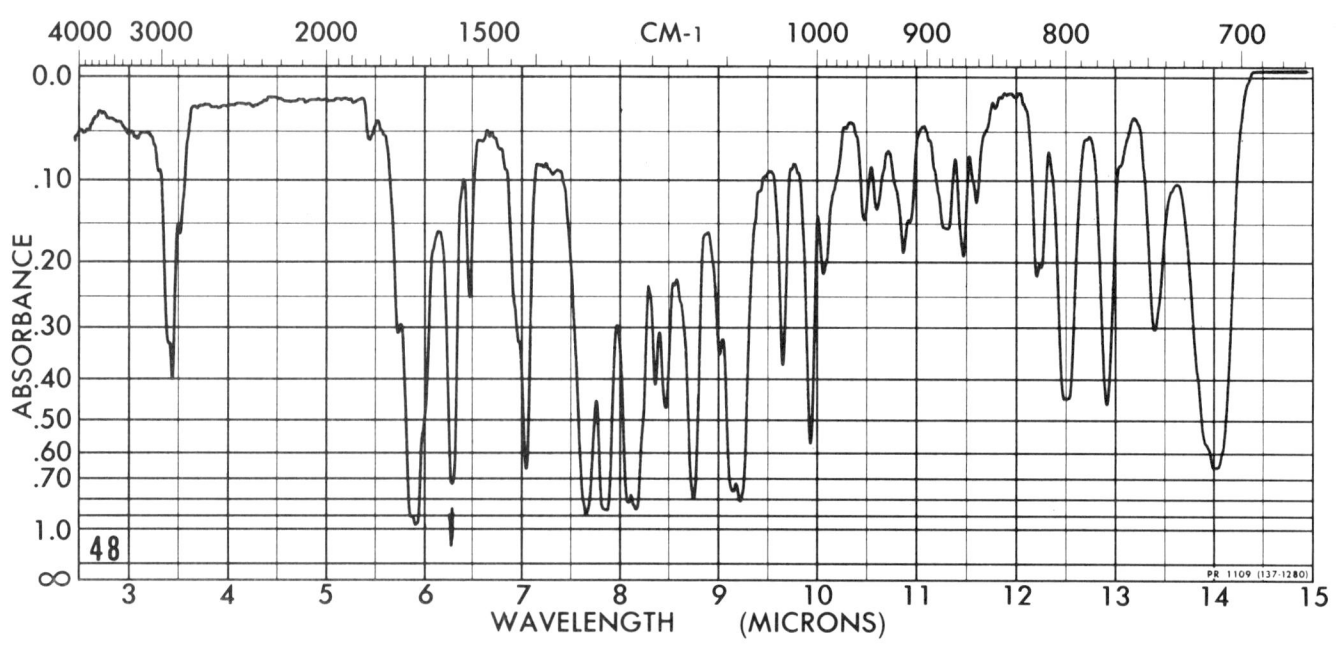

48
IRR. 3.58

48
4000Hz (200)ppm 3000 2000 1000 0
2000 (100) 1500 1000 500 0
1000 (50) 750 250 0
800 (40) 600 200 0
600 (30) 450 150 0
PK. HT. PPM
26 144.3
31 142.6
26 141.2
31 139.5
8 138.2
11 53.9
27 52.4
31 52.2
31 50.6
42 50.5
8 48.8
PK. HT. PPM
6 163.6
10 160.0
52 143.5
58 140.4
15 138.1
49 71.3
54 59.7
60 51.5
30 51.3
200 190 180 170 160 150 140 130 120 110 100 90 80 70 60 50 40 30 20 10 0
δ(PPM)

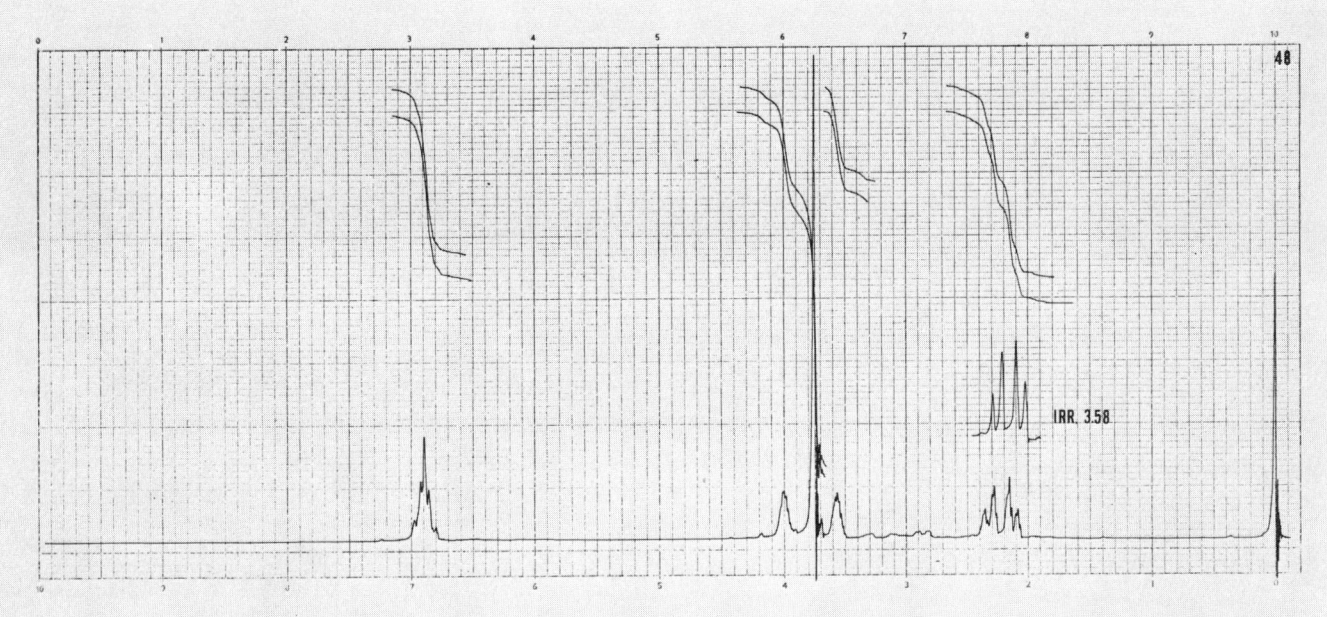

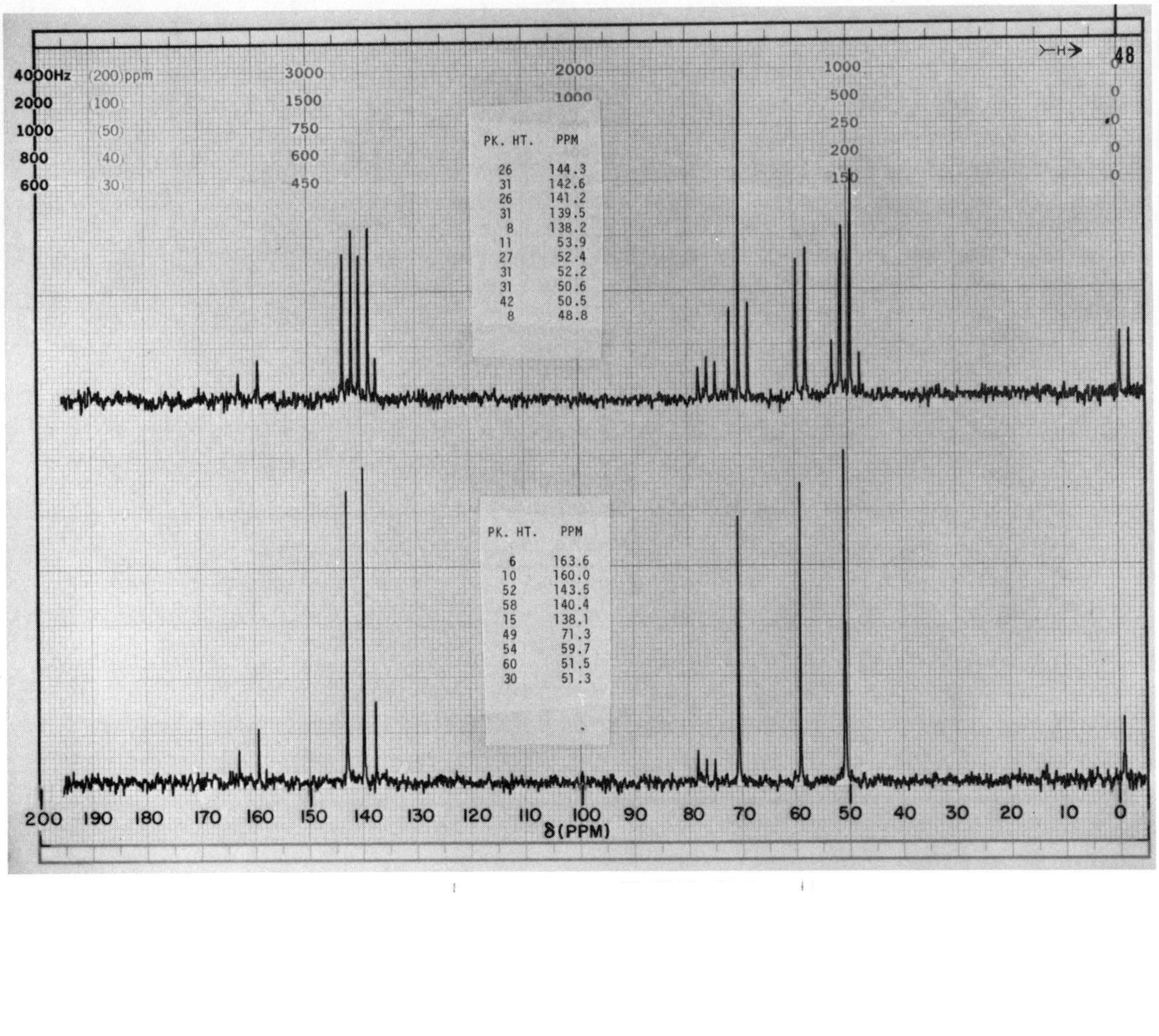

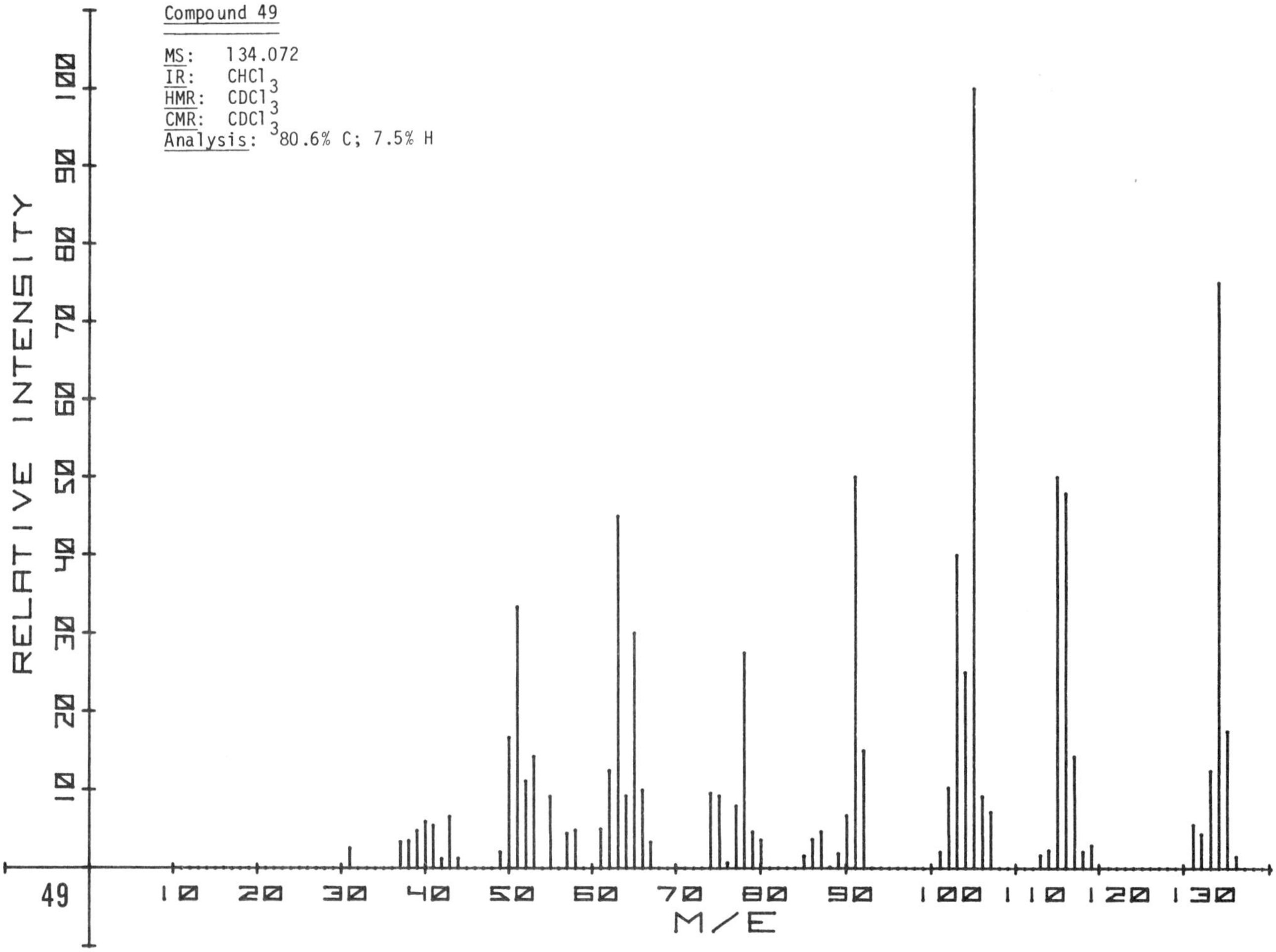

Compound 49
MS: 134.072
IR: CHCl3
HMR: CDCl3
CMR: CDCl3
Analysis: 80.6% C; 7.5% H
RELATIVE INTENSITY
10 20 30 40 50 60 70 80 90 100
49
M/E
10 20 30 40 50 60 70 80 90 100 110 120 130

4000 3000 2000 1500 CM-1 1000 900 800 700
ABSORBANCE
0.0 .10 .20 .30 .40 .50 .60 .70 1.0 ∞
49
3 4 5 6 7 8 9 10 11 12 13 14 15
WAVELENGTH (MICRONS)

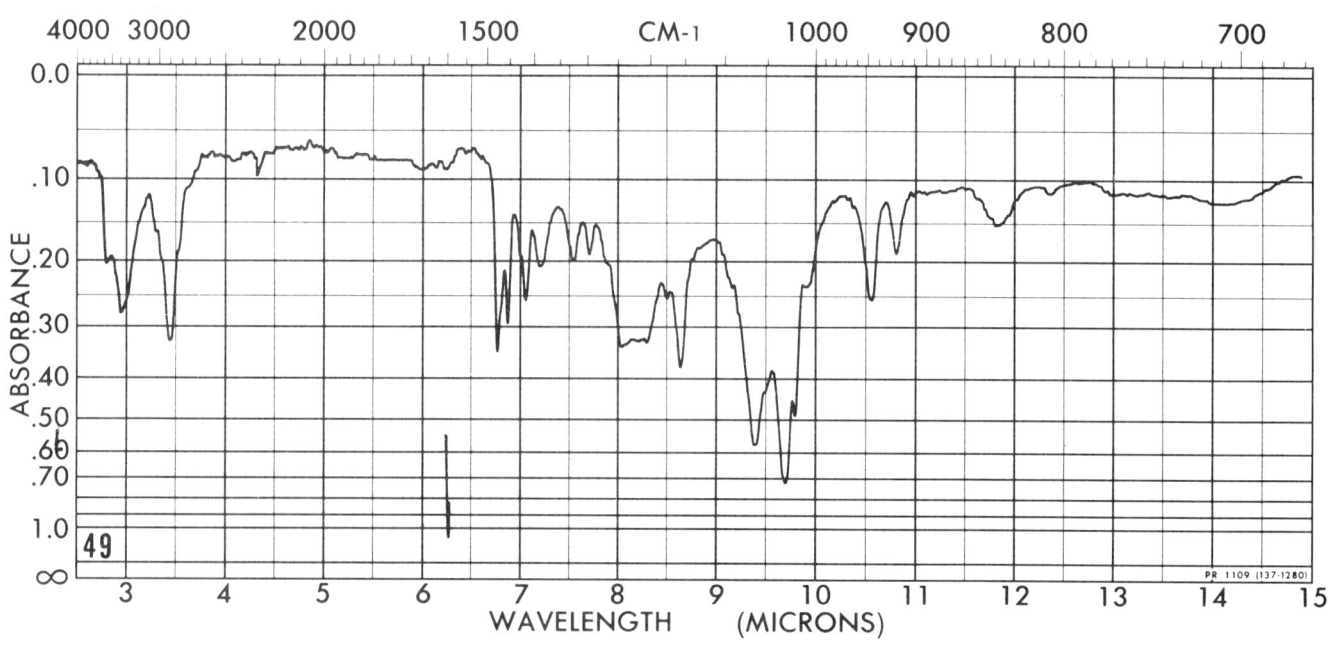

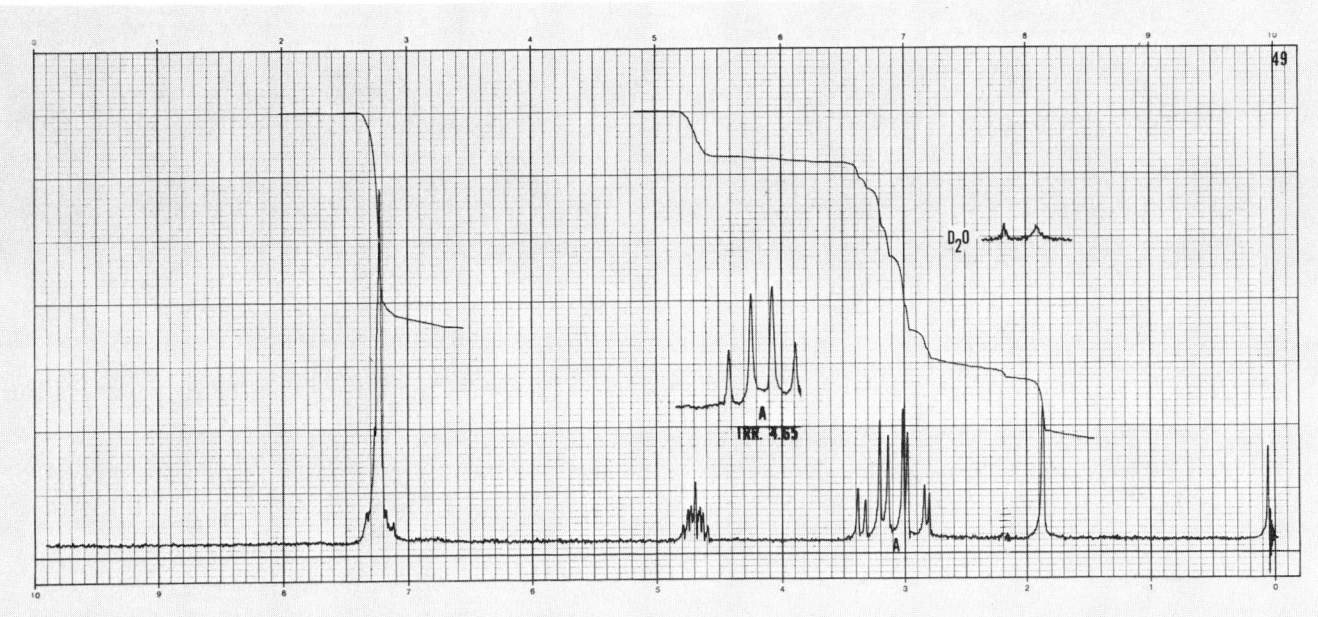

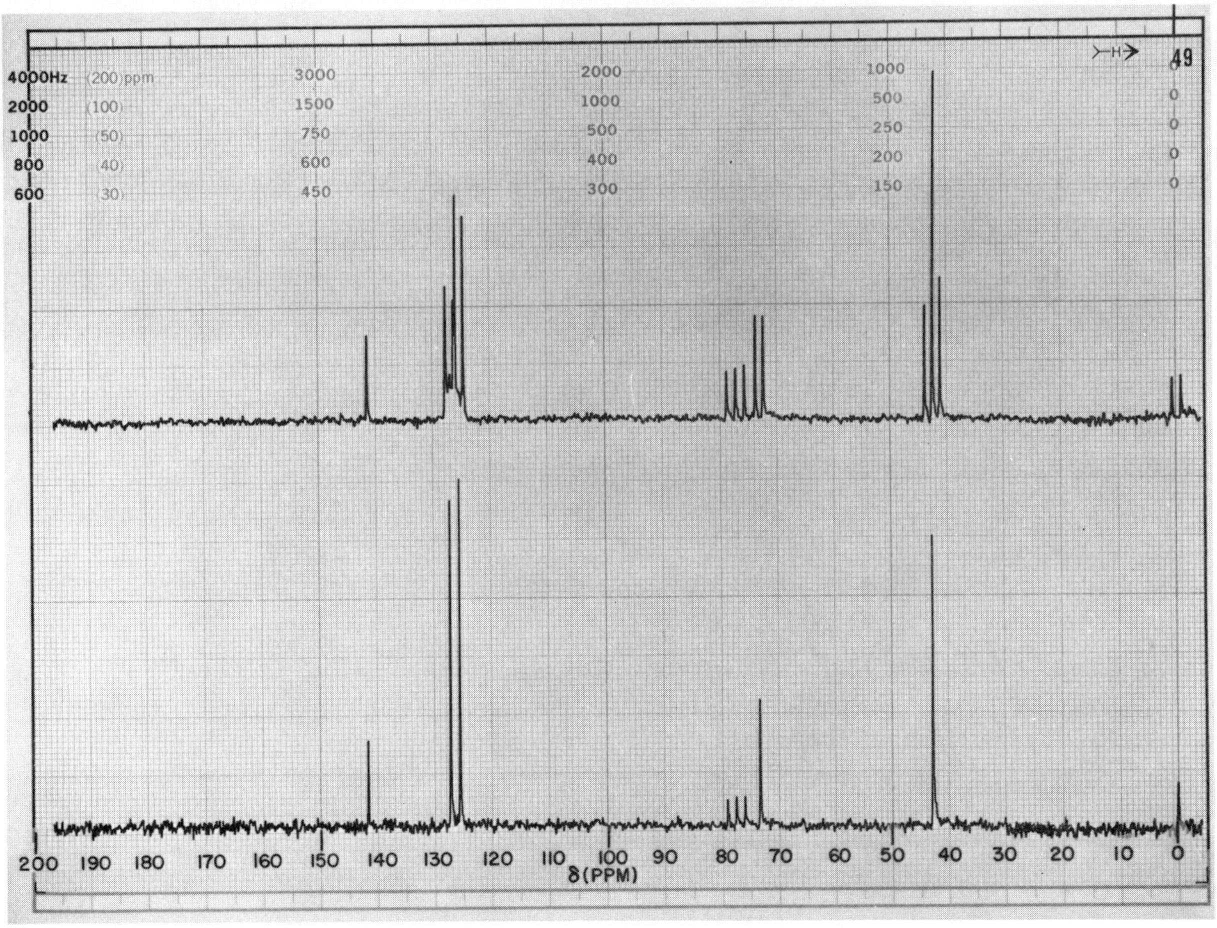

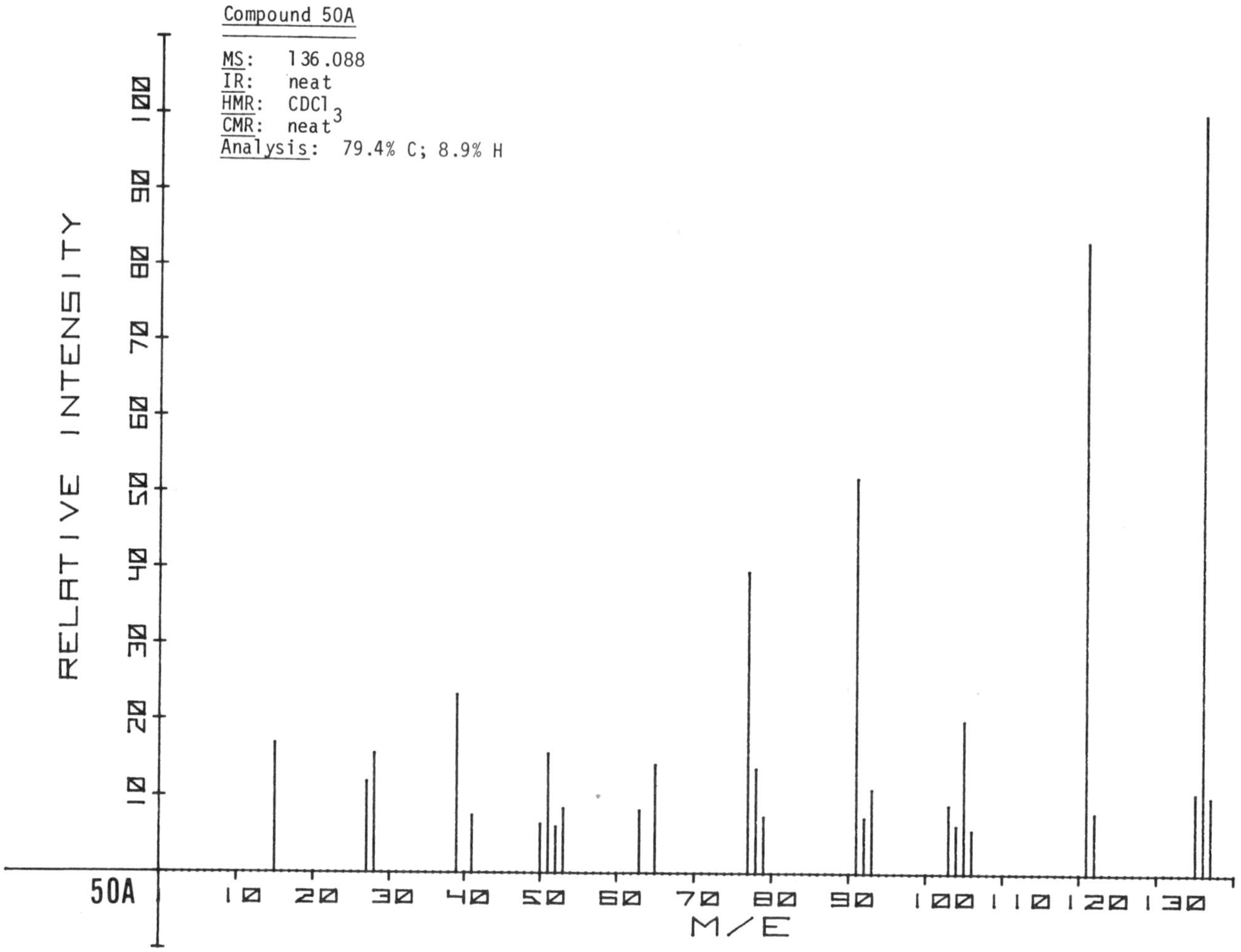

Compound 50A
MS: 136.088
IR: neat
HMR: CDCl3
CMR: neat
Analysis: 79.4% C; 8.9% H
RELATIVE INTENSITY
10 20 30 40 50 60 70 80 90 100
50A
10 20 30 40 50 60 70 80 90 100 110 120 130
M/E

4000 3000 2000 1500 CM-1 1000 900 800 700
ABSORBANCE
0.0 .10 .20 .30 .40 .50 .60 .70 1.0 ∞
50A
3 4 5 6 7 8 9 10 11 12 13 14 15
WAVELENGTH (MICRONS)
PR 1109 (137-1280)

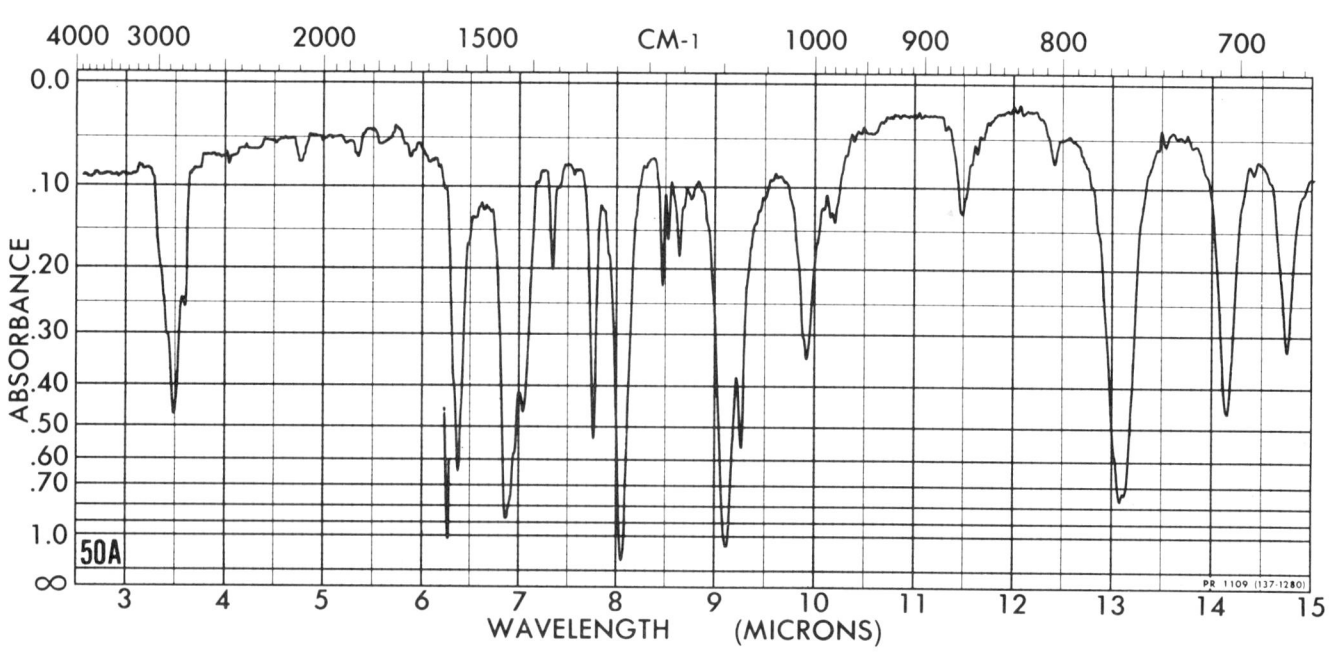

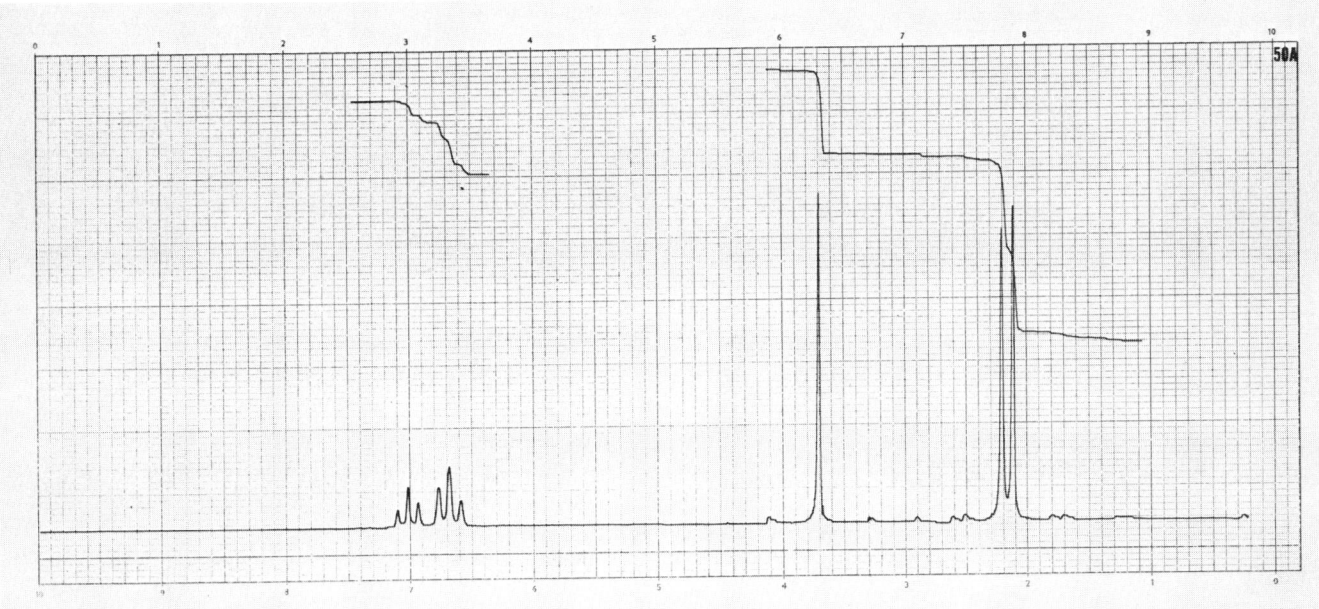

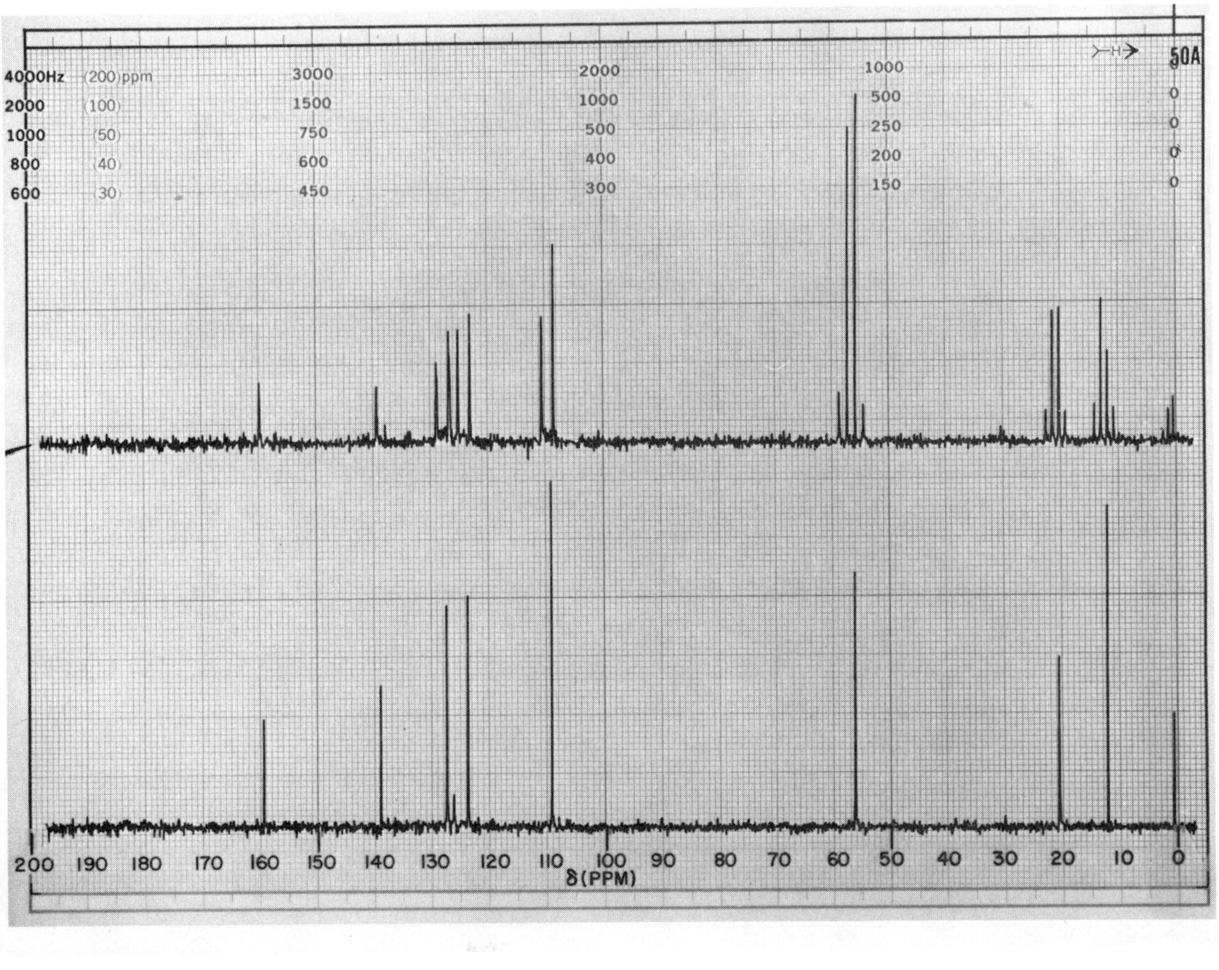

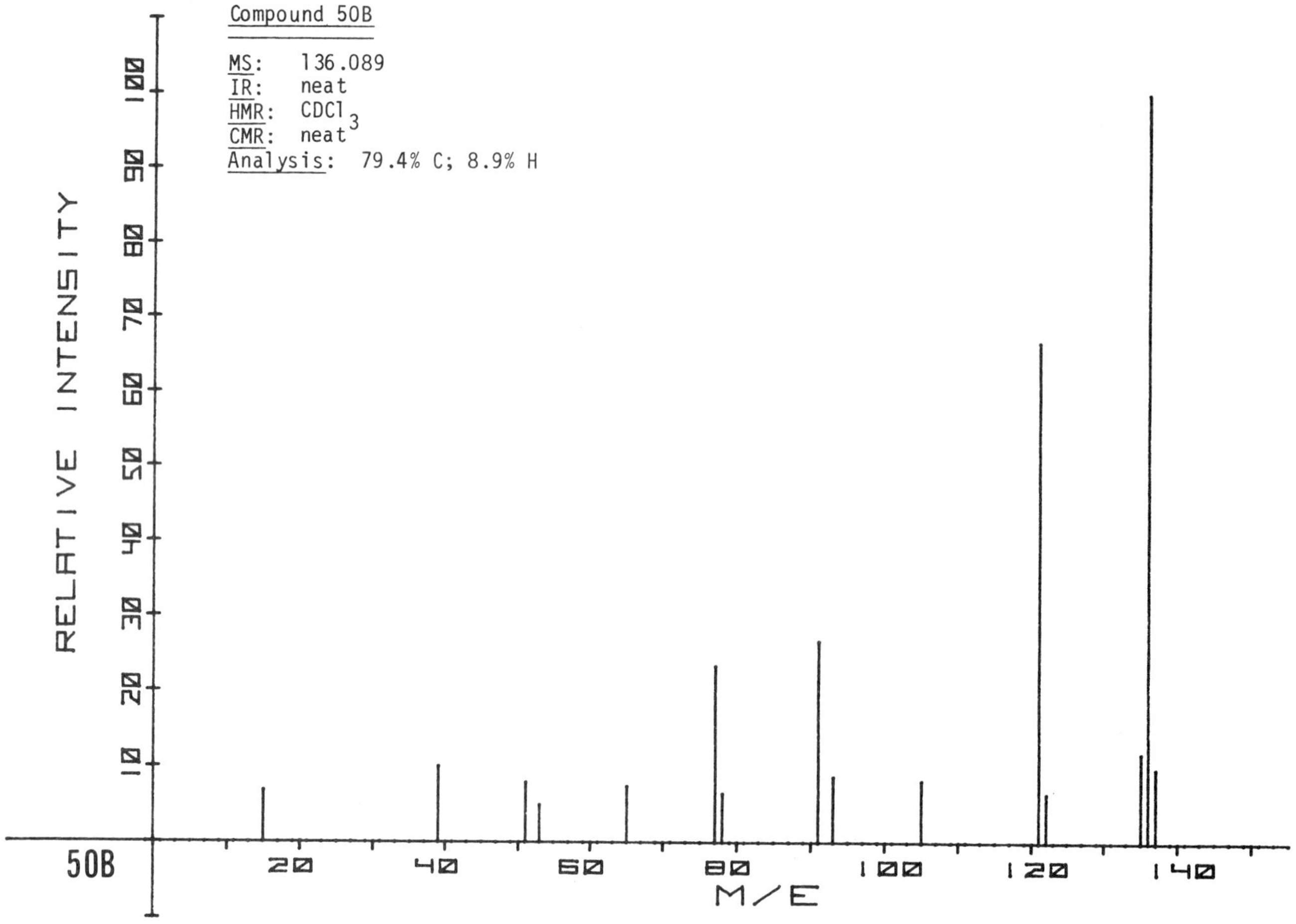

Compound 50B
MS: 136.089
IR: neat
HMR: CDCl3
CMR: neat
Analysis: 79.4% C; 8.9% H
RELATIVE INTENSITY
10 20 30 40 50 60 70 80 90 100
50B
20 40 60 80 100 120 140
M/E

4000 3000 2000 1500 CM-1 1000 900 800 700
ABSORBANCE
0.0 .10 .20 .30 .40 .50 .60 .70 1.0 ∞
50B
3 4 5 6 7 8 9 10 11 12 13 14 15
WAVELENGTH (MICRONS)

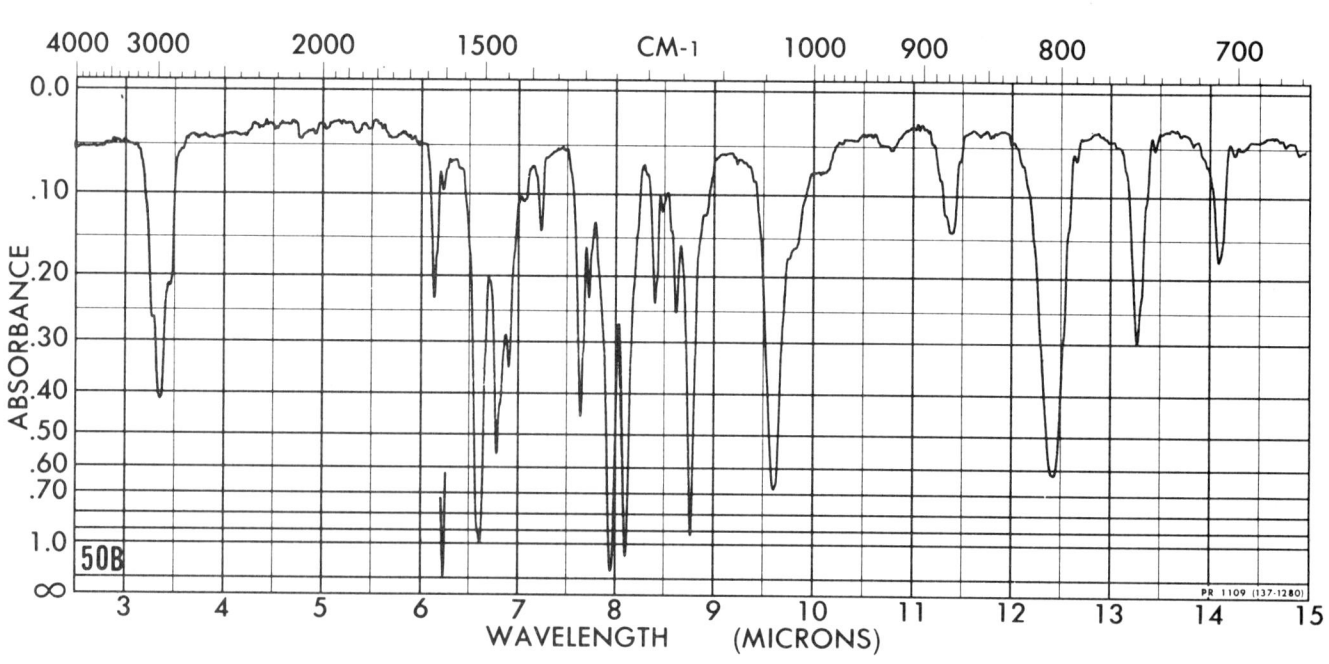

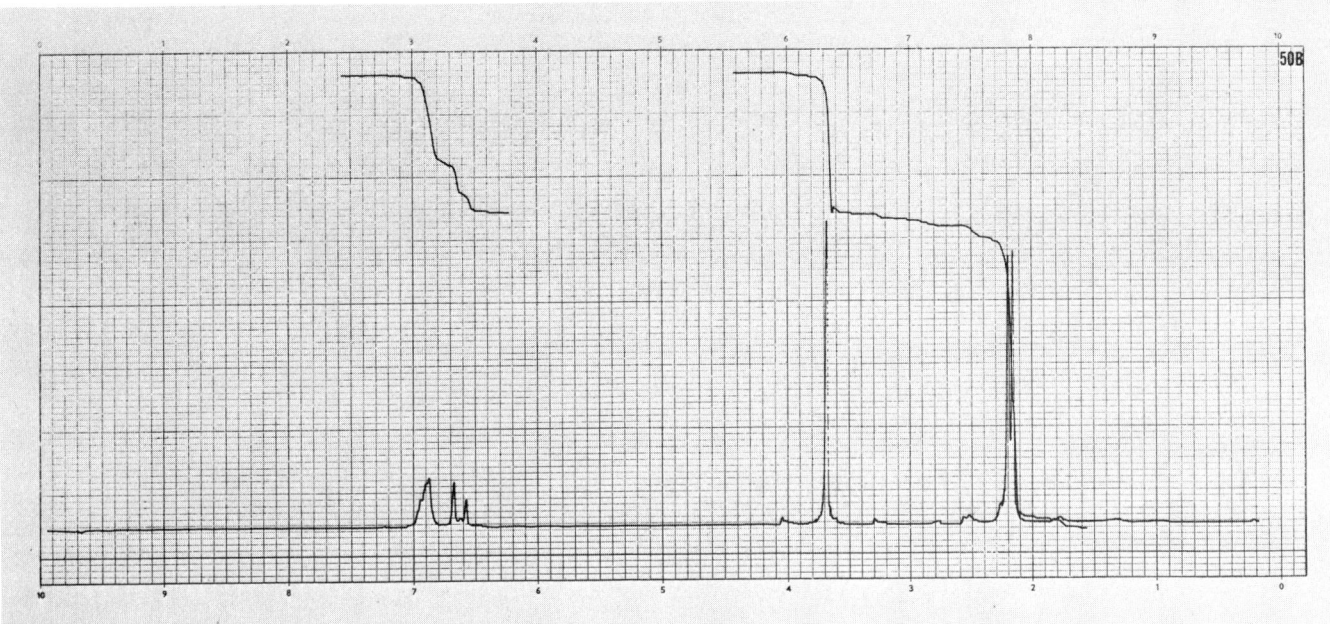

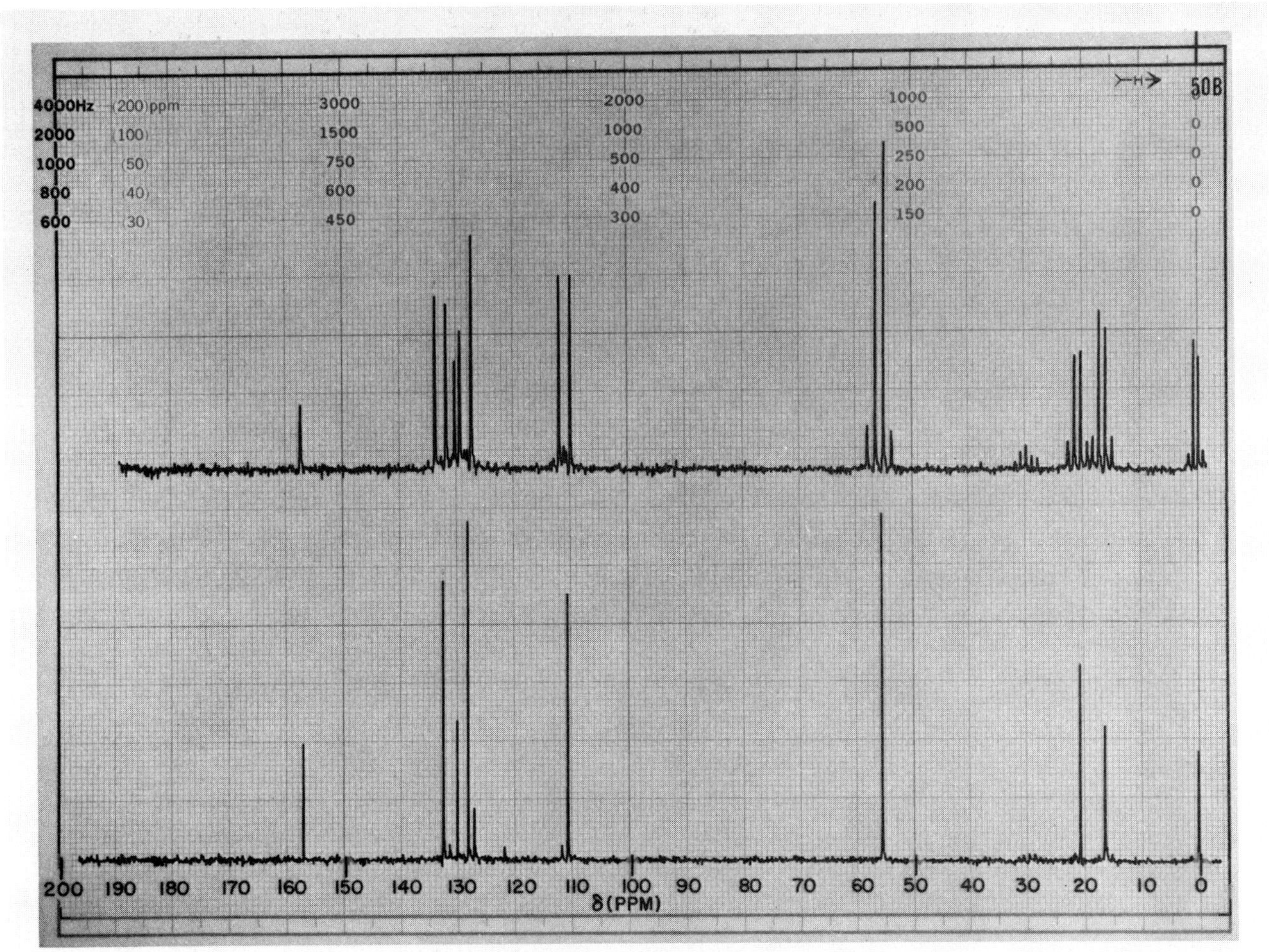

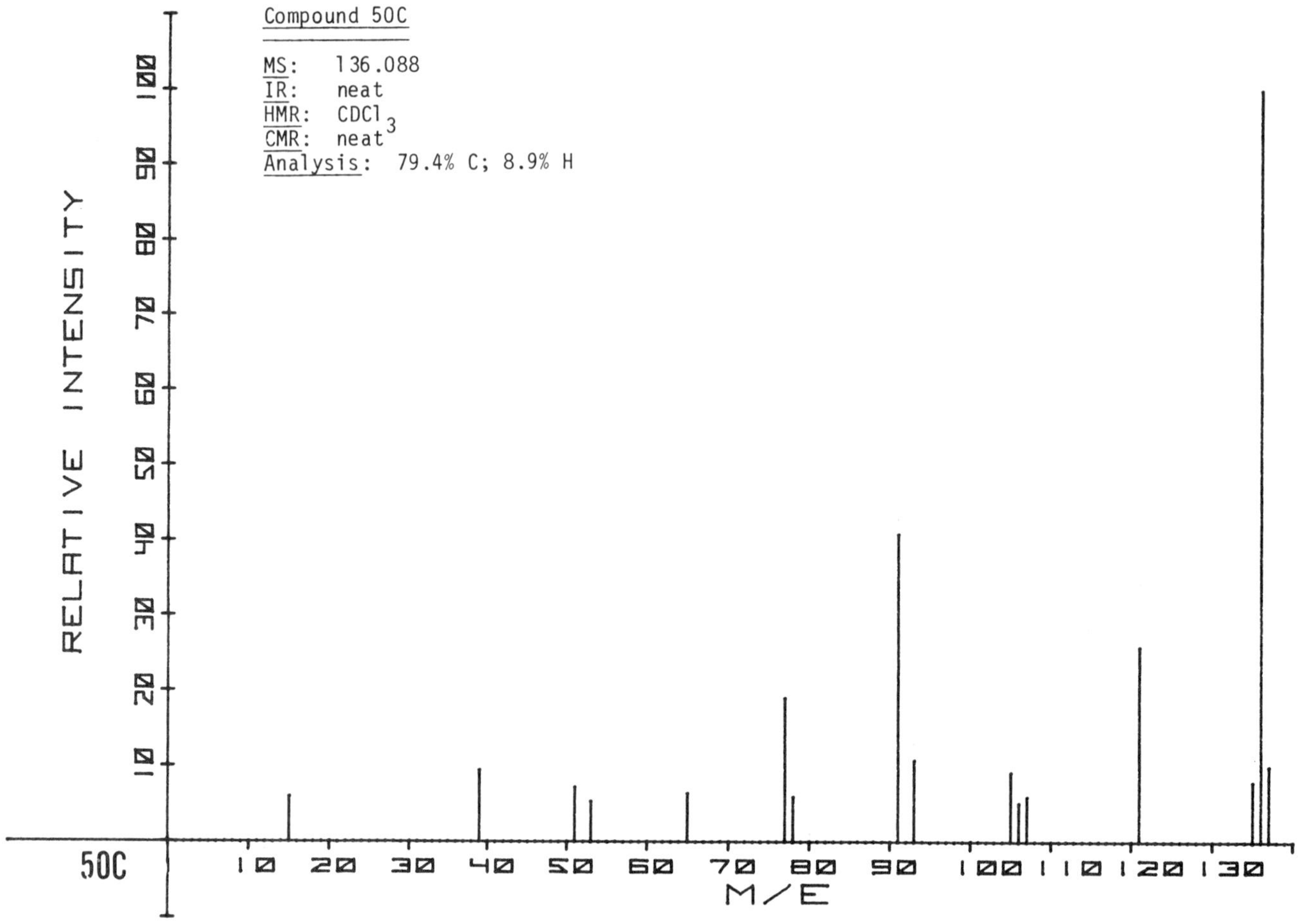

Compound 50C
MS: 136.088
IR: neat
HMR: CDCl3
CMR: neat
Analysis: 79.4% C; 8.9% H
RELATIVE INTENSITY
10 20 30 40 50 60 70 80 90 100
50C
10 20 30 40 50 60 70 80 90 100 110 120 130
M/E

4000 3000 2000 1500 CM-1 1000 900 800 700
0.0 .10 .20 .30 .40 .50 .60 .70 1.0 ∞
ABSORBANCE
50C
3 4 5 6 7 8 9 10 11 12 13 14 15
WAVELENGTH (MICRONS)

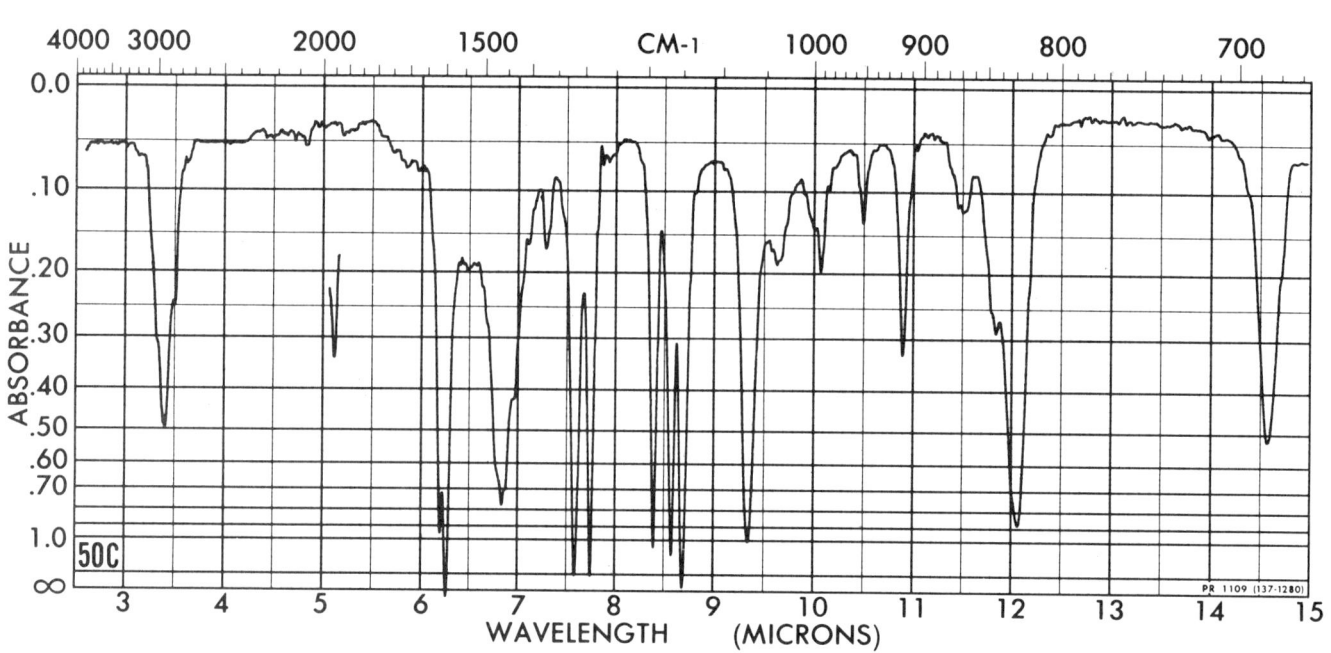

50C
10
9
8
7
6
5
4
3
2
1
0

4000Hz (200)ppm 3000 2000 1000
2000 (100) 1500 1000 500
1000 (50) 750 500 250
800 (40) 600 400 200
600 (30) 450 300 150
50C
200 190 180 170 160 150 140 130 120 110 100 90 80 70 60 50 40 30 20 10 0
δ(PPM)

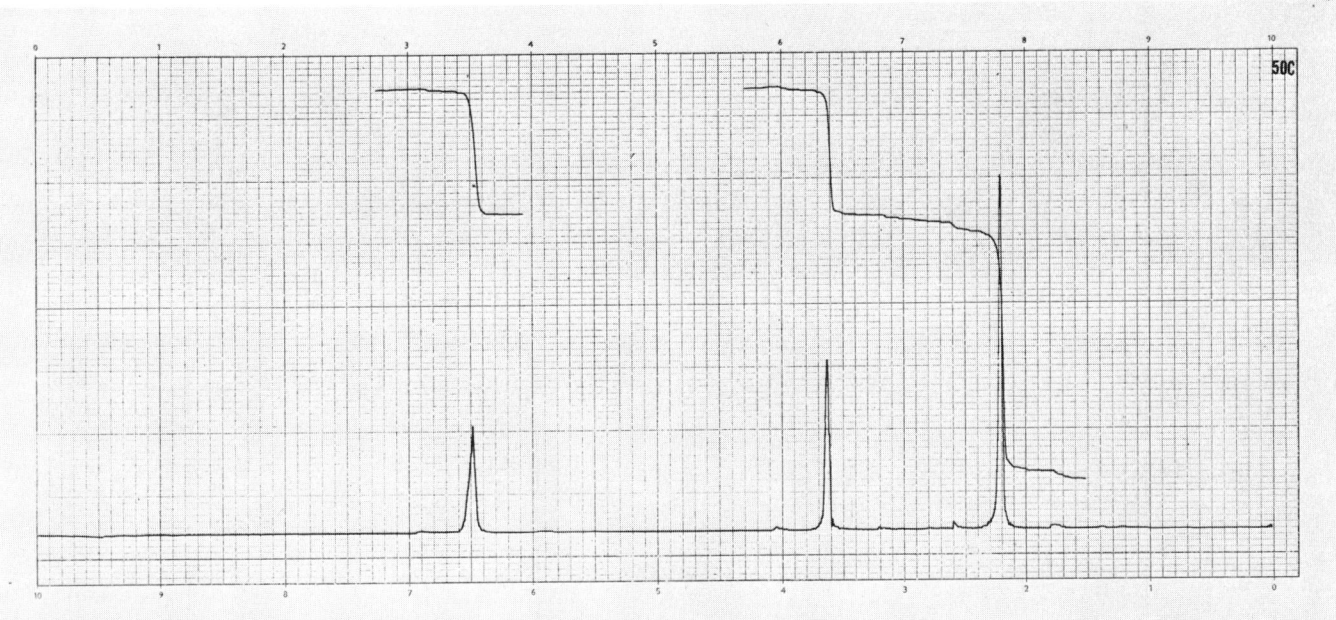

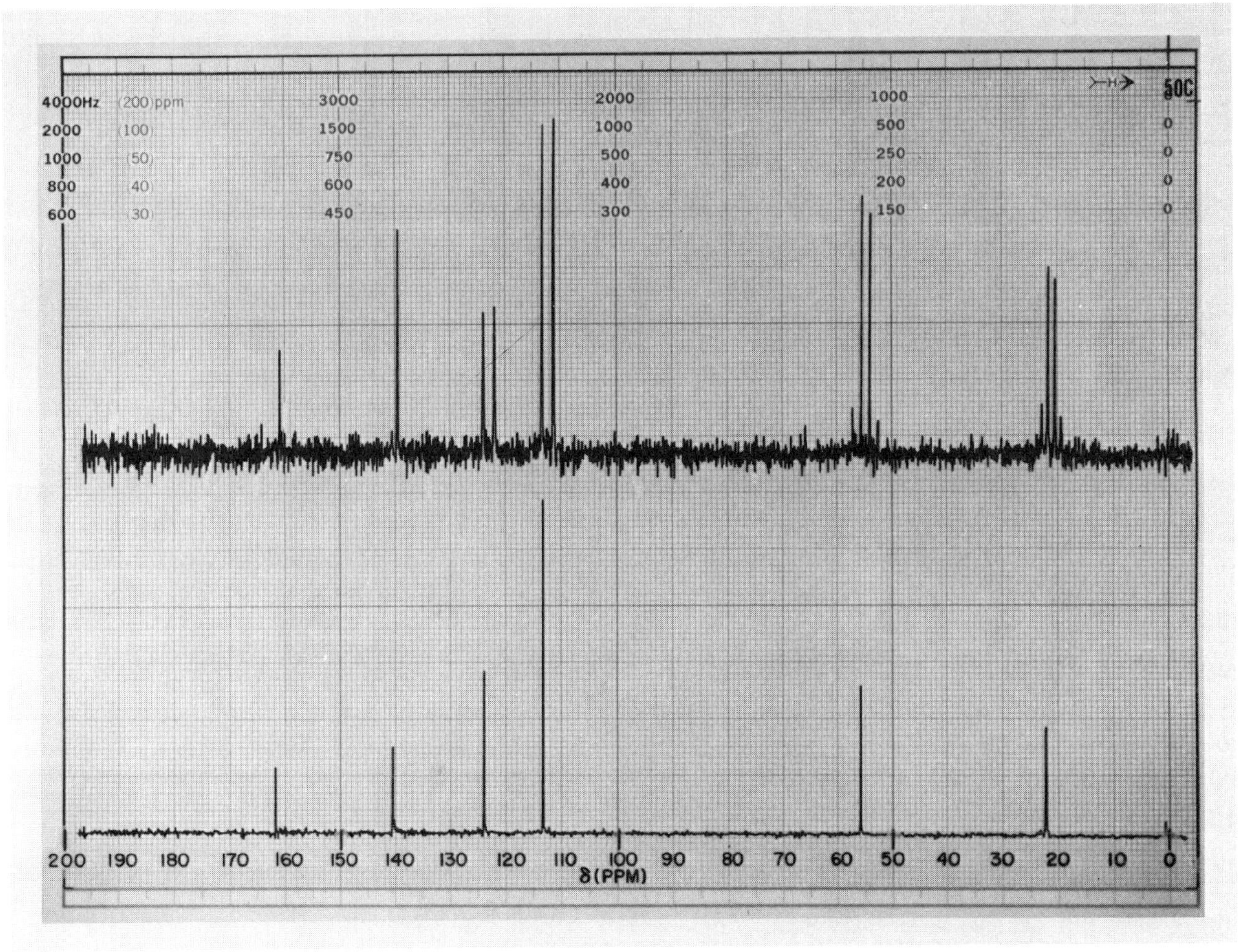

142 COMPOUND 50D

Compound 50D
MS: 136.090
IR: neat
HMR: CDCl$_3$
CMR: neat
Analysis: 79.4% C; 8.9% H

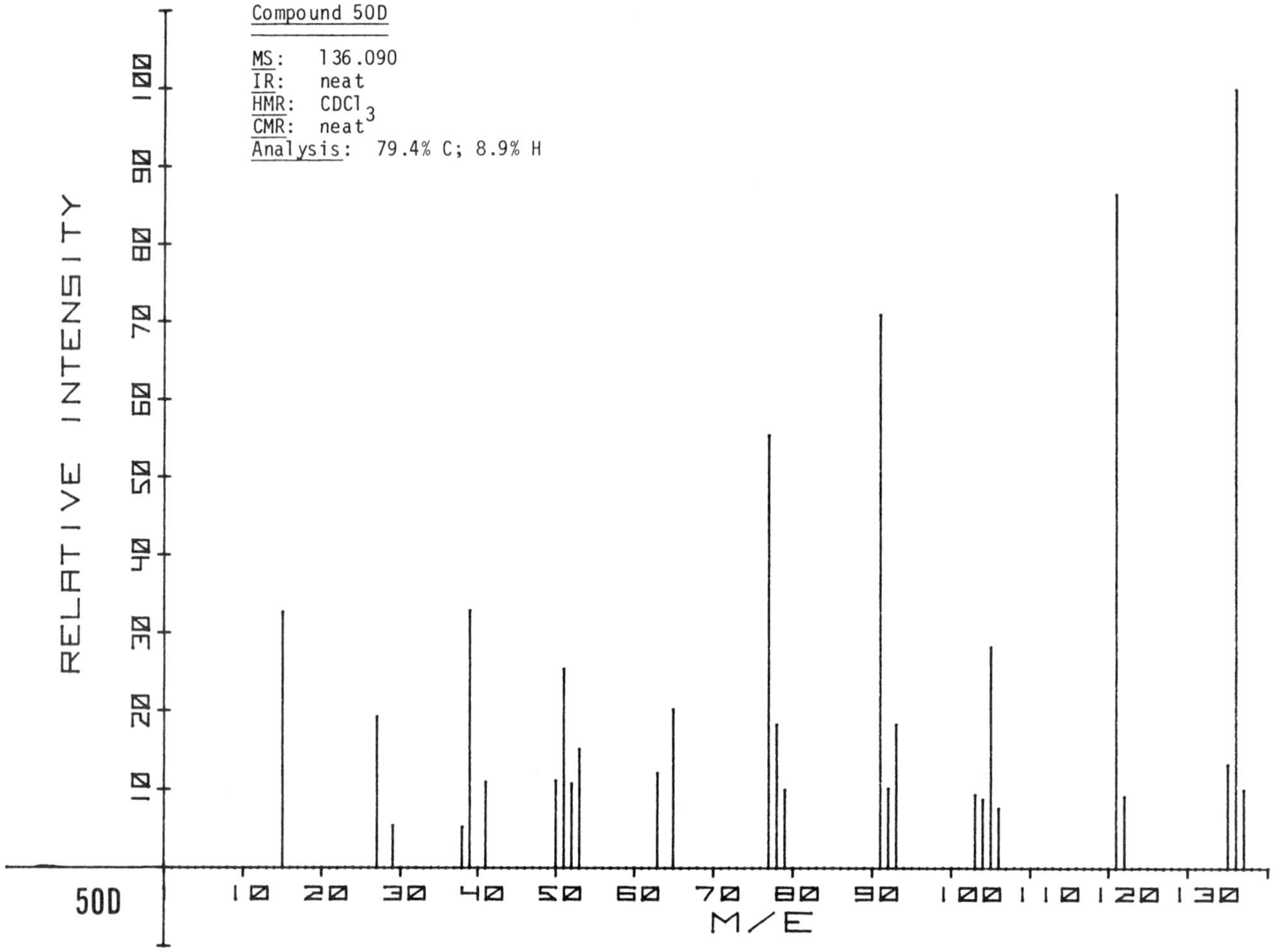

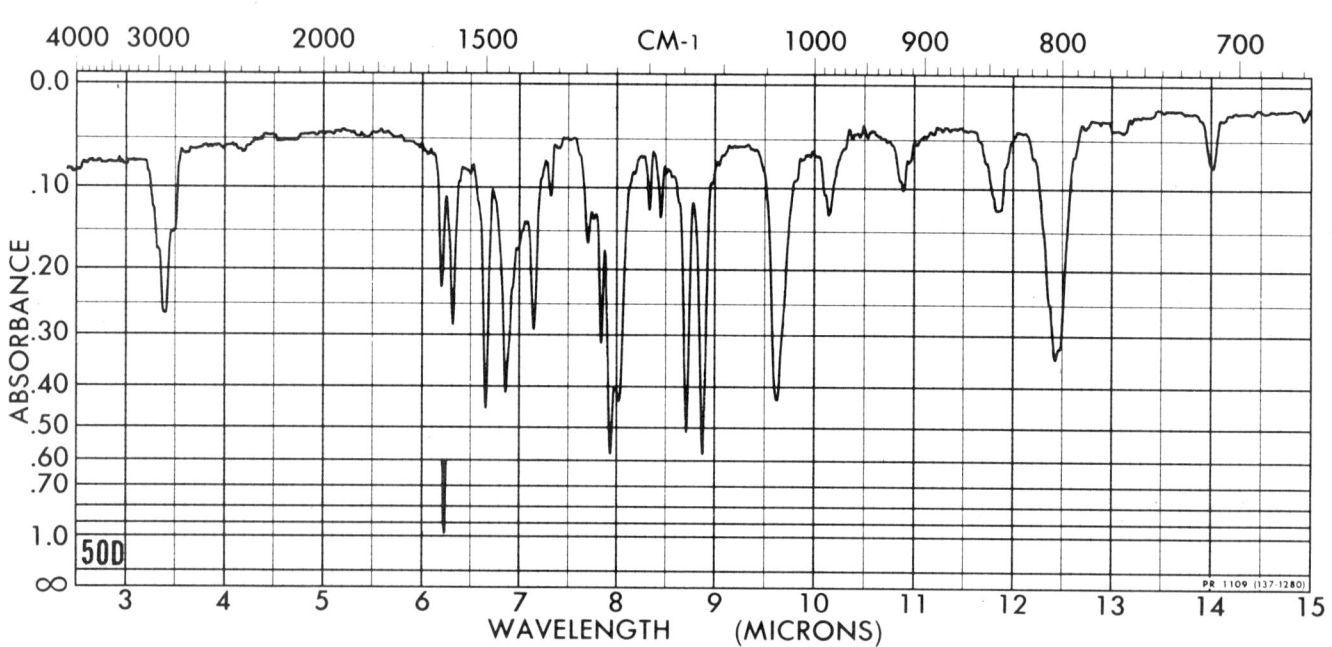

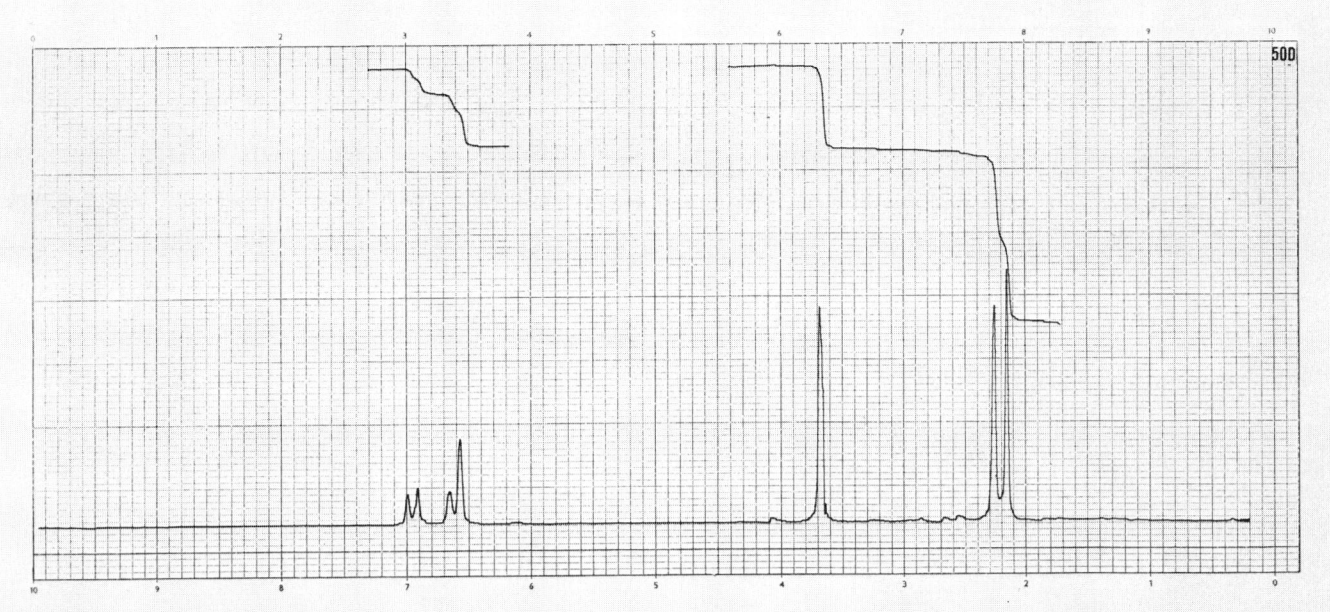

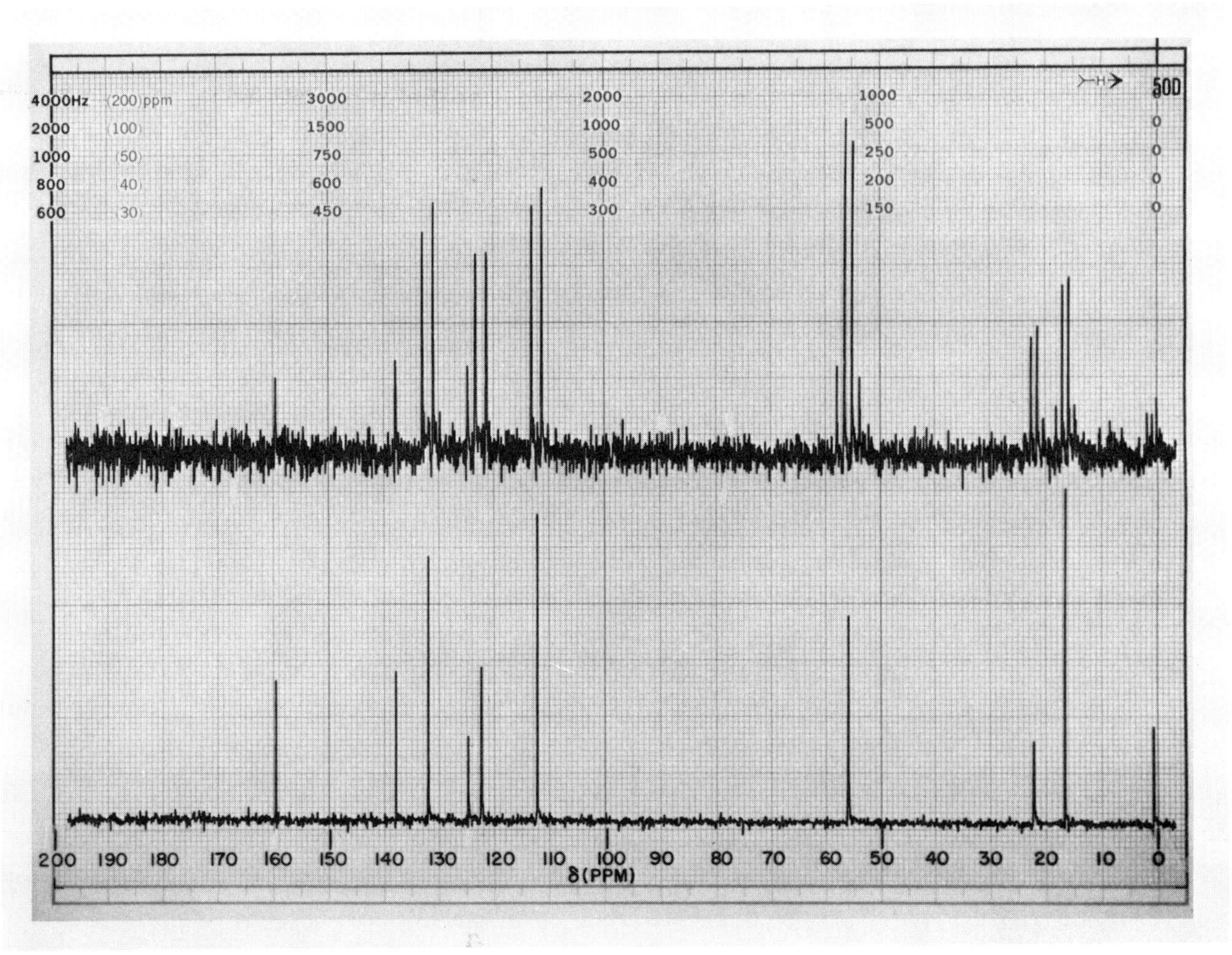

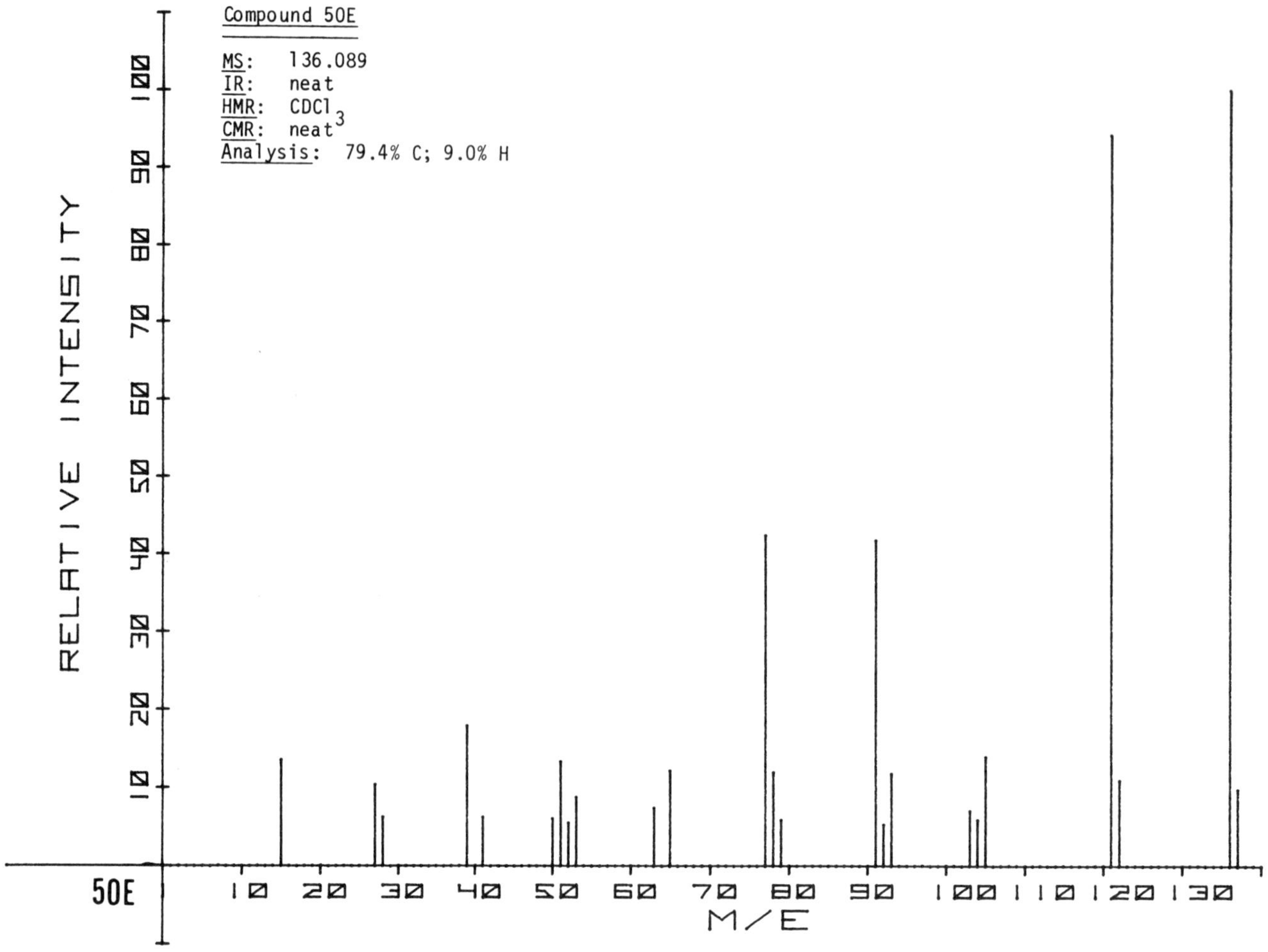

Compound 50E
MS: 136.089
IR: neat
HMR: $CDCl_3$
CMR: neat
Analysis: 79.4% C; 9.0% H
RELATIVE INTENSITY
50E
M/E

CM-1
ABSORBANCE
50E
WAVELENGTH (MICRONS)
PR 1109 (137-1280)

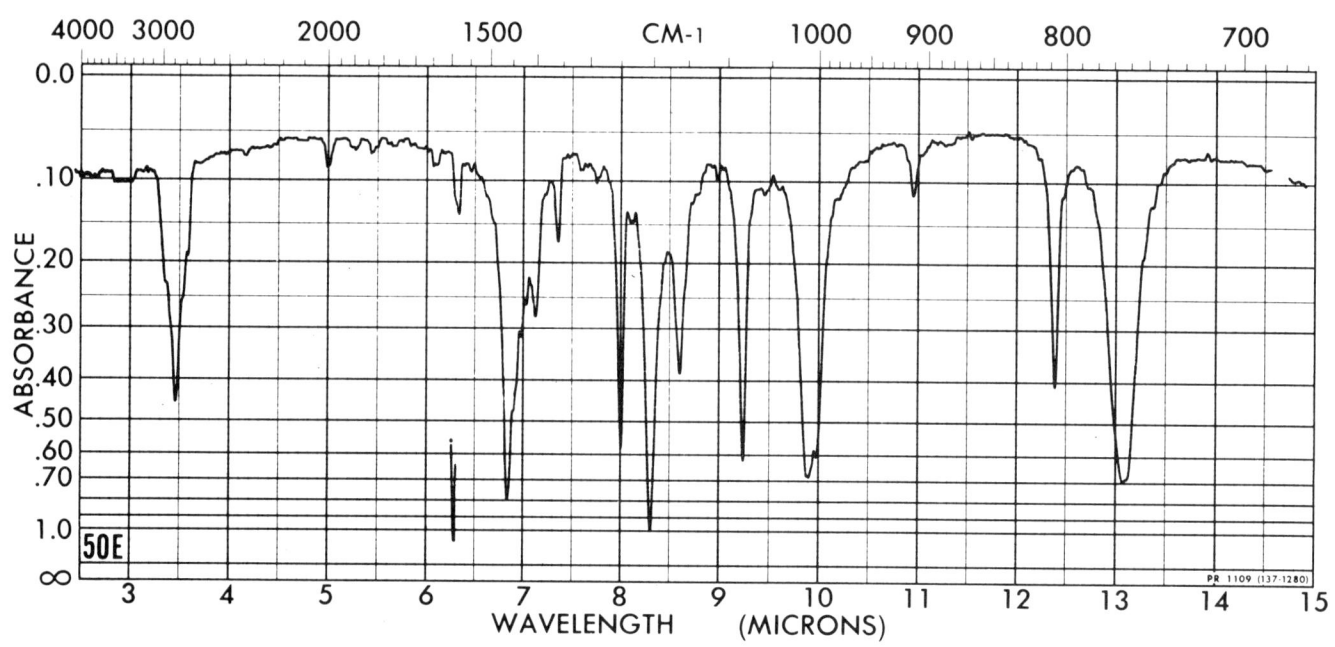

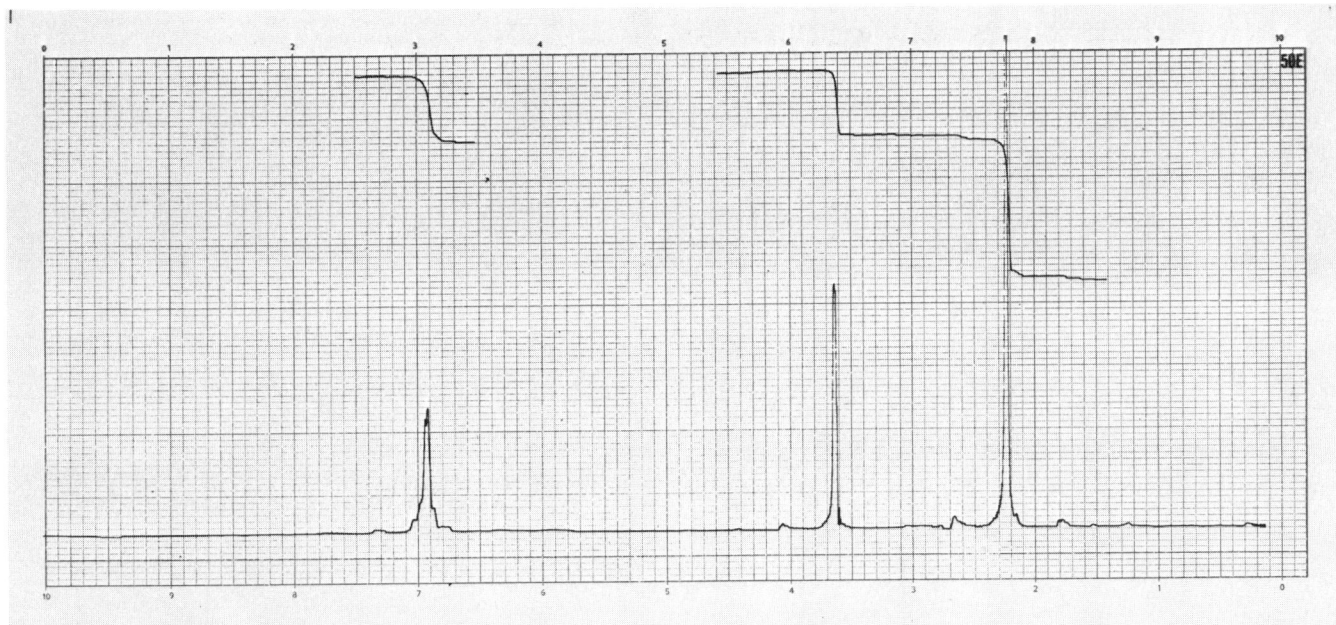

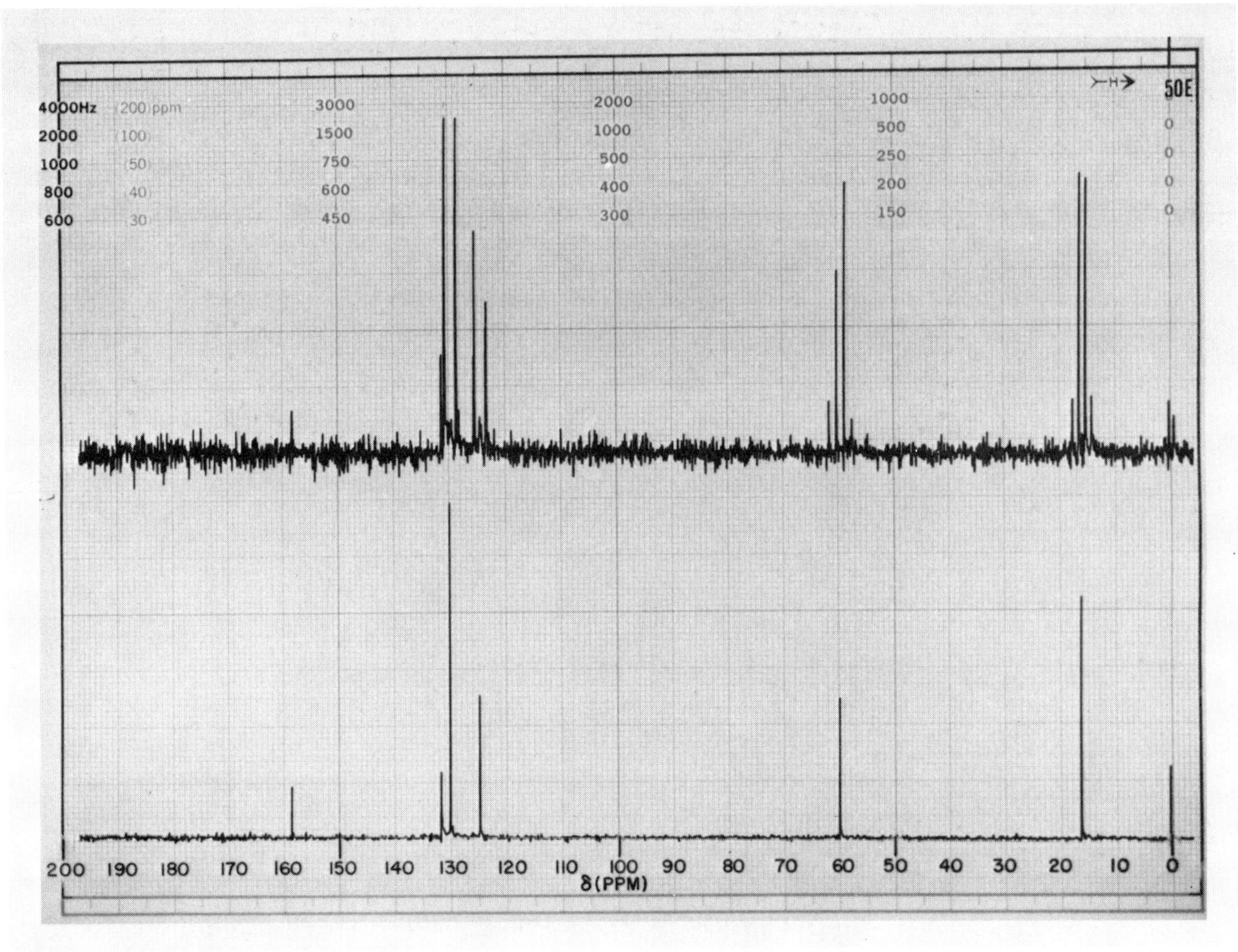

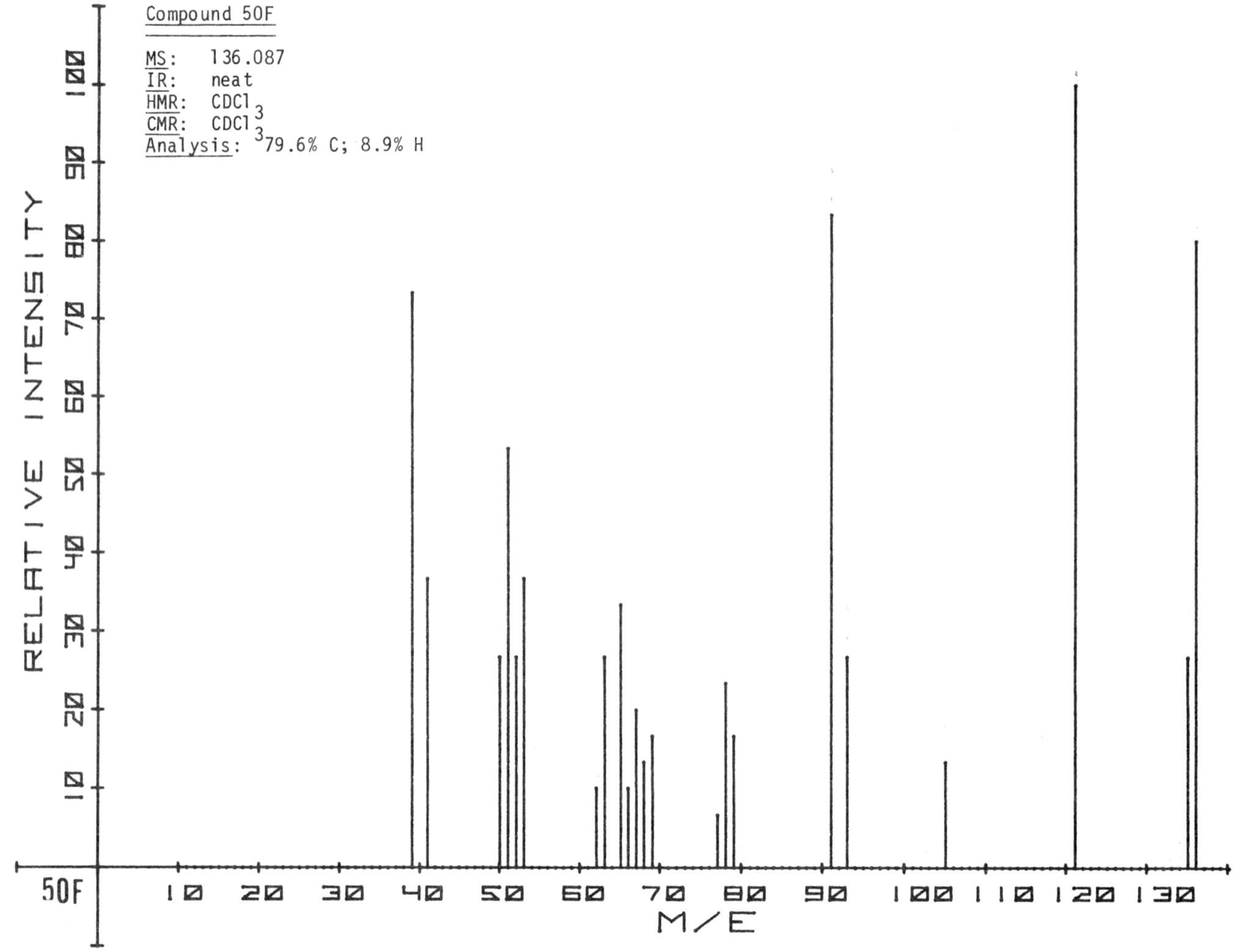

Compound 50F
MS: 136.087
IR: neat
HMR: CDCl3
CMR: CDCl3
Analysis: 79.6% C; 8.9% H
RELATIVE INTENSITY
50F
M/E

CM-1
ABSORBANCE
50F
WAVELENGTH (MICRONS)
PR 1109 (137-1280)

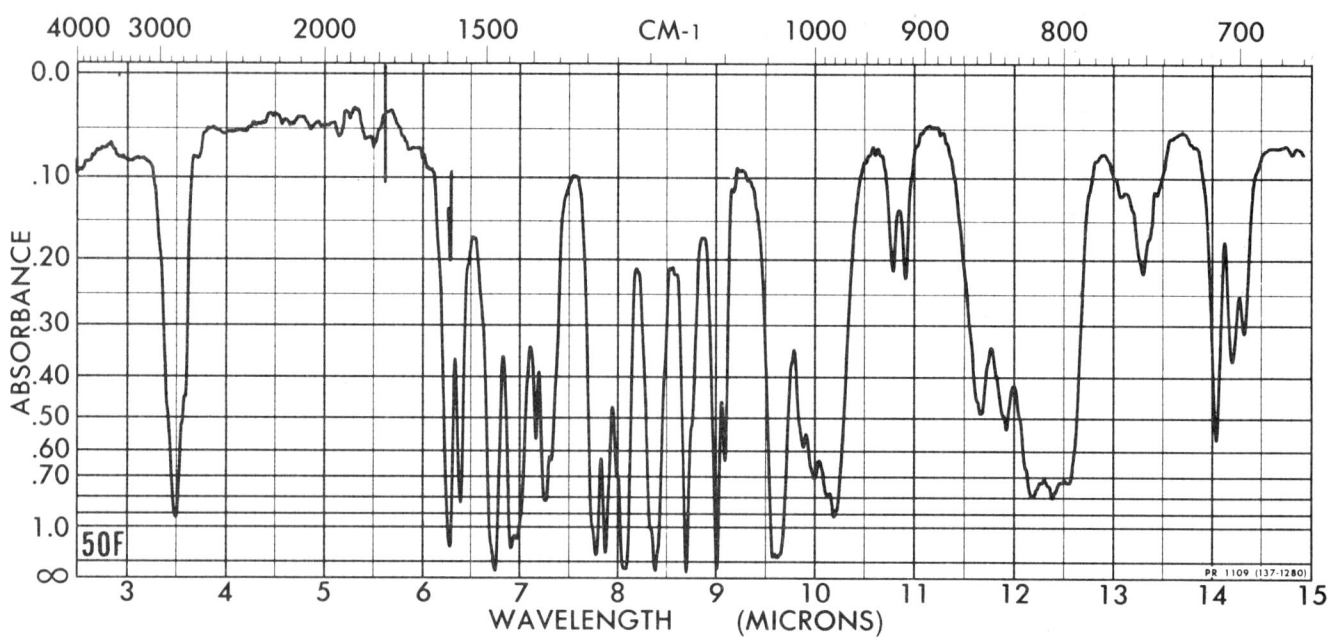

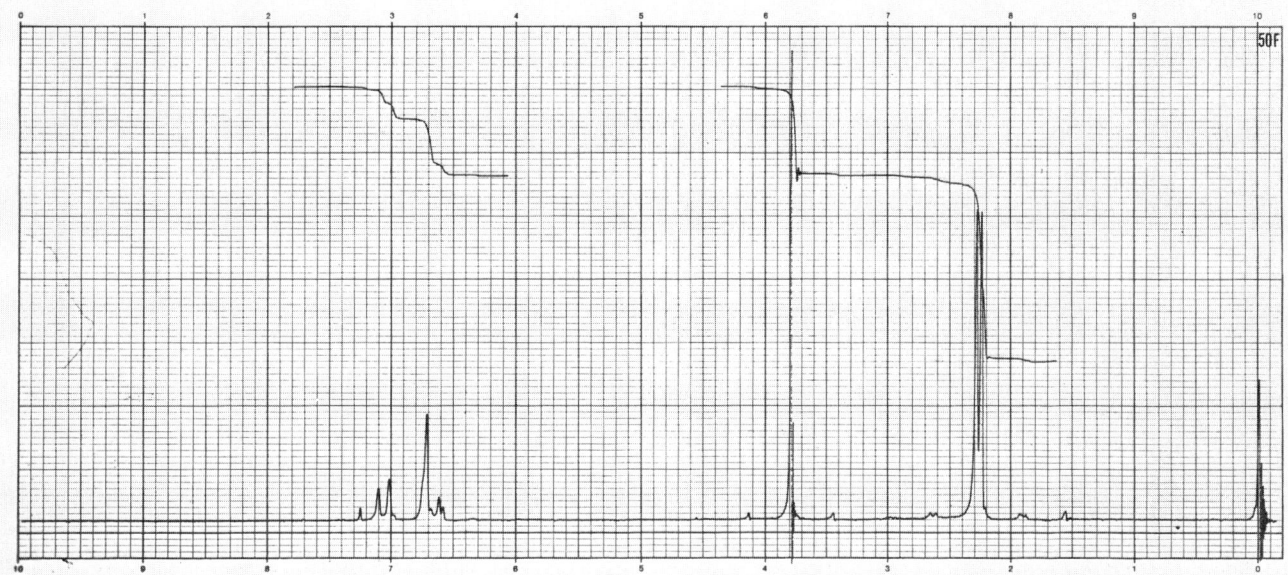

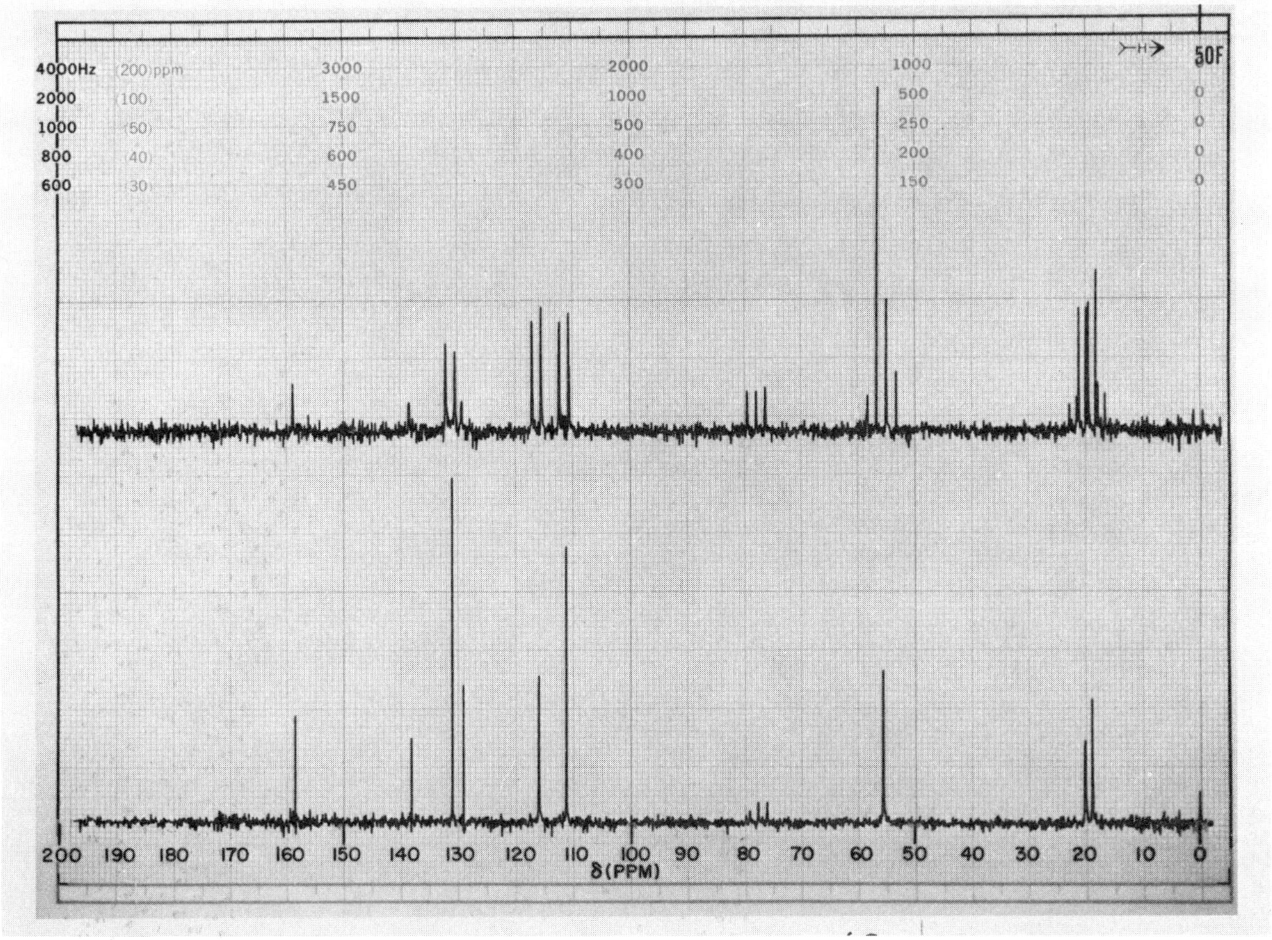

COMPOUND 51

Compound 51
MS: 168.078
IR: melt
HMR: CDCl_3
CMR: CDCl_3
Analysis: 64.3% C; 7.3% H

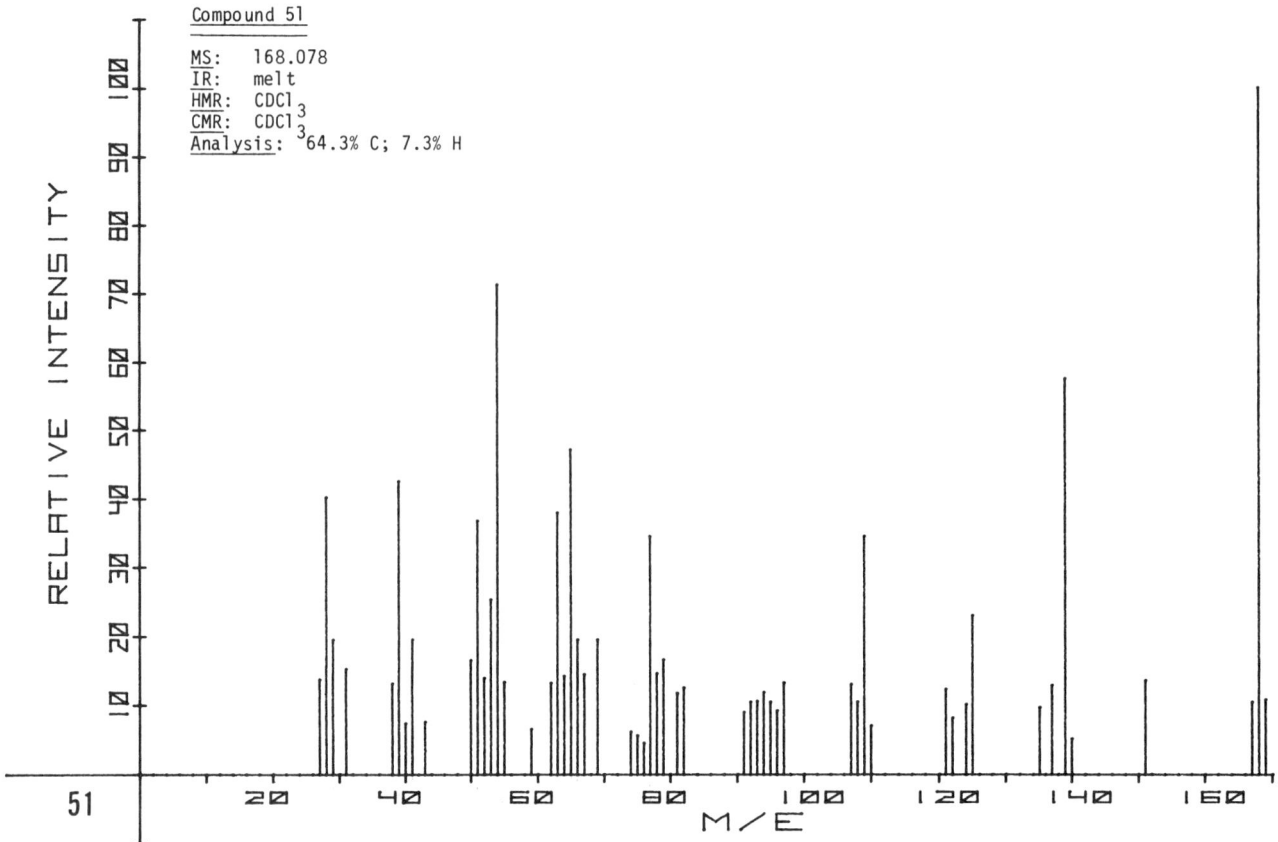

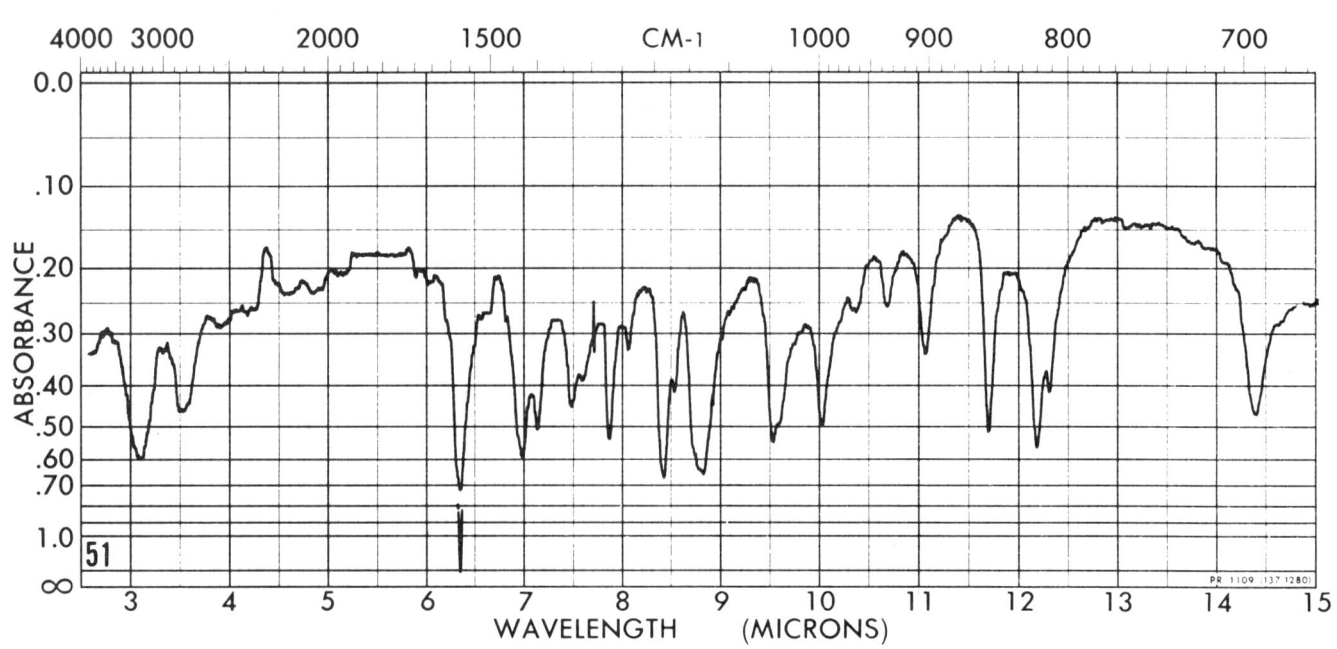

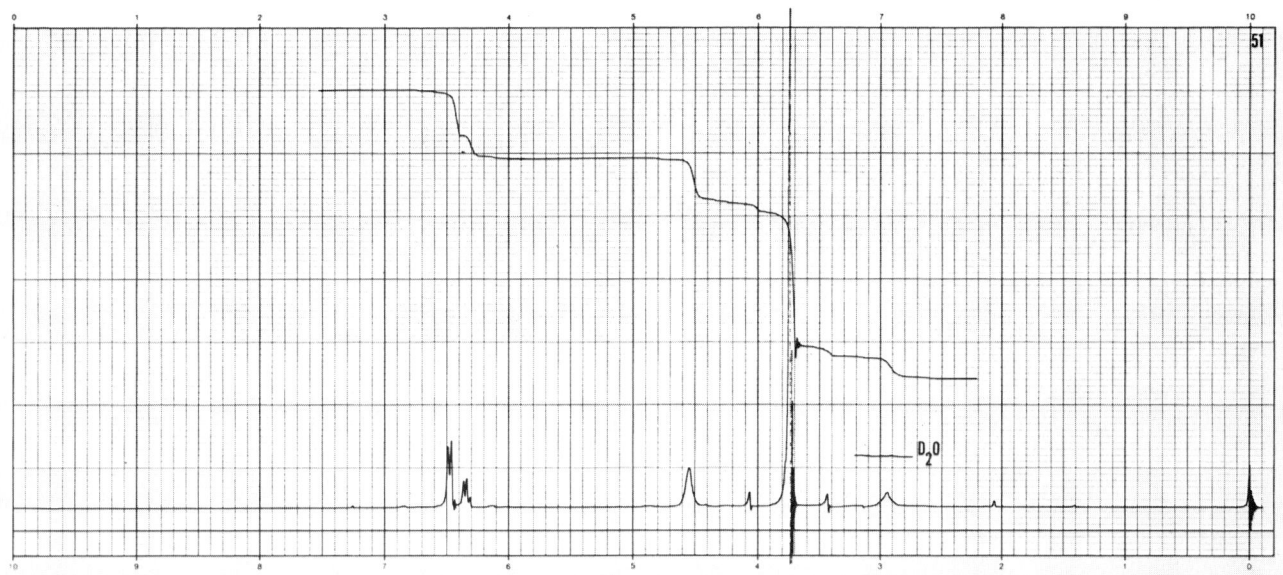

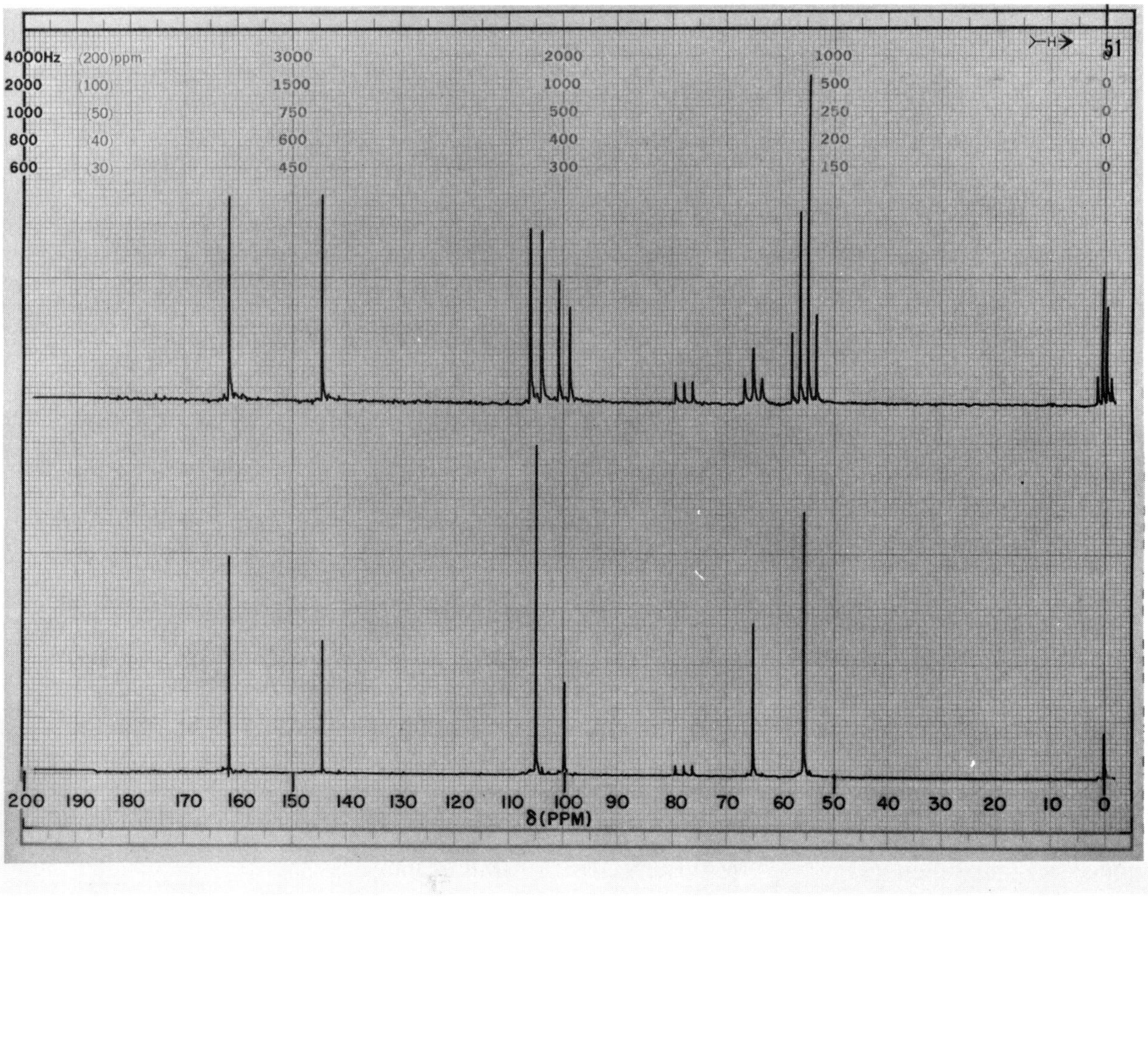

COMPOUND 52

Compound 52

MS: 135.105
IR: neat
HMR: CDCl$_3$
CMR: CDCl$_3$
Analysis: 80.0% C; 9.8% H; 10.4% N

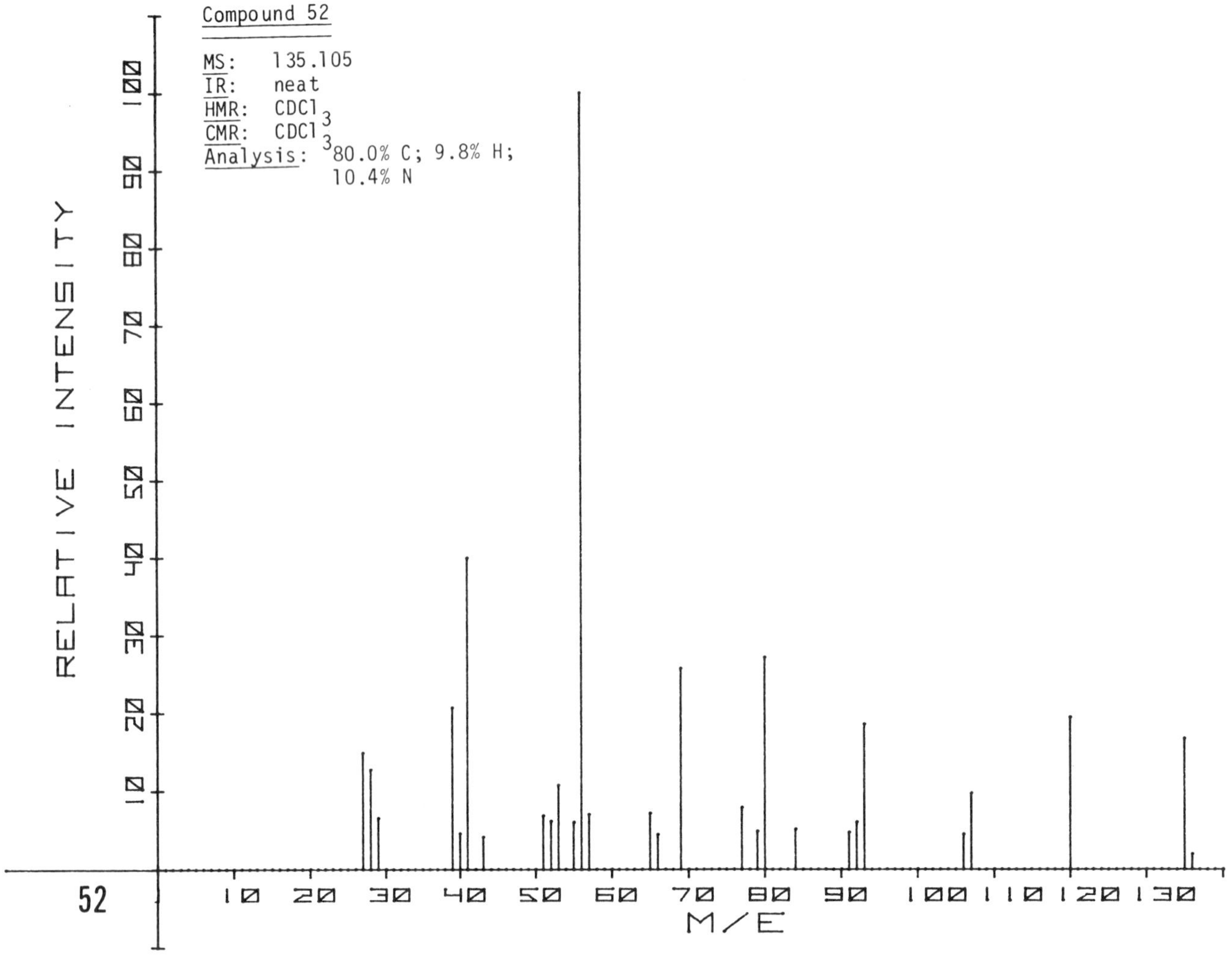

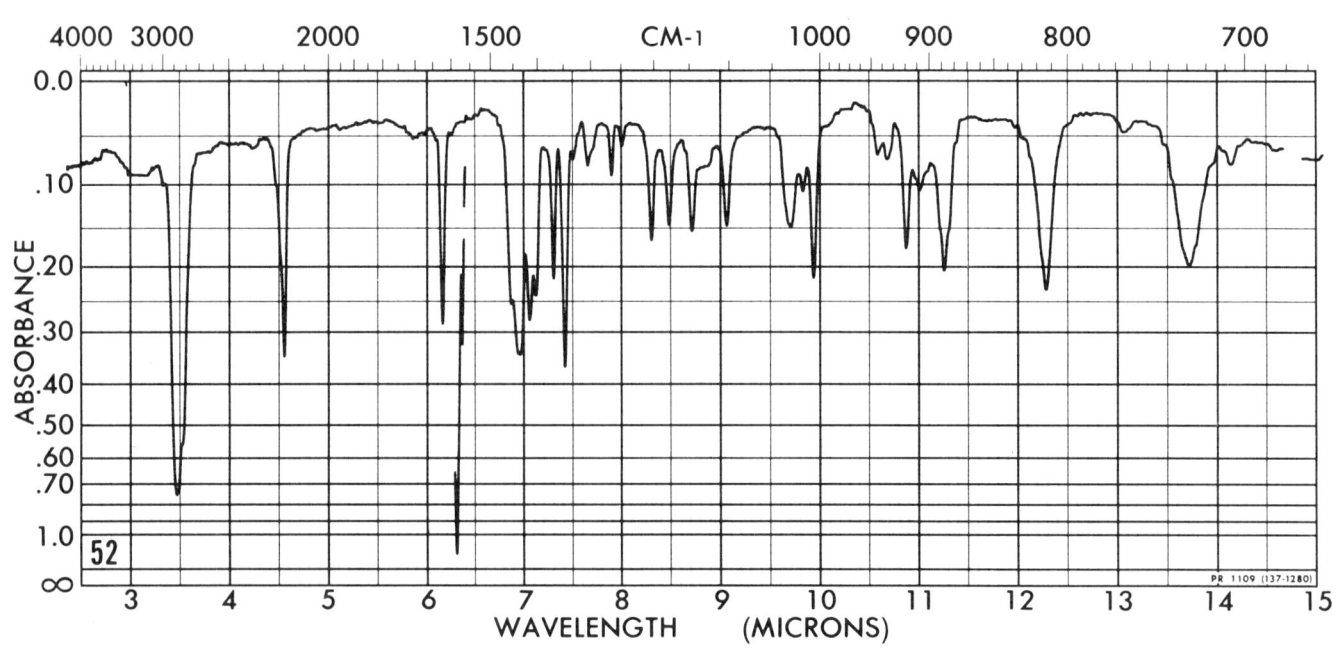

52
IRR. 2.25
4X
IRR. 2.25
4X

52
4000Hz (200)ppm 3000 2000 1000
2000 (100) 1500 1000 500
1000 750 500 250
800 600 400 200
600 450 300 150
Pk. Ht. PPM
22 144.6
24 143.1
12 119.2
25 111.4
27 39.3
24 34.1
17 29.0
60 28.3
35 27.9
59 27.5
15 26.8
19 24.6
Pk. Ht. PPM
17 143.9
4 119.3
8 111.4
26 39.4
24 34.1
60 27.9
26 24.6
200 190 180 170 160 150 140 130 120 110 100 90 80 70 60 50 40 30 20 10 0
δ(PPM)

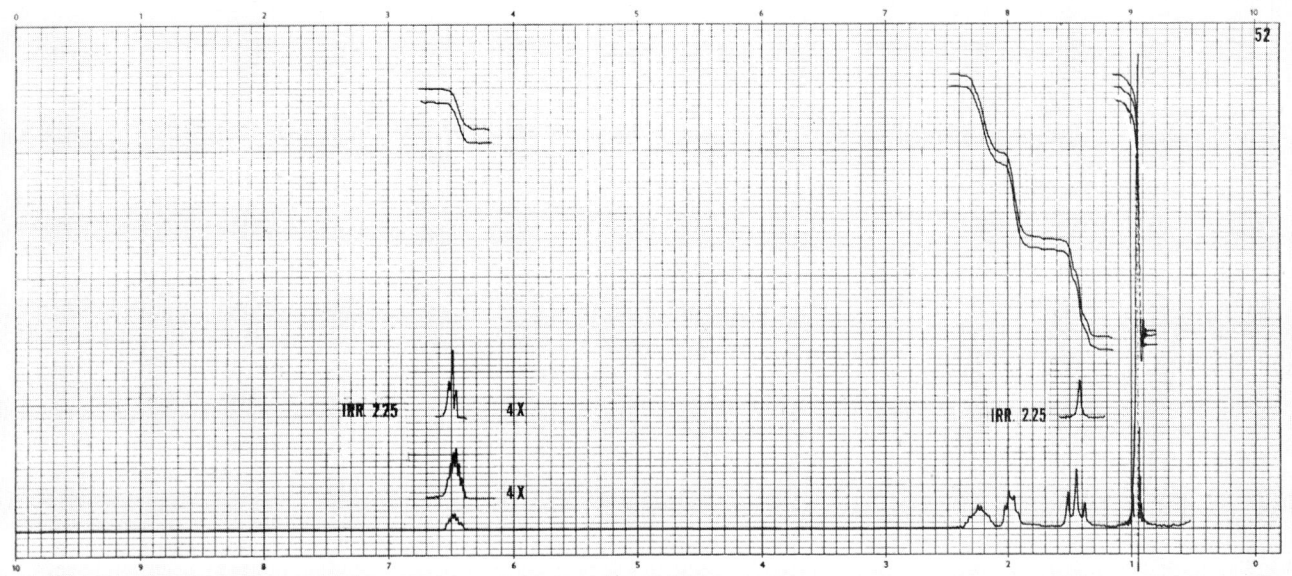

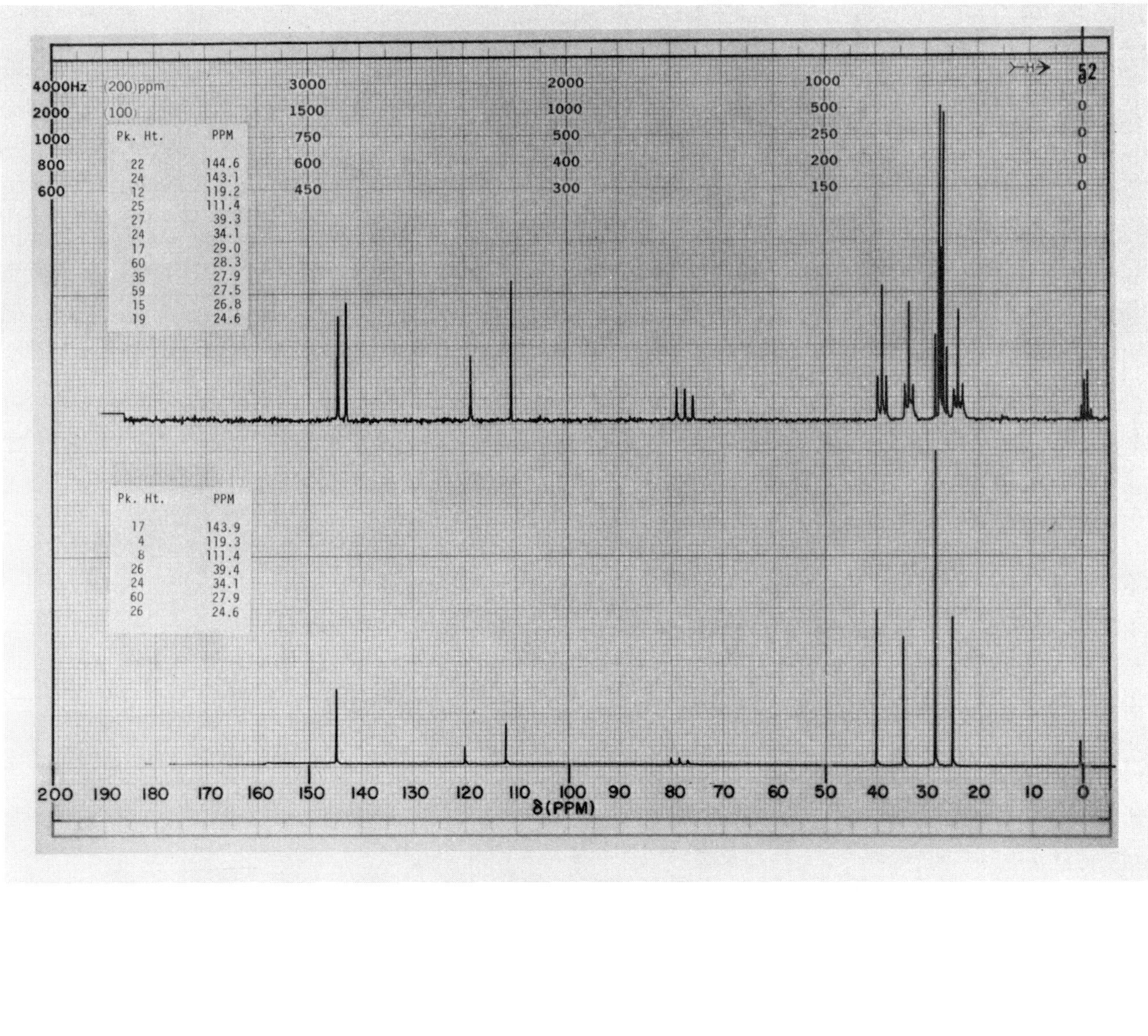

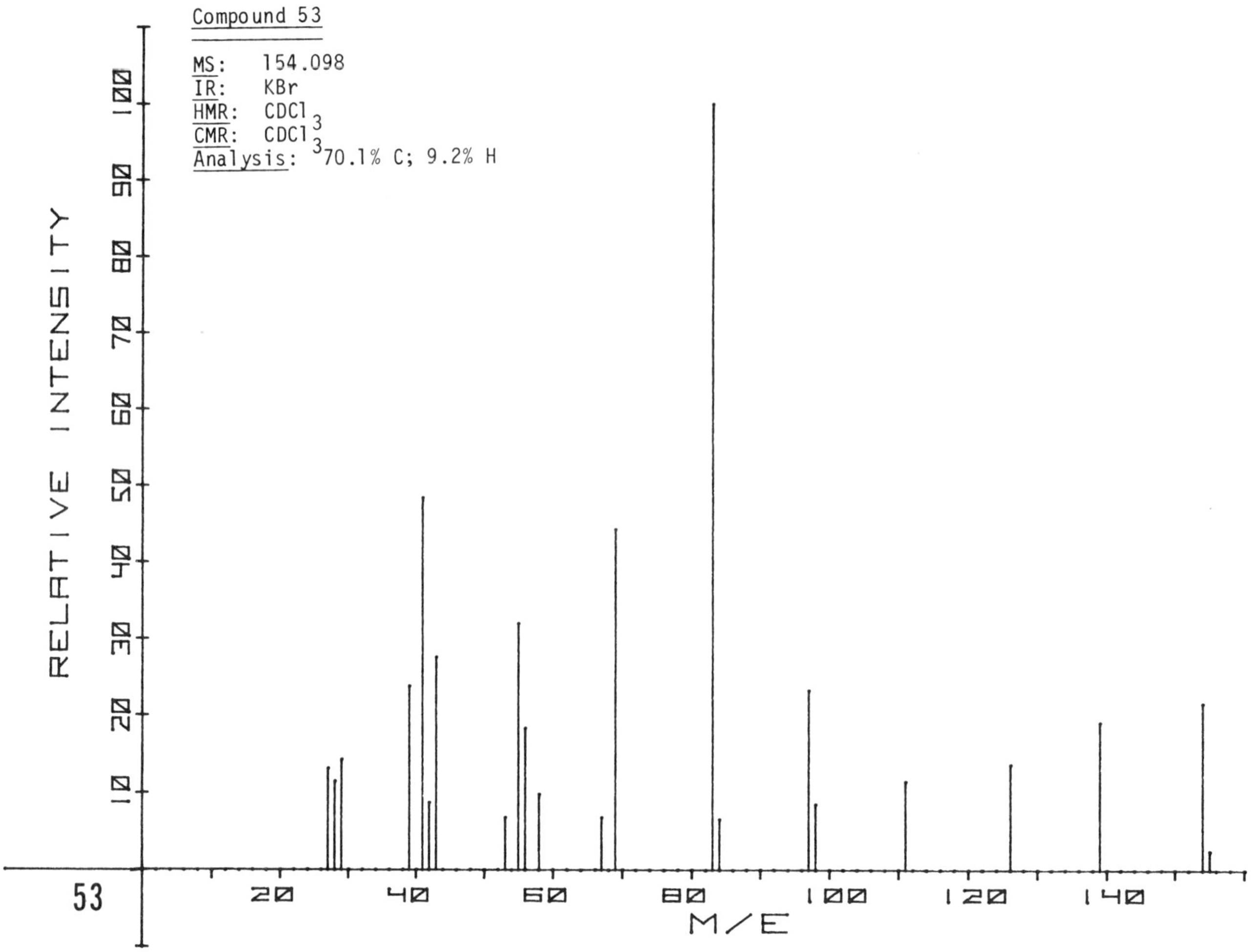

Compound 53
MS: 154.098
IR: KBr
HMR: CDCl3
CMR: CDCl3
Analysis: 70.1% C; 9.2% H
RELATIVE INTENSITY
10 20 30 40 50 60 70 80 90 100
20 40 60 80 100 120 140
M/E
53

4000 3000 2000 1500 CM-1 1000 900 800 700
ABSORBANCE
0.0 .10 .20 .30 .40 .50 .60 .70 1.0 ∞
53
3 4 5 6 7 8 9 10 11 12 13 14 15
WAVELENGTH (MICRONS)

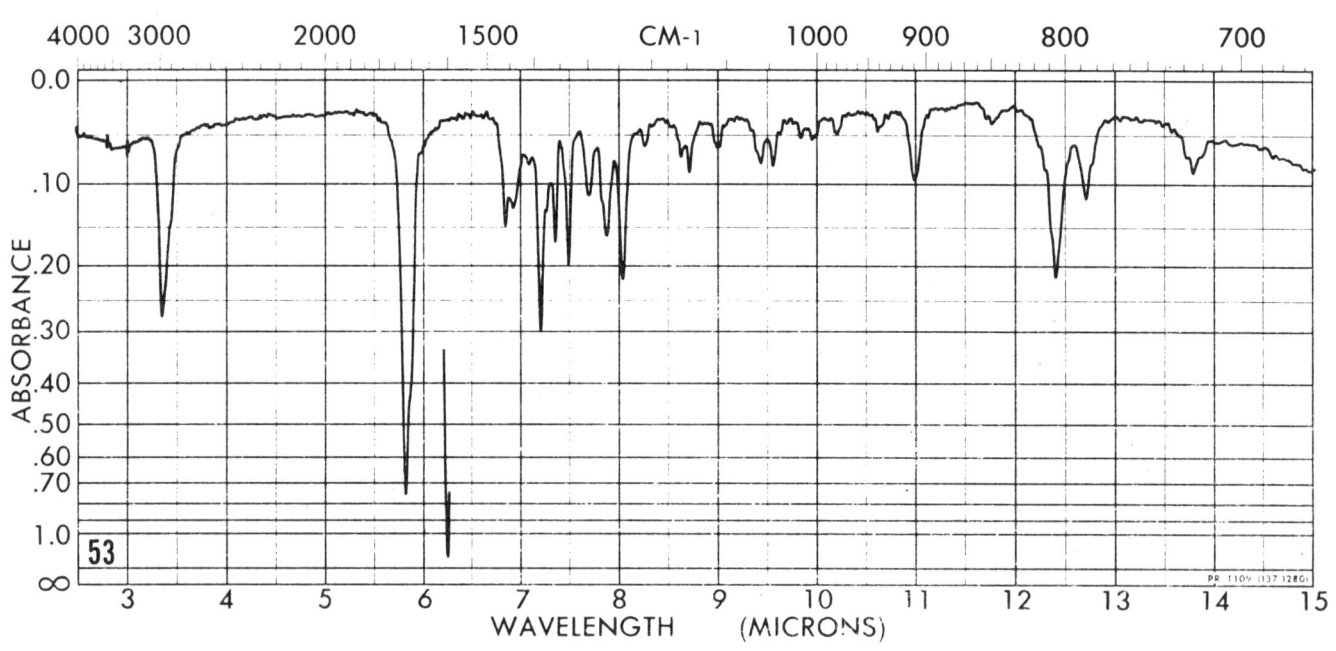

53

53
4000Hz (200)ppm 3000 2000 1000 0
2000 (100) 1500 500 0
1000 (50) 750 250 0
800 (40) 600 200 0
600 (30) 450 150 0
PK. HT. PPM
22 63.9
42 62.2
46 60.2
8 33.0
37 31.5
42 30.0
34 28.5
34 27.0
14 26.2
11 25.4
60 24.7
60 23.1
16 21.6
PK. HT. PPM
25 207.1
26 63.9
43 61.2
56 47.9
60 42.7
43 35.9
52 30.7
48 27.8
51 23.9
207.1
200 190 180 170 160 150 140 130 120 110 100 90 80 70 60 50 40 30 20 10 0
δ(PPM)

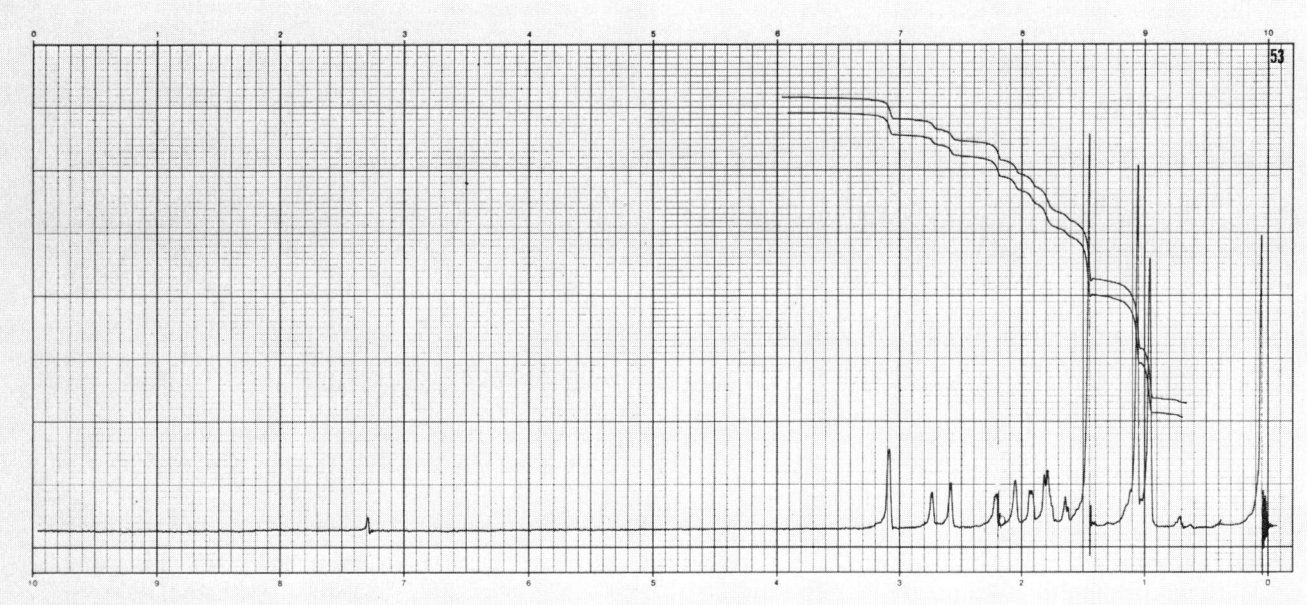

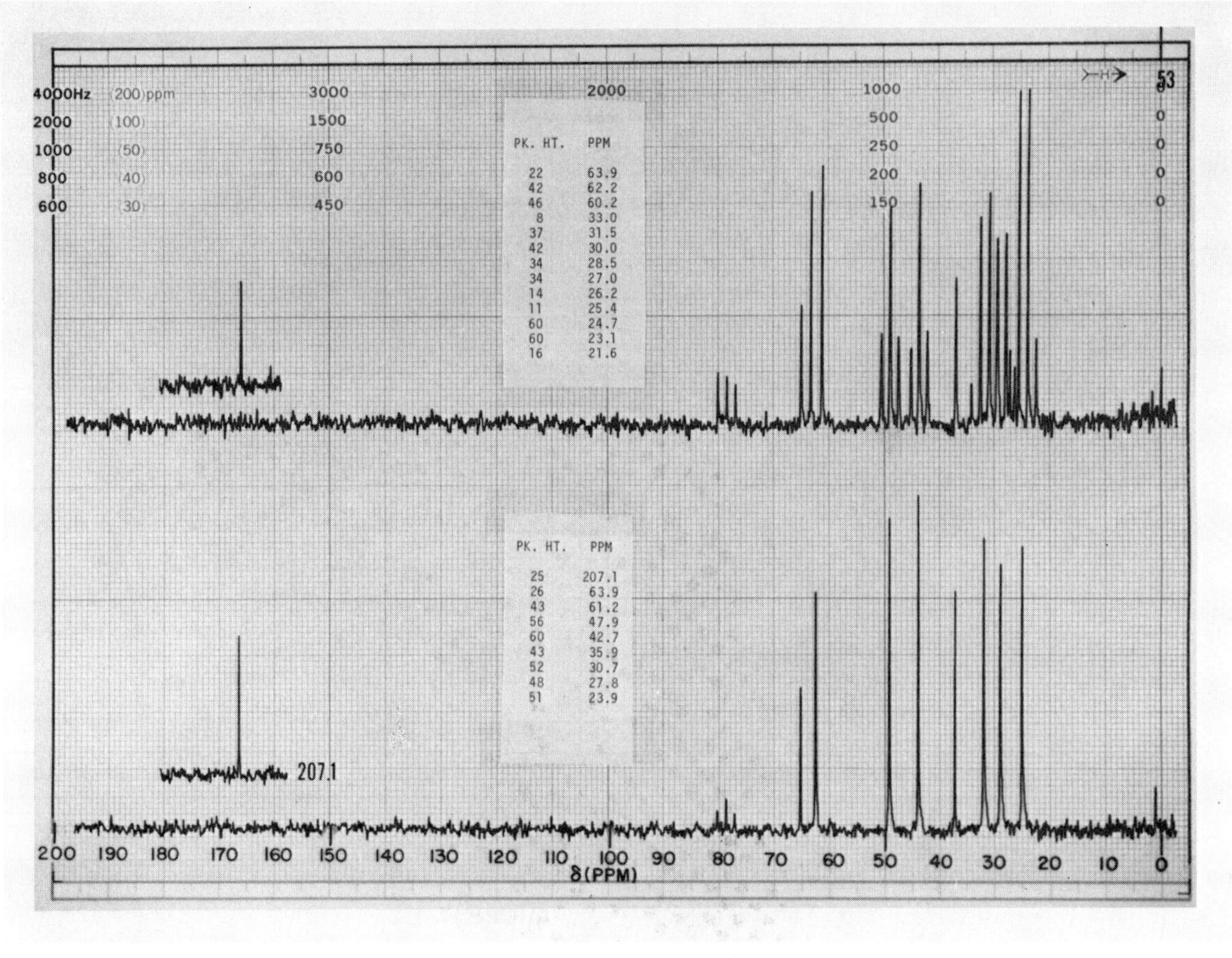

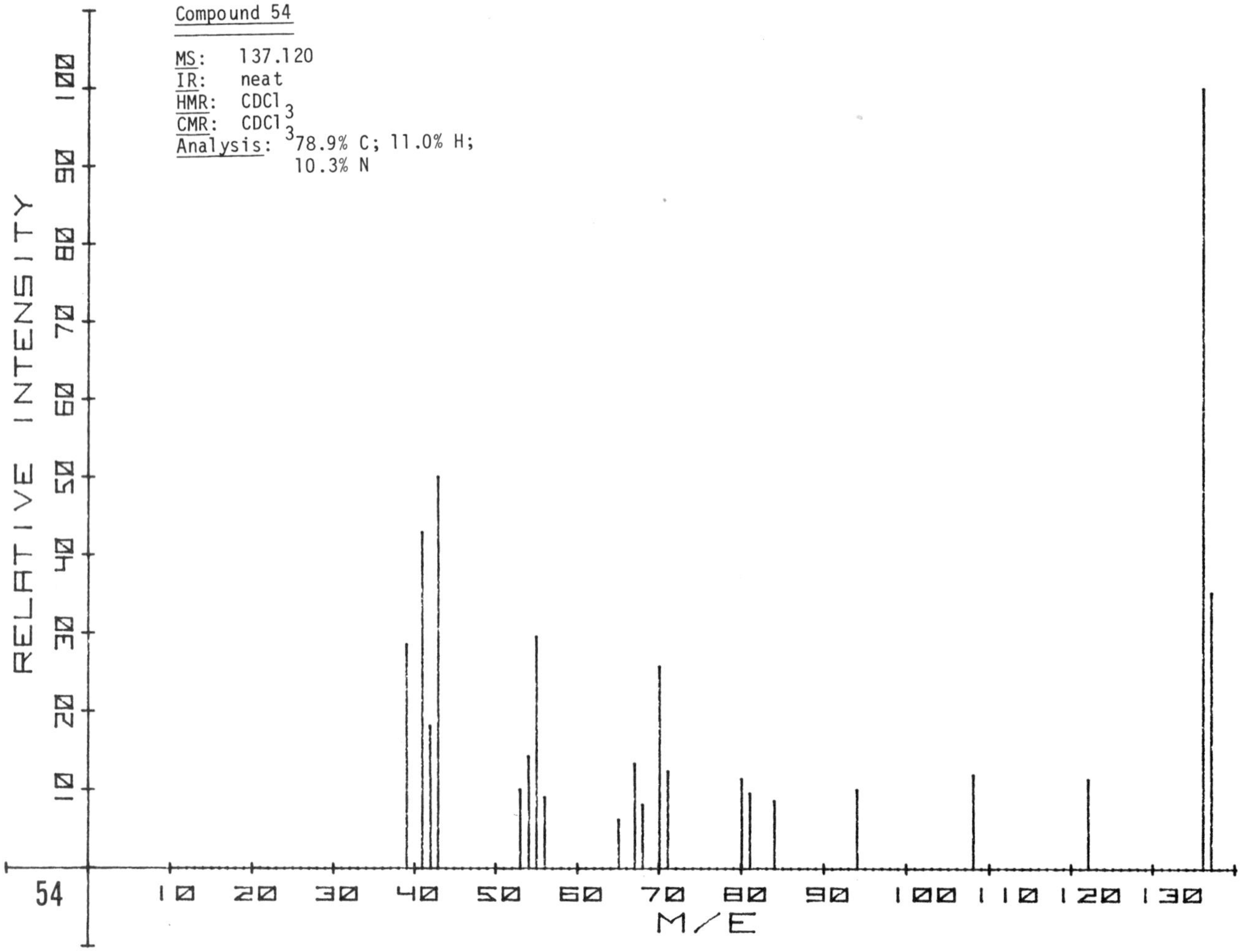

Compound 54
MS: 137.120
IR: neat
HMR: CDCl3
CMR: CDCl3
Analysis: 78.9% C; 11.0% H; 10.3% N
RELATIVE INTENSITY
10 20 30 40 50 60 70 80 90 100
54
10 20 30 40 50 60 70 80 90 100 110 120 130
M/E

4000 3000 2000 1500 CM-1 1000 900 800 700
ABSORBANCE
0.0 .10 .20 .30 .40 .50 .60 .70 1.0 ∞
54
PR 1109 (137-1280)
3 4 5 6 7 8 9 10 11 12 13 14 15
WAVELENGTH (MICRONS)

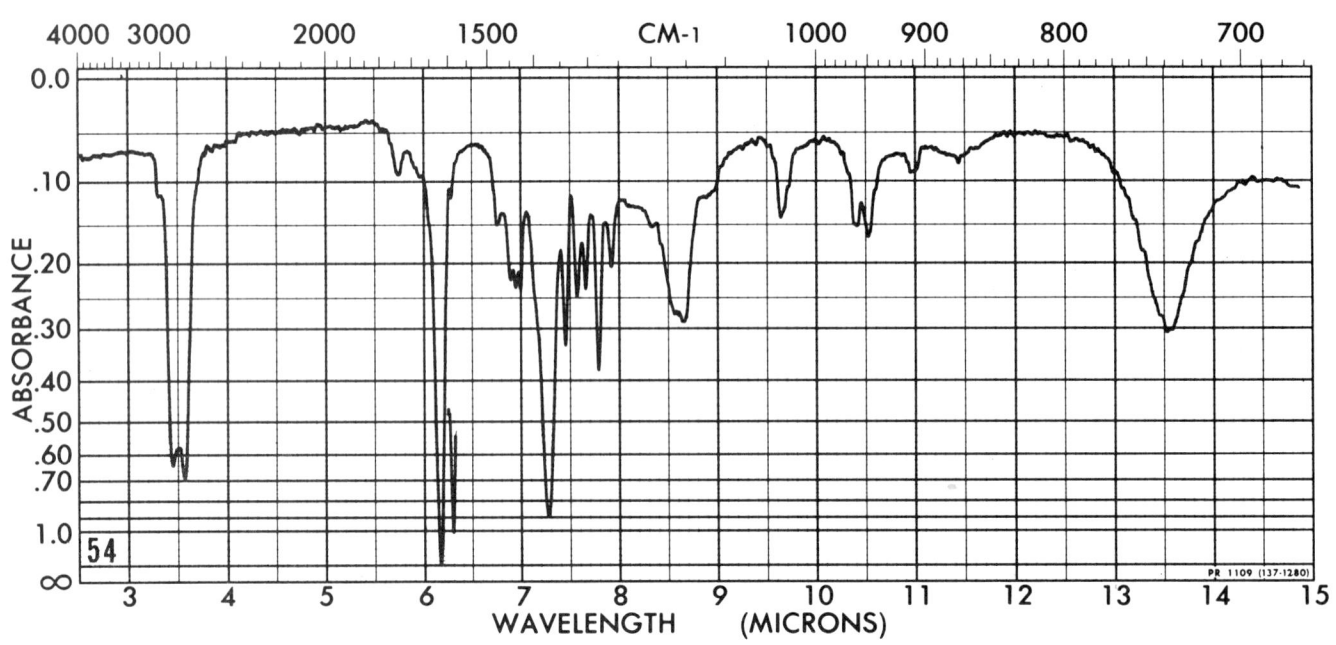

54
IRR. 1.81
IRR. 3.0
A
A

54
4000Hz (200)ppm
2000 (100)
1000 (50)
800 (40)
600 (30)
2000 1000 500 400 300
1000 500 250 200 150
PK. HT. PPM
6 29.6
17 27.7
17 26.6
17 25.7
15 25.5
15 25.3
60 24.6
19 23.5
21 23.2
22 22.7
8 21.6
9 21.2
PK. HT. PPM
5 143.3
17 93.7
45 47.5
18 27.7
60 24.7
20 23.5
26 23.2
200 190 180 170 160 150 140 130 120 110 100 90 80 70 60 50 40 30 20 10 0
δ(PPM)

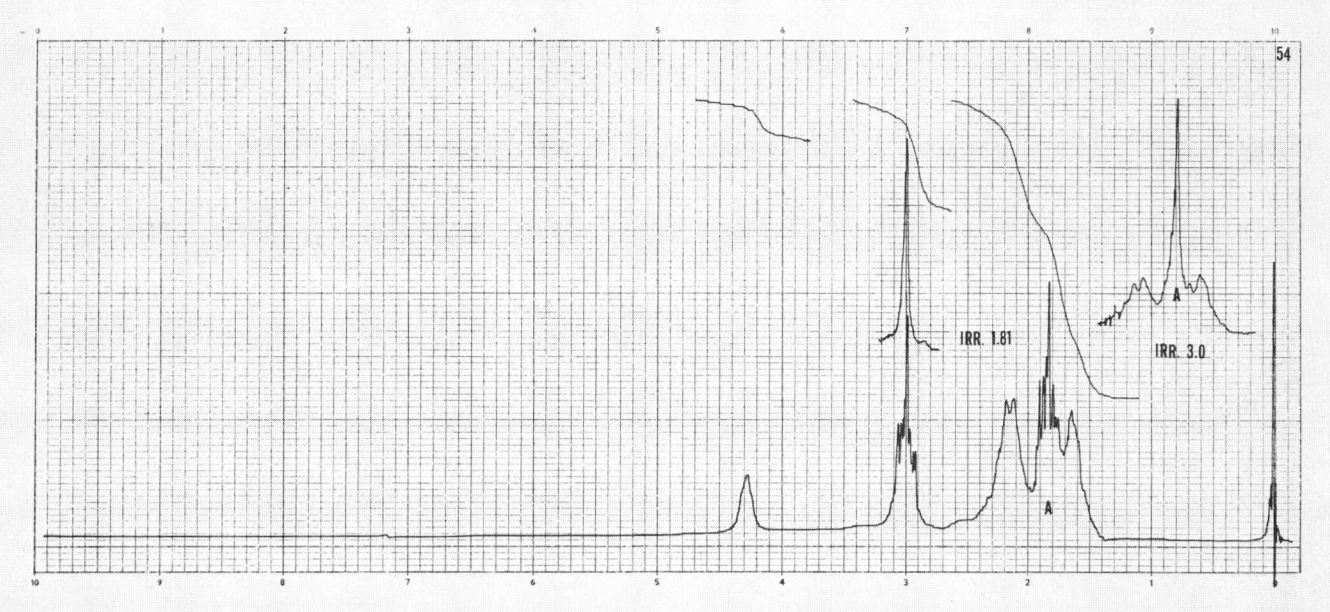

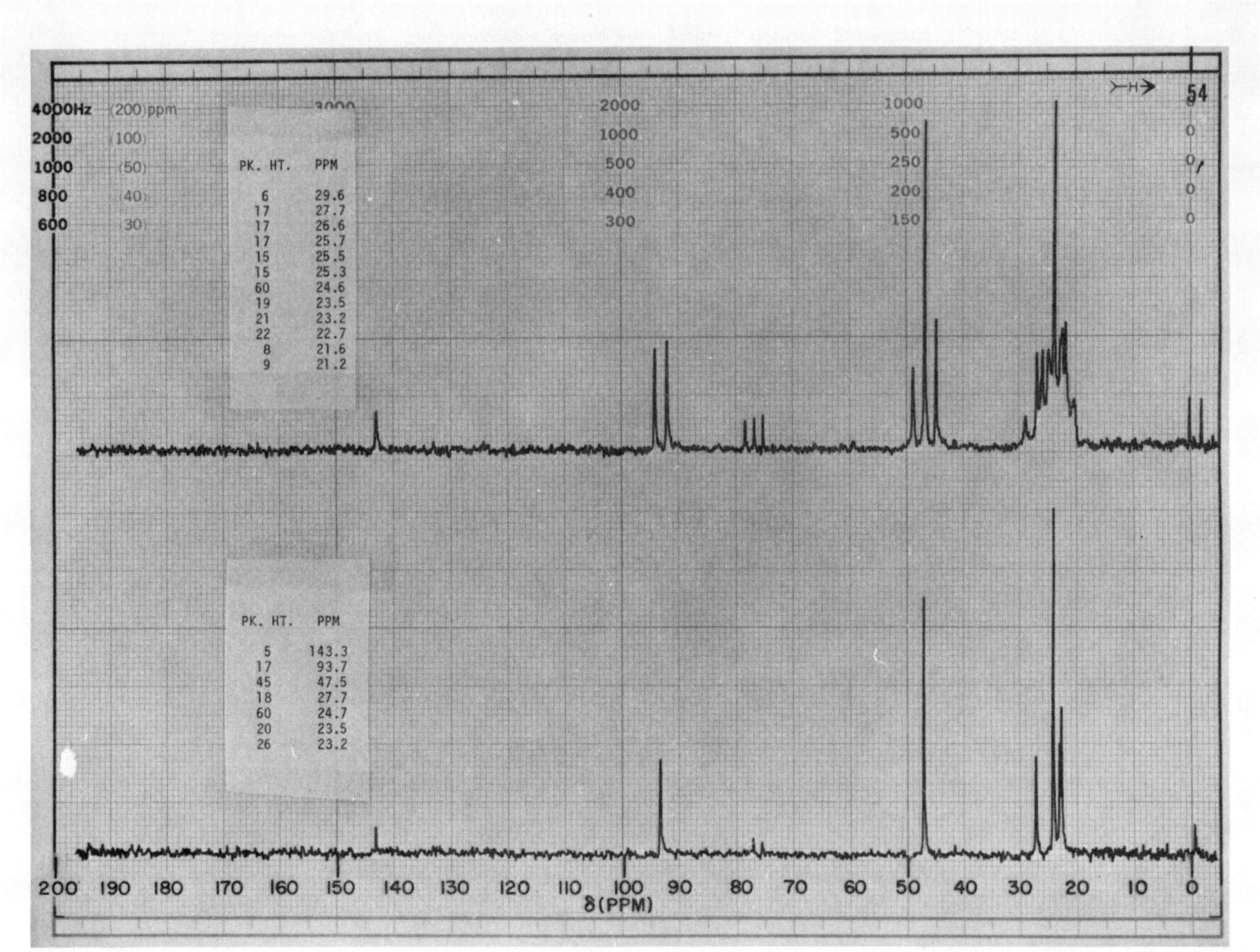

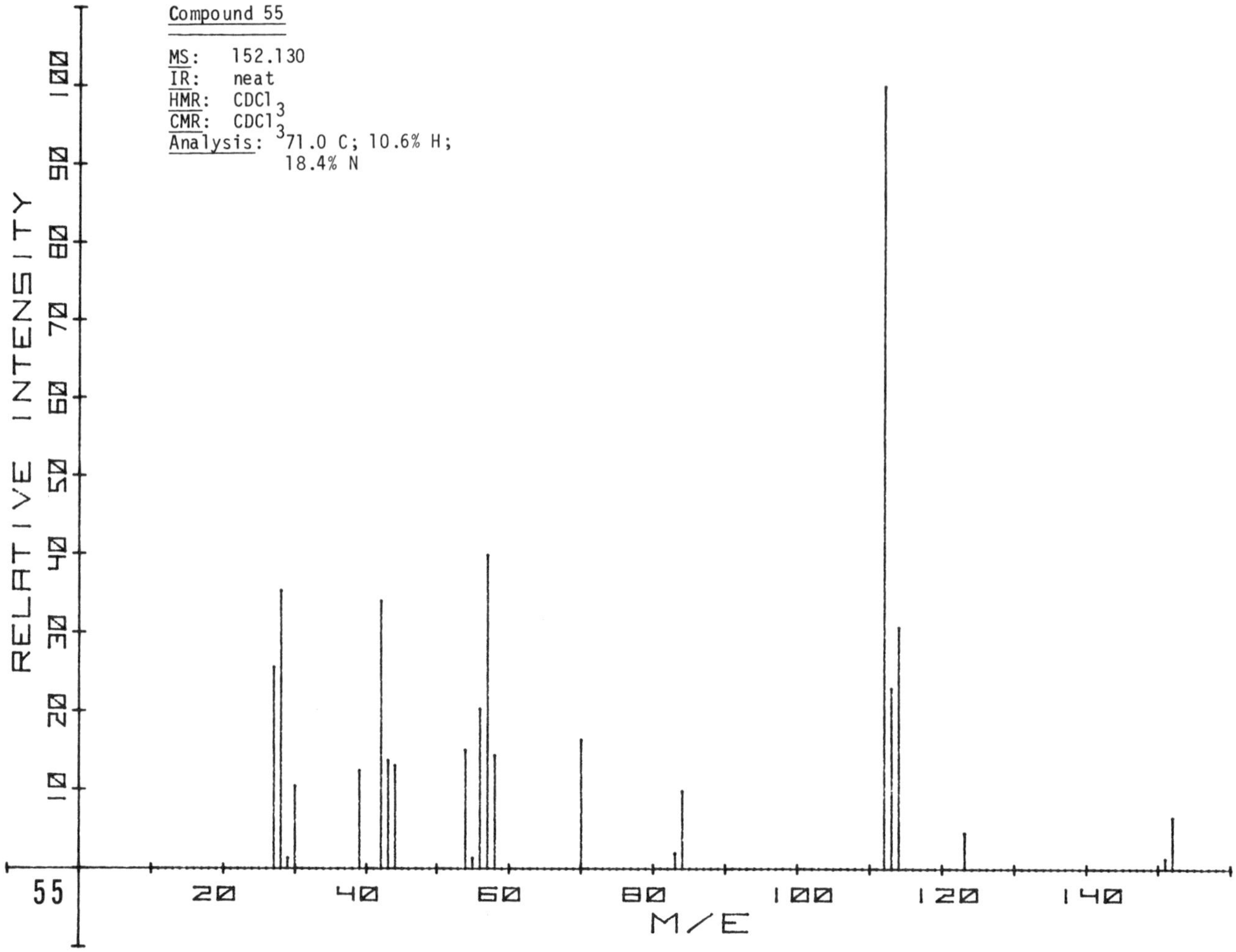

Compound 55
MS: 152.130
IR: neat
HMR: CDCl3
CMR: CDCl3
Analysis: 71.0 C; 10.6% H; 18.4% N
RELATIVE INTENSITY
10
20
30
40
50
60
70
80
90
100
55
20
40
60
80
100
120
140
M/E

4000
3000
2000
1500
CM-1
1000
900
800
700
0.0
.10
.20
.30
.40
.50
.60
.70
1.0
∞
ABSORBANCE
55
3
4
5
6
7
8
9
10
11
12
13
14
15
WAVELENGTH (MICRONS)

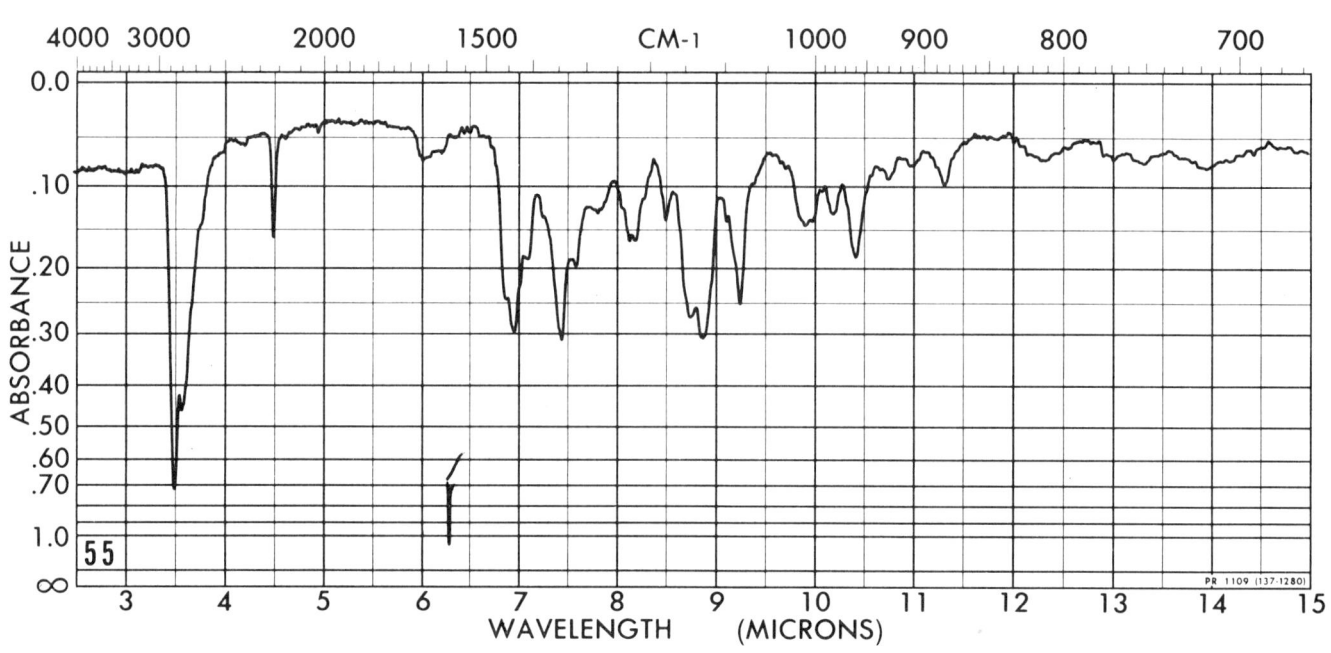

55

55
4000Hz (200)ppm 3000 2000 1000
2000 (100) 1500 1000 500 0
1000 (50) 750 500 250 0
800 (40) 600 400 200 0
600 (30) 450 300 150 0
Pk. Ht. PPM
11 119.2
17 56.8
13 55.6
54 54.9
33 53.6
31 52.9
11 51.6
17 30.5
18 29.1
60 28.5
42 27.0
25 26.5
15 24.9
19 18.6
45 16.6
15 14.5
Pk. Ht. PPM
6 119.2
60 54.9
34 53.6
57 28.5
54 27.0
30 16.6
200 190 180 170 160 150 140 130 120 110 100 90 80 70 60 50 40 30 20 10 0
δ(PPM)

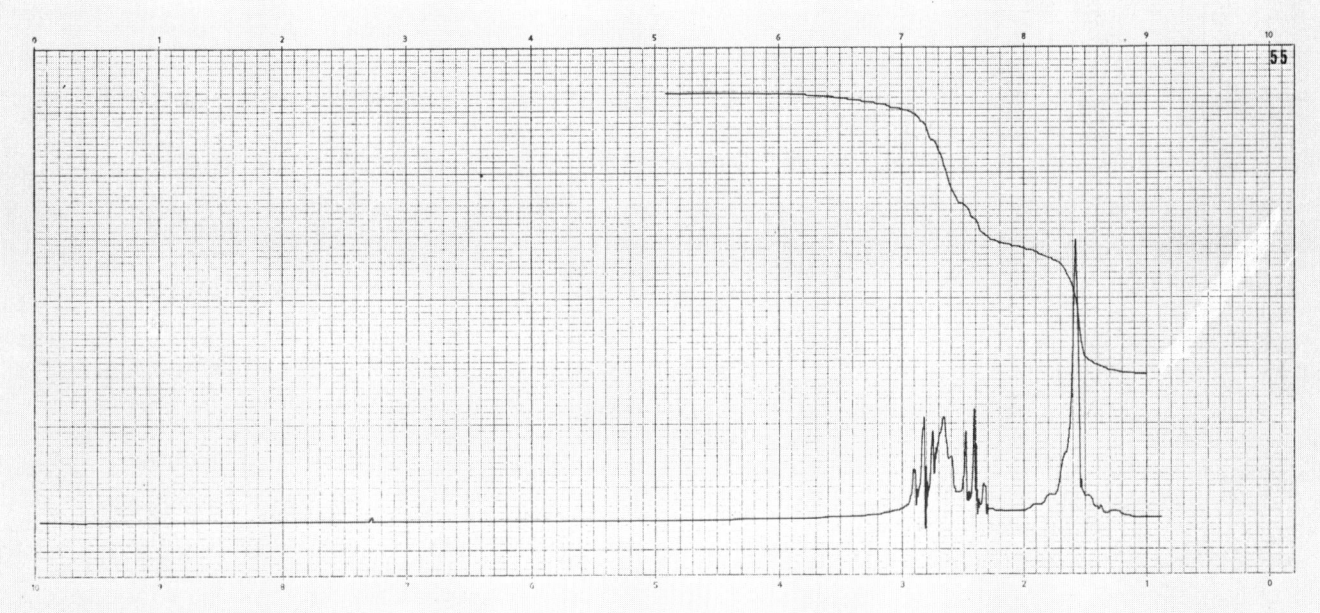

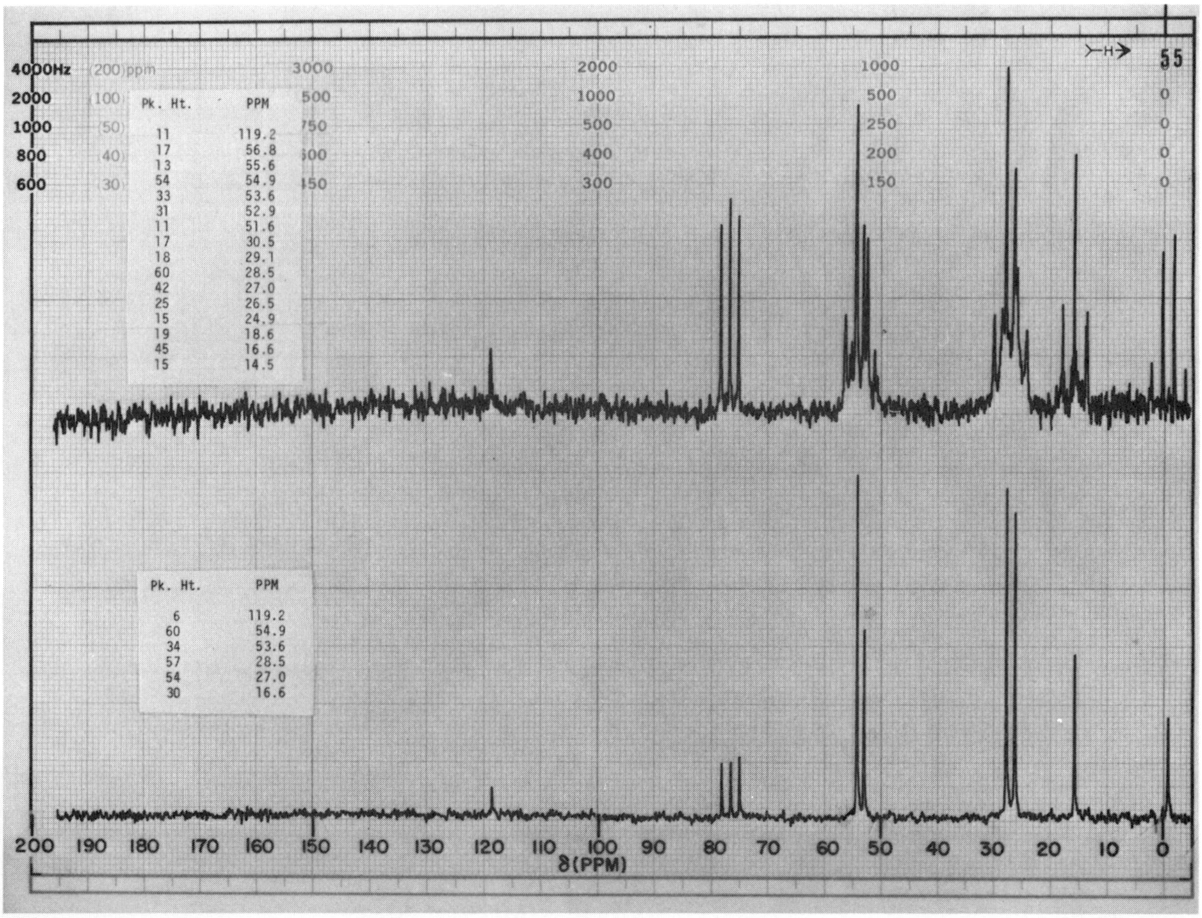

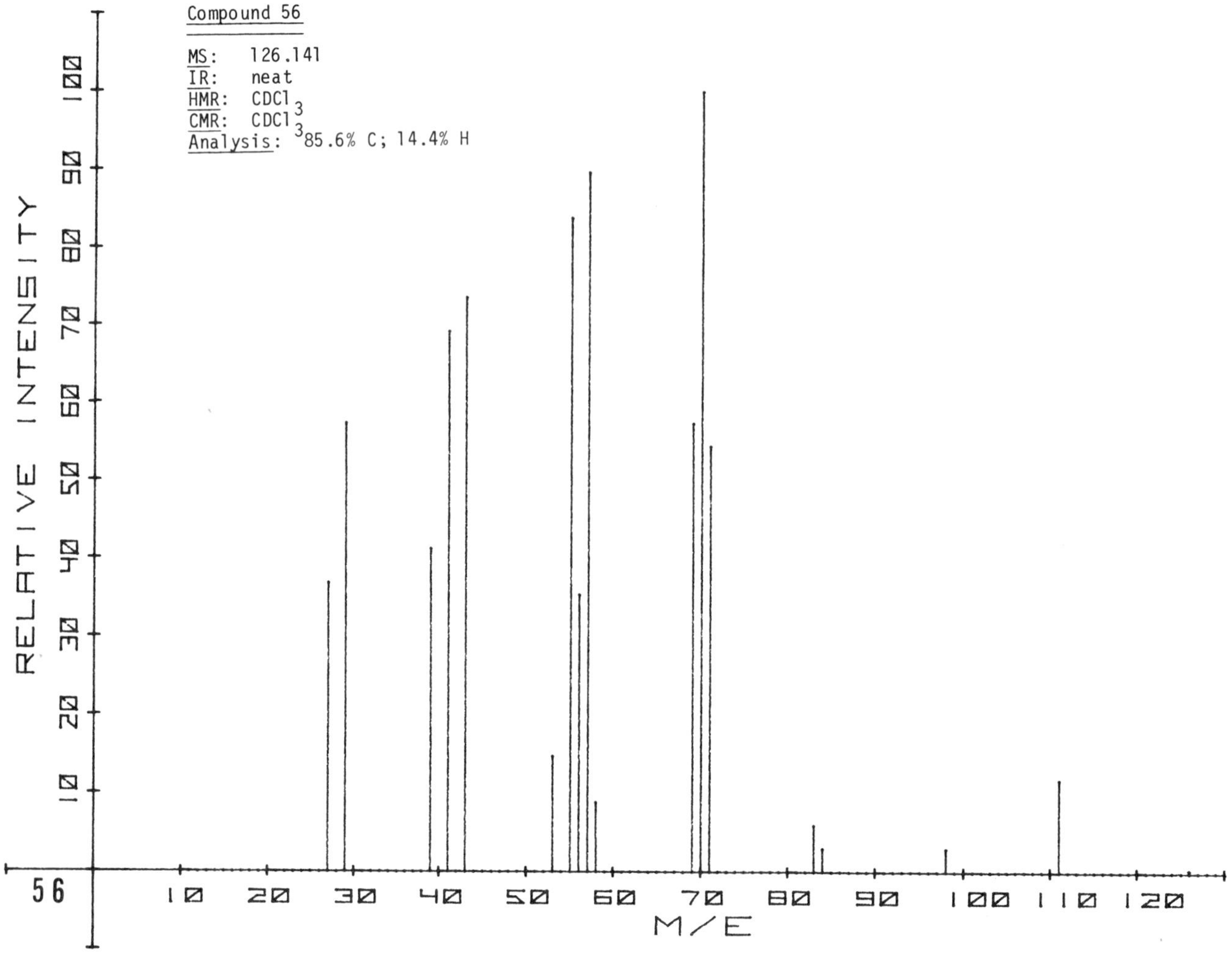

Compound 56
MS: 126.141
IR: neat
HMR: CDCl3
CMR: CDCl3
Analysis: 85.6% C; 14.4% H
RELATIVE INTENSITY
M/E
56

CM-1
ABSORBANCE
WAVELENGTH (MICRONS)
56

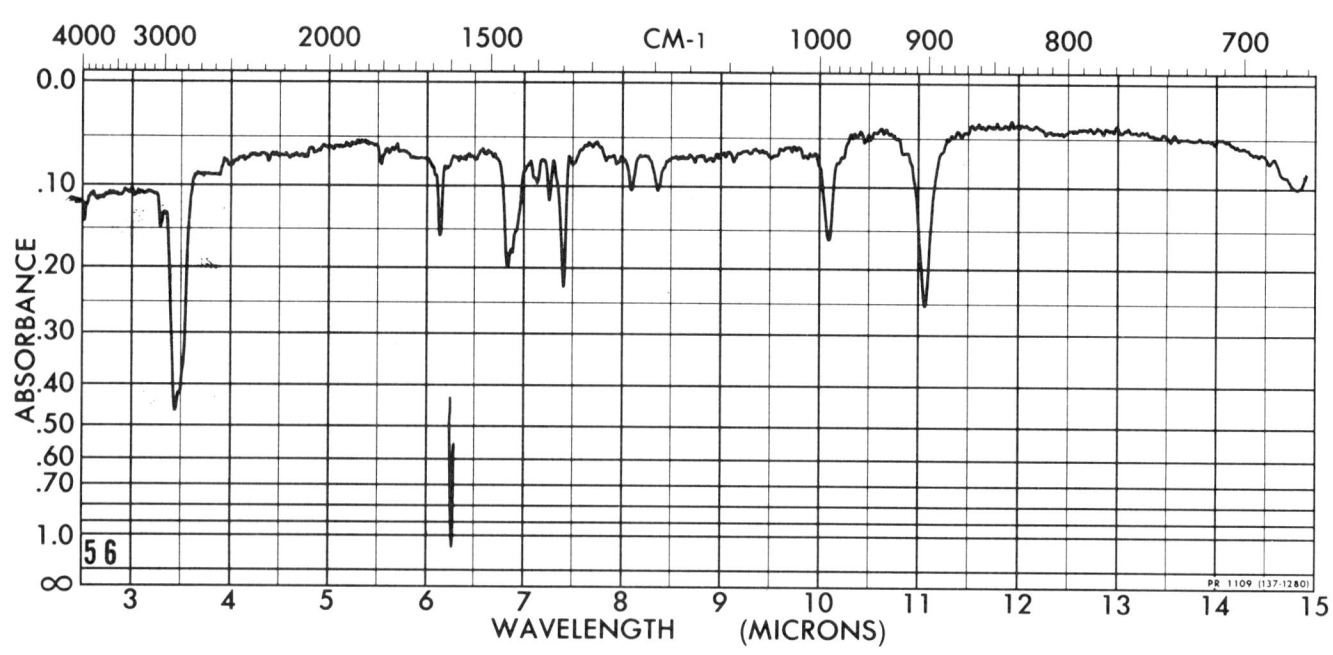

56
IRR. 5.77
IRR. 2.29
A
B
IRR. 2.29
A
B

56
4000Hz (200)ppm 3000 2000 1000
2000 (100) 1500 1000 500 0
1000 (50) 750 500 250 0
800 (40) 400 200 0
600 (30) 300 150 0
PK. HT. PPM
12 35.6
13 34.5
13 32.3
13 31.2
58 30.9
60 29.5
16 28.1
PK. HT. PPM
18 147.0
24 111.5
20 51.0
19 35.0
15 31.2
60 30.2
22 23.5
200 190 180 170 160 150 140 130 120 110 100 90 80 70 60 50 40 30 20 10 0
δ(PPM)

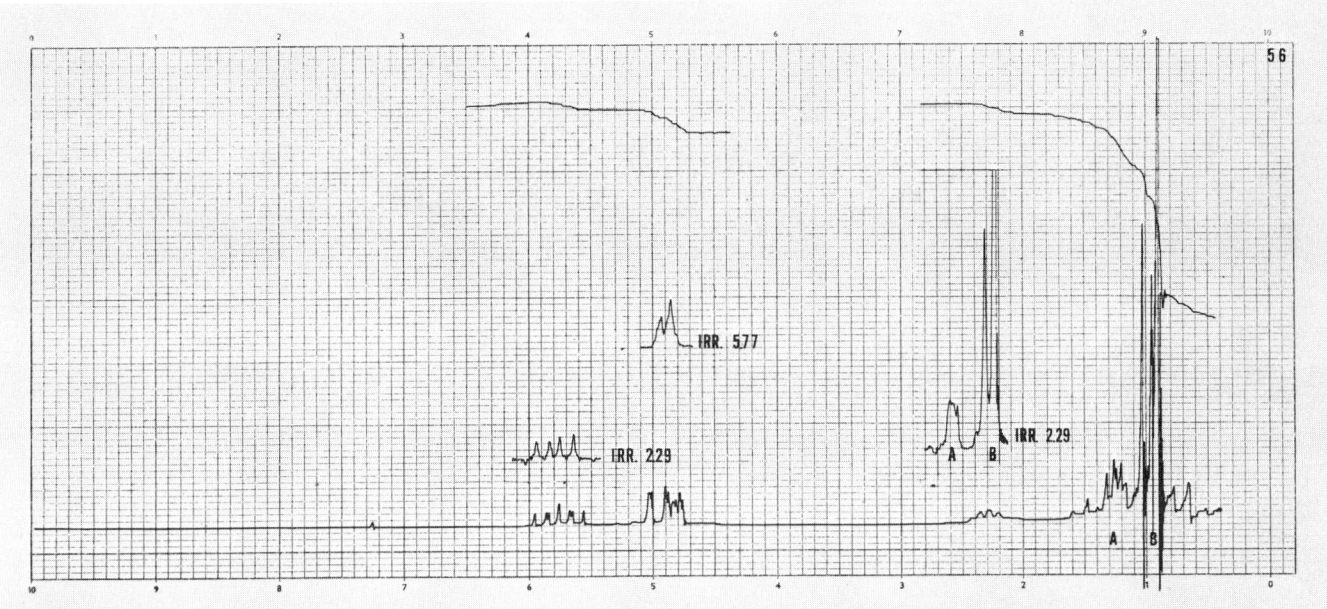

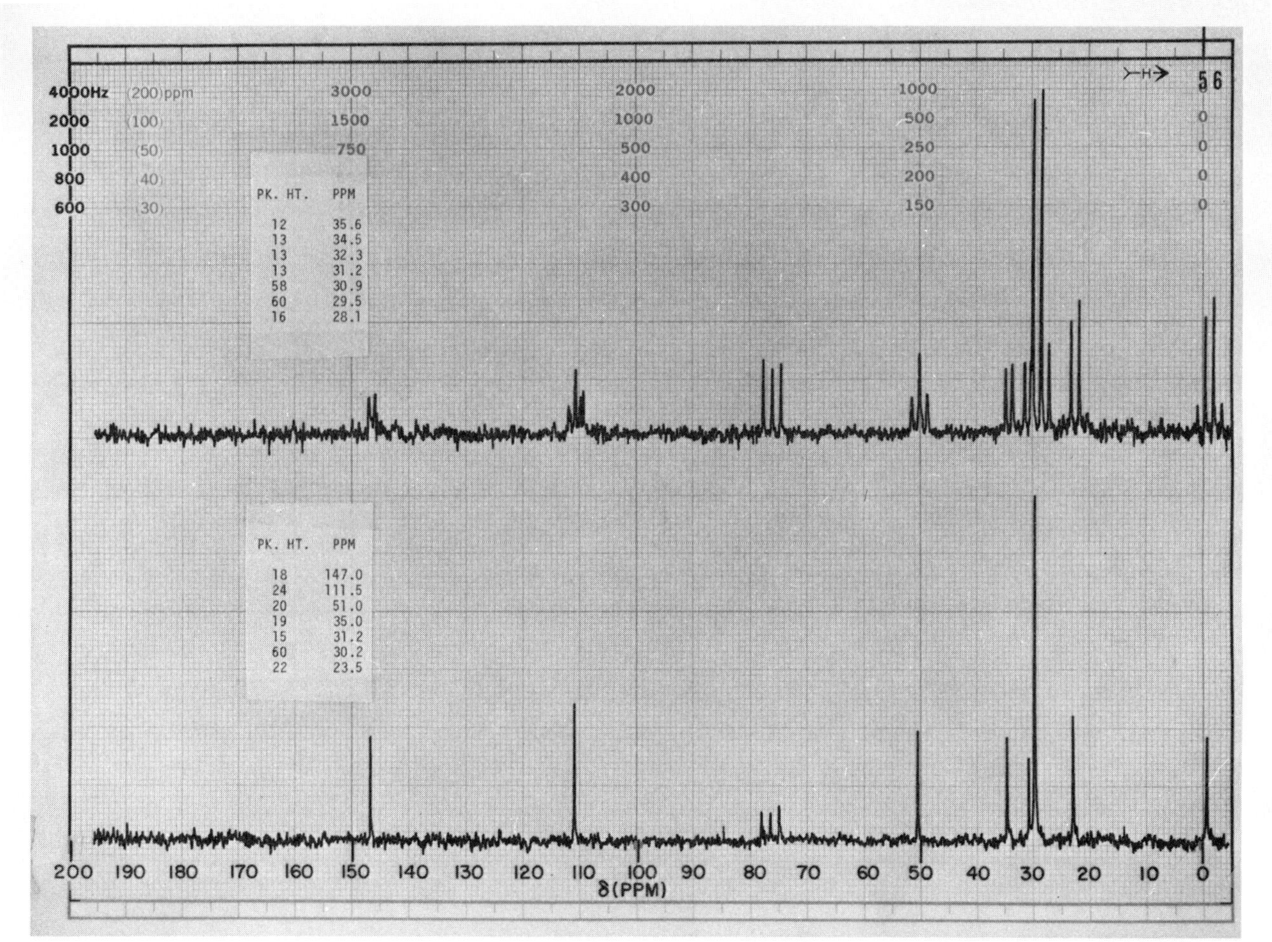

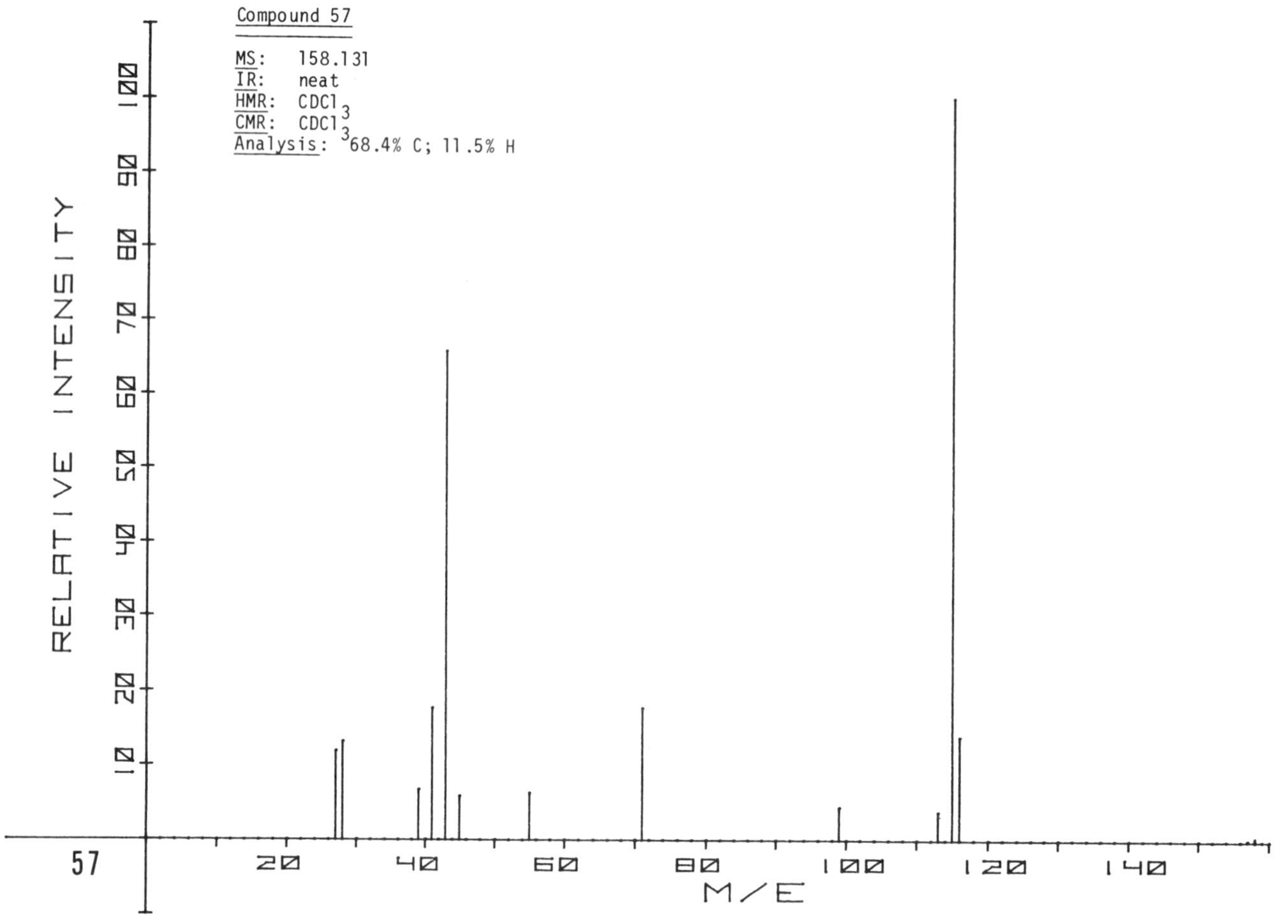

Compound 57
MS: 158.131
IR: neat
HMR: CDCl3
CMR: CDCl3
Analysis: 68.4% C; 11.5% H
RELATIVE INTENSITY
10 20 30 40 50 60 70 80 90 100
57
20 40 60 80 100 120 140
M/E

4000 3000 2000 1500 CM-1 1000 900 800 700
0.0 .10 .20 .30 .40 .50 .60 .70 1.0 ∞
ABSORBANCE
57
3 4 5 6 7 8 9 10 11 12 13 14 15
WAVELENGTH (MICRONS)

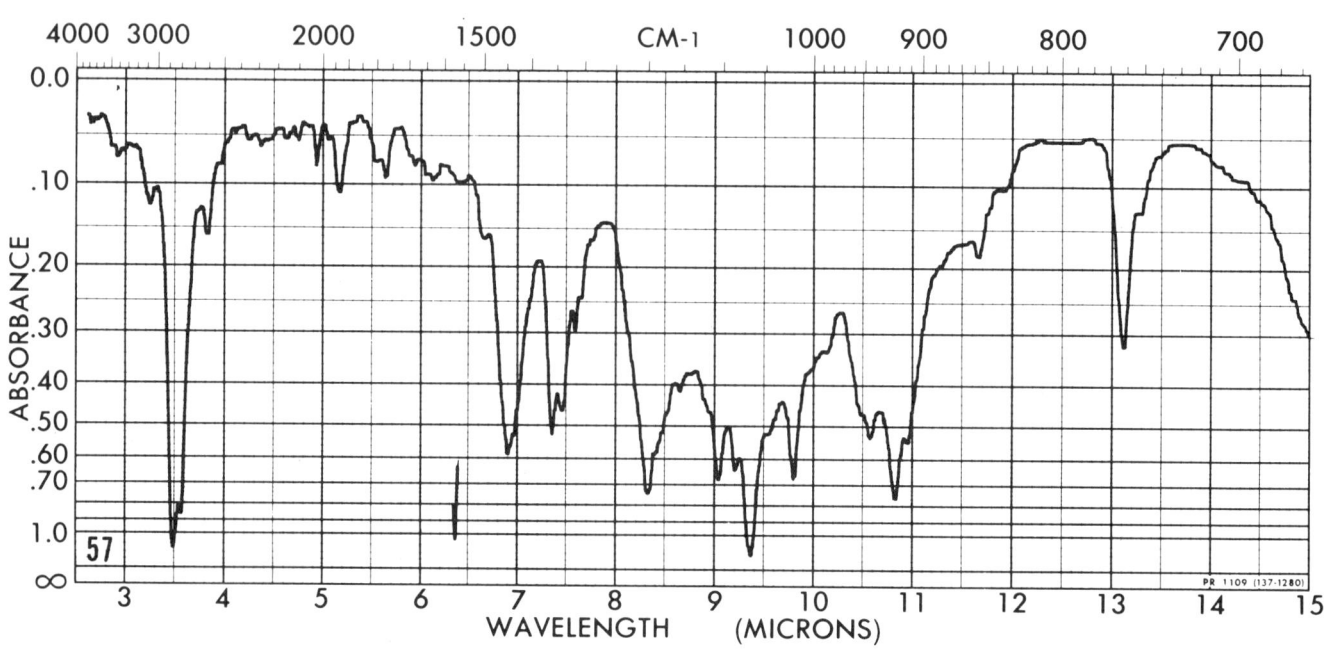

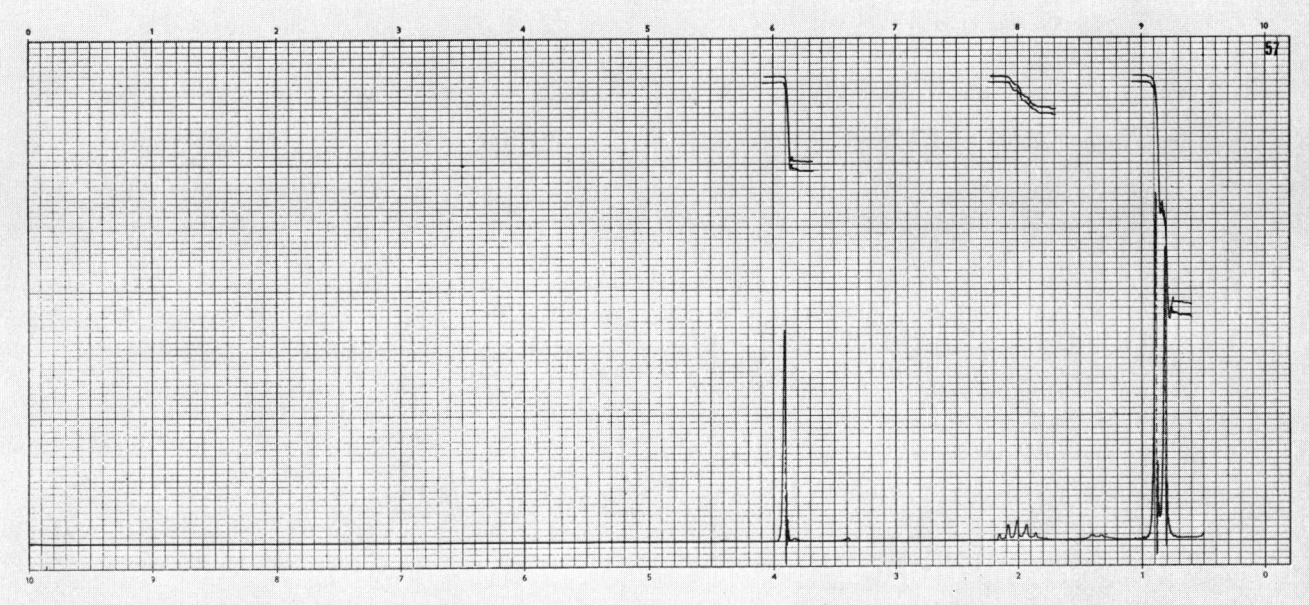

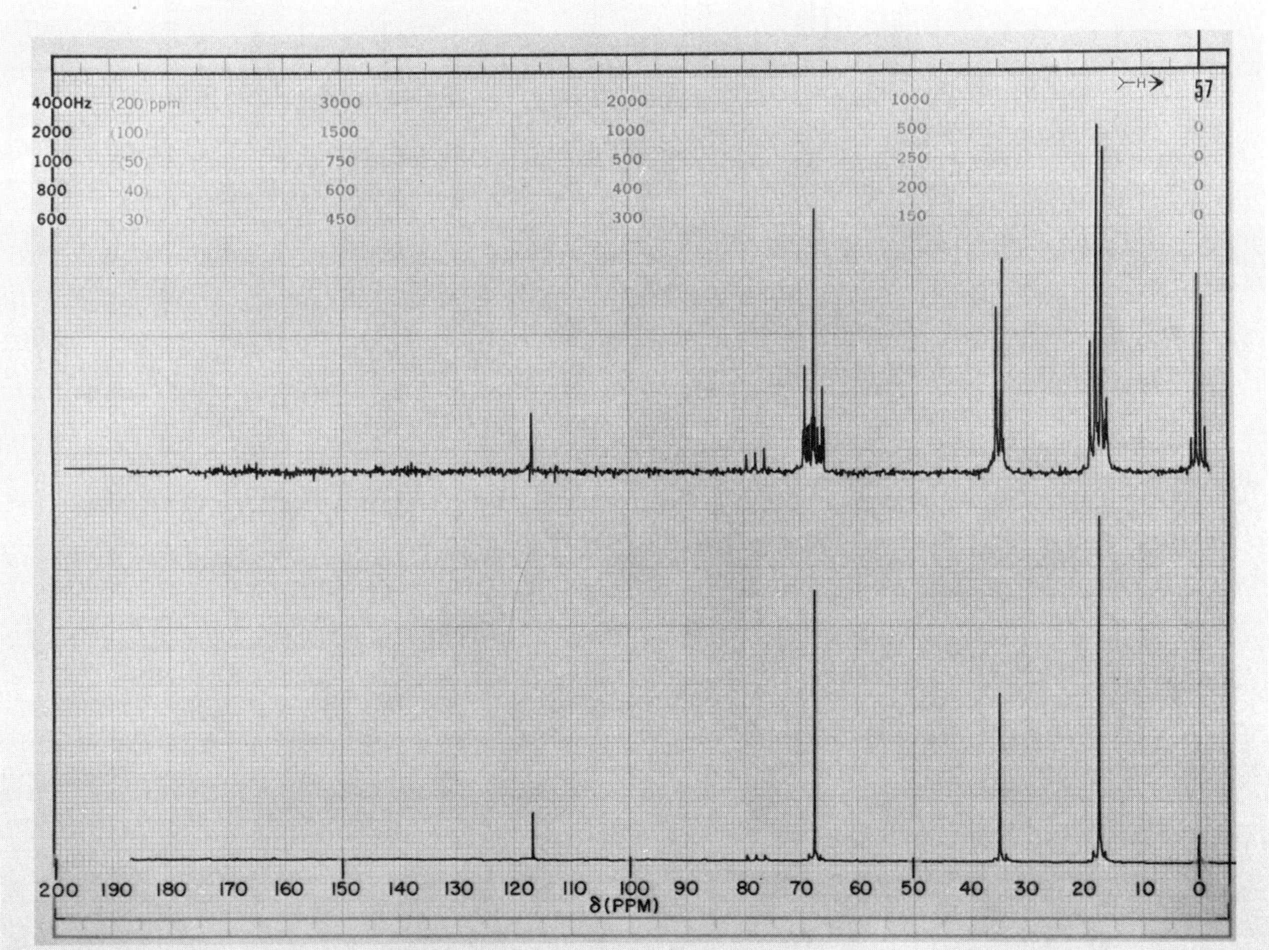

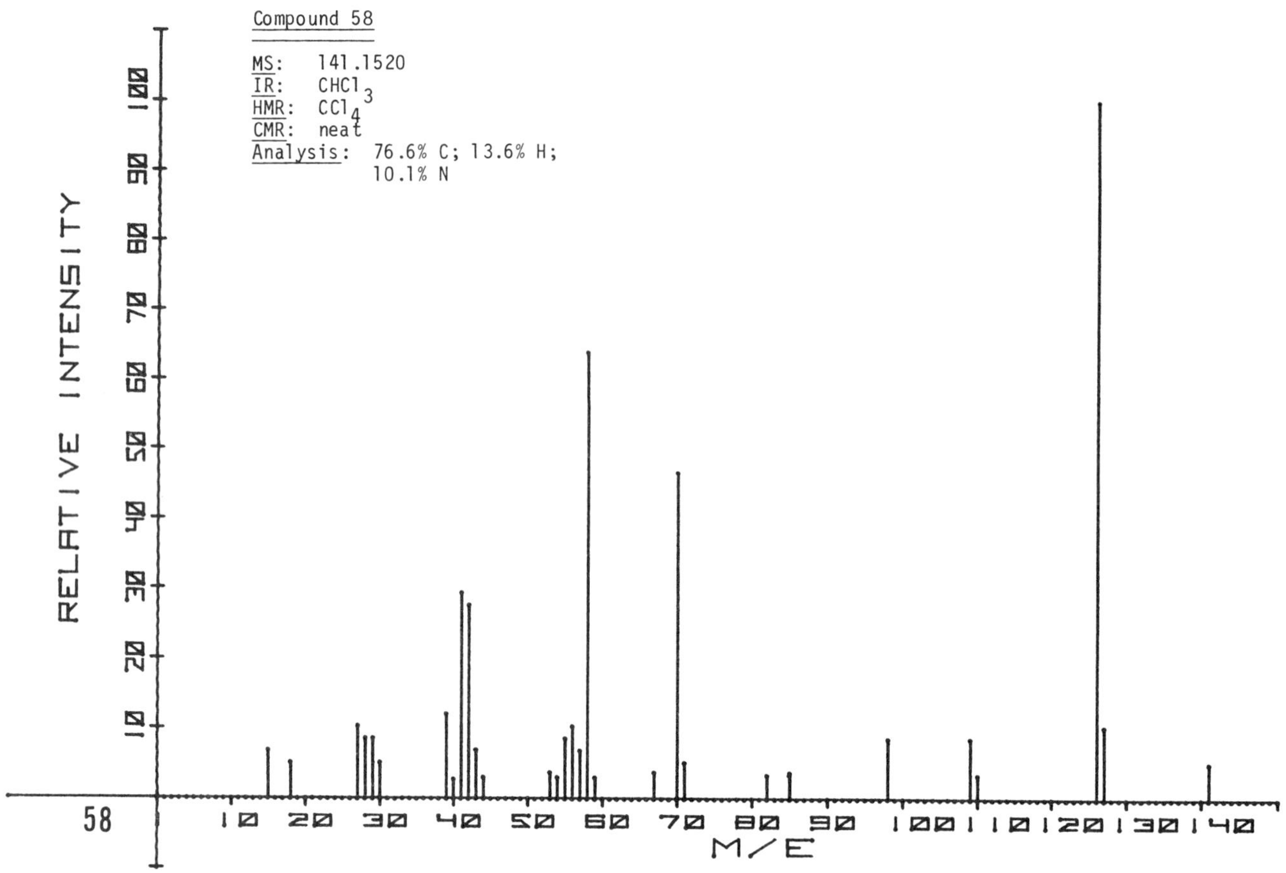

Compound 58
MS: 141.1520
IR: $CHCl_3$
HMR: CCl_4
CMR: neat
Analysis: 76.6% C; 13.6% H; 10.1% N
RELATIVE INTENSITY
M/E
58

CM-1
ABSORBANCE
WAVELENGTH (MICRONS)
58

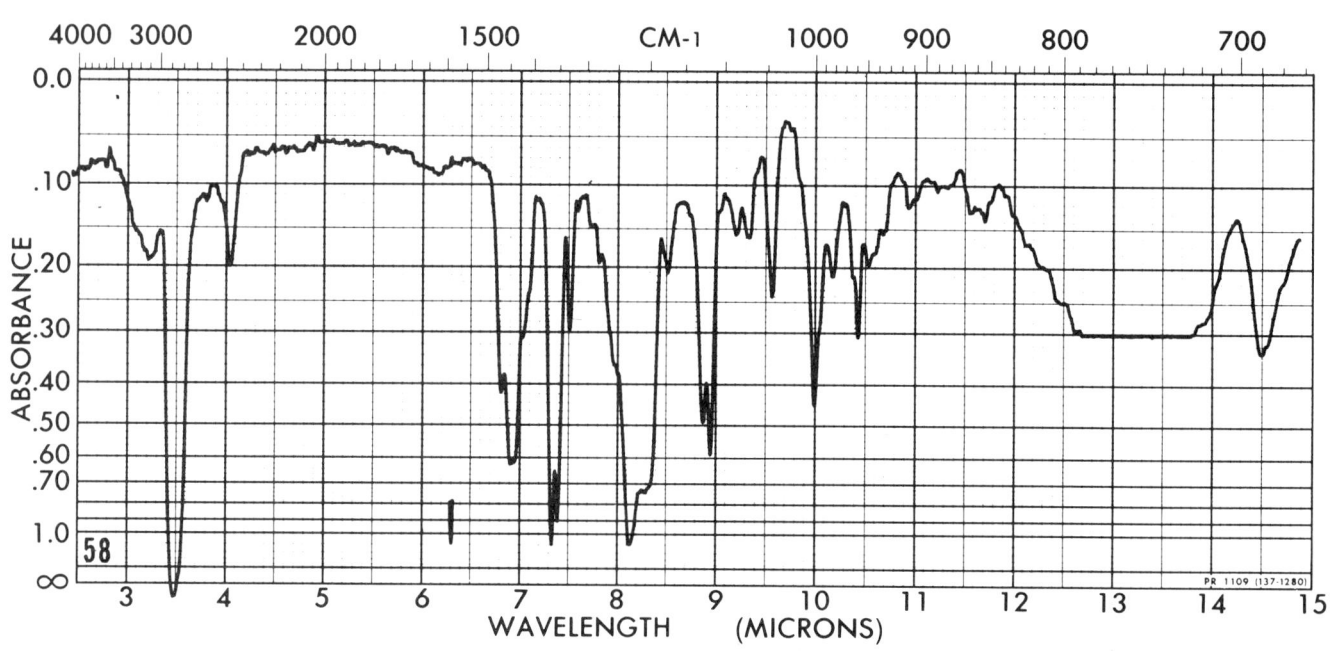

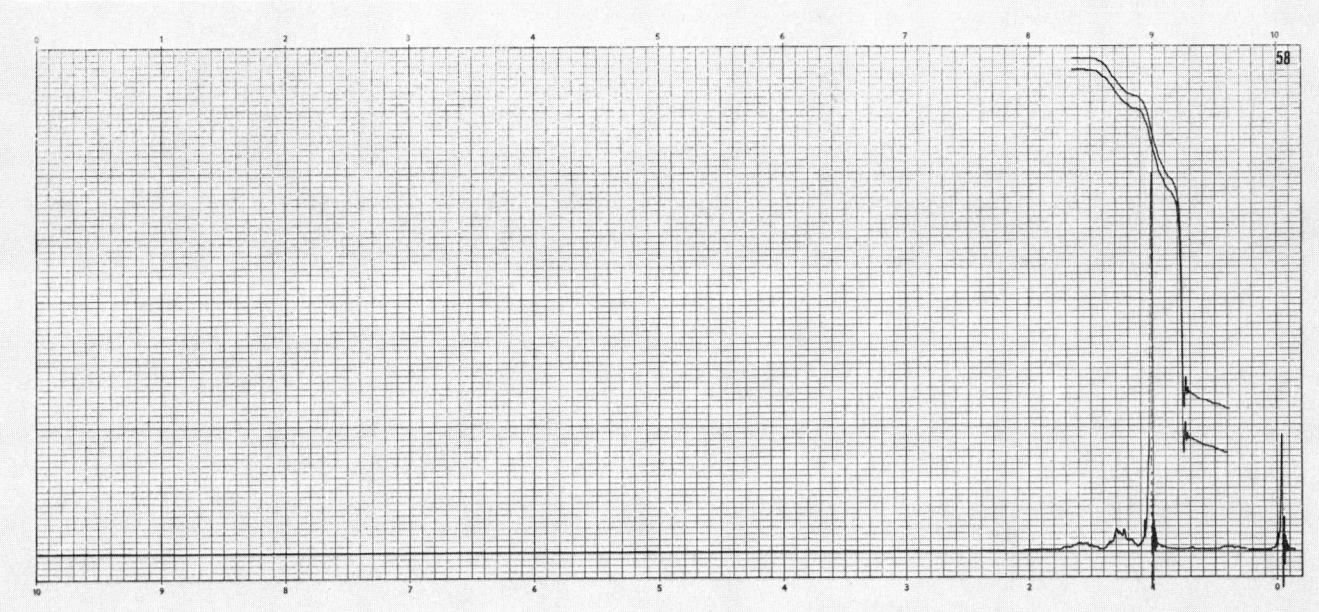

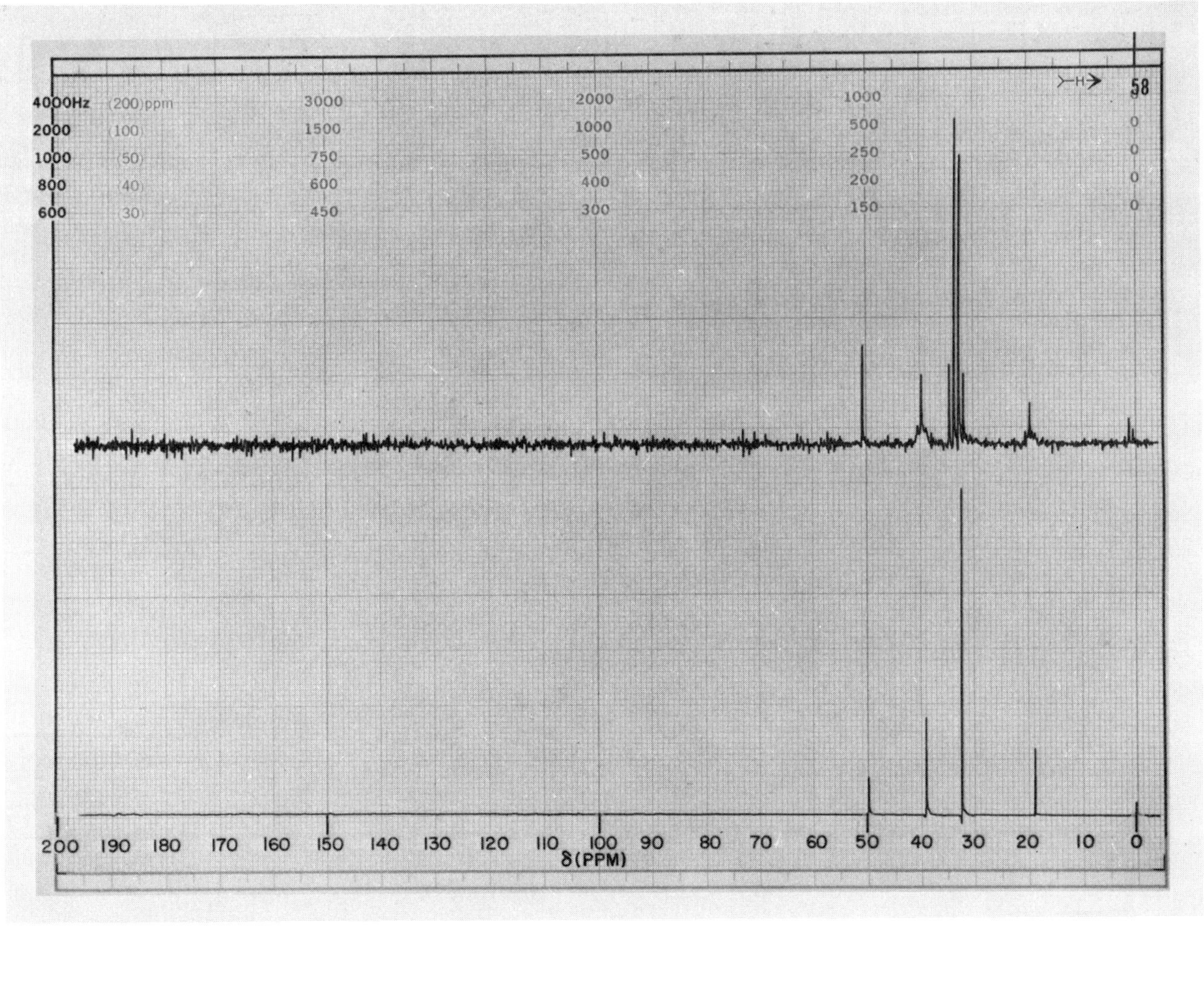

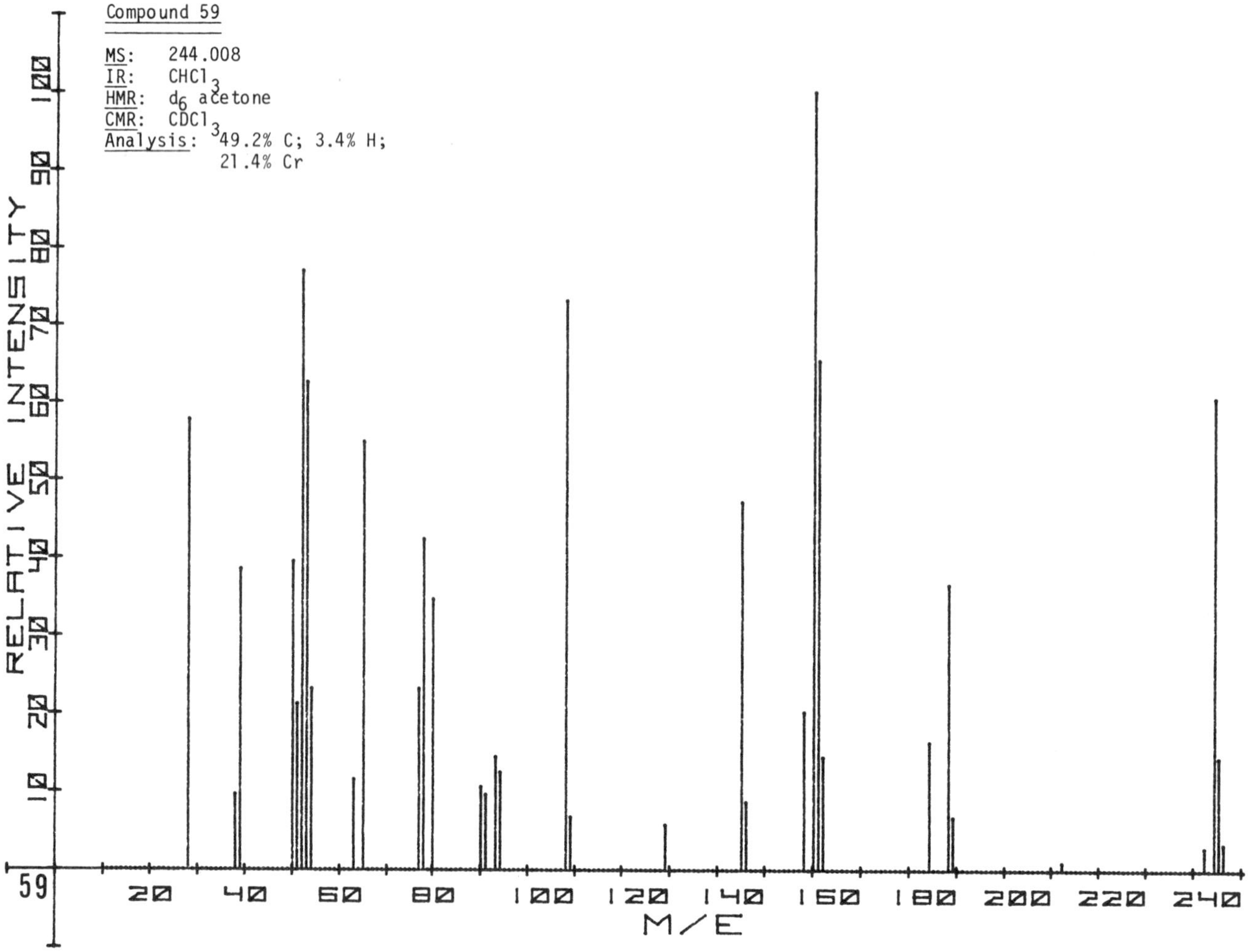

Compound 59
MS: 244.008
IR: CHCl3
HMR: d6 acetone
CMR: CDCl3
Analysis: 49.2% C; 3.4% H; 21.4% Cr
RELATIVE INTENSITY
10
20
30
40
50
60
70
80
90
100
59
20
40
60
80
100
120
140
160
180
200
220
240
M/E

4000
3000
2000
1500
CM-1
1000
900
800
700
ABSORBANCE
0.0
.10
.20
.30
.40
.50
.60
.70
1.0
∞
59
3
4
5
6
7
8
9
10
11
12
13
14
15
WAVELENGTH (MICRONS)

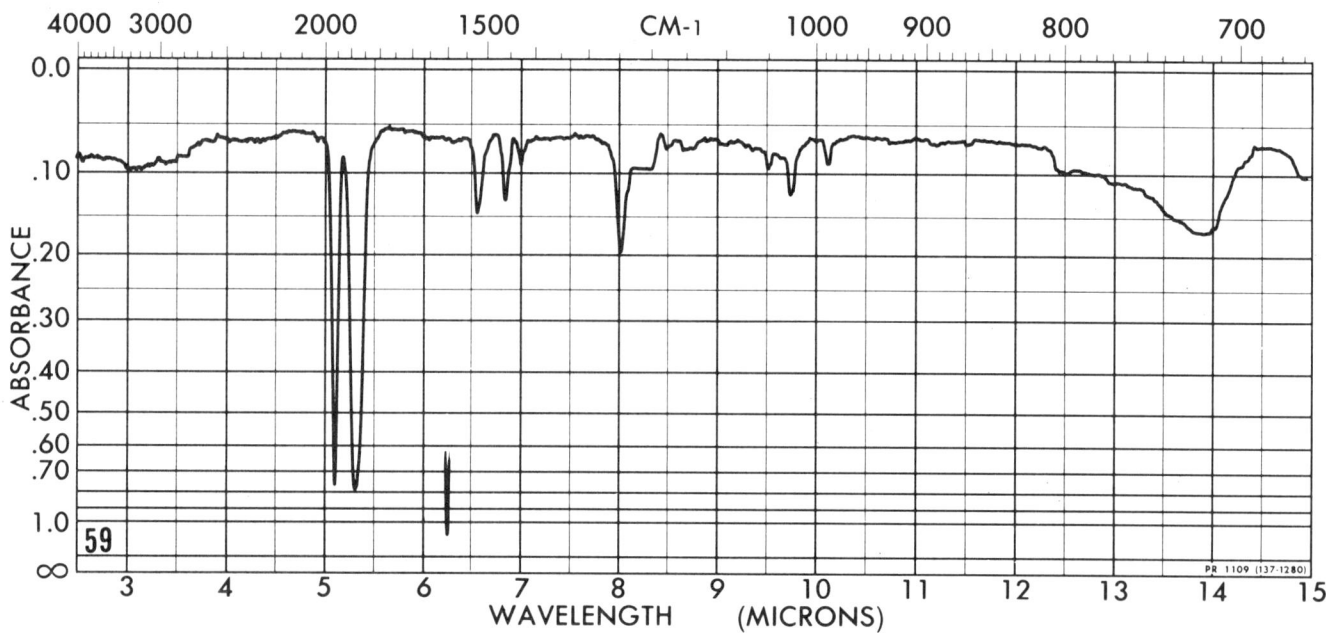

59

59
4000Hz (200)ppm 3000 2000 1000
2000 (100) 1500 1000 500
1000 (50) 750 500 250
800 (40) 400 200
600 (30) 300 150
PK. HT. PPM
60 79.2
50 77.4
7 58.8
39 56.3
24 54.8
6 52.3
PK. HT. PPM
13 233.3
7 143.4
60 95.2
28 85.6
57 78.3
22 55.6
233.3
200 190 180 170 160 150 140 130 120 110 100 90 80 70 60 50 40 30 20 10 0
δ(PPM)

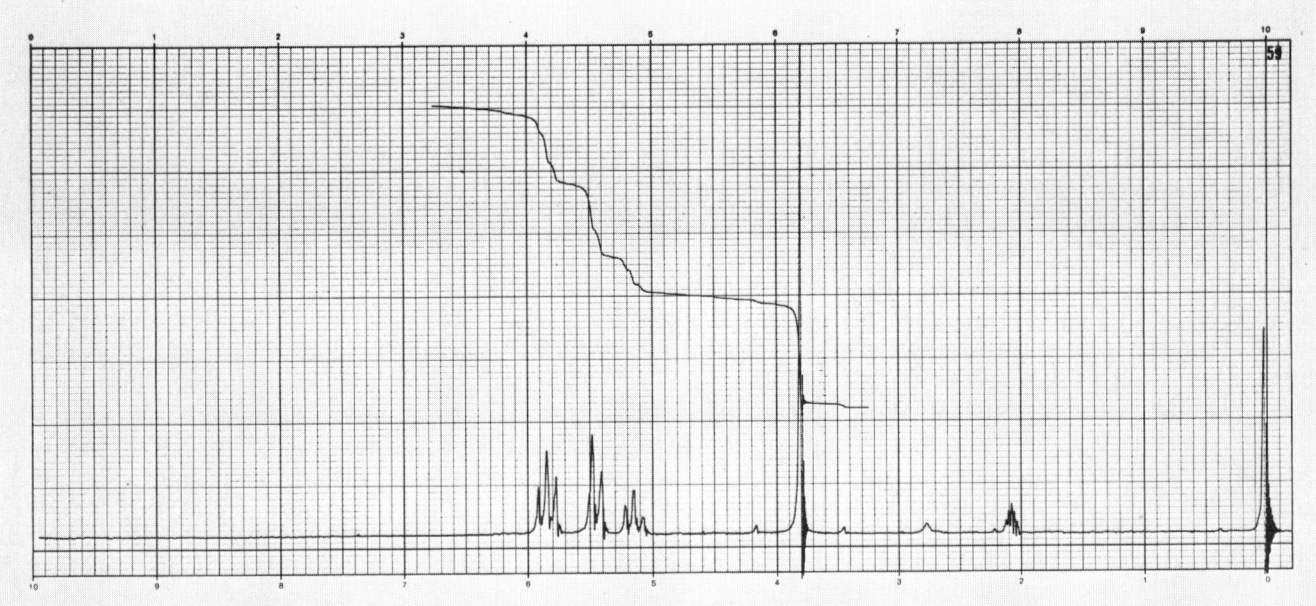

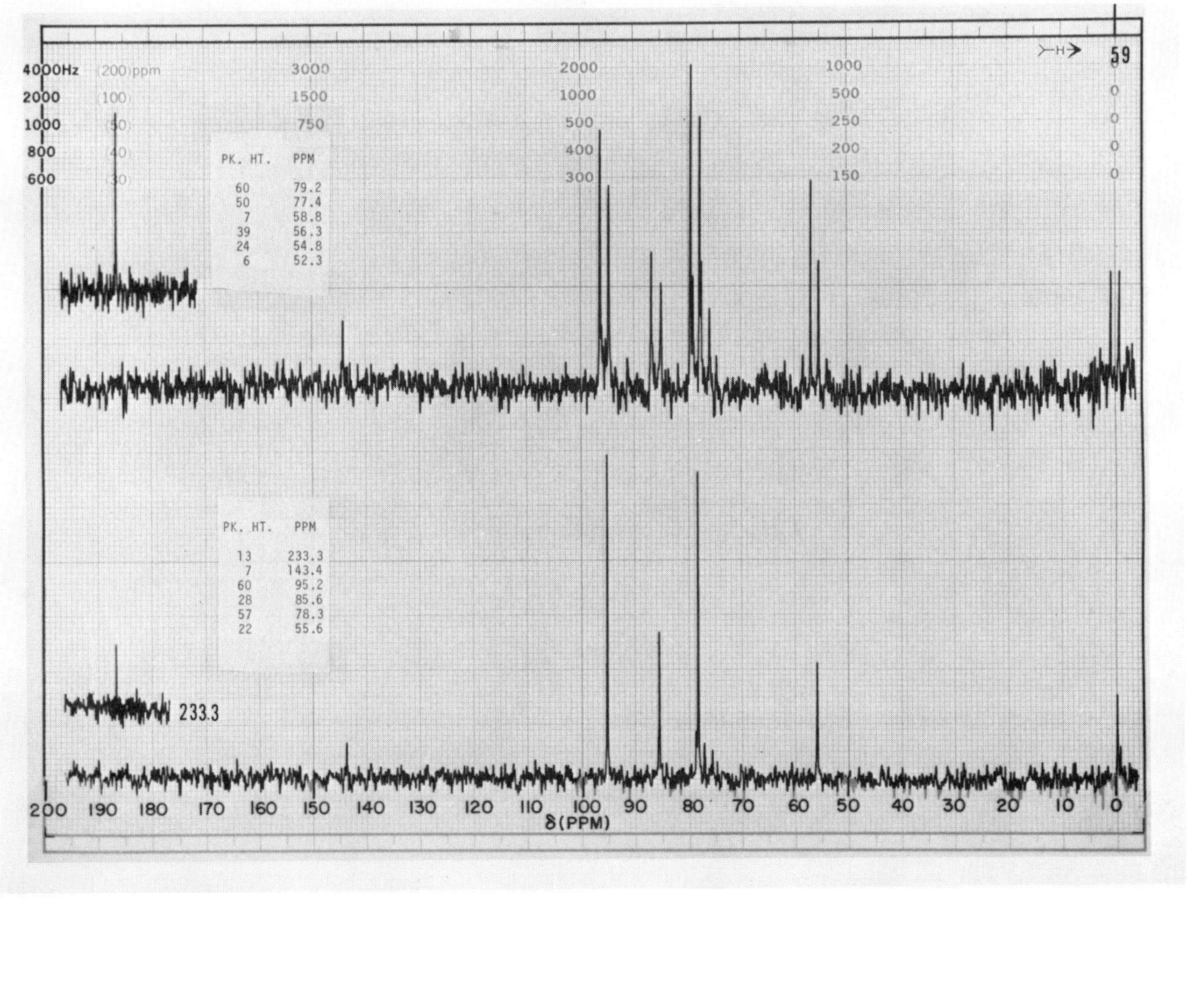

COMPOUND 60A

Compound 60A

MS: 160.053
IR: KBr
HMR: d_6 acetone
CMR: d_6 DMSO
Analysis: 75.0% C; 5.1% H

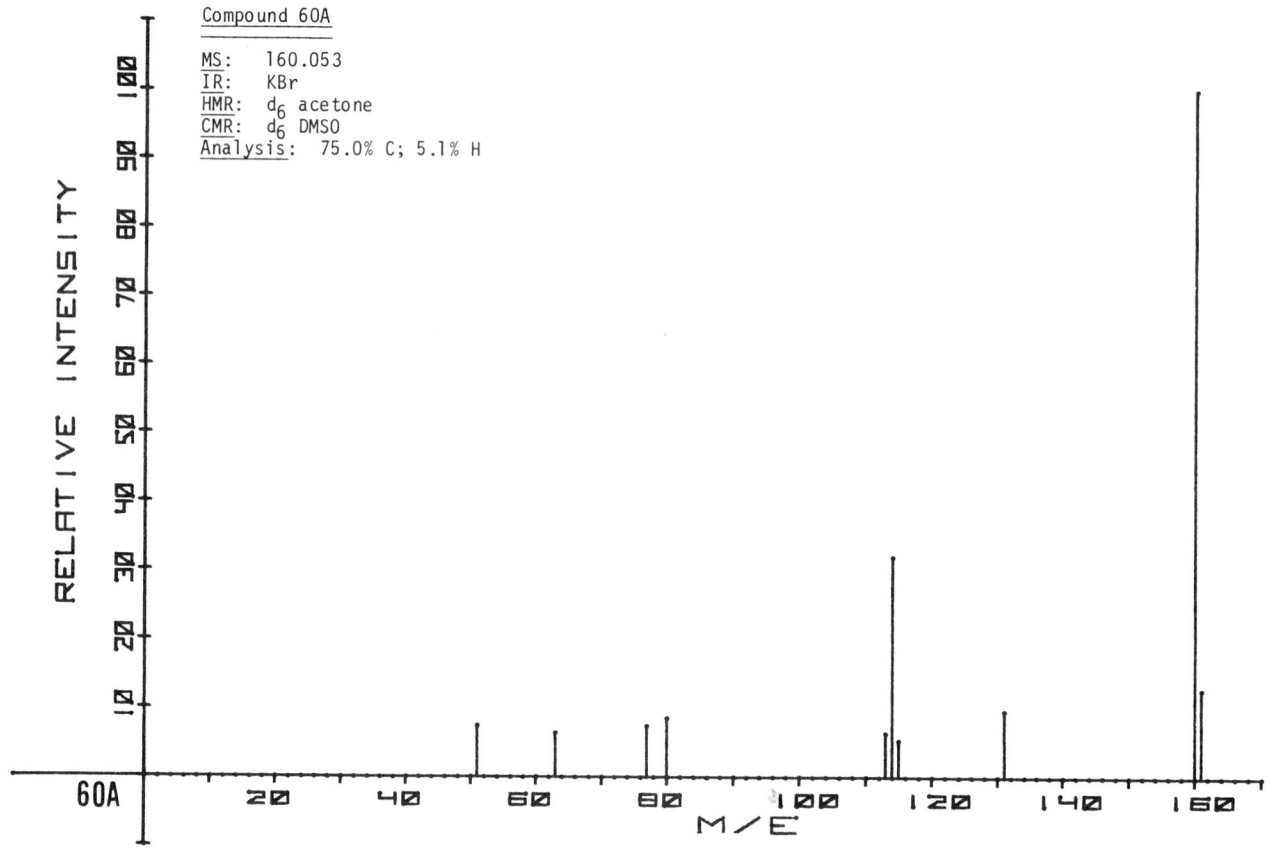

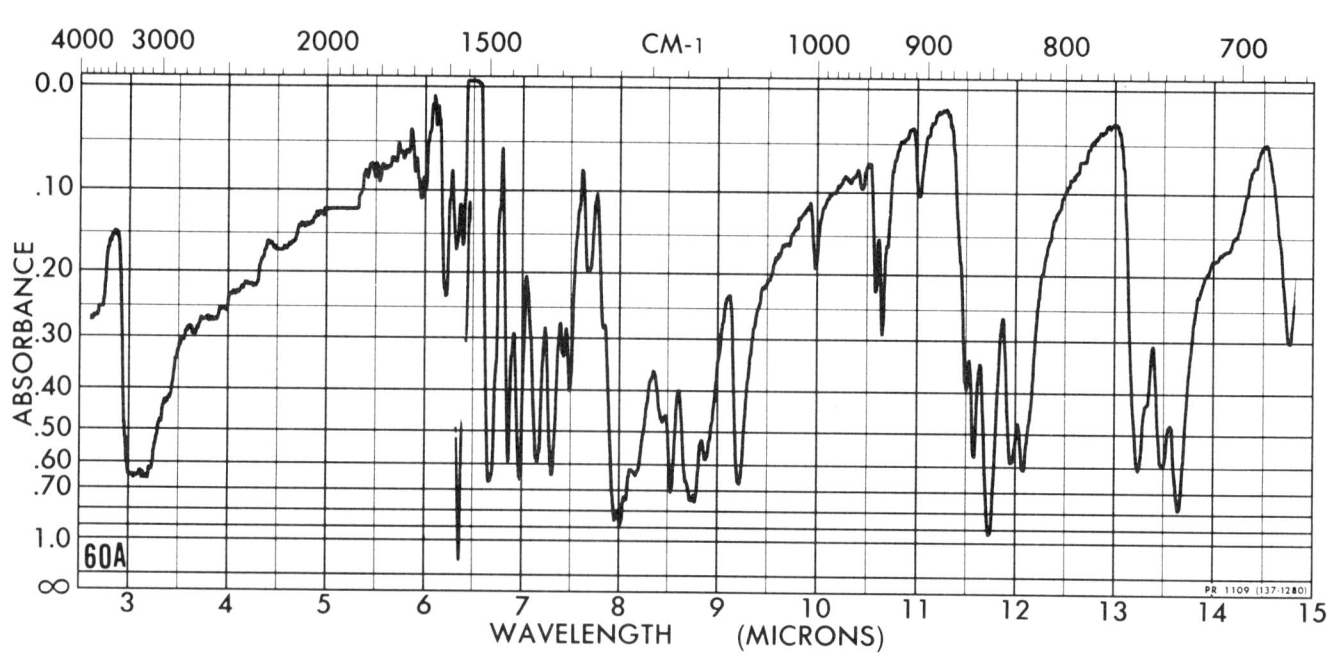

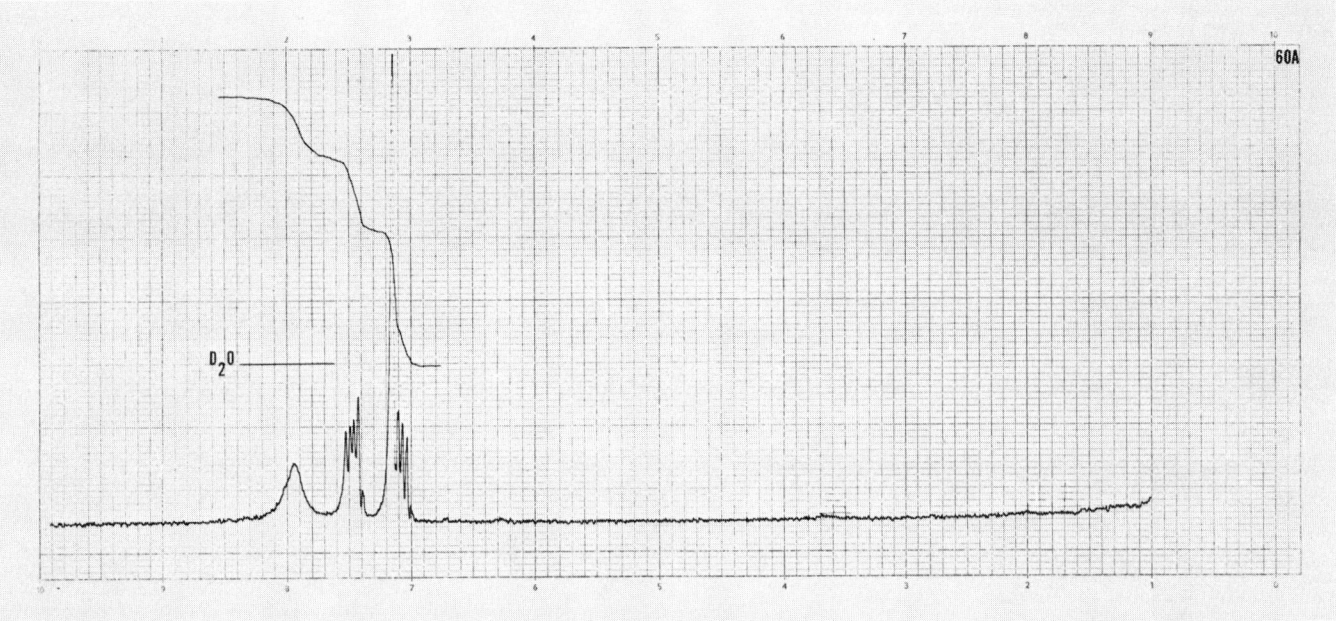

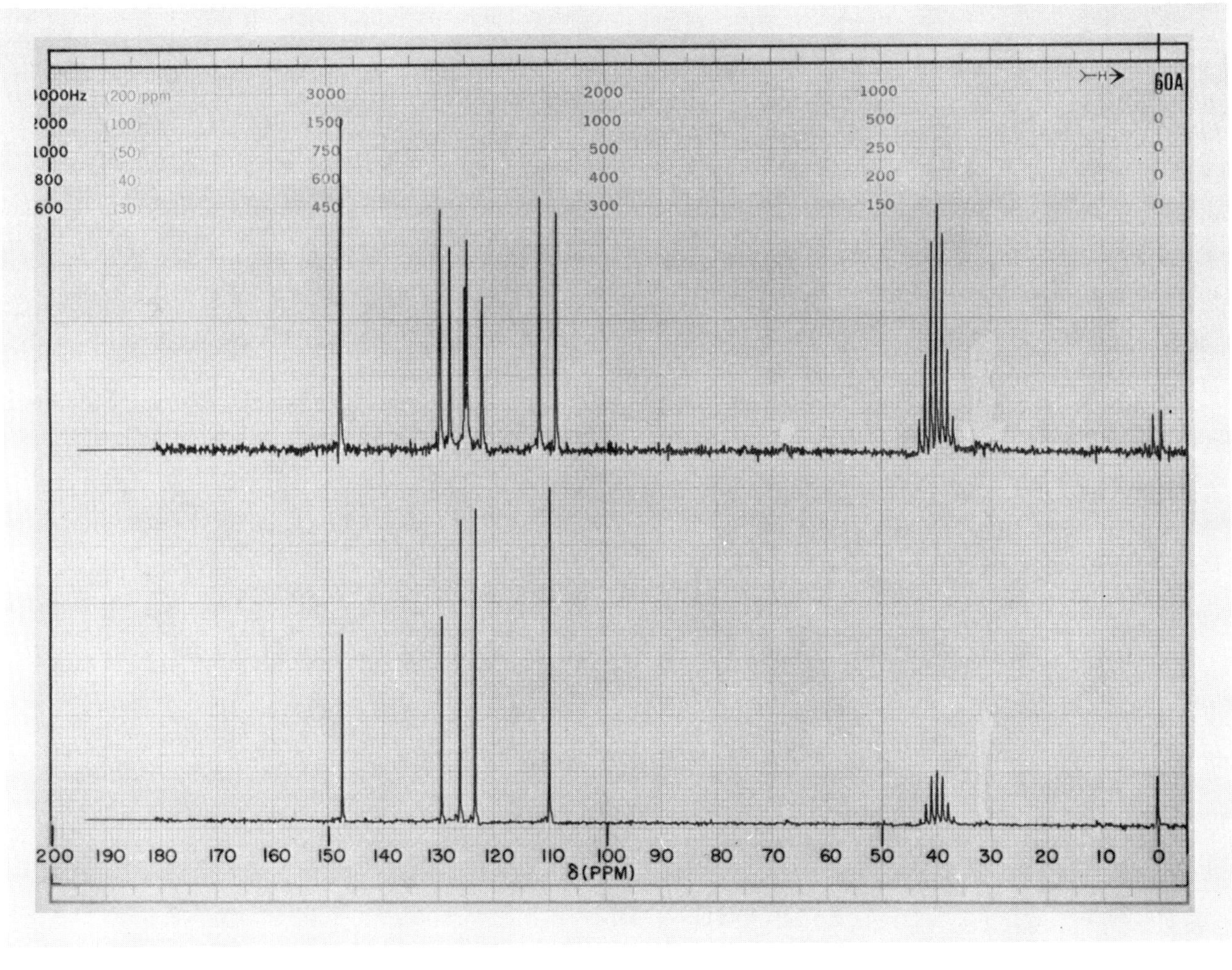

COMPOUND 60B

Compound 60B

MS: 160.052
IR: KBr
HMR: d6 acetone
CMR: d6 DMSO
Analysis: 75.1% C; 5.0% H

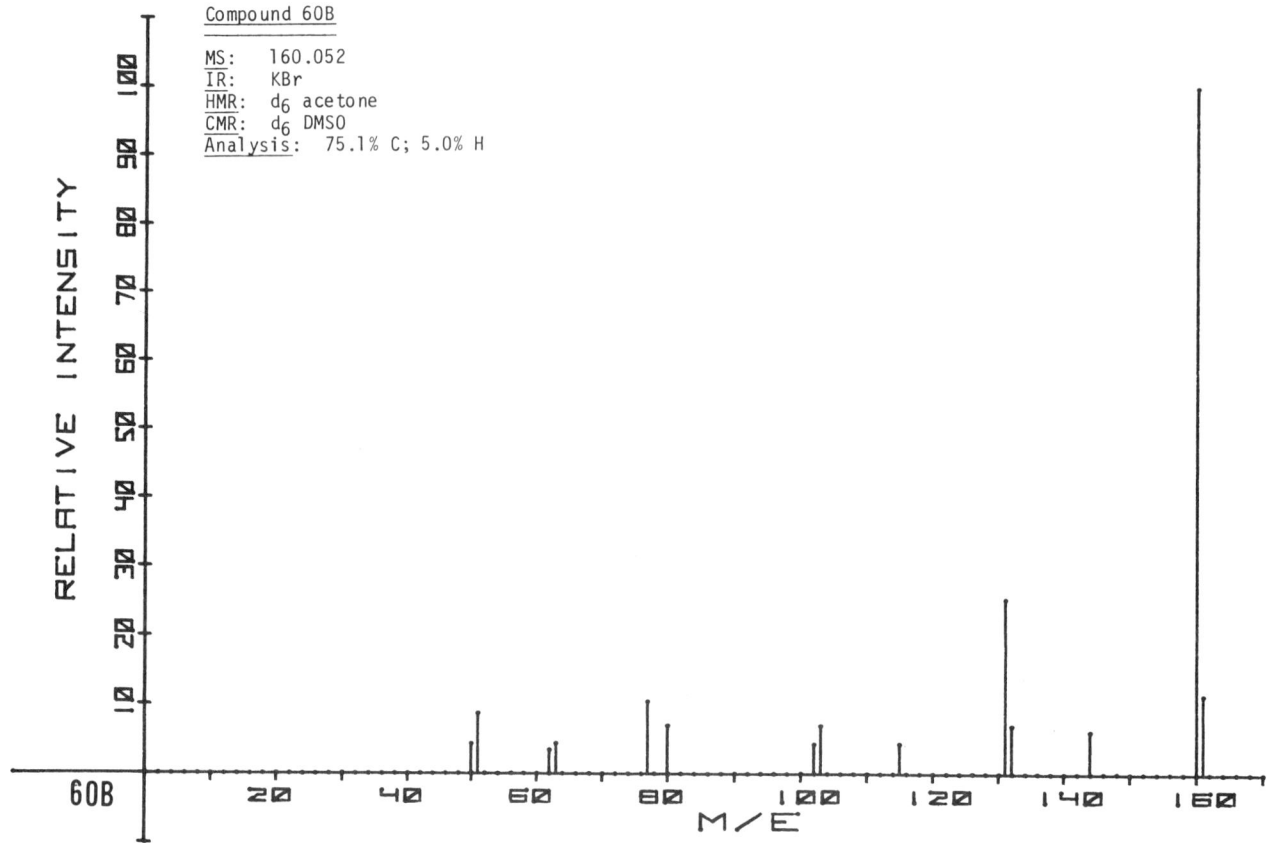

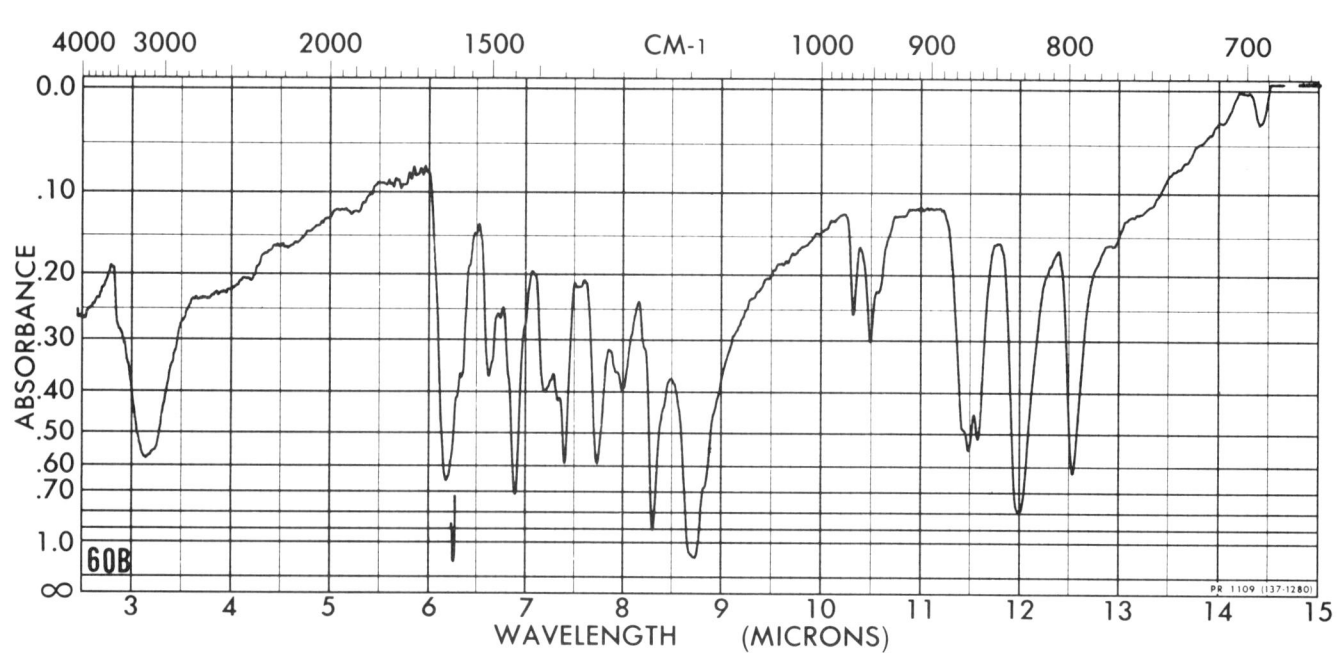

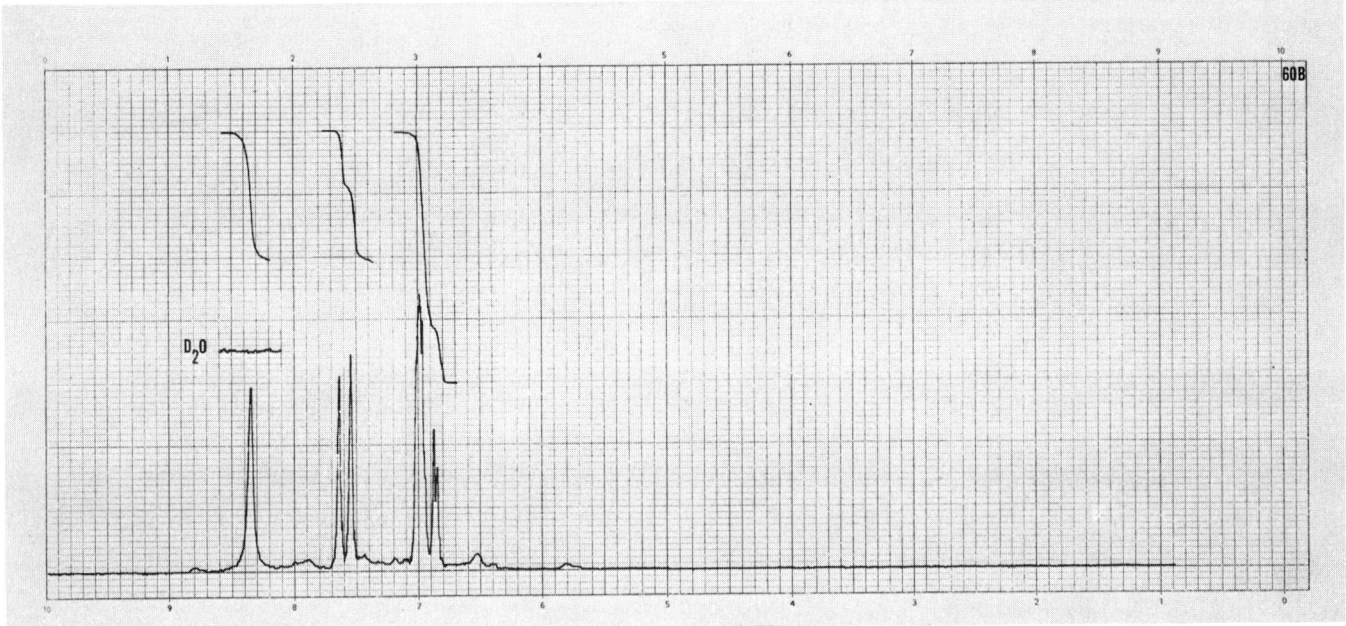

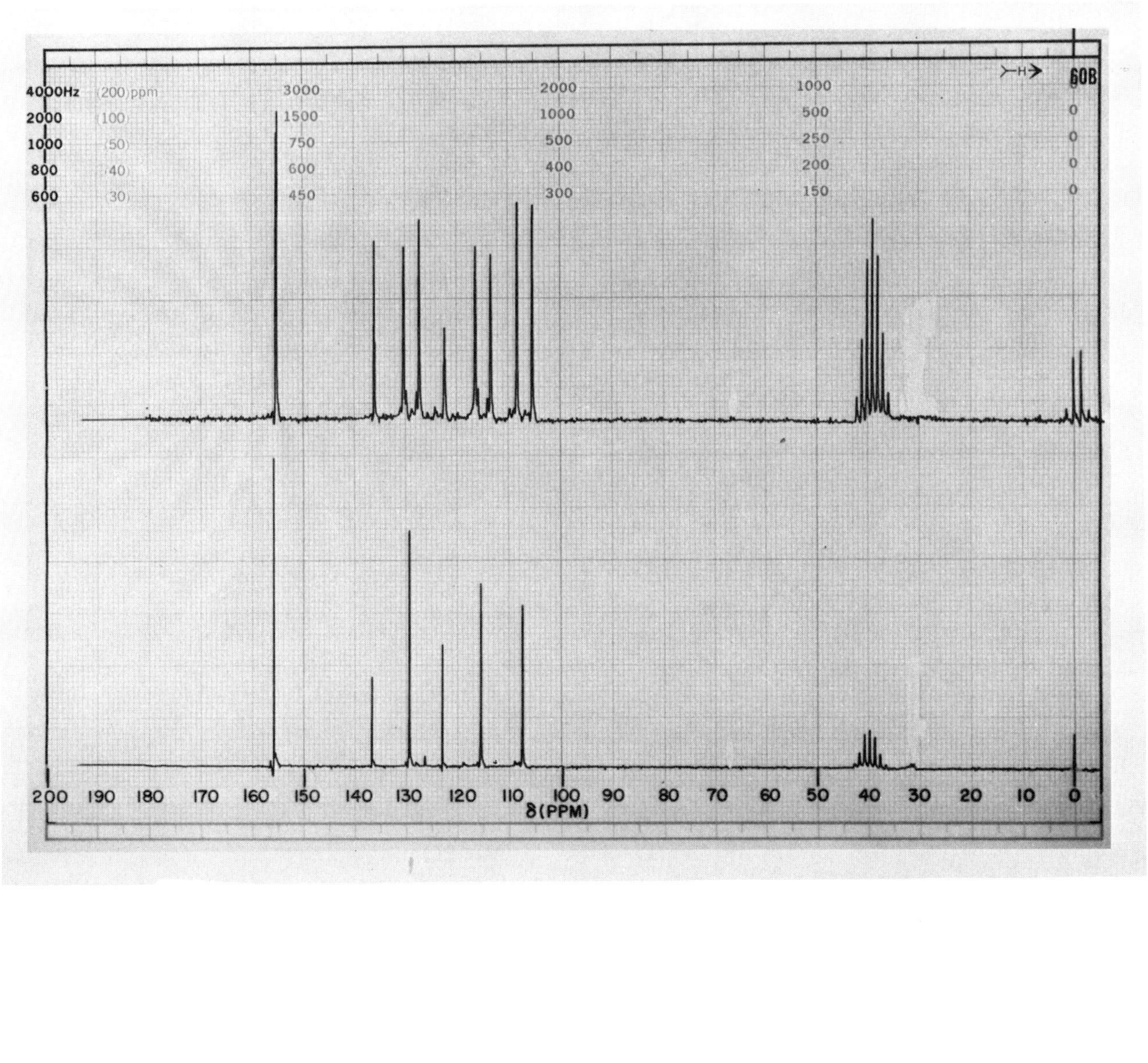

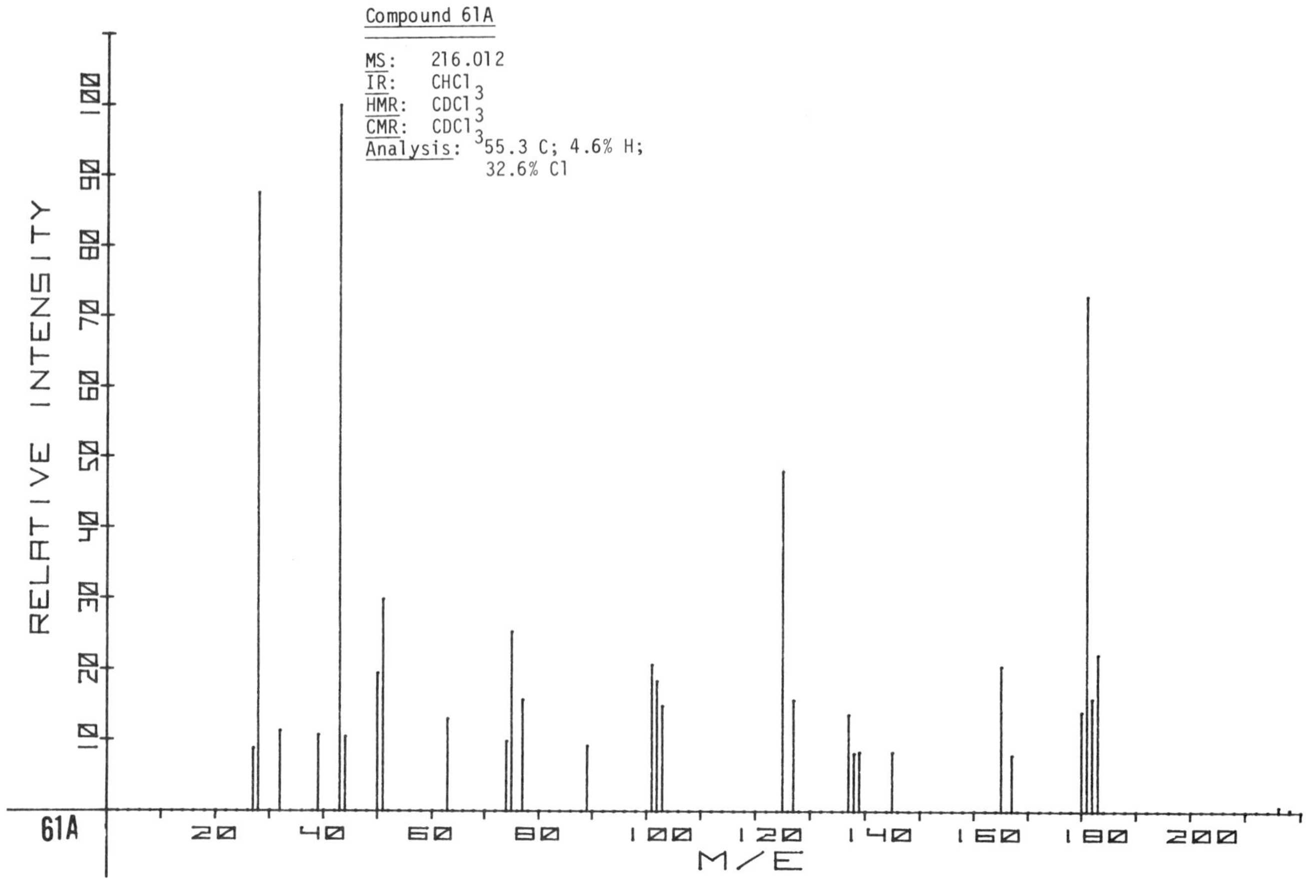

Compound 61A
MS: 216.012
IR: CHCl3
HMR: CDCl3
CMR: CDCl3
Analysis: 55.3 C; 4.6% H; 32.6% Cl
RELATIVE INTENSITY
10 20 30 40 50 60 70 80 90 100
61A
20 40 60 80 100 120 140 160 180 200
M/E

4000 3000 2000 1500 CM-1 1000 900 800 700
0.0 .10 .20 .30 .40 .50 .60 .70 1.0 ∞
ABSORBANCE
61A
3 4 5 6 7 8 9 10 11 12 13 14 15
WAVELENGTH (MICRONS)
PR 1109 (137-1280)

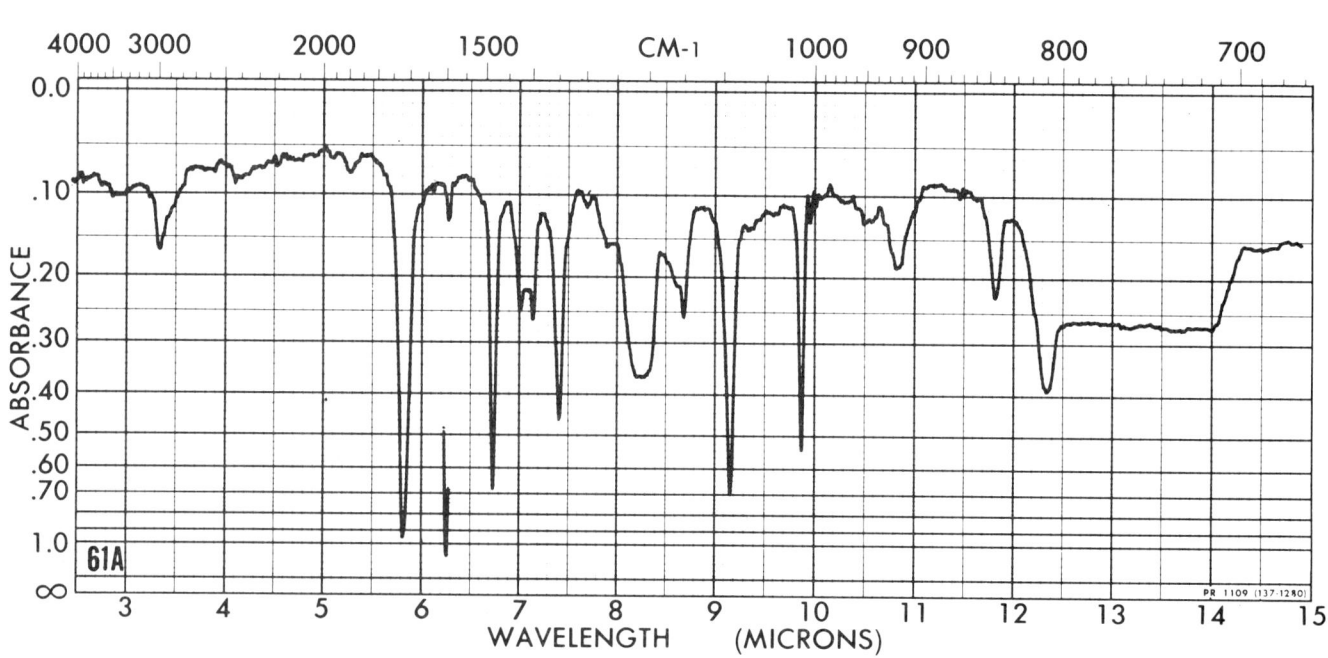

61A
IRR. 7.15
IRR. 4.25

61A
PK. HT. PPM
21 135.1
17 132.5
37 132.0
60 129.7
38 129.6
44 127.2
PK. HT. PPM
11 201.5
28 135.1
25 132.5
60 130.8
44 128.5
21 63.5
17 38.5
9 26.3
201.5
δ(PPM)

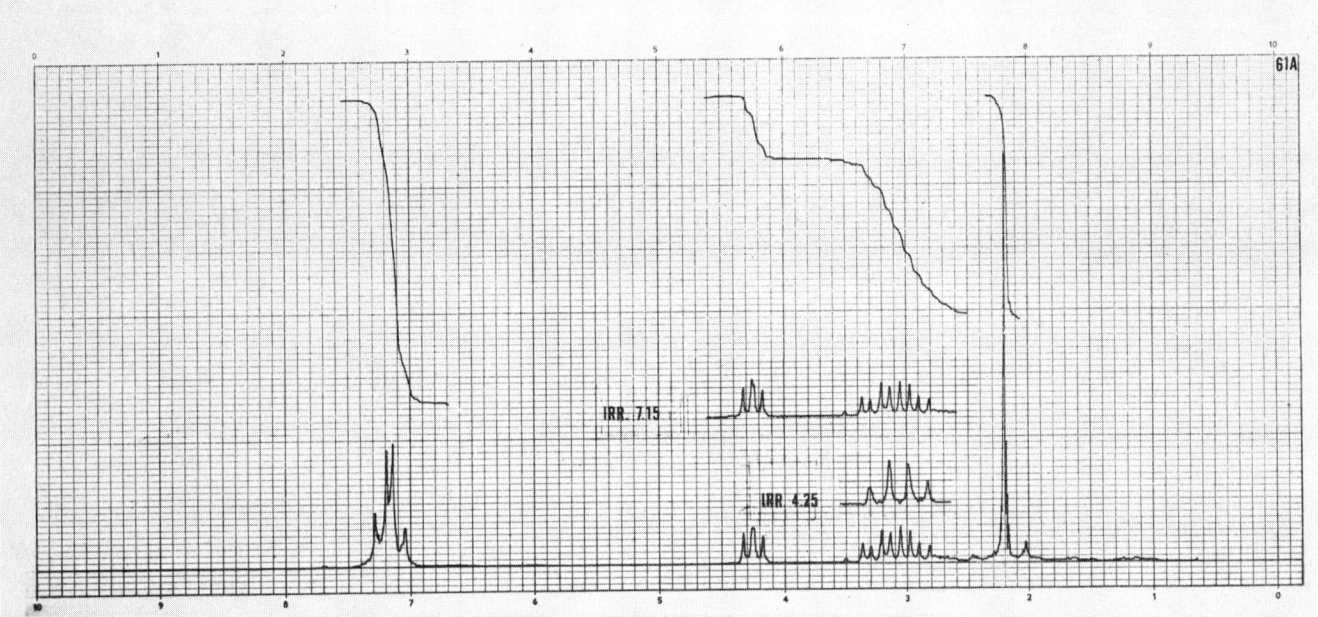

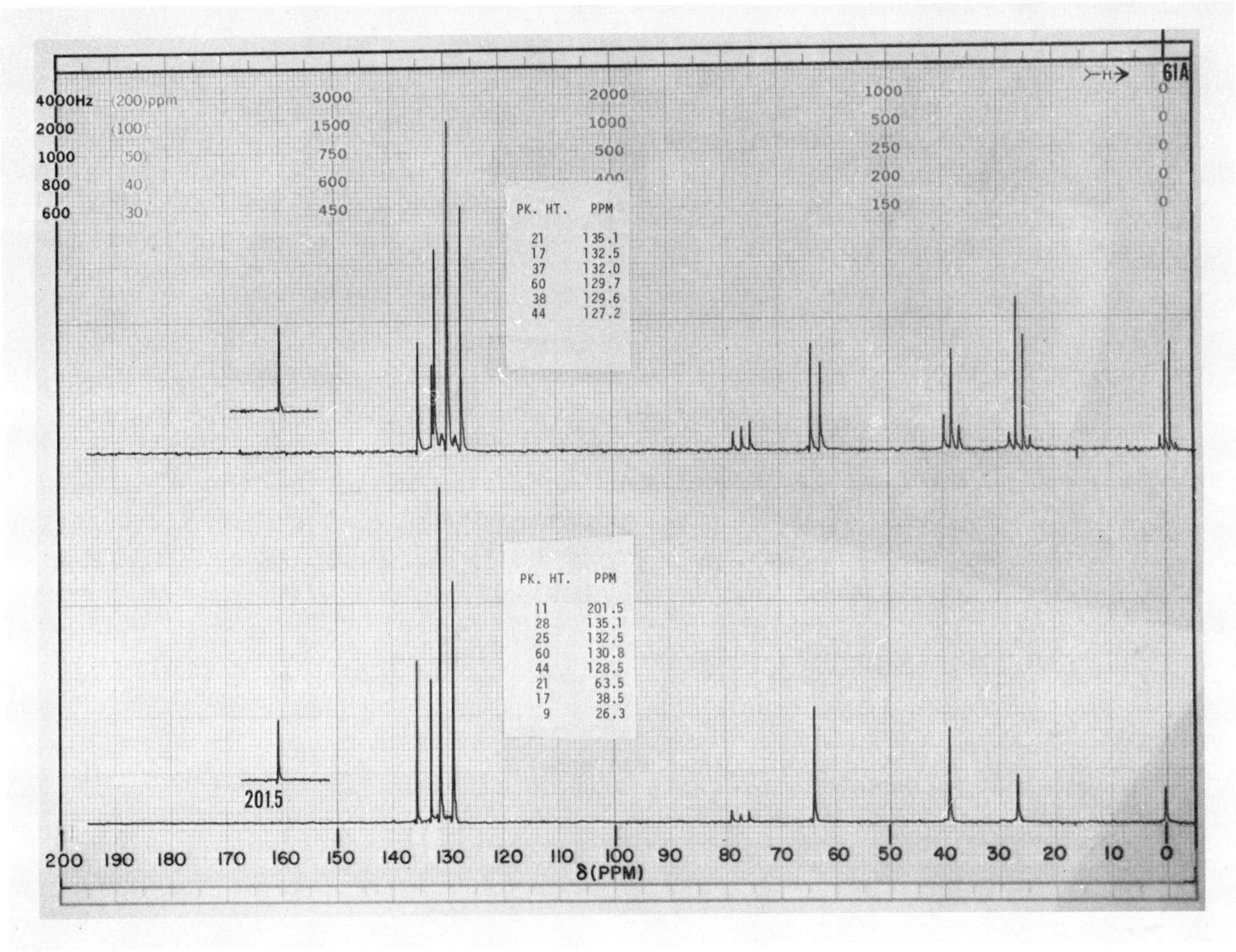

COMPOUND 61B

Compound 61B
MS: 216.012
IR: CHCl$_3$
HMR: CDCl$_3$
CMR: CDCl$_3$
Analysis: 55.3% C; 4.7% H; 32.7% Cl

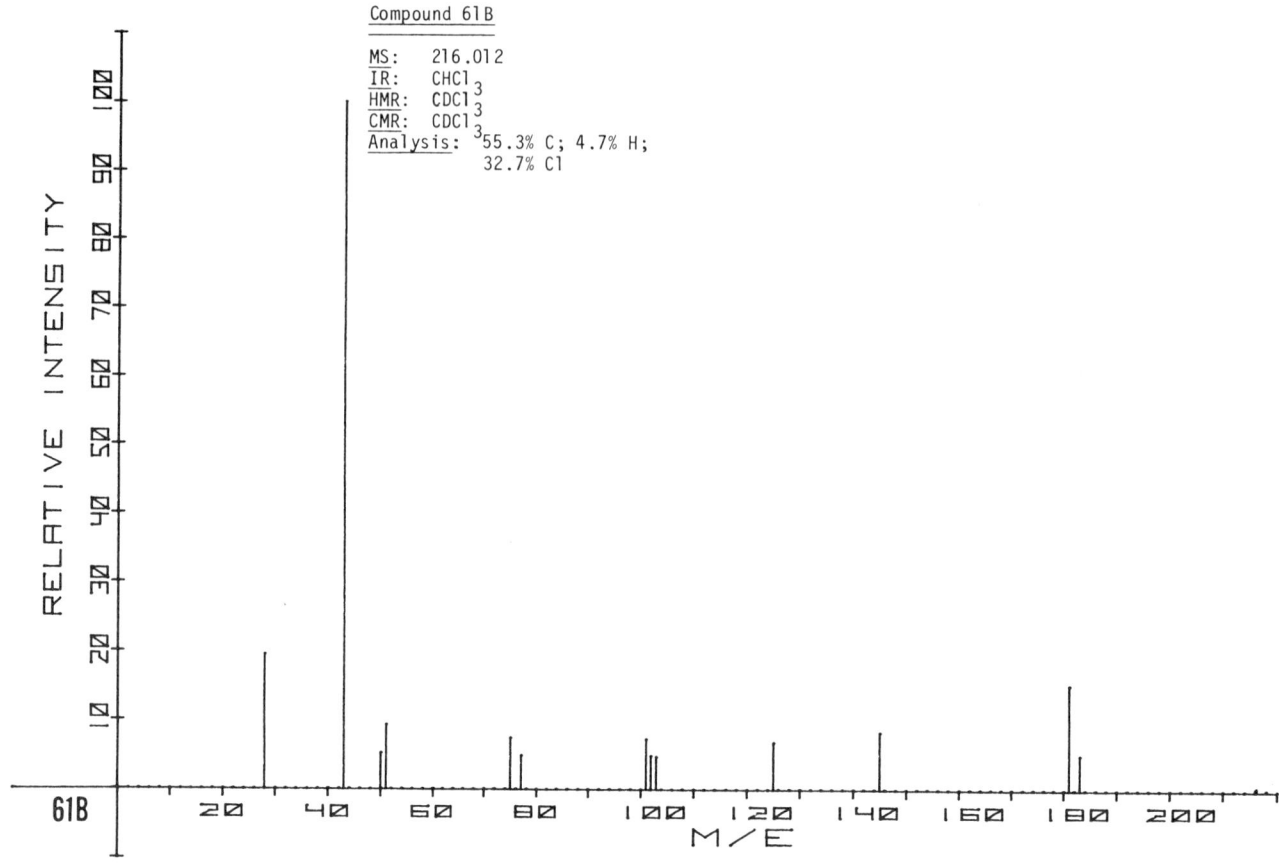

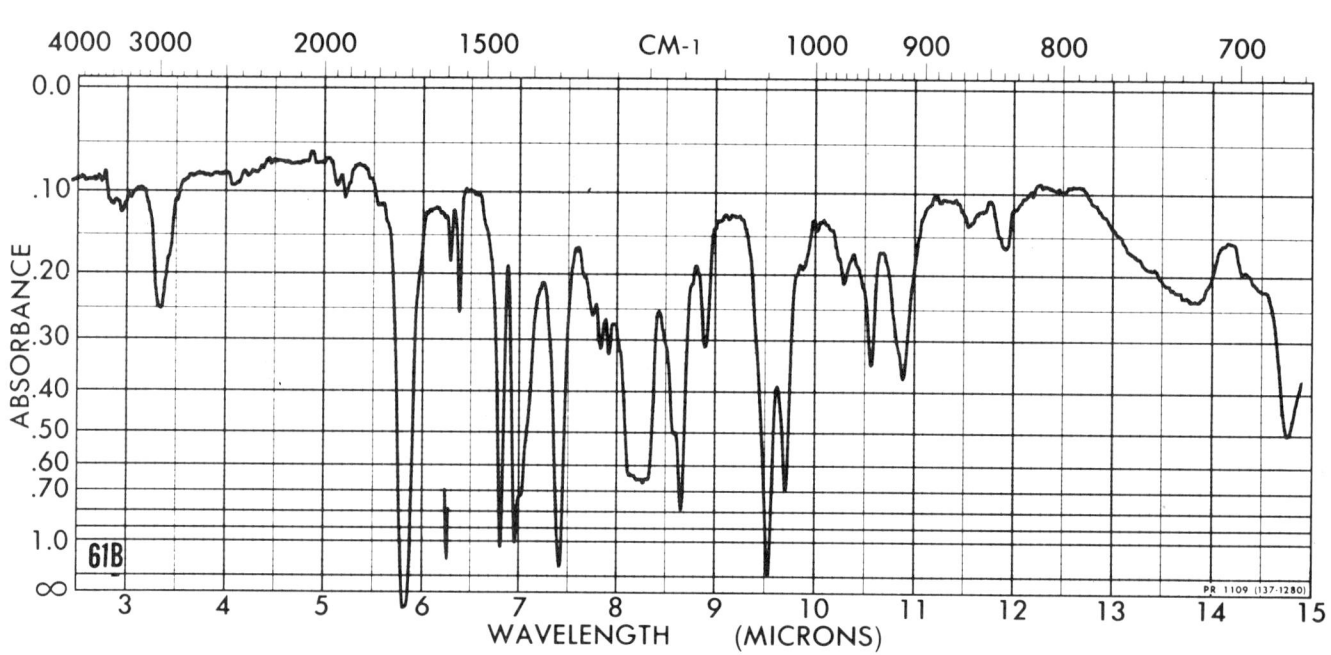

61B
IRR. 7.20
IRR. 4.47

61B
4000Hz (200)ppm 3000 2000 1000 0
2000 (100) 1500 1000 500 0
1000 (50) 750 500 250 0
800 (40) 600 200 0
600 (30) 450 150 0
PK. HT. PPM
25 134.1
10 133.9
29 133.3
40 130.7
33 129.9
62 128.1
24 127.4
25 125.6
PK. HT. PPM
43 200.9
57 134.1
42 133.9
60 132.1
37 129.4
36 128.6
60 126.8
36 61.7
26 37.0
16 26.3
200.9
δ (PPM)

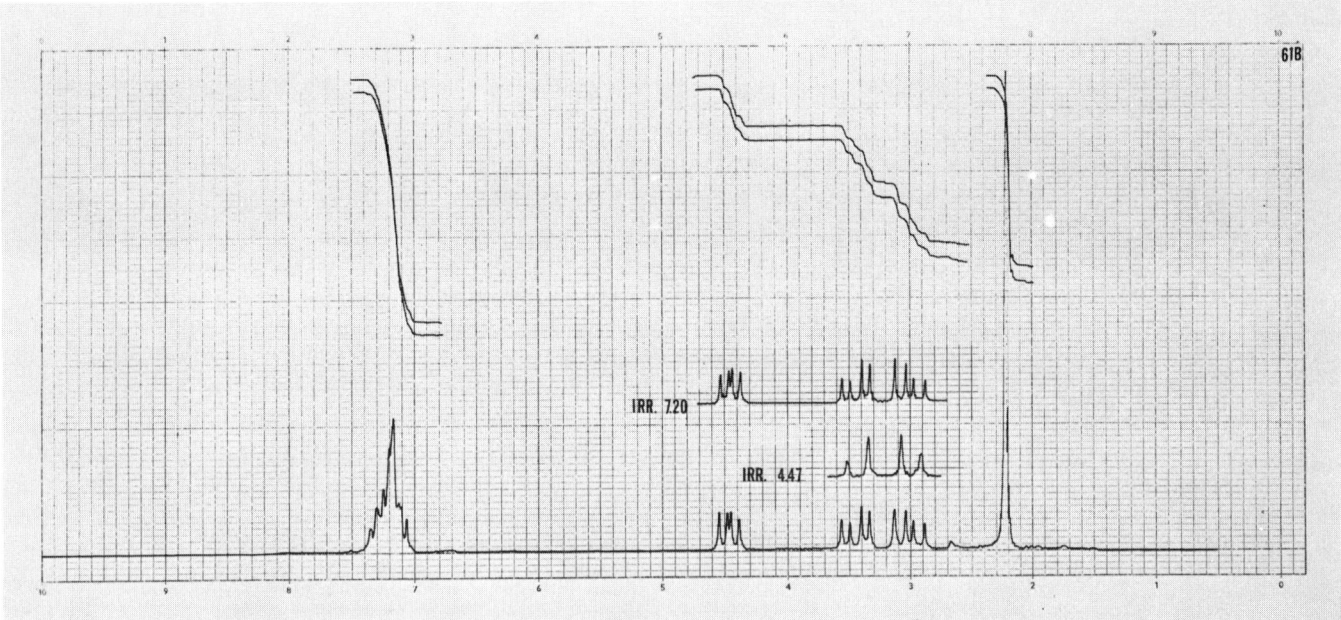

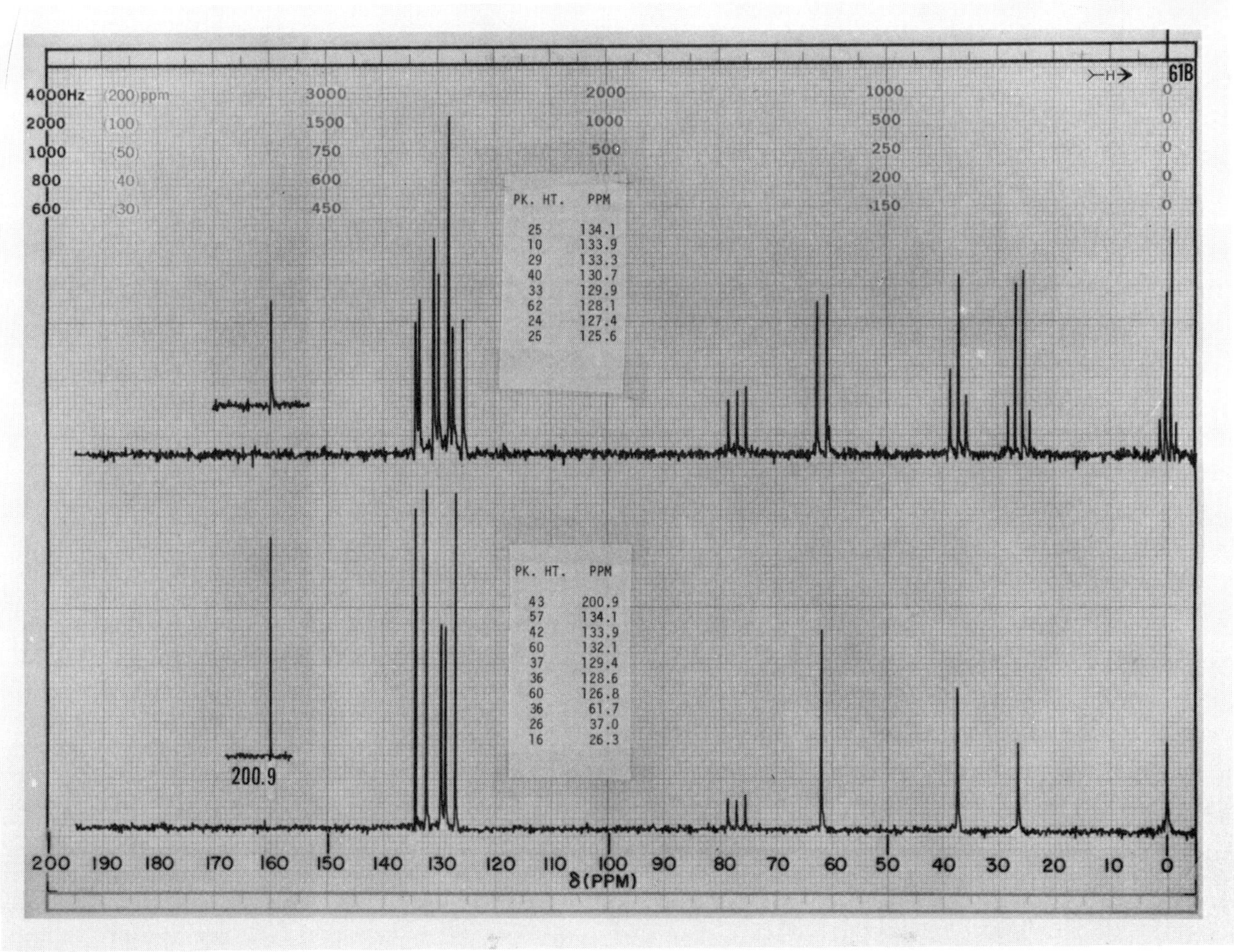

COMPOUND 61C

Compound 61C

MS: 216.011
IR: CHCl$_3$
HMR: CDCl$_3$
CMR: CDCl$_3$
Analysis: 55.3% C; 4.6% H; 32.7% Cl

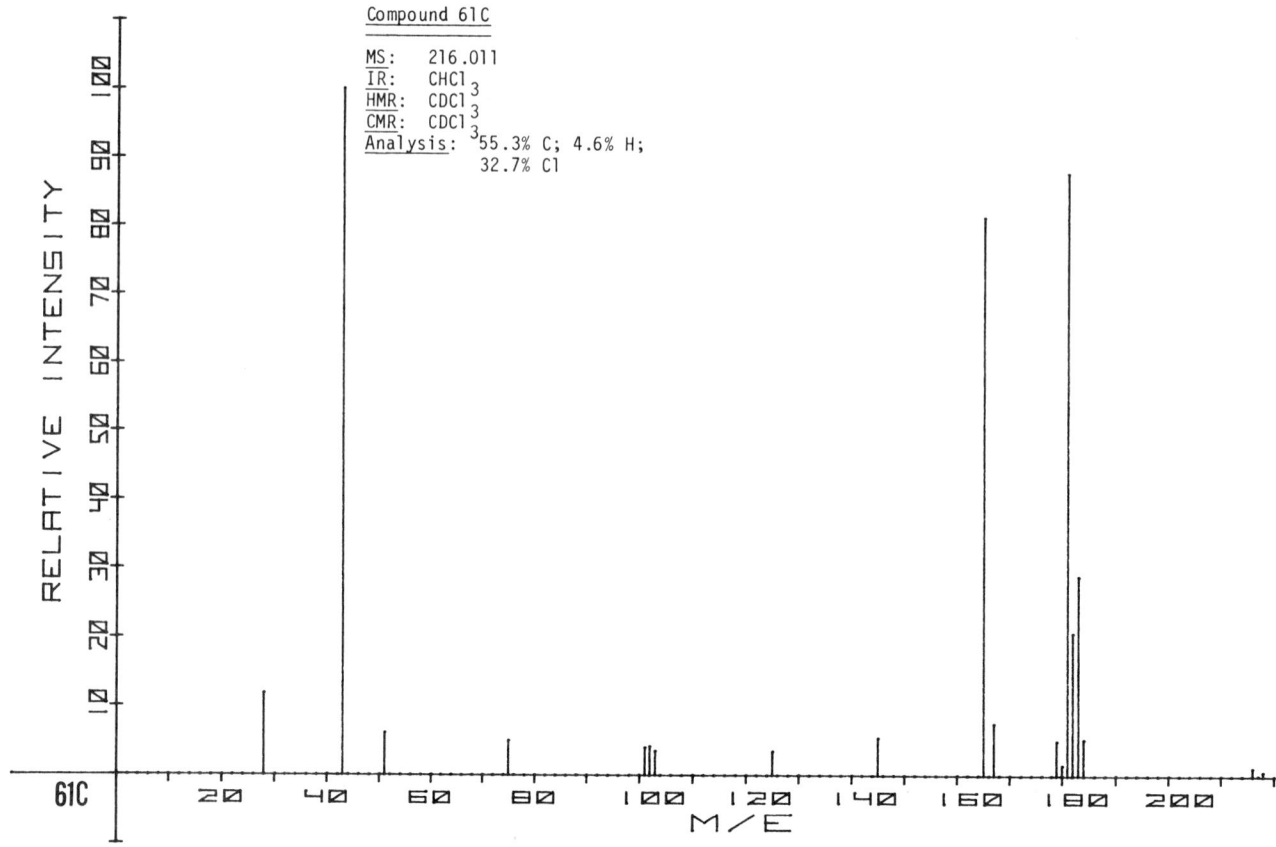

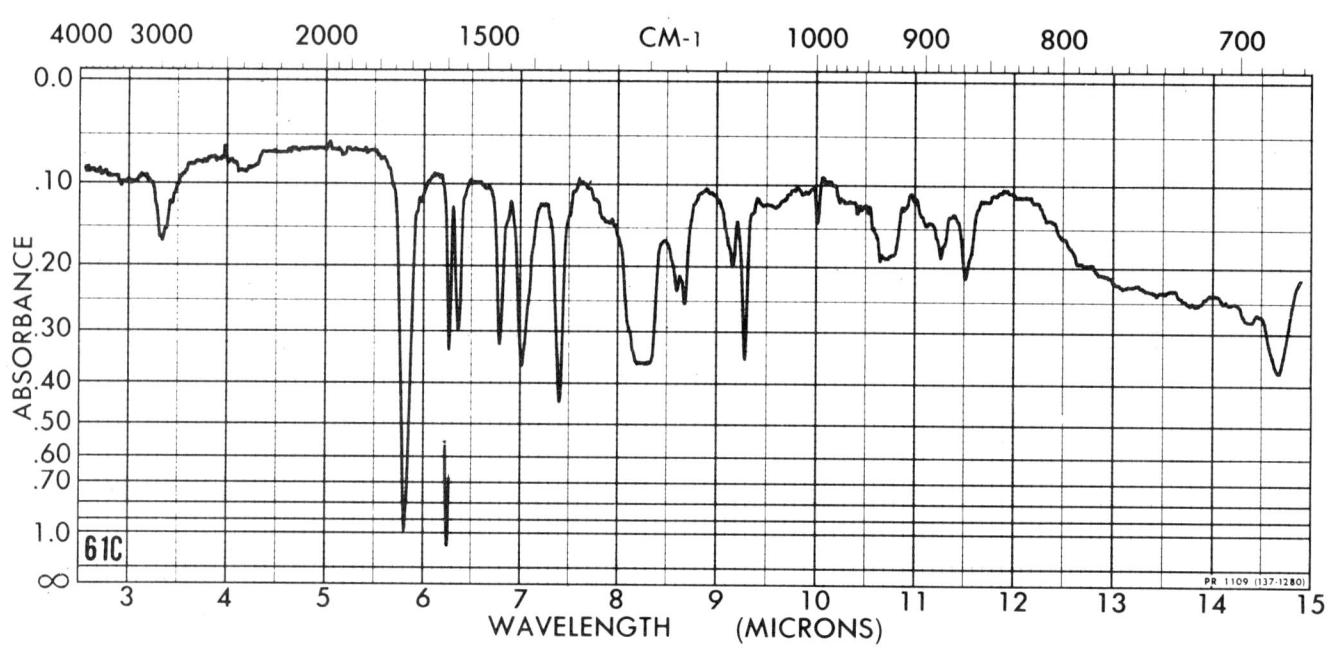

61C
IRR. 7.15
IRR. 4.28

61C
4000Hz (200)ppm 3000 2000 1000 0
2000 (100) 1500 1000 500 0
1000 (50) 750 500 250 0
800 (40) 600 200 0
600 (30) 450 150 0
PK. HT. PPM
45 138.7
37 133.8
37 131.1
38 130.5
33 128.9
52 128.4
36 128.0
28 126.4
32 125.8
PK. HT. PPM
20 201.3
60 138.7
53 133.8
44 129.8
48 129.3
47 127.7
40 127.1
28 63.3
22 38.7
16 26.3
201.3
δ(PPM)

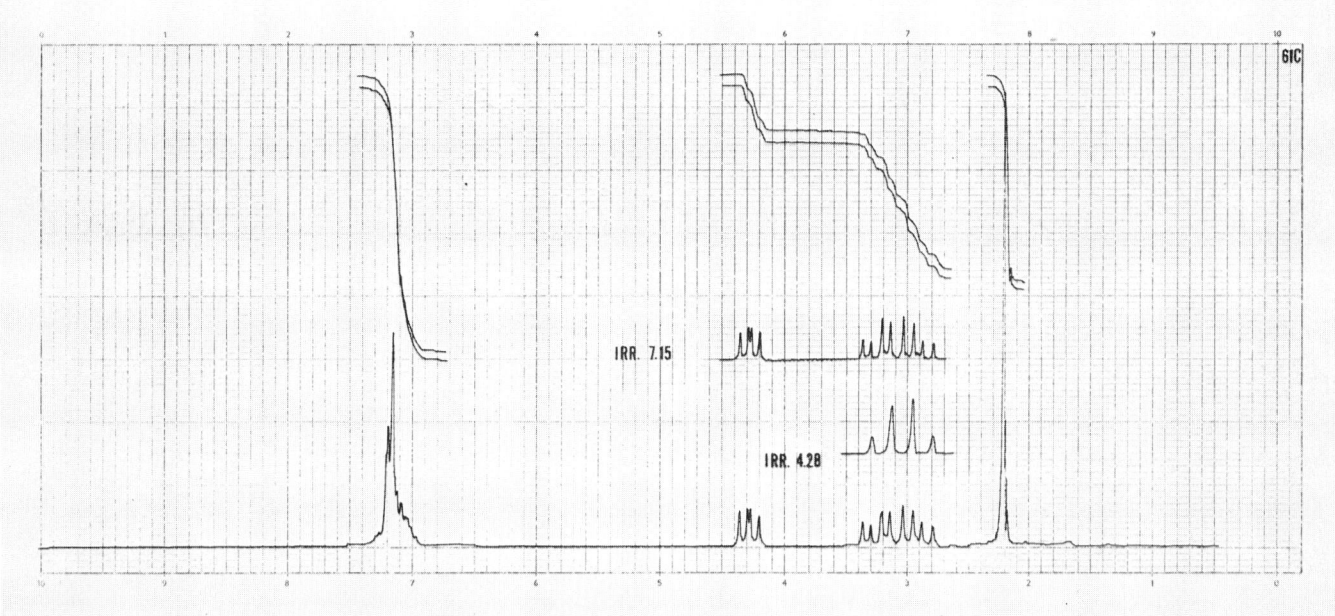

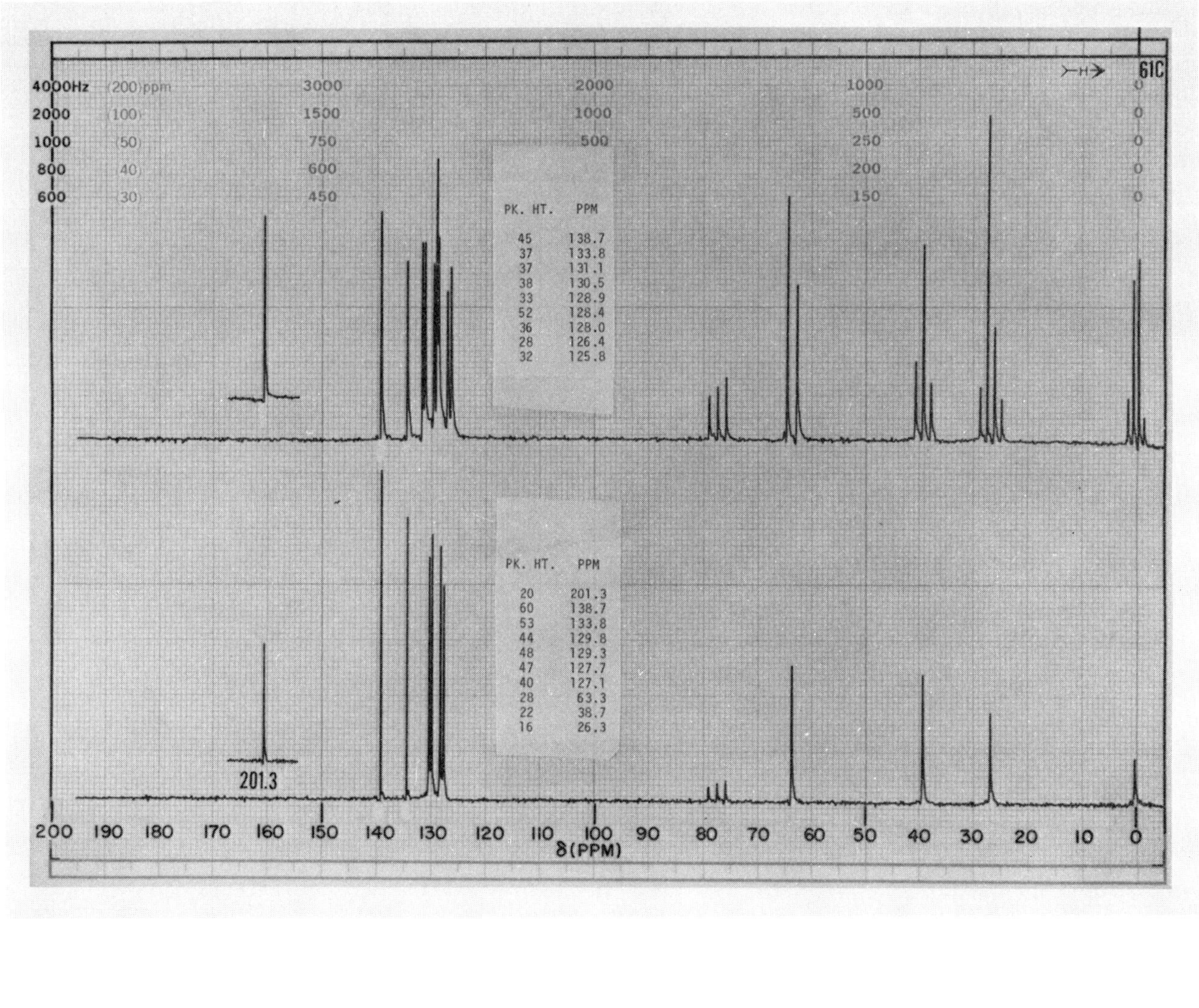

COMPOUND 62

Compound 62
MS: 162.068
IR: KBr
HMR: CDCl$_3$
CMR: CDCl$_3$
Analysis: 74.1% C; 6.2% H

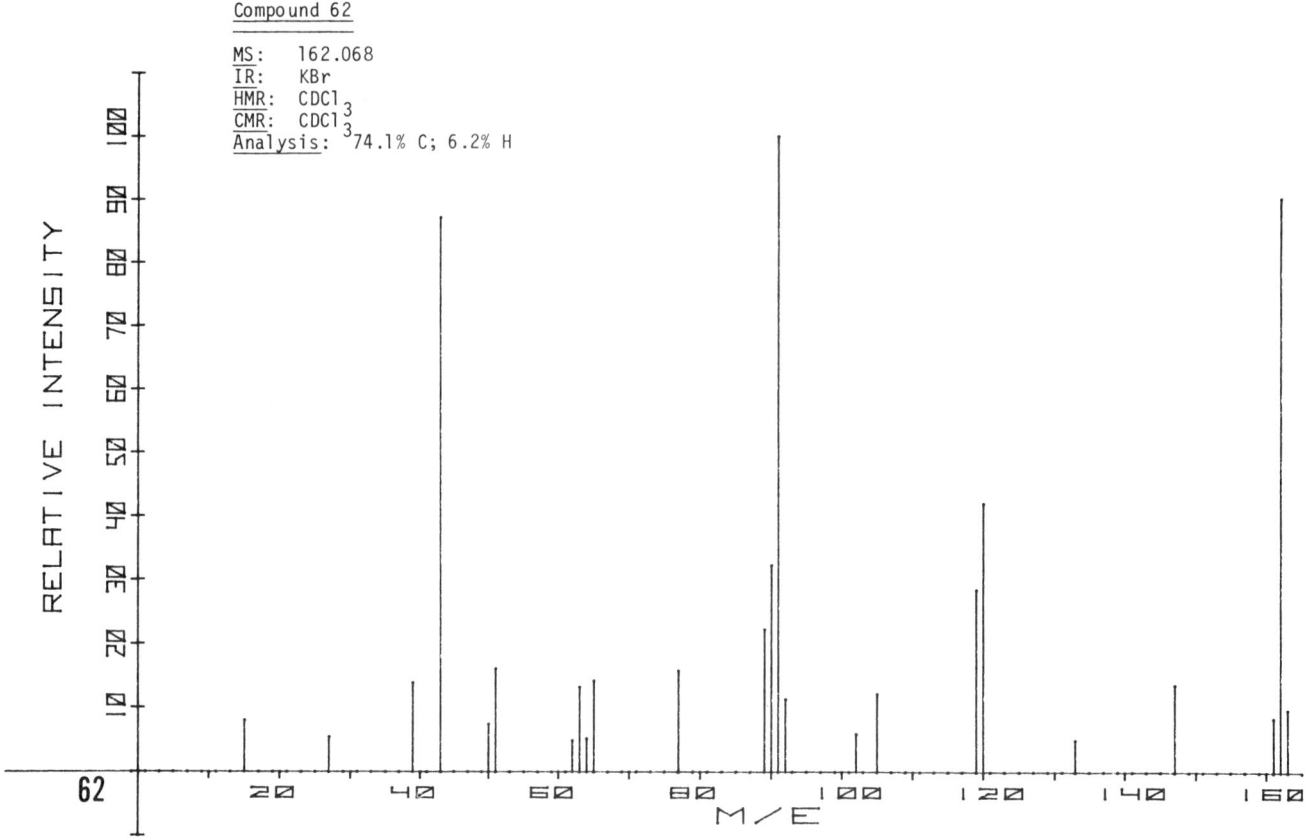

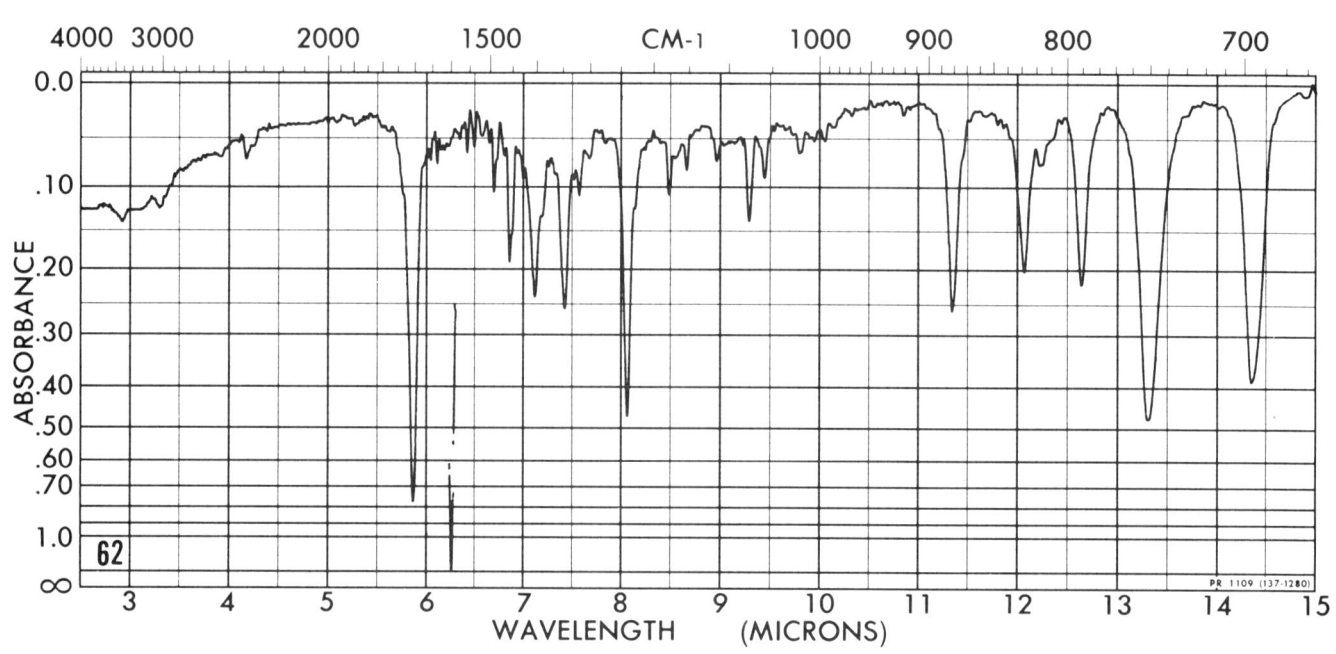

62

62
PK. HT. PPM
8 204.0
11 135.2
26 129.9
37 129.5
28 128.1
38 127.9
60 126.5
60 124.9
6 27.1
21 25.5
23 23.9
6 22.3
PK. HT. PPM
5 204.0
10 135.1
30 129.0
49 128.7
60 125.7
14 63.4
17 57.7
8 24.7
204.0
δ(PPM)

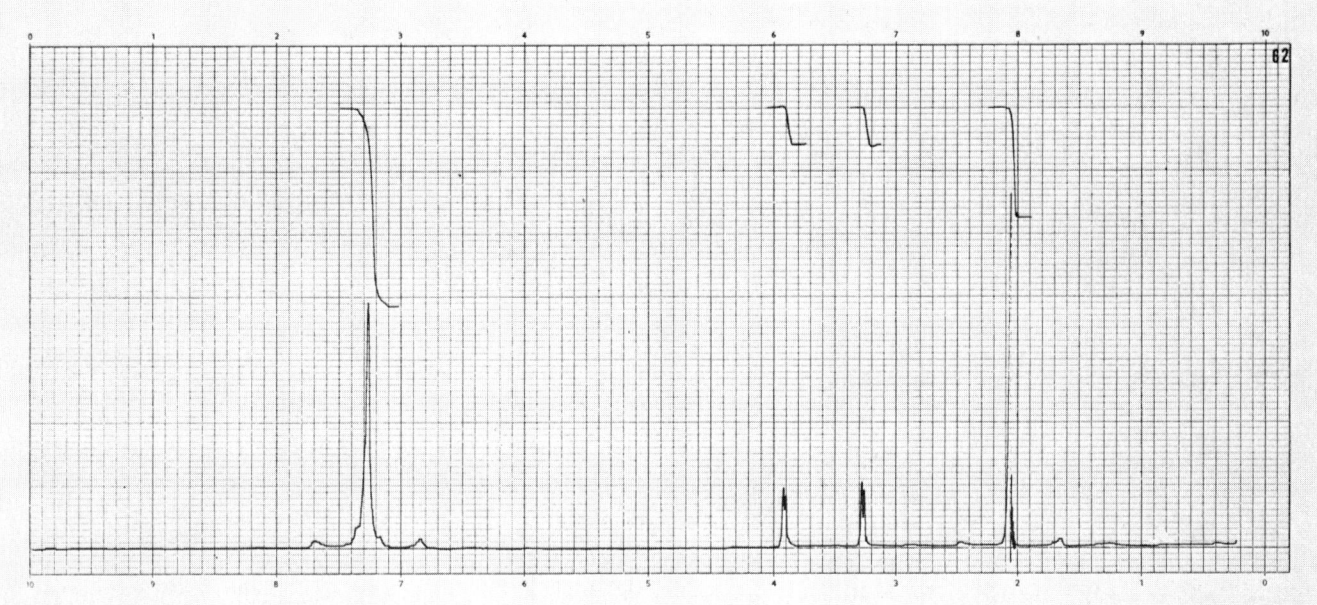

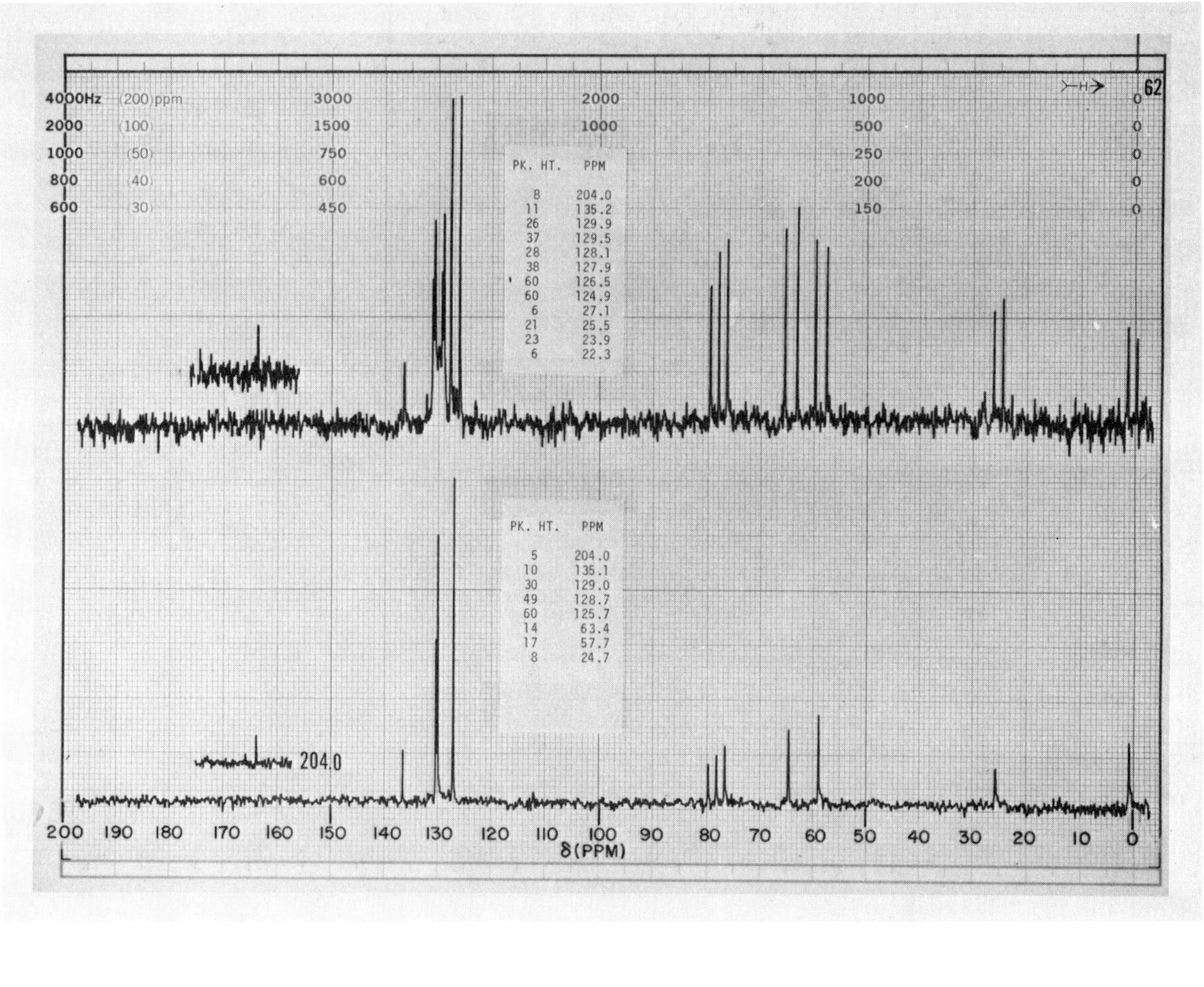

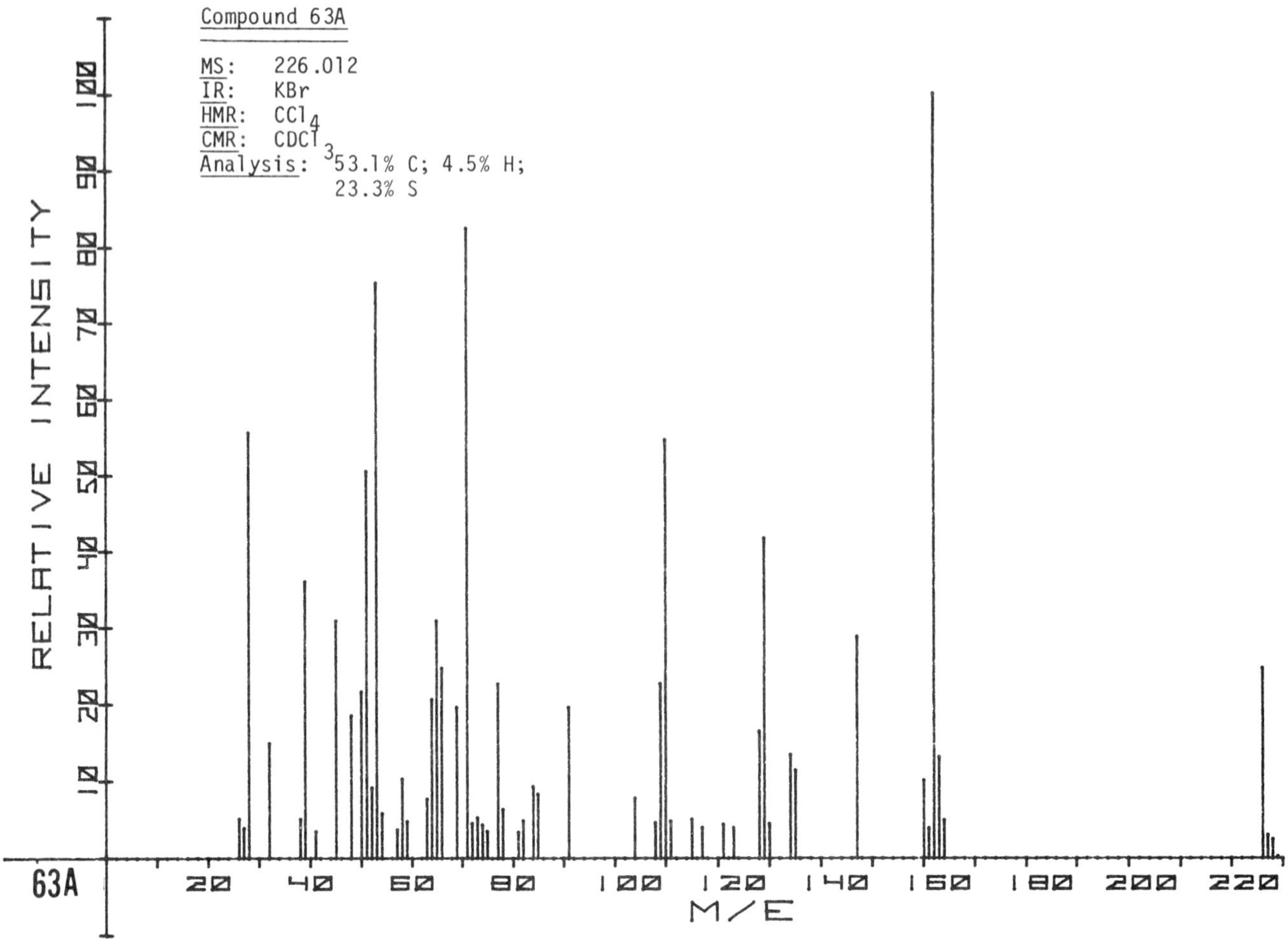

Compound 63A
MS: 226.012
IR: KBr
HMR: CCl_4
CMR: $CDCl_3$
Analysis: 53.1% C; 4.5% H; 23.3% S
RELATIVE INTENSITY
10
20
30
40
50
60
70
80
90
100
63A
20
40
60
80
100
120
140
160
180
200
220
M/E

4000 3000 2000 1500 CM-1 1000 900 800 700
0.0
.10
.20
.30
.40
.50
.60
.70
1.0
∞
ABSORBANCE
63A
3 4 5 6 7 8 9 10 11 12 13 14 15
WAVELENGTH (MICRONS)

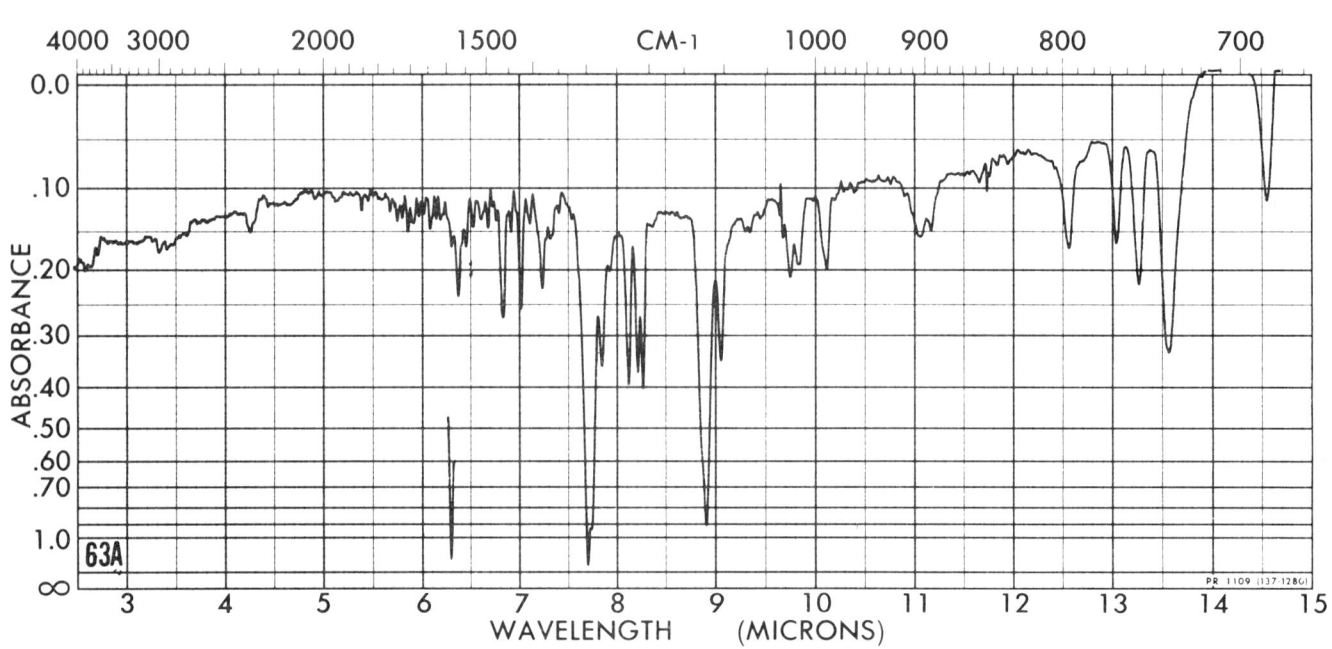

63A

63A
4000Hz (200)ppm 3000 2000 1000 0
2000 (100) 1500 1000 500 0
1000 (50) 500 250 0
800 (40) 400 200 0
600 (30) 300 150 0
PK. HT. PPM
60 133.9
52 132.0
23 130.5
18 129.8
18 129.7
21 128.7
15 127.9
19 121.4
25 119.7
PK. HT. PPM
31 132.9
17 131.9
60 129.6
32 128.9
16 120.5
22 57.7
26 57.3
200 190 180 170 160 150 140 130 120 110 100 90 80 70 60 50 40 30 20 10 0
δ(PPM)

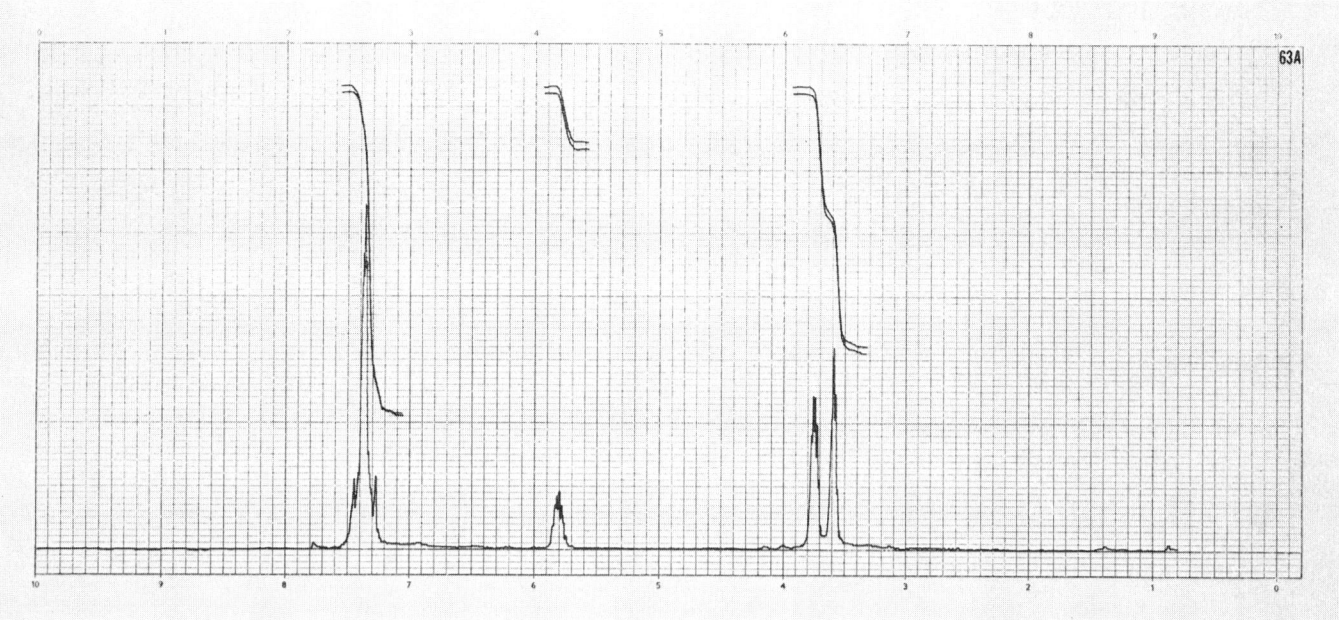

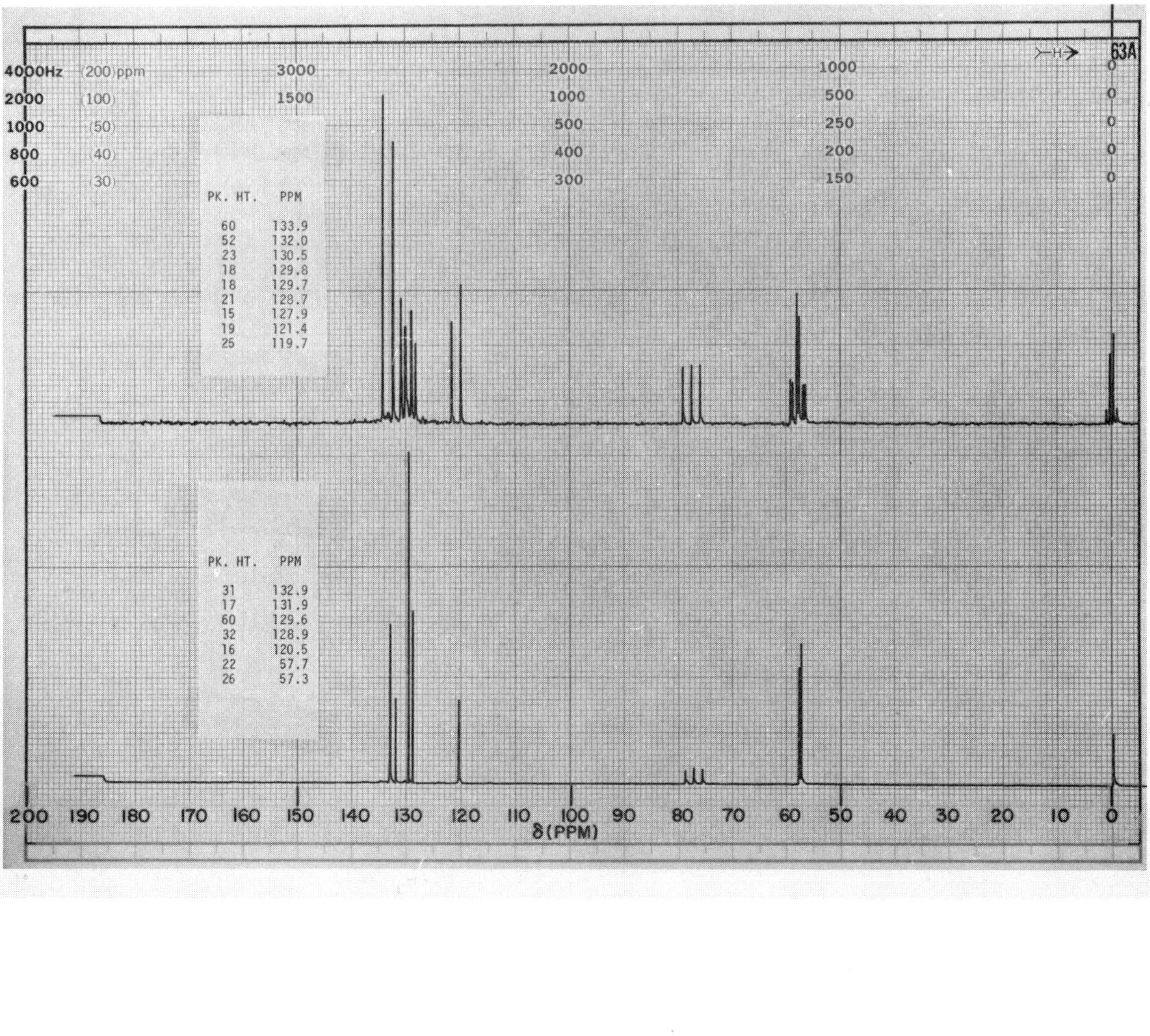

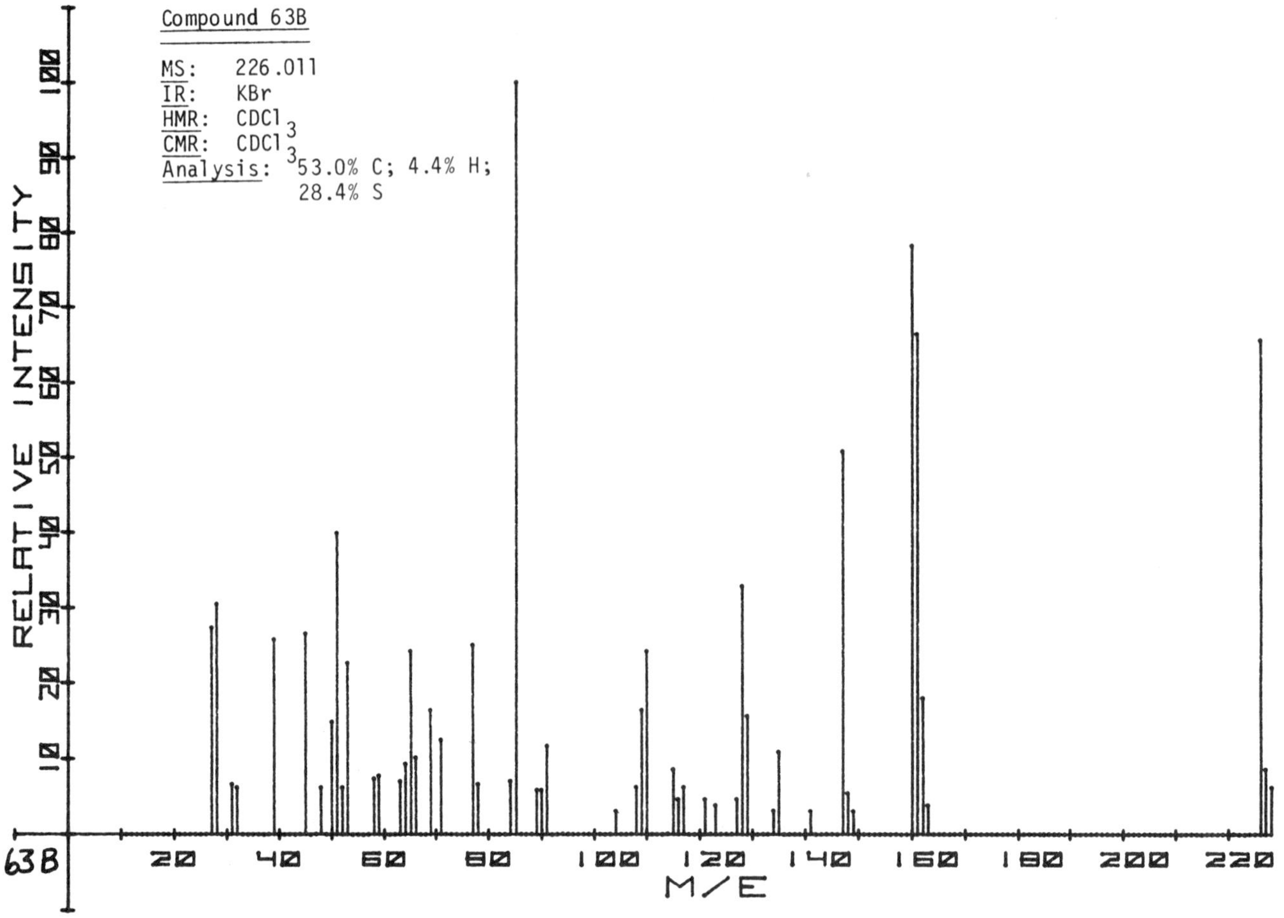

Compound 63B
MS: 226.011
IR: KBr
HMR: CDCl3
CMR: CDCl3
Analysis: 53.0% C; 4.4% H; 28.4% S
RELATIVE INTENSITY
M/E
63B

CM-1
ABSORBANCE
WAVELENGTH (MICRONS)
63B

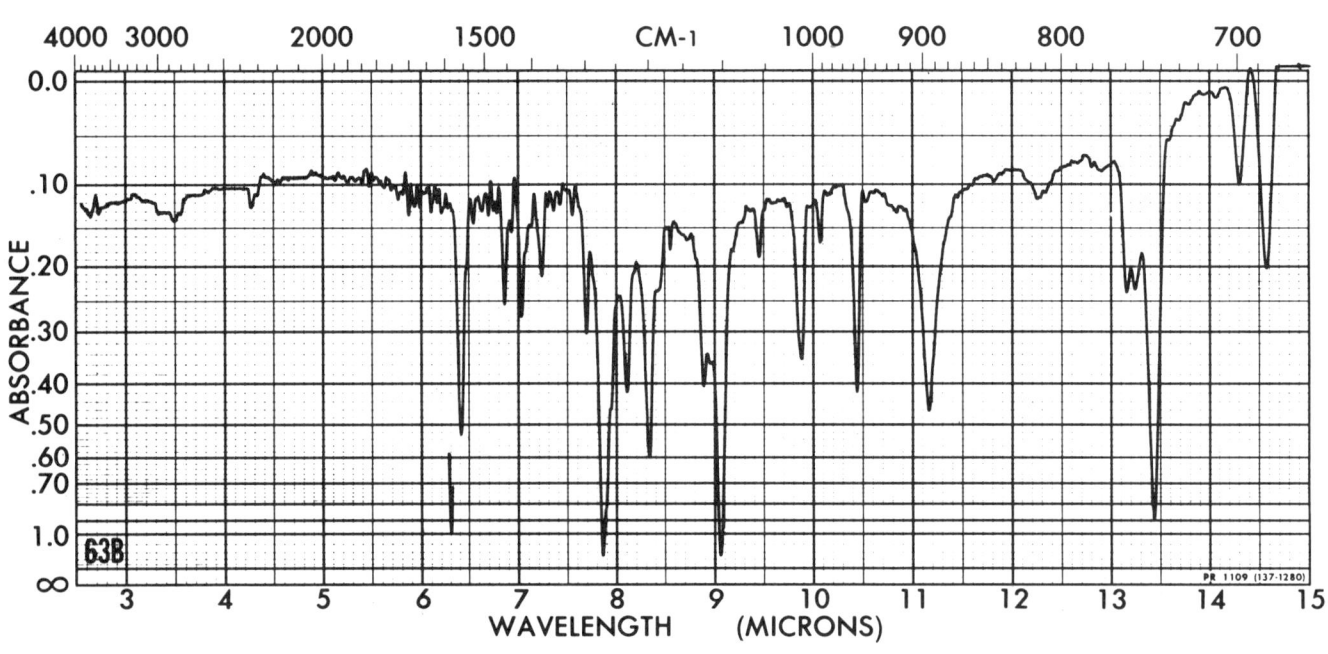

63B

4000Hz (200)ppm 3000 2000 1000 0
2000 (100) 1000 500 0
1000 (50) 500 250 0
800 (40) 400 200 0
600 (30) 300 150 0
63B
PK. HT. PPM
11 155.7
60 135.0
35 130.6
59 130.1
11 128.0
41 120.5
28 50.3
24 29.0
PK. HT. PPM
60 135.9
36 134.0
19 131.5
32 131.0
5 130.1
16 129.6
26 129.1
11 128.0
200 190 180 170 160 150 140 130 120 110 100 90 80 70 60 50 40 30 20 10 0
δ(PPM)

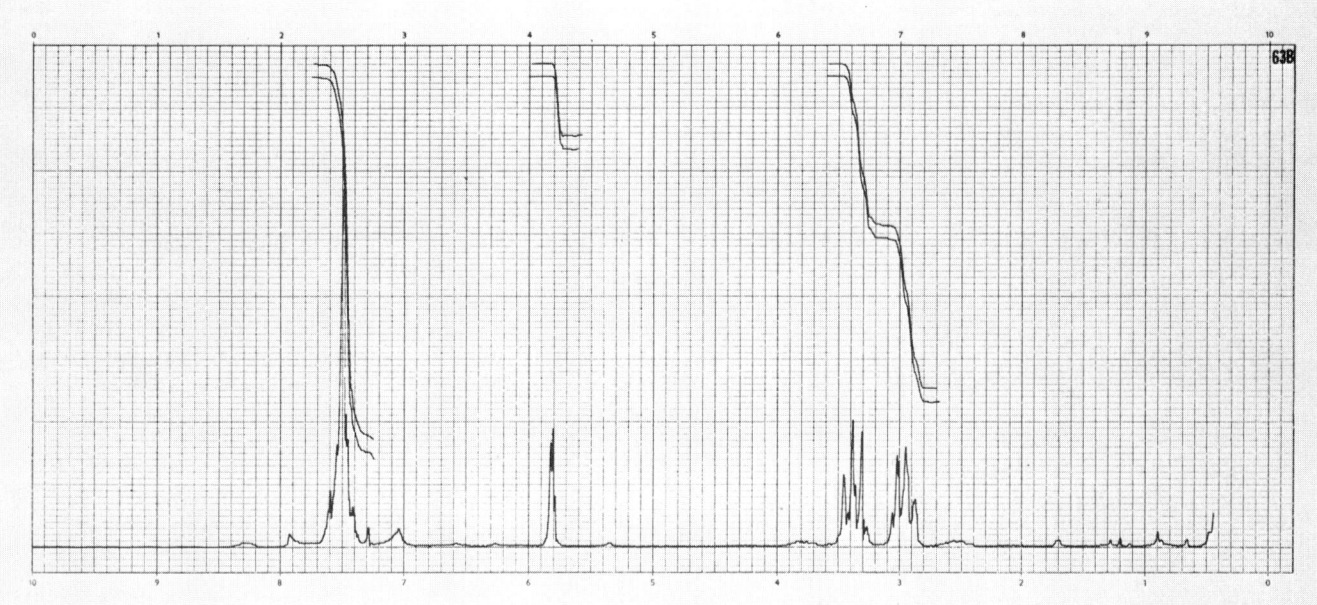

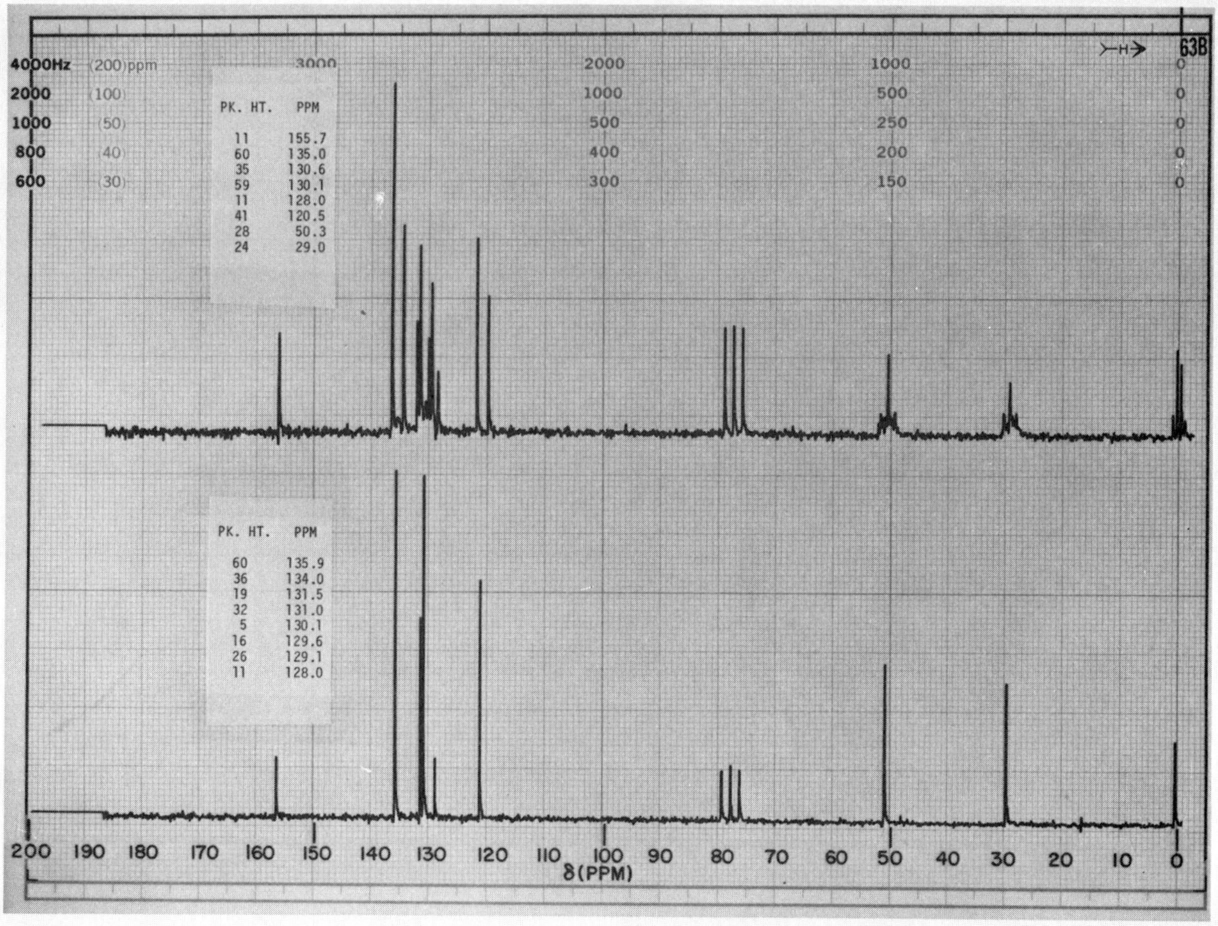

COMPOUND 64

Compound 64

MS: 132.094
IR: neat
HMR: CDCl₃
CMR: CDCl₃
Analysis: 90.9% C; 9.1% H

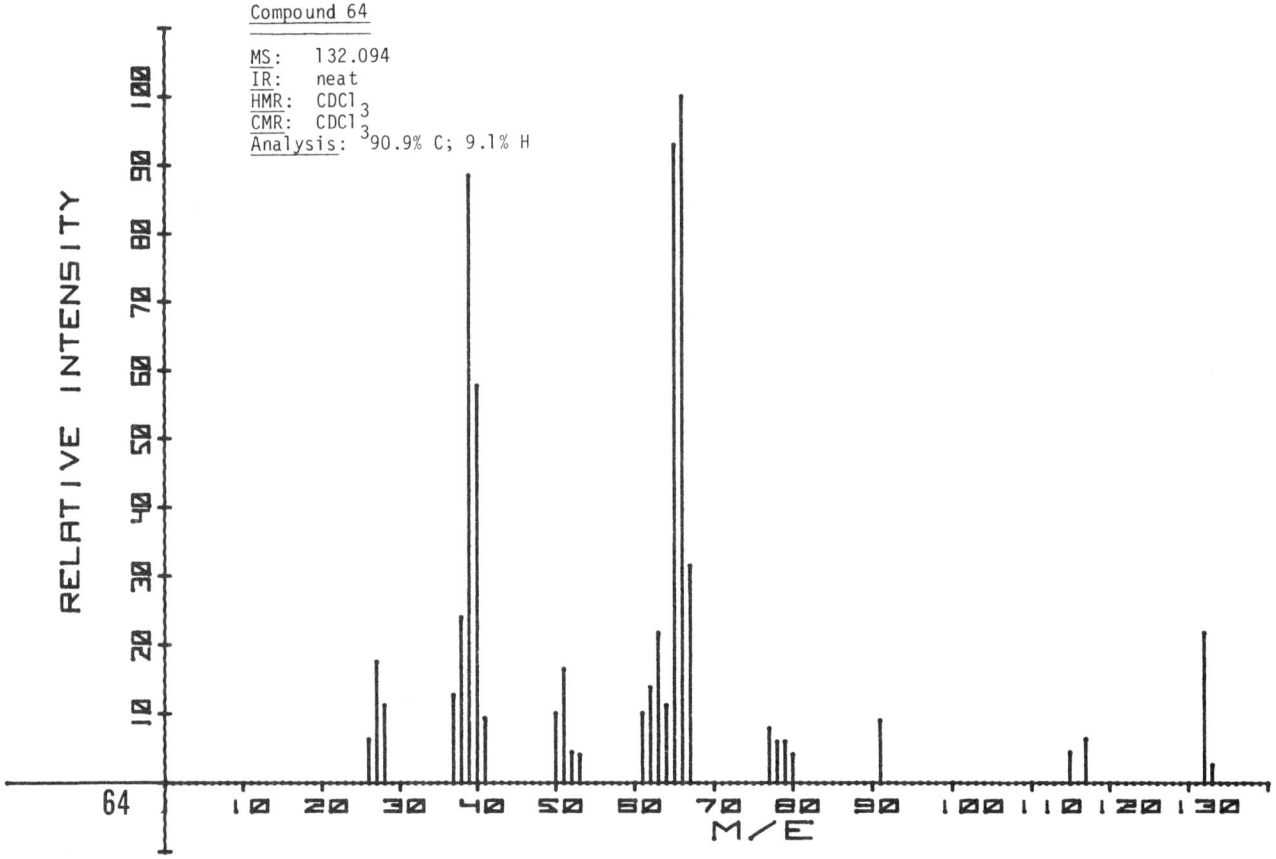

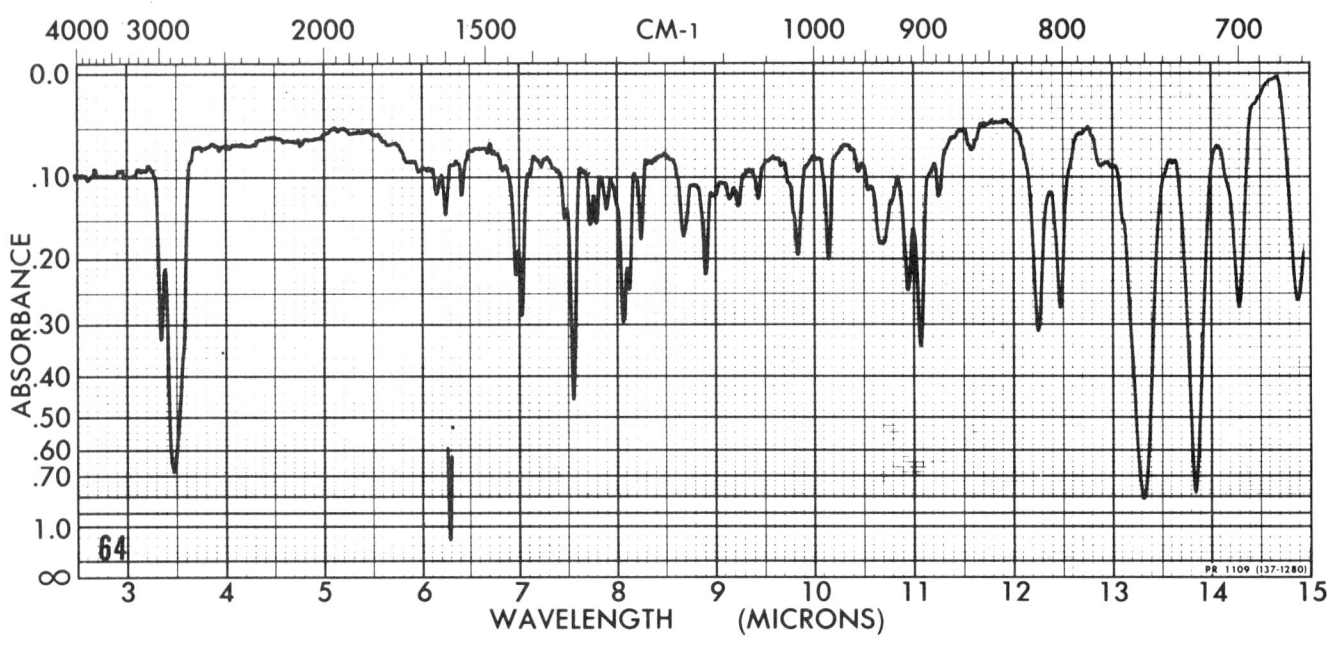

64
IRR. 3.18
IRR. 2.80

4000Hz (200)ppm 3000 2000 1000 0
2000 (100) 1500 1000 500 0
1000 (50) 500 250 0
800 (40) 400 200 0
600 (30) 300 150 0
64
PK. HT. PPM
28 137.2
27 134.8
26 133.5
36 133.1
37 133.0
26 131.0
36 130.9
41 130.8
19 47.2
18 46.2
20 45.5
18 44.5
PK. HT. PPM
46 136.0
47 132.2
60 132.0
49 131.9
43 54.9
39 50.4
34 46.3
34 45.3
39 41.3
47 34.7
200 190 180 170 160 150 140 130 120 110 100 90 80 70 60 50 40 30 20 10 0
δ(PPM)

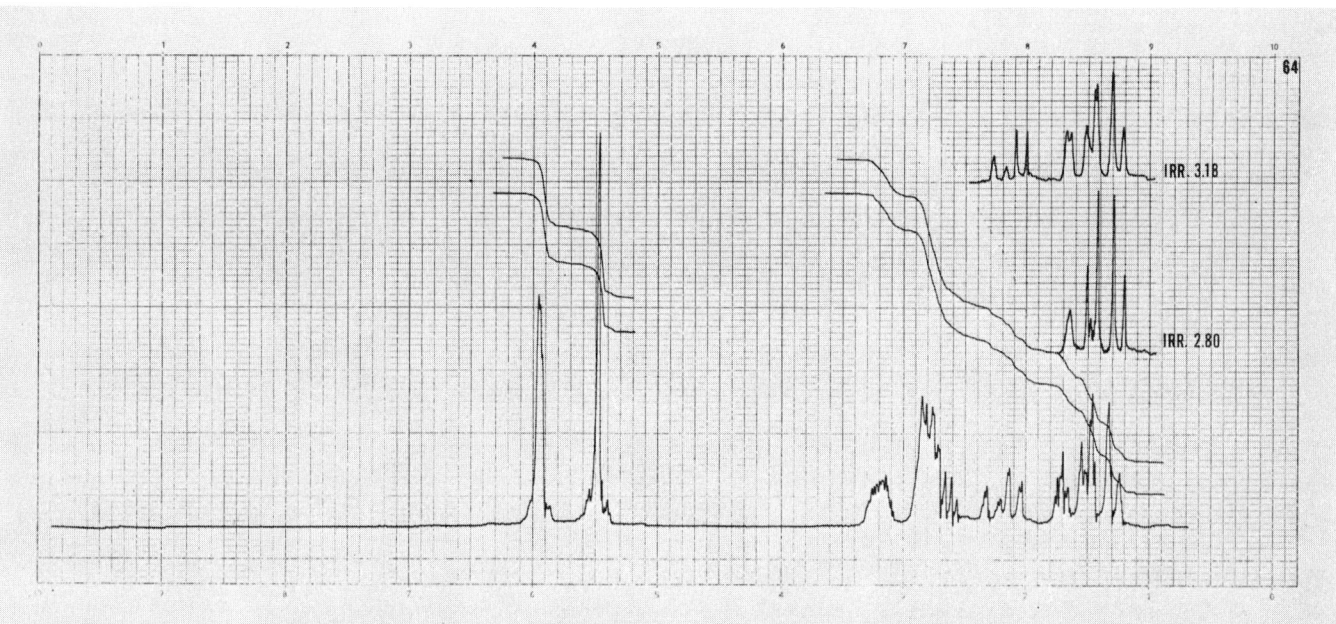

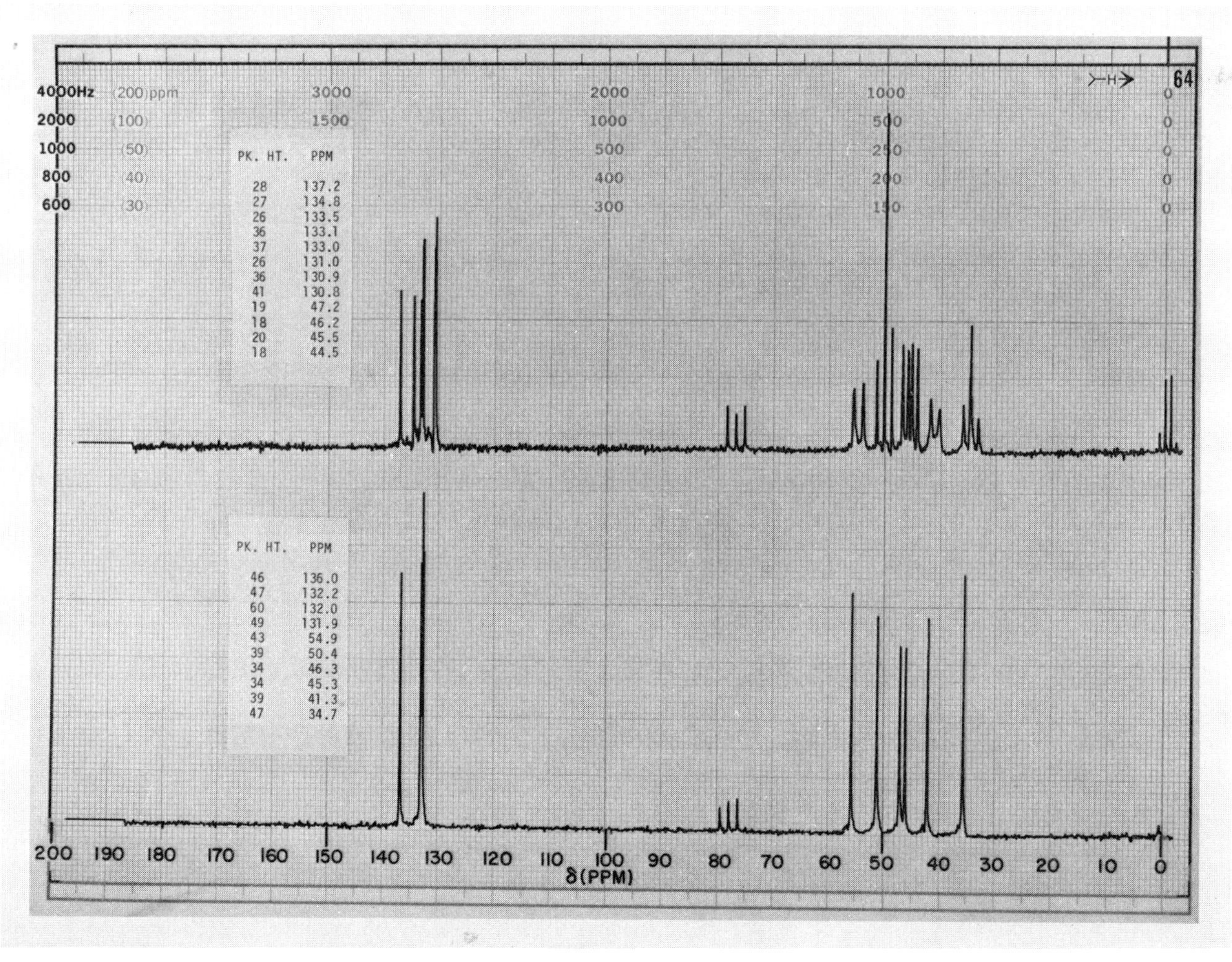

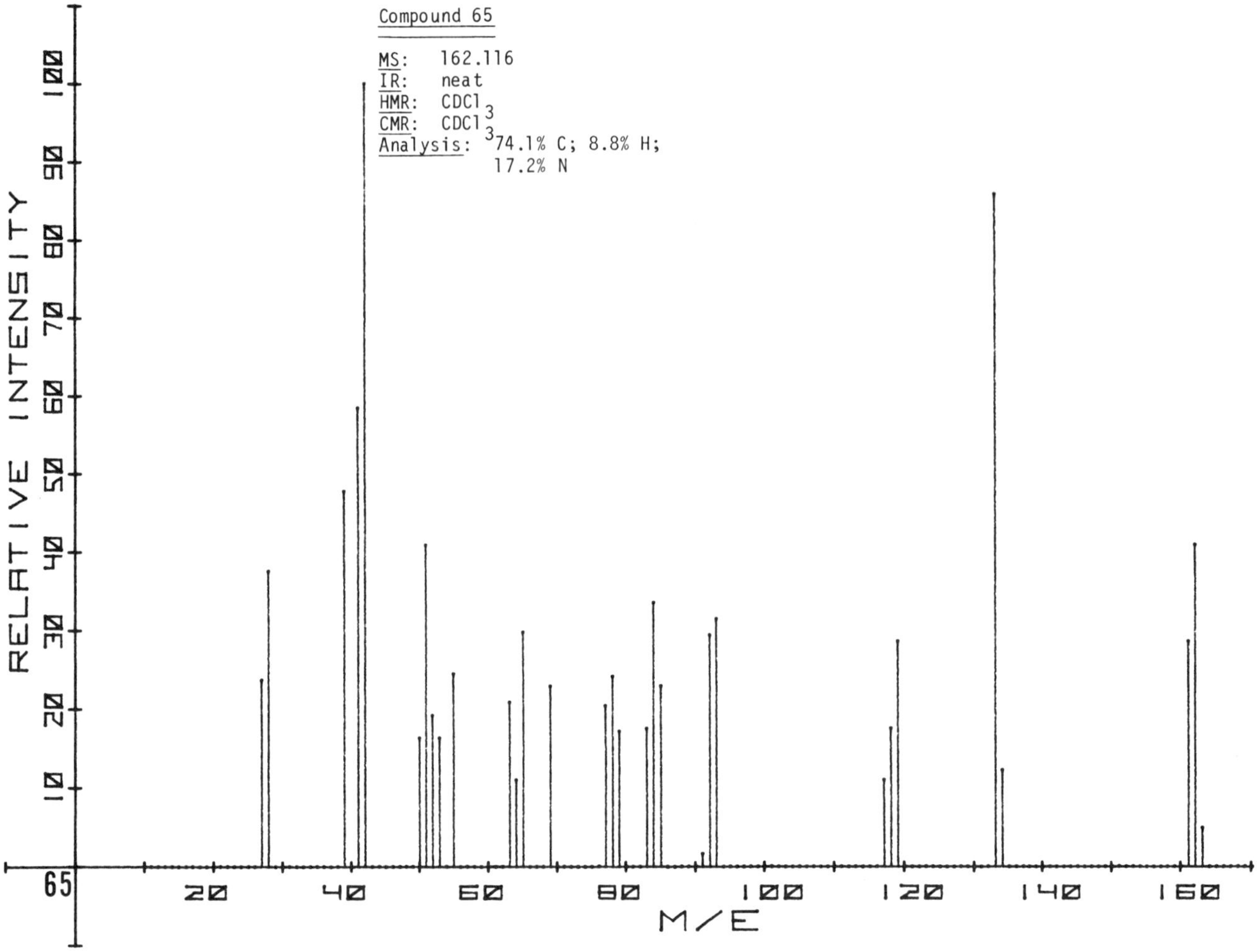

Compound 65
MS: 162.116
IR: neat
HMR: CDCl3
CMR: CDCl3
Analysis: 74.1% C; 8.8% H; 17.2% N
RELATIVE INTENSITY
M/E
65

CM-1
ABSORBANCE
WAVELENGTH (MICRONS)
65
PR 1109 (137-1280)

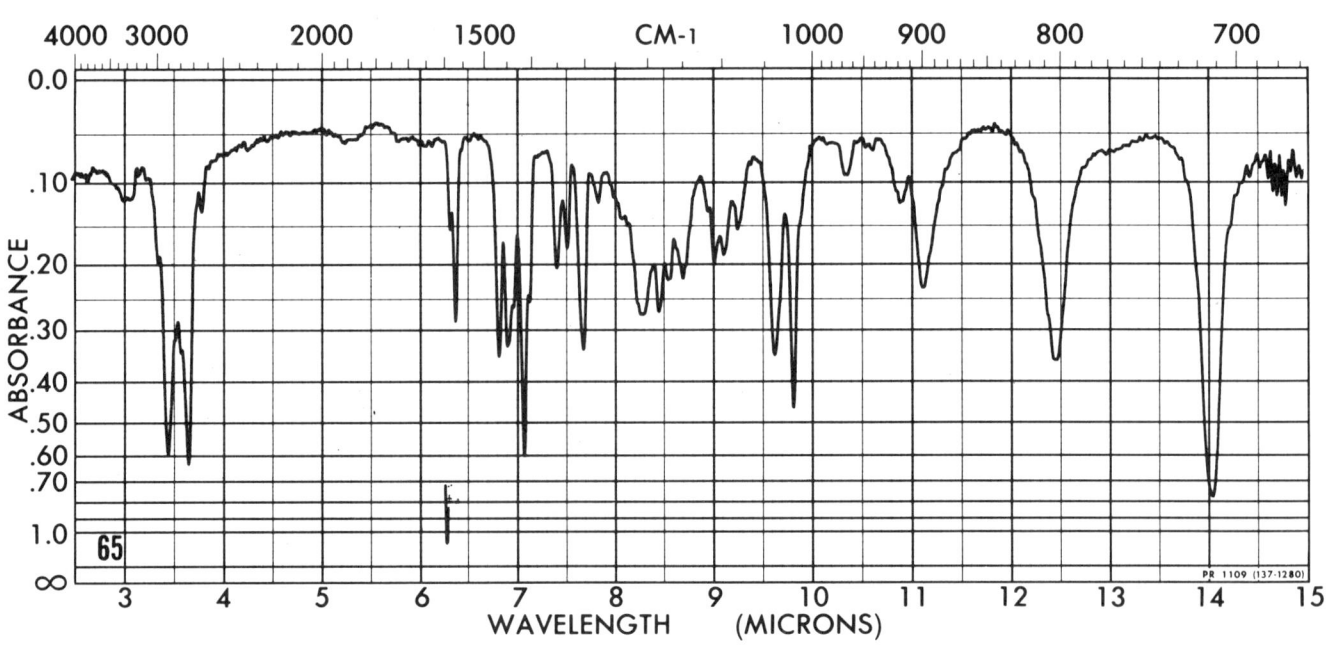

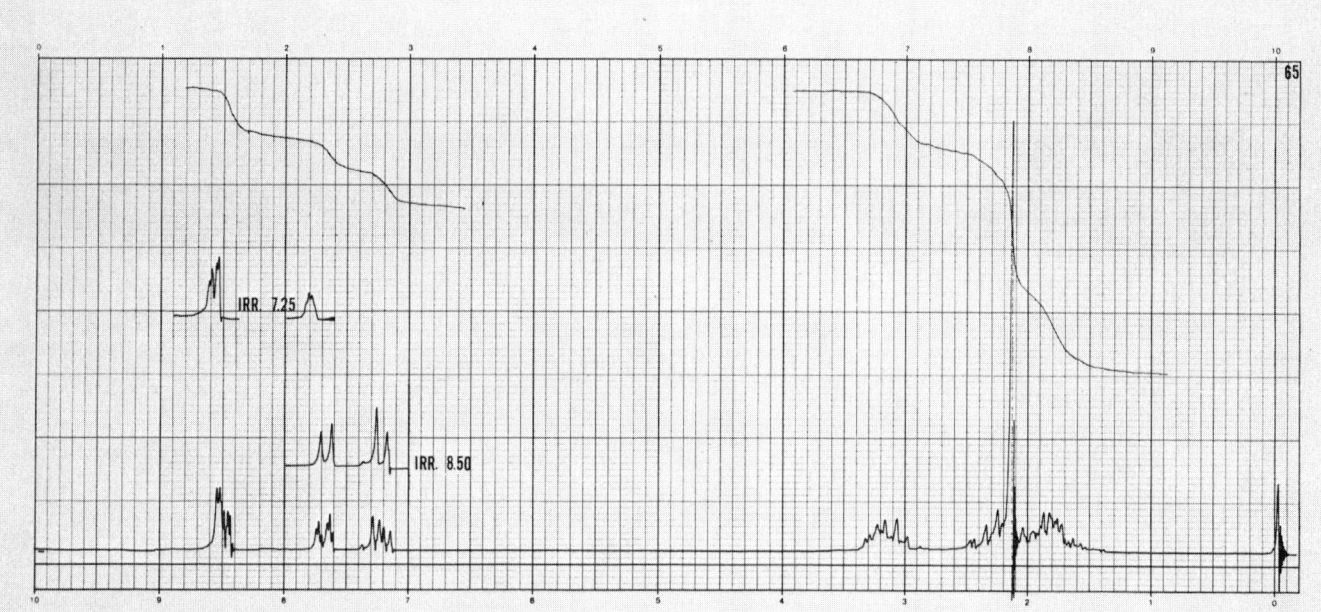

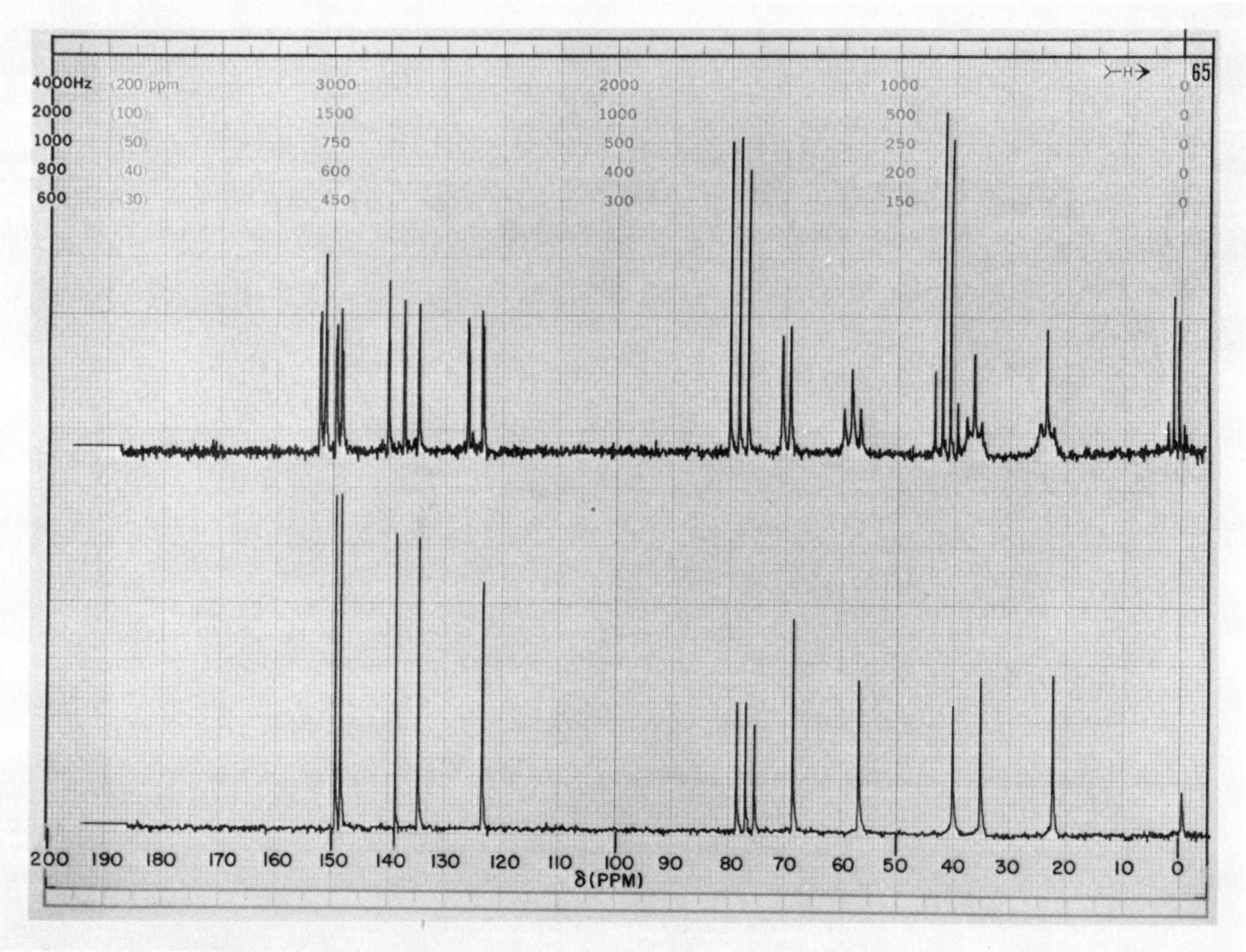

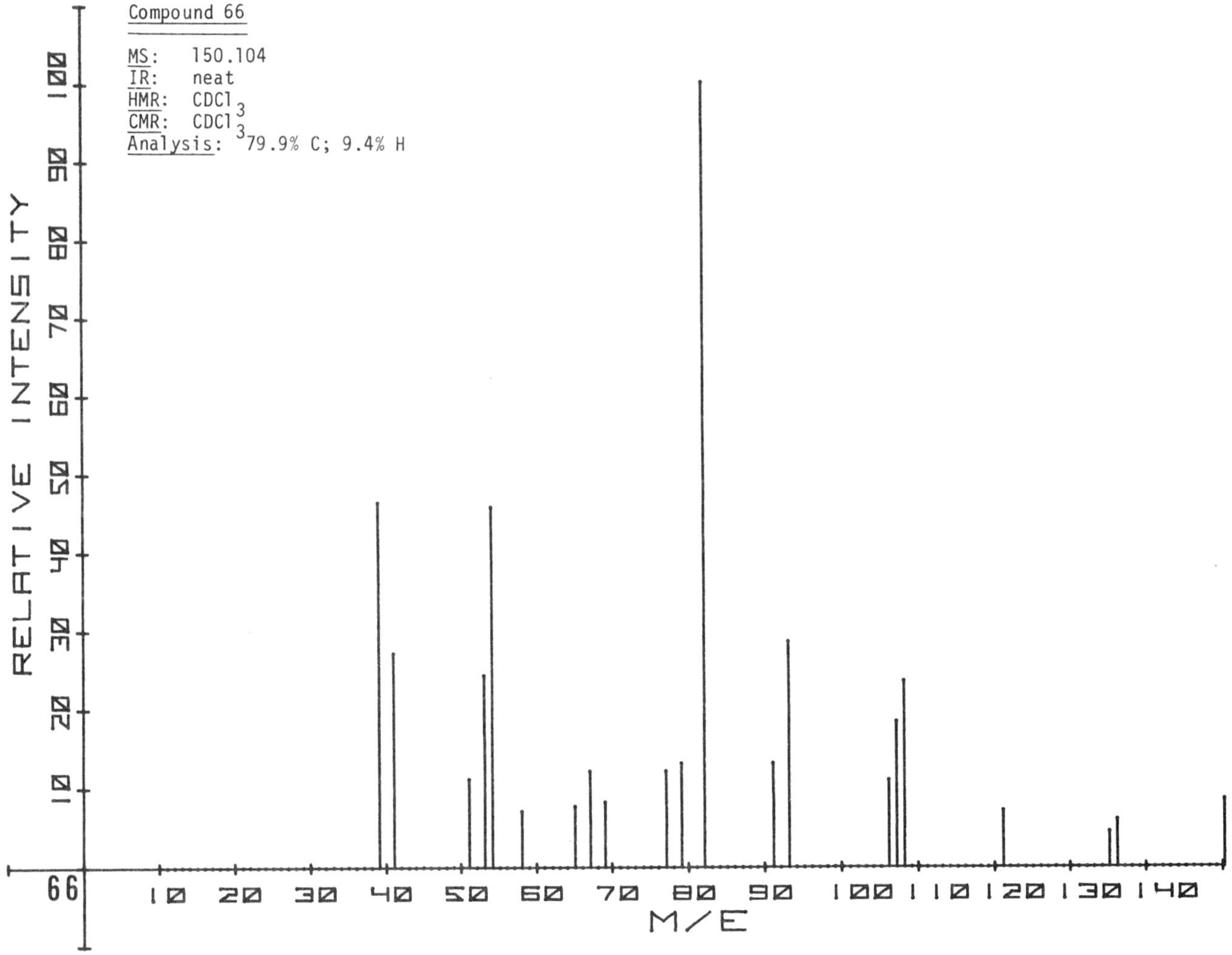

COMPOUND 66

Compound 66

MS: 150.104
IR: neat
HMR: CDCl$_3$
CMR: CDCl$_3$
Analysis: 79.9% C; 9.4% H

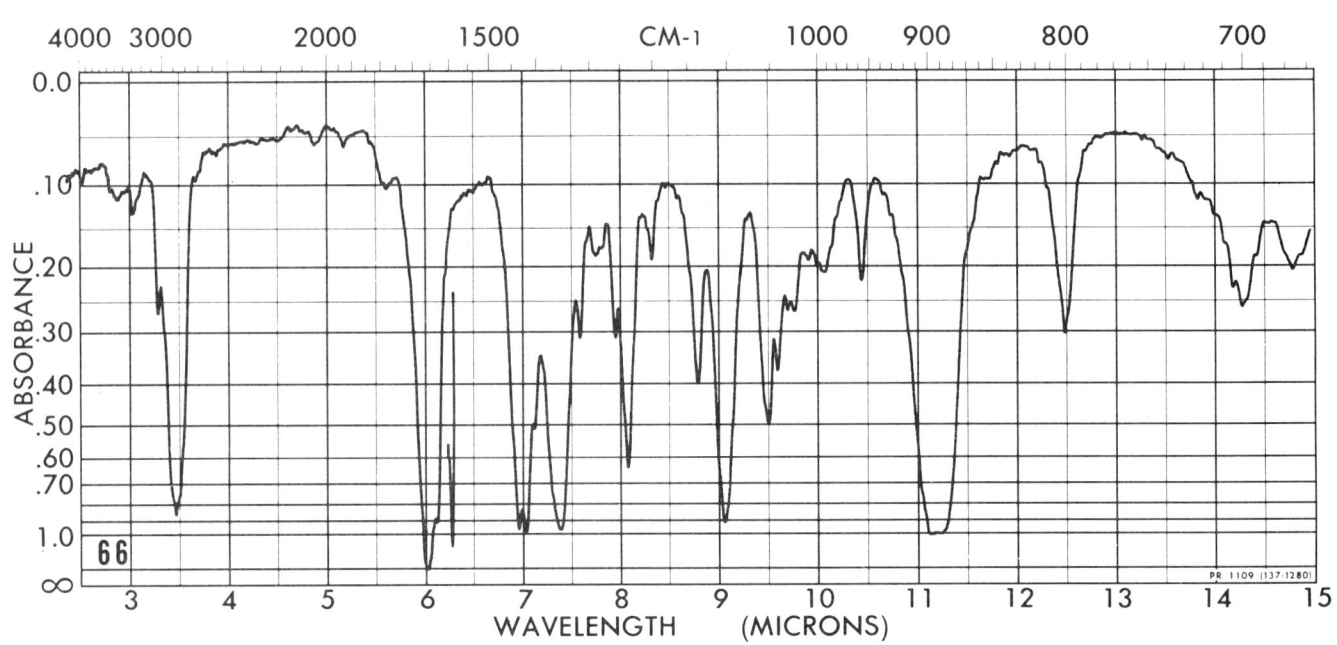

66
IRR. 2.68
IRR. 1.78

66
4000Hz (200)ppm 3000 2000 1000 0
2000 (100) 1500 1000 500 0
1000 (50) 750 500 250 0
800 (40) 600 200 0
600 (30) 450 150 0
PK. HT. PPM
19 146.7
38 145.3
40 143.4
18 45.1
49 43.3
18 41.6
23 41.4
PK. HT. PPM
11 199.3
19 146.7
51 144.4
19 135.4
60 110.5
48 43.2
56 42.5
46 31.3
31 20.5
29 15.6
199.3
200 190 180 170 160 150 140 130 120 110 100 90 80 70 60 50 40 30 20 10 0
δ(PPM)

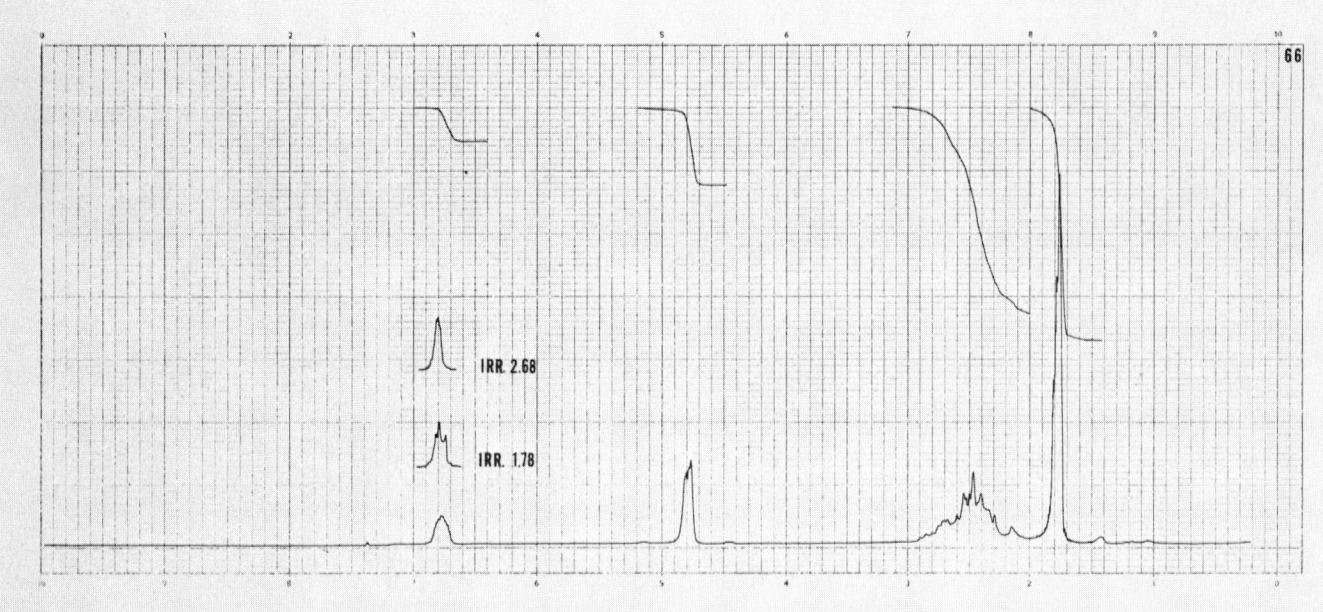

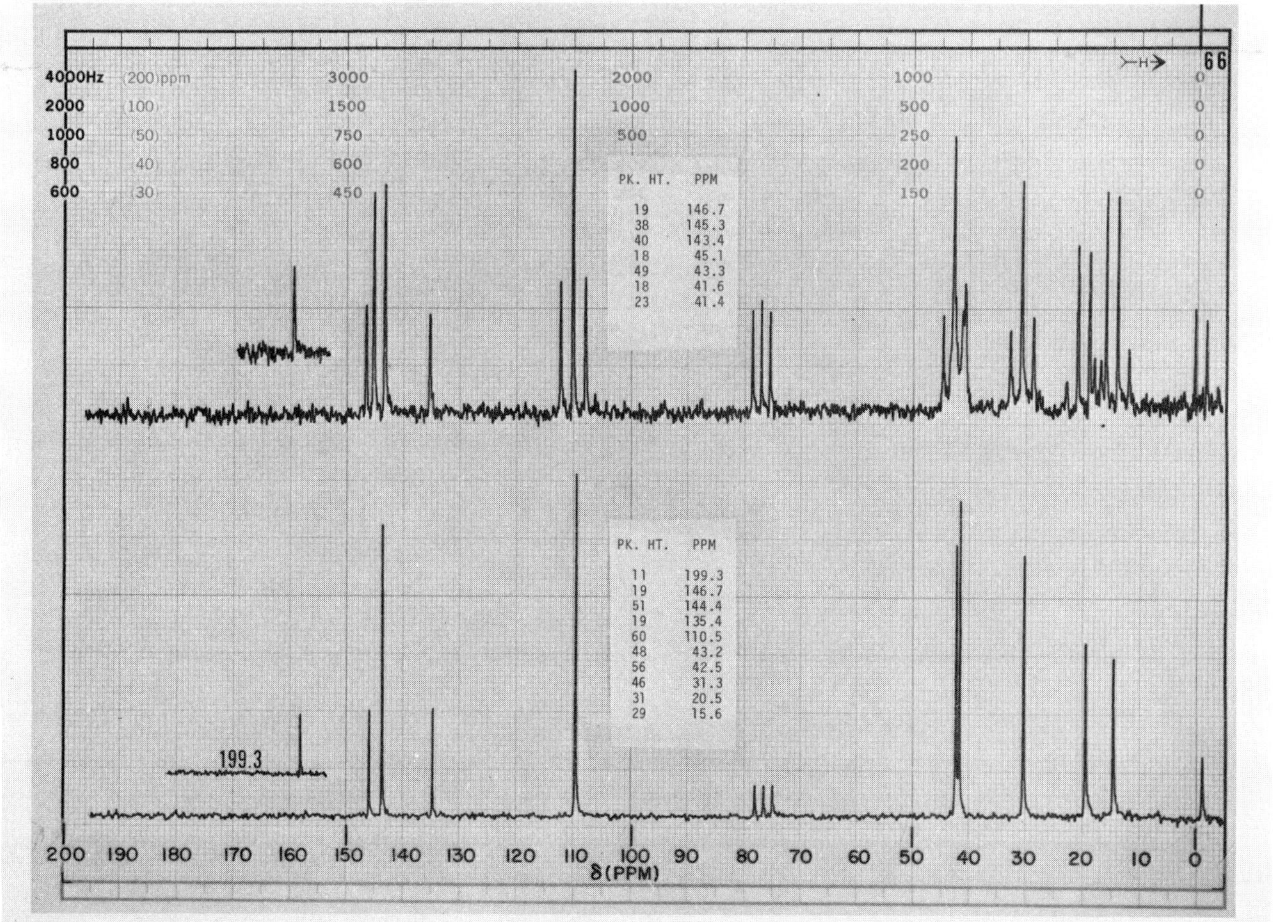

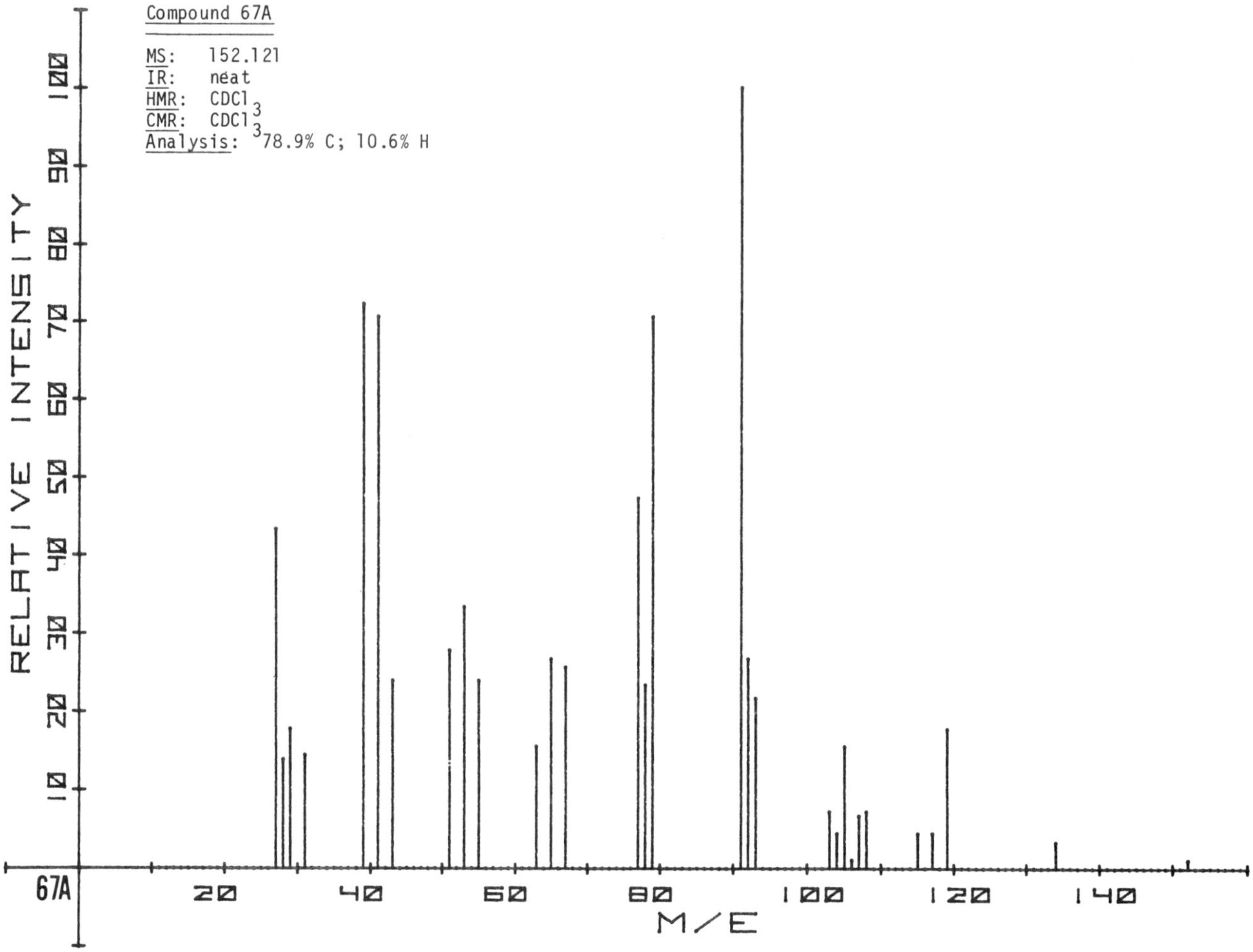

Compound 67A
MS: 152.121
IR: neat
HMR: CDCl3
CMR: CDCl3
Analysis: 78.9% C; 10.6% H
RELATIVE INTENSITY
M/E
67A

CM-1
ABSORBANCE
WAVELENGTH (MICRONS)
67A

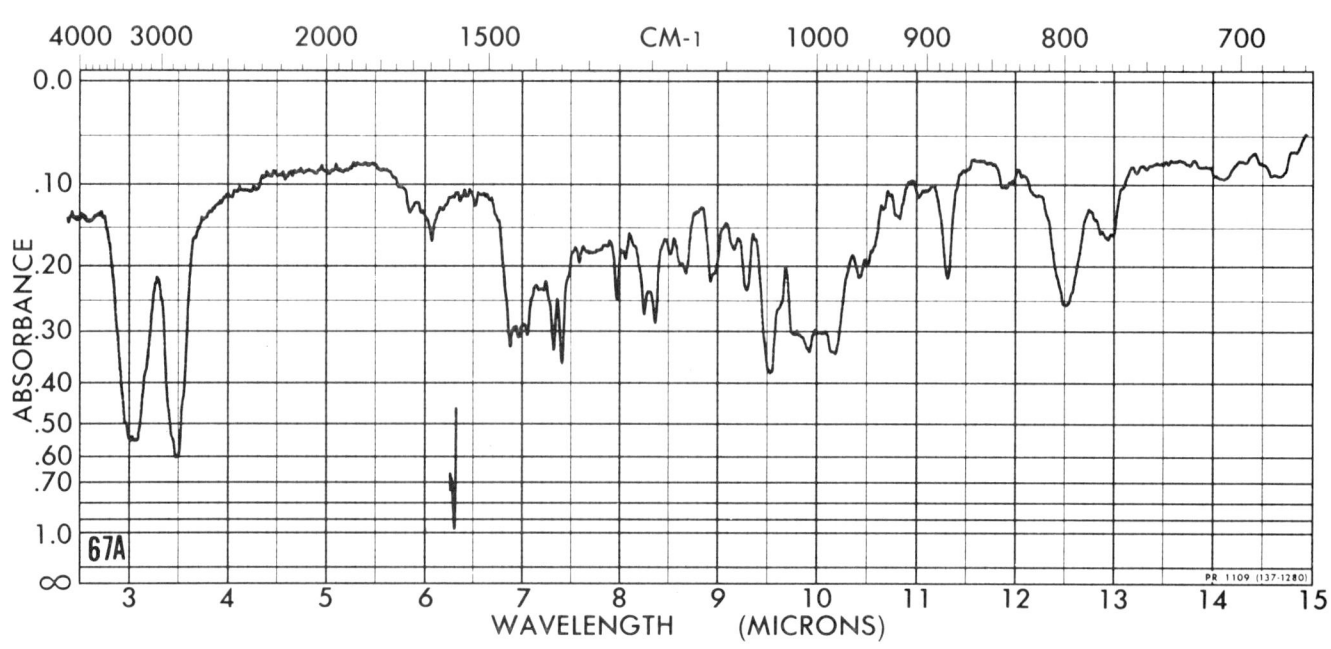

67A
IRR. 5.50
D2O

67A
4000Hz 200 ppm 3000 2000 1000 0
2000 (100) 1500 1000 500 0
1000 750 500 250 0
800 600 400 200 0
600 450 300 150 0
PK. HT. PPM
24 44.2
24 42.6
25 41.8
25 40.2
17 38.0
16 33.2
25 32.5
31 31.6
60 31.2
18 30.1
26 29.9
PK. HT. PPM
28 147.9
60 117.6
52 65.7
46 43.4
60 41.0
31 37.9
45 31.6
57 31.2
39 26.2
38 21.1
200 190 180 170 160 150 140 130 120 110 100 90 80 70 60 50 40 30 20 10 0
δ(PPM)

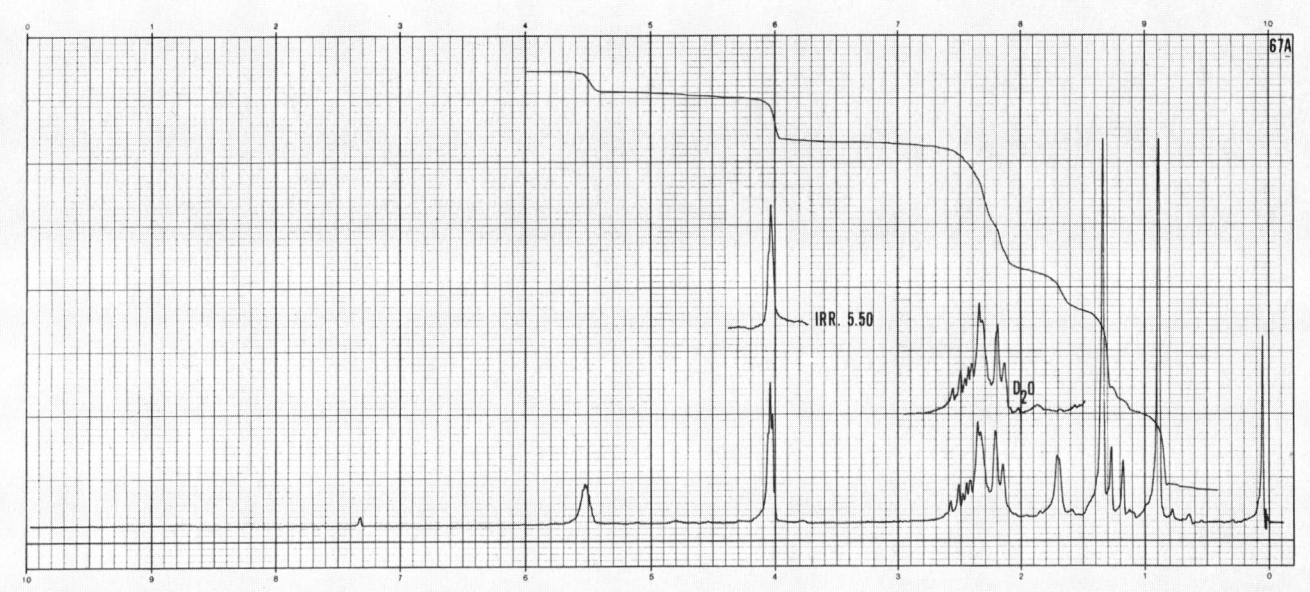

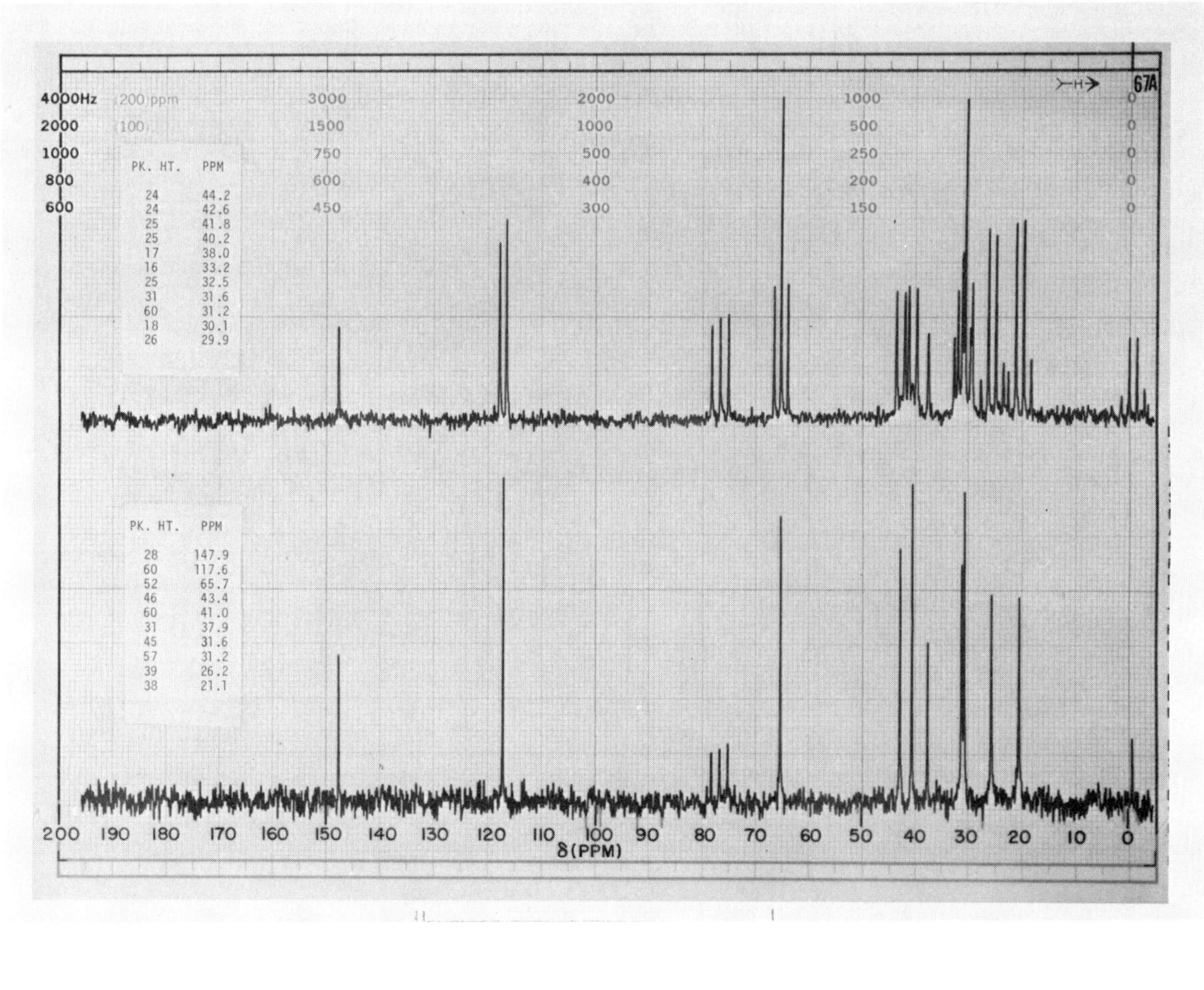

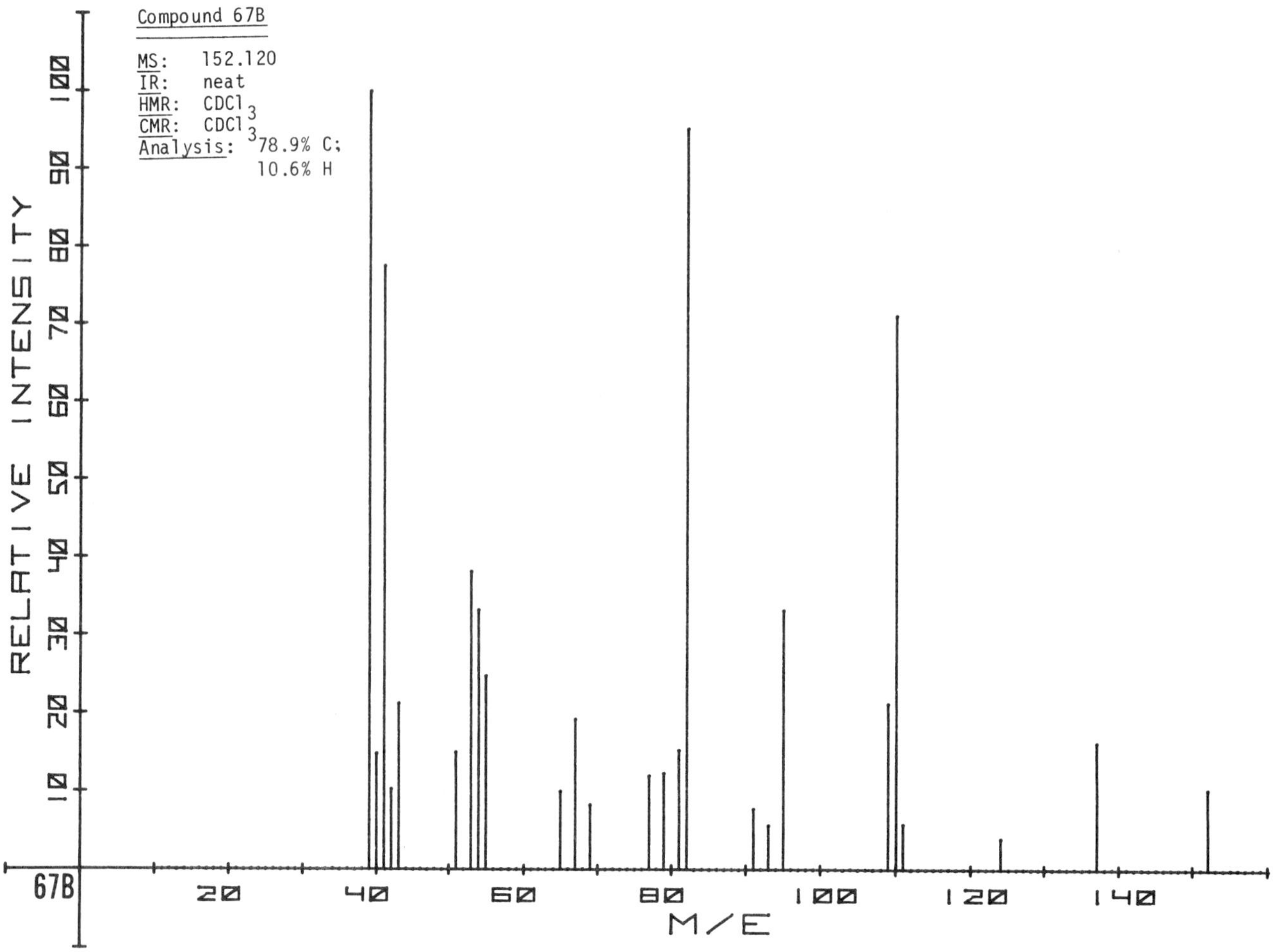

Compound 67B
MS: 152.120
IR: neat
HMR: CDCl3
CMR: CDCl3
Analysis: 78.9% C; 10.6% H
RELATIVE INTENSITY
M/E
67B

CM-1
ABSORBANCE
WAVELENGTH (MICRONS)
67B

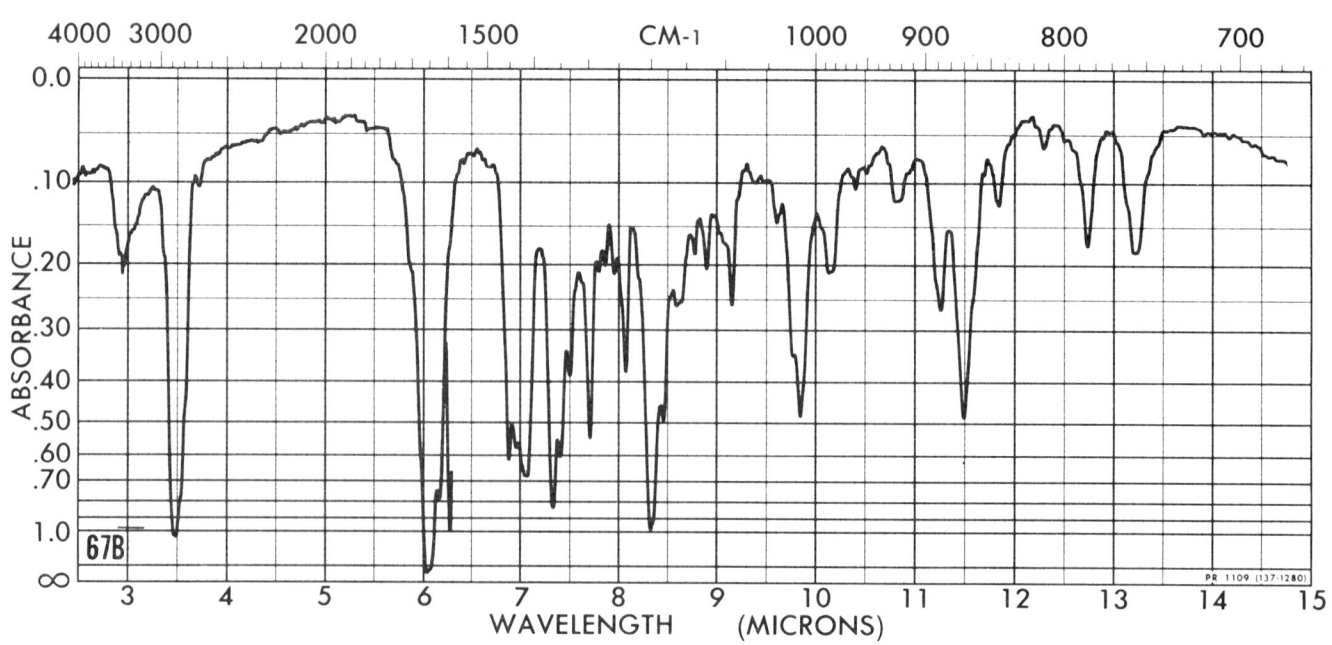

67B
A
IRR. 2.4
IRR. 1.95
A

67B
4000Hz (200)ppm 3000 1000 0
2000 (100) 1500 500 0
1000 (50) 750 250 0
800 (40) 600 200 0
600 (30) 450 150 0
PK. HT. PPM
11 32.6
35 30.4
16 28.2
15 27.4
40 27.0
44 25.1
40 24.9
13 24.5
13 24.0
58 23.0
13 22.2
43 21.8
25 20.7
60 19.6
45 17.5
16 15.3
PK. HT. PPM
15 201.1
15 161.0
60 126.8
49 51.6
55 30.4
56 25.9
32 24.0
54 23.1
41 20.7
38 18.6
201.1
200 190 180 170 160 150 140 130 120 110 100 90 80 70 60 50 40 30 20 10 0
δ(PPM)

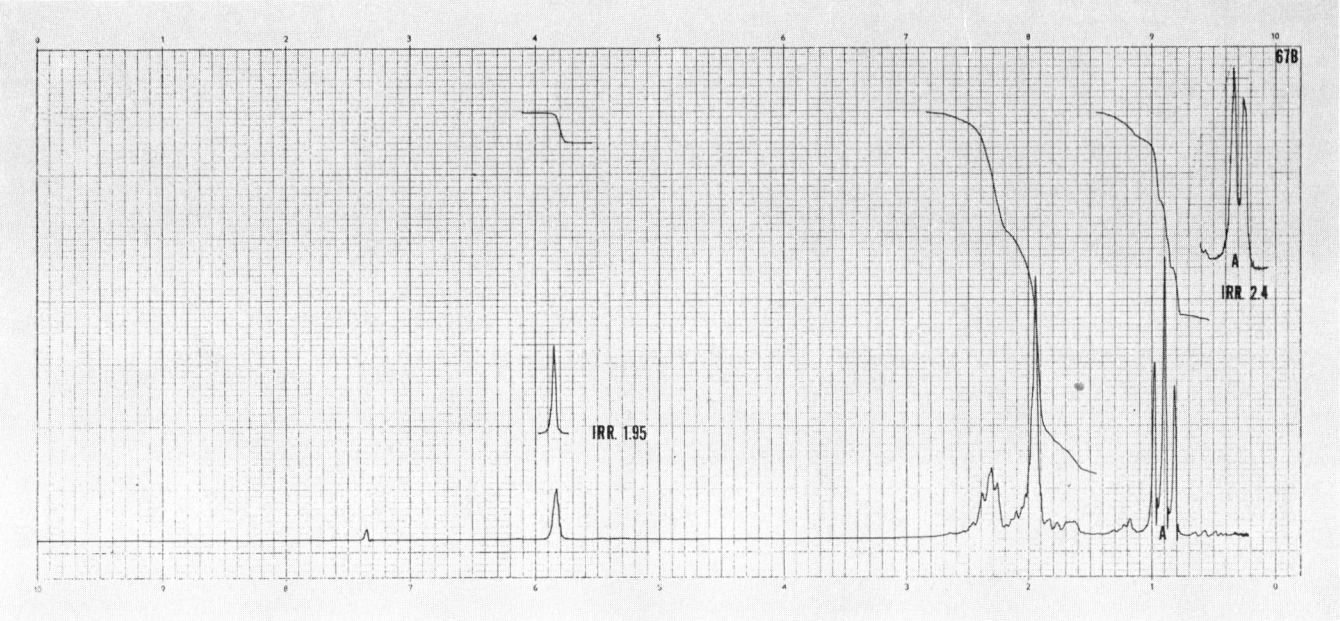

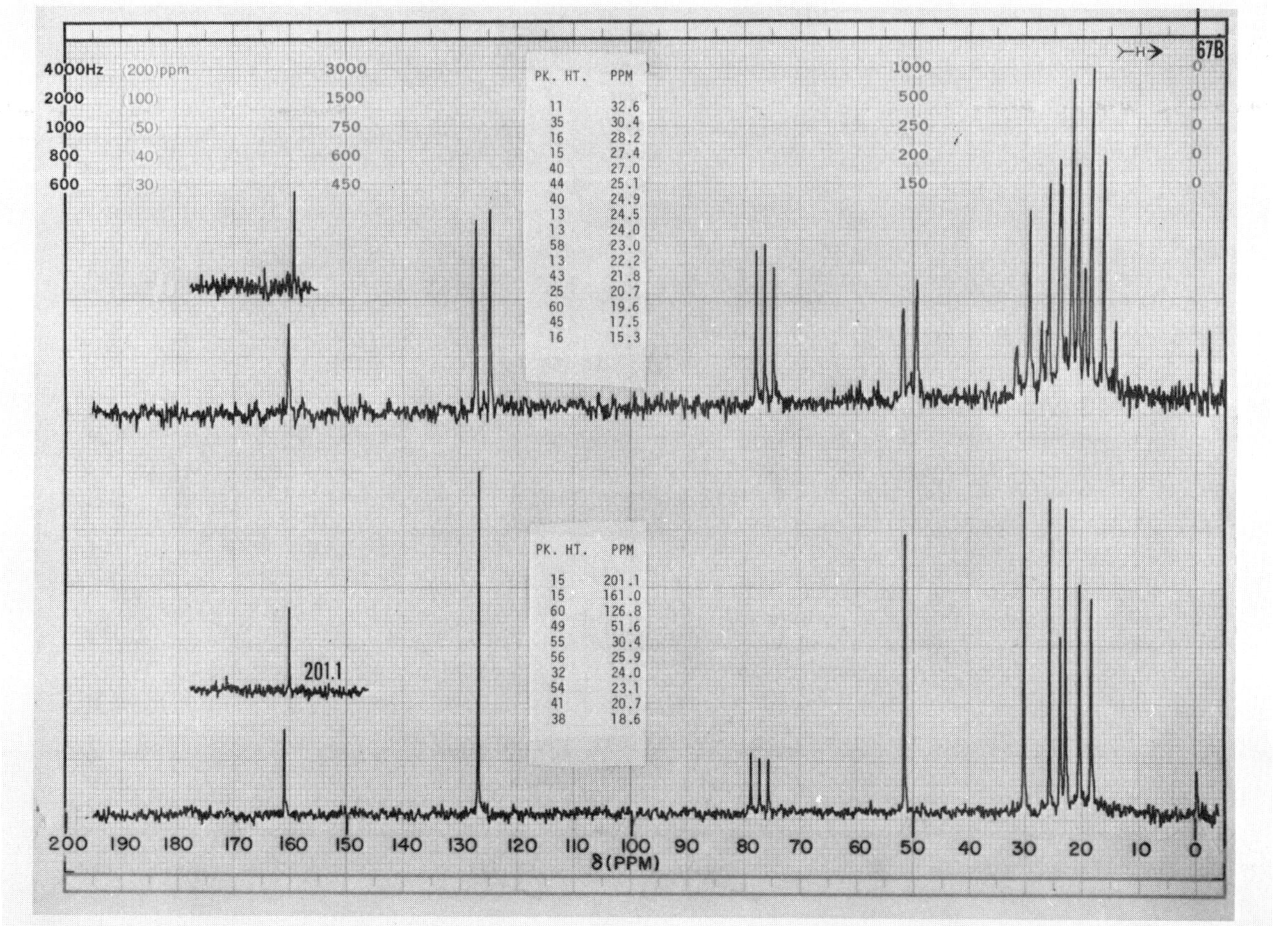

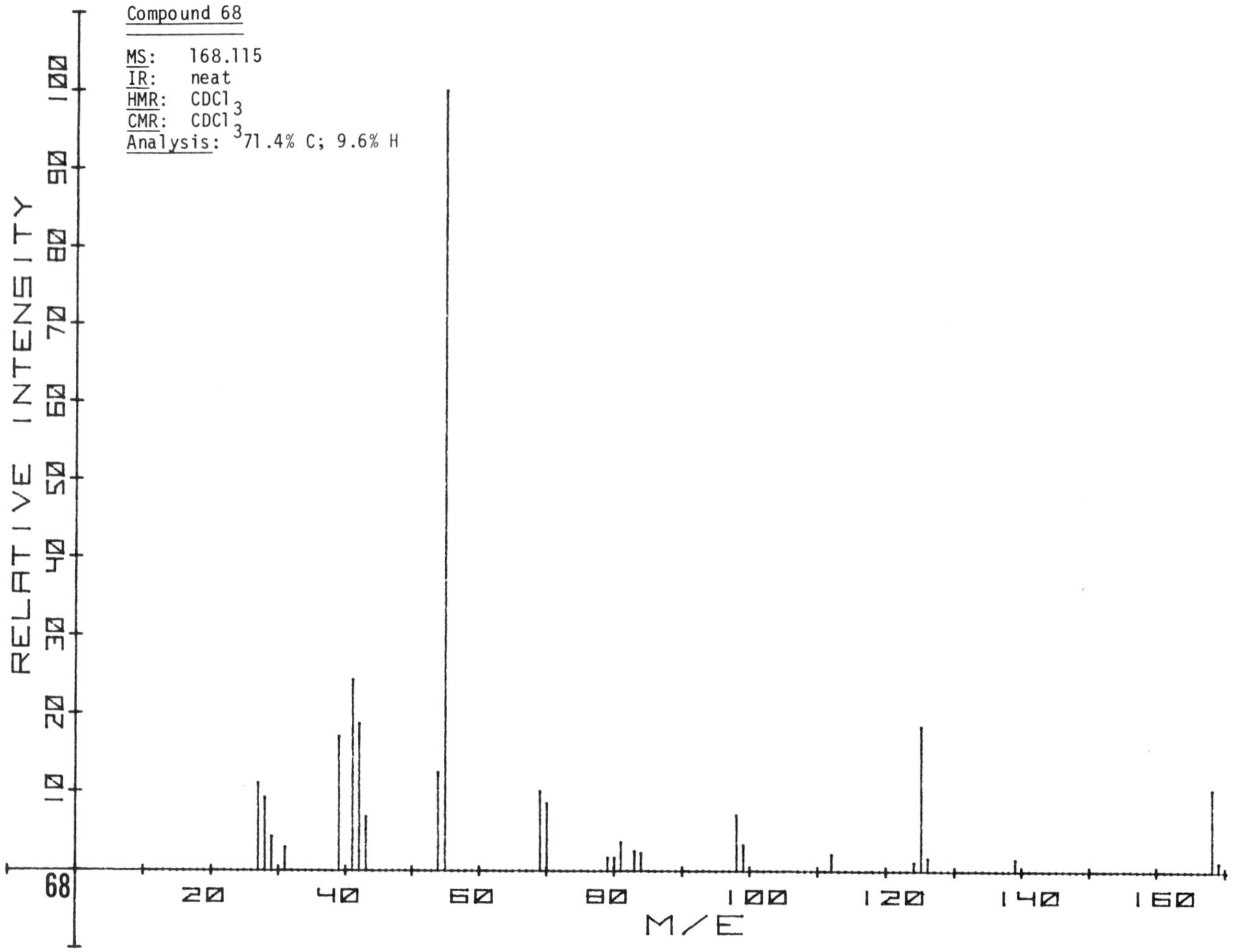

Compound 68
MS: 168.115
IR: neat
HMR: CDCl3
CMR: CDCl3
Analysis: 71.4% C; 9.6% H
RELATIVE INTENSITY
10 20 30 40 50 60 70 80 90 100
68
20 40 60 80 100 120 140 160
M/E

4000 3000 2000 1500 CM-1 1000 900 800 700
ABSORBANCE
0.0 .10 .20 .30 .40 .50 .60 .70 1.0 ∞
68
3 4 5 6 7 8 9 10 11 12 13 14 15
WAVELENGTH (MICRONS)

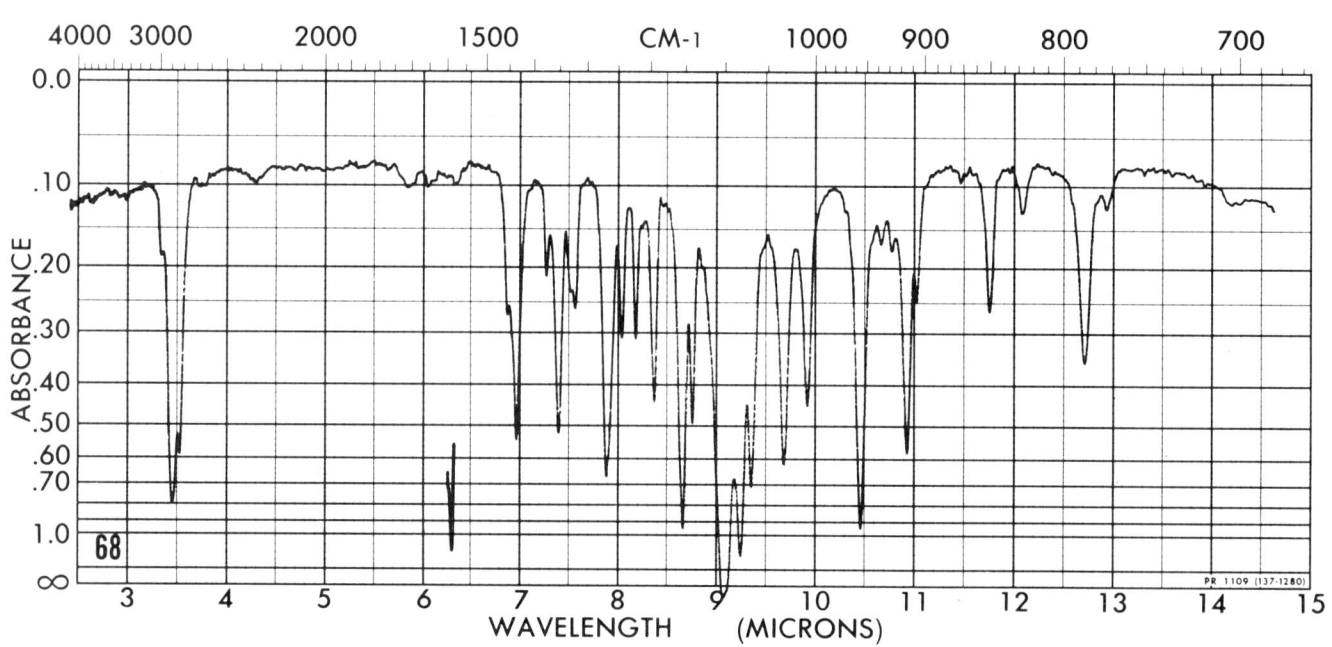

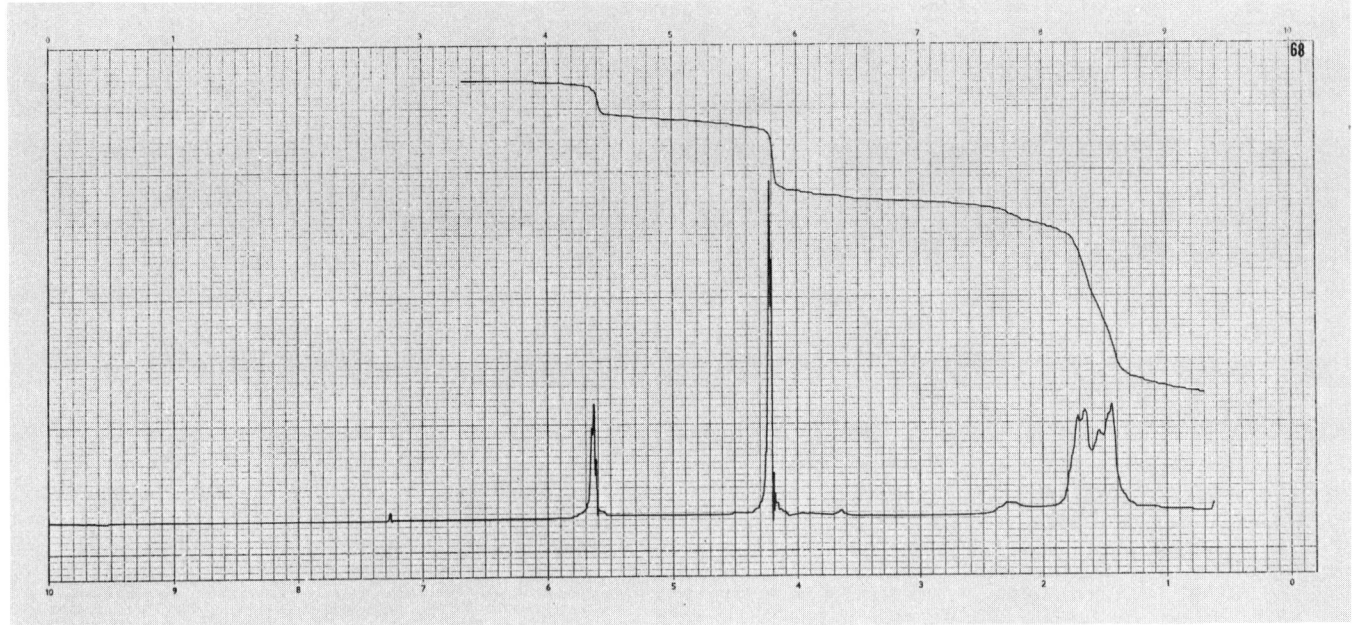

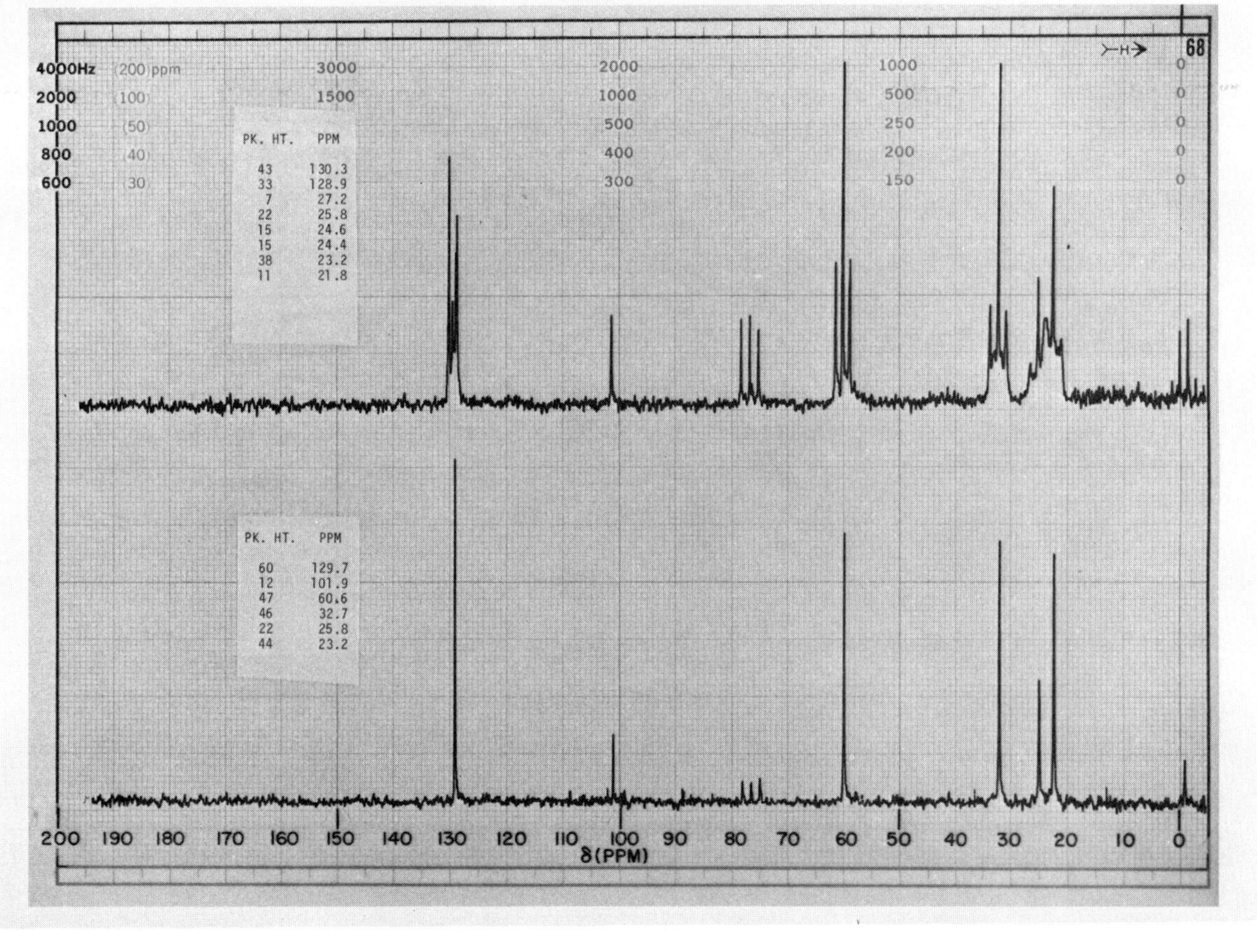

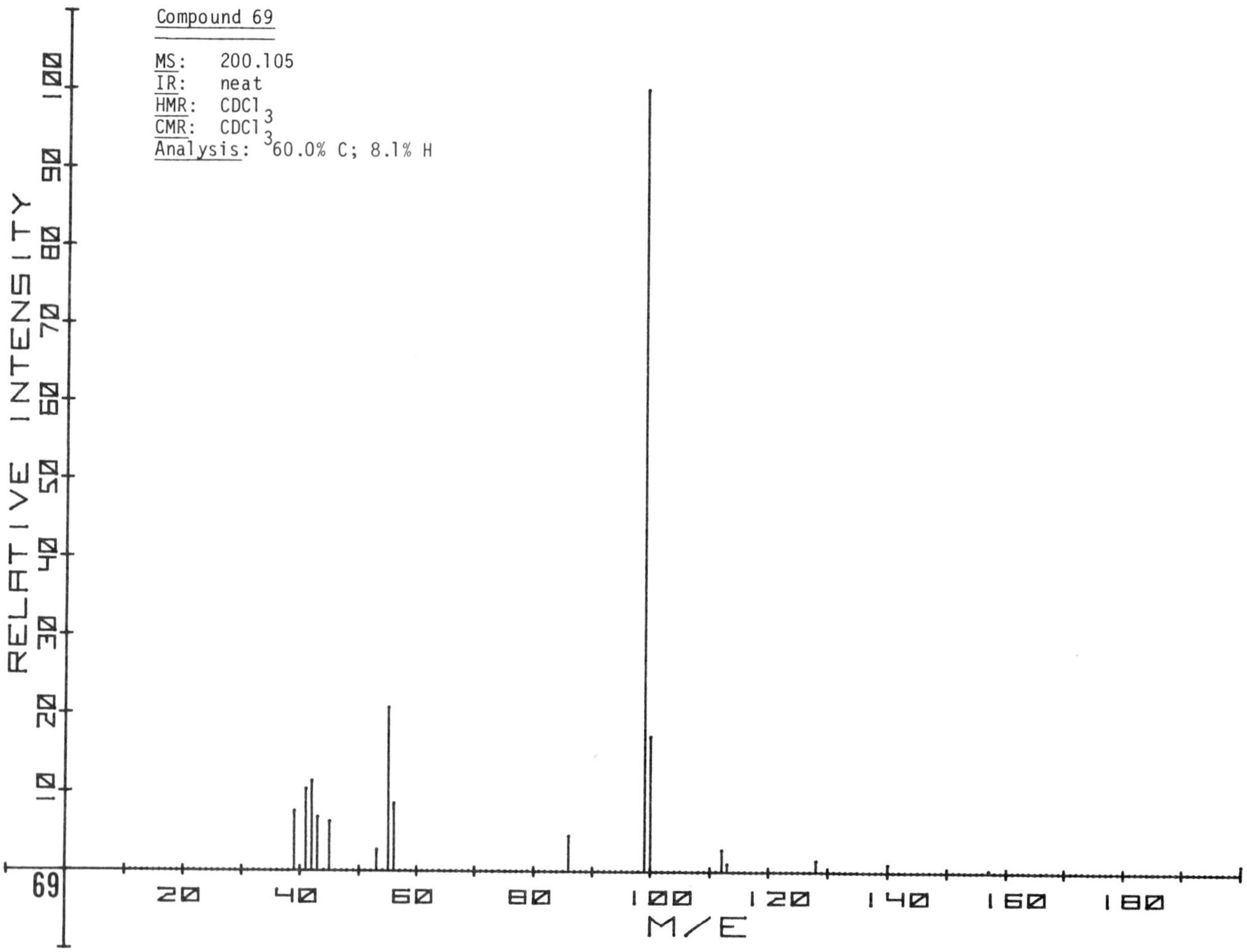

Compound 69
MS: 200.105
IR: neat
HMR: CDCl3
CMR: CDCl3
Analysis: 60.0% C; 8.1% H
RELATIVE INTENSITY
M/E
69

CM-1
ABSORBANCE
WAVELENGTH (MICRONS)
69
PR 1109 (137-1280)

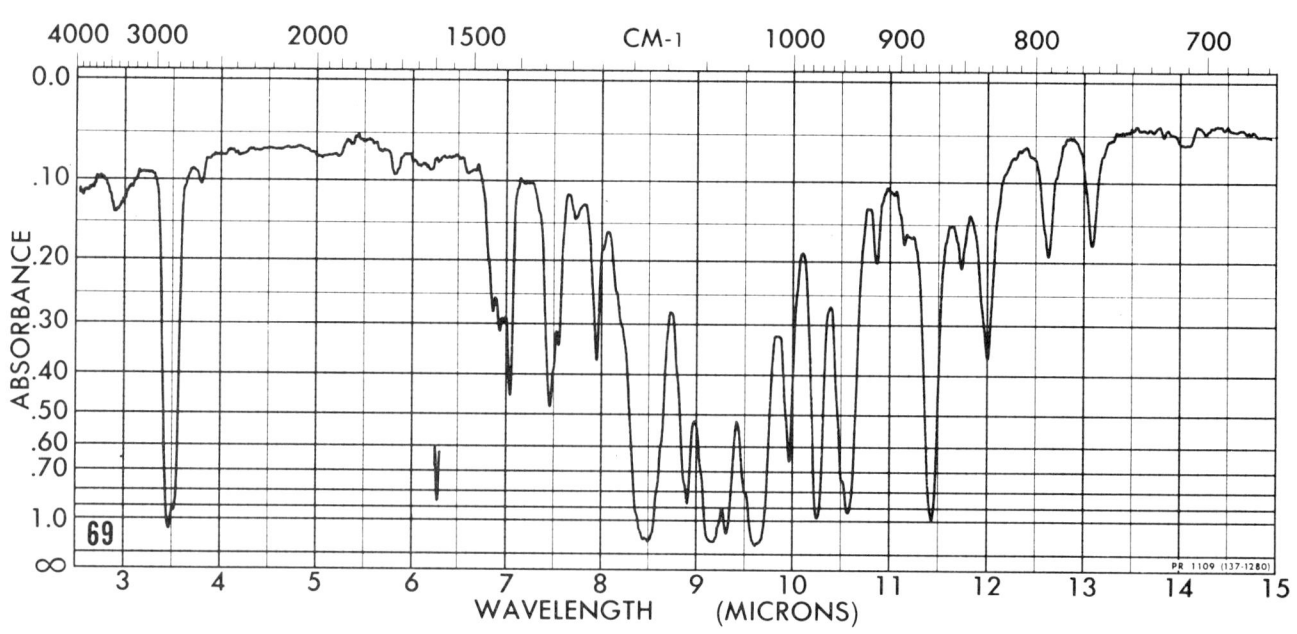

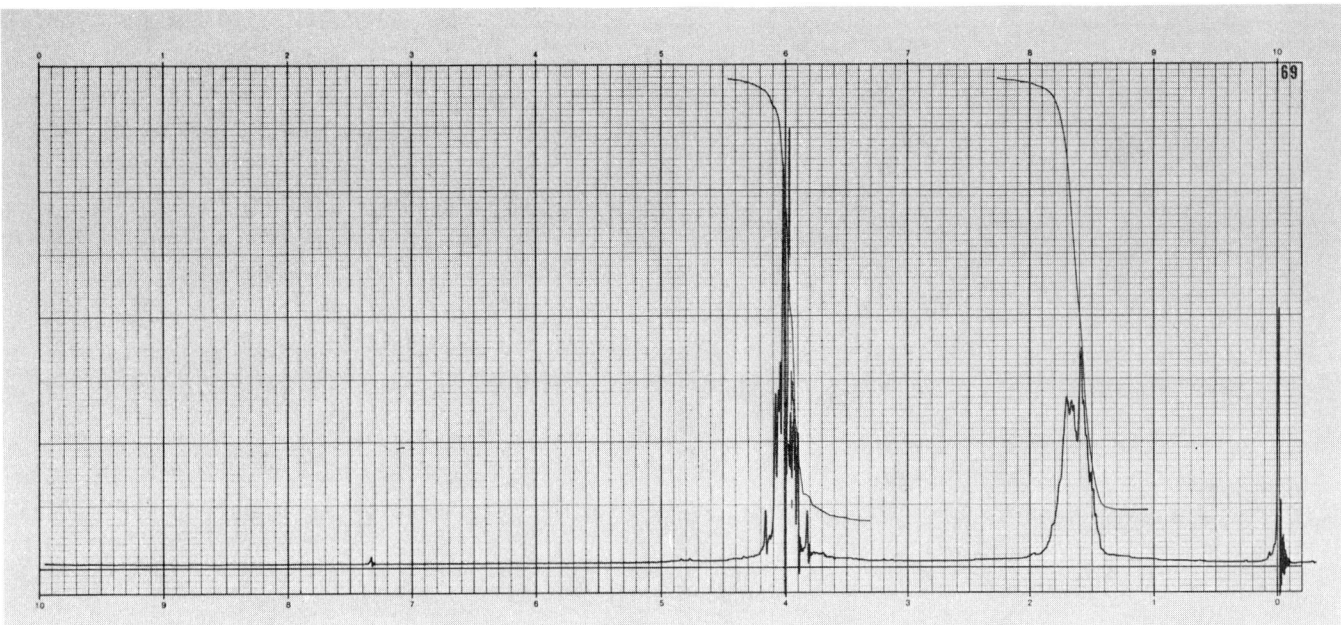

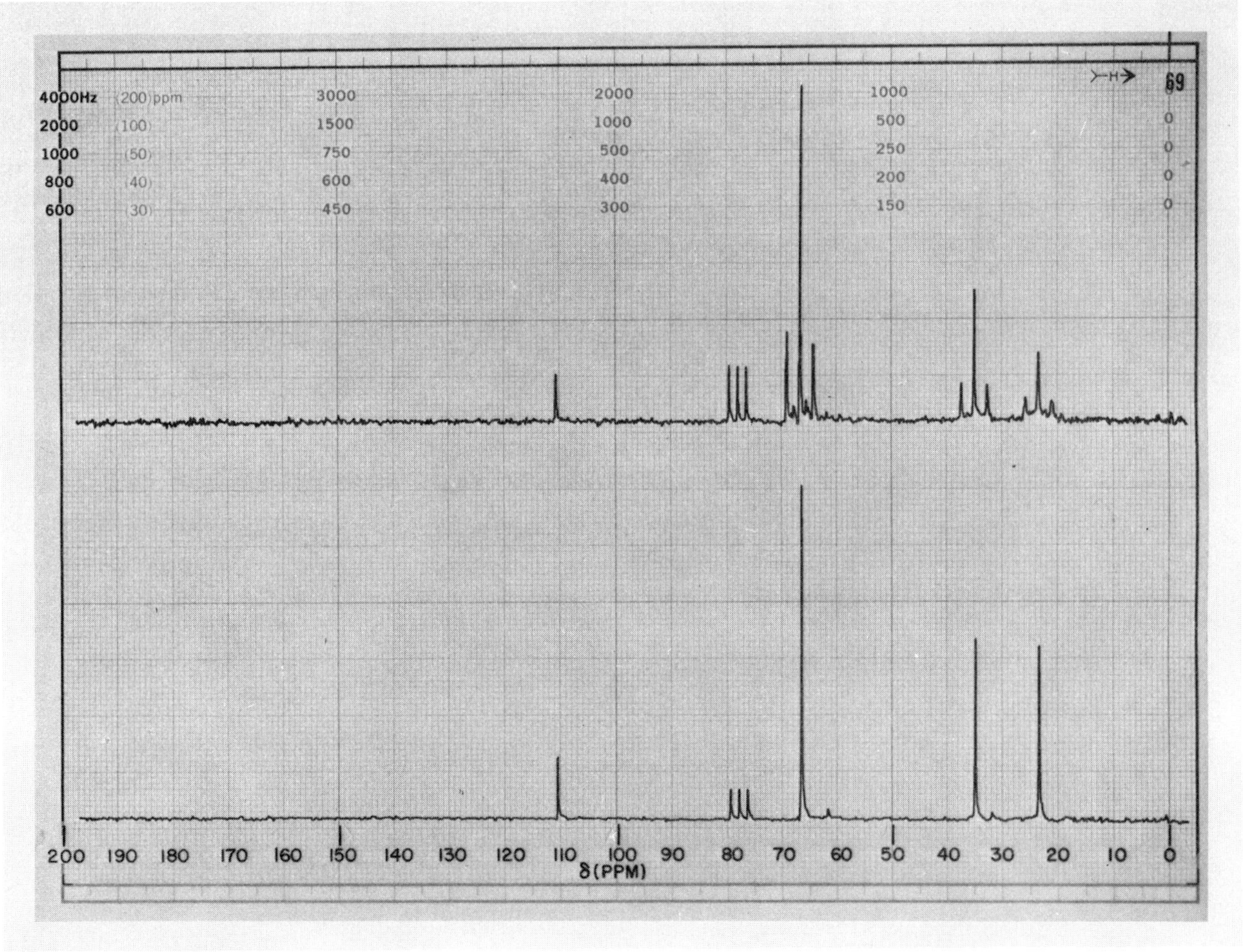

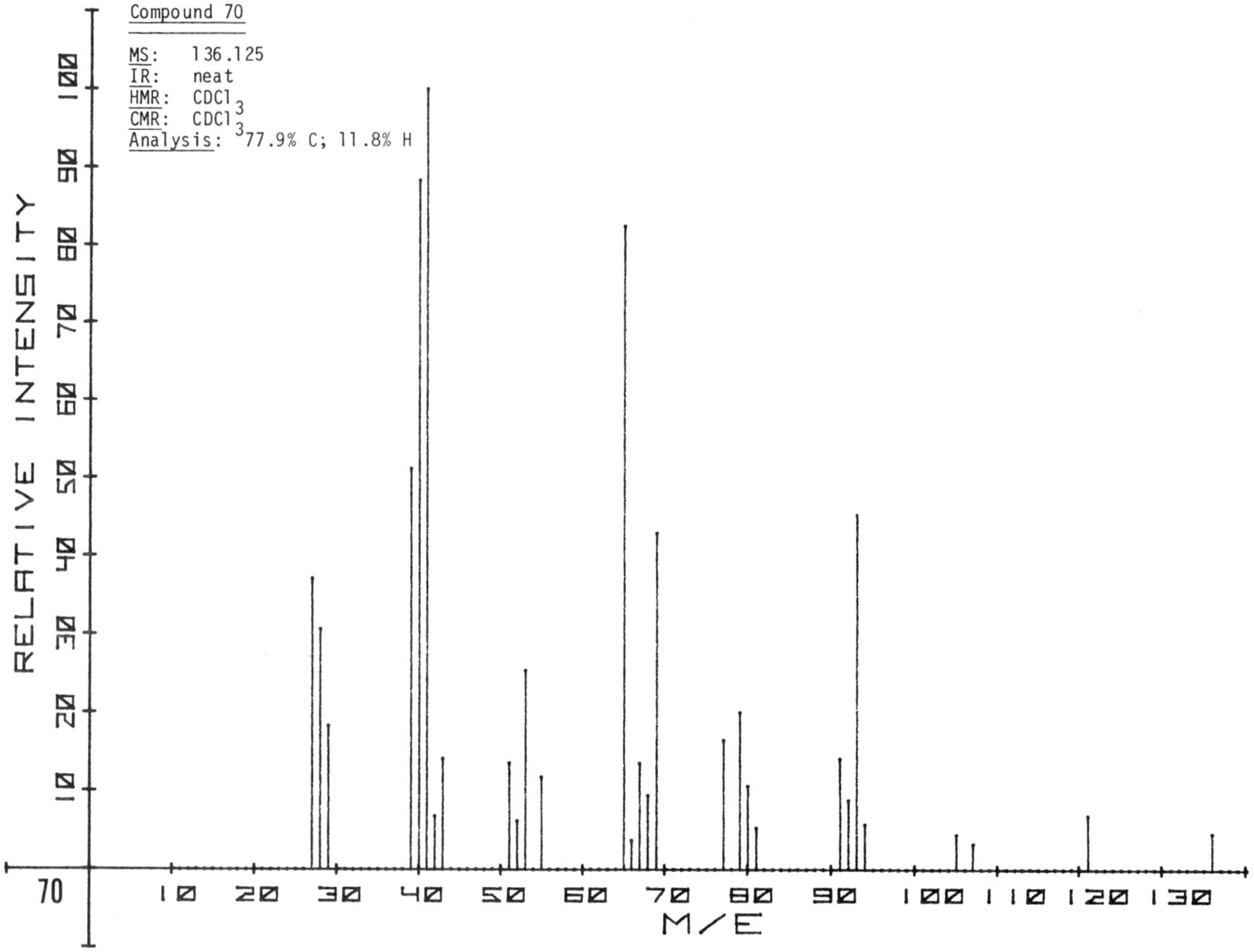

Compound 70
MS: 136.125
IR: neat
HMR: $CDCl_3$
CMR: $CDCl_3$
Analysis: 77.9% C; 11.8% H
RELATIVE INTENSITY
M/E
70

CM-1
ABSORBANCE
WAVELENGTH (MICRONS)
70
PR 1109 (137-1280)

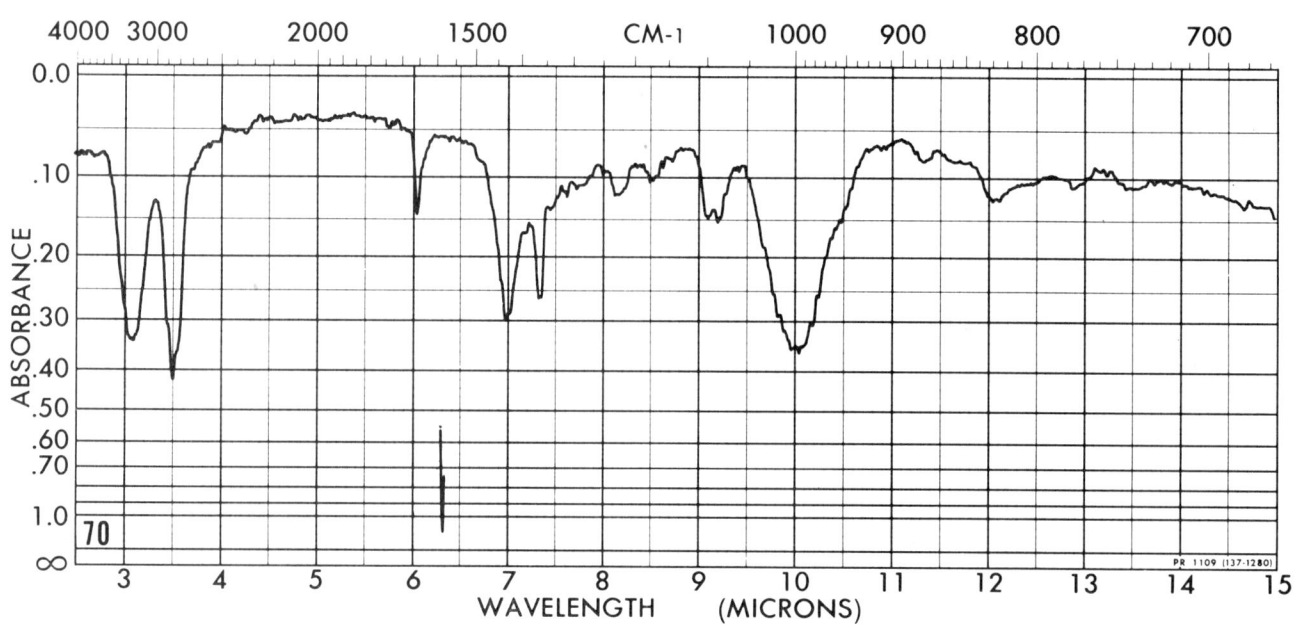

70
D2O
A
IRR. 2.10
IRR. 5.45
IRR. 4.20
A

4000Hz (200)ppm
2000 (100)
1000 (50)
800 (40)
600 (30)
PK. HT. PPM
31 125.1
41 124.8
40 123.0
44 122.8
16 28.6
60 26.6
48 24.6
12 22.6
6 20.7
8 19.2
20 18.7
22 17.2
21 16.7
23 15.2
8 14.7
9 13.2
2000 1000 500 400 300
1000 500 250 200 150
70
PK. HT. PPM
23 138.9
20 131.6
50 124.1
60 123.8
50 59.1
48 39.6
43 26.5
35 25.6
19 17.7
21 16.2
200 190 180 170 160 150 140 130 120 110 100 90 80 70 60 50 40 30 20 10 0
δ(PPM)

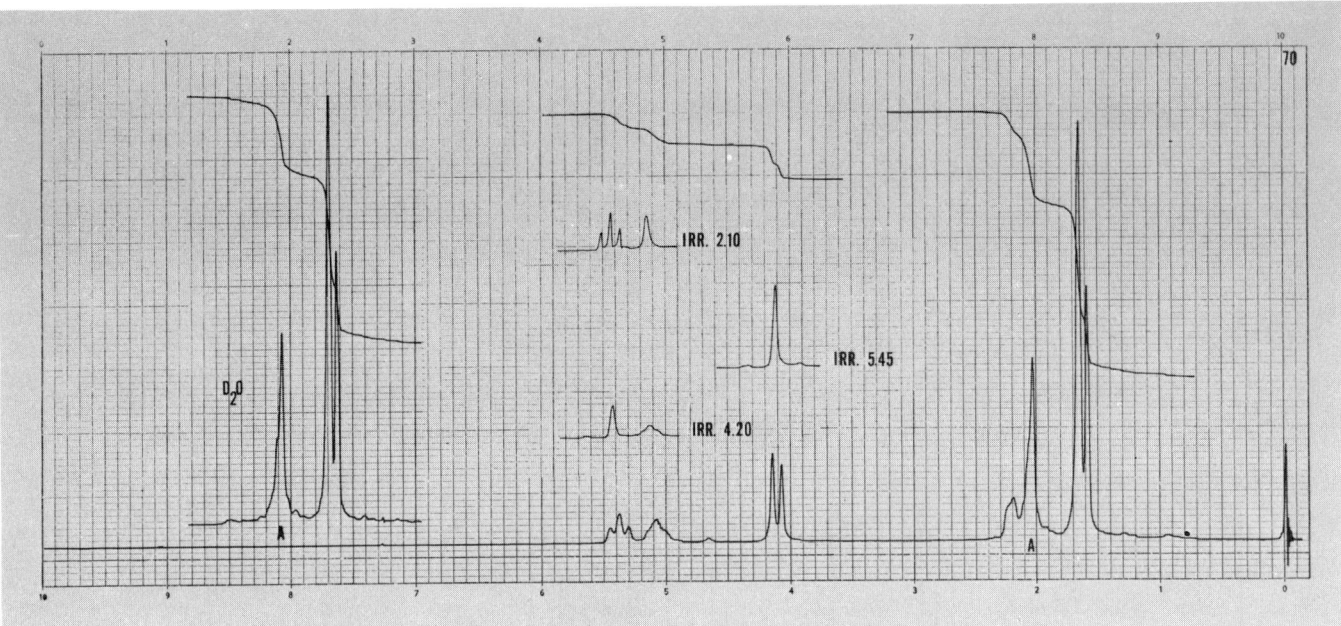

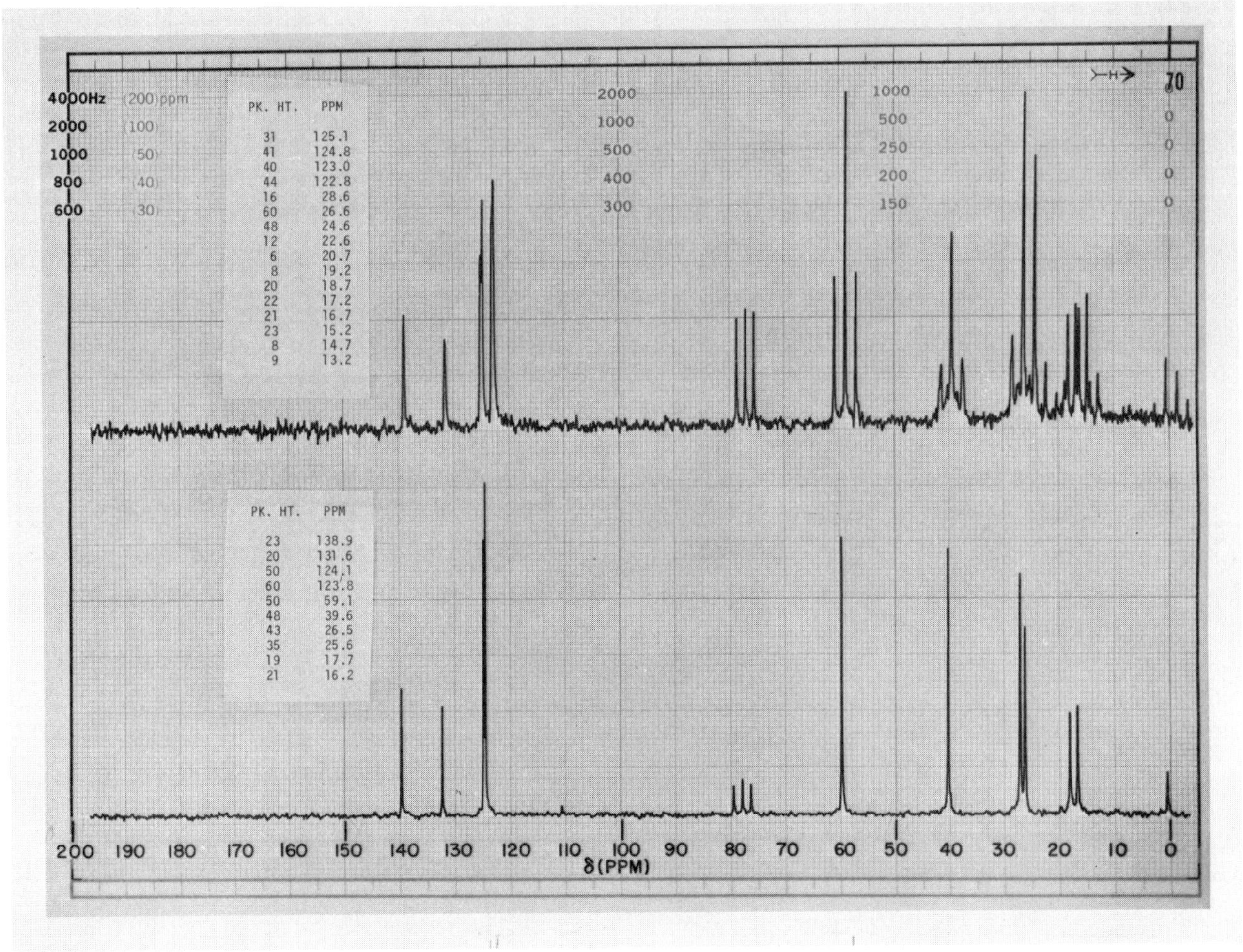

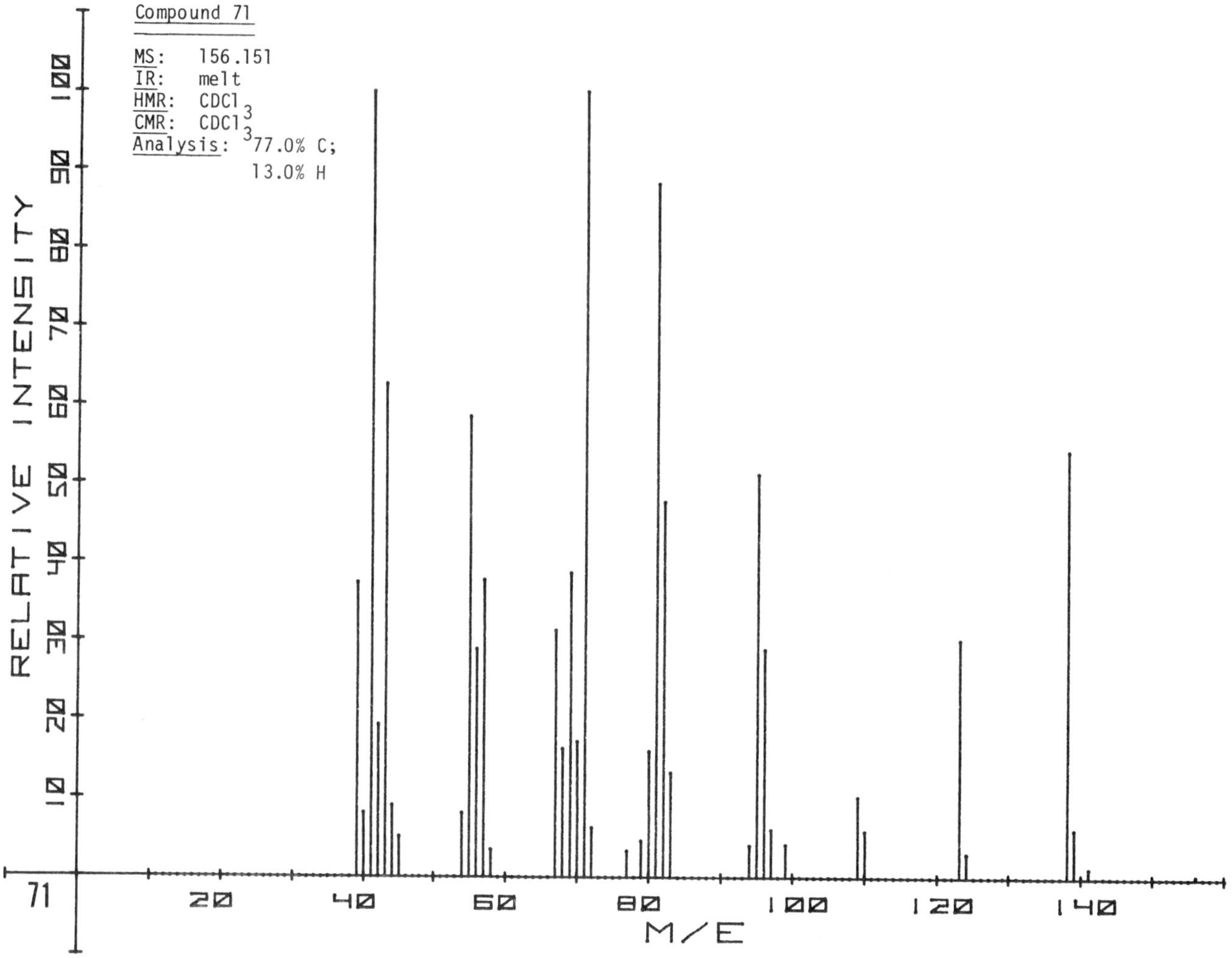

Compound 71
MS: 156.151
IR: melt
HMR: CDCl3
CMR: CDCl3
Analysis: 77.0% C; 13.0% H
RELATIVE INTENSITY
71
M/E

CM-1
ABSORBANCE
71
WAVELENGTH (MICRONS)

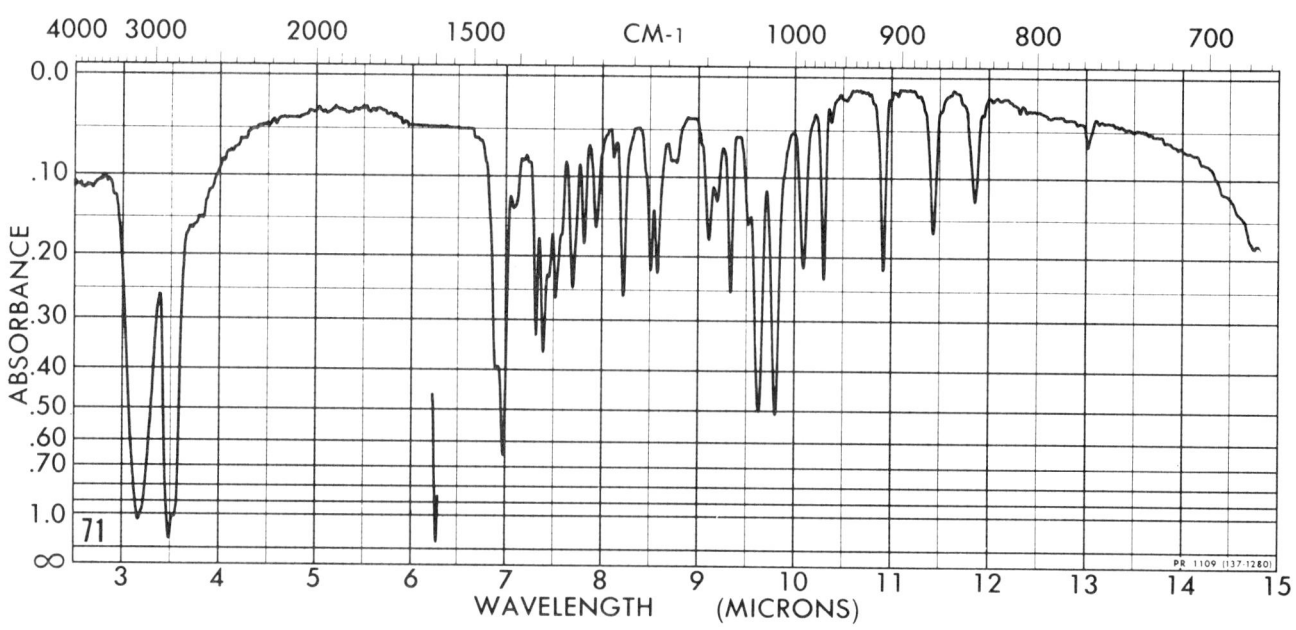

71
D2O

71
4000Hz (200)ppm
2000 (100)
1000 (50)
800 (40)
600 (30)
PK. HT. PPM
11 36.5
23 34.8
15 33.1
20 32.4
21 31.1
60 26.3
57 25.1
21 24.7
23 24.5
37 23.3
54 23.1
56 21.8
50 21.5
49 20.4
17 20.0
14 18.9
15 18.4
46 16.9
49 15.4
13 13.9
PK. HT. PPM
49 71.3
58 50.1
48 45.2
47 34.8
55 31.8
60 25.7
43 23.3
40 22.3
40 21.1
42 16.1
δ(PPM)

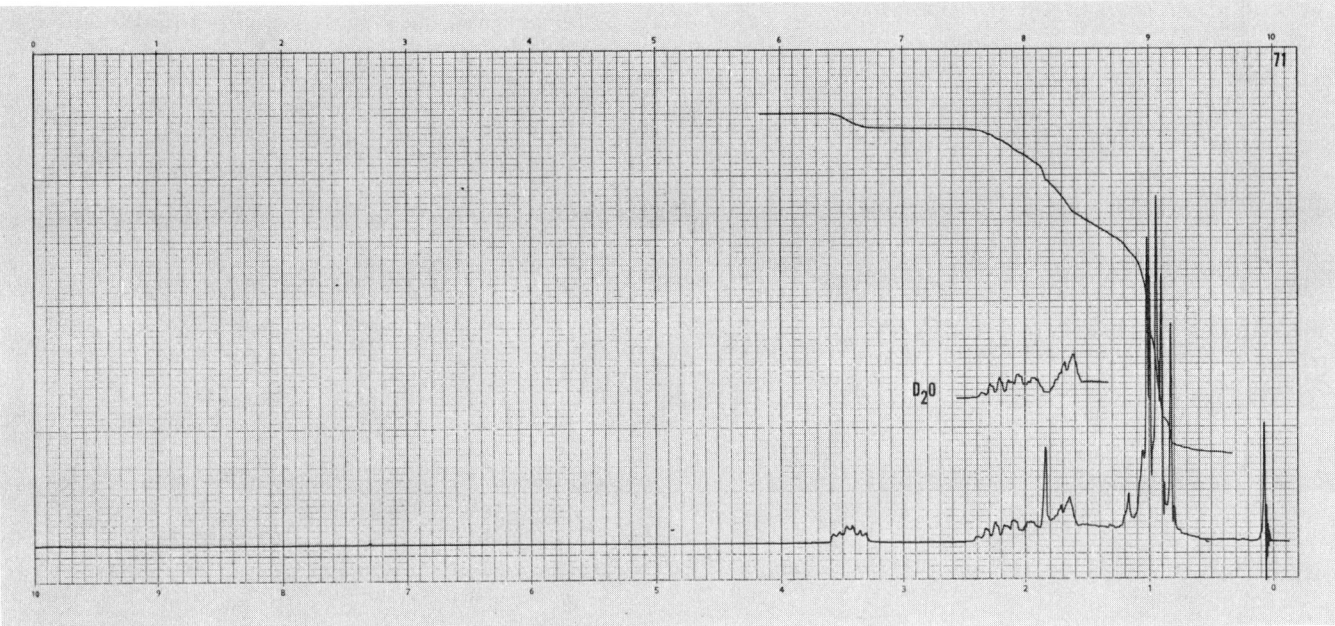

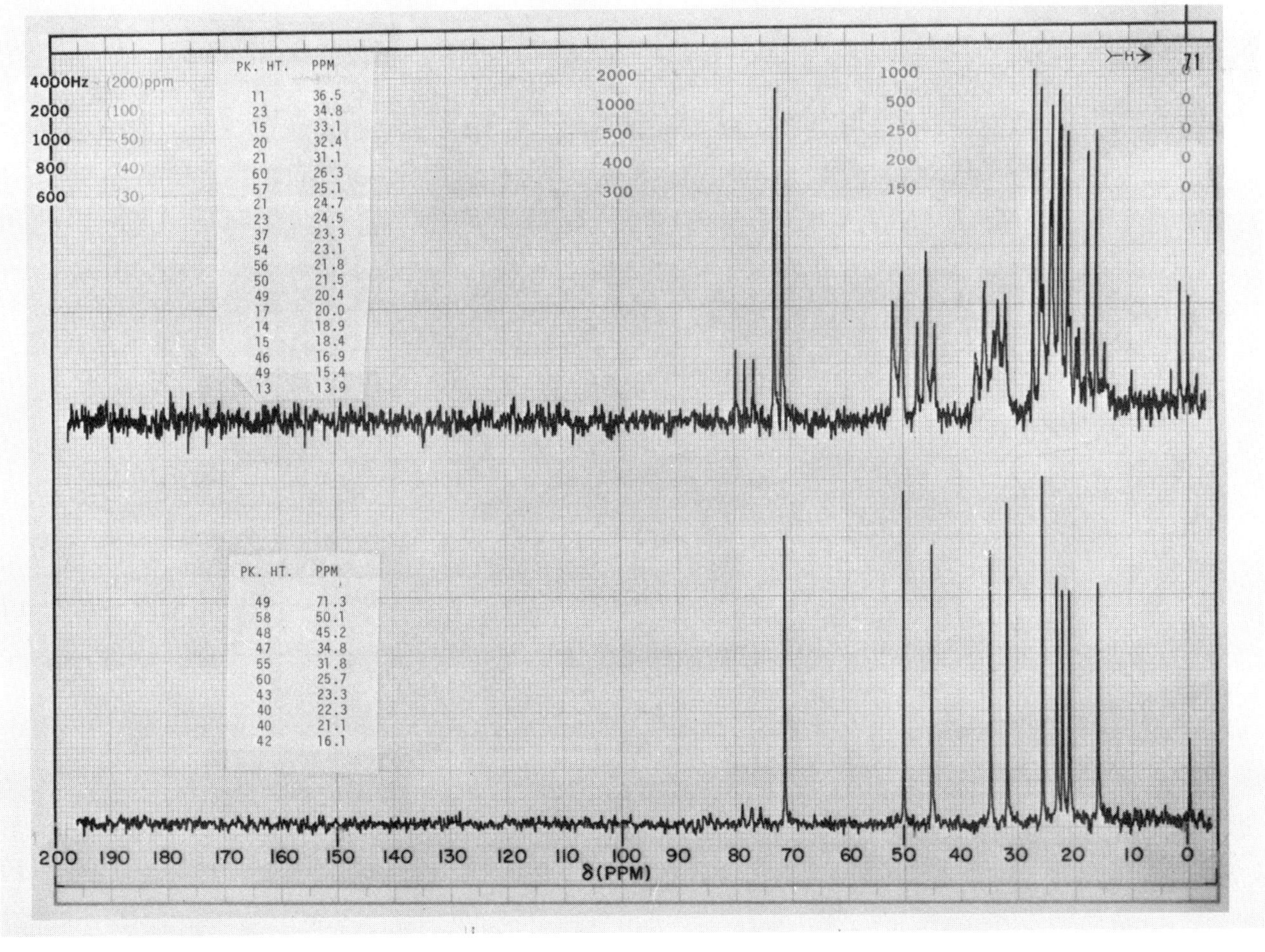

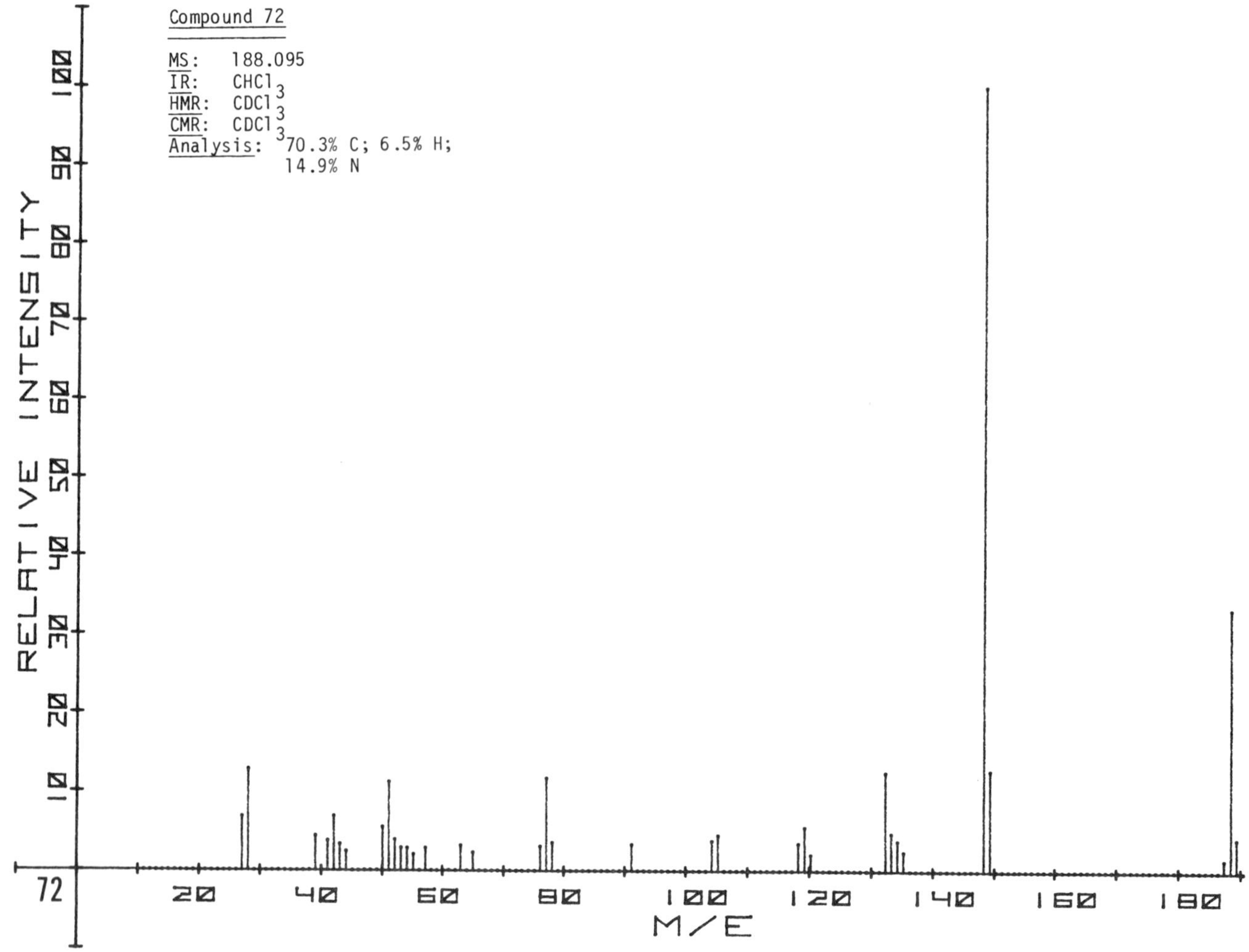

Compound 72
MS: 188.095
IR: CHCl3
HMR: CDCl3
CMR: CDCl3
Analysis: 70.3% C; 6.5% H; 14.9% N
RELATIVE INTENSITY
10 20 30 40 50 60 70 80 90 100
72
20 40 60 80 100 120 140 160 180
M/E

4000 3000 2000 1500 CM-1 1000 900 800 700
0.0 .10 .20 .30 .40 .50 .60 .70 1.0 ∞
ABSORBANCE
72
3 4 5 6 7 8 9 10 11 12 13 14 15
WAVELENGTH (MICRONS)

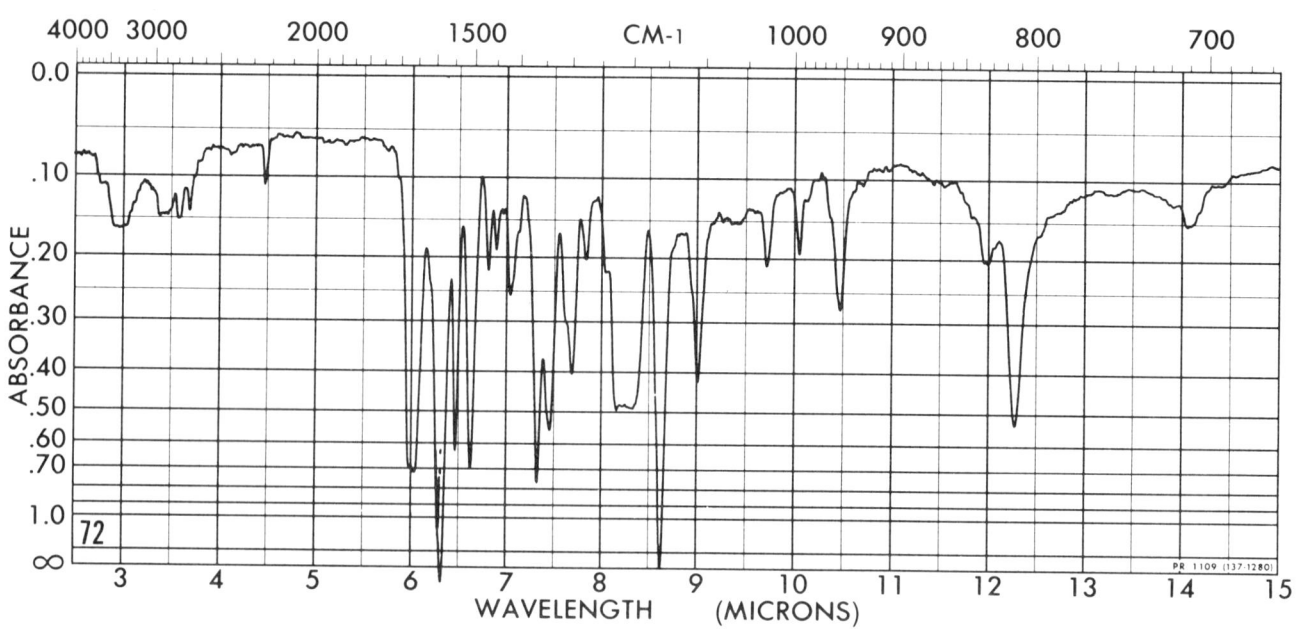

72
IRR. 3.85
IRR. 2.67

72
4000Hz (200)ppm 3000 2000 1000
2000 (100) 1500 1000 500
1000 (50) 750 500 250
800 (40) 600 400 200
600 (30) 450 300 150
200 190 180 170 160 150 140 130 120 110 100 90 80 70 60 50 40 30 20 10 0
δ(PPM)

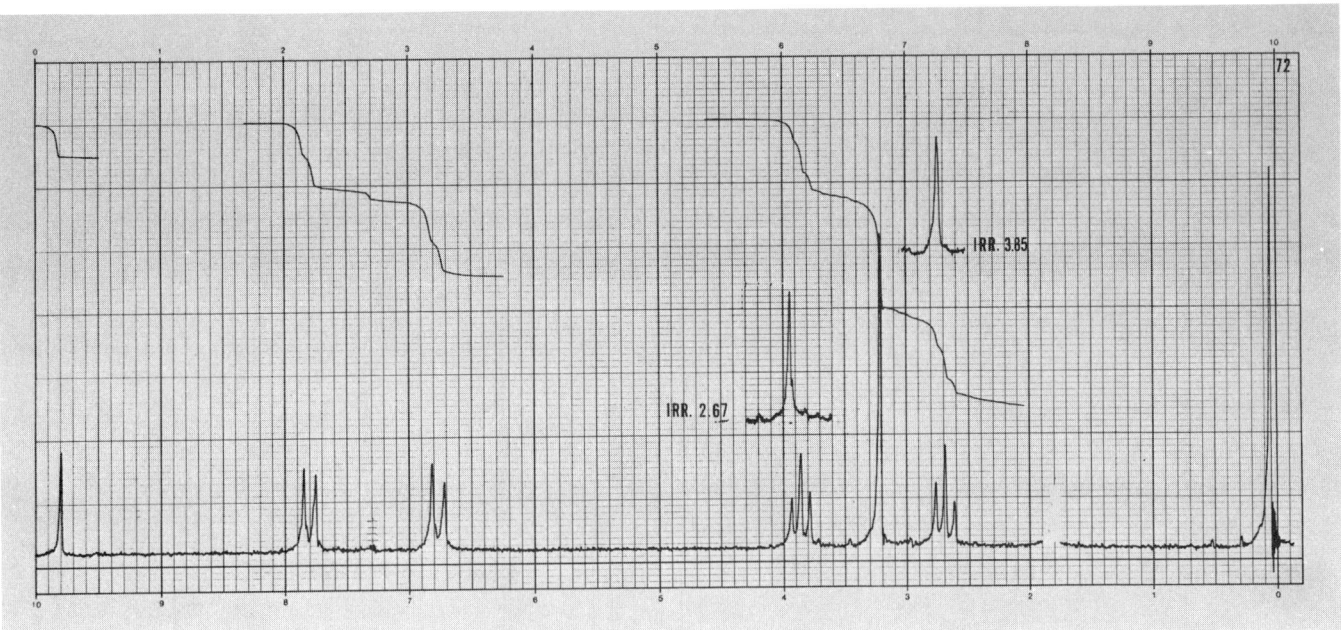

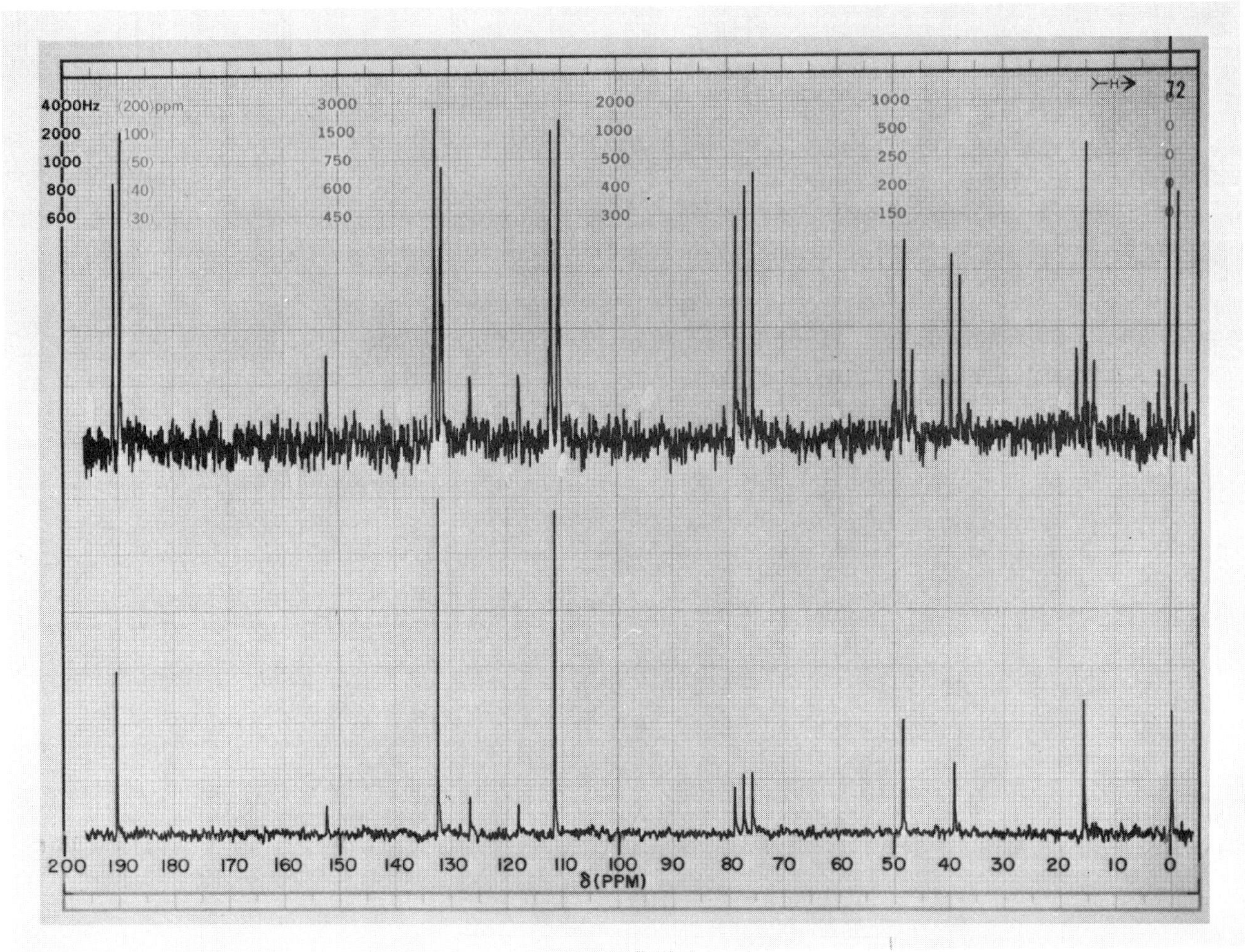

COMPOUND 73

Compound 73
MS: 178.100
IR: KBr
HMR: CDCl$_3$
CMR: CDCl$_3$
Analysis: 74.2% C; 8.0% H

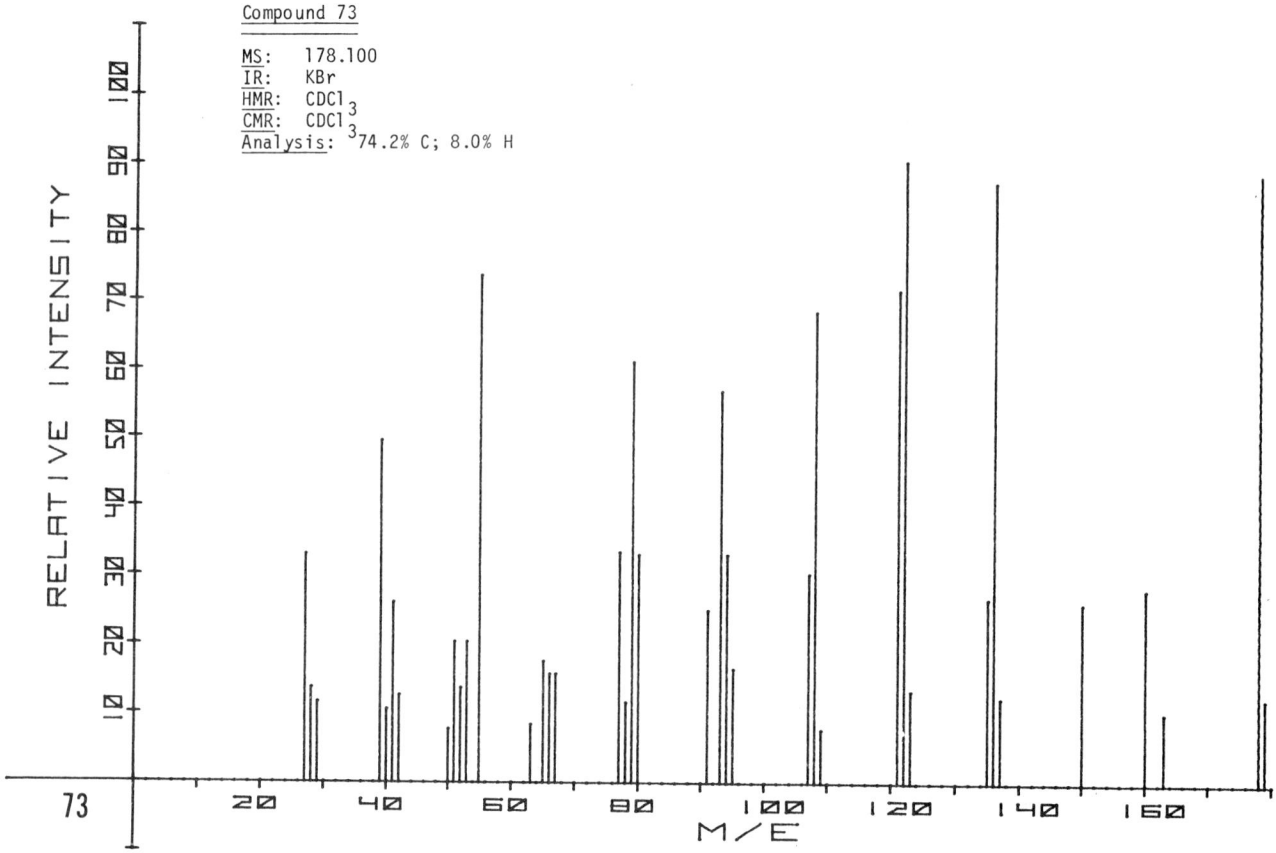

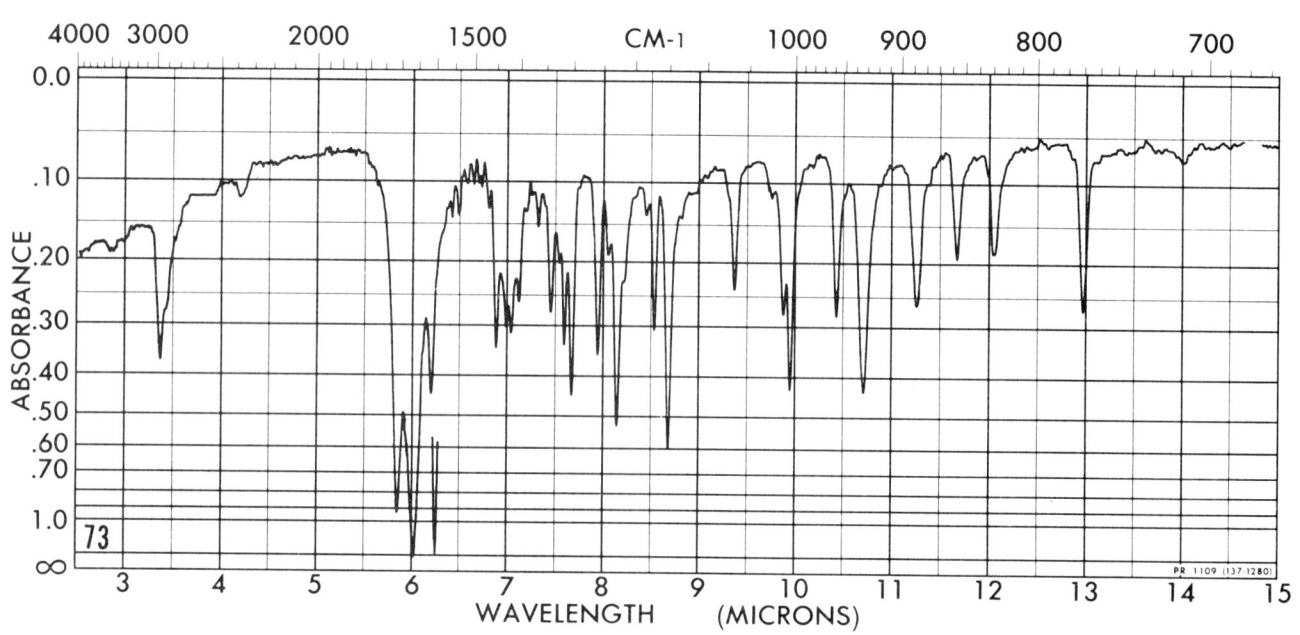

73

4000Hz (200)ppm
2000 (100)
1000 (50)
800 (40)
600 (30)
2000 1000 500 400 300
1000 500 250 200 150
73
PK. HT. PPM
13 34.4
32 33.6
17 32.8
13 32.6
28 31.6
14 30.7
17 30.5
34 29.7
13 28.9
13 24.4
11 23.7
54 23.6
23 22.9
58 22.8
16 22.0
PK. HT. PPM
14 210.8
21 197.9
26 165.9
60 125.6
27 50.6
49 37.6
52 33.6
45 31.7
46 29.7
38 23.2
57 22.9
210.8
200 190 180 170 160 150 140 130 120 110 100 90 80 70 60 50 40 30 20 10 0
δ(PPM)

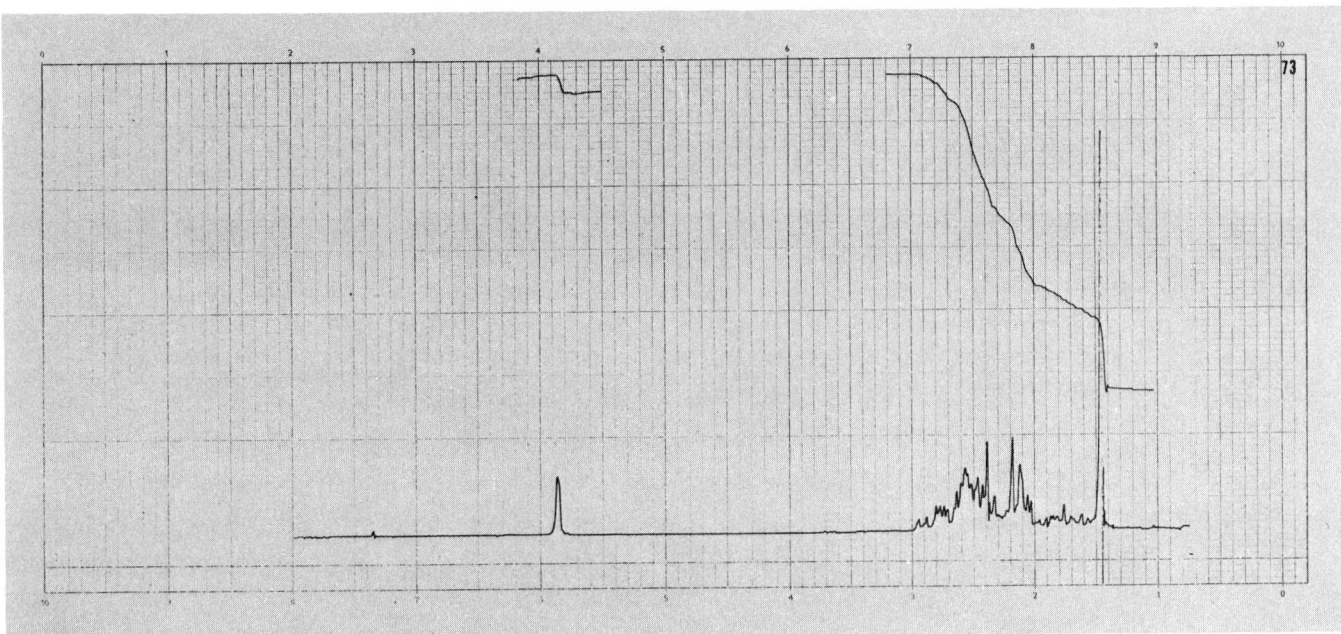

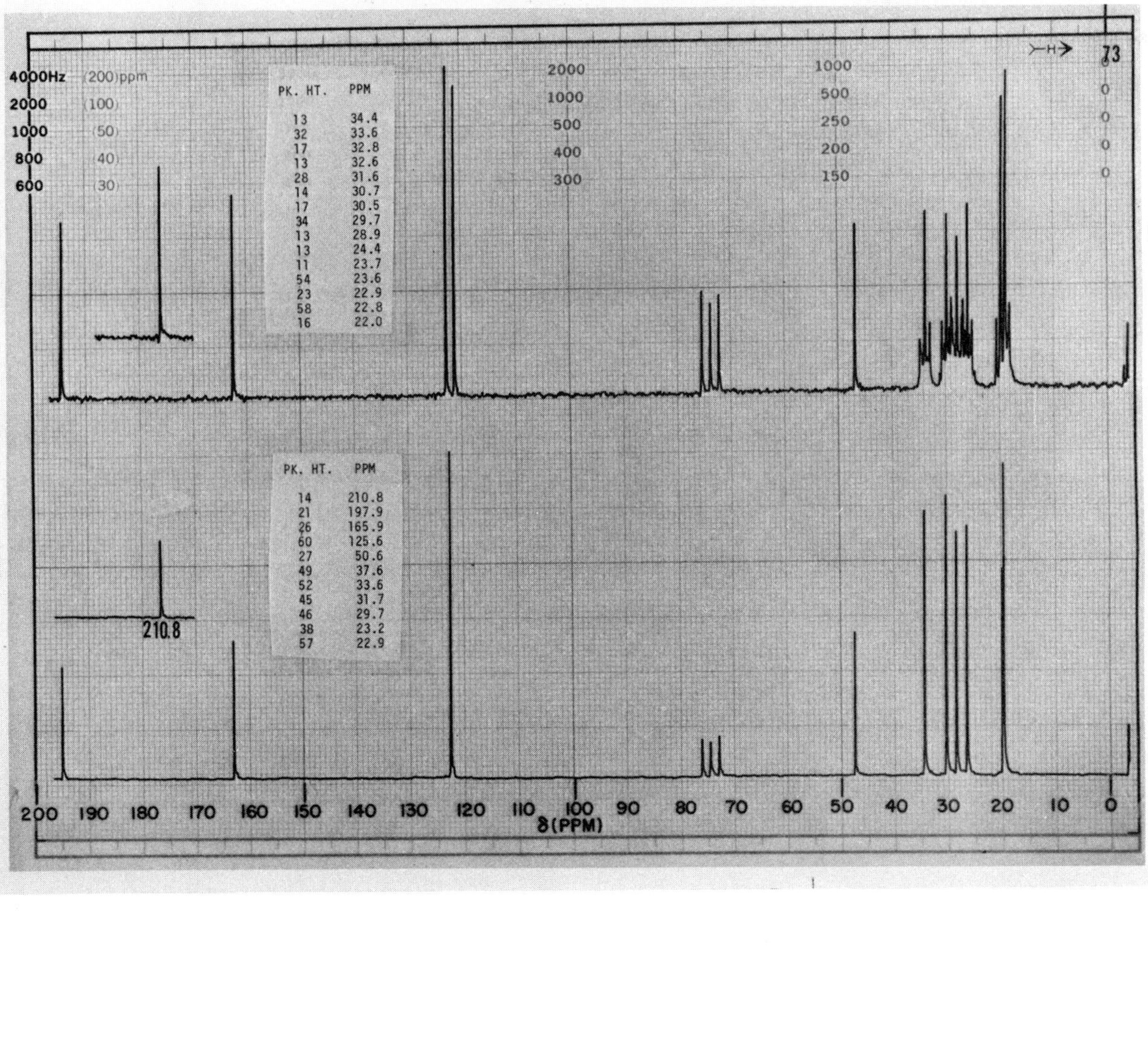

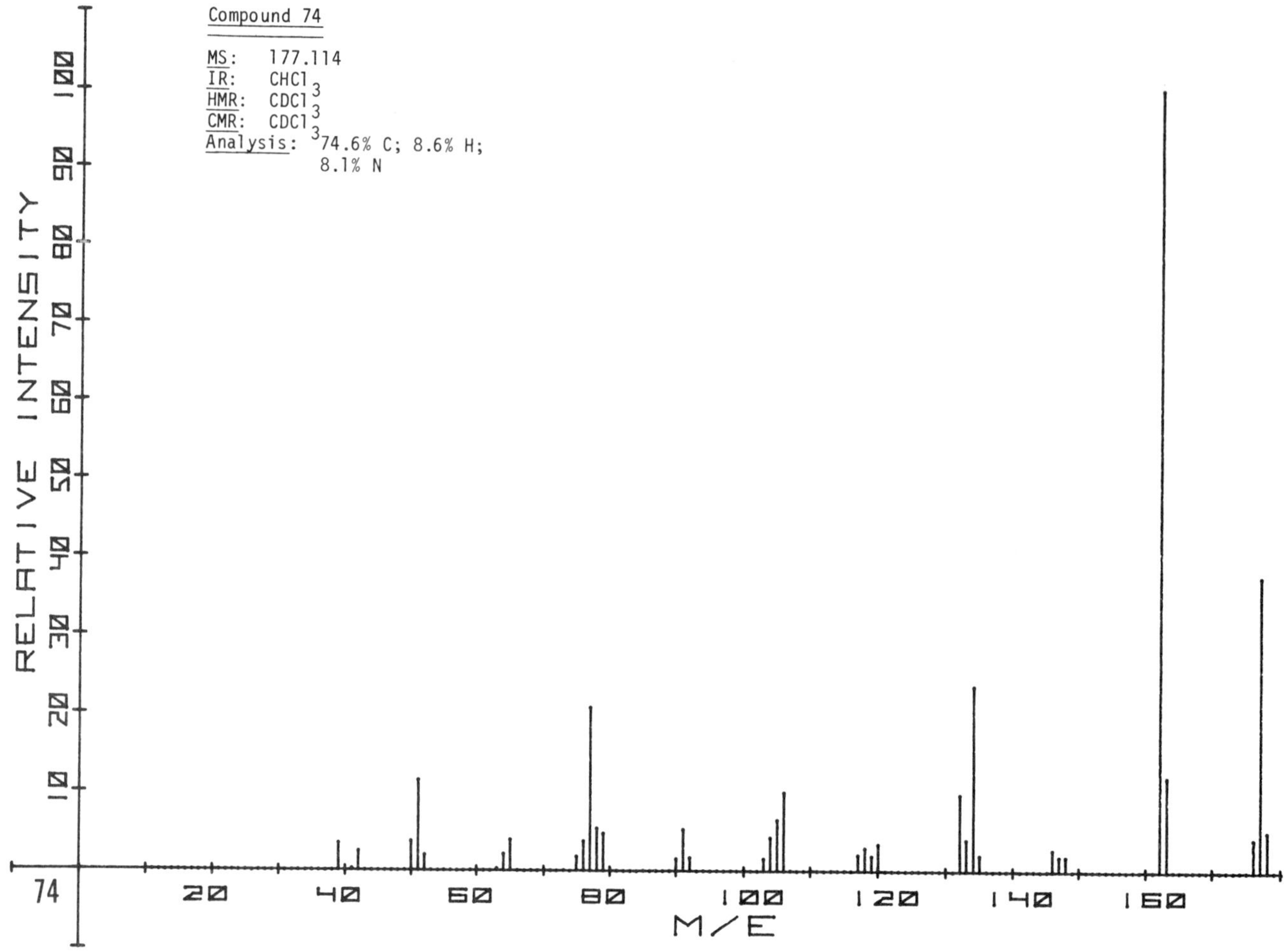

COMPOUND 74

Compound 74
MS: 177.114
IR: CHCl$_3$
HMR: CDCl$_3$
CMR: CDCl$_3$
Analysis: 74.6% C; 8.6% H; 8.1% N

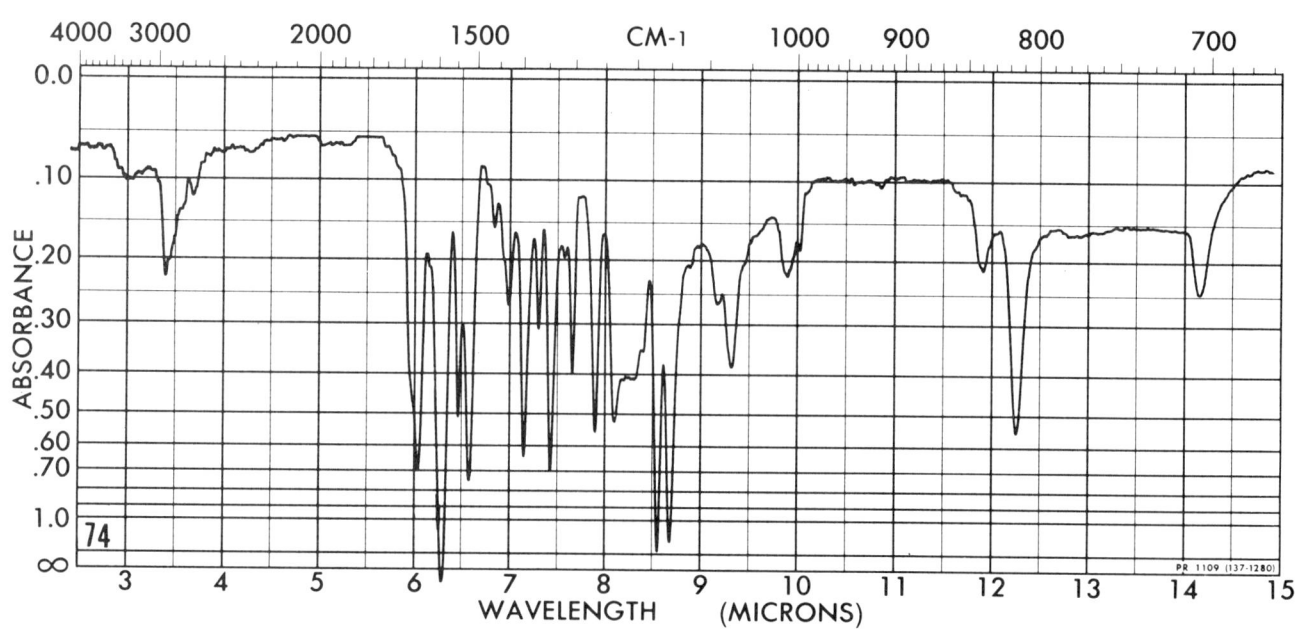

74
10
9
8
7
6
5
4
3
2
1
0

74
4000Hz (200)ppm 3000 2000 1000 0
2000 (100) 1500 1000 500 0
1000 (50) 750 500 250 0
800 (40) 600 400 200 0
600 (30) 450 300 150 0
200 190 180 170 160 150 140 130 120 110 100 90 80 70 60 50 40 30 20 10 0
δ(PPM)

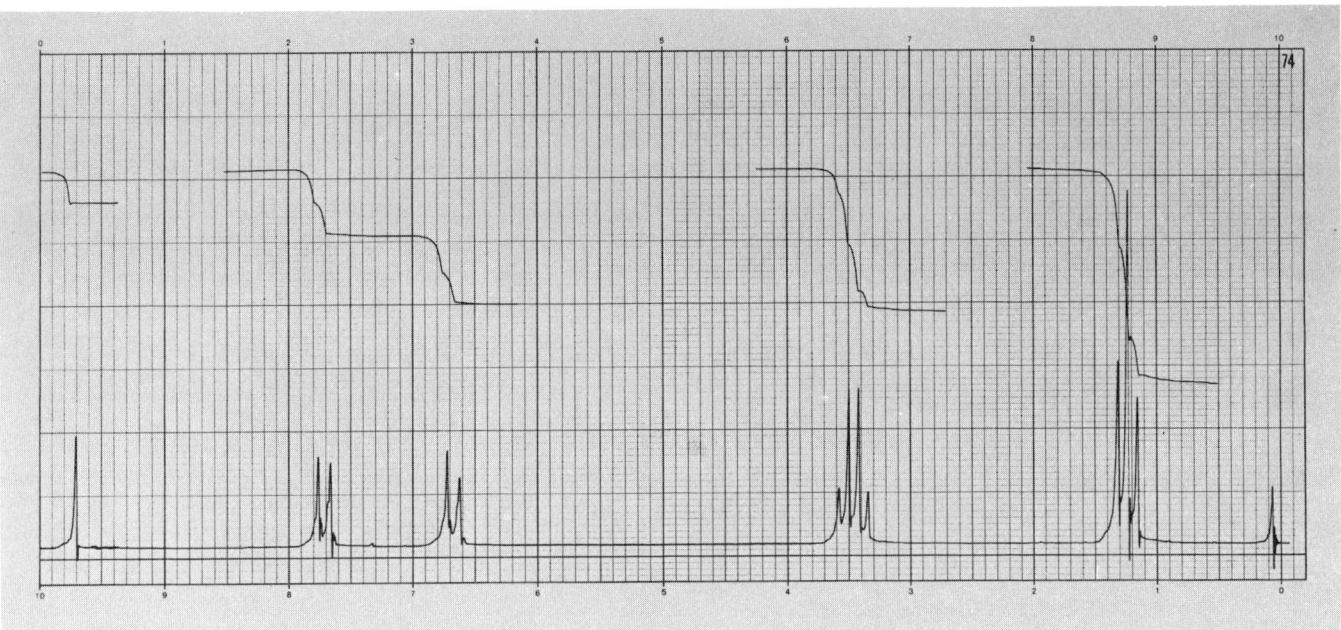

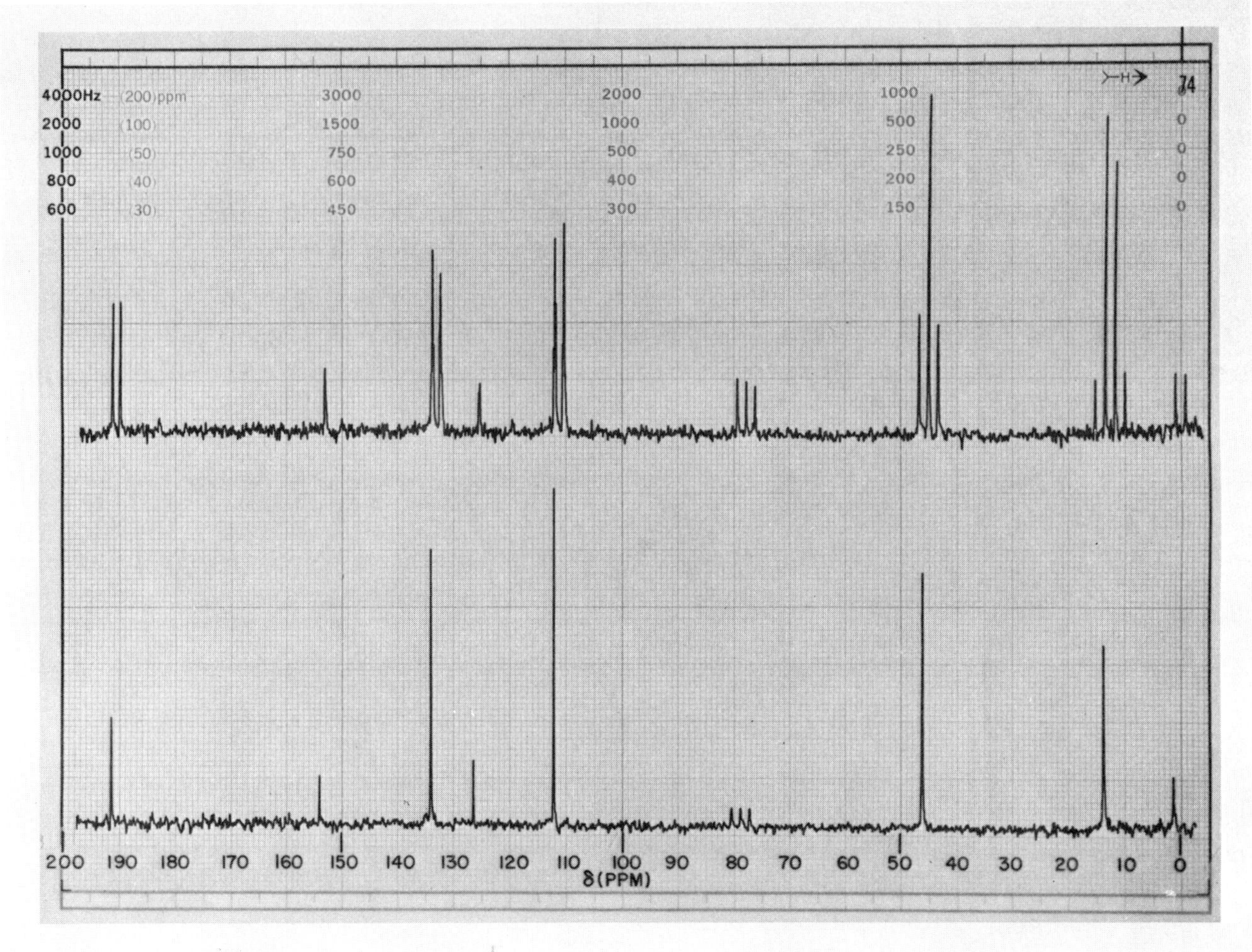

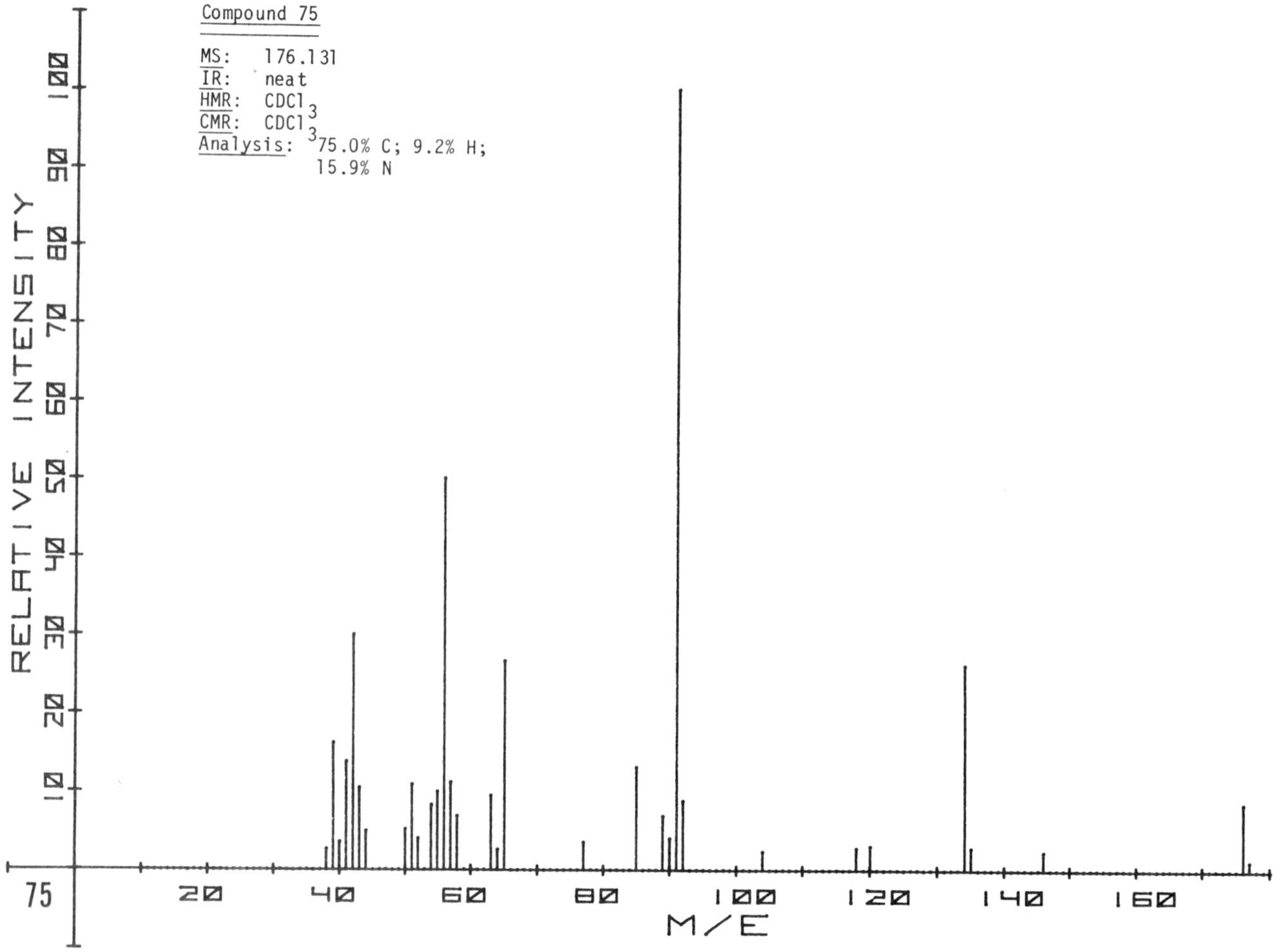

Compound 75
MS: 176.131
IR: neat
HMR: CDCl3
CMR: CDCl3
Analysis: 75.0% C; 9.2% H; 15.9% N
RELATIVE INTENSITY
10 20 30 40 50 60 70 80 90 100
75
20 40 60 80 100 120 140 160
M/E

4000 3000 2000 1500 CM-1 1000 900 800 700
0.0 .10 .20 .30 .40 .50 .60 .70 1.0 ∞
ABSORBANCE
75
3 4 5 6 7 8 9 10 11 12 13 14 15
WAVELENGTH (MICRONS)

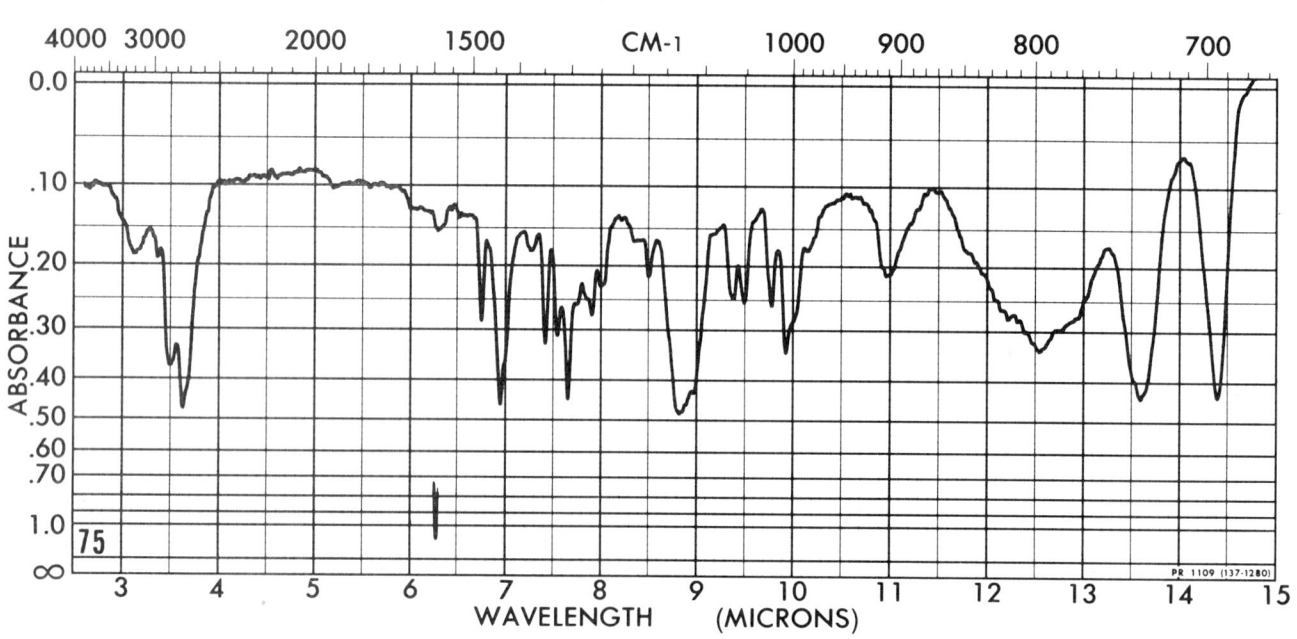

75
D_2O

75
PK. HT. PPM
56 129.7
38 128.7
60 128.3
33 127.5
40 127.3
20 126.1
PK. HT. PPM
11 138.2
49 129.0
60 128.0
32 126.8
23 63.6
39 54.5
40 46.1
δ(PPM)

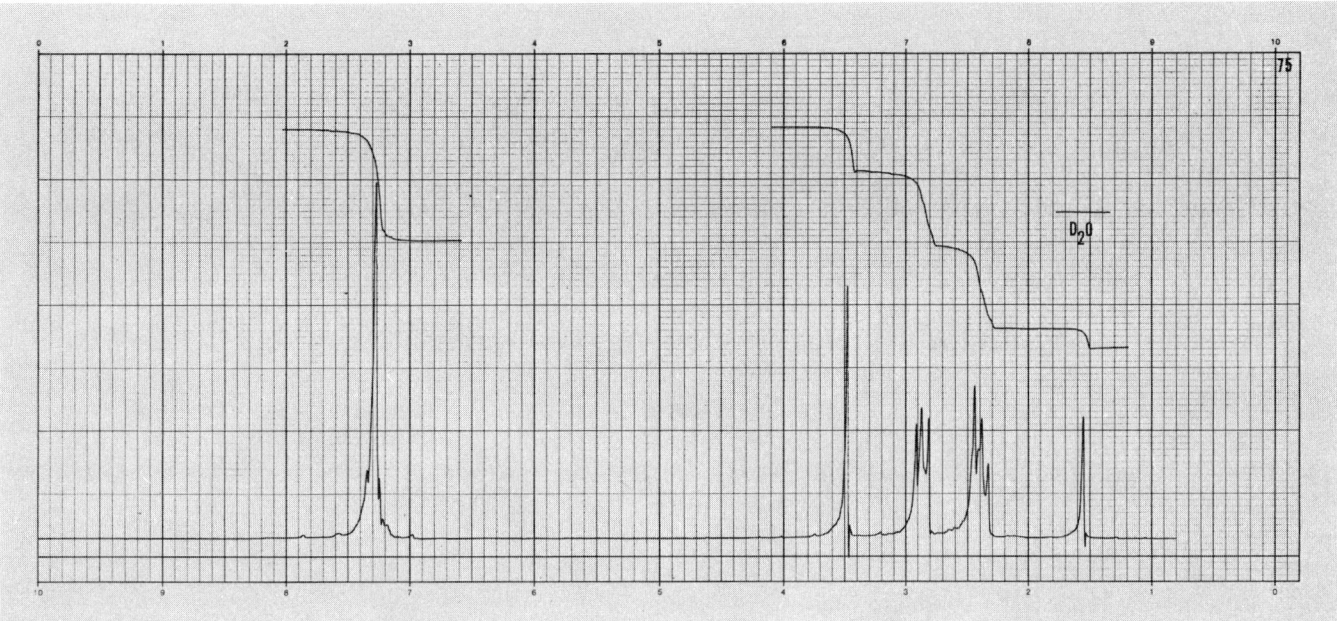

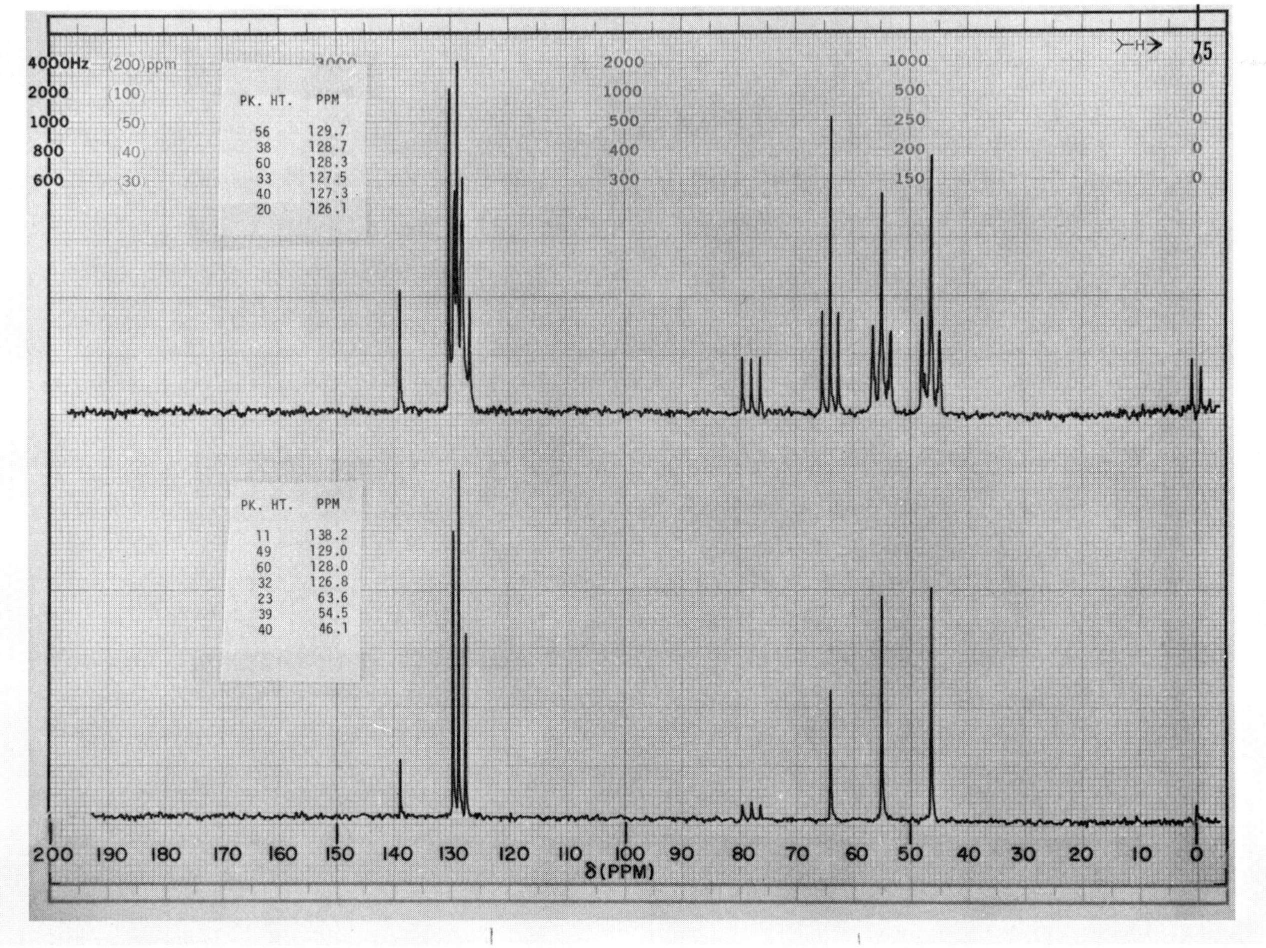

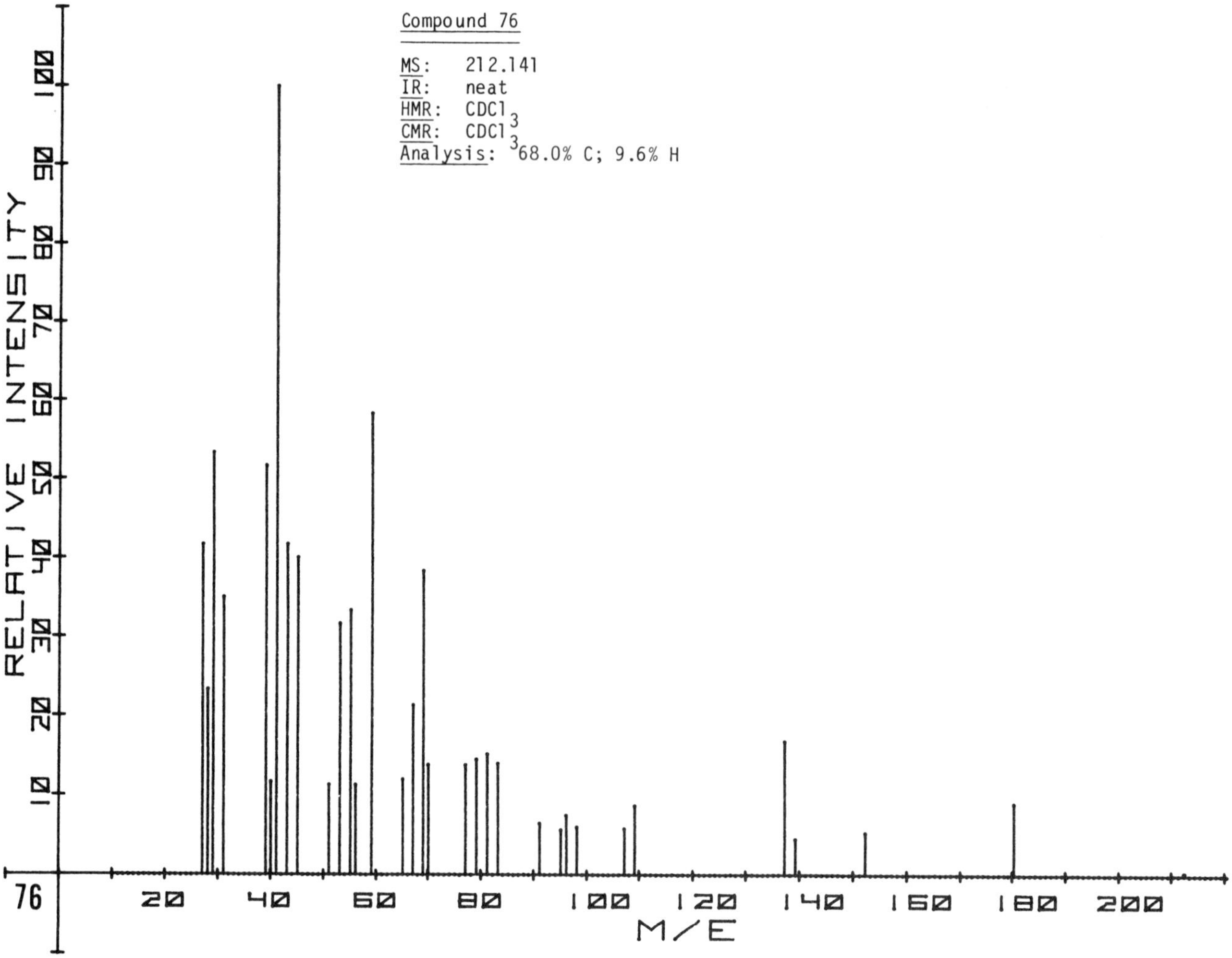

Compound 76
MS: 212.141
IR: neat
HMR: CDCl3
CMR: CDCl3
Analysis: 68.0% C; 9.6% H
RELATIVE INTENSITY
10 20 30 40 50 60 70 80 90 100
76
20 40 60 80 100 120 140 160 180 200
M/E

4000 3000 2000 1500 CM-1 1000 900 800 700
ABSORBANCE
0.0 .10 .20 .30 .40 .50 .60 .70 1.0 ∞
76
3 4 5 6 7 8 9 10 11 12 13 14 15
WAVELENGTH (MICRONS)

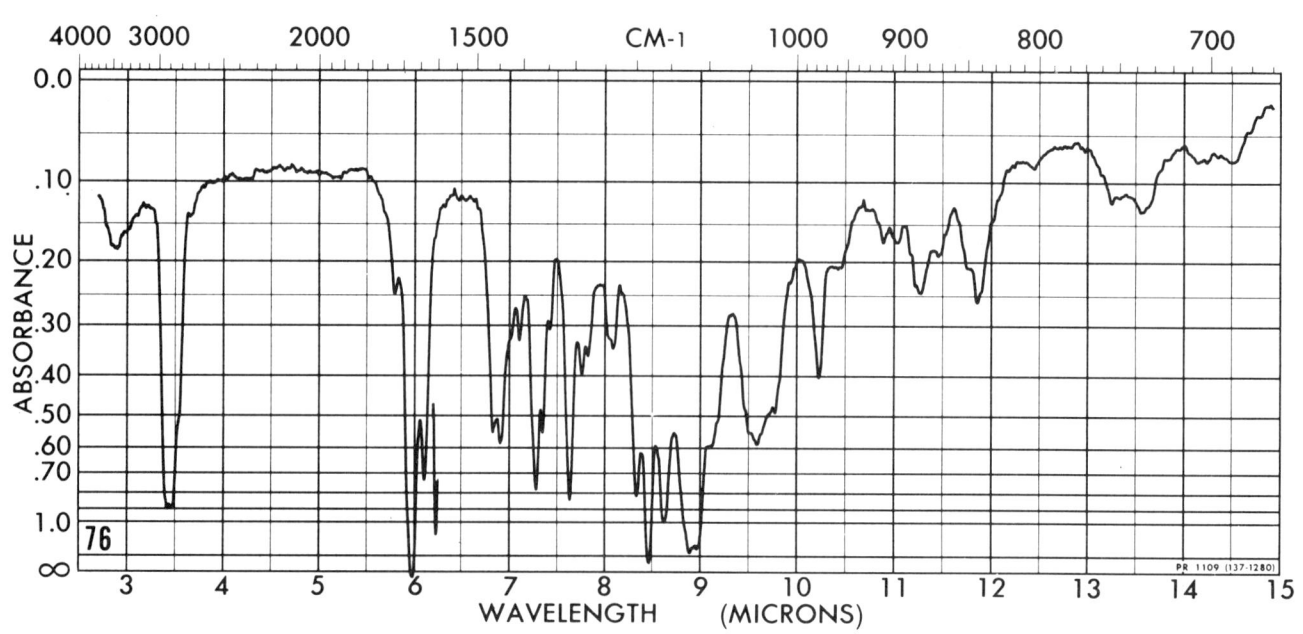

76

76
4000Hz (200)ppm 3000 2000 1000
2000 1500 1000 500
1000 750 500 250
800 600 400 200
600 450 300 150
Pk. Ht. PPM
17 194.2
8 147.2
14 142.9
32 71.9
49 70.7
34 59.8
37 57.5
36 52.2
32 45.5
29 33.0
54 29.2
60 27.0
31 18.8
34 16.6
Pk. Ht. PPM
11 194.2
8 147.3
10 142.9
31 71.9
37 70.7
24 58.7
36 52.2
35 45.5
25 33.0
60 28.1
18 17.7
200 190 180 170 160 150 140 130 120 110 100 90 80 70 60 50 40 30 20 10 0
δ(PPM)

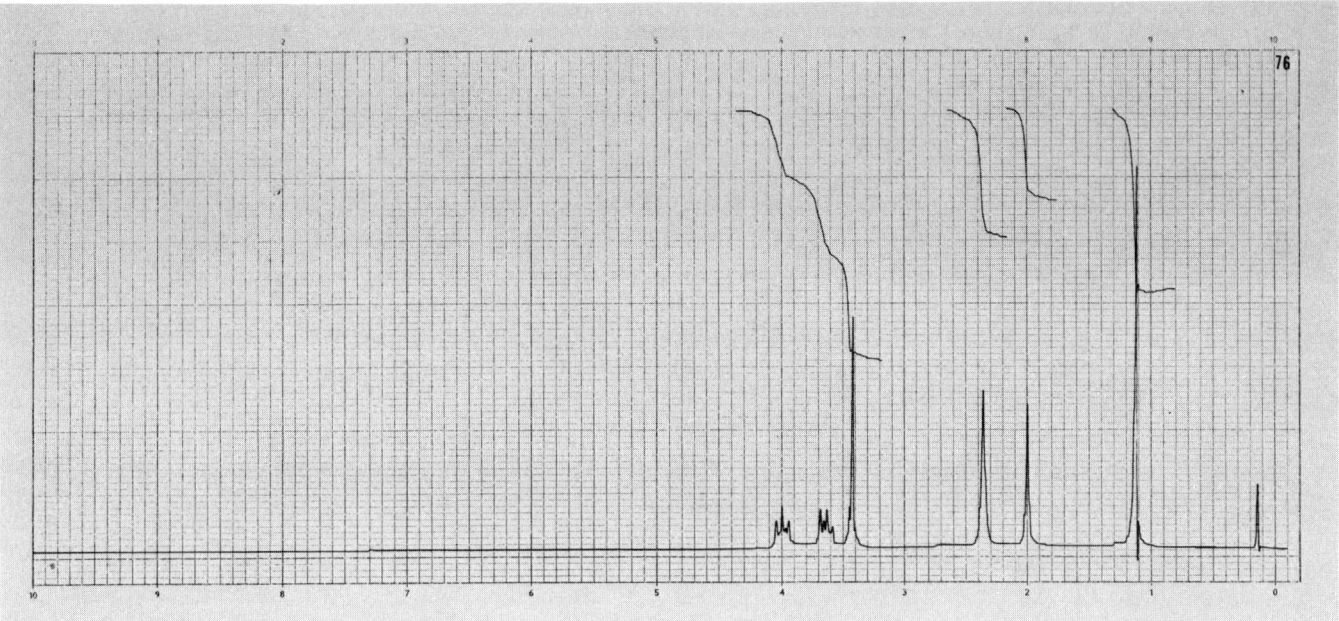

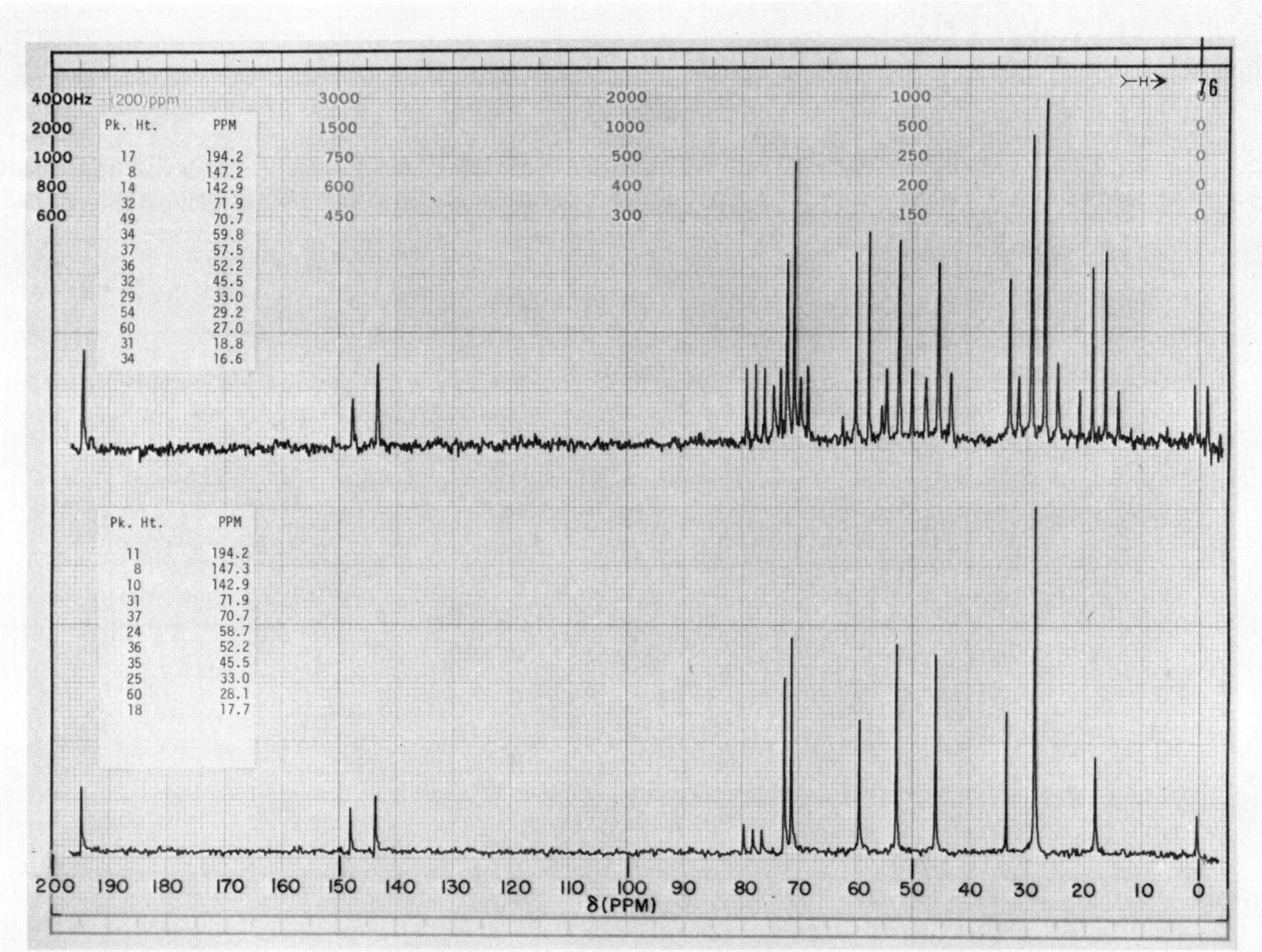

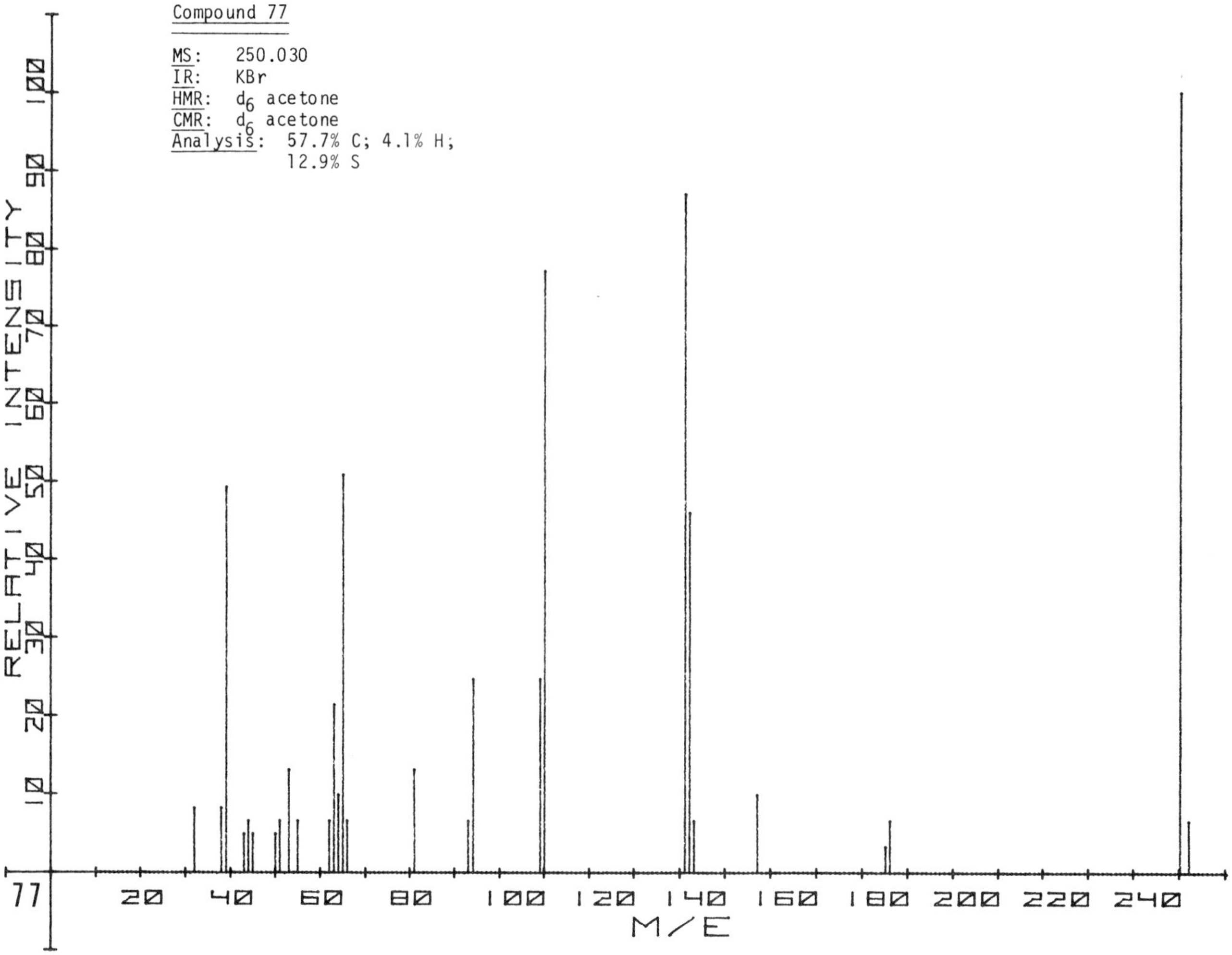

Compound 77
MS: 250.030
IR: KBr
HMR: d_6 acetone
CMR: d_6 acetone
Analysis: 57.7% C; 4.1% H; 12.9% S
RELATIVE INTENSITY
10 20 30 40 50 60 70 80 90 100
77
20 40 60 80 100 120 140 160 180 200 220 240
M/E

4000 3000 2000 1500 CM-1 1000 900 800 700
ABSORBANCE
0.0 .10 .20 .30 .40 .50 .60 .70 1.0 ∞
77
3 4 5 6 7 8 9 10 11 12 13 14 15
WAVELENGTH (MICRONS)

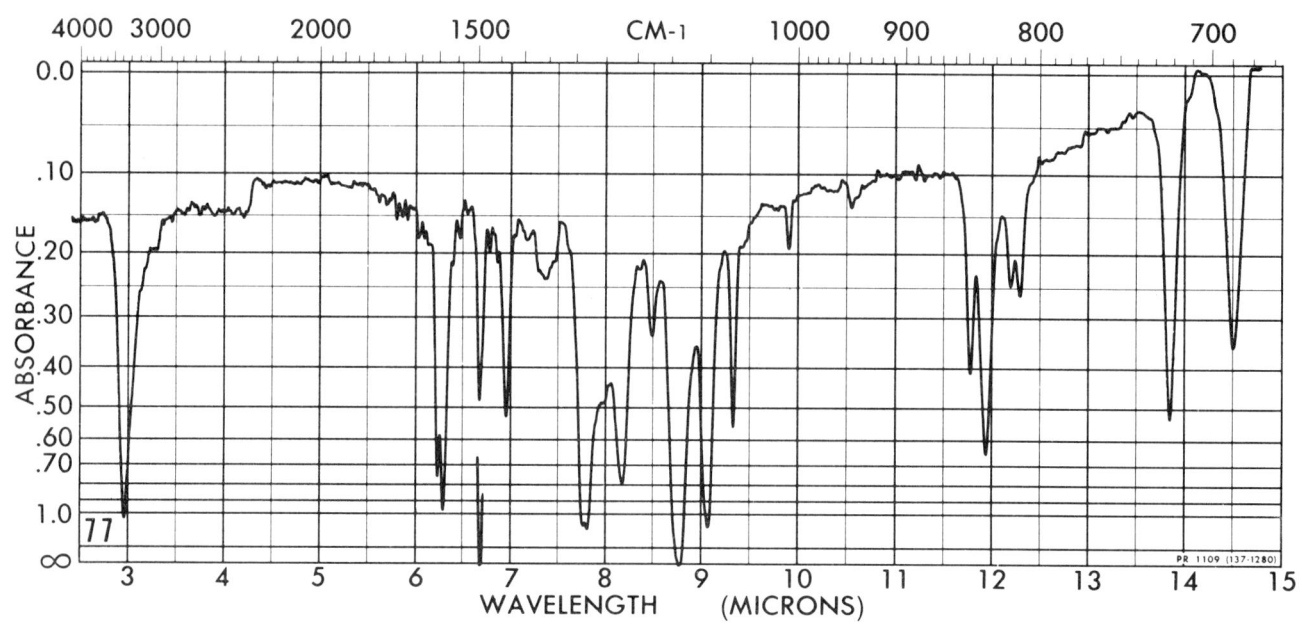

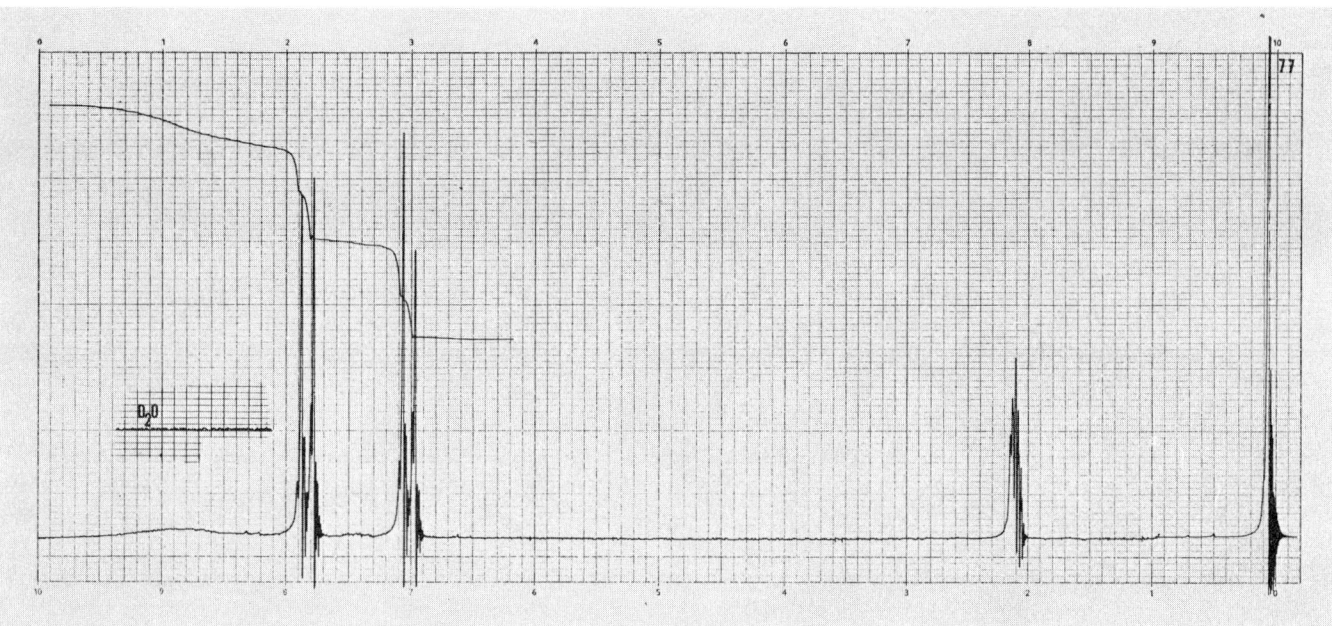

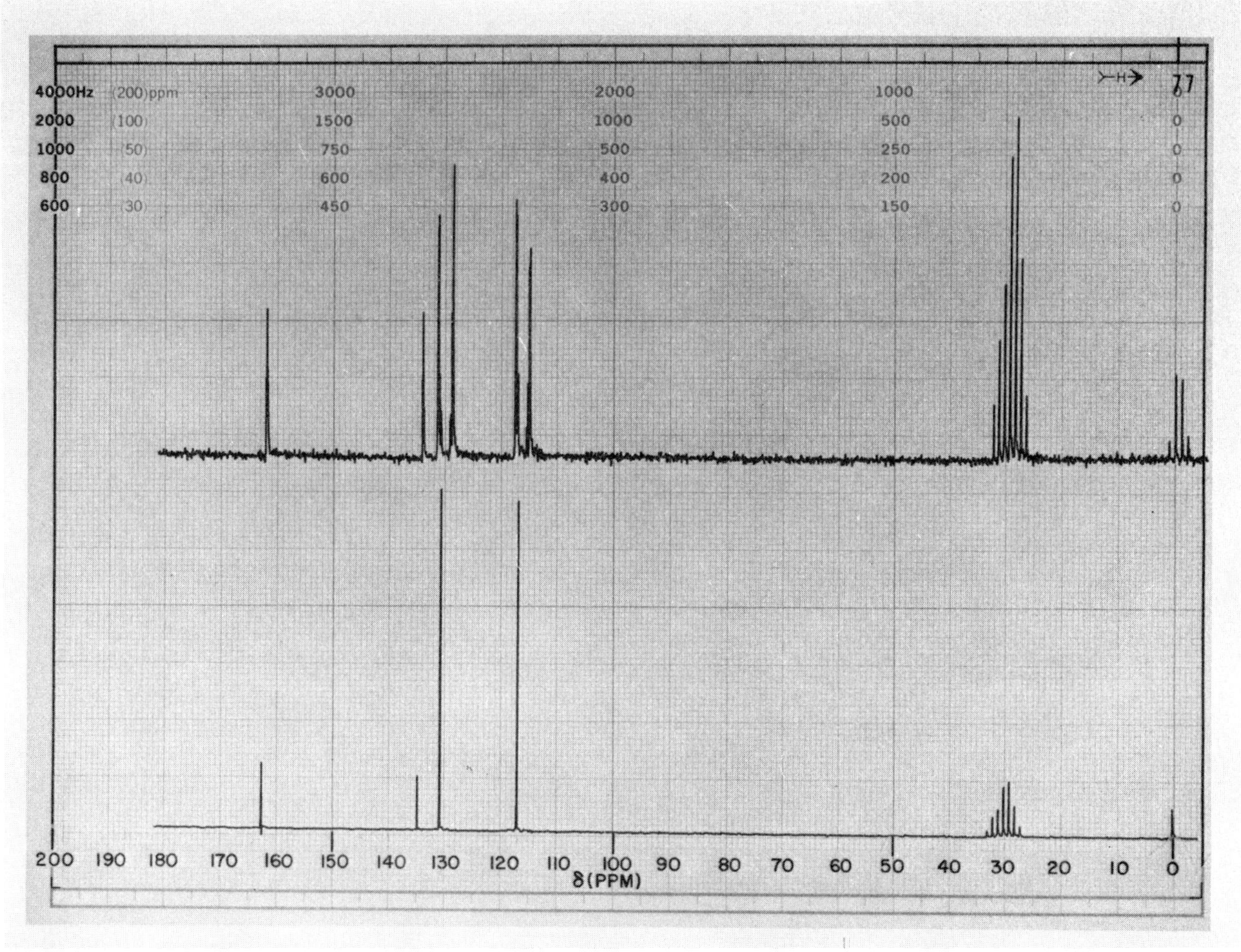

COMPOUND 78

Compound 78
MS: 190.082
IR: neat
HMR: CDCl$_3$
CMR: CDCl$_3$
Analysis: 75.9% C; 7.5% H; 16.9% S

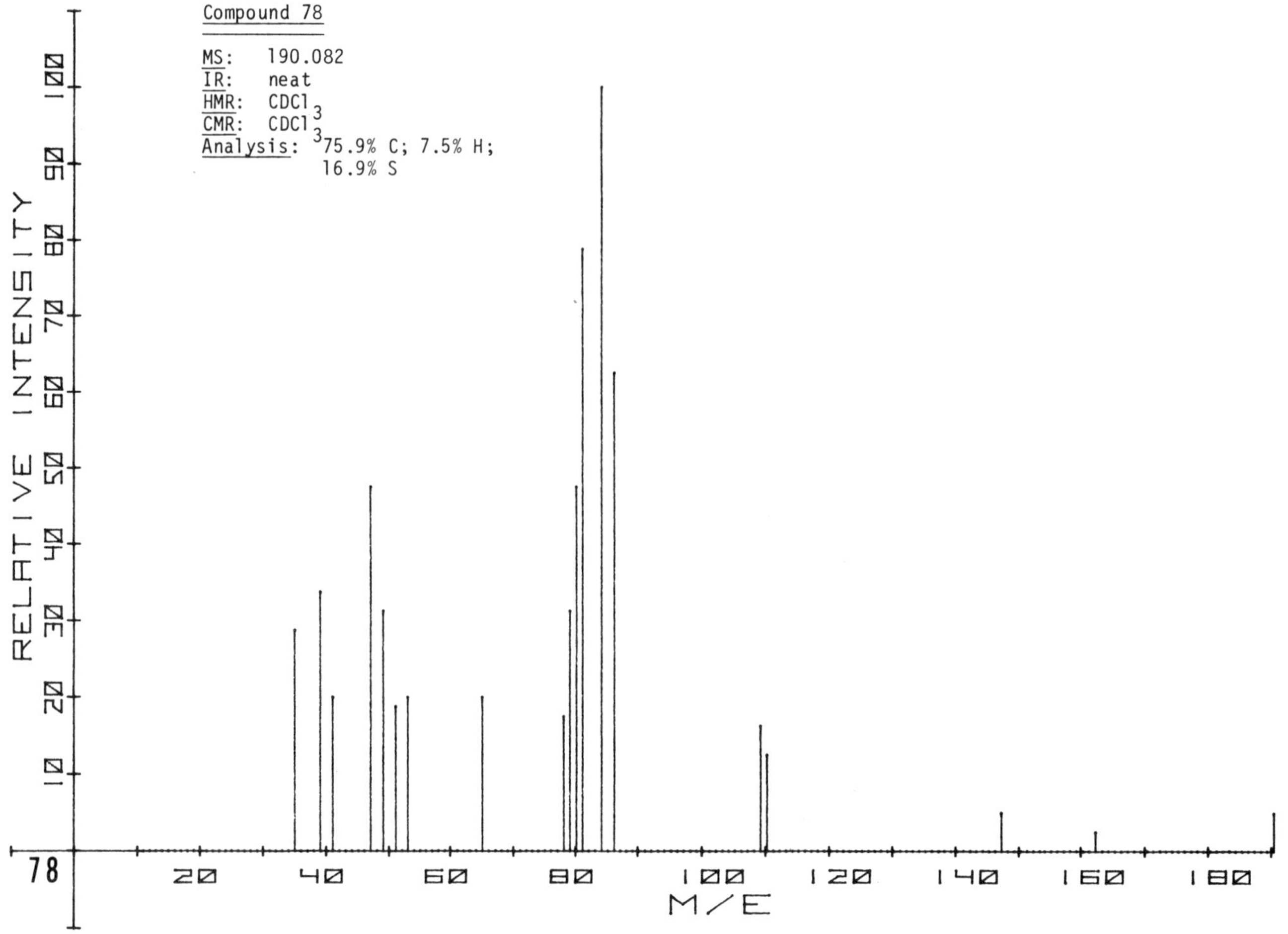

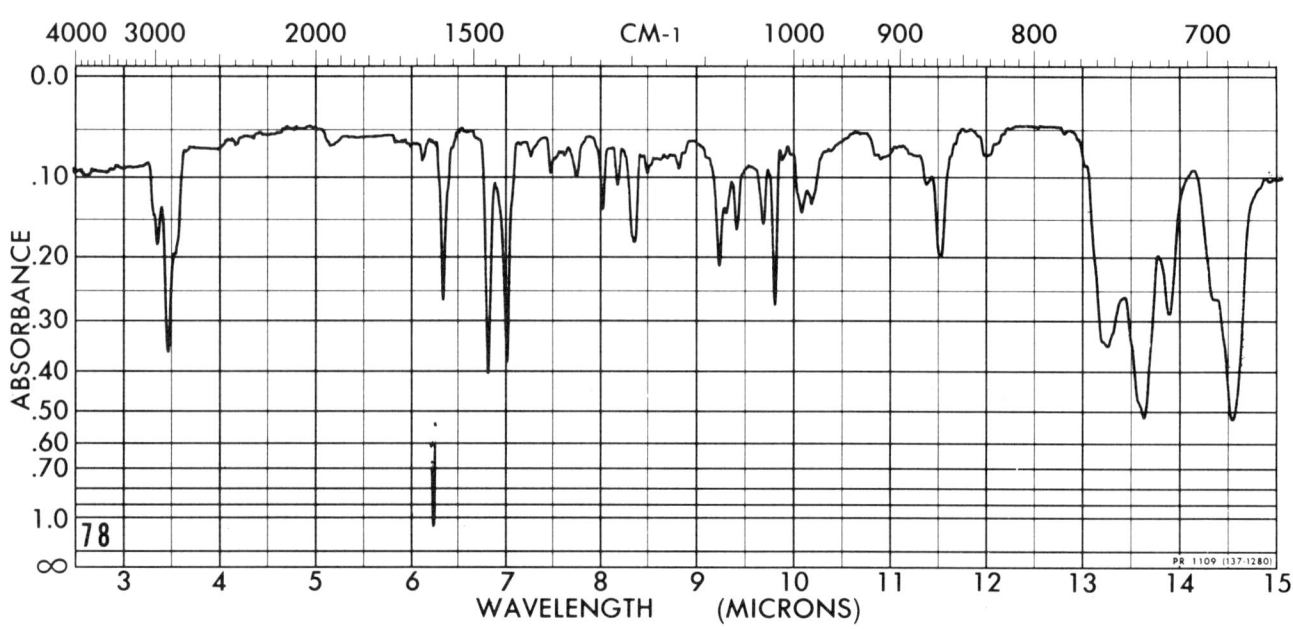

78

78
4000Hz (200)ppm 3000 2000 1000 0
2000 (100) 1500 1000 500 0
1000 (50) 500 250 0
800 (40) 400 200 0
600 (30) 300 150 0
PK. HT. PPM
10 136.0
60 132.2
17 131.0
53 130.3
37 129.8
19 129.5
44 127.8
23 127.5
14 126.2
27 125.5
PK. HT. PPM
3 136.0
60 131.3
26 130.2
43 128.8
26 127.0
34 126.5
15 43.9
22 28.9
24 25.0
25 19.5
200 190 180 170 160 150 140 130 120 110 100 90 80 70 60 50 40 30 20 10 0
δ(PPM)

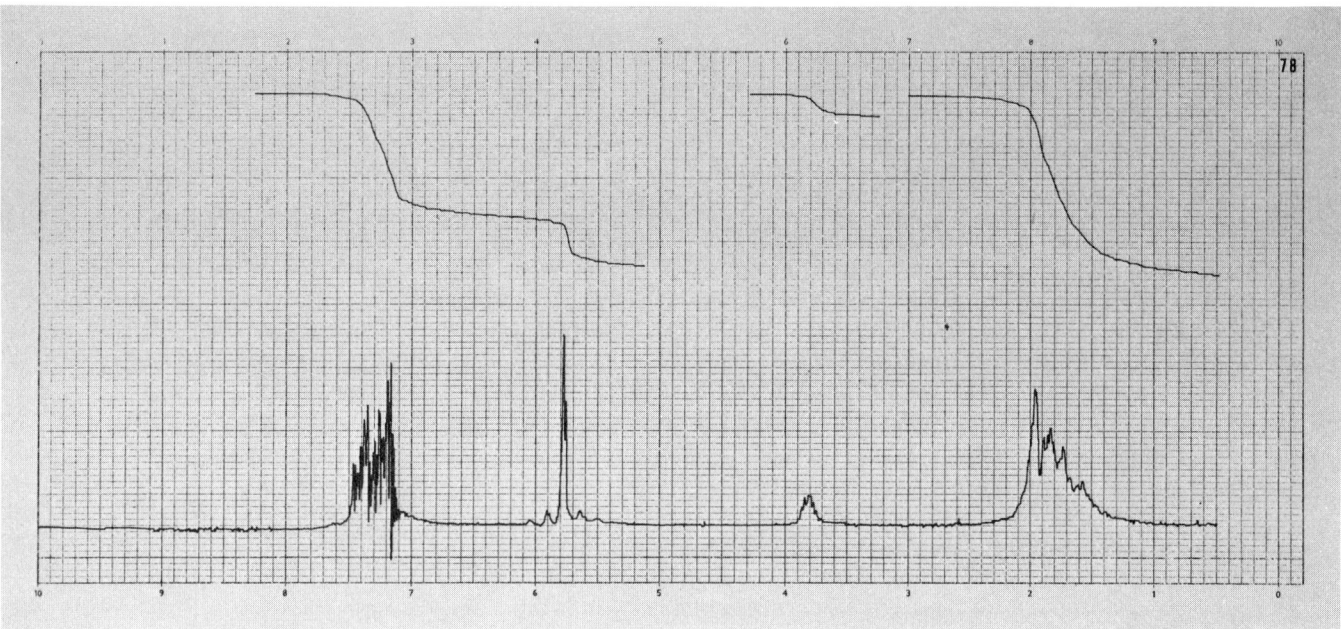

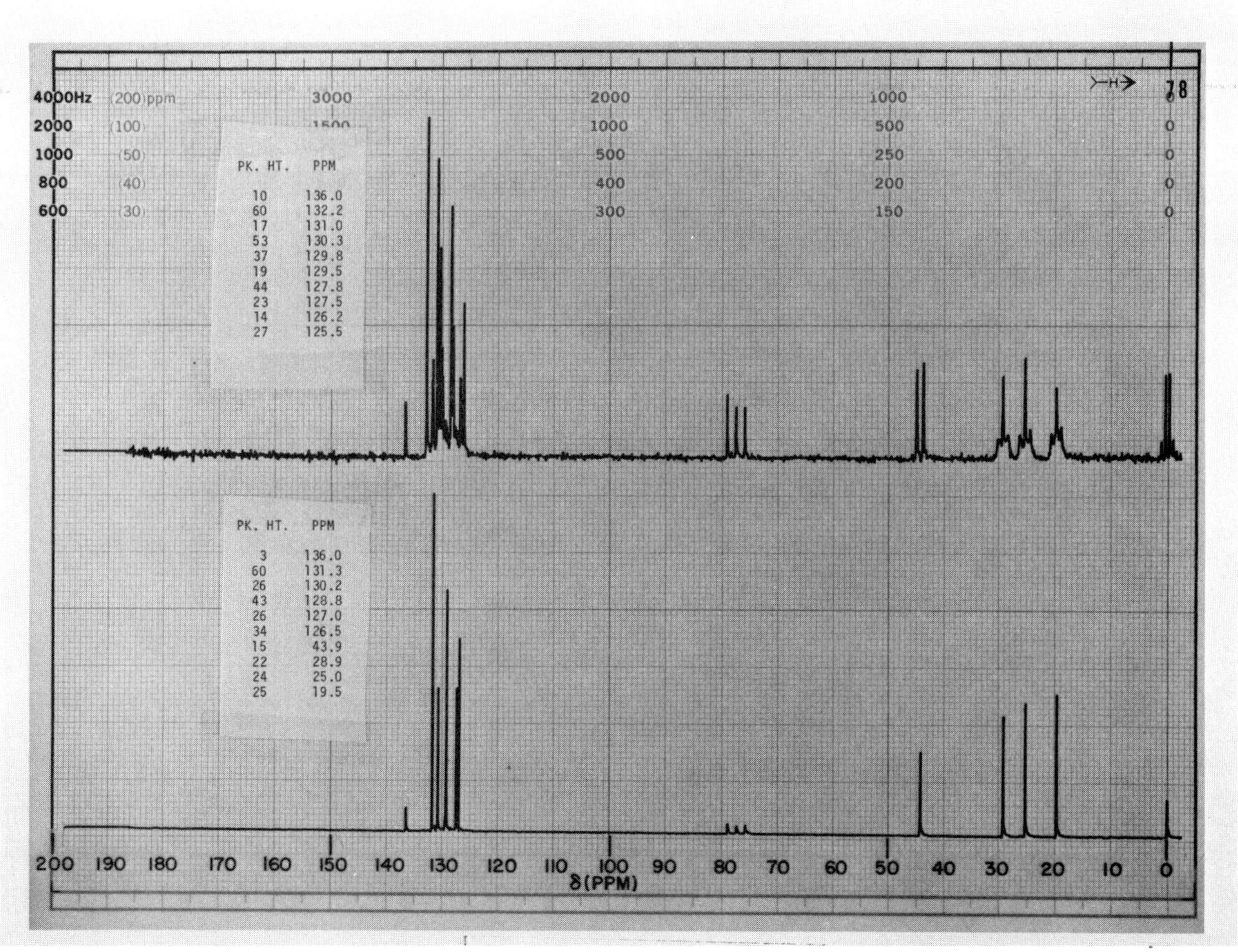

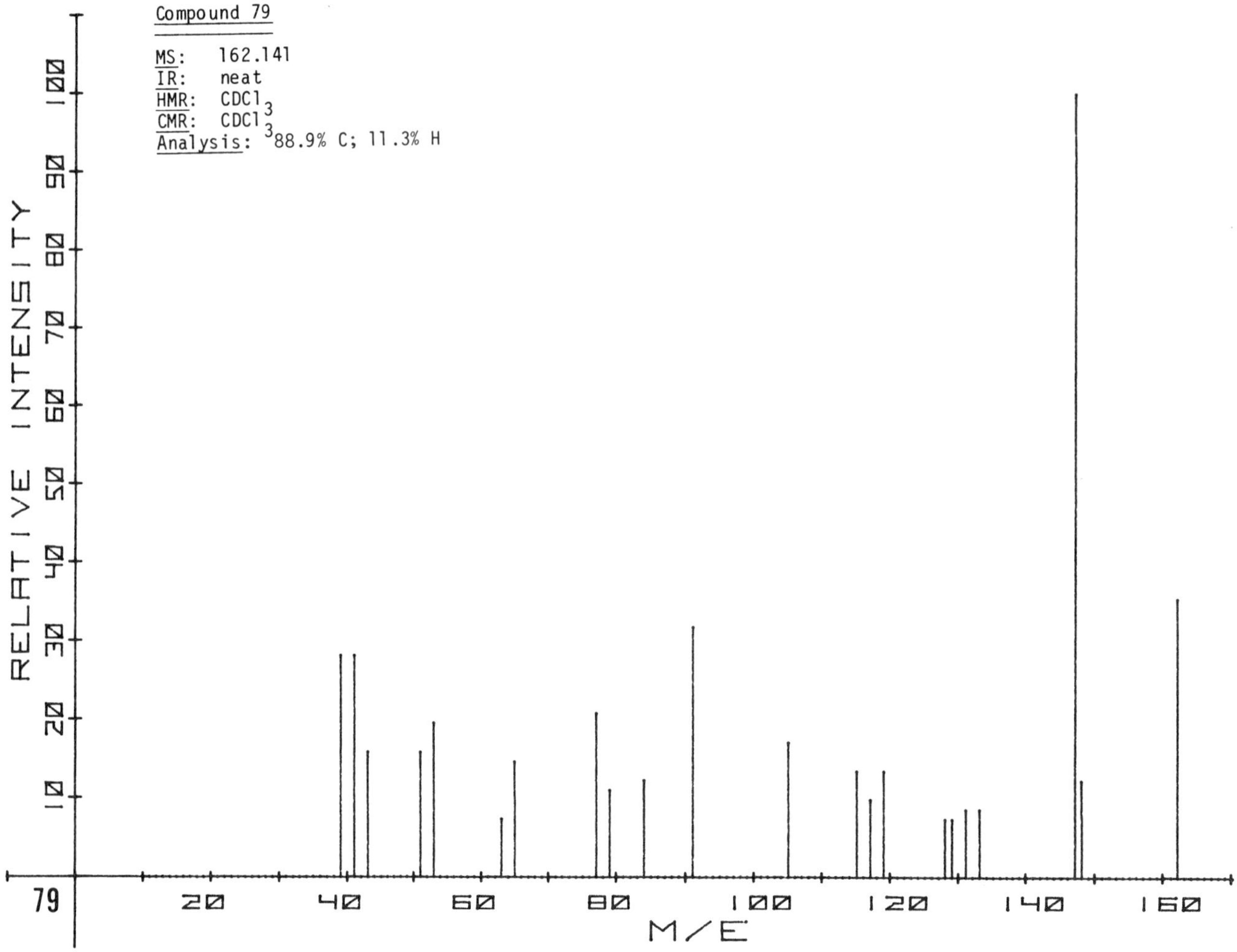

Compound 79
MS: 162.141
IR: neat
HMR: $CDCl_3$
CMR: $CDCl_3$
Analysis: 88.9% C; 11.3% H
RELATIVE INTENSITY
10 20 30 40 50 60 70 80 90 100
79
20 40 60 80 100 120 140 160
M/E

4000 3000 2000 1500 CM-1 1000 900 800 700
ABSORBANCE
0.0 .10 .20 .30 .40 .50 .60 .70 1.0 ∞
79
3 4 5 6 7 8 9 10 11 12 13 14 15
WAVELENGTH (MICRONS)

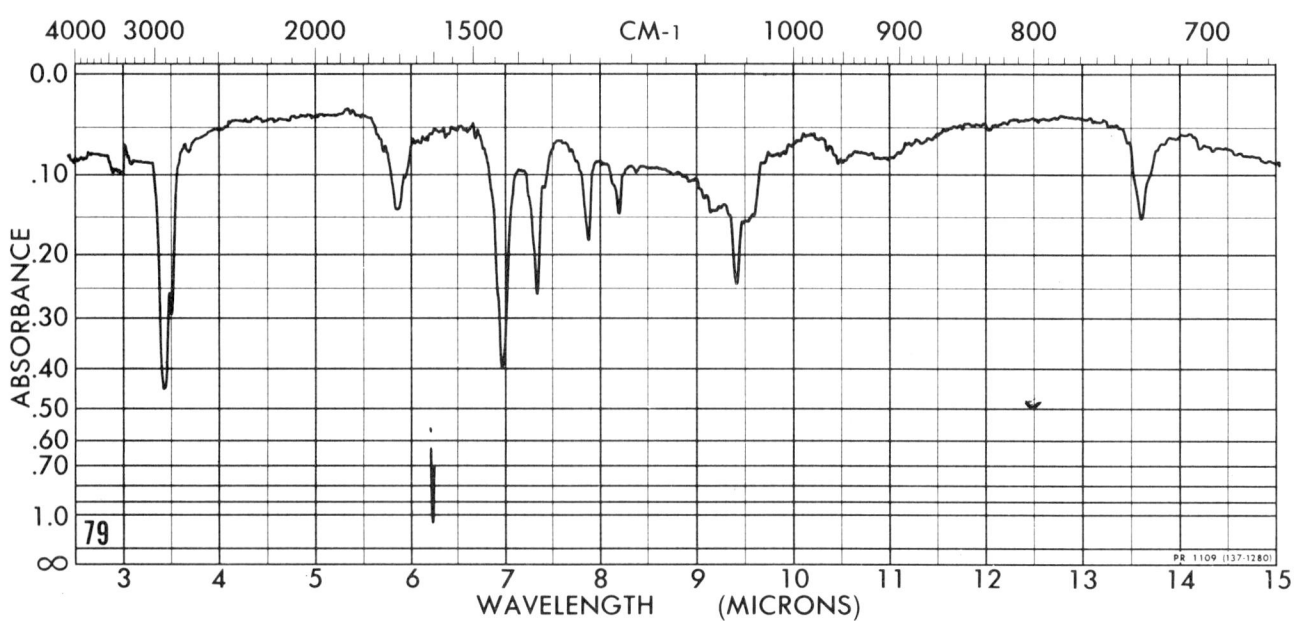

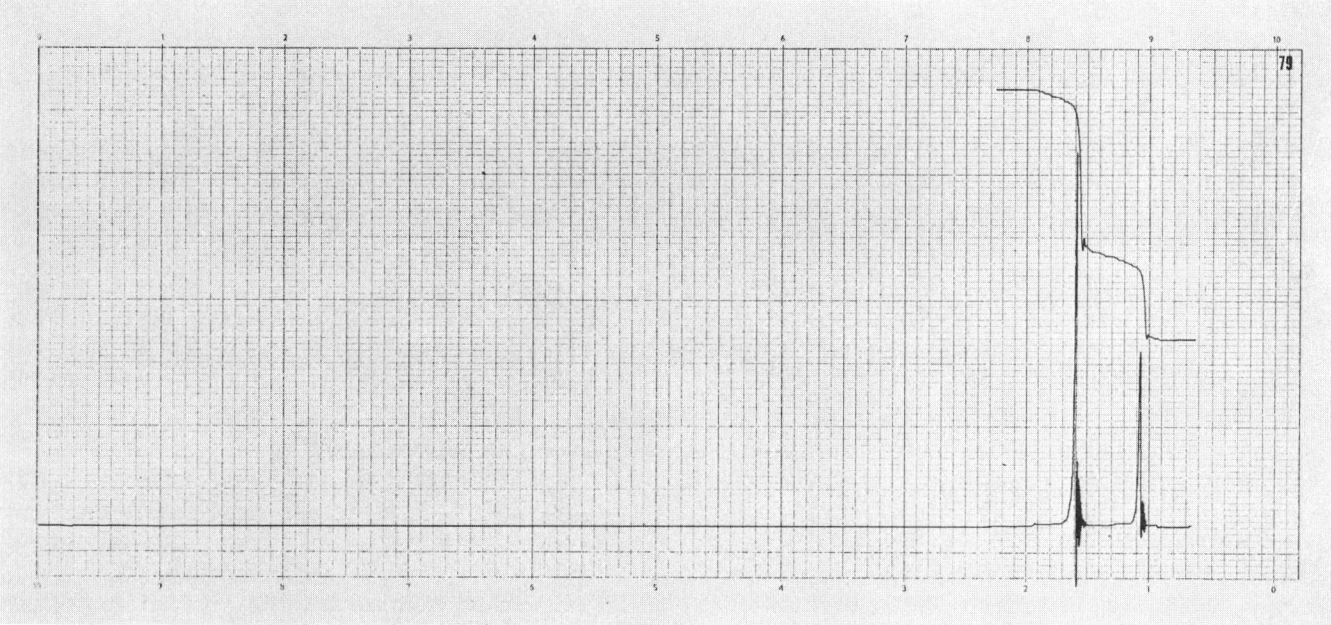

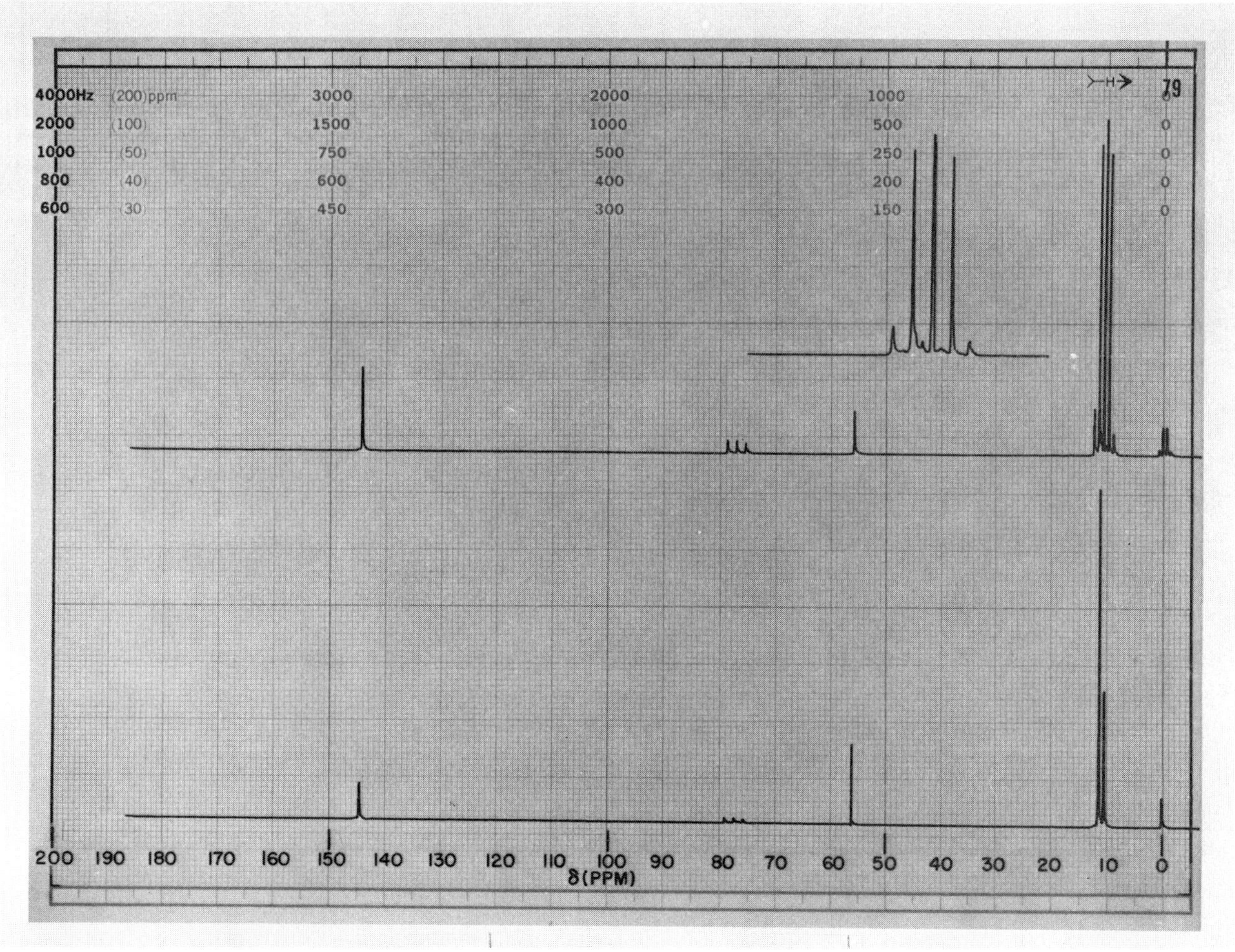

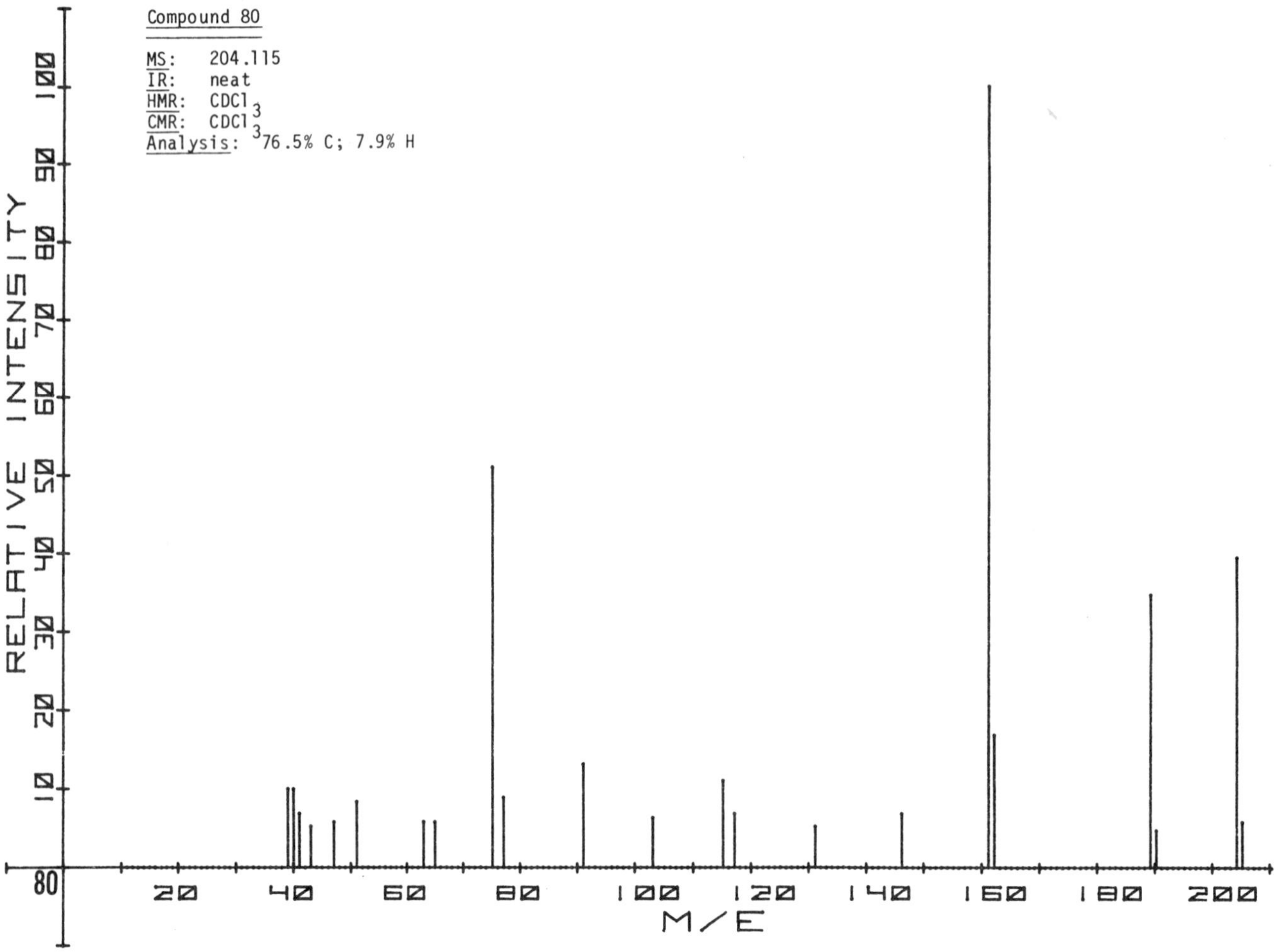

Compound 80
MS: 204.115
IR: neat
HMR: CDCl3
CMR: CDCl3
Analysis: 76.5% C; 7.9% H
RELATIVE INTENSITY
M/E
80

CM-1
ABSORBANCE
WAVELENGTH (MICRONS)
80

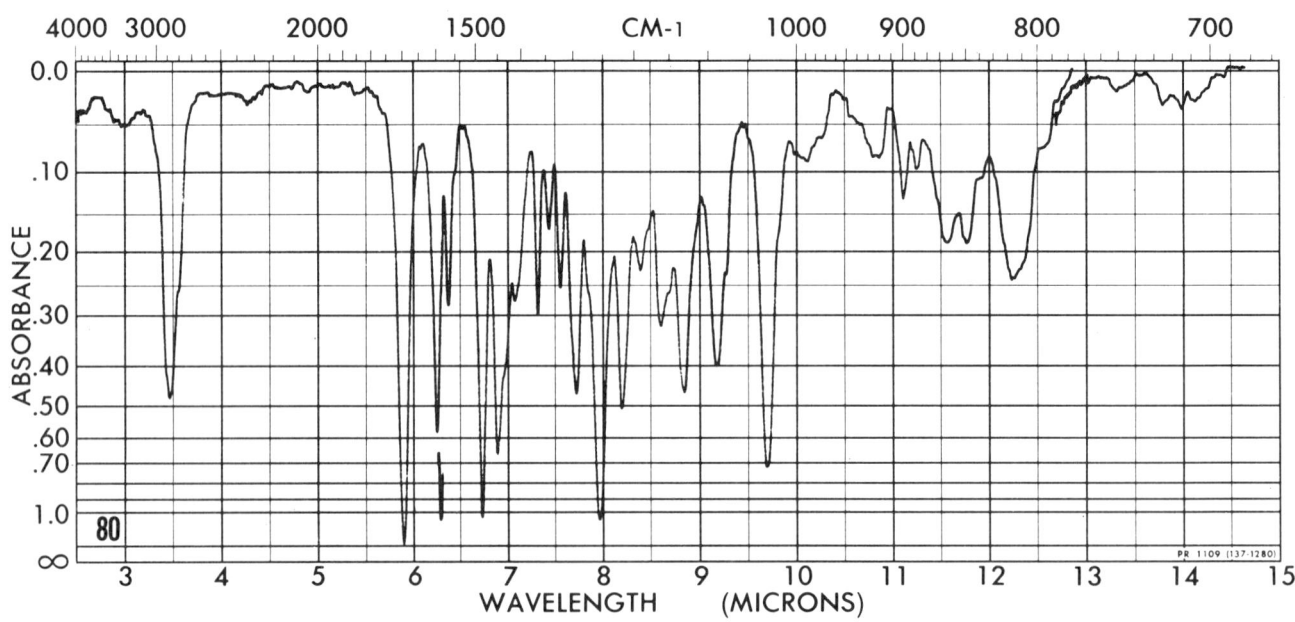

80

4000Hz (200)ppm
2000 (100)
1000 (50)
800 (40)
600 (30)
2000 1000 500 400 300
1000 500 250 200 150
80
PK. HT. PPM
48 113.8
33 113.5
47 112.5
27 112.1
7 30.0
14 29.3
30 28.7
60 27.8
12 27.4
55 26.2
11 24.7
27.8
PK. HT. PPM
10 214.3
16 158.0
24 136.4
20 135.6
58 127.2
57 113.2
60 112.8
29 55.1
28 47.1
40 37.0
42 28.7
60 27.0
214.3
200 190 180 170 160 150 140 130 120 110 100 90 80 70 60 50 40 30 20 10 0
δ(PPM)

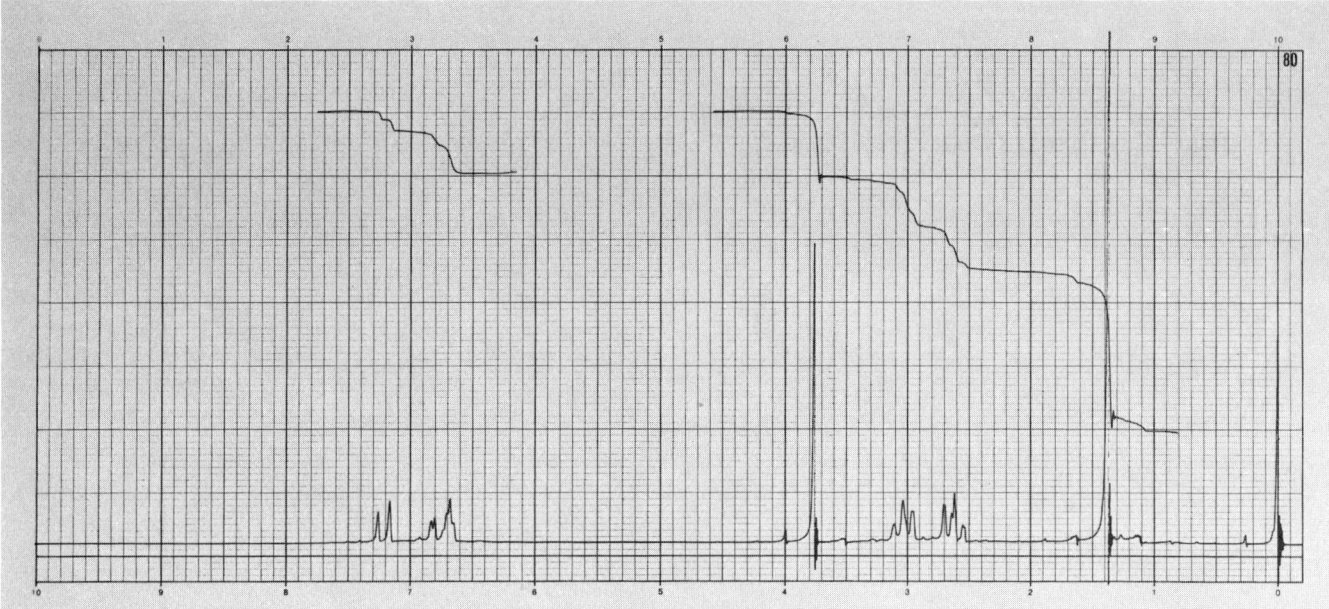

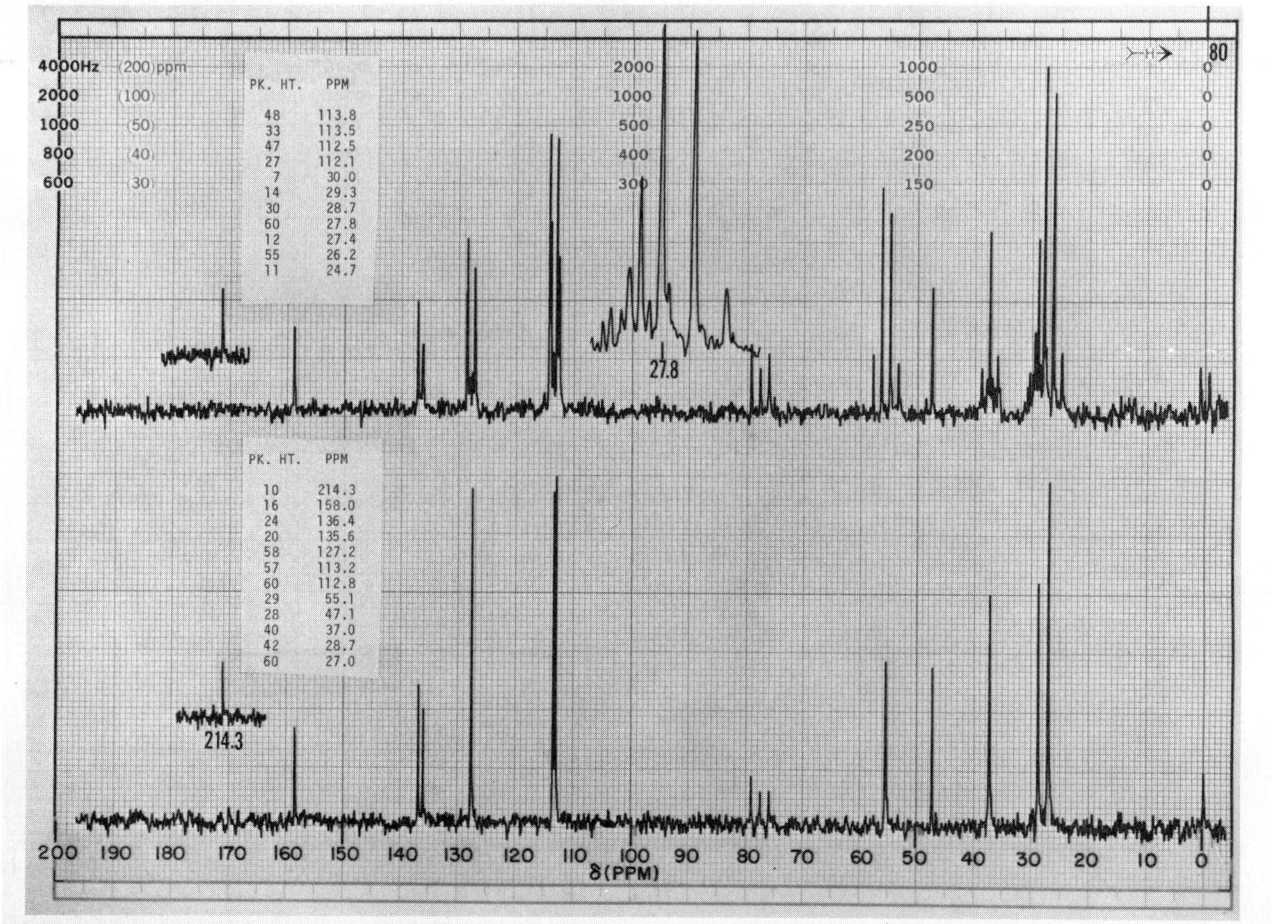

COMPOUND 81

Compound 81
MS: 192.151
IR: neat
HMR: CDCl$_3$
CMR: neat
Analysis: 81.3% C; 10.4% H

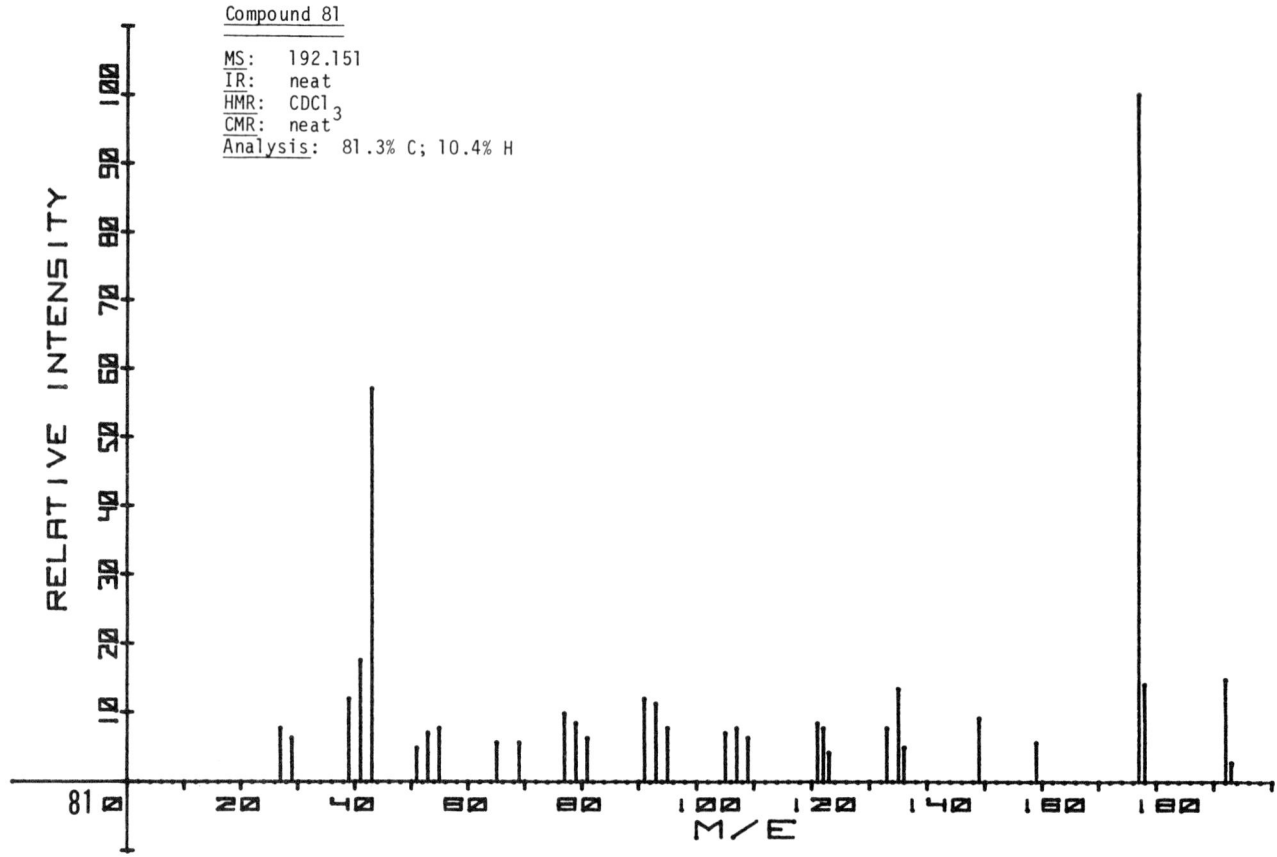

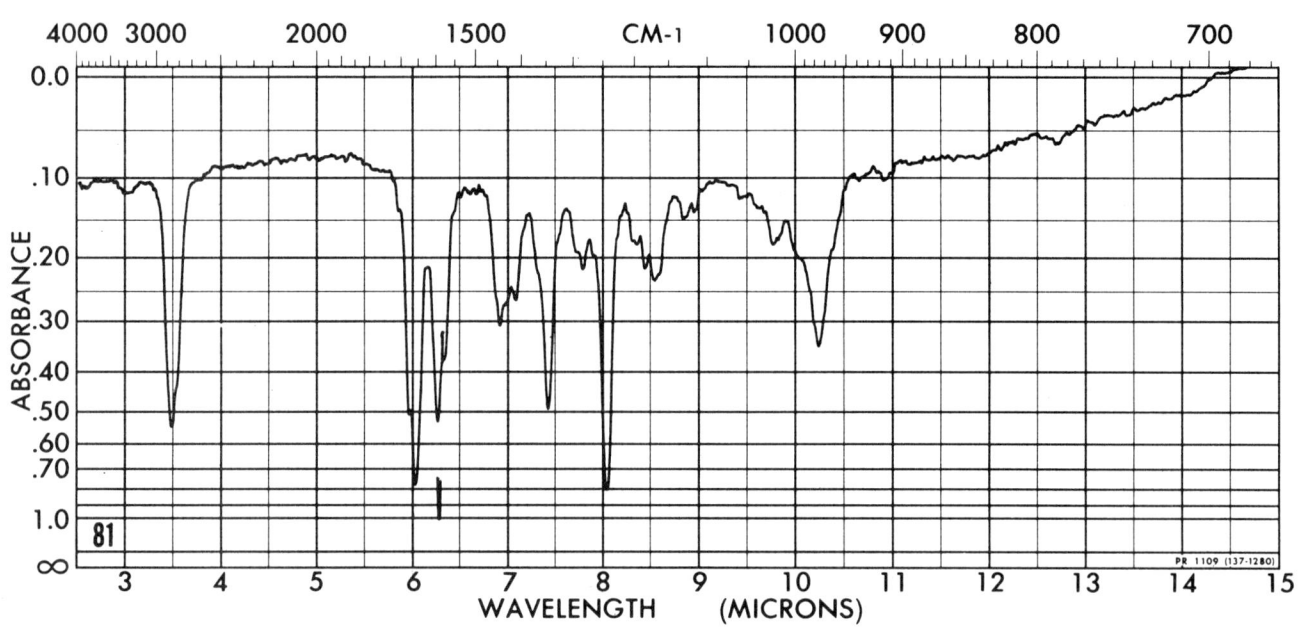

81
IRR. 6.10

4000Hz 3000 2000 1000 0
2000 1500 1000 500 0
1000 750 500 250 0
800 600 400 200 0
600 450 300 150 0
81
PK. HT. PPM
4 195.8
22 142.5
17 141.0
17 136.2
23 134.7
23 132.9
17 131.4
15 39.9
43 34.1
25 33.5
14 30.0
55 29.2
60 28.4
15 27.6
27 27.4
27 26.5
26 22.0
24 21.1
15 19.2
PK. HT. PPM
10 195.9
21 141.7
15 136.2
17 134.5
17 132.2
15 39.9
31 34.1
15 33.5
60 28.8
15 27.0
15 21.6
26 19.2
200 190 180 170 160 150 140 130 120 110 100 90 80 70 60 50 40 30 20 10 0
δ(PPM)

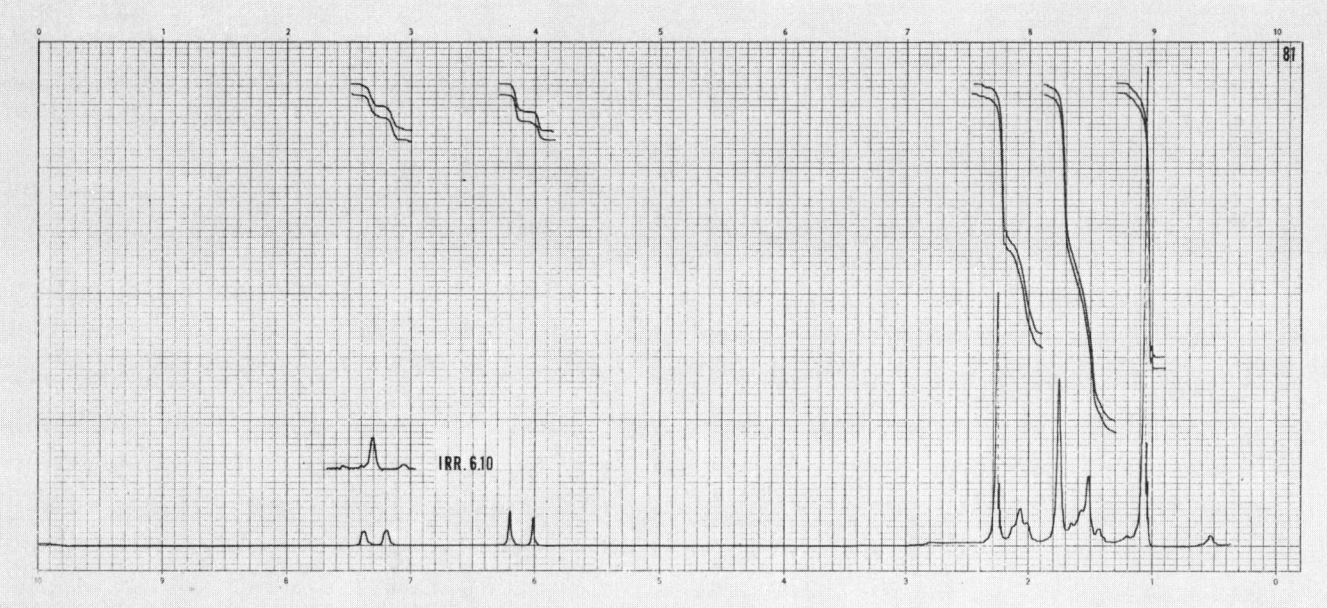

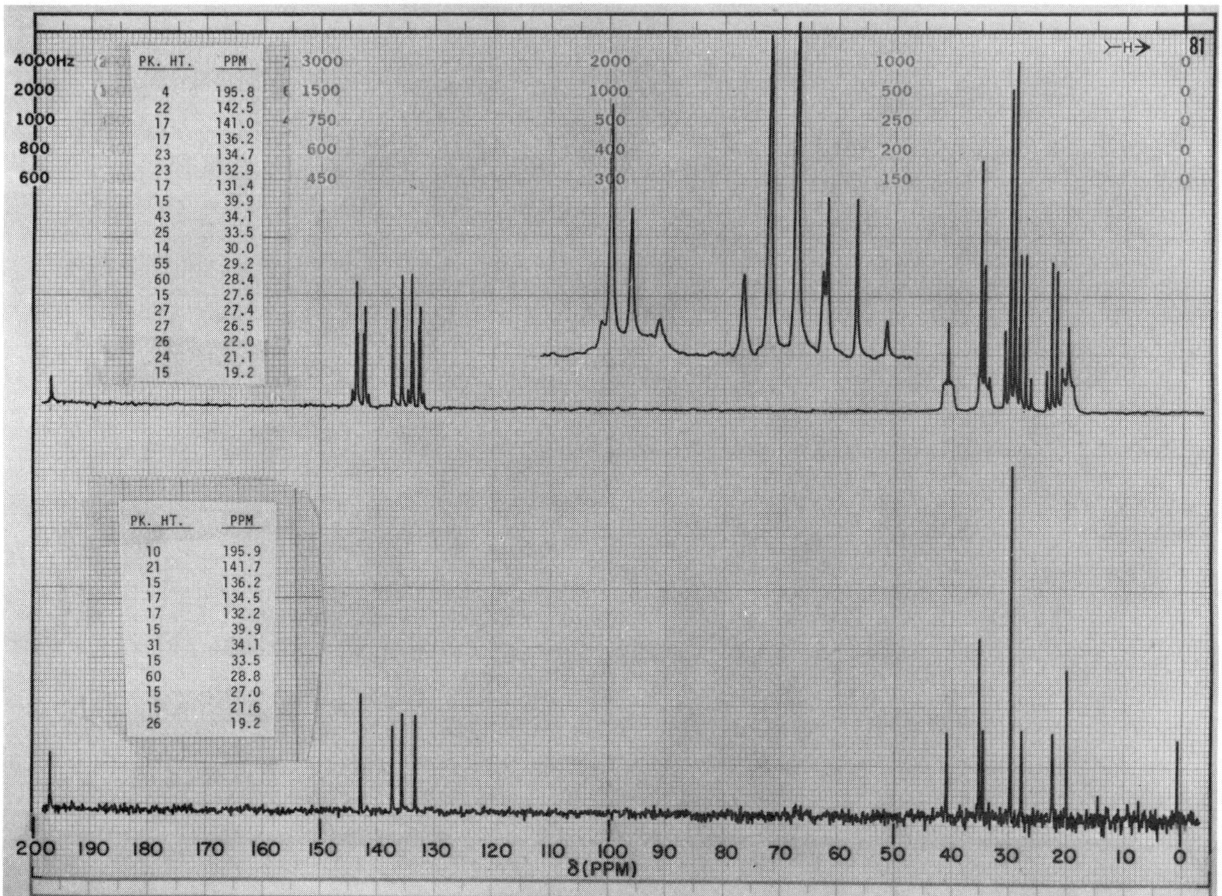

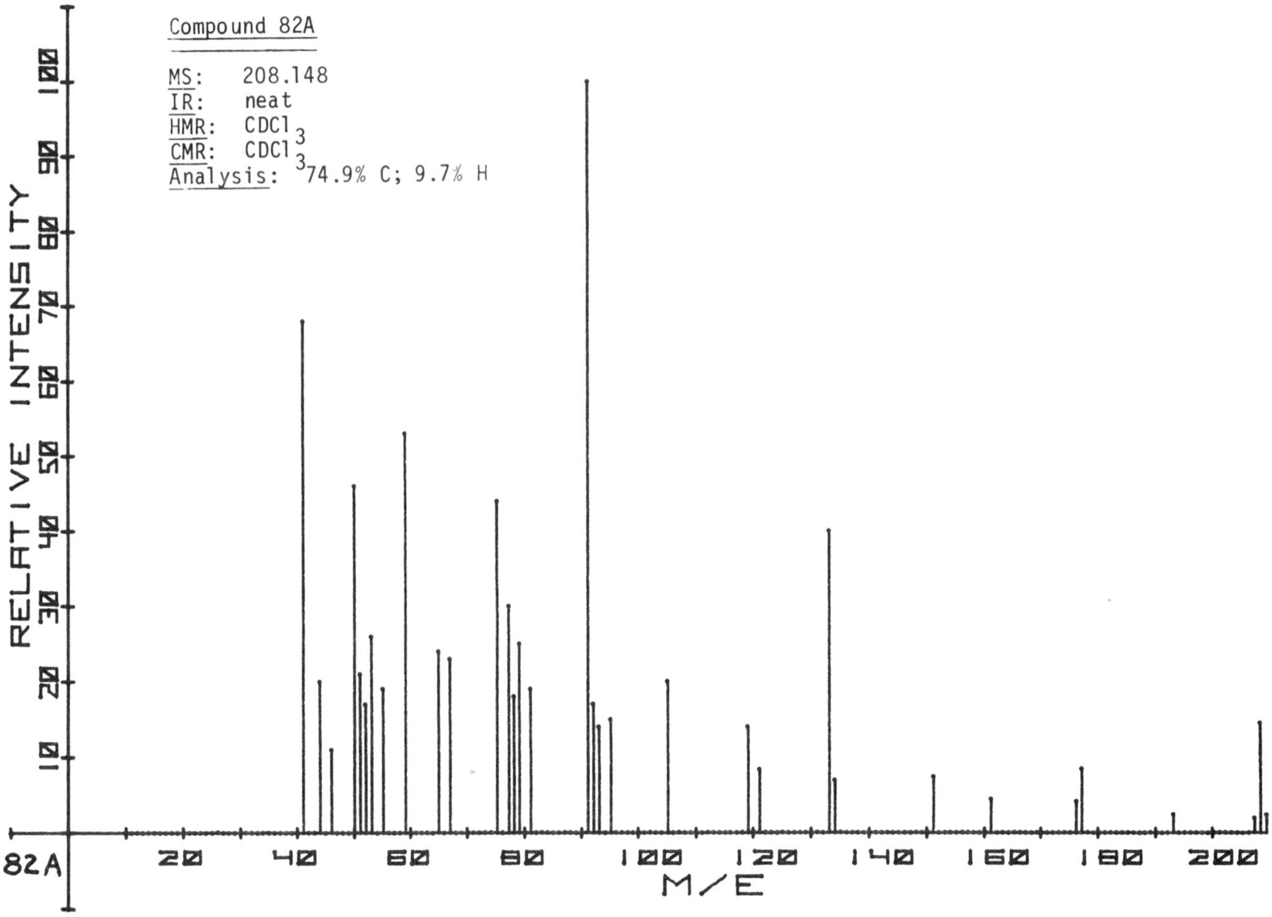

Compound 82A
MS: 208.148
IR: neat
HMR: CDCl3
CMR: CDCl3
Analysis: 74.9% C; 9.7% H
RELATIVE INTENSITY
10 20 30 40 50 60 70 80 90 100
82A
20 40 60 80 100 120 140 160 180 200
M/E

4000 3000 2000 1500 CM-1 1000 900 800 700
0.0
.10
.20
.30
.40
.50
.60
.70
1.0
∞
ABSORBANCE
82A
PR 1109 (137-1280)
3 4 5 6 7 8 9 10 11 12 13 14 15
WAVELENGTH (MICRONS)

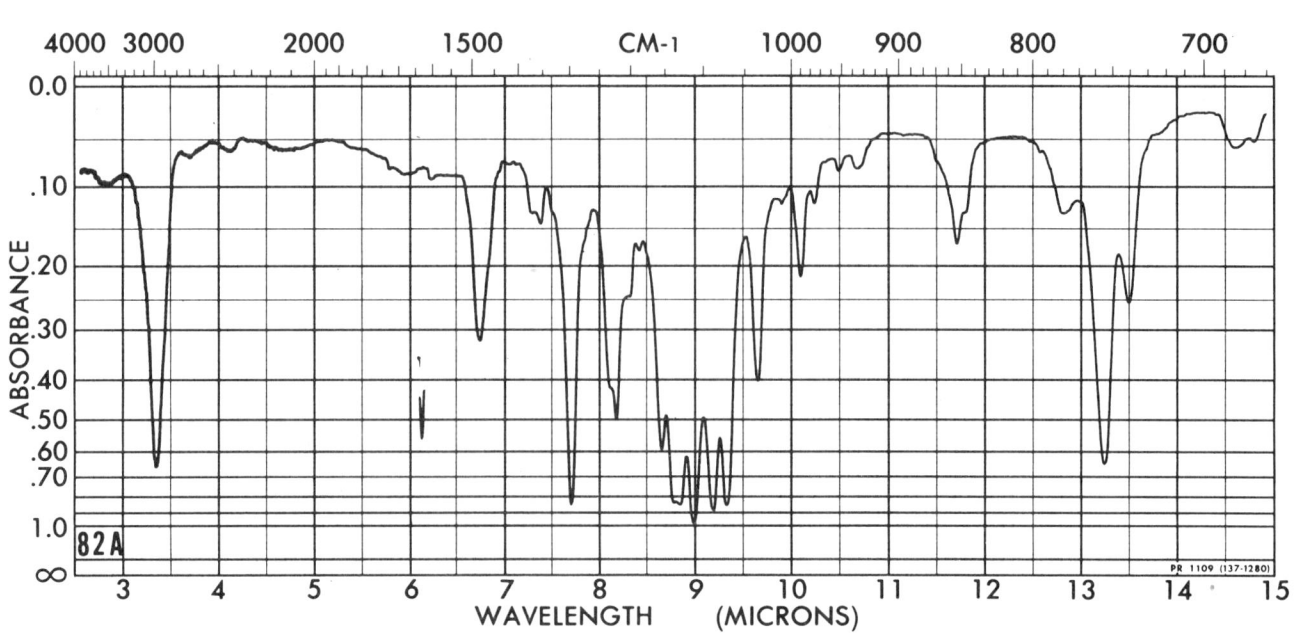

82A

82A
4000Hz (200)ppm
2000 (100)
1000 (50)
800 (40)
600 (30)
2000 1000
1000 500
500 250
400 200
300 150
PK. HT. PPM
7 53.6
26 52.0
6 51.4
31 50.4
40 50.2
25 49.8
5 48.8
36 48.6
32 48.2
5 46.6
8 23.5
31 22.4
15 21.3
20 20.3
7 19.3
PK. HT. PPM
52 132.5
60 119.0
14 51.2
46 49.4
17 49.0
49 36.8
33 22.4
31 20.3
200 190 180 170 160 150 140 130 120 110 100 90 80 70 60 50 40 30 20 10 0
δ(PPM)

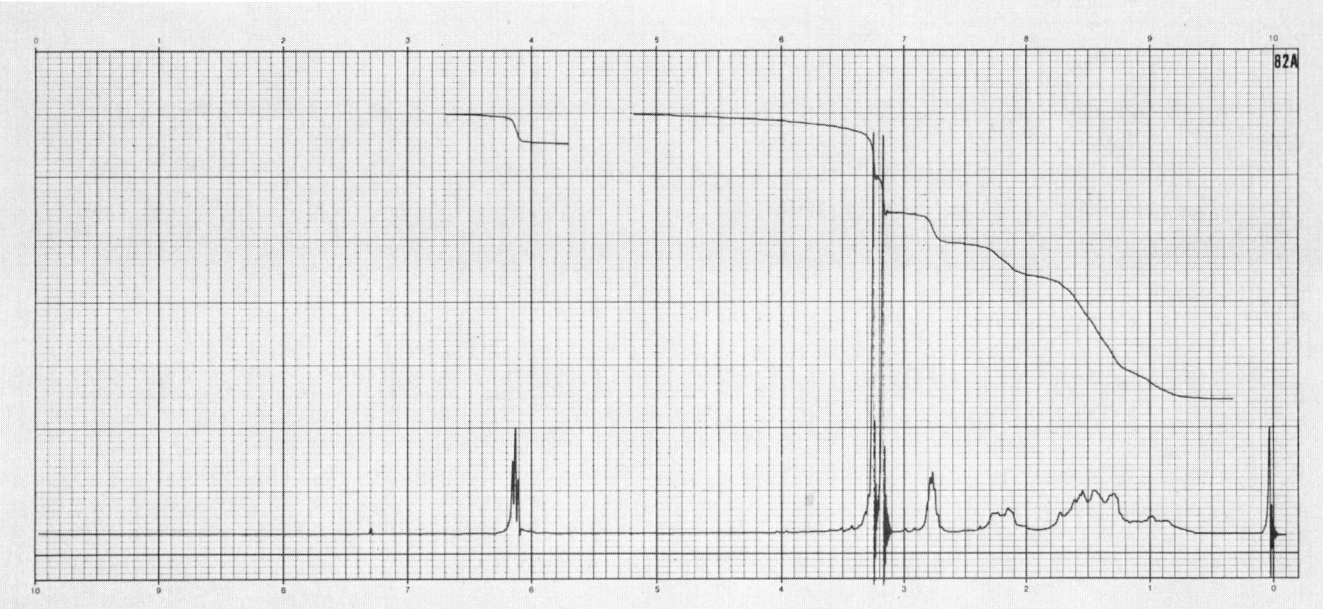

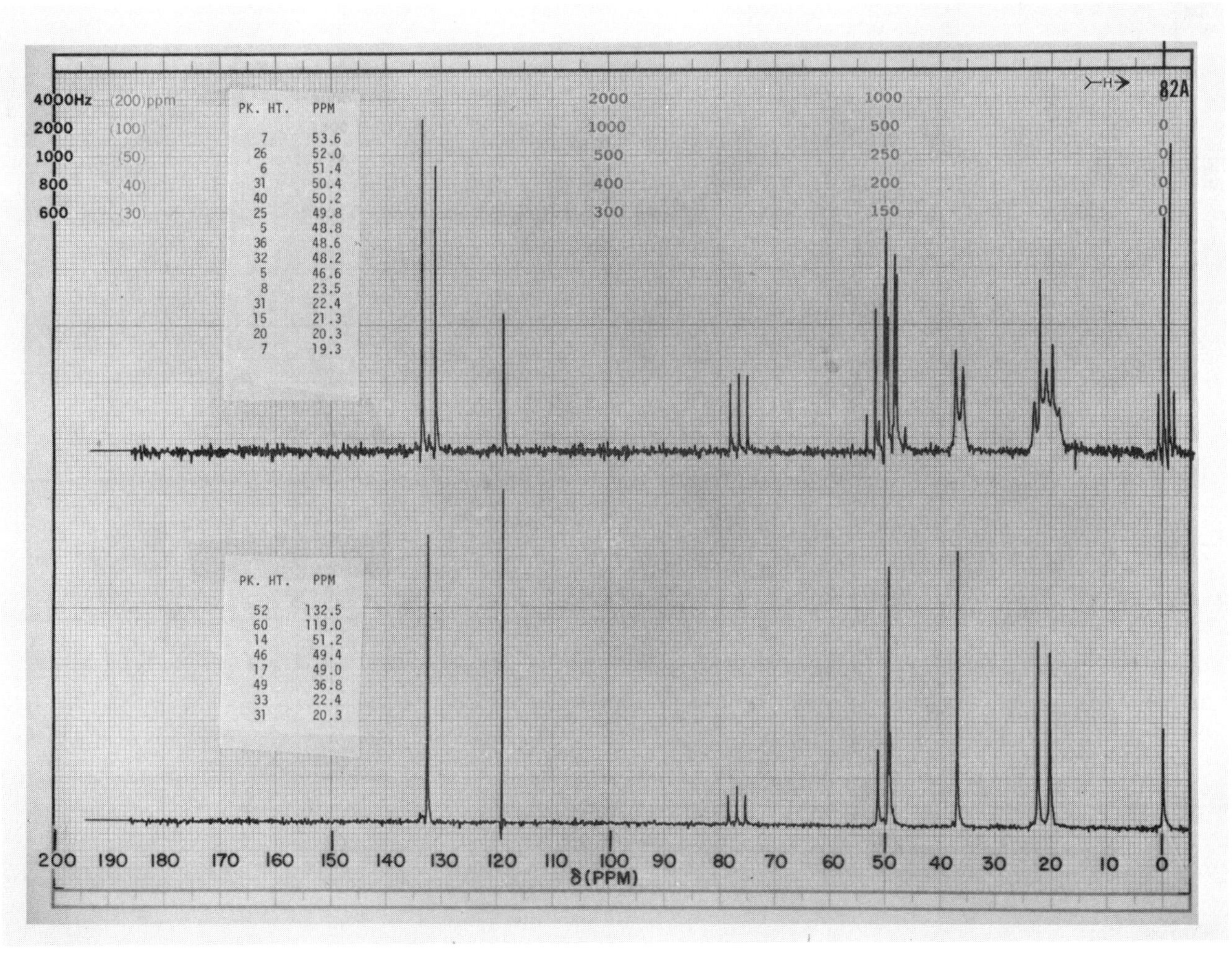

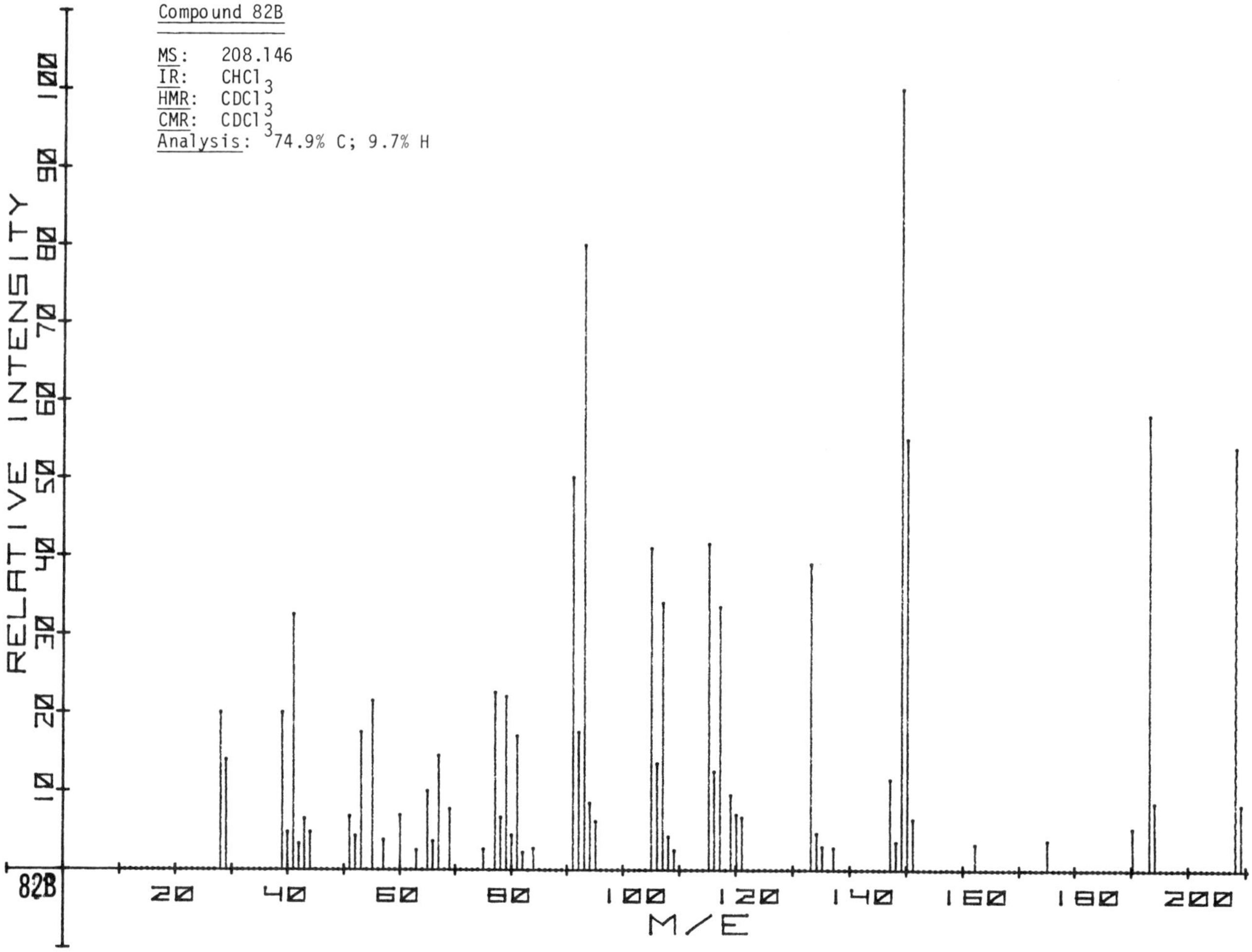

Compound 82B
MS: 208.146
IR: CHCl3
HMR: CDCl3
CMR: CDCl3
Analysis: 74.9% C; 9.7% H
RELATIVE INTENSITY
10 20 30 40 50 60 70 80 90 100
82B
20 40 60 80 100 120 140 160 180 200
M/E

4000 3000 2000 1500 CM-1 1000 900 800 700
0.0 .10 .20 .30 .40 .50 .60 .70 1.0 ∞
ABSORBANCE
82B
3 4 5 6 7 8 9 10 11 12 13 14 15
WAVELENGTH (MICRONS)

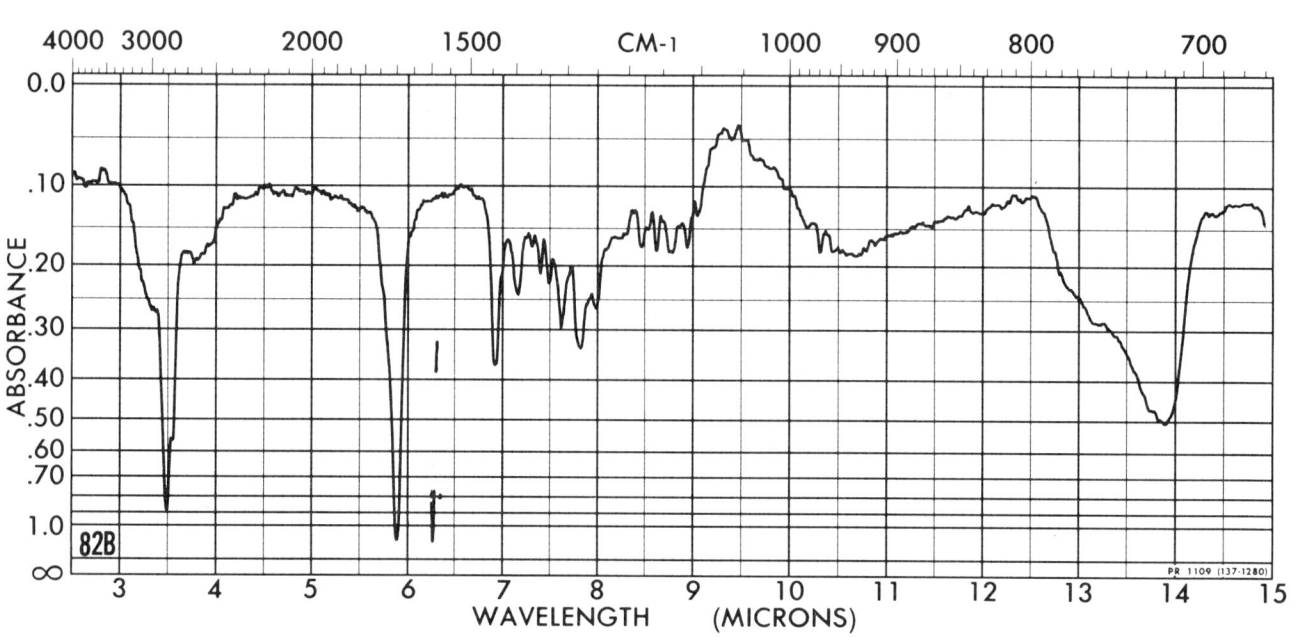

82B
D2O
OFFSET 3 PPM

82B
4000Hz (200)ppm
2000 (100)
1000 (50)
800 (40)
600 (30)
2000
1000
500
400
300
1000
500
250
200
150
PK. HT. PPM
12 50.6
26 49.8
32 49.3
60 48.5
17 47.9
29 47.1
27 33.5
12 33.0
43 31.6
35 30.6
49 30.2
48 29.9
19 28.8
51 28.5
PK. HT. PPM
16 178.6
24 49.3
24 48.5
52 43.7
47 41.6
25 35.9
19 33.5
20 30.9
29 30.6
60 29.2
200 190 180 170 160 150 140 130 120 110 100 90 80 70 60 50 40 30 20 10 0
δ(PPM)

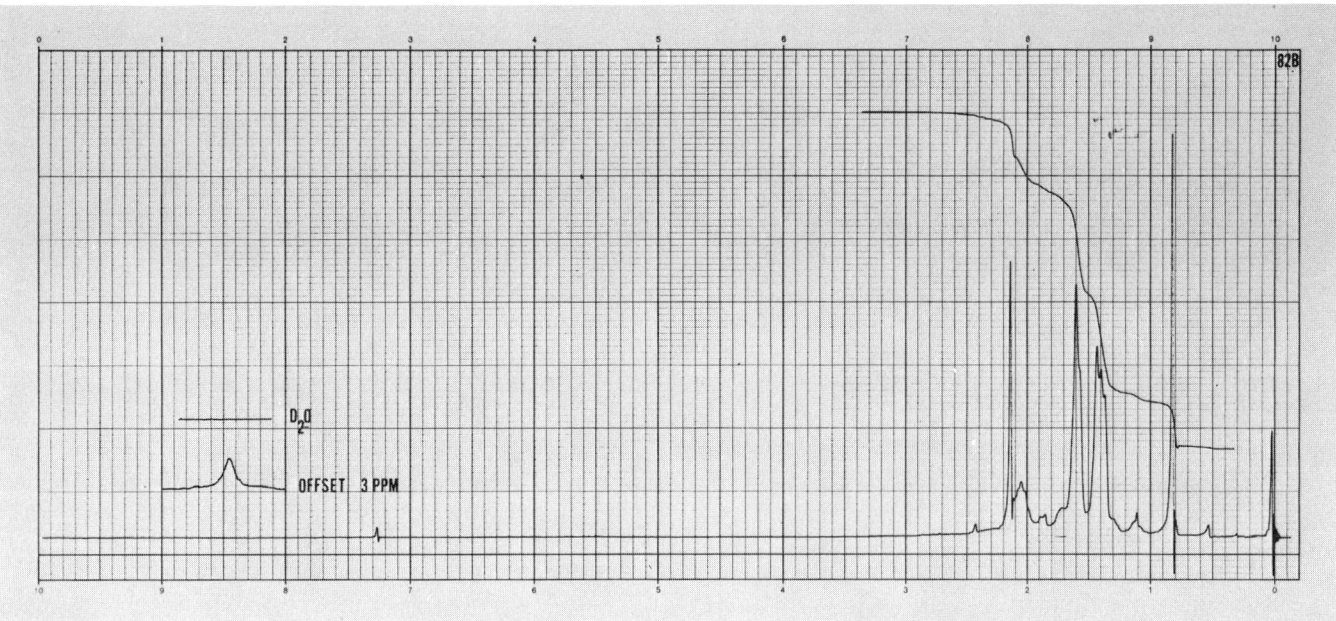

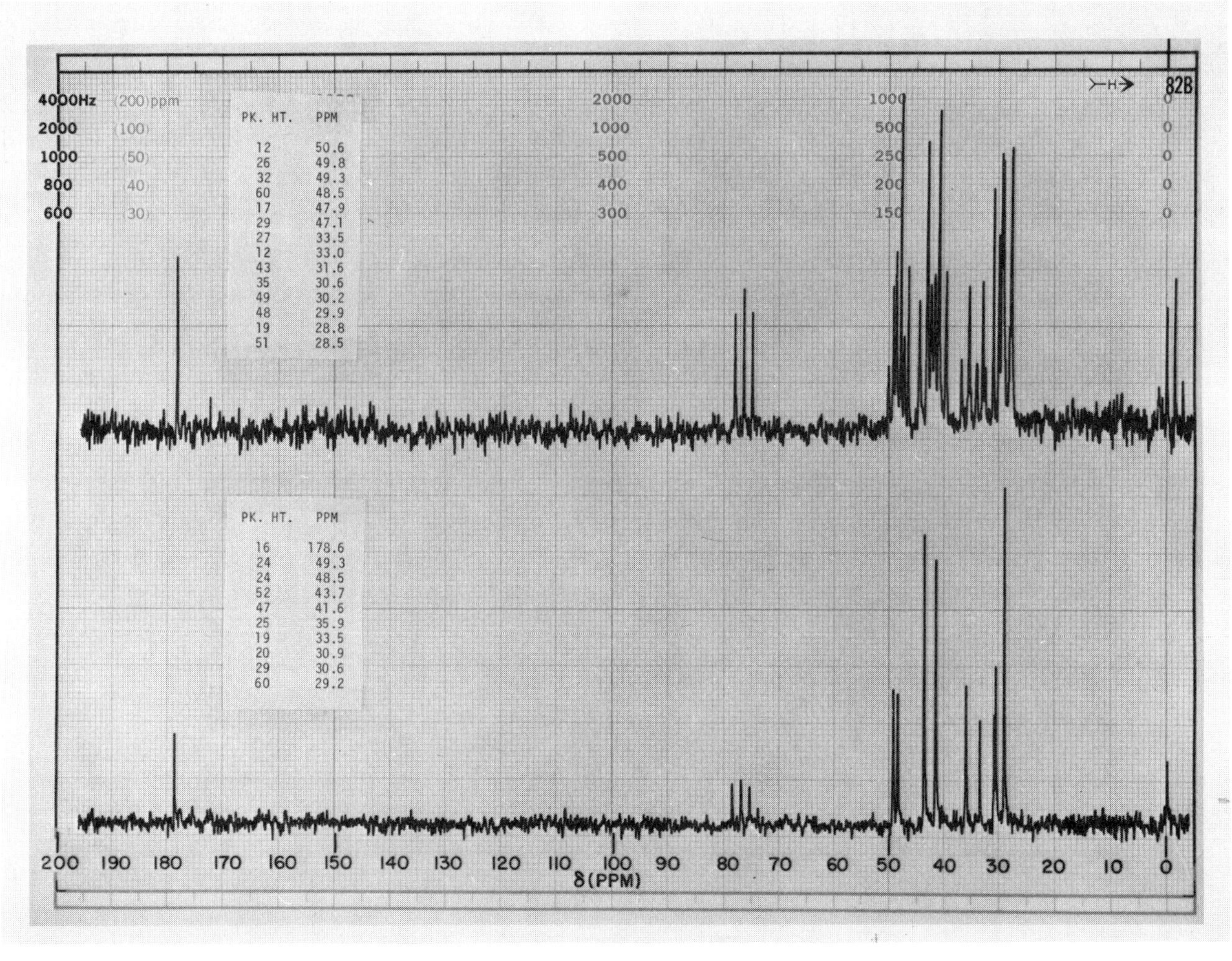

COMPOUND 83

Compound 83
MS: 214.147
IR: melt
HMR: CCl$_4$
CMR: d$_6$ acetone
Analysis: 78.4% C; 8.5% H; 13.1% N

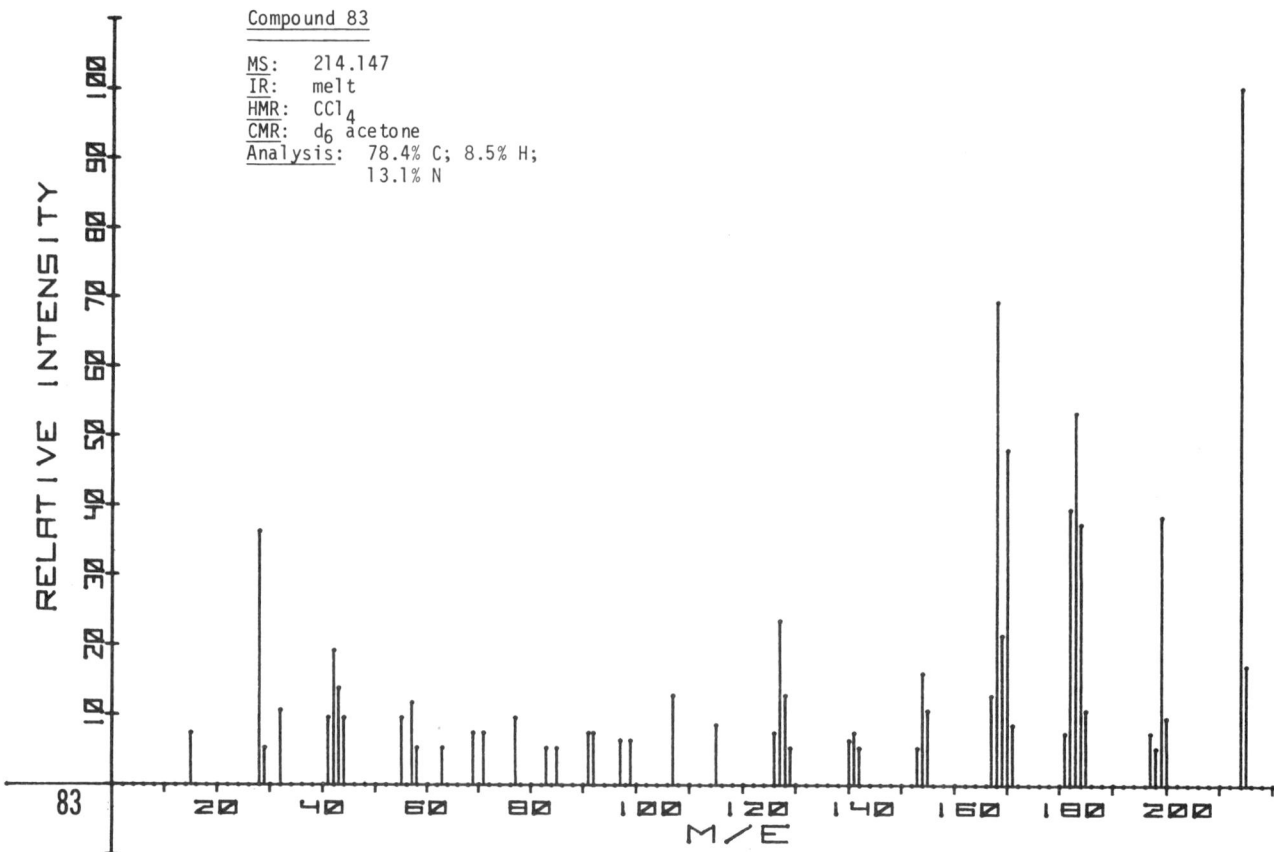

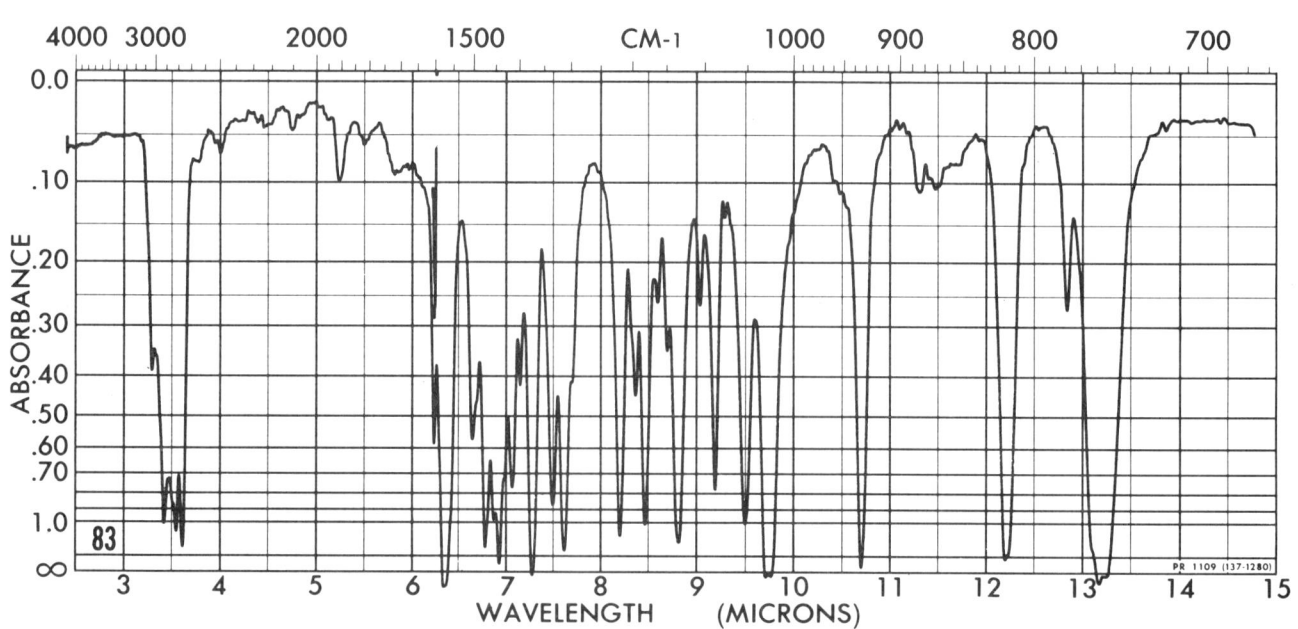

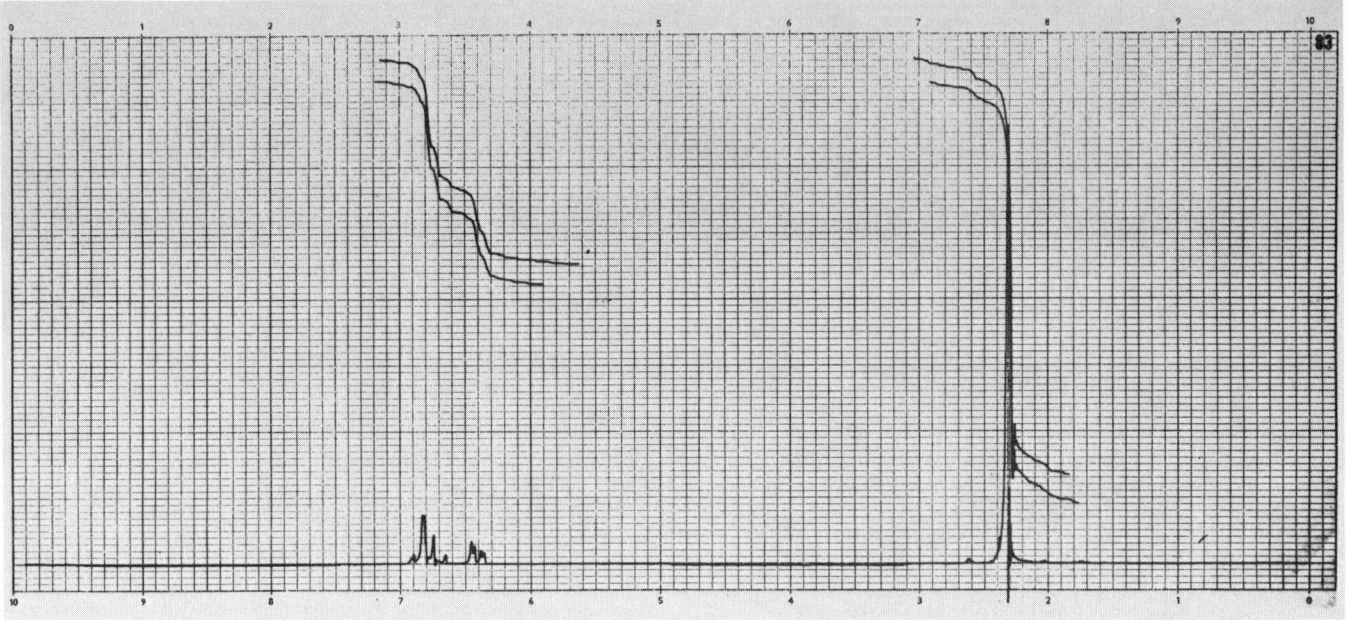

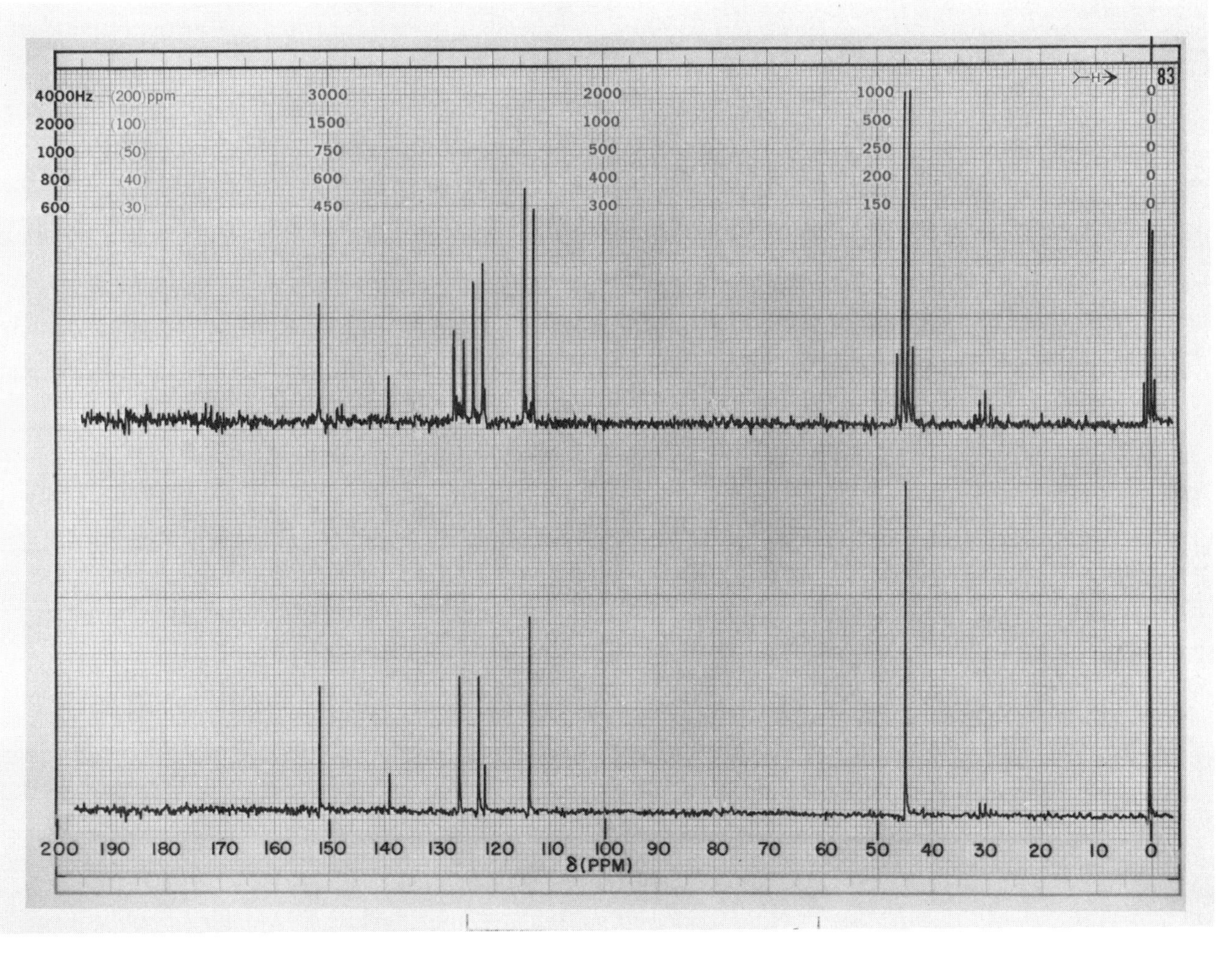

COMPOUND 84

Compound 84

MS: 266.115
IR: KBr
HMR: CDCl_3
CMR: CDCl_3
Analysis: 63.3% C; 6.9% H

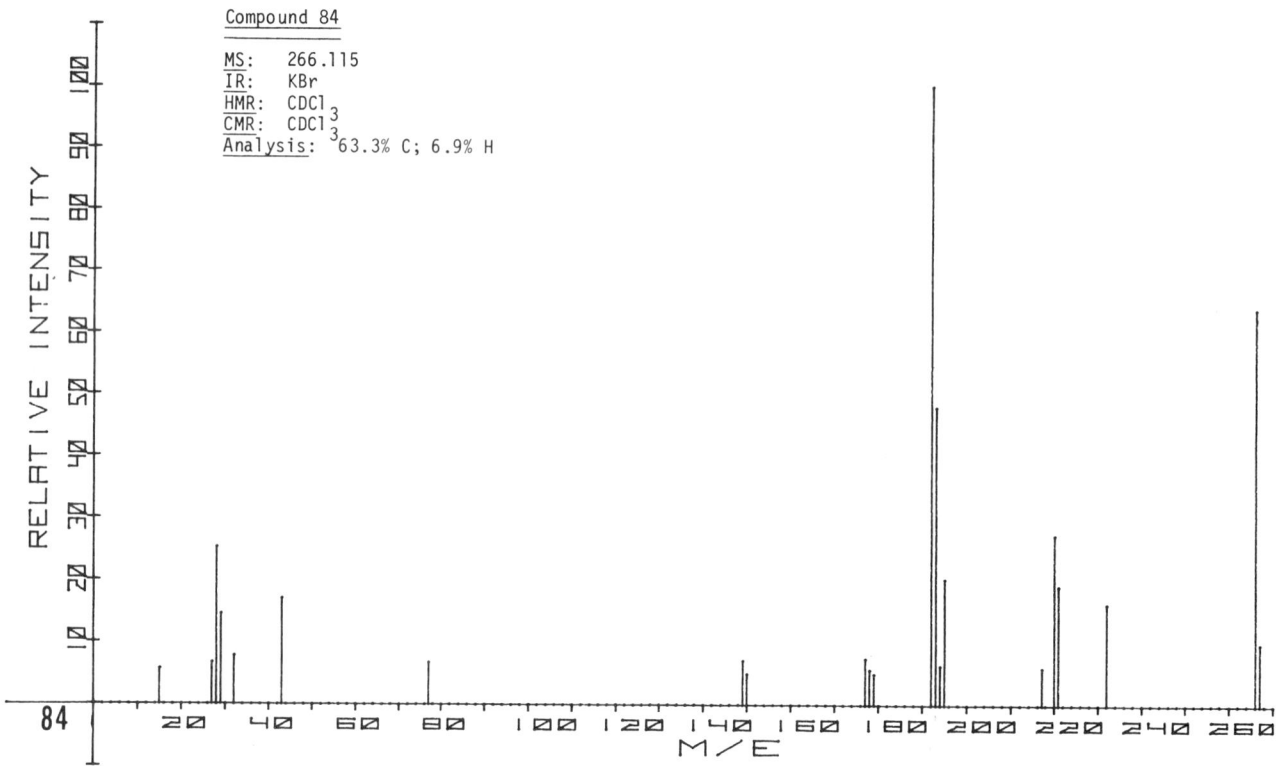

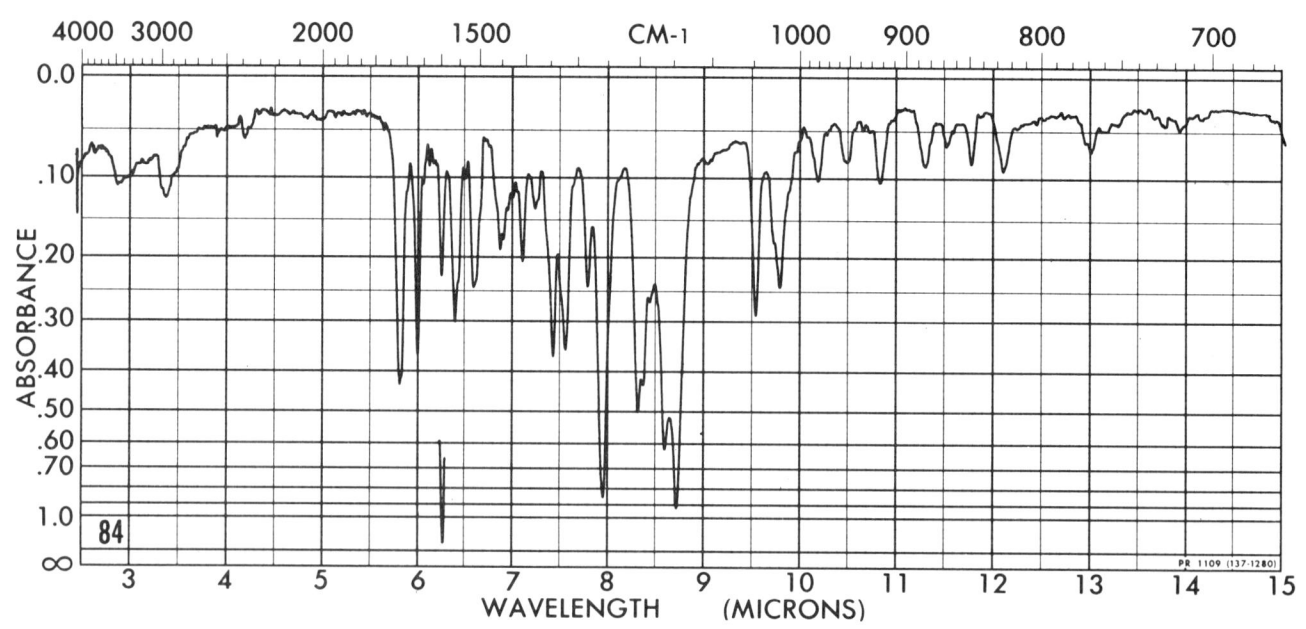

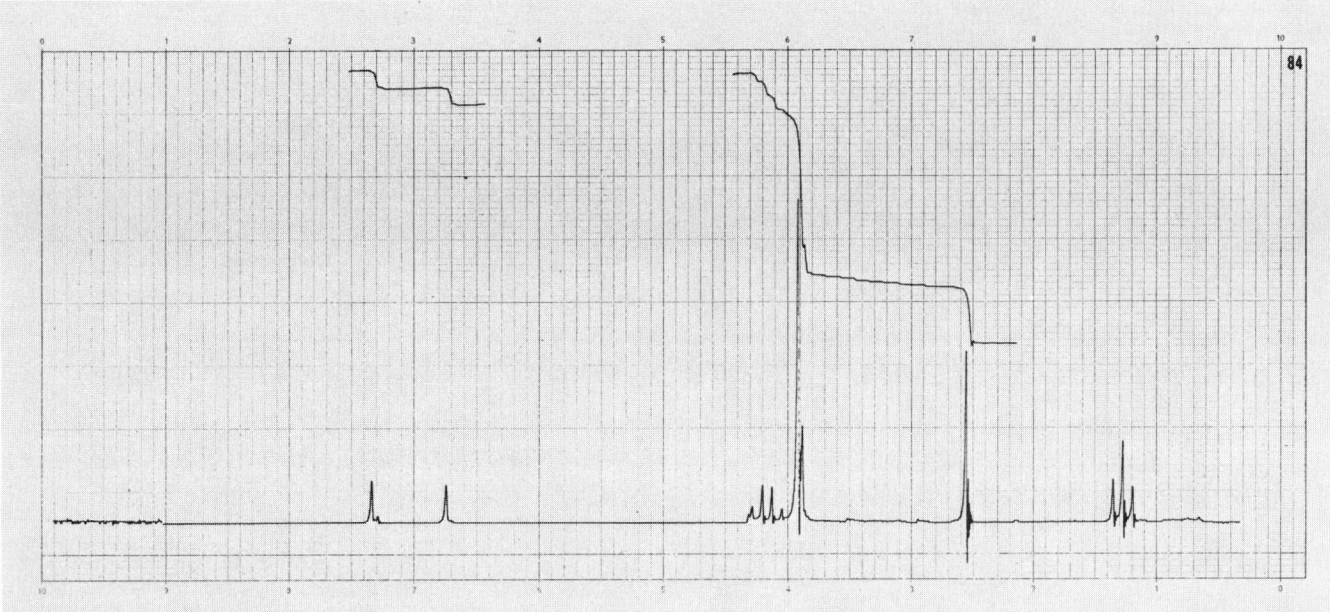

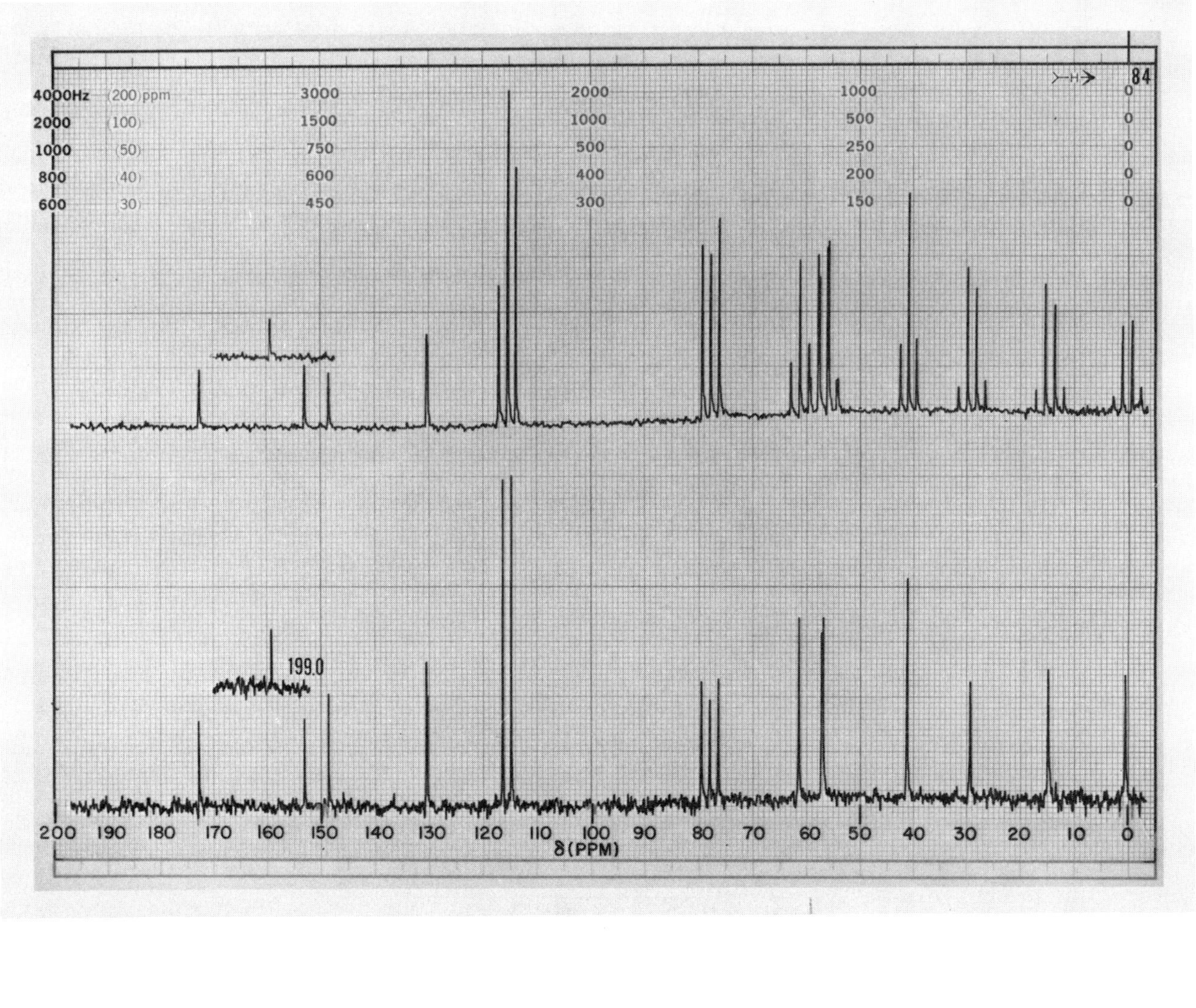

COMPOUND 85

Compound 85

MS: 206.167
IR: neat
HMR: CDCl₃
CMR: CDCl₃
Analysis: 81.6% C; 10.9% H

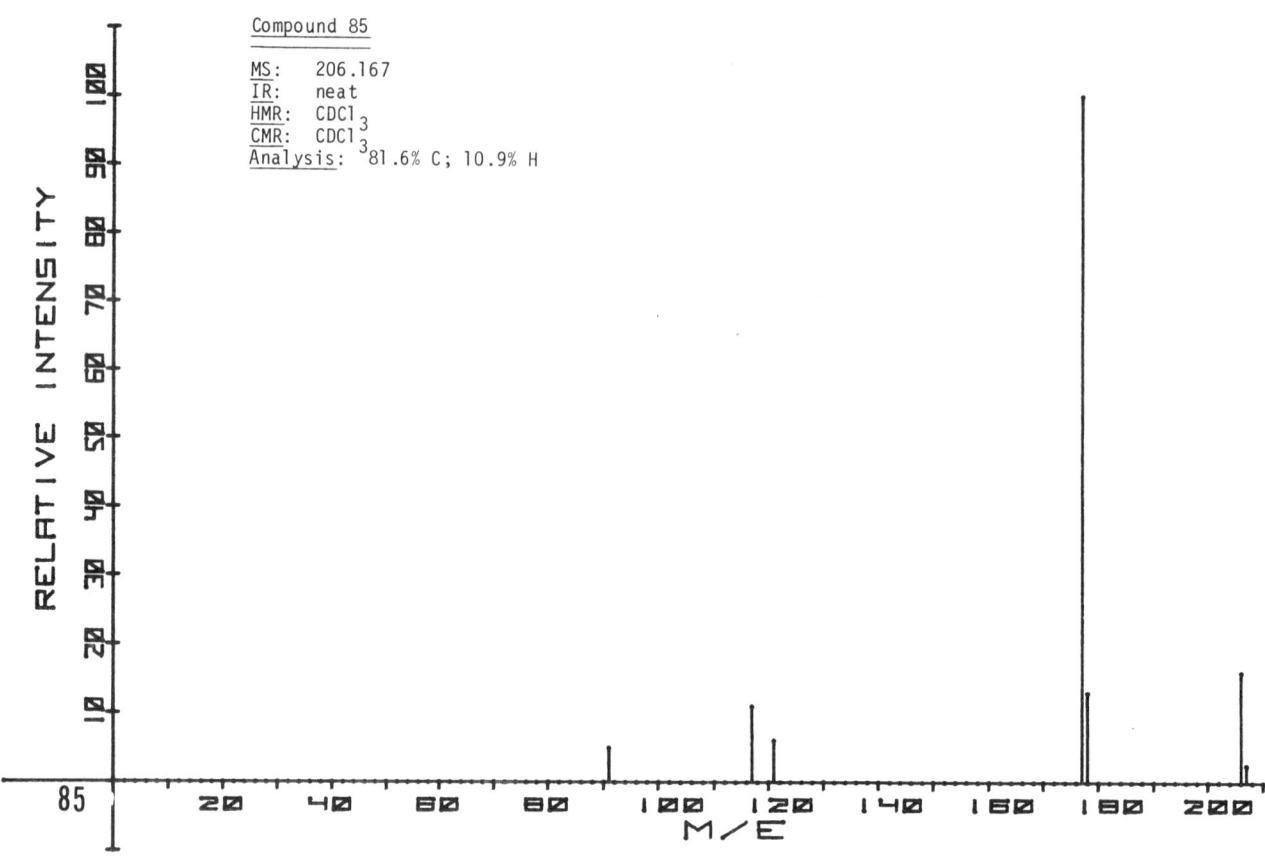

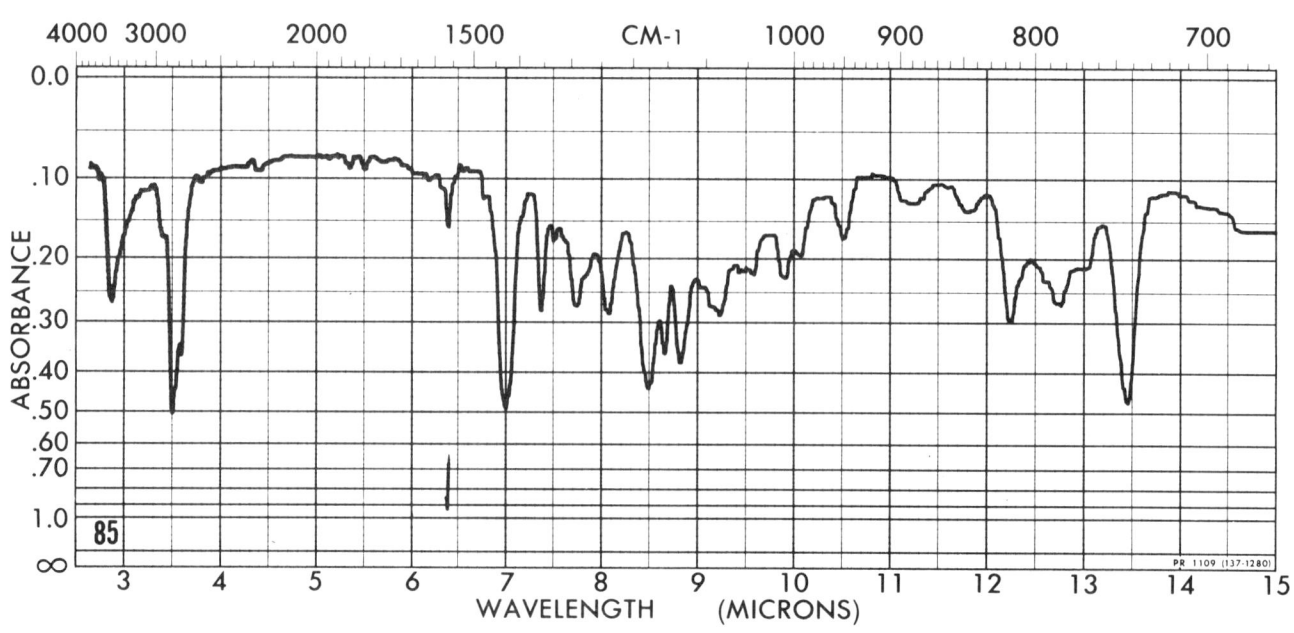

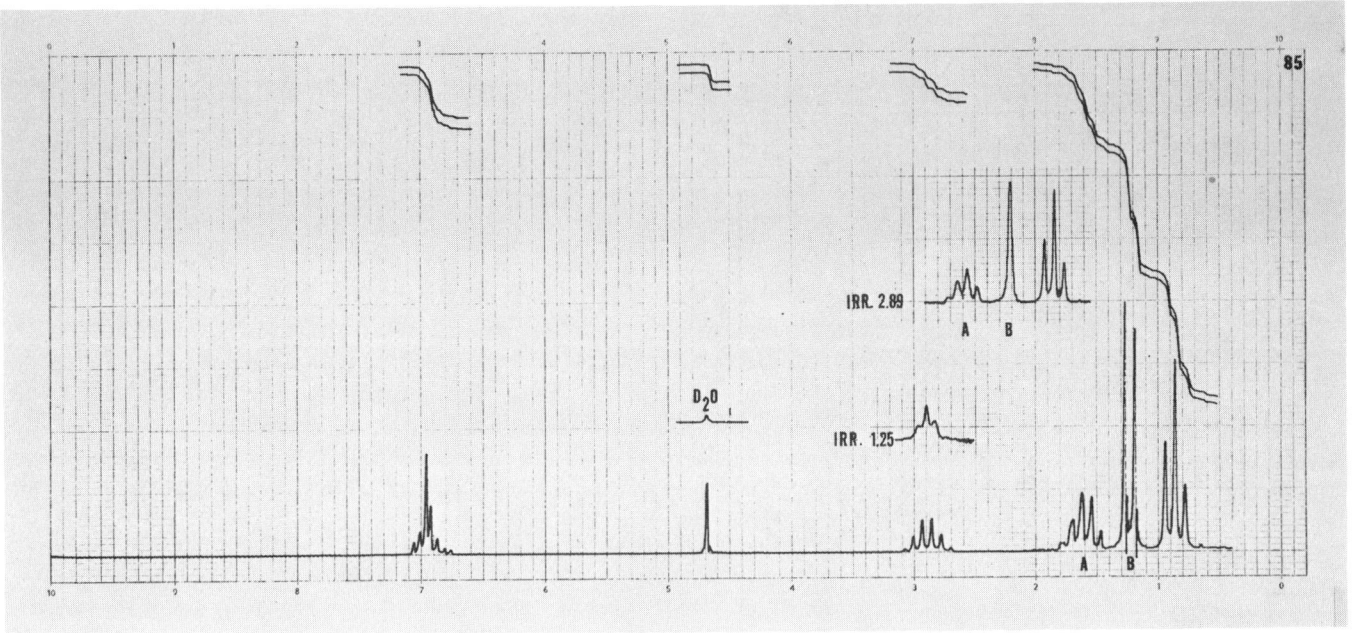

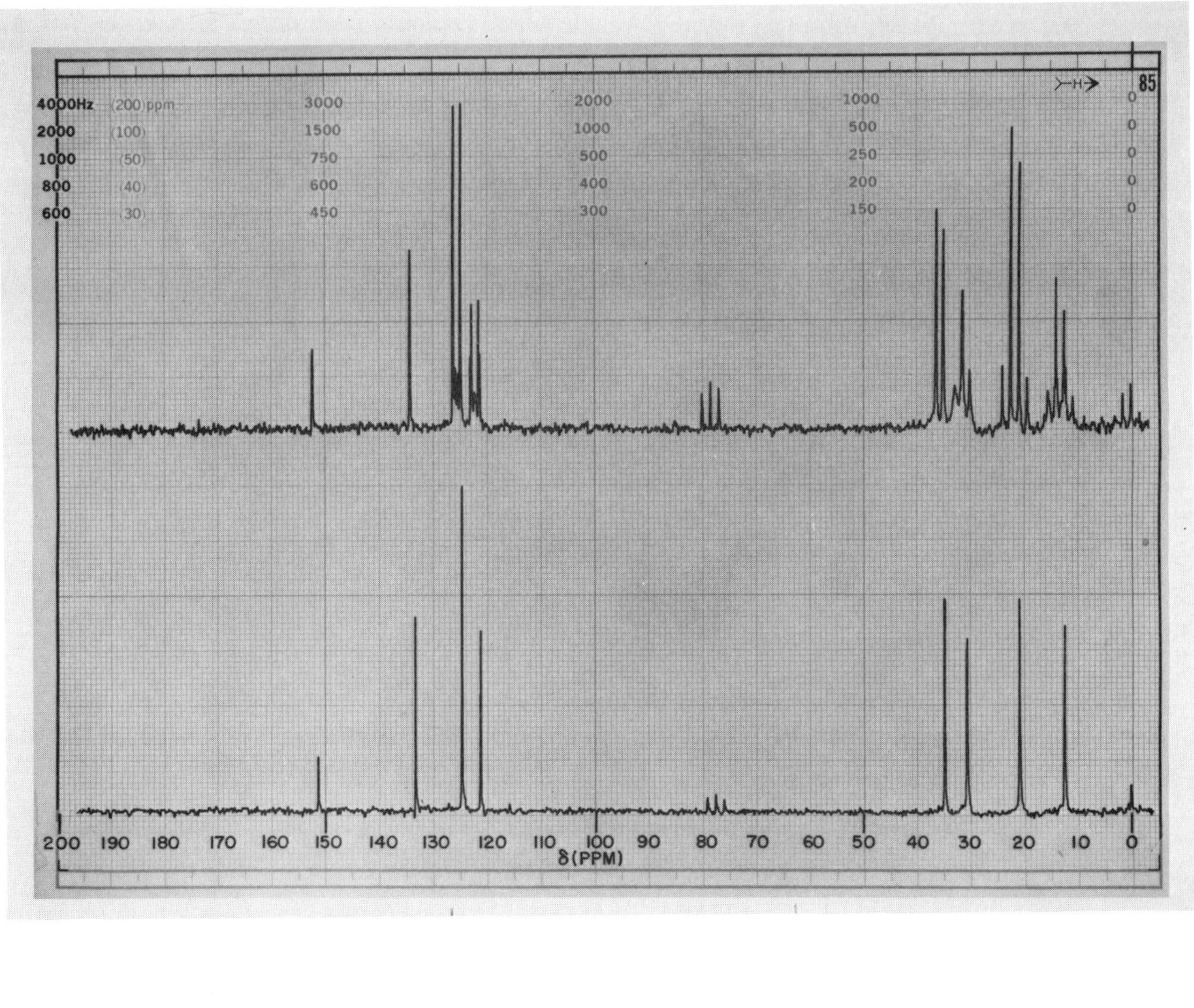

COMPOUND 86

Compound 86

MS: 251.095
IR: neat
HMR: CDCl$_3$
CMR: CDCl$_3$
Analysis: 76.6% C; 5.3% H; 5.6% N

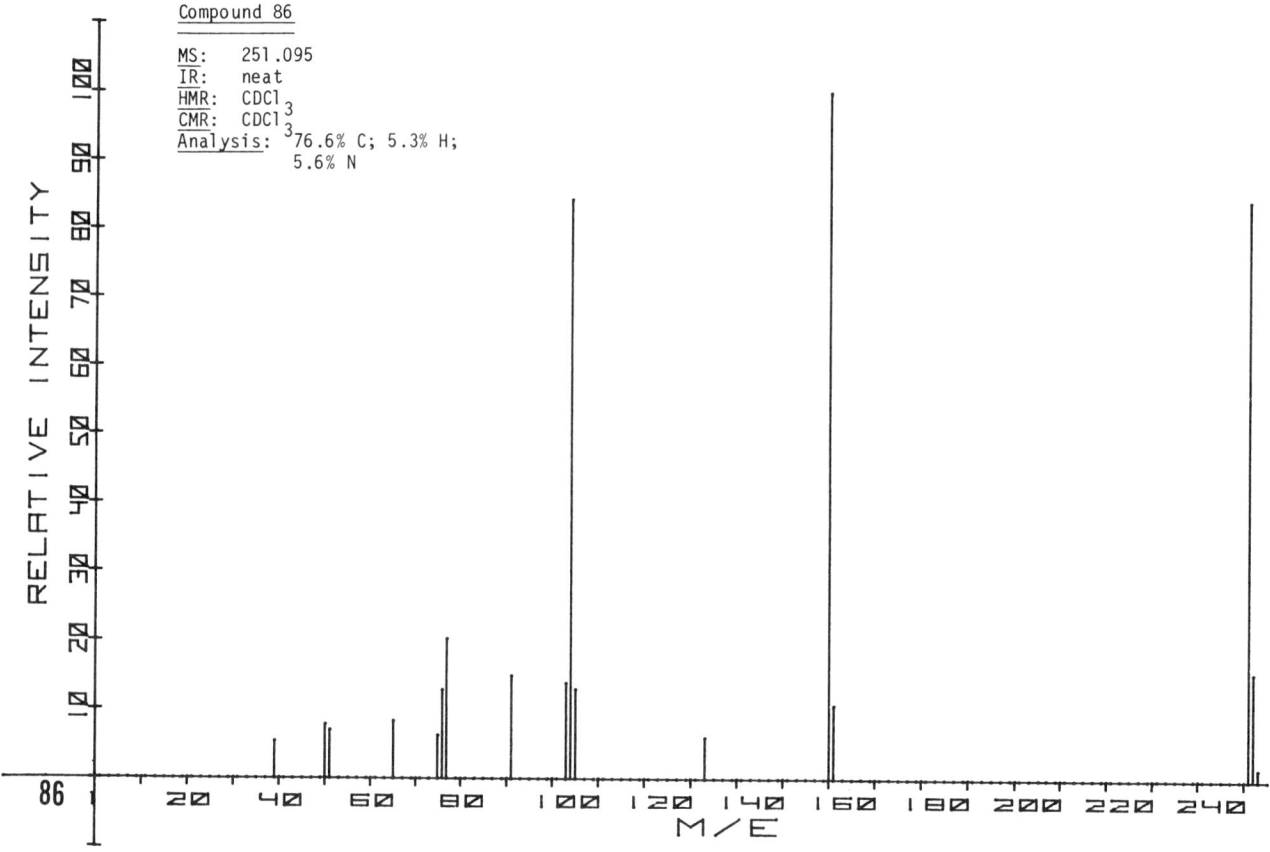

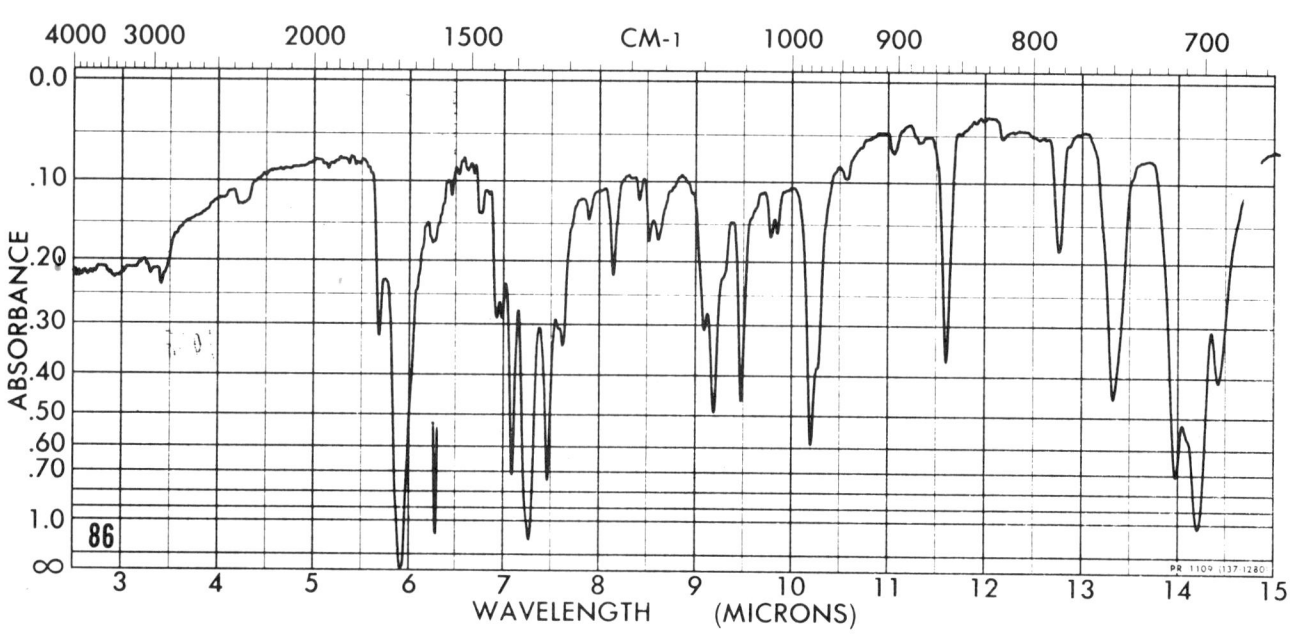

86

86
4000Hz (200)ppm
2000 (100)
1000 (50)
800 (40)
600 (30)
2000
1000
500
400
300
1000
500
250
200
150
PK. HT. PPM
18 138.1
49 134.8
40 132.8
24 132.1
47 129.7
36 129.5
44 127.9
52 127.6
19 125.6
53 124.1
60 122.1
PK. HT. PPM
11 168.0
13 138.1
39 133.8
13 132.1
47 128.8
60 128.5
21 126.6
45 123.1
24 39.2
22 34.6
200 190 180 170 160 150 140 130 120 110 100 90 80 70 60 50 40 30 20 10 0
δ(PPM)

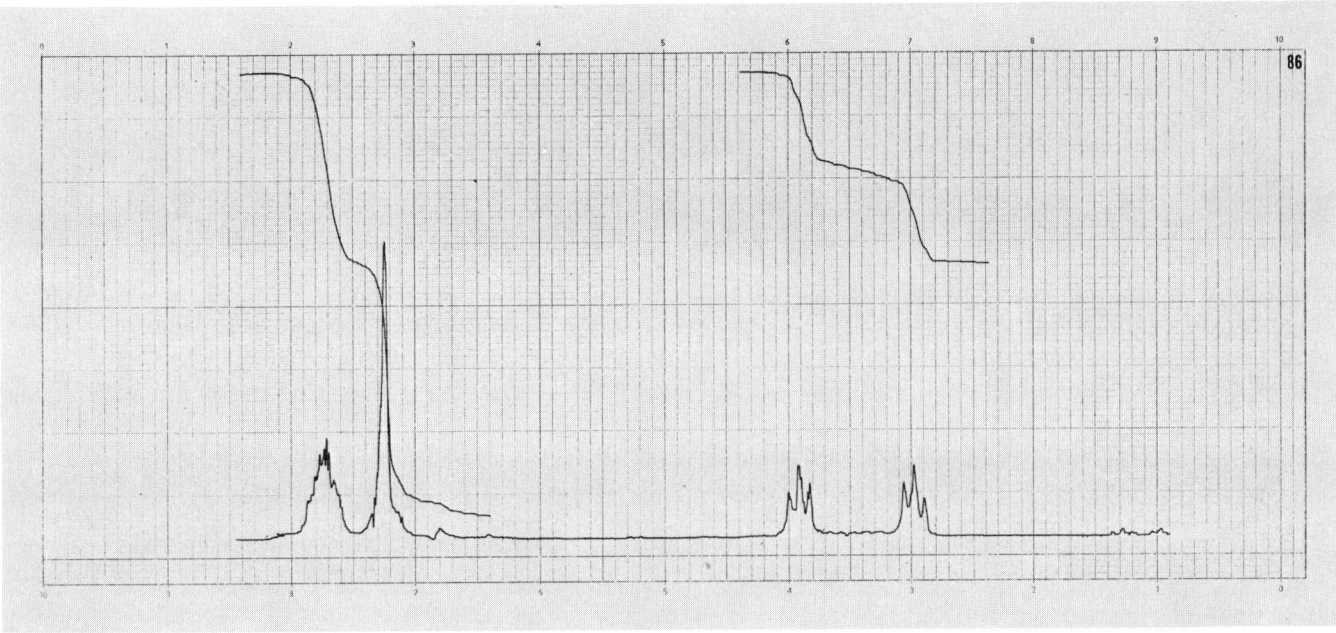

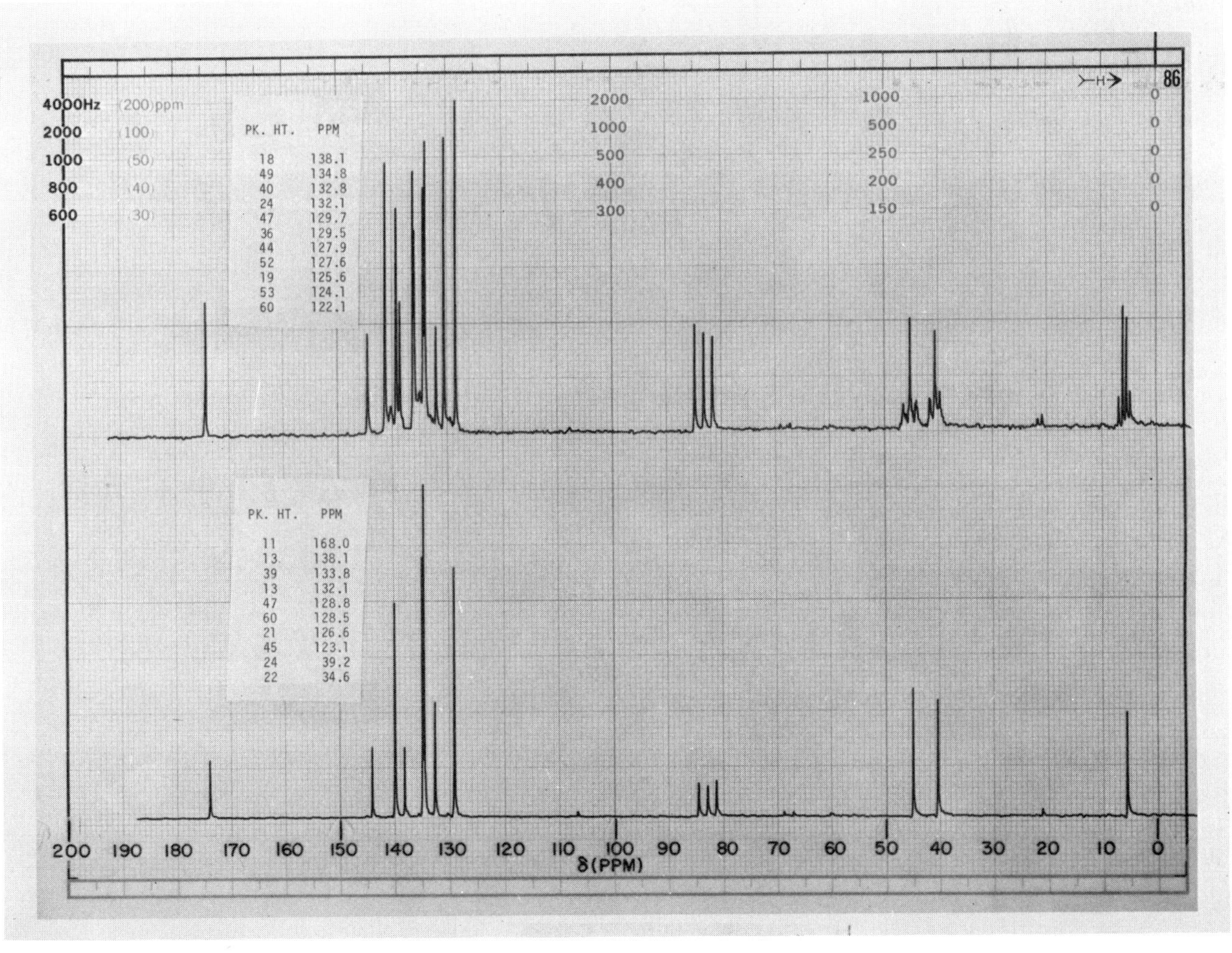

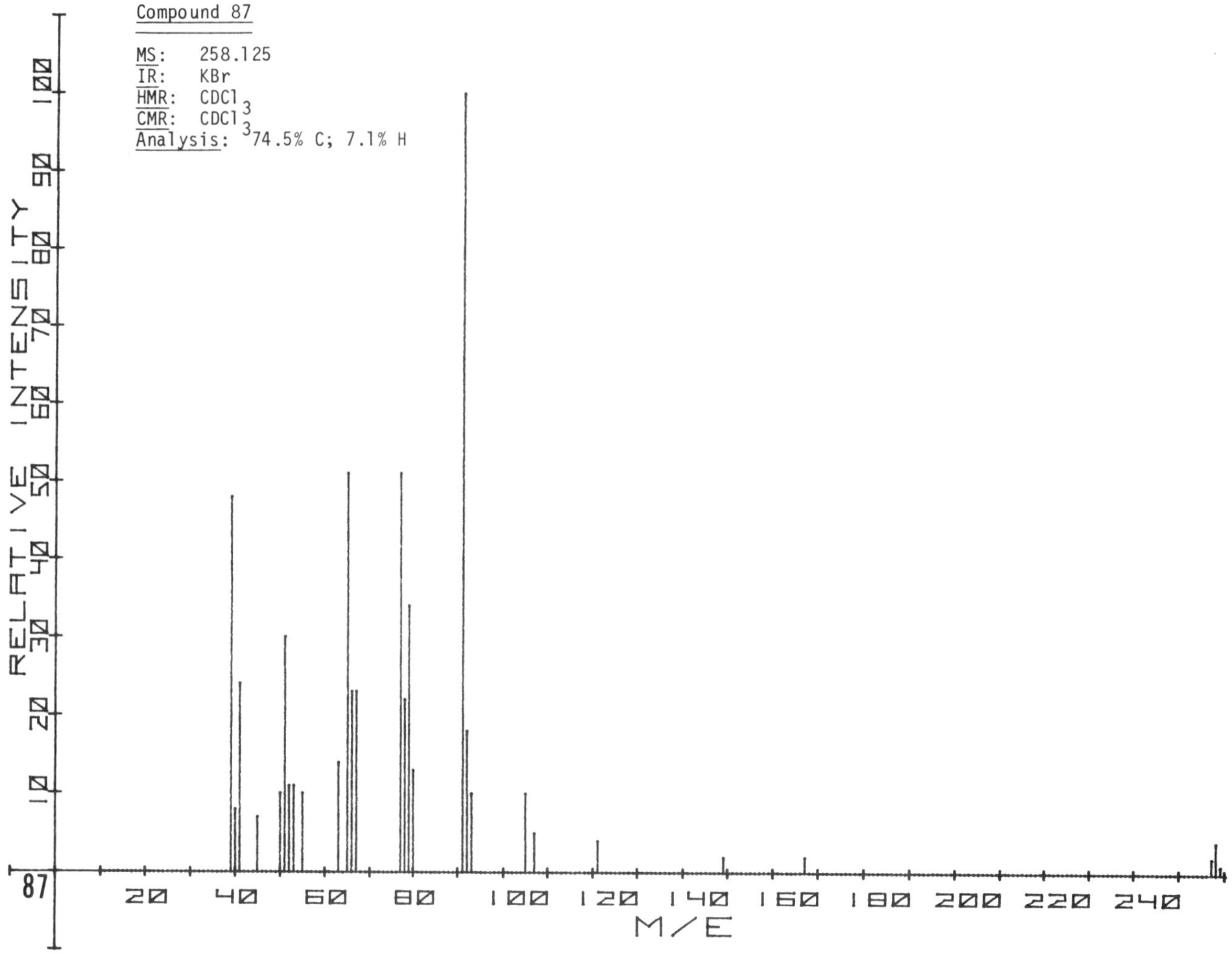

Compound 87
MS: 258.125
IR: KBr
HMR: CDCl3
CMR: CDCl3
Analysis: 74.5% C; 7.1% H
RELATIVE INTENSITY
10 20 30 40 50 60 70 80 90 100
87
20 40 60 80 100 120 140 160 180 200 220 240
M/E

4000 3000 2000 1500 CM-1 1000 900 800 700
0.0 .10 .20 .30 .40 .50 .60 .70 1.0 ∞
ABSORBANCE
87
3 4 5 6 7 8 9 10 11 12 13 14 15
WAVELENGTH (MICRONS)

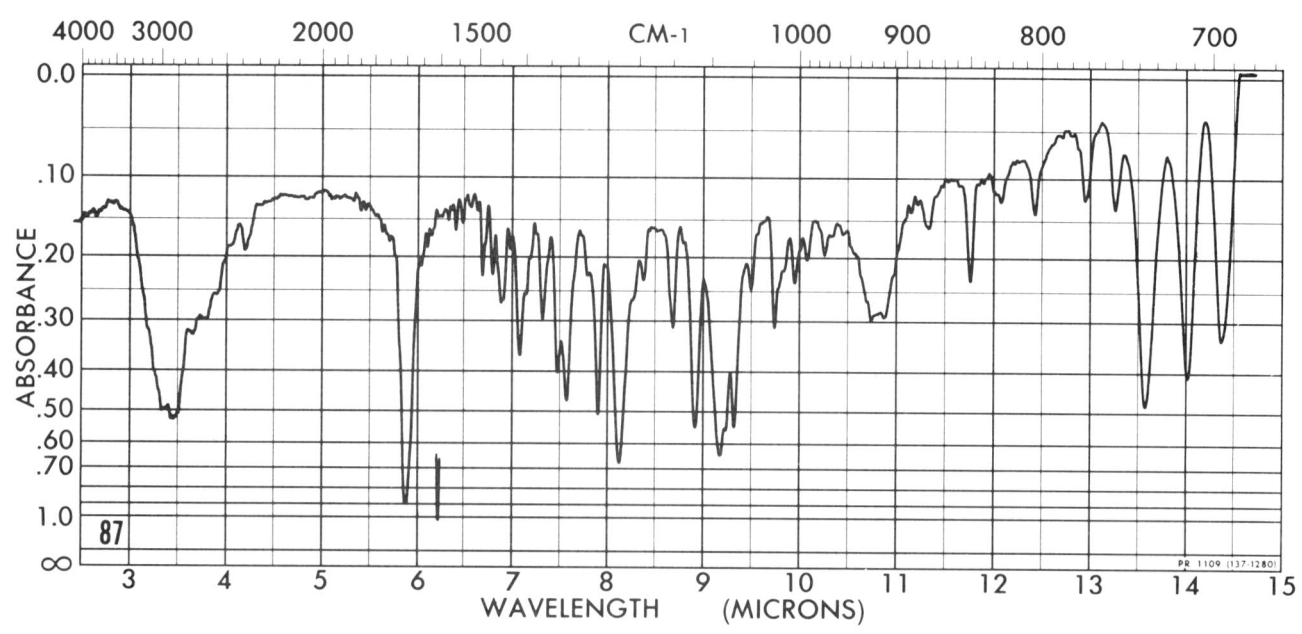

87
IRR. 2.12
A
IRR. 2.85
D2O
IRR. 3.20
IRR. 2.85
A
10 9 8 7 6 5 4 3 2 1 0

87
4000Hz (200) ppm
2000 (100)
1000 (50)
800 (40)
600 (30)
2000 1000 500 400 300
1000 500 250 200 150
PK. HT. PPM
12 138.5
17 135.9
17 134.2
19 130.7
23 129.3
21 128.9
60 128.4
31 127.3
45 126.6
14 45.0
16 44.1
14 43.8
10 43.1
PK. HT. PPM
21 180.3
5 138.5
20 135.1
18 129.8
42 128.3
60 127.5
16 73.0
10 68.3
15 60.8
17 47.4
17 44.4
16 43.6
10 29.8
200 190 180 170 160 150 140 130 120 110 100 90 80 70 60 50 40 30 20 10 0
δ(PPM)

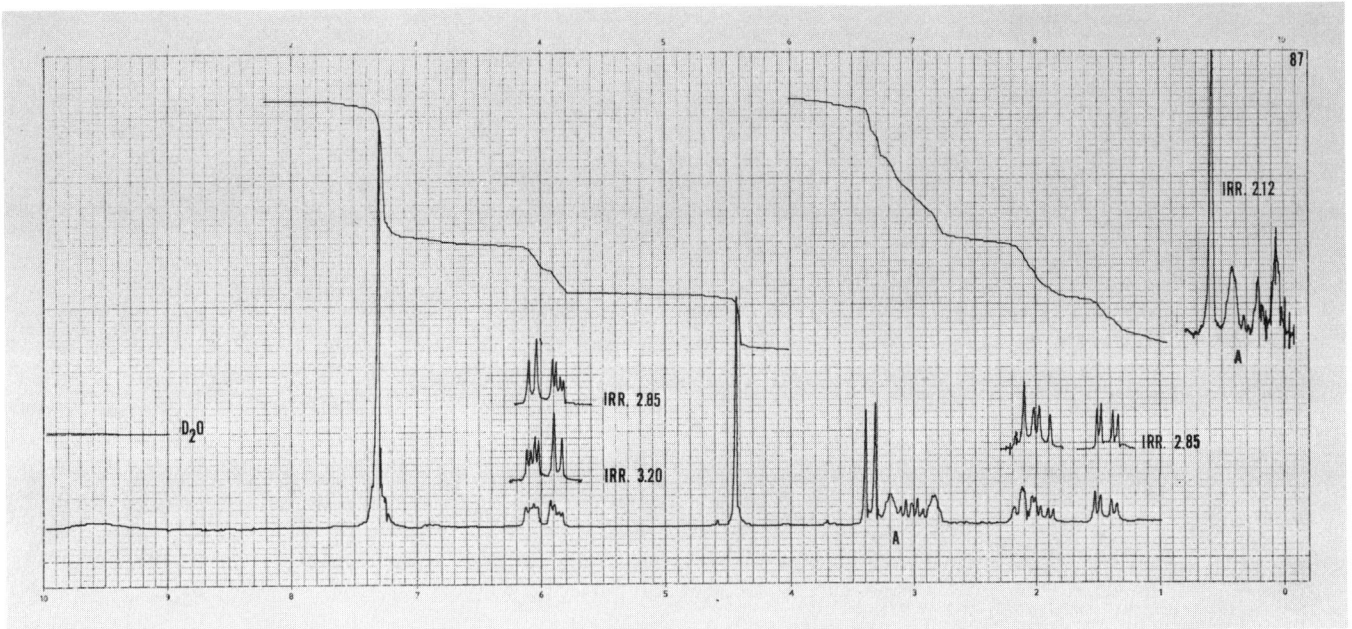

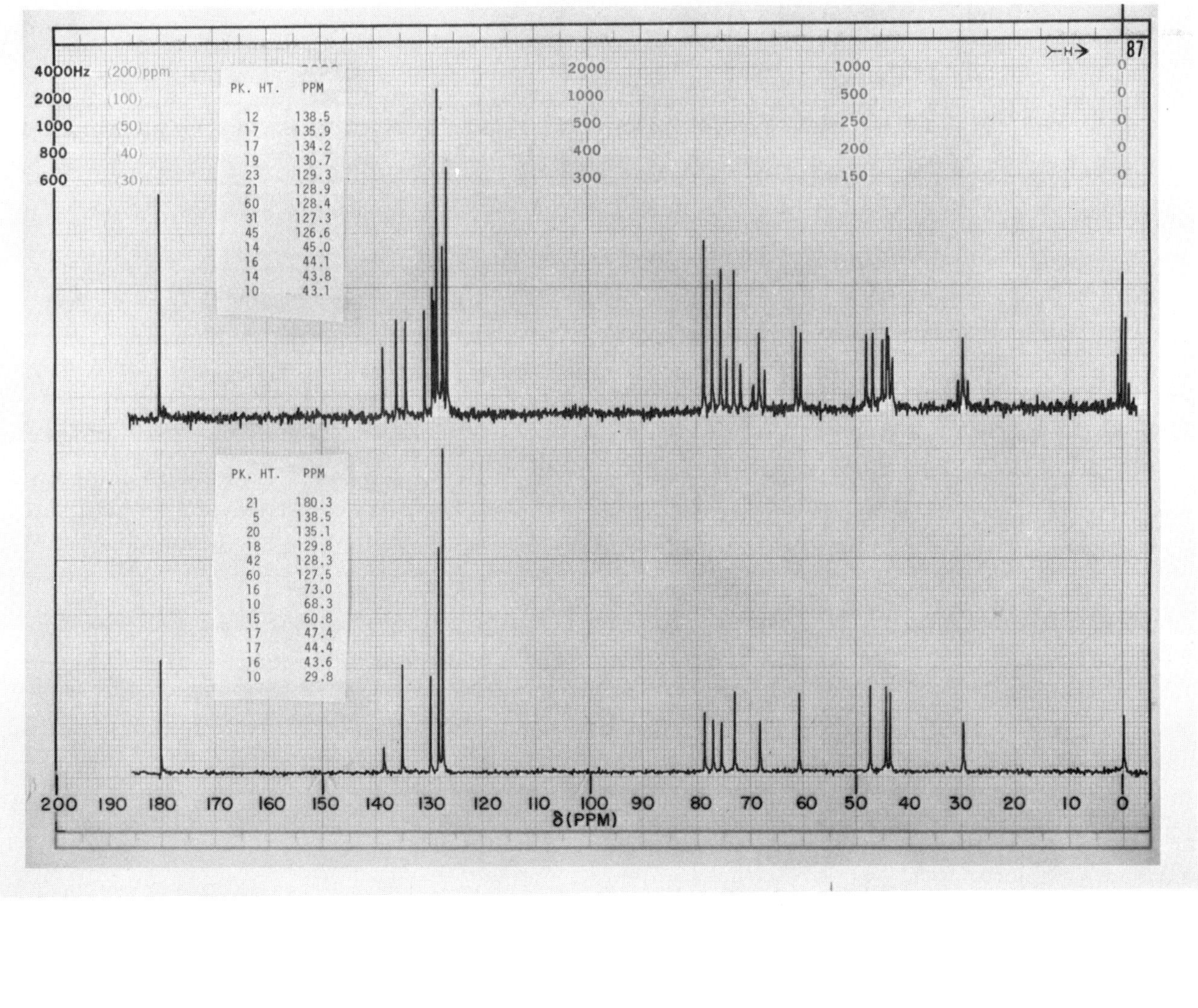

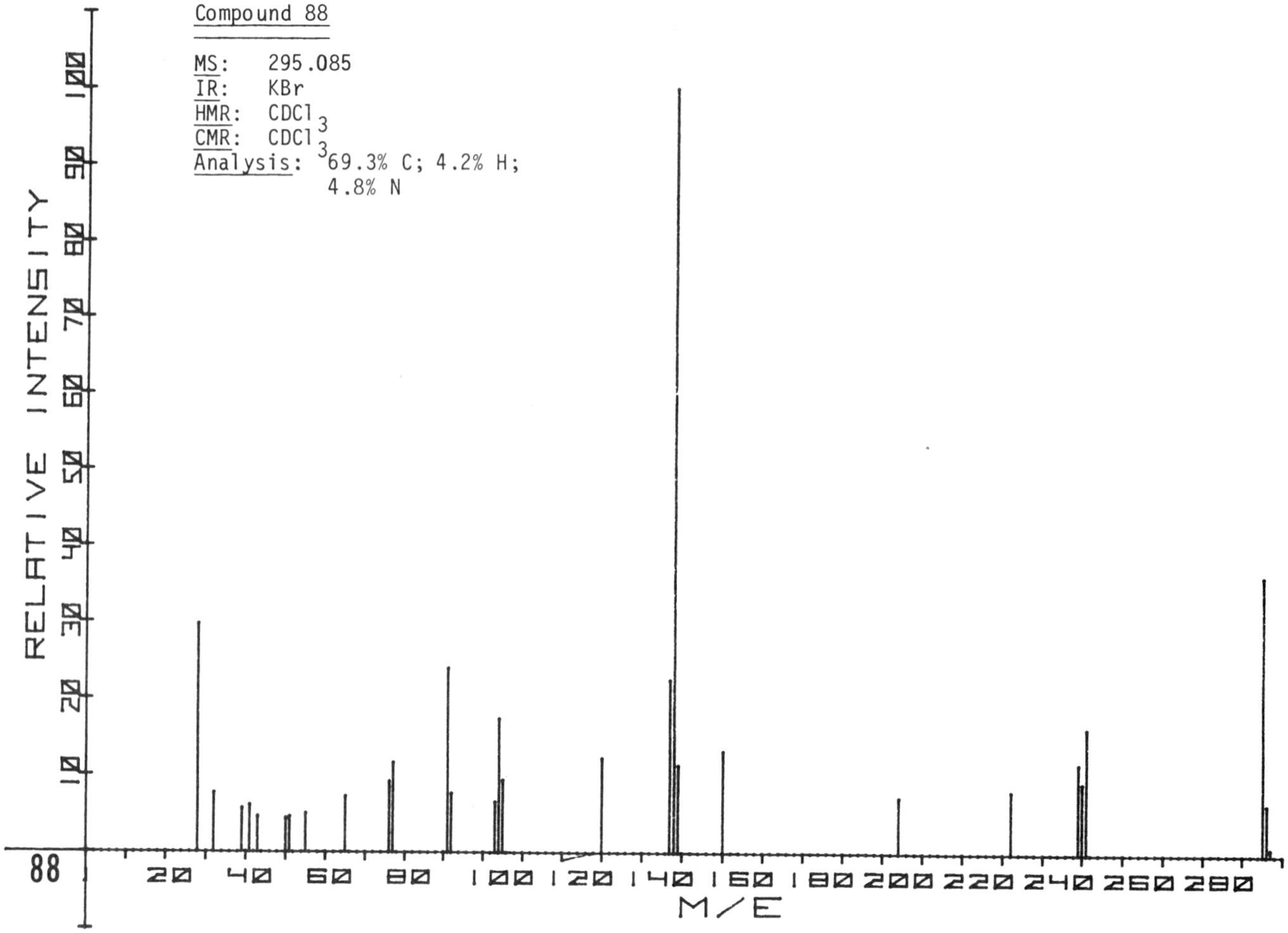

Compound 88
MS: 295.085
IR: KBr
HMR: CDCl3
CMR: CDCl3
Analysis: 69.3% C; 4.2% H; 4.8% N
RELATIVE INTENSITY
10 20 30 40 50 60 70 80 90 100
88
20 40 60 80 100 120 140 160 180 200 220 240 260 280
M/E

4000 3000 2000 1500 CM-1 1000 900 800 700
0.0 .10 .20 .30 .40 .50 .60 .70 1.0 ∞
ABSORBANCE
88
3 4 5 6 7 8 9 10 11 12 13 14 15
WAVELENGTH (MICRONS)

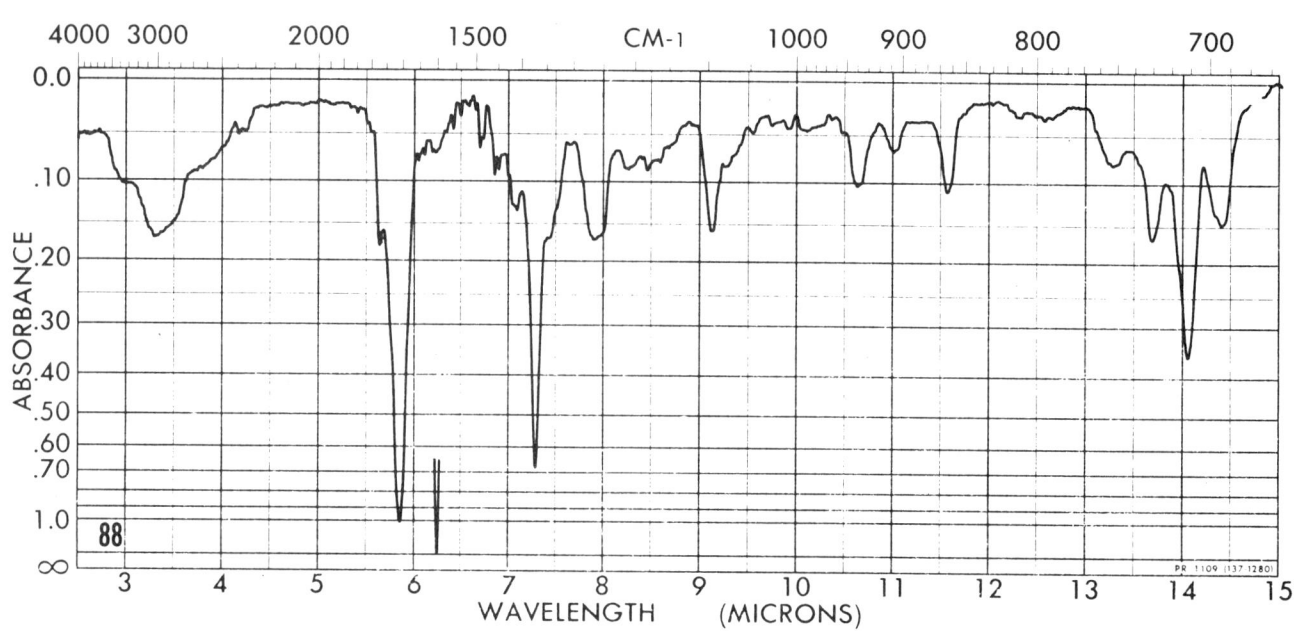

88
D2O
OFFSET 2 PPM

88
4000Hz (200)ppm
2000 (100)
1000 (50)
800 (40)
600 (30)
2000
1000
500
400
300
1000
500
250
200
150
PK. HT. PPM
19 136.5
29 135.2
24 133.1
30 131.6
34 129.8
24 129.6
46 127.9
24 127.7
12 126.0
30 124.6
44 122.6
PK. HT. PPM
19 174.7
9 167.5
8 136.5
45 134.2
28 131.6
46 128.9
59 128.6
19 127.0
45 123.6
15 53.1
13 34.4
200 190 180 170 160 150 140 130 120 110 100 90 80 70 60 50 40 30 20 10 0
δ(PPM)

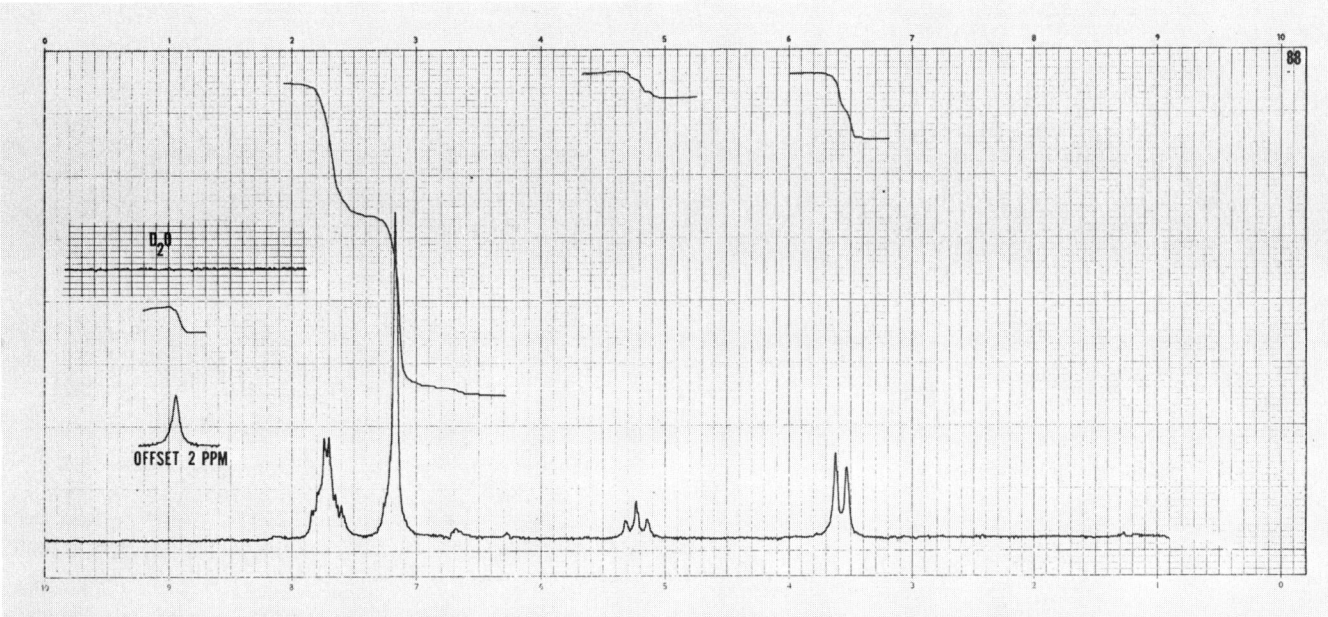

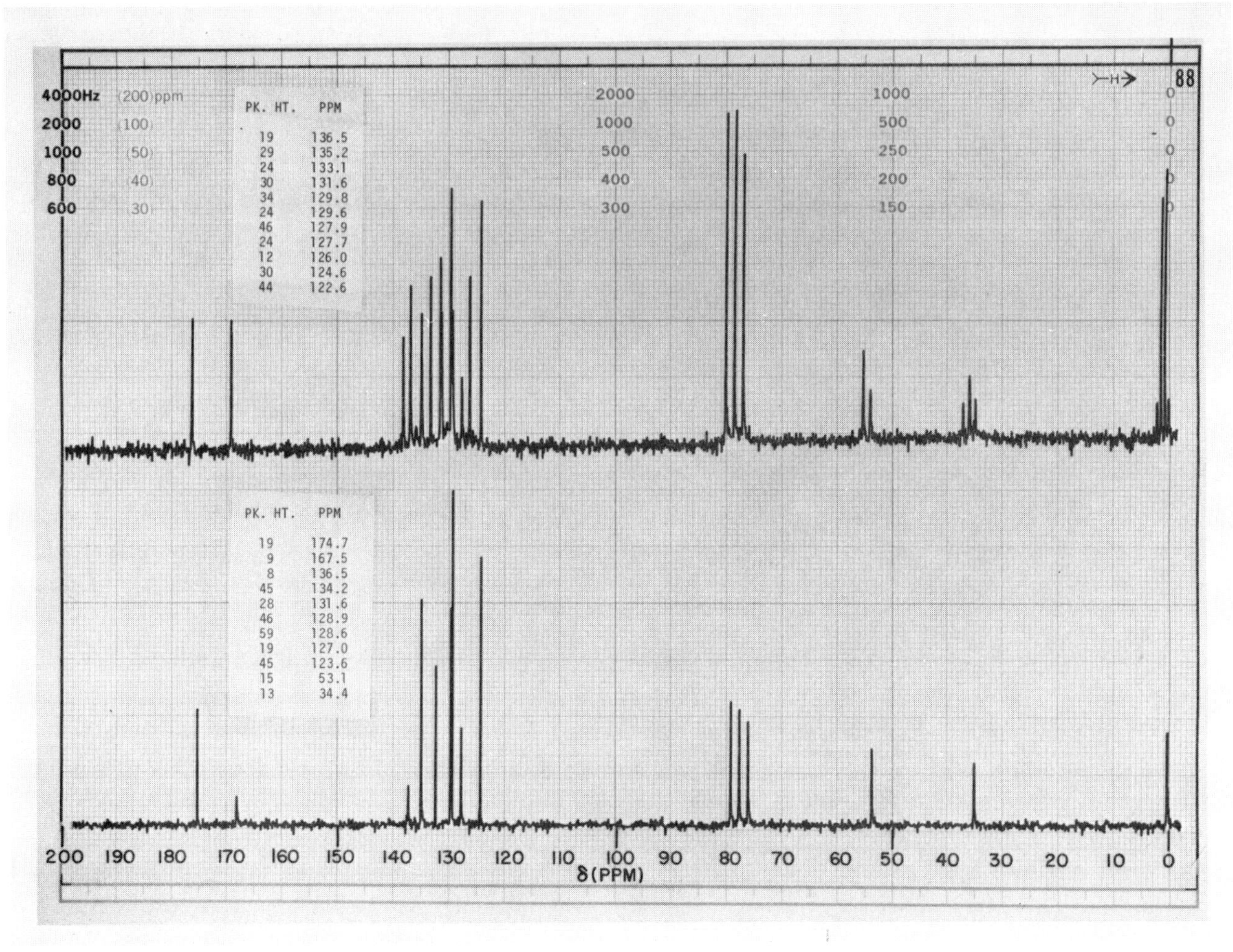

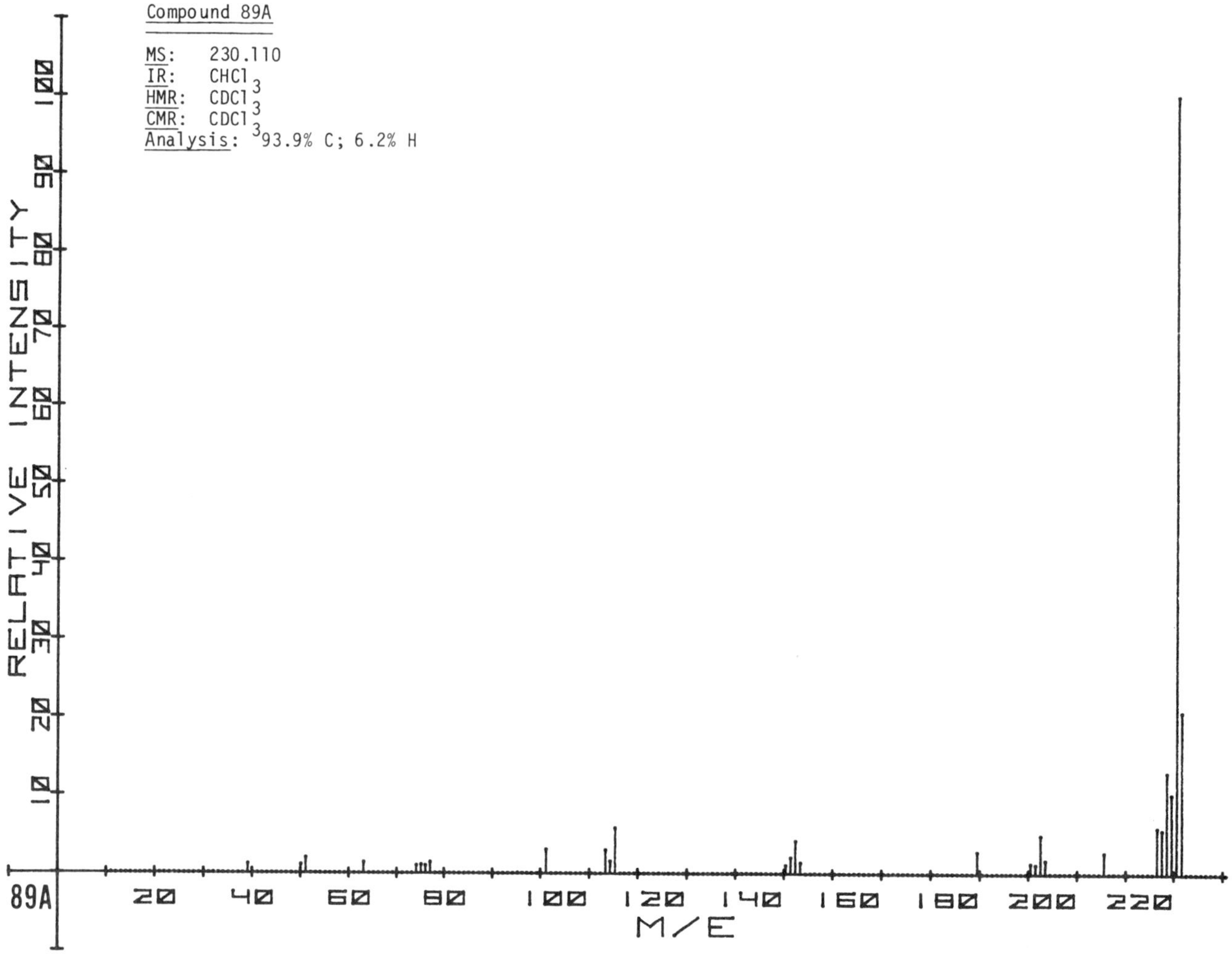

Compound 89A
MS: 230.110
IR: $CHCl_3$
HMR: $CDCl_3$
CMR: $CDCl_3$
Analysis: 93.9% C; 6.2% H
RELATIVE INTENSITY
10 20 30 40 50 60 70 80 90 100
20 40 60 80 100 120 140 160 180 200 220
M/E
89A

4000 3000 2000 1500 CM-1 1000 900 800 700
ABSORBANCE
0.0 .10 .20 .30 .40 .50 .60 .70 1.0 ∞
3 4 5 6 7 8 9 10 11 12 13 14 15
WAVELENGTH (MICRONS)
89A
PR 1109 (137-1280)

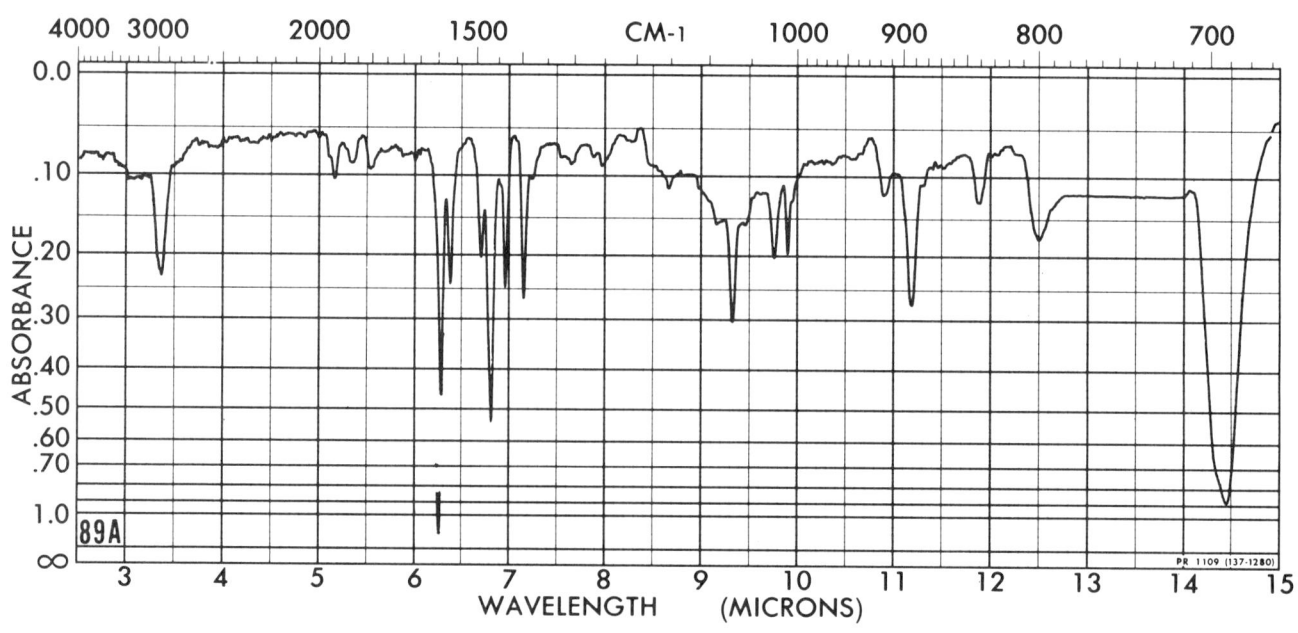

89A

89A
4000Hz (200)ppm 3000 2000 1000
2000 (100) 1000 500
1000 (50) 500 250
800 (40) 400 200
600 (30) 300 150
PK. HT. PPM
12 129.7
23 129.5
18 128.6
39 128.1
51 127.9
18 127.7
52 126.7
60 126.6
14 126.3
34 125.4
PK. HT. PPM
9 141.8
8 141.2
13 129.1
60 128.8
32 127.3
46 127.2
7 127.0
32 126.1
200 190 180 170 160 150 140 130 120 110 100 90 80 70 60 50 40 30 20 10 0
δ(PPM)

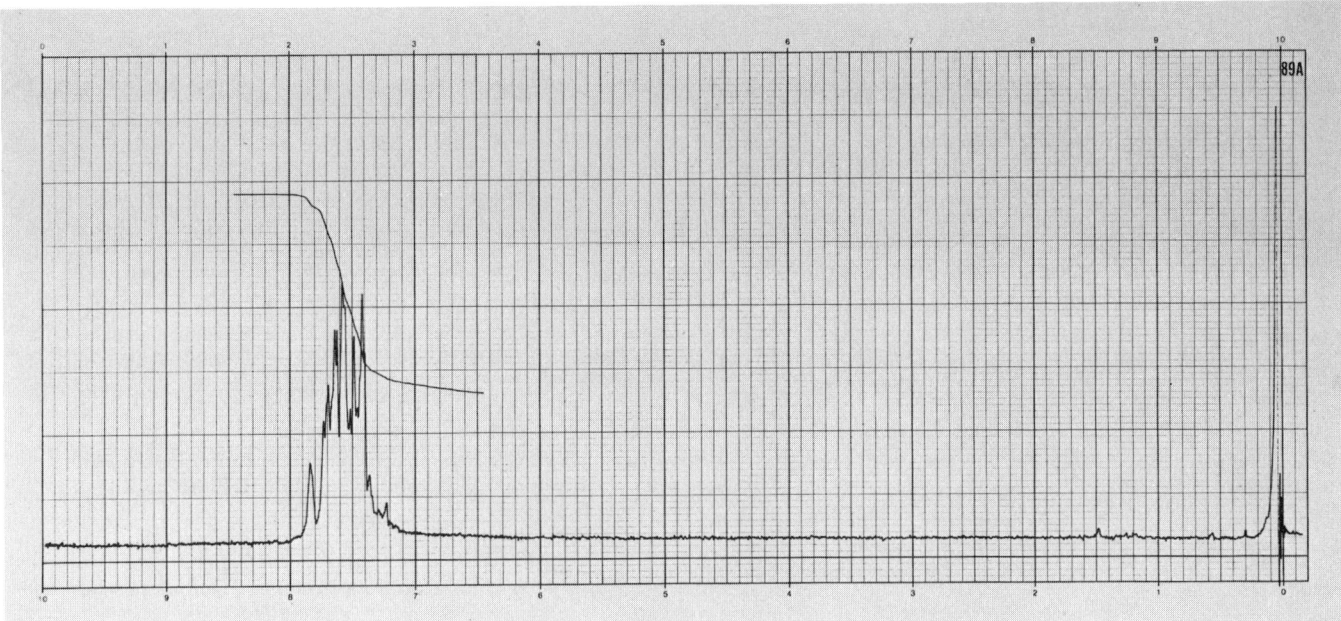

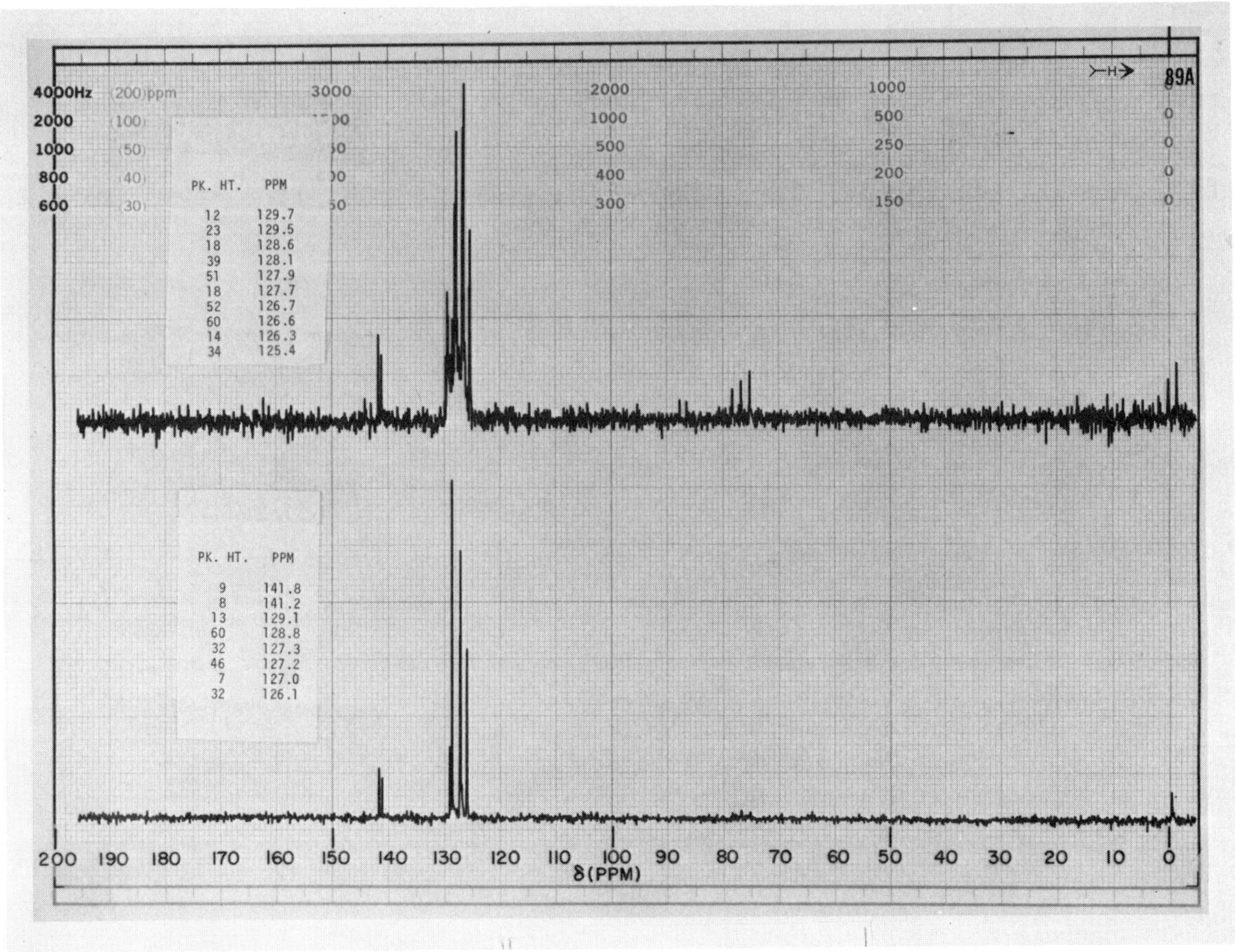

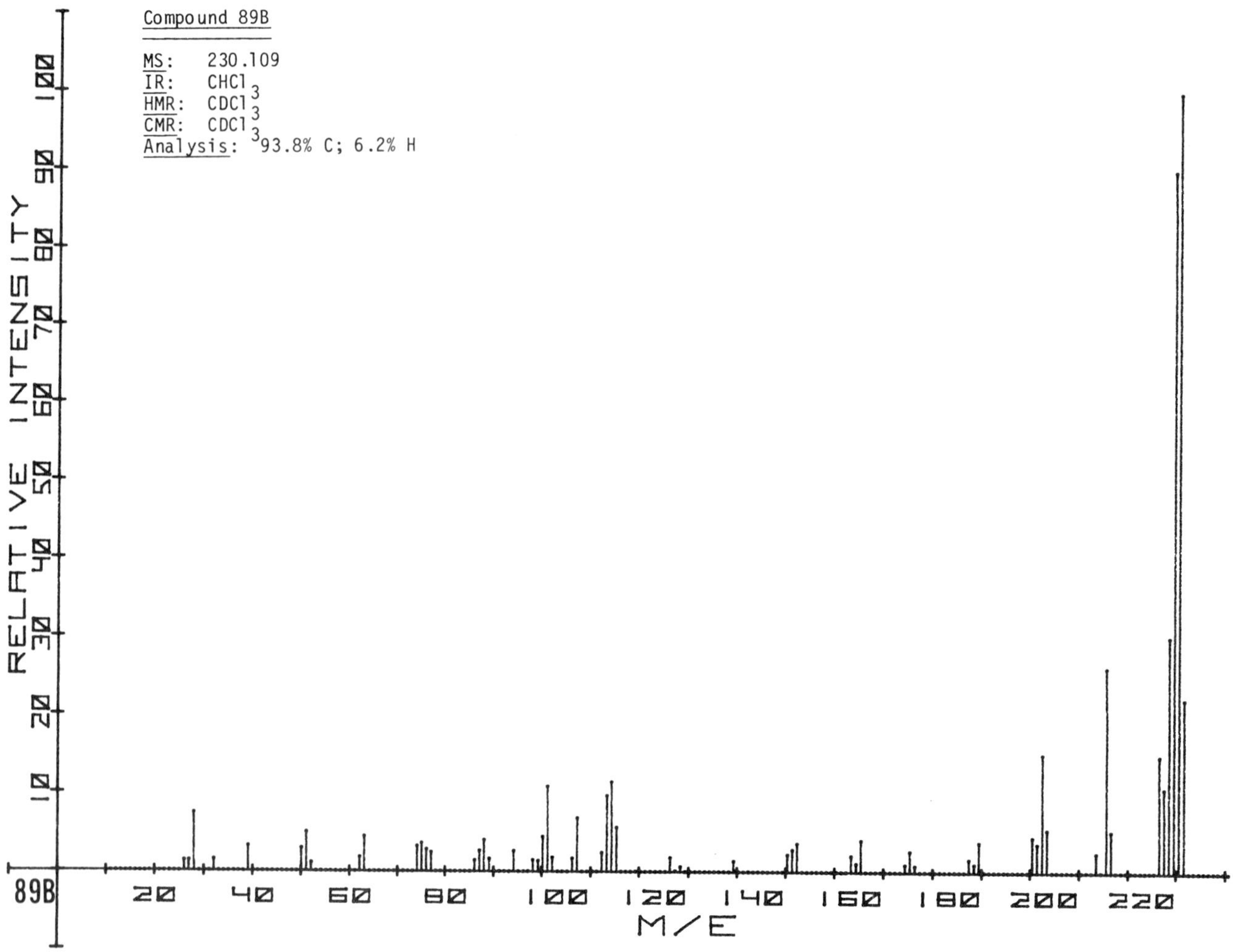

Compound 89B
MS: 230.109
IR: CHCl3
HMR: CDCl3
CMR: CDCl3
Analysis: 93.8% C; 6.2% H
RELATIVE INTENSITY
89B
M/E

CM-1
ABSORBANCE
89B
WAVELENGTH (MICRONS)

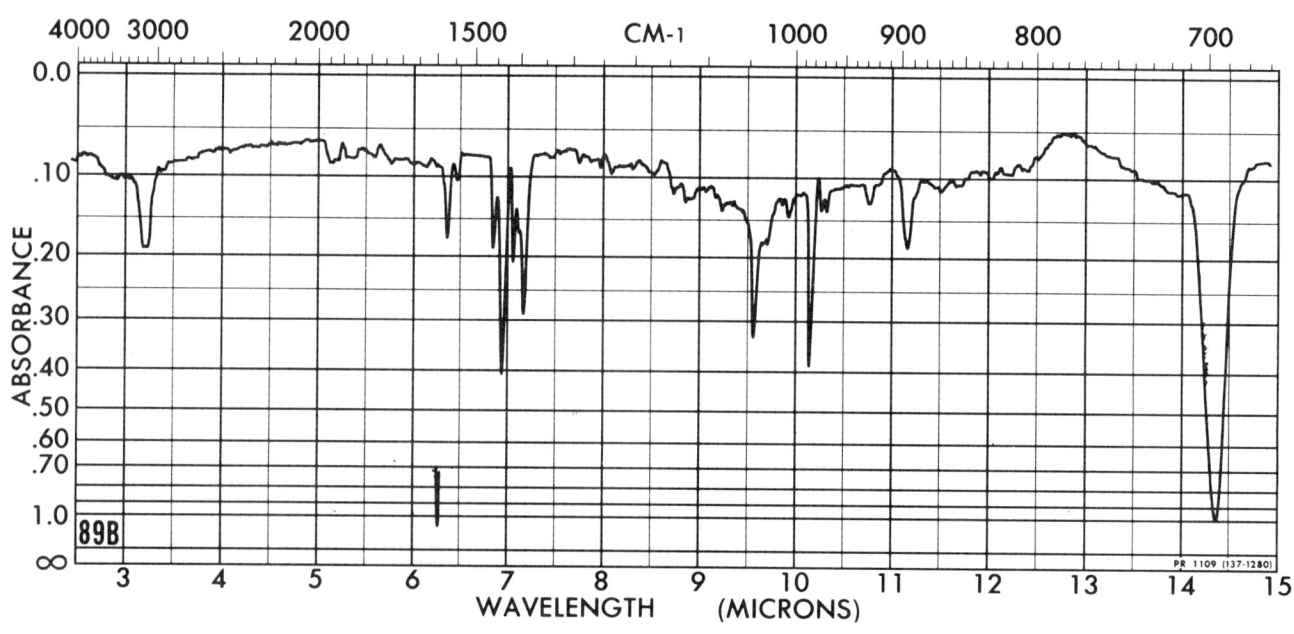

89B

4000Hz (200)ppm
2000 (100)
1000 (50)
800 (40)
600 (30)
2000
1000
500
400
300
1000
500
250
200
150
0
89B
PK. HT. PPM
18 131.2
54 130.5
24 129.9
60 129.1
22 128.4
16 128.0
34 127.1
26 127.0
17 126.7
16 125.6
PK. HT. PPM
9 141.5
8 140.6
19 130.5
60 129.8
55 127.8
30 127.4
29 126.3
200 190 180 170 160 150 140 130 120 110 100 90 80 70 60 50 40 30 20 10 0
δ(PPM)

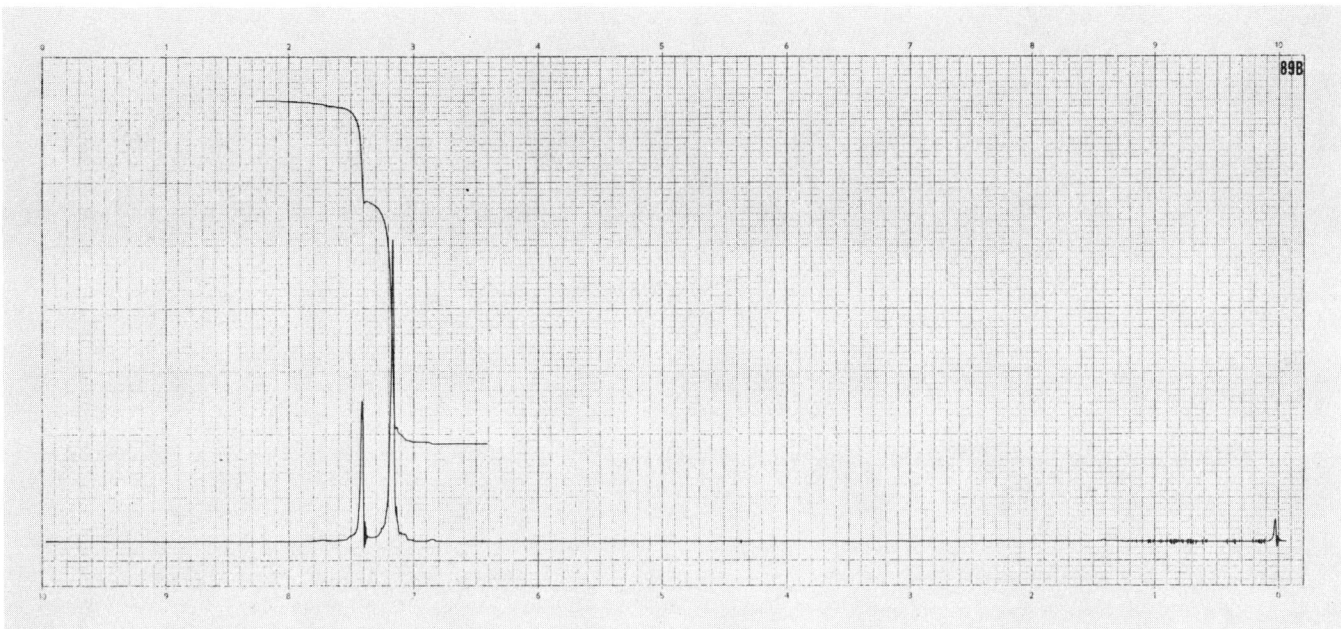

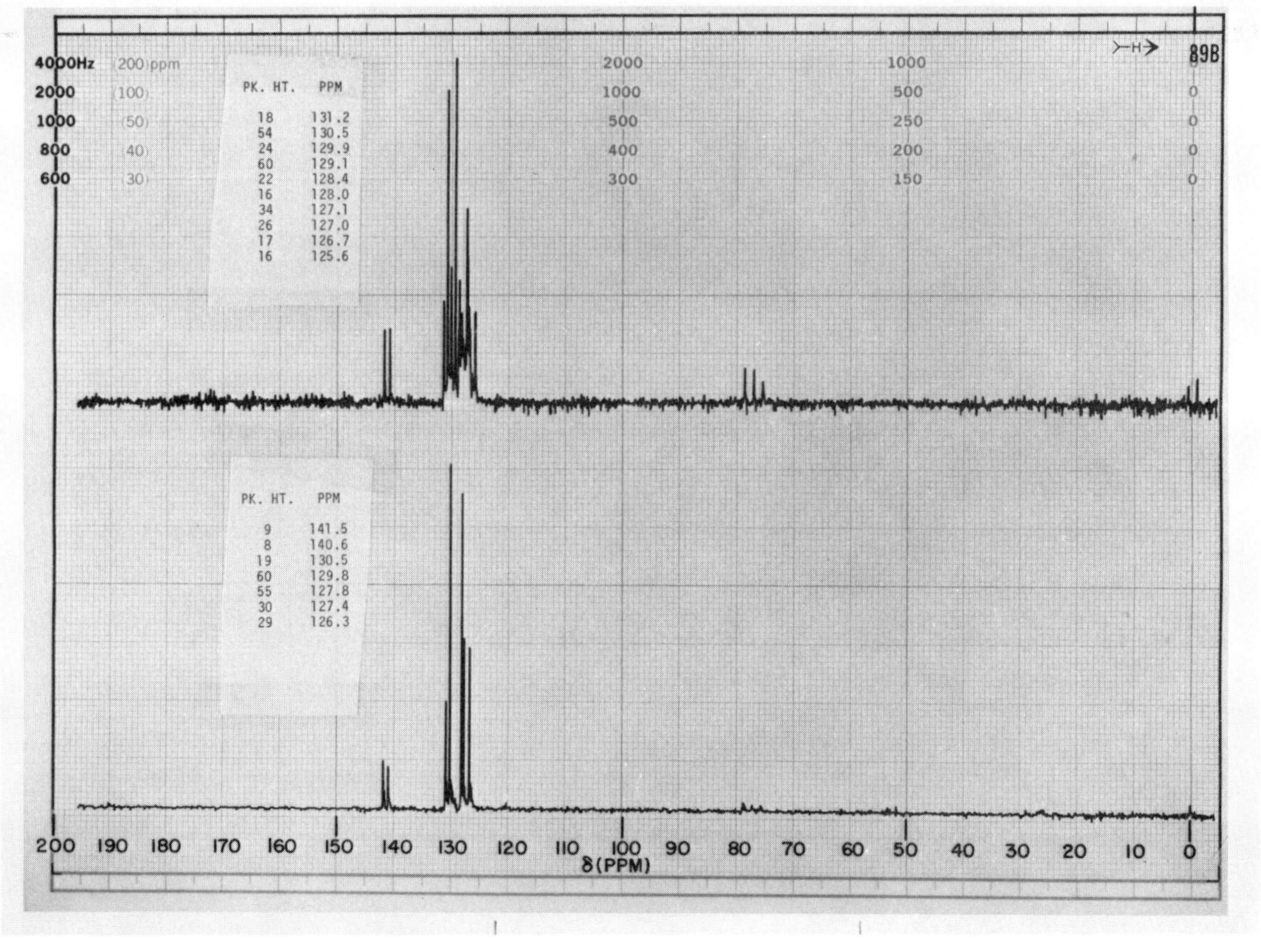

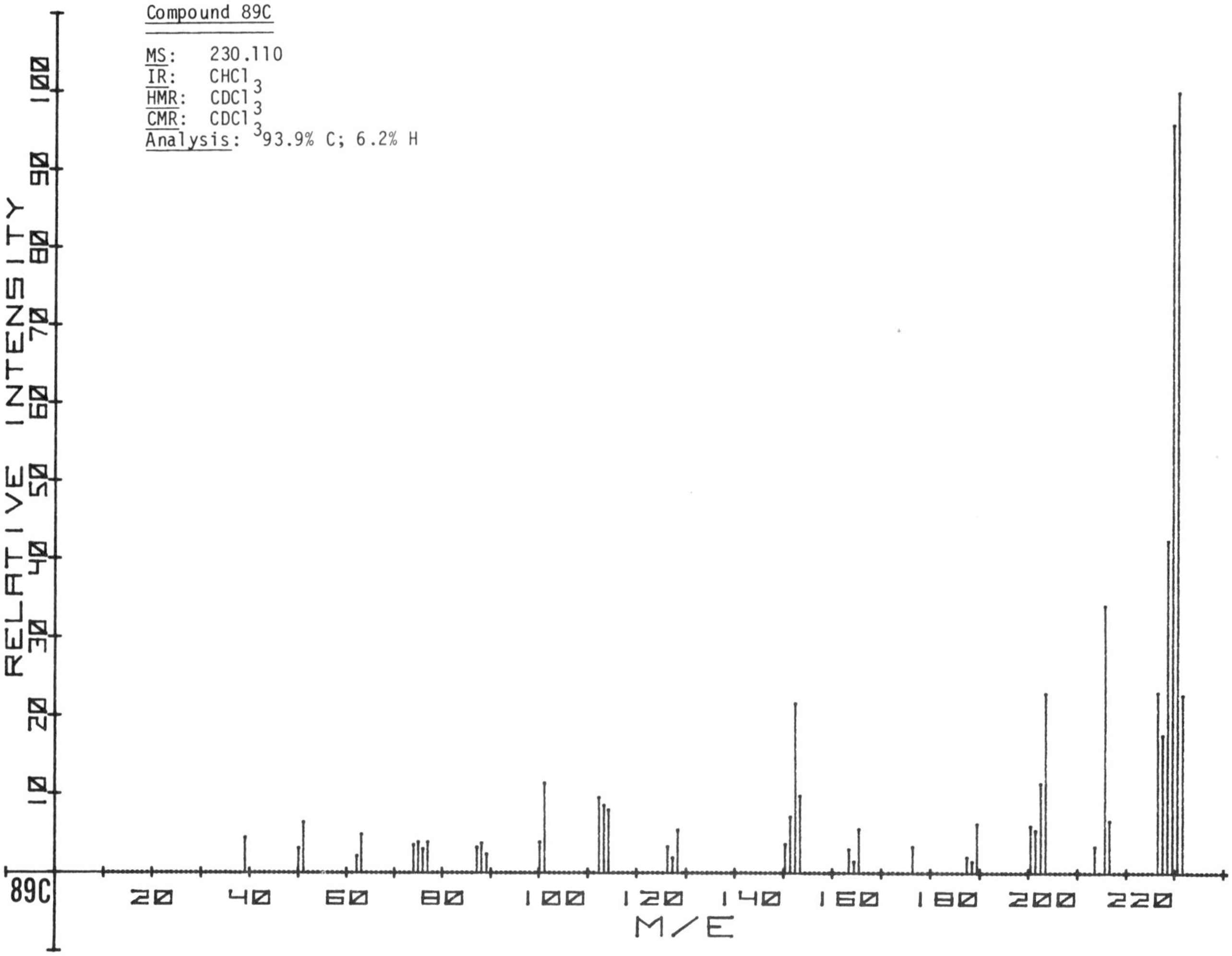

Compound 89C
MS: 230.110
IR: CHCl3
HMR: CDCl3
CMR: CDCl3
Analysis: 93.9% C; 6.2% H
RELATIVE INTENSITY
10 20 30 40 50 60 70 80 90 100
89C
20 40 60 80 100 120 140 160 180 200 220
M/E

4000 3000 2000 1500 CM-1 1000 900 800 700
ABSORBANCE
0.0 .10 .20 .30 .40 .50 .60 .70 1.0 ∞
89C
3 4 5 6 7 8 9 10 11 12 13 14 15
WAVELENGTH (MICRONS)

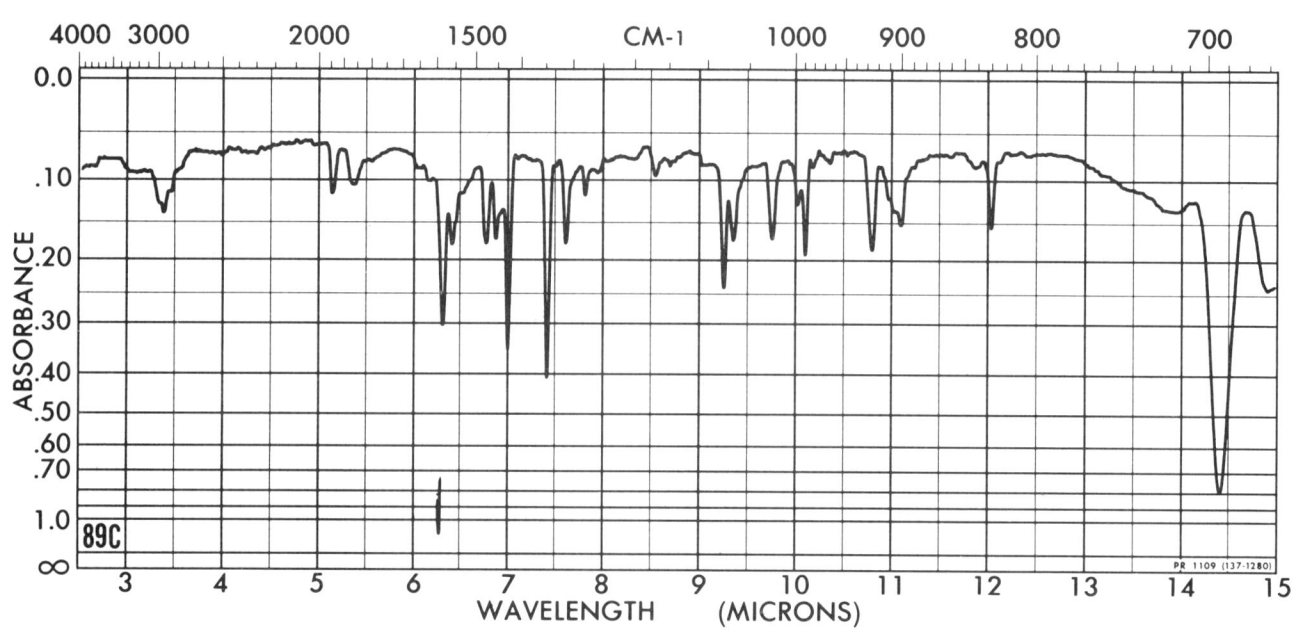

89C

89C
4000Hz (200)ppm 3000 2000 1000
2000 (100) 1500 1000 500 0
1000 (50) 750 500 250 0
800 (40) 600 400 200 0
600 (30) 450 300 150 0
PK. HT. PPM
4 151.8
3 144.0
13 141.3
22 133.2
60 132.8
23 131.5
57 131.3
18 129.5
31 128.5
25 127.8
27 126.8
27 125.2
26 123.4
PK. HT. PPM
2 151.8
3 144.0
9 141.3
30 132.3
59 132.1
28 128.6
60 127.7
24 124.3
200 190 180 170 160 150 140 130 120 110 100 90 80 70 60 50 40 30 20 10 0
δ(PPM)

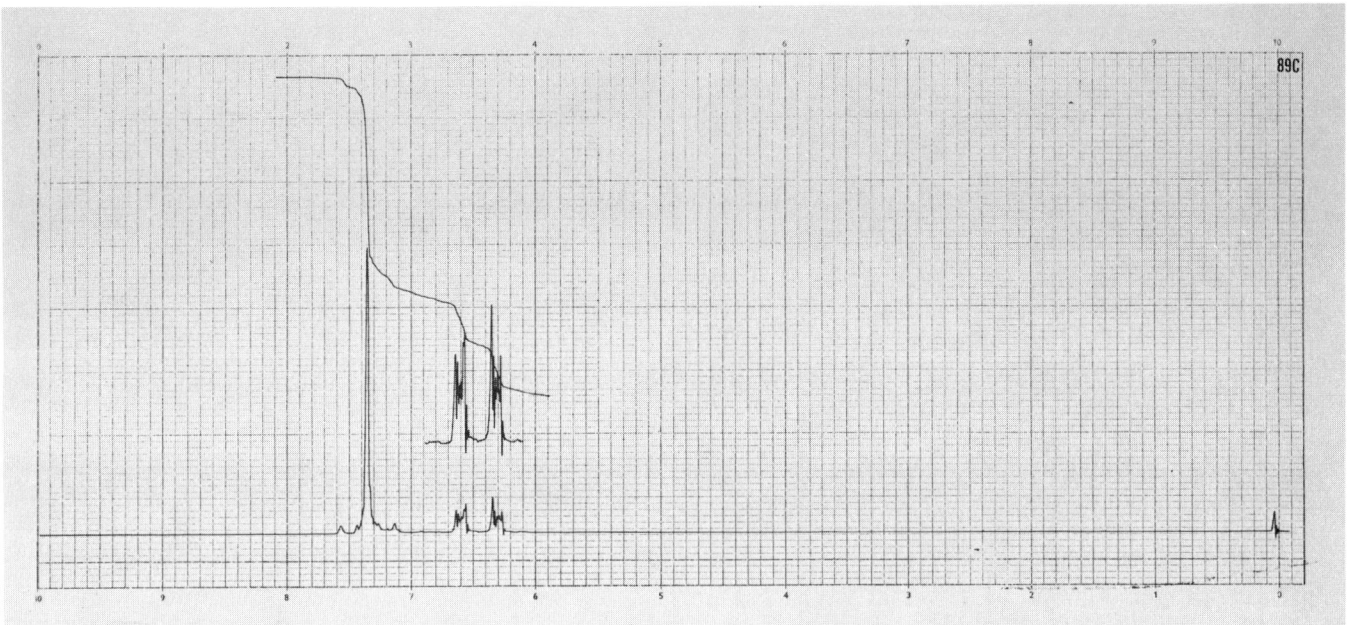

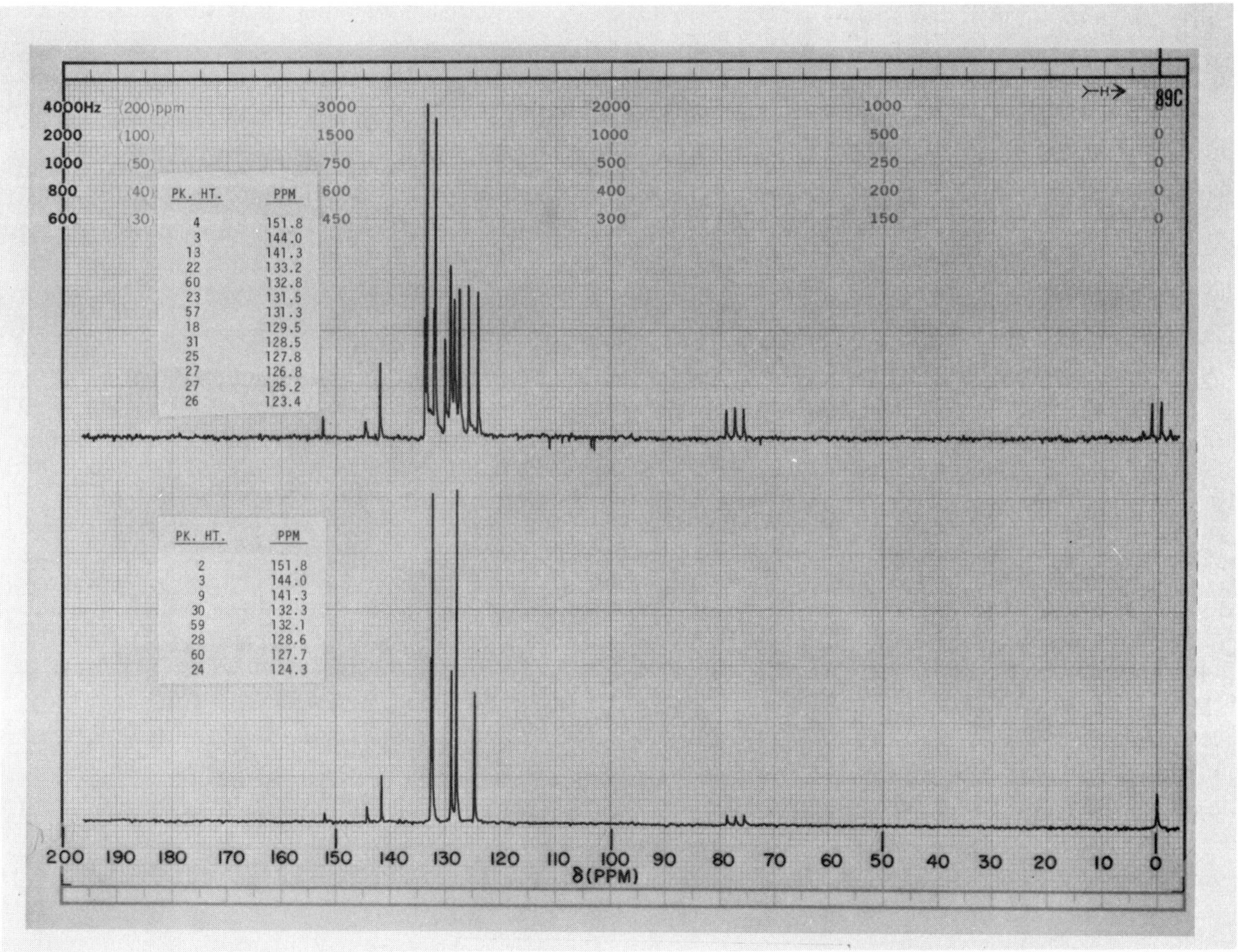

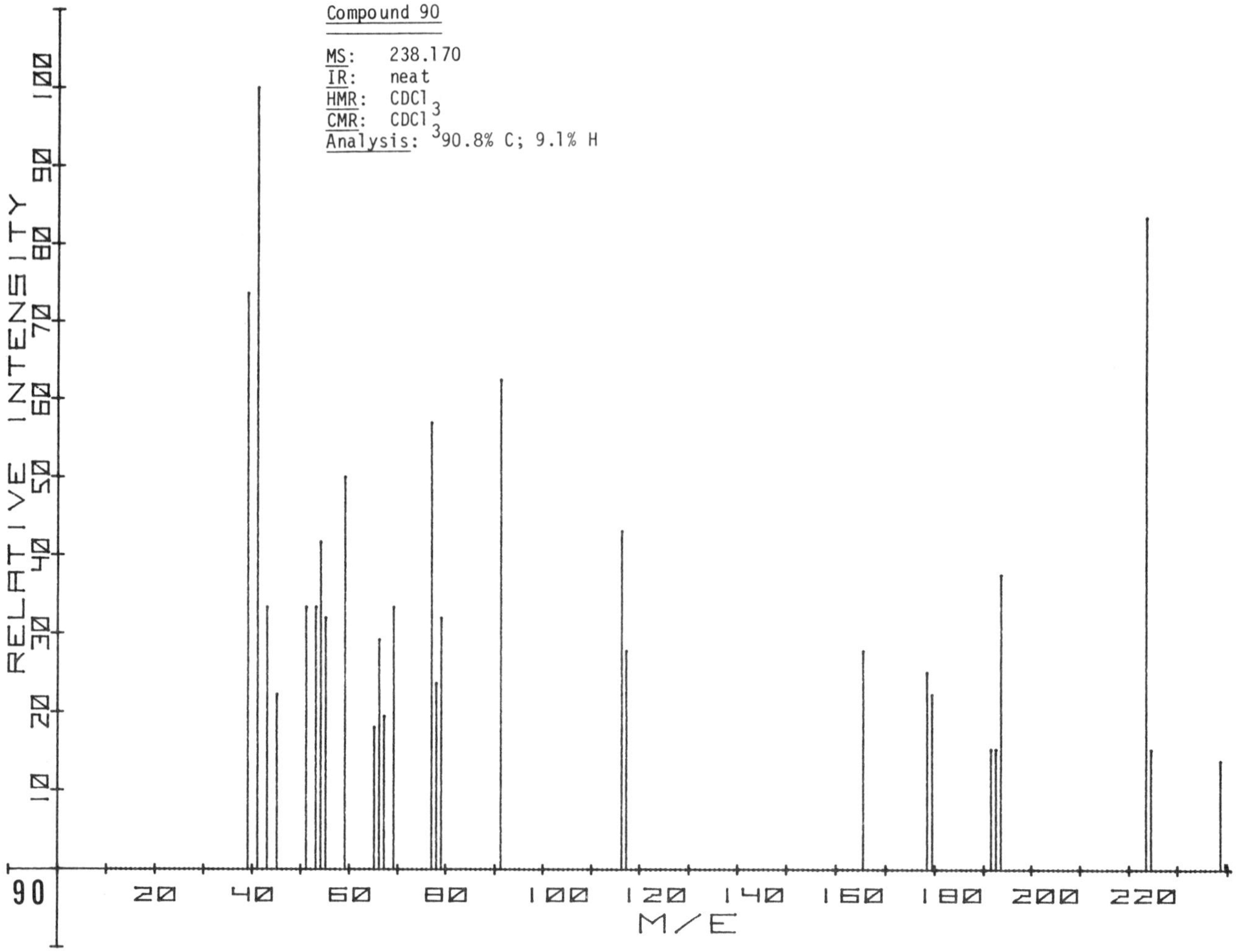

Compound 90
MS: 238.170
IR: neat
HMR: CDCl3
CMR: CDCl3
Analysis: 90.8% C; 9.1% H
RELATIVE INTENSITY
10 20 30 40 50 60 70 80 90 100
90
20 40 60 80 100 120 140 160 180 200 220
M/E

4000 3000 2000 1500 CM-1 1000 900 800 700
0.0 .10 .20 .30 .40 .50 .60 .70 1.0 ∞
ABSORBANCE
90
3 4 5 6 7 8 9 10 11 12 13 14 15
WAVELENGTH (MICRONS)

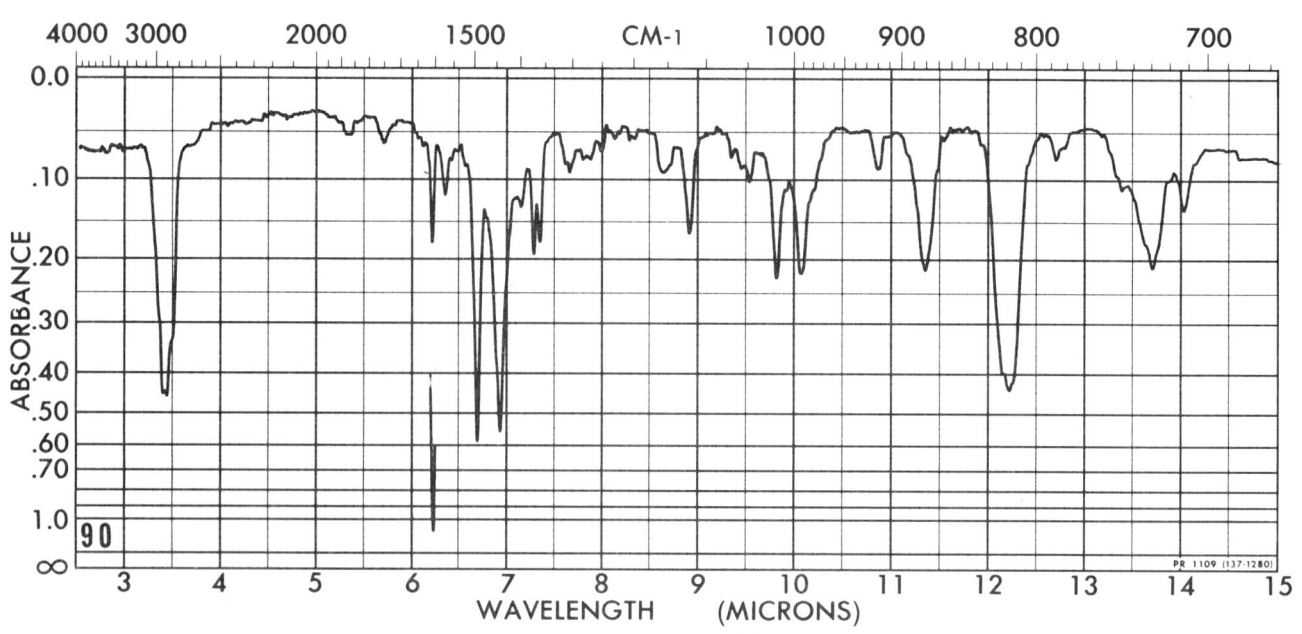

90
10X

4000Hz (200)ppm
2000 (100)
1000 (50)
800 (40)
600 (30)
2000
1000
500
400
300
1000
500
250
200
150
PK. HT. PPM
2 23.4
11 22.2
11 21.0
9 20.7
6 20.2
31 19.6
38 19.1
29 18.5
34 18.0
6 17.4
4 16.9
PK. HT. PPM
25 143.8
34 135.4
51 133.0
24 129.2
22 128.6
24 124.6
10 43.7
5 21.6
14 19.1
14 18.5
90
200 190 180 170 160 150 140 130 120 110 100 90 80 70 60 50 40 30 20 10 0
δ(PPM)

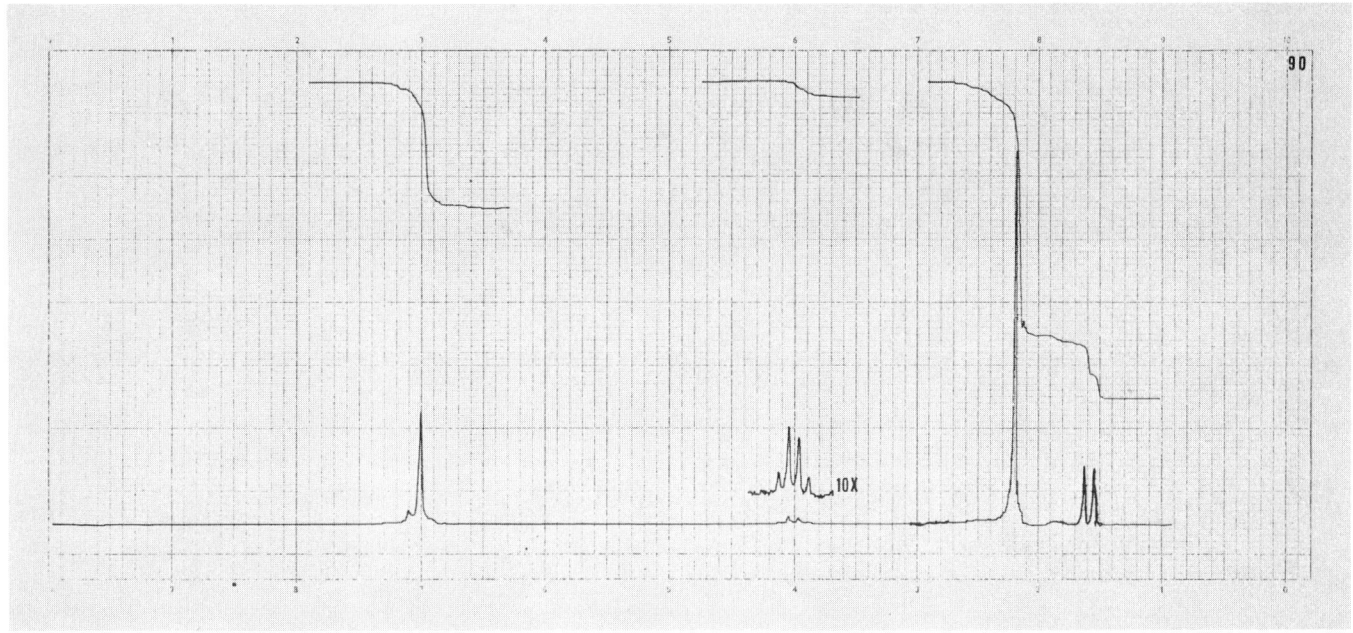

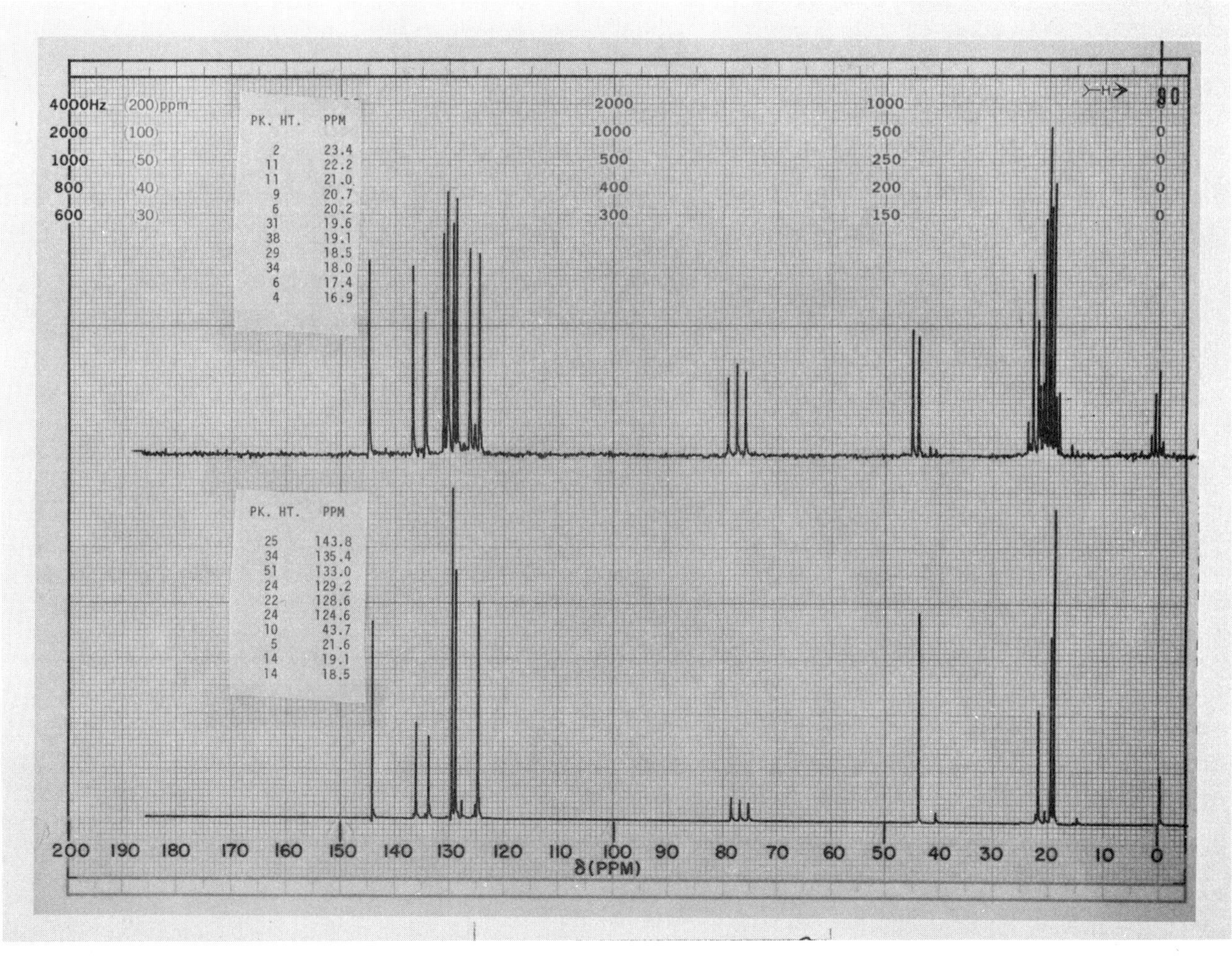

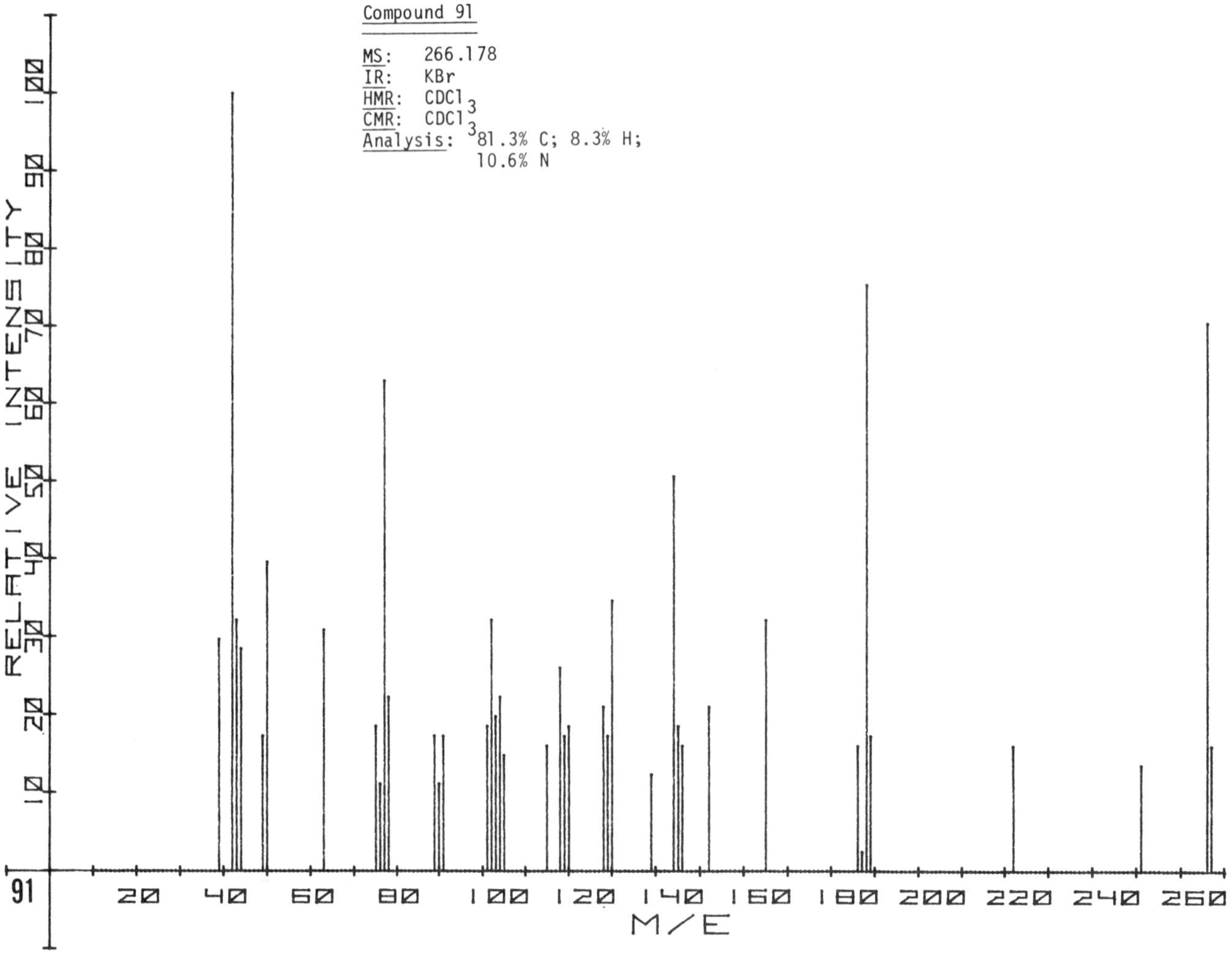

Compound 91
MS: 266.178
IR: KBr
HMR: CDCl3
CMR: CDCl3
Analysis: 81.3% C; 8.3% H; 10.6% N
RELATIVE INTENSITY
10 20 30 40 50 60 70 80 90 100
91
20 40 60 80 100 120 140 160 180 200 220 240 260
M/E

4000 3000 2000 1500 CM-1 1000 900 800 700
0.0 .10 .20 .30 .40 .50 .60 .70 1.0 ∞
ABSORBANCE
91
PR 1109 (137-1280)
3 4 5 6 7 8 9 10 11 12 13 14 15
WAVELENGTH (MICRONS)

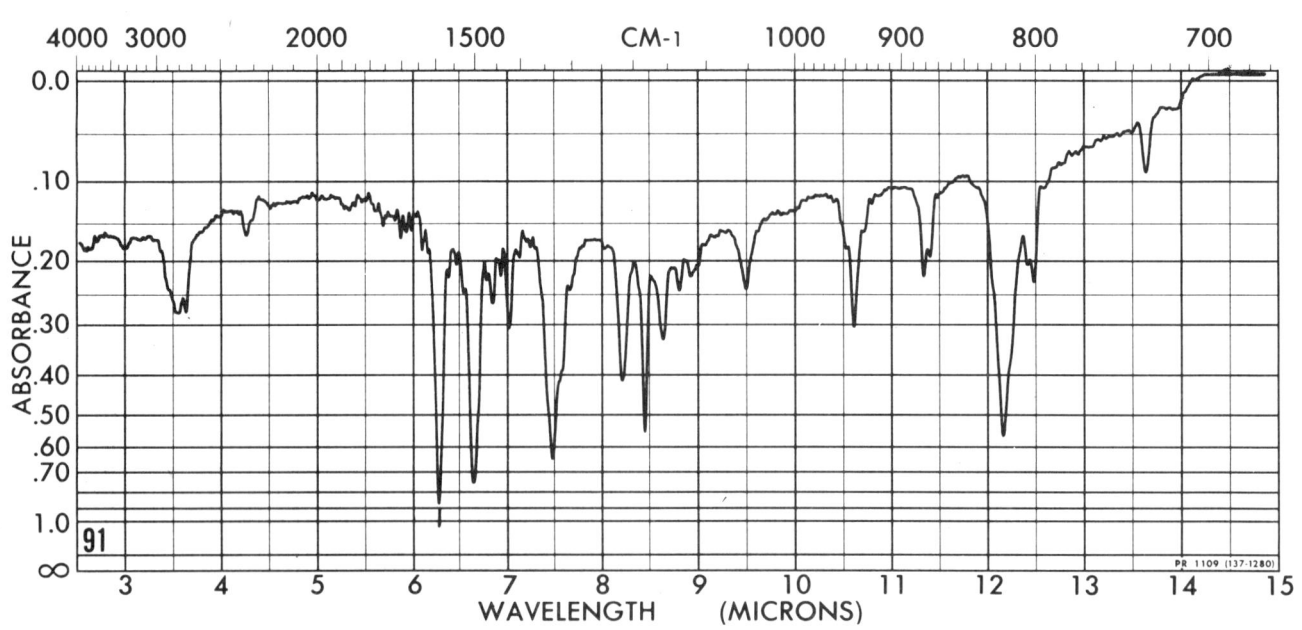

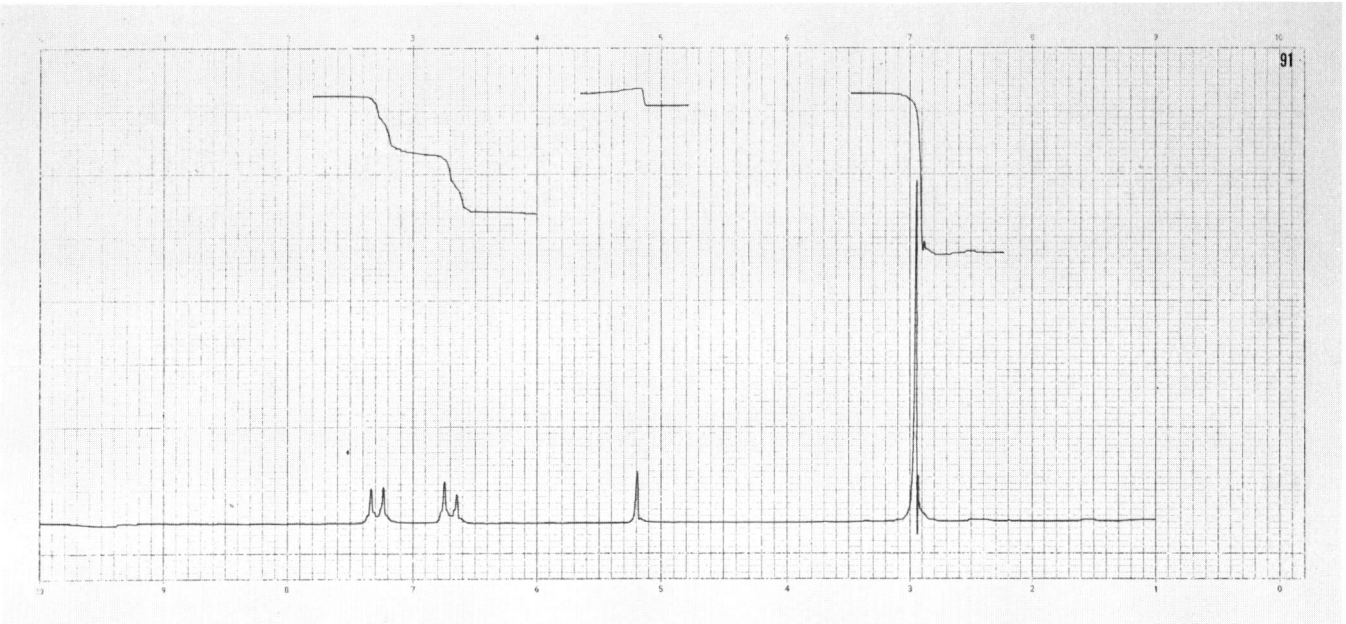

91

91
4000Hz (200)ppm
2000 (100)
1000 (50)
800 (40)
600 (30)
2000 1000
1000 500
500 250
400 200
300 150
PK. HT. PPM
10 130.5
30 130.2
25 128.3
29 112.9
24 111.1
3 110.5
24 109.0
5 107.5
PK. HT. PPM
13 150.1
7 149.7
4 130.5
60 129.2
58 112.0
8 108.9
20 40.5
200 190 180 170 160 150 140 130 120 110 100 90 80 70 60 50 40 30 20 10 0
δ(PPM)

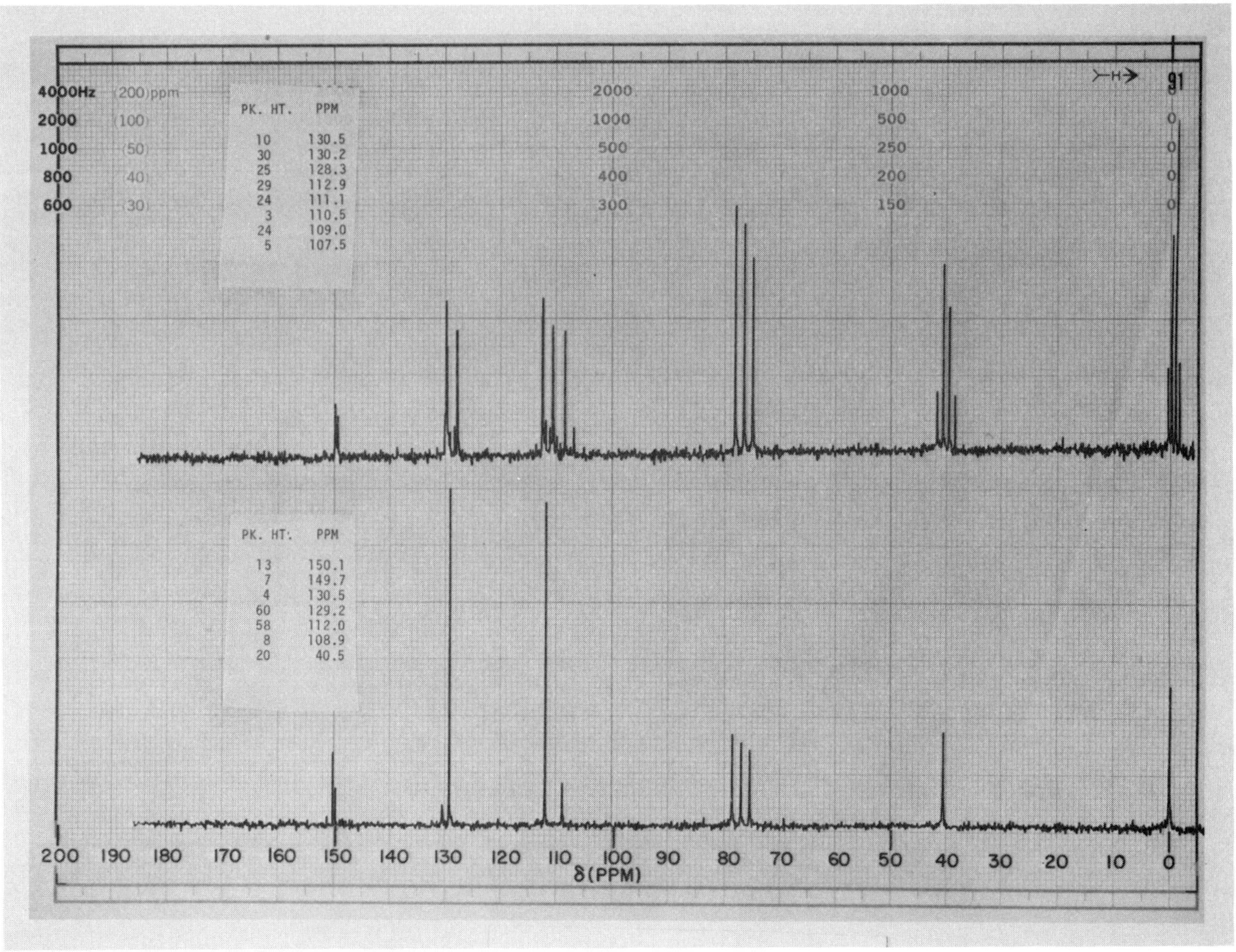

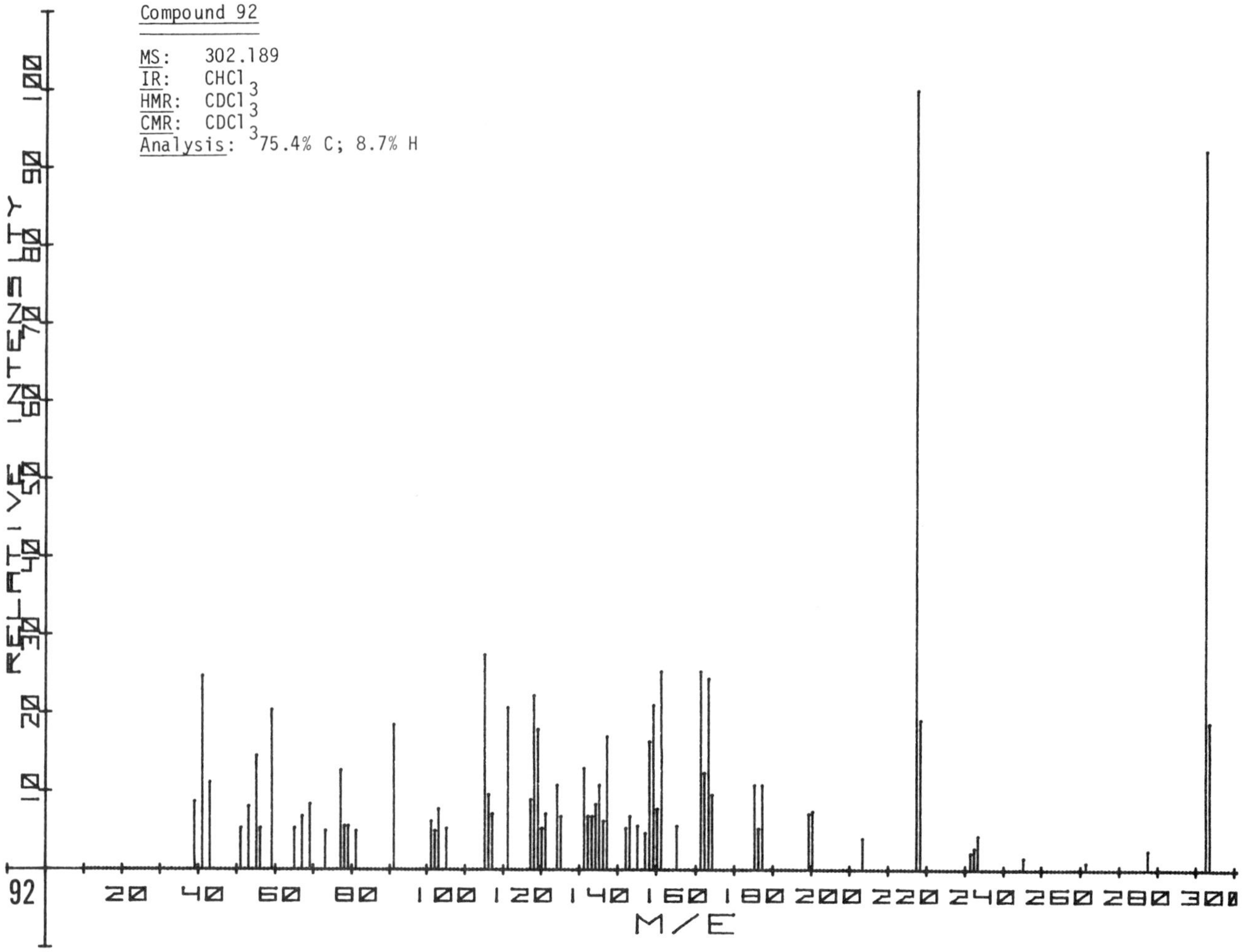

Compound 92
MS: 302.189
IR: CHCl3
HMR: CDCl3
CMR: CDCl3
Analysis: 75.4% C; 8.7% H
RELATIVE INTENSITY
92
M/E
20 40 60 80 100 120 140 160 180 200 220 240 260 280 300
10 20 30 40 50 60 70 80 90 100

4000 3000 2000 1500 CM-1 1000 900 800 700
ABSORBANCE
0.0 .10 .20 .30 .40 .50 .60 .70 1.0 ∞
92
3 4 5 6 7 8 9 10 11 12 13 14 15
WAVELENGTH (MICRONS)
PR 1109 (137-1280)

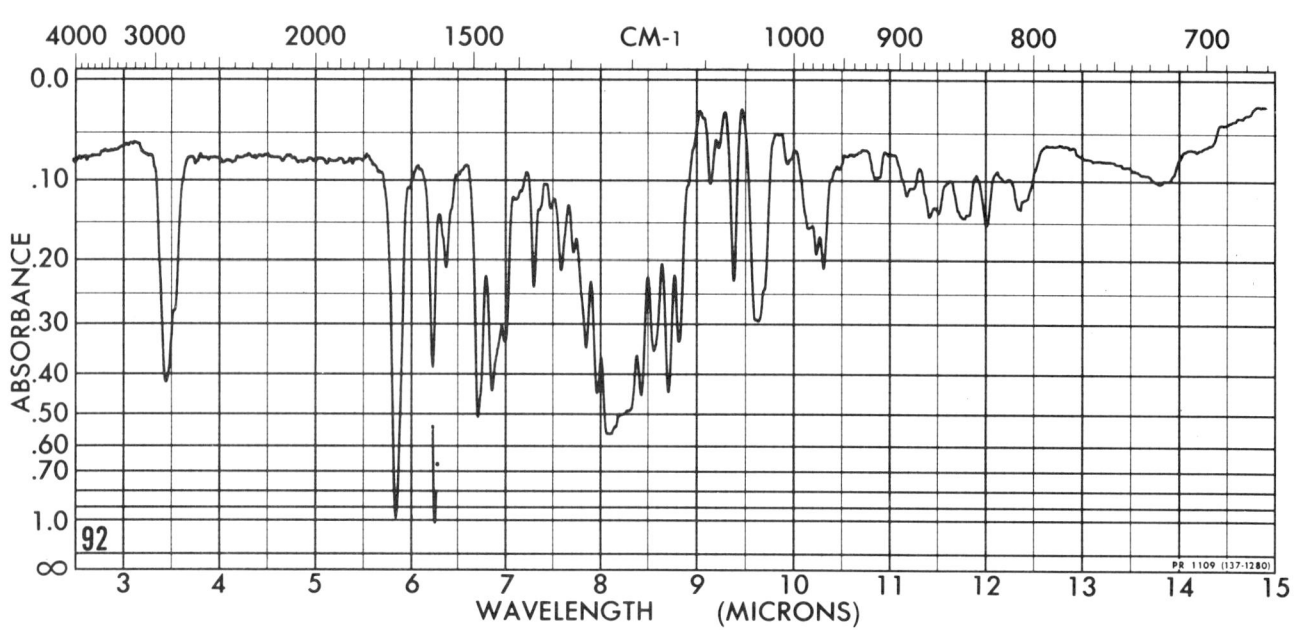

92

4000Hz (200)ppm 3000
2000 (100) 1500
1000 (50) 750
800 (40) 600
600 (30) 450
1000
500
250
200
150
92
PK. HT. PPM
29 109.6
53 109.3
26 108.4
60 108.2
6 53.7
41 53.0
36 52.3
6 51.6
26 51.0
26 50.4
8 49.9
36 49.1
27 48.3
6 47.5
12 37.9
26 37.3
13 36.6
62 36.4
12 36.1
24 35.5
14 34.9
13 19.4
29 18.9
14 18.4
14 18.5
33 17.8
11 17.4
PK. HT. PPM
22 175.3
23 155.6
17 146.9
58 127.5
25 125.1
57 109.0
58 108.8
47 52.6
41 50.7
44 48.7
36 41.7
43 37.2
27 36.4
51 35.4
42 29.0
43 26.2
43 20.6
38 18.9
38 17.8
200 190 180 170 160 150 140 130 120 110 100 90 80 70 60 50 40 30 20 10 0
δ(PPM)

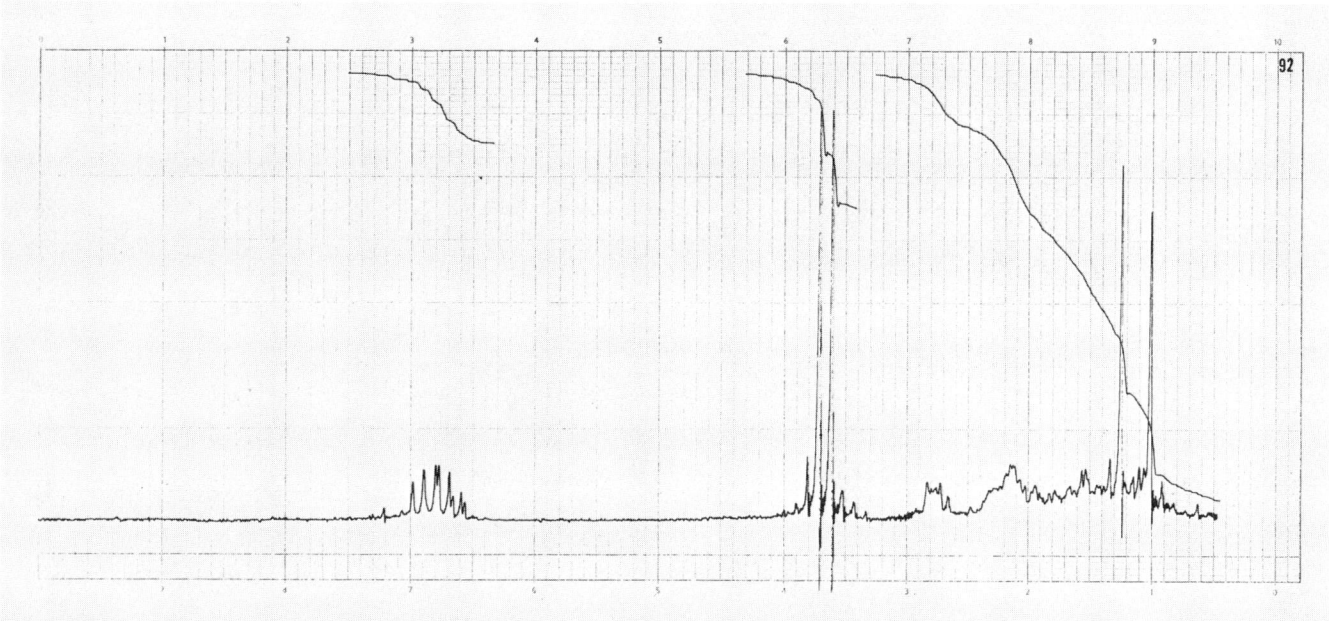

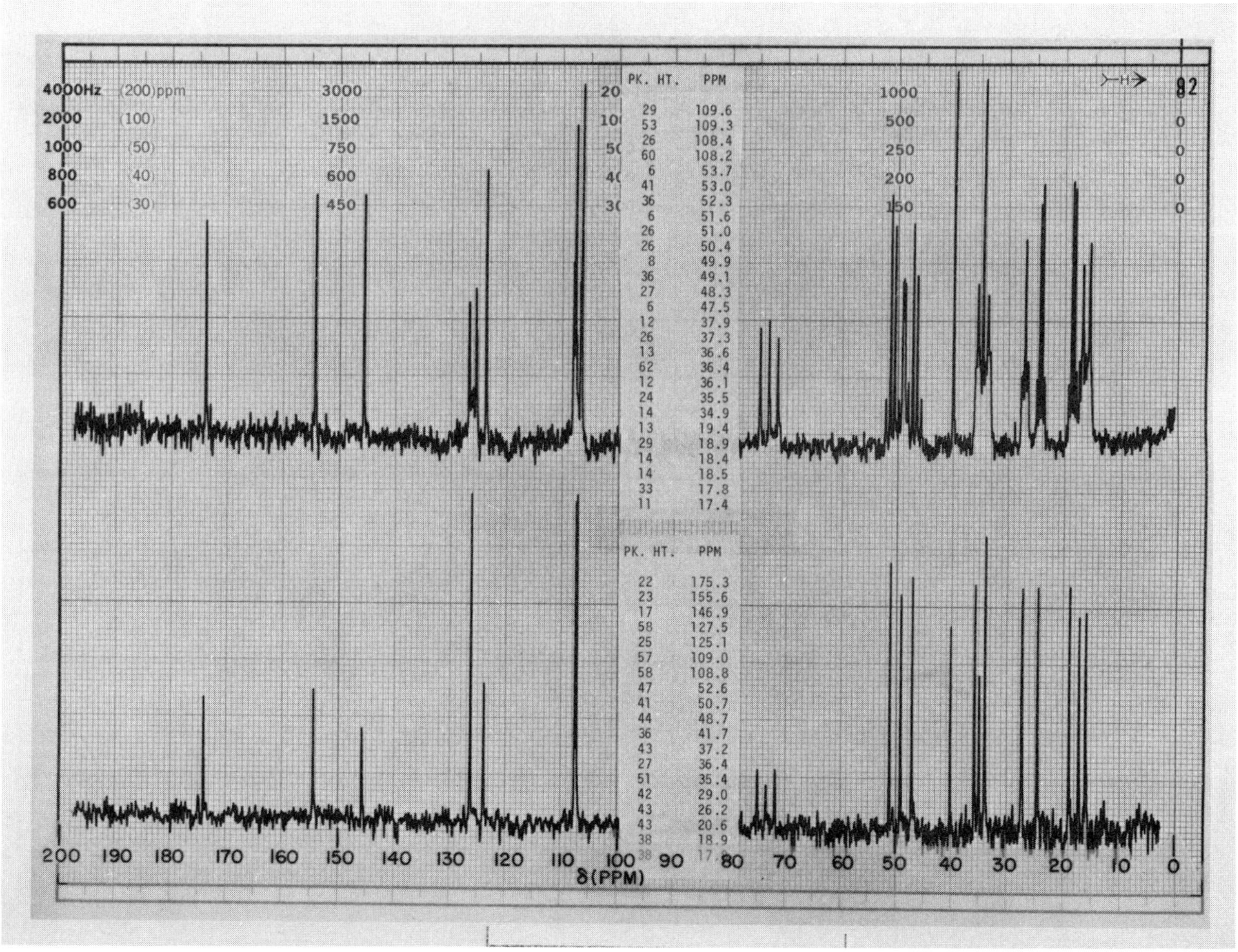

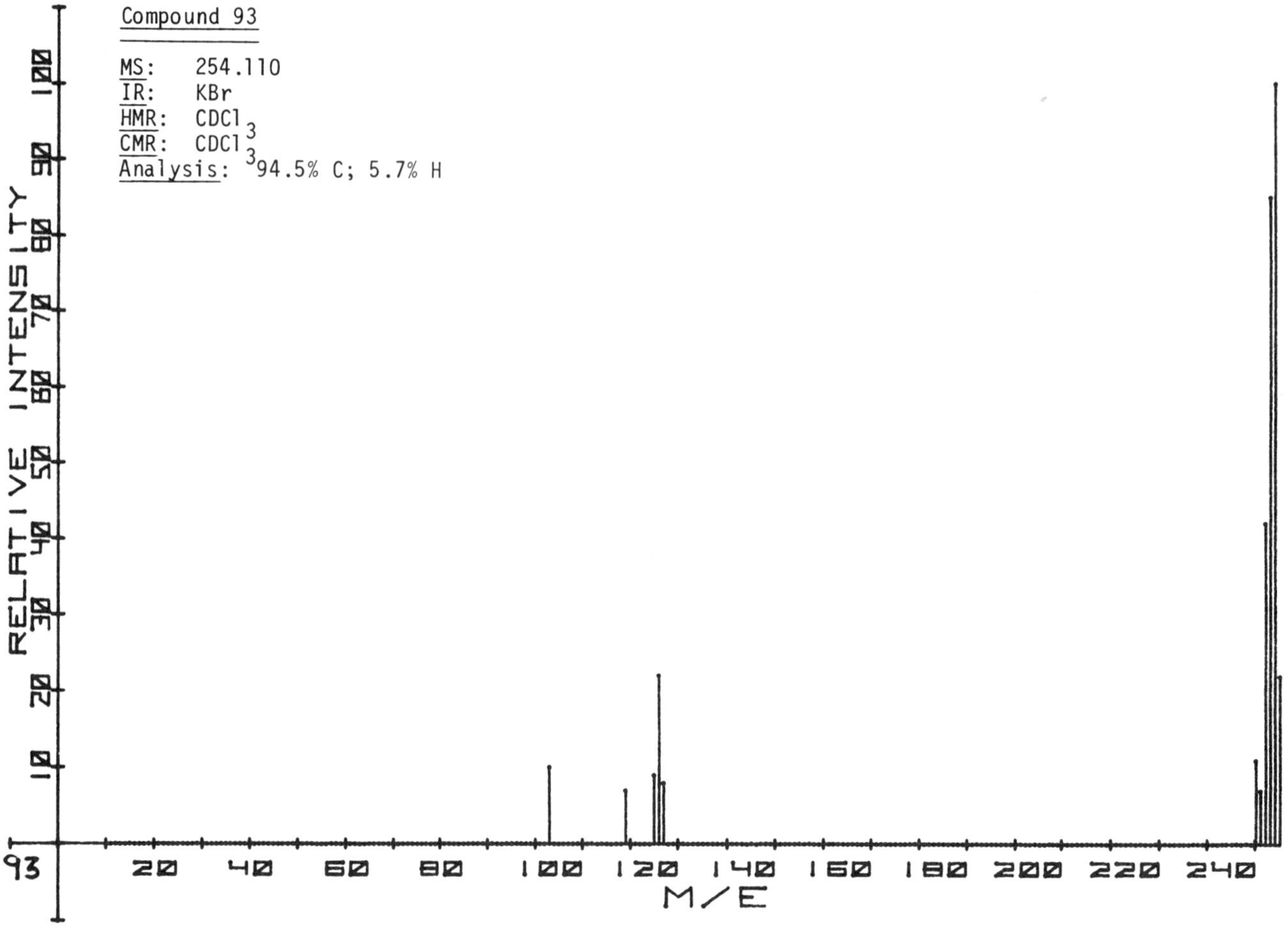

Compound 93
MS: 254.110
IR: KBr
HMR: CDCl3
CMR: CDCl3
Analysis: 94.5% C; 5.7% H
RELATIVE INTENSITY
10 20 30 40 50 60 70 80 90 100
93
20 40 60 80 100 120 140 160 180 200 220 240
M/E

4000 3000 2000 1500 CM-1 1000 900 800 700
0.0 .10 .20 .30 .40 .50 .60 .70 1.0 ∞
ABSORBANCE
93
3 4 5 6 7 8 9 10 11 12 13 14 15
WAVELENGTH (MICRONS)

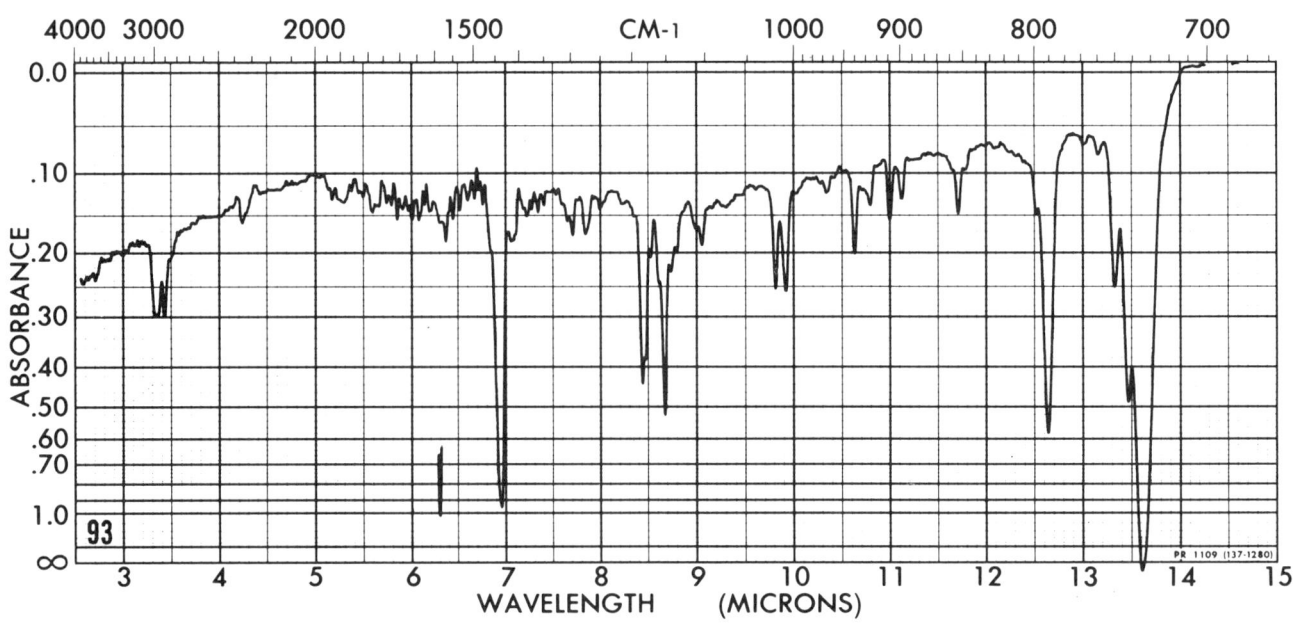

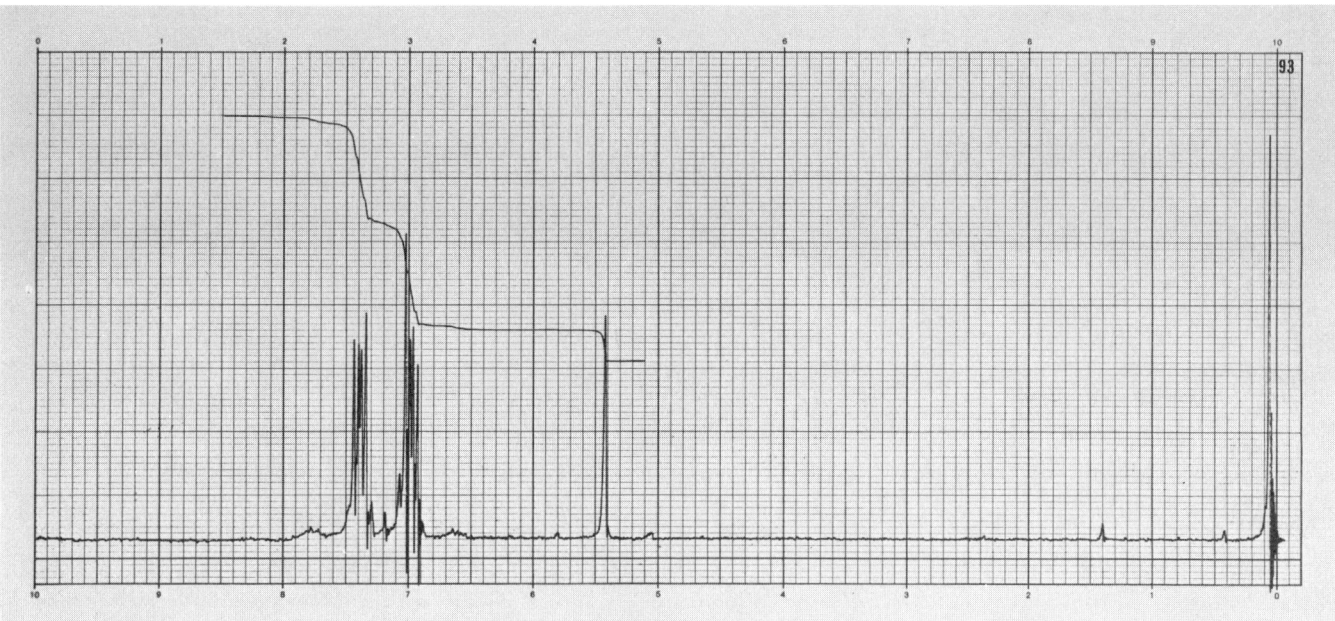

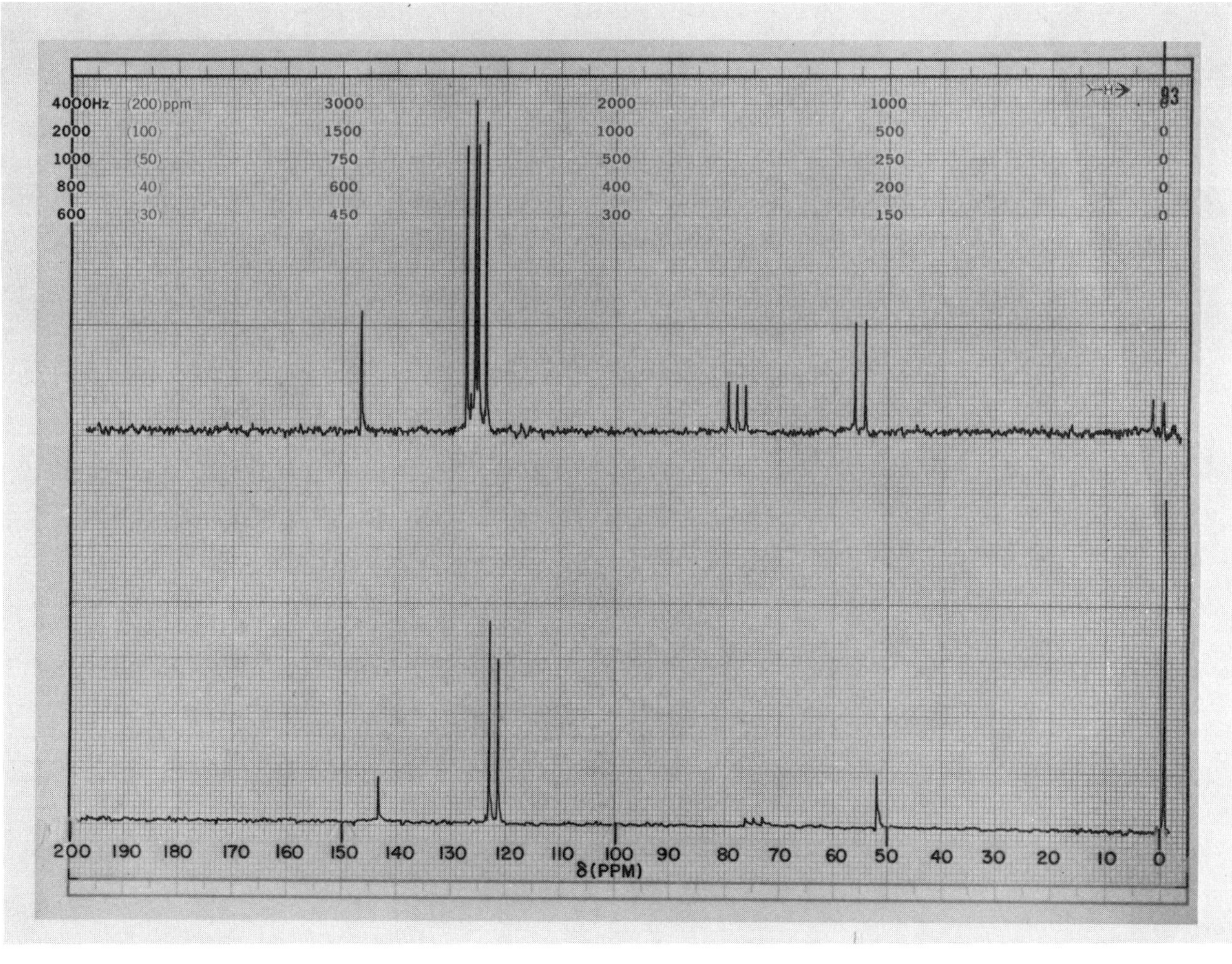

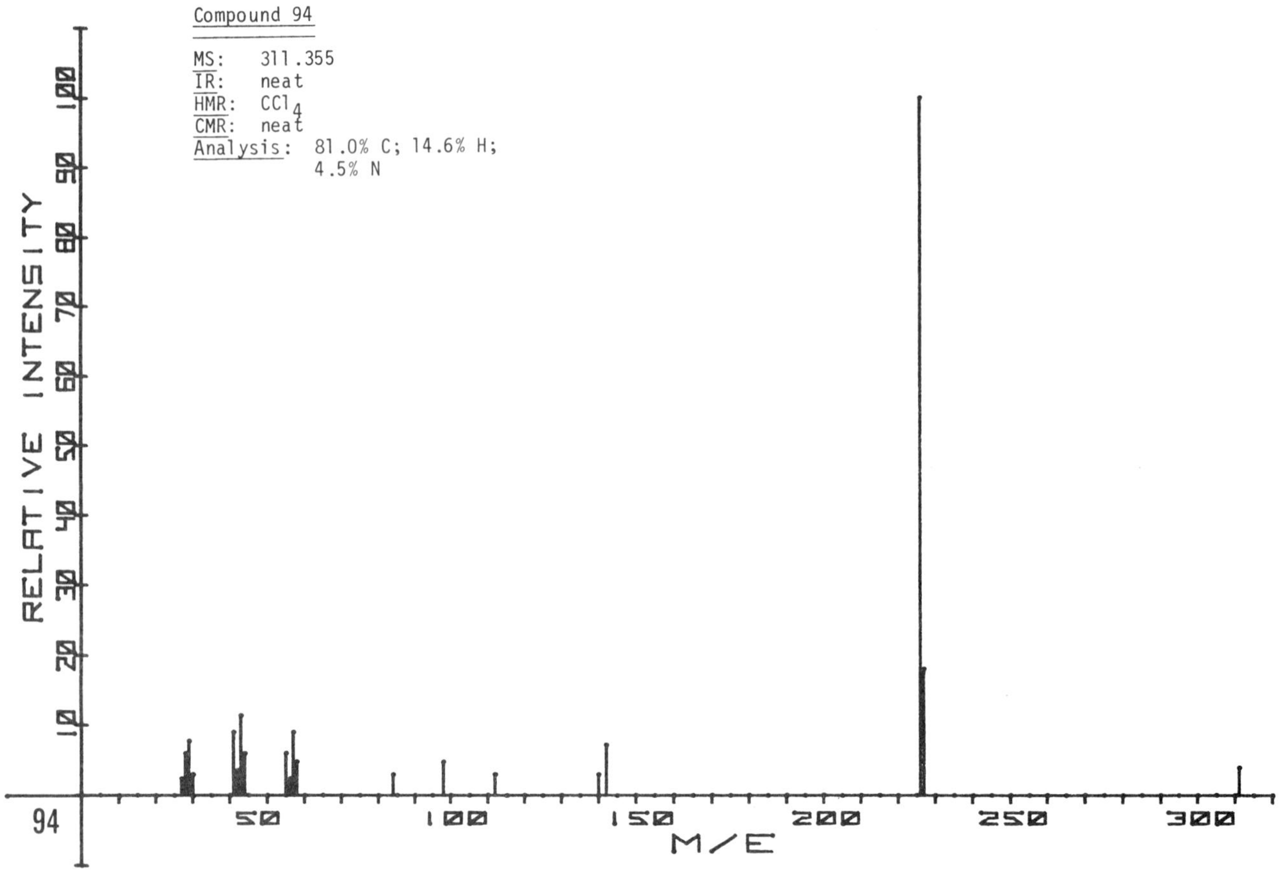

Compound 94
MS: 311.355
IR: neat
HMR: CCl_4
CMR: neat
Analysis: 81.0% C; 14.6% H; 4.5% N
RELATIVE INTENSITY
M/E
94

CM-1
ABSORBANCE
WAVELENGTH (MICRONS)
94

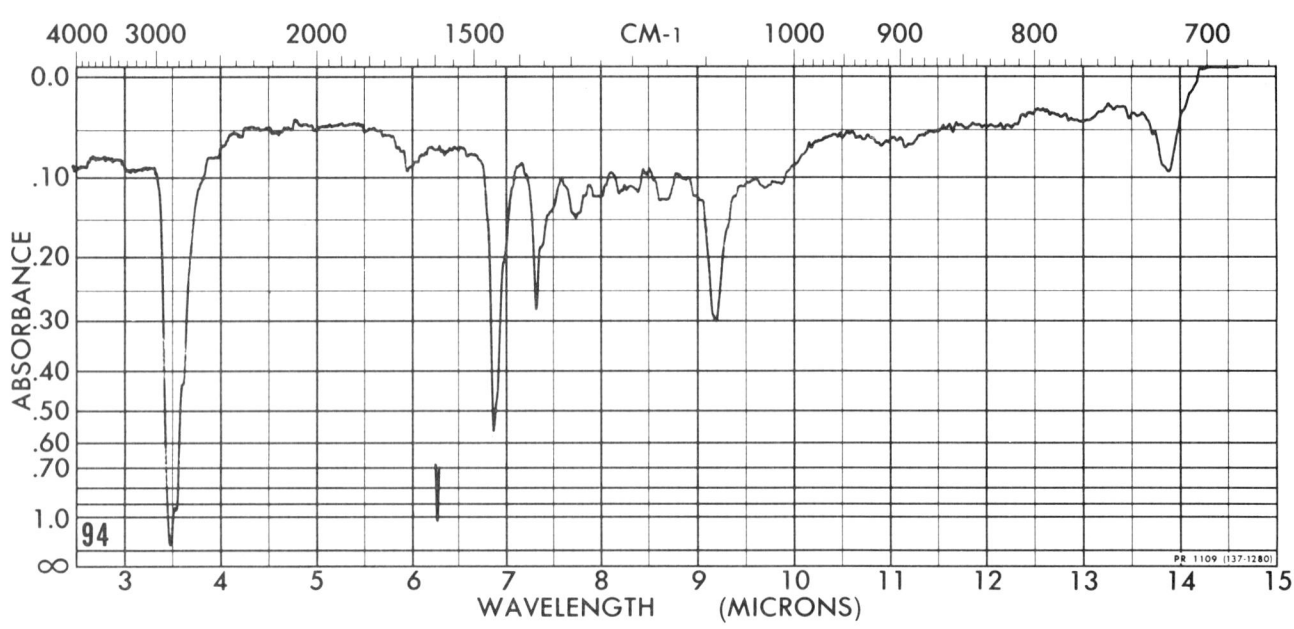

94
10
9
8
7
6
5
4
3
2
1
0

4000Hz (200)ppm
2000 (100)
1000 (50)
800 (40)
600 (30)
PK. HT. PPM
44 29.7
29 28.8
60 28.0
58 27.8
24 27.2
2000
1000
500
400
300
1000
500
250
200
150
94
PK. HT. PPM
46 54.6
59 32.4
57 29.8
48 28.0
60 27.8
55 23.0
48 14.2
200 190 180 170 160 150 140 130 120 110 100 90 80 70 60 50 40 30 20 10 0
δ(PPM)

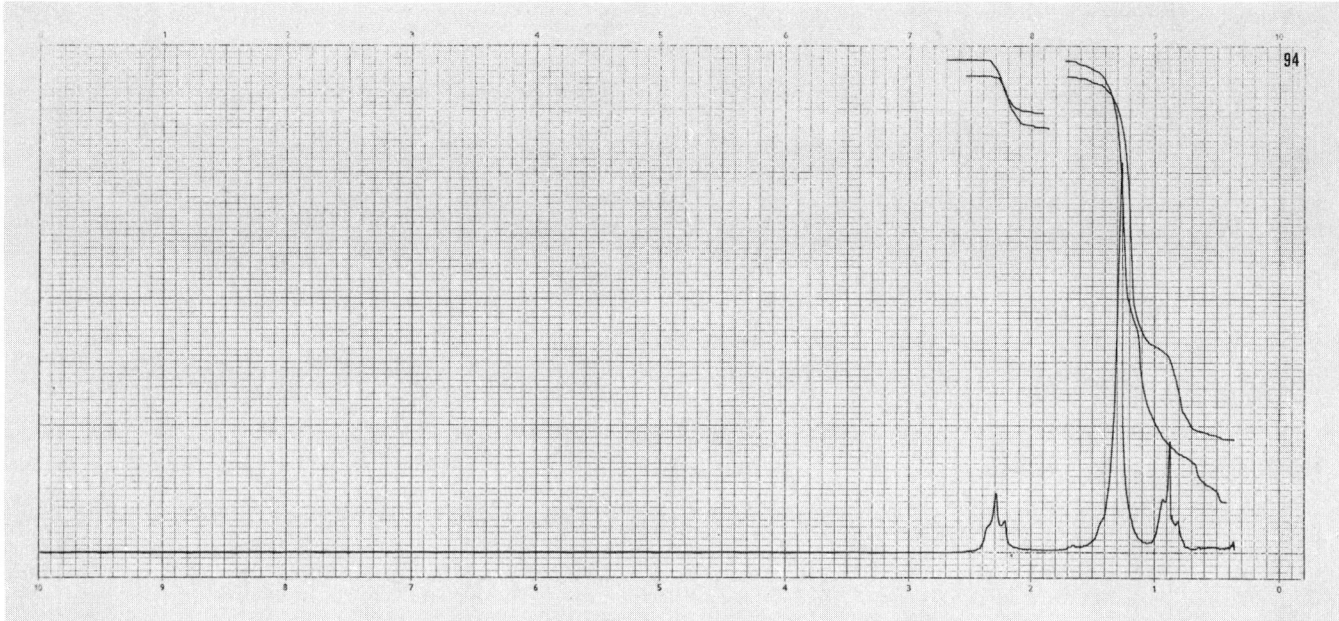

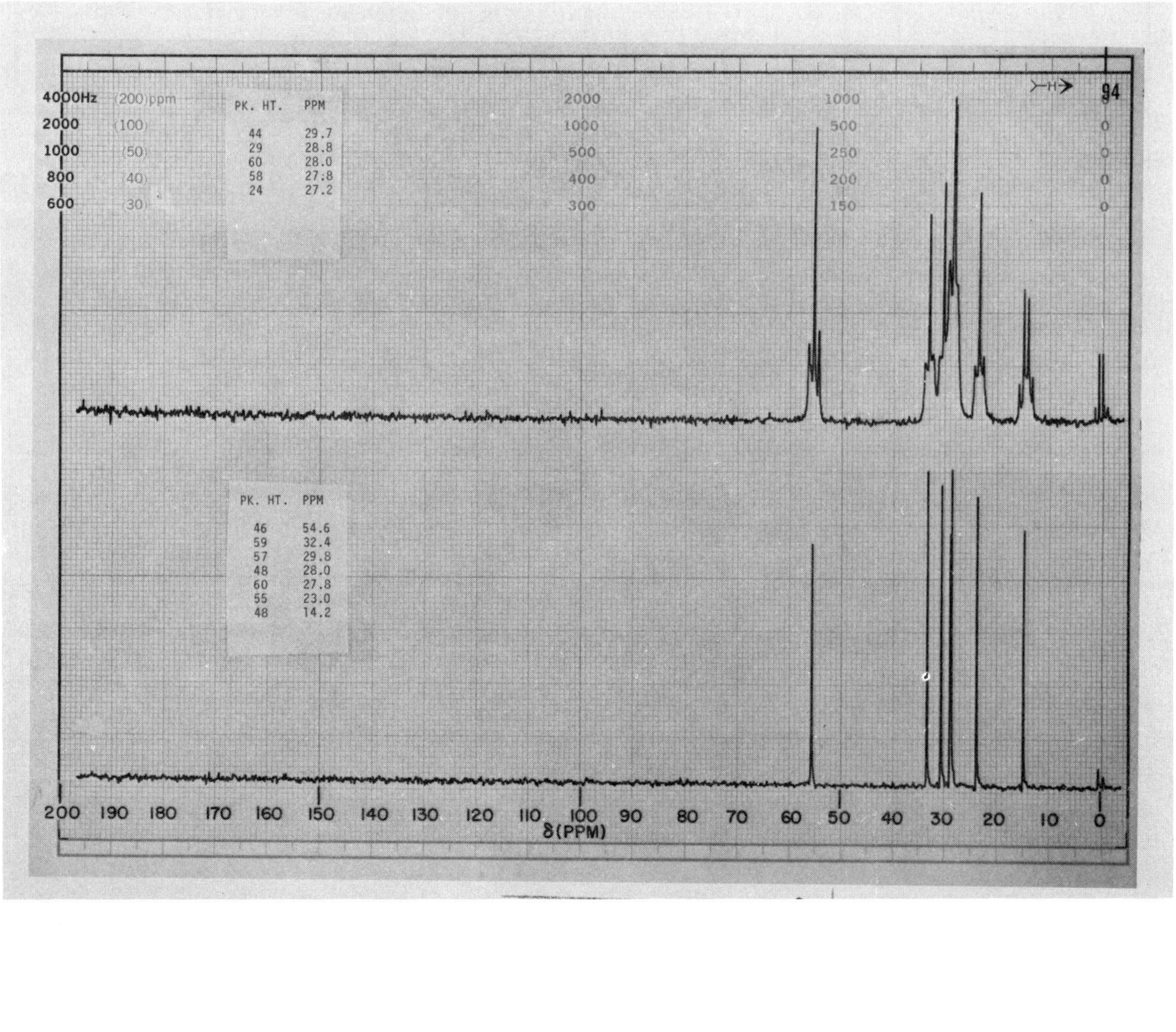

COMPOUND 95

Compound 95

MS: 346.216
IR: CHCl$_3$
HMR: CDCl$_3$
CMR: CDCl$_3$
Analysis: 76.3% C; 7.5% H; 16.2% N

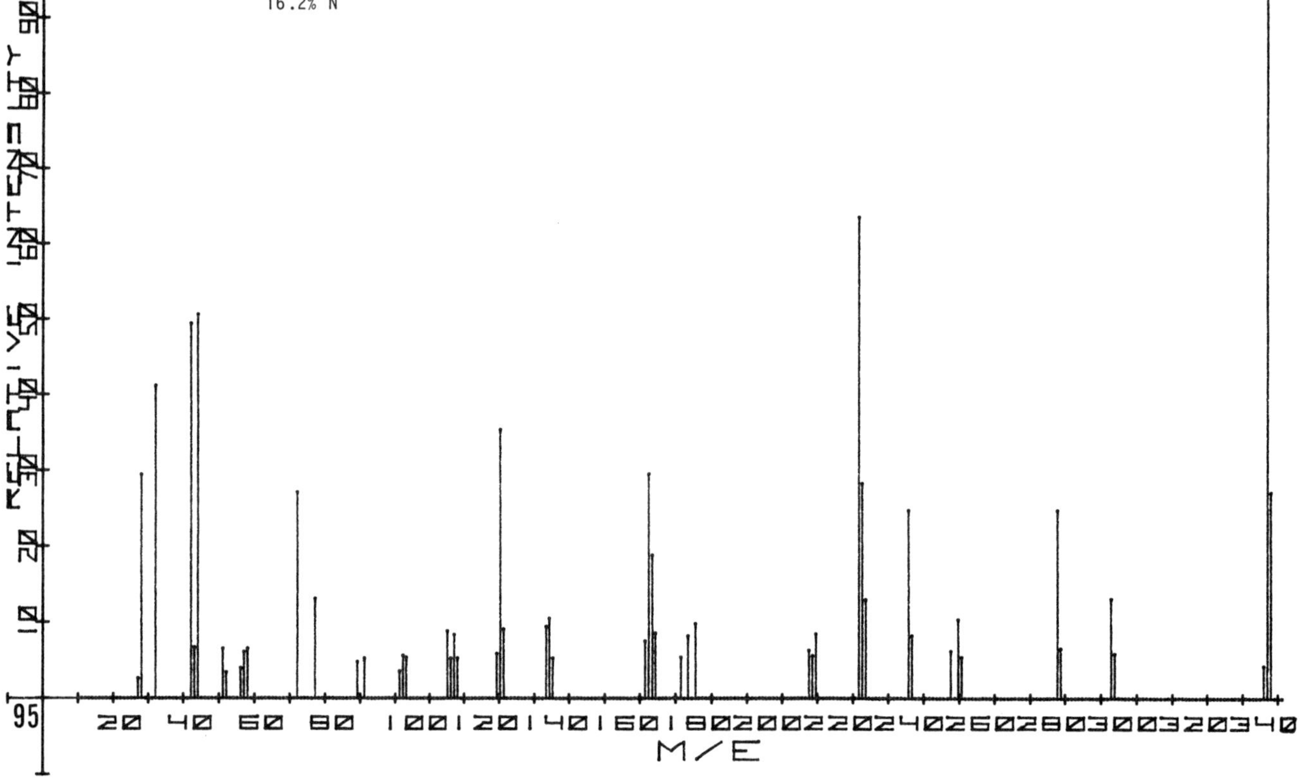

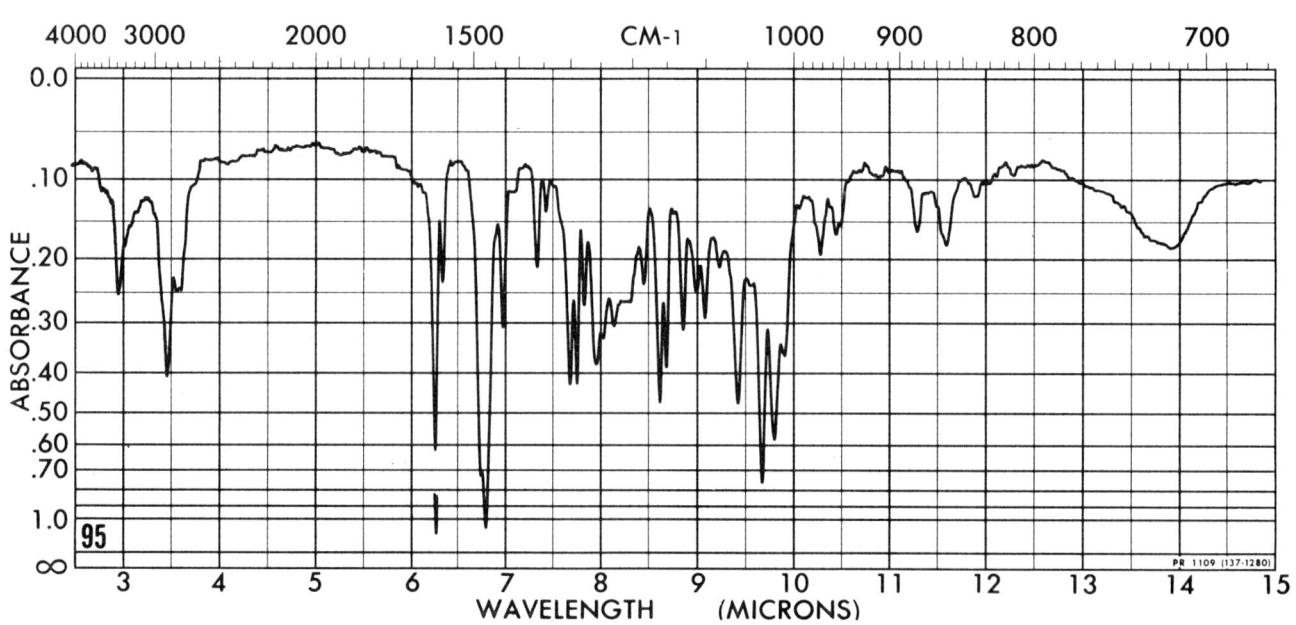

95
D2O

95
4000Hz (200)ppm
2000 (50)
1000 (40)
800 (30)
600
3000
2000
1000
PK. HT. PPM
42 145.4
44 126.5
43 125.1
52 124.4
43 116.4
41 112.0
28 71.1
24 46.5
15 42.6
60 36.0
18 31.8
PK. HT. PPM
9 127.1
12 125.9
40 125.1
13 125.0
14 123.8
200 190 180 170 160 150 140 130 120 110 100 90 80 70 60 50 40 30 20 10 0
δ(PPM)

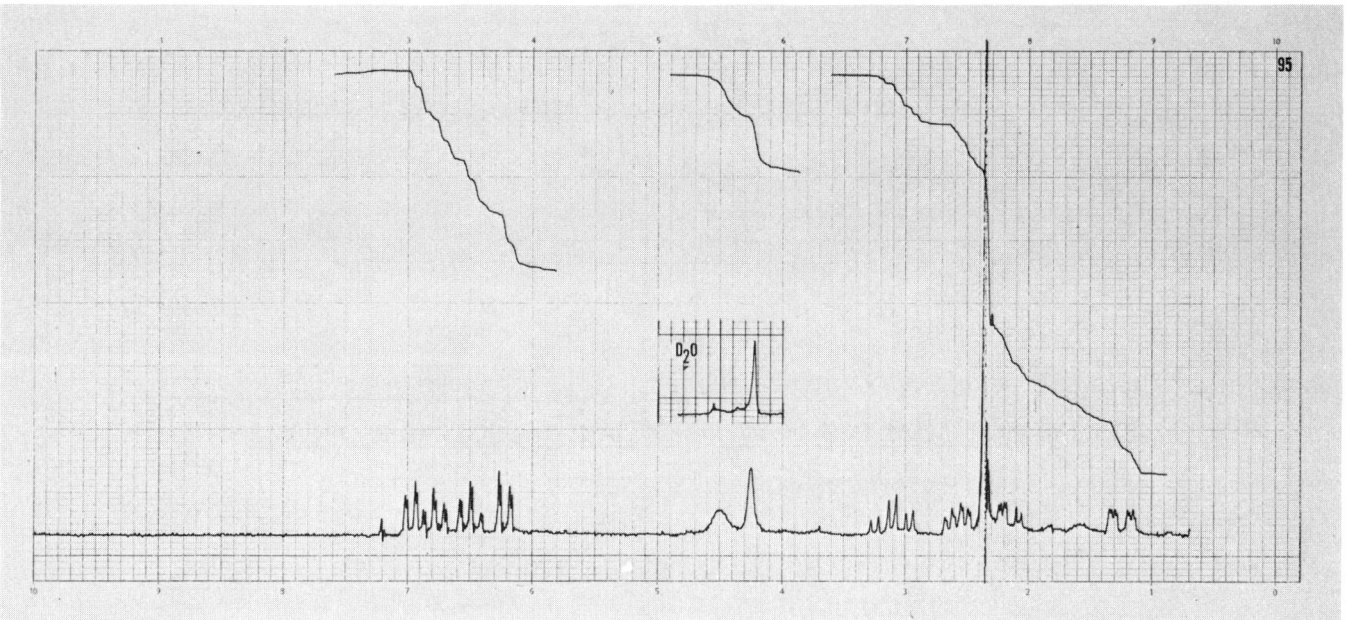

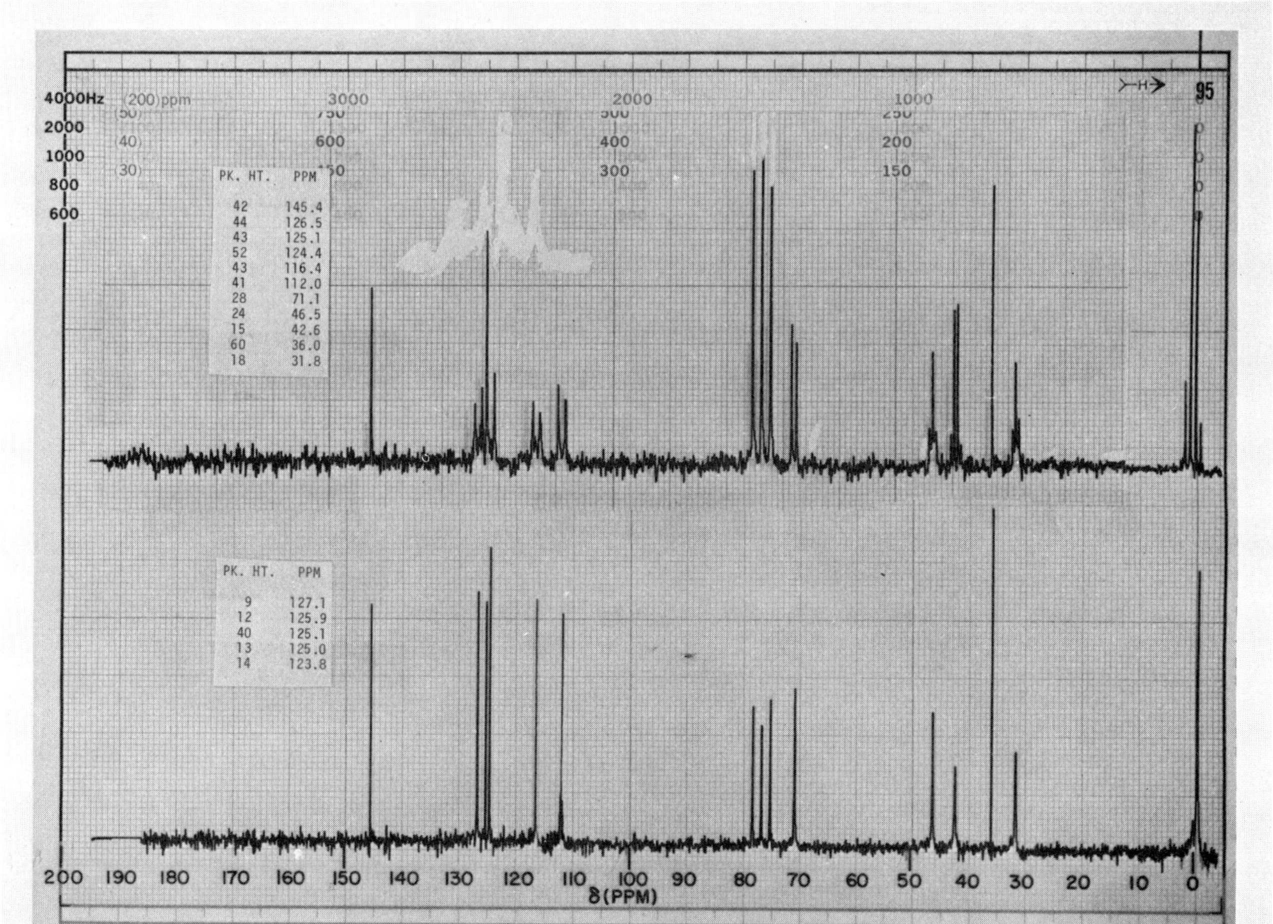

APPENDIX I

GENERAL PROTOCOL FOR SOLVING ORGANIC SPECTRAL PROBLEMS

I. DATA COLLECTION

A. <u>Formula</u>

1. Determination of the empirical formula from the combustion analysis.

$$\%A = \frac{(\text{number of A atoms/compound}) \times (\text{atomic weight of A})}{(\text{molecular weight of the compound})} \times 100\%$$

or $\quad n_A$ = number of A atoms/compound $= \dfrac{(\%A) \times (M.W. \text{ compound})}{(100\%) \times (A.W. \text{ of A})} \quad (1)$

Since the molecular weight is not known, the exact formula of the unknown cannot be unambiguously determined. (Note that the highest m/e peak in the mass spectrum is not necessarily the molecular ion, but may be a fragment ion.)

The combustion data can be used to determine the combustion ratio $(C_w H_x N_y O_z)$ of elements present in the unknown:

$$C_w H_x N_y O_z \cdot (M.W. \text{ Compound}) = \text{Exact Formula}$$

where $\quad w$ = carbon subratio $= \dfrac{(\%C)}{(100\%) \times (12)} \quad (2)$

$\quad x$ = hydrogen subratio $= \dfrac{(\%H)}{(100\%) \times (1)} \quad (3)$

$\quad y$ = nitrogen subratio $= \dfrac{(\%N)}{(100\%) \times (14)} \quad (4)$

It should be noted that all elements (C, H, N, S, P, halogens, metals) <u>except for oxygen</u> are determined by specific analysis. The oxygen content of an unknown is usually deduced by subtraction.

$$z = \text{oxygen subratio} = \frac{(100\% - \%C - \%H - \%N)}{(100\%) \times (16)} \quad (5)$$

The combustion ratio is then divided by the lowest subratio present to provide an empirical formula where the subscripts w, x, y, z, are small whole numbers.

Example: Unknown 18 has a combustion analysis of 62.6%C, 11.4%H, and 12.3%N.

Applying Equations (2)-(5) produces the combustion ratio:

$$C_{0.0522}H_{0.113}N_{0.0087}O_{0.0086} \times (\text{M.W. Compound}) = \text{Exact Formula} \quad (6)$$

Dividing by 0.0086 provides the empirical formula:

$$C_{6.07}H_{13.26}N_{1.01}O_{1.00} = C_6H_{13}NO$$

2. Determination of the exact formula.

   a. From the combustion ratio and the mass spectral "molecular weight:"

   As can be seen in Equation (1), the exact formula of the unknown can be computed from the combustion analysis--<u>provided the molecular weight can be determined</u>.

   i. As a first approximation, <u>assume</u> that the highest m/e ion in the mass spectrum is the molecular ion.

   ii. Calculate the tentative exact formula by multiplying the combustion ratio in Equation (6) by the highest m/e value from the mass spectrum.

   Example: Unknown 18 has m/e 115 as the highest ion. This yields $C_6H_{13}NO$ as the tentative exact formula.

   iii. Verify that the proposed formula is consistent with the indicated molecular weight.

   $$(6 \times 12) + (13 \times 1) + (14 \times 1) + (16 \times 1) = 115$$

   iv. Subject the highest m/e ion to the criteria for molecular ions. Two criteria must be met for the highest m/e ion to qualify as the molecular ion: (1) The ion must be an "odd-electron" ion; and (2) It must be capable of yielding important high-mass fragments through the loss of logical neutral species.* <u>It is important to realize that an</u>

---

*For an in-depth explanation of molecular ion determination see, F. W. McLafferty, "Interpretation of Mass Spectra," 2nd edition, Chapter 3, (W. A. Benjamin, New York, 1973).

ion can meet both of these criteria and still not be the molecular ion.

An odd-electron ion is an ion which contains an uneven number of electrons. It can arise either from loss of a single electron in the ionization process (to generate the molecular ion) or via a two-bond fragmentation process from the molecular ion. A one-bond fragmentation process from a molecular ion yields "even-electron" ions in which the outer shell electrons are paired.*

These concepts can be demonstrated with isopropanol 1. Ionization of 1 by loss of a single electron yields the molecular ion 2 which is an odd-electron ion. Loss of a methyl radical (a logical neutral loss) via one-bond cleavage produces oxonium ion 3, an even-electron ion. Alternatively, a two-bond cleavage process (with the logical loss of the neutral species, H_2O) produces the radical-cation 4, the same odd-electron ion that would be produced by ionization of propene 5.

The unsaturation number equation can be used to determine whether a given formula is an odd-electron or even-electron ion:

$$U.N. = w - 1/2\,x + 1/2\,y + 0z + 1 \qquad (7)$$

---

*For an in-depth explanation of molecular ion determination see, F. W. McLafferty, "Interpretation of Mass Spectra," 2nd edition, Chapter 3, (W. A. Benjamin, New York, 1973).

where   w is the number of C atoms,
x is the number of H atoms,
y is the number of N atoms,
and   z is the number of O atoms for a given formula.

The unsaturation number (U.N.) will be a whole number for odd-electron ions and a fraction for even-electron ions.*

<u>Example</u>: Ion <u>2</u> (C_3H_8O); U.N. = 0 ∴ an OE ion;
whereas Ion <u>3</u> (C_2H_5O); U.N. = 1/2 ∴ an EE ion.

The ion $C_6H_{13}NO$ (unknown 18) has U.N. = 6 - 13/2 + 1/2 + 0 + 1 = 1 is an odd-electron ion and <u>may</u> be the molecular ion. After the structure of this ion is further defined, it can be tested to see if it is capable of yielding other high mass ions in the mass spectrum by logical neutral losses.

  b. Determine the elemental composition of the highest m/e ion from the exact mass data.

The exact mass for any ion in the mass spectrum can be accurately determined (±.003 AMU). It is therefore possible to determine the elemental composition for any peak in the mass spectrum by using tables,** or by simply adding up the individual masses of all the elements in the proposed formula. The peak which yields the most information by this process is clearly the highest m/e ion.

<u>Example</u>: Unknown 18, from Beynon's Table under mass 115:

		Calculated**	Found
<u>6</u>	$C_4H_{11}N_4$	115.0980	115.098
<u>7</u>	$C_6H_{13}NO$	115.0997	
<u>8</u>	$C_5H_{13}N_3$	115.1109	

In this instance it is easy to see that Formula <u>7</u> is the only one which fits the data. Formula <u>6</u> is an even-electron ion (U.N. = 1.5). Formula <u>8</u> is an odd-electron ion (U.N. = 1) but it is outside the commonly accepted error limit (±.003 AMU) and also contains no oxygen, which has been inferred to be present by combustion analysis. We could have alternatively chosen to calculate the exact mass for $C_6H_{13}NO$ and to compare the calculated value with the experimental value of 115.098.

---

*For an in-depth explanation of molecular ion determination see, F. W. McLafferty, "Interpretation of Mass Spectra," 2nd edition, Chapter 3, (W. A. Benjamin, New York, 1973).

**J. H. Beynon, A. E. Williams, "Mass and Abundance Tables for Use in Mass Spectrometry," (Elsevier, New York, 1963).

		Exact Mass* of Element	
6	x	12.00000	= 72.00000
13	x	1.007825	= 13.10172
1	x	14.00307	= 14.00307
1	x	15.99491	= 15.99491
			115.100 ± .001

B. **Unsaturation**

1. Determine the unsaturation number (U.N.) of the unknown by applying Equation (7).

   a. For any $C_wH_xN_yO_z$ compound the unsaturation number is given by the formula:

   $$U.N. = w - 1/2\,x + 1/2\,y + 0z + 1 \qquad (7)$$

   b. Compounds containing additional elements are solved by the same formula where the additional elements are included according to their valance. (*i.e.* w = tetravalent elements, x = monovalent elements, y = trivalent elements, and z = divalent elements.)

   Example: $C_{18}H_{32}BrClFINO_3PS_2Si_2$ would be calculated as follows:

   $$U.N. = (18C + 2Si) - 1/2(32H + 1Cl + 1Br + 1F + 1I) + 1/2(1N + 1P) + 0(3O + 2S) + 1$$

   $$U.N. = 20 - 36/2 + 2/2 + 1 = 4$$

2. Delineate the types of unsaturation present in the unknown by examination of IR, ^{13}C-NMR, and ^{1}H-NMR.

   Unsaturations are either multiple bonds X=Y (1 unsaturation), X≡Y (two unsaturations), or rings (one unsaturation per ring).

   For Unknown 18 ($C_6H_{13}NO$ = one unsaturation)

   a. The IR shows the absence of:

      i. C=O (5.5-6.2 μ).
      ii. C=N (6.0-6.5 μ).
      iii. N=O (6.4-6.5 μ).

---

*R. C. Weast Ed., "Handbook of Chemistry and Physics," 52nd. *Ed.* (The Chemical Rubber Co., Cleveland, Ohio).

b. The ^{13}C-NMR shows the absence of:

   i. C=O (180-250 ppm δ).
   ii. C=C (120-180 ppm δ).
   iii. C=N (120-160 ppm δ).

c. The ^{1}H-NMR lacks olefinic C-H absorption (5-8 δ). Therefore the unsaturation present is most probably a ring.

C. Functional Groups and Part Structure

1. Examine the IR for easily identified functional groups.

   a. H-C≡C, H-O, H-NR$_2$, H$_2$NR (2.5-3.2 μ).
   b. C≡C, C≡N (4.42-4.7 μ).
   c. C=O (5.5-6.4 μ); try to delineate what type of carbonyl group(s) is(are) present.

   Example: Unknown 18. The unknown lacks absorption in the 2.8-6.5 μ range; therefore, the nitrogen and oxygen atoms are both fully substituted with carbon atoms (No N-H or O-H in IR) or they might be bonded together.

2. Examine the ^{13}C-NMR (CMR).

   a. Determine the number of unique carbon types observable in the fully-decoupled CMR (lower scan). Compare that number with the number of carbons known to be present in the formula. If there are fewer observable carbon types than the total number of carbons present, there is most probably at least one element of symmetry in the compound.

   Example: Unknown 18. There are four unique carbon types but six carbons in the formula; therefore, the compound probably has some symmetry.

   b. Examine the off-resonance decoupled CMR spectrum (upper scan) to determine the number of hydrogen atoms on each carbon unit. Tabulate the carbon units present and note their general chemical shift area.

   Example: Unknown 18

   i. 3 CH$_2$ (moderately deshielded). The implication which can be drawn from these chemical shifts is that these three carbons are each attached to a single electronegative (deshielding) element.

   ii. 1 CH$_3$ (upfield). This shift is consistent for a C-CH$_3$ group.

   It is easy to see by simple inspection that the 66.9 δ peak is a triplet (i.e. a CH$_2$ group) and that the 11.8 δ peak is a quartet (i.e. a CH$_3$ group). The overlapping nature of the off-resonance spectrum for the remaining two carbon atoms appears somewhat complex at first glance. The following diagram (Figure I) is determined by examination of the computer tabulated shift positions.

Figure I

It can now be seen that the pattern is a simple set of overlapping triplets (2 CH_2's). It is helpful to remember that in more complicated overlapping patterns singlets and triplets will have approximately the same chemical shift for their center peaks, while doublets and quartets will be symmetrically centered about the original chemical shifts. (See Figure II.)

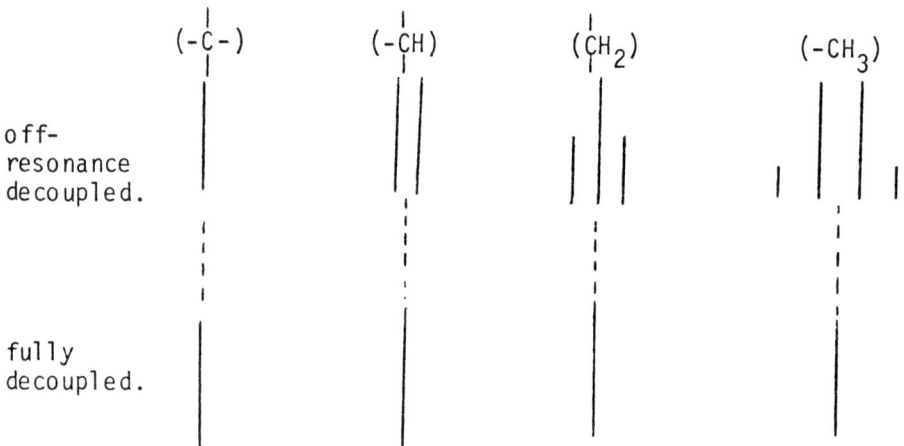

Figure II

3. Examine the ^{1}H-NMR (HMR).

   a. Determine the number of hydrogen atoms at each resonance position by calculating the ratio of the integration heights.

Example: Unknown 18

  i. <u>A</u> 4H, 3.80 δ, m
  ii. <u>B</u> 6H, 2.55 δ, m
  iii. <u>C</u> 3H, 1.18 δ, t (J=7Hz)

 b. Analyze any coupling information present.

Example: Unknown 18

The only easily identifiable coupling pattern is the 3H triplet (J=7Hz) at 1.18 δ. (Therefore the CH<u>$_2$-CH$_3$</u> group is present.)

 c. Analyze any other experiments such as D_2O additions or homonuclear decoupling.

Example: Unknown 18

Irradiation of the 6H (six hydrogen) multiplet at 2.55 δ simplifies both the 4H multiplet at 3.8 δ and 3H triplet at 1.18 δ to singlets. Irradiation of the 3.8 δ multiplet causes the 2.55 δ multiplet to collapse to a quartet (J=7Hz) and an overlapping singlet. This experiment reveals these facts:

  i. The group $X-CH_2-CH_3$ is present. Since the decoupling experiment showed the CH_2 moiety to be coupled <u>only</u> to a CH_3 group, X is not a carbon atom containing additional protons.

  ii. Two $Y-CH_2-CH_2-Z$ groups are present in the molecule (Y and Z having no hydrogens). Since there are four hydrogens at 3.8 δ which are only coupled to four hydrogens at 2.55 δ, there are two sets of CH_2-CH_2 groups. As has been previously shown, the remaining two hydrogens at 2.55 δ are coupled to the methyl group.

II. DEDUCTION OF THE STRUCTURAL FORMULA

 A. <u>Summarize the data which is most supported by all the spectra</u>.

Example: Unknown 18

  1. $C_6H_{13}NO$, one unsaturation (Ring).
  2. CH_3-CH_2-X (CMR + HMR).
  3. 2 $Y-CH_2CH_2-Z$ (CMR + HMR).
  4. 3 CH_2's with electronegative groups and one CH_2-CH_3 group.

 B. <u>Utilize the CMR and HMR evidence to determine the presence of any symmetry elements</u>.

Example: Unknown 18

1. Since the formula shows there are six carbon atoms and the CMR indicates only four unique types of carbon, there must be some symmetry element present. For the purposes of analysis, it is assumed that accidental equivalences of ^{13}C carbon types do not occur. (This assumption is valid in over 95% of the simple organic compounds which the authors have examined.)

2. It has already been shown from the HMR decoupling experiments that two $Y-CH_2-CH_2-Z$ groups are present. Since the other two unique carbon types are the $X-CH_2CH_3$ group, this is in accord with the need for four carbon types.

C. Assemble the structure by piecing together the part structures.

$C_6H_{13}NO$

$- 2\ YCH_2CH_2Z$ (Y and Z must be N and O)

leaves C_2H_5

Since the $-CH_2CH_3$ group must also be attached to a heteroatom and the oxygen's two valances have already been used to generate the two YCH_2CH_2Z groups, it is possible to deduce these part structures.

plane of symmetry

III. CHECK THE ASSIGNMENT TO MAKE SURE IT IS CONSISTENT WITH ALL THE DATA!

A. IR

No OH, NH, olefins, or C=O.

B. CMR

The molecule has four carbon types (3 CH_2 + 1 CH_3) with reasonable chemical shifts and a plane of symmetry which makes the two OCH_2CH_2N groups identical.

C. HMR

1. Isolated ethyl group. Calculation* of the expected chemical shift of the CH_2 group attached to nitrogen may be done as follows:

   base: CH_2 1.20 δ

             α -NR_2 +1.33
                            2.53 δ calculated; 2.55 δ observed.

2. 2 OCH_2-CH_2N groups, calculated for the OCH_2 group:

   base: CH_2 1.20 δ

             α -OR   +2.35

             β -NR_2 + .13
                            3.68 δ calculated; 3.80 δ observed.

3. HMR decoupling experiments are consistent as previously indicated.

D. Mass Spectrum

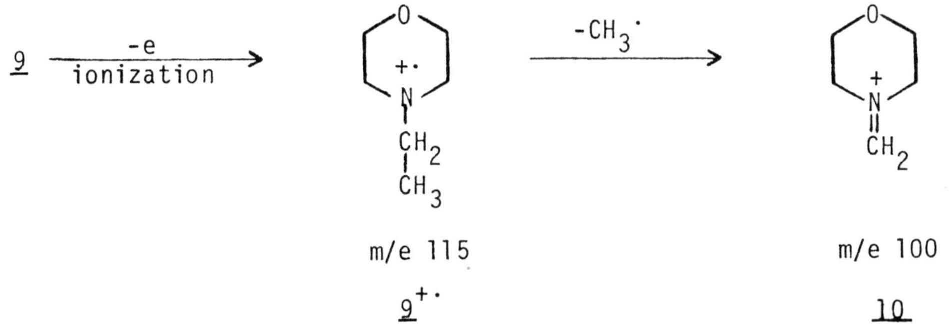

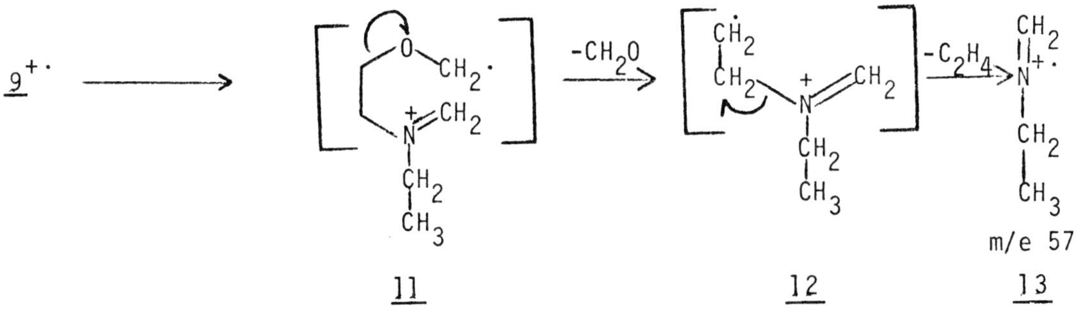

*See Appendix III.

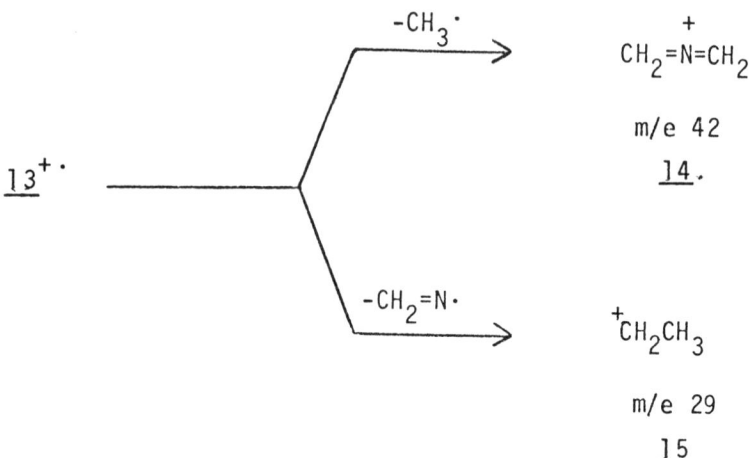

APPENDIX II

SOLVED PROBLEMS

This section contains ten unknowns which have been worked by the general method of Appendix I. It should be noted that in several instances the data does not allow an unambiguous structural assignment to be made. It is important when solving spectral problems, both in course work and in research, to realize the scope <u>and limitation</u> of the methods involved. In those cases where general spectral methods do not allow a clearcut differentiation to be made between several possible structures, it is necessary to undertake additional spectral or chemical experiments to fully establish the structure in question.

UNKNOWN 14B

I. DATA COLLECTION (UNKNOWN 14B)

    A. <u>Formula</u>

        1. Empirical formula: $C_6H_{11}BrO_2$.

        2. Exact formula: Same.

    B. <u>Unsaturation</u>

        1. U.N. = 1.

        2. Type of unsaturation.

            a. The IR shows a C=O at 5.75μ (consistent with ester).

b. The CMR shows 170 δ singlet (consistent with an ester C=O).

c. The HMR shows no olefinic hydrogens but has a low field (~4δ) quartet consistent with an O-ethyl group of an ester.

C. <u>Functional Groups and Part Structure</u>

1. IR : Ester.

2. CMR.

   a. Six unique carbons; therefore no symmetry.

   b. Carbon units.

   i. CO_2R.
   ii. 2 CH_2 (one low field, one high field).
   iii. 1 CH (low field).
   iv. 2 CH_3.

3. HMR.

   a. Hydrogen units.

   i.   <u>A</u> 3H, q (J=7.5Hz), t (J=7.5Hz), 4.20 δ.
   ii. <u>B</u> 2H, m 2.05 δ.
   iii. <u>C</u> 3H, t (J=7.5Hz), 1.30 δ.
   iv. <u>D</u> 3H, t (J=7.5Hz), 1.03 δ.

   b. Decoupled HMR.

   i. Decoupling at <u>C</u> changes <u>A</u> to a triplet and overlapping singlet; irradiation at <u>B</u> changes <u>A</u> to a quartet and overlapping singlet; irradiation of <u>B</u> changes <u>D</u> to a singlet.

   ii. These part structures can be assigned.

$$CH_3-CH_2-O-\overset{O}{\underset{\|}{C}}-$$

   <u>C</u>   quartet portion of <u>A</u>

$$CH_3-CH_2-\underset{X}{CH}-Y$$ (X + Y, bearing no vicinal hydrogens)

   <u>D</u>   <u>B</u>   triplet portion of <u>A</u>

II. DEDUCTION OF THE STRUCTURAL FORMULA

Simple assembledge of the parts from the HMR decoupling experiment gives α bromo ester <u>1</u> as the only possible structure.

$$CH_3-CH_2-\underset{\underset{Br}{|}}{\overset{\overset{H}{|}}{C}}-\overset{\overset{O}{\|}}{C}-OC_2H_5$$

<u>1</u>

III. CHECK THE ASSIGNMENT

A. <u>IR</u> : Ester C=O.

B. <u>CMR</u> : Consistent with the assigned structure.

C. <u>HMR</u> :

1. The structure is consistent with the decoupling.

2. Calculation* of the chemical shift of the α-Bromomethine:

   base   1.55 δ

   α −Br   +2.20

   α −CO_2R +0.95

   ─────────

   4.70 δ calculated <u>vs.</u> 4.20 δ observed

D. <u>MS</u> : (Note the ∼1:1 intensity of ions bearing Bromine).

1. 194,196 (M^+, ^{79}Br and ^{81}Br).

2. 149,151 ($M^+ - \cdot OC_2H_5$).

3. 115 ($M^+ - [^{79}Br$ and $^{81}Br]$).

4. 121,123 ($M^+ - \cdot CO_2Et$).

5. 166,168 ($M^+ - CH_2=CH_2$ by McLafferty rearrangement).

6. 29 ($C_2H_5^+$).

*See Appendix III.

UNKNOWN 30

I. DATA COLLECTION (UNKNOWN 30)

A. Formula

1. Empirical formula: $C_7H_{14}O_3$.

2. Exact formula: Cannot be solved from the M.S. since m/e 103 is an even-electron ion. The three possible elemental compositions within the acceptable error limits for this ion are: CH_7N_6 (103.073); $C_3H_9N_3O$ (103.074); and $C_5H_{11}O_2$ (103.076). Since there is no nitrogen in the unknown, the last choice is the correct one.

B. Unsaturation

1. U.N. = 1.

2. Type of unsaturation.

   a. No C=O is present in the IR.
   b. No olefin or C=O carbons are present in the CMR.
   c. No olefinic hydrogens are present in the HMR.
   d. The unsaturation is therefore a ring.

C. Functional Groups and Part Structure

1. IR: shows no OH or C=O groups.

2. CMR.

   a. There are six unique carbon types; therefore, two carbons are assumed equivalent by symmetry.

   b. Carbon units.

      i. 2 CH (one very low field; probably bears more than one deshielding group).

      ii. 3 CH_2.

      iii. 1 CH_3.

3. HMR.

   a. Hydrogen units.

      i. <u>A</u> 1H, d (J=6Hz), 4.15 δ.
      ii. <u>B</u> 4H, m, 3.55 δ.
      iii. <u>C</u> 1H, "q", 2.89 δ.
      iv. <u>D</u> 2H, d (J=4Hz), 2.57 δ.
      v. <u>E</u> 6H, dt, 1.20 δ.

b. Decoupled HMR.

1. Irradiation at <u>A</u> causes the "q" of <u>C</u> to sharpen to a t with the same coupling as D (~4 Hz); therefore, there is a single hydrogen which is flanked by a CH_2 group and a CH group where W, X, Y, Z are insulators.

$$W-CH_2-\underset{\underset{\underline{D}}{\uparrow}}{C}H-\underset{\underset{\underline{C}}{\uparrow}}{\overset{\overset{X}{|}}{C}}H-\underset{\underset{\underline{A}}{\uparrow}}{\overset{\overset{H}{|}}{C}}\overset{Y}{\underset{Z}{<}}$$

2. Further irradiation caused no other changes. The complex pattern <u>B</u> and <u>E</u> is therefore insulated from the rest of the spectrum and is only coupled <u>B</u> to <u>E</u>.

II. DEDUCTION OF THE STRUCTURAL FORMULA

A. <u>Summarize the Data</u>

1. Formula: $C_7H_{14}O_3$--one unsaturation (ring).

2. W-CH$_2$-CH-CHYZ; W, X, Y, Z are insulators bearing no vicinal hydrogens.
   |
   X

3. 4H multiplet coupled to a 6H multiplet.

B. <u>Determine the Symmetry</u>. Two equivalent CH_3 groups by CMR.

C. <u>Assemble the Structure</u>

1. W-CH$_2$-CH-CH(Y,Z) + C_4H_{10} + O + O + O.
   |
   X

2. Since there are only three insulating units (the three oxygen atoms) and W, X, Y, & Z must be used in this way, one of the three oxygens must serve as a bridge to provide the ring unsaturation.

3. The remaining C_4H_{10} must be two non-equivalent CH_2CH_3 groups attached to the additional oxygen atoms.

4. There are five reasonable structures possible.

a. The bridging oxygen is W=X.

$$CH_2\underset{O}{\overset{\diagdown \diagup}{-\!\!\!-\!\!\!-}}CHCH(OC_2H_5)_2$$

<u>1</u>

b. The bridging oxygen is W=Y or W=Z.

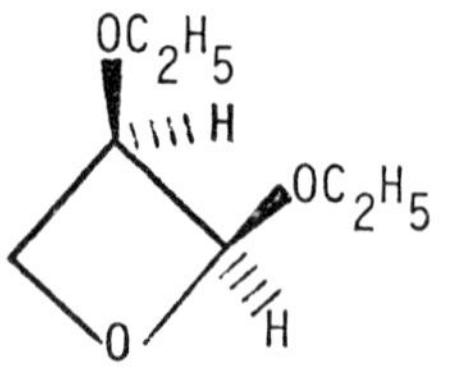

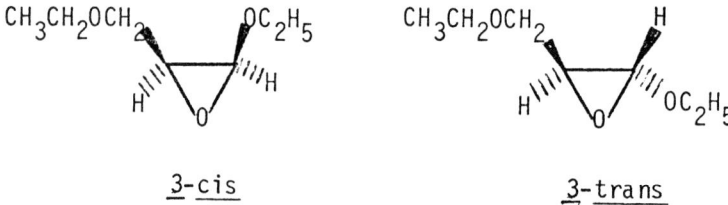

2-trans 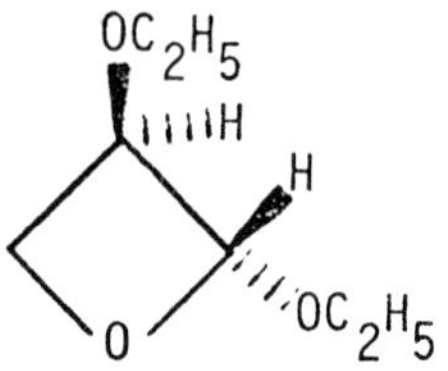   2-cis

c. The bridging oxygen is X=Y or X=Z.

3-cis   3-trans

d. Bridging Y and Z generates aldehyde 4. This structure has an aldehyde carbonyl and is not consistent with the IR, CMR or HMR data.

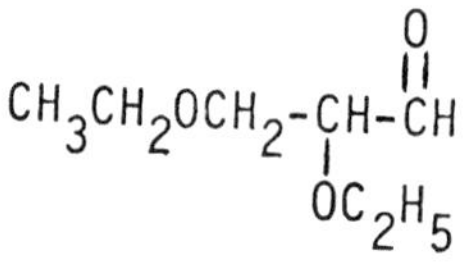

$$CH_3CH_2OCH_2-\underset{OC_2H_5}{\underset{|}{CH}}-\overset{O}{\overset{\|}{CH}}$$

4

5. All five possible structures have one feature which does not fit the CMR data - they all have <u>seven</u> unique carbon atoms. Structure 1 might appear at first glance to have only five unique carbon atoms; however, examination of Newmann projection 1' shows that there is no rotational operation which can make the two ethyl groups equivalent.

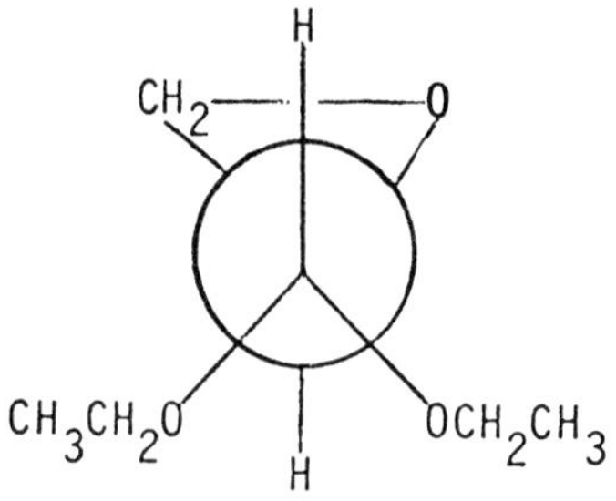

1'

6. Since the decoupling experiment is clear and these five structures are the only ones possible which accommodate all the rest of the data, this must be a case of accidental equivalence of the remote methyl carbons of O-ethyl groups. On this basis it is tempting to opt for structure <u>1</u> because the methyl groups of this structure are in the most similar environment. However, since <u>all</u> the structures bear two O-ethyl groups, it is not inconceivable that any of the other possibilities might still be correct.

7. The mass spectrum is most helpful in assigning the correct structure. The base peak at m/e 103 was found to be $C_5H_{11}O_2$. The only structure which can easily afford this unit is <u>1</u>. Note that the m/e 47 ion (C_2H_7O) could be rationalized from all structures and is not useful in deciding between them.

III. CHECK THE ASSIGNMENT

A. <u>IR</u>: Large 8.7–10μ CHOR and CH_2OR absorptions.

B. <u>CMR</u>: The low field carbon is the one bearing the two oxygen atoms. The accidental equivalence is acceptable in terms of the <u>almost</u>-symmetrical structure of <u>1</u>.

C. <u>HMR</u>: The CH_2 protons of the O-ethyl group are not equivalent and split each other. There are two such situations which overlap to generate the complicated pattern observed.

D. <u>MS</u>: See above.

<u>UNKNOWNS 41A,B</u>

I. DATA COLLECTION (UNKNOWNS 41 A,B)

A. <u>Formula</u>

1. Empirical formula: Both 41A,B have the same combustion analysis. Solving the combustion ratios shows the empirical formula to be $C_8H_{14}O_2$.

2. Exact formula.

   a. The combustion ratio and the mass spectral "molecular weight" yield the formula $C_8H_{14}O_2$ (odd-electron ion and = m/e <u>148</u>).

   b. Determine the elemental composition of the highest m/e ion in the mass spectrum.

Formula	Calculated Exact Mass $\pm$.003 of 142.098*
<u>1</u> $C_4H_{10}N_6$	142.0966
<u>2</u> $C_6H_{12}N_3O$	142.0980
<u>3</u> $C_8H_{14}O_2$	142.0998

   Formulas <u>1</u>,<u>2</u> can be eliminated since there is no nitrogen in the unknown (none found in the analysis); note also that <u>2</u> is an even-electron ion.

B. <u>Unsaturation</u>

   1. U.N. = 2.

   2. Type of unsaturation.

      a. The IR shows a C=O for each of the unknowns.

      b. The CMR shows three sp^2-type carbons for each unknown [~170 δ (C=O) 120 δ & 135 δ (C=C)].

      c. The HMR shows a 2H multiplet at ~5.5 δ indicative of two hydrogens on a double bond.

      d. Each unknown, therefore, has one C=C and one C=O to account for its two unsaturations.

C. <u>Functional Groups and Part Structure</u>

   1. IR.

      a. No OH.
      b. C=O at 5.77 μ (ester?)

   2. CMR.

      a. Both 41A and 41B have eight unique carbon types.

      b. Carbon units. Note that this accounts for all carbons and hydrogens present in the molecule. (It also accounts for both oxygens if the carbonyl is an ester.)

         i. 1 C=O (no H).
         ii. 2 CH (sp^2=olefin).
         iii. 2 CH_3.
         iv. 3 CH_2.

*J. H. Beynon, A. E. Williams, "Mass and Abundance Tables for Use in Mass Spectrometry," (Elsevier, New York, 1963).

3. HMR.

    a. Hydrogen units.

        i. A 2H, m, 5.5 δ.
        ii. B 2H, q (J=7Hz), 4.2 δ.
        iii. C 2H, d (J=6Hz), 3.1 δ.
        iv. D 2H, m, 2.15 δ.
        v. E 3H, t (J=7Hz), 1.25 δ.
        vi. F 3H, t (J=7Hz), 0.98 δ.

    b. Decoupled HMR.

        i. There is an isolated X-CH$_2$-CH$_3$ group where X is an electronegative element (B + E).

        ii. There is a CH$_2$CH$_3$ group (D + F) which is further coupled to the olefinic group at 5.5 δ.

        iii. There is a CH$_2$ group (C) which is only coupled to the olefinic protons at 5.5 δ.

II. DEDUCTION OF THE STRUCTURAL FORMULA

    A. Summarize the Data (Both unknowns)

        1. C$_8$H$_{18}$O$_2$ (two unsaturations: CH=CH, C=O).

        2. The carbonyl unit must be an ethyl ester because:

            a. Chemical shift of C=O from CMR.

            b. Chemical shift $\overset{\overset{O}{\|}}{C}$-OCH$_2$-CH$_3$ from HMR.

            c. IR 5.77 μ.

        3. There is a CH$_3$CH$_2$CH= unit.

        4. There is a =HC-CH$_2$-X unit where X has no further hydrogens.

    B. Determine the Symmetry. None.

    C. Assemble the Structure. There is only one possible solution.

CH$_3$CH$_2$O-$\overset{\overset{O}{\|}}{C}$-CH$_2$-CH=CHCH$_2CH_3$ for both 41A and 41B. The question is which is trans and which is cis? The IR of 41A has a 10.3μ C=C stretch typical of trans double bonds. Unfortunately, isomer 41A does not show the corresponding 14μ IR stretch often seen for cis double bonds. A further piece of evidence can be obtained from the CMR. In molecules where there are two carbons on one side of a double bond (cisoid), these carbons are mutually shielded (moved upfield)

relative to the same carbons of the <u>transoid</u> isomer.* Therefore, since compound
<u>41A</u> has its allylic methylene groups at higher field, it must be the <u>cis</u> isomer.

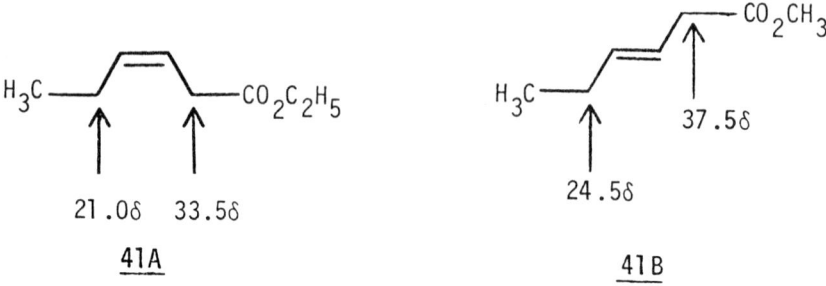

Note that this effect is particularly useful in assigning the stereochemistry of
trisubstituted olefins.

III. CHECK THE ASSIGNMENT

A. <u>IR</u>: Ester 41B <u>trans</u> olefin.

B. <u>CMR</u>: Consistent.

C. <u>HMR</u>

1. The decoupling is consistent with the assigned structure. The multiplet at
2.15 δ (D) arises as a result of similar coupling constants to both the olefin
and the methyl group (Jad ≅ Jfd).

$$CH_{3f}-CH_{2d}-CH_a=CH_a-CH_2\overset{O}{\overset{\|}{C}}OCH_2CH_3$$

<u>F</u>   <u>D</u>   <u>A</u>   <u>C</u>   <u>B</u> <u>E</u>

*N. K. Wilson, J. B. Stothers, Topics in Stereochemistry, <u>8</u>, 1 (1974).

2. Calculated* chemical shift positions.

<u>D</u>		<u>C</u>		<u>B</u>	
base	1.20	base	1.20	base	1.20
α-C=C	+.75	α-C=C	+.75	α-C̈-O (C=O)	+2.98
		α-CO₂R	+1.05		

1.95 δ calculated   3.00 δ calculated   4.18 δ calculated
(2.05 δ observed)   (3.10 δ observed)   (4.15 δ observed)

D. <u>MS</u>:

$$CH_3-CH_2-CH=CHCH_2 \overset{O}{\underset{\|}{C}} -O-CH_2-CH_3$$

            69 | 29

UNKNOWN 53

I. DATA COLLECTION (UNKNOWN 53)

   A. <u>Formula</u>

      1. Empirical formula: $C_9H_{14}O_2$.
      2. Exact formula: Same.

   B. <u>Unsaturation</u>

      1. U.N. = 3.

      2. Types of unsaturation.

         a. CMR: ketone carbon 207 PPM, no olefin carbons; possible C≡C-H at 61 and 64 PPM? (normal C≡C range 65-90 PPM**).

---

*See Appendix III.

**"Carbon-13 Nuclear Magnetic Resonance for Organic Chemists," G. C. Levy, G. L. Nelson, p. 71, (Wiley-Interscience, New York, 1972).

b. IR: 5.80μ - saturated ketone, no H-C≡C-, ∴ not an acetylene at 61, 64 δ in the CMR.

c. HMR: no olefinic H.

d. Since there are nine unique carbons in the CMR and there is only a single ketone, the remaining two units of unsaturation must both be rings.

C. Functional Groups and Part Structure

1. IR: ketone carbonyl, No OH; therefore the remaining oxygen atom must be an ether type.

2. CMR.

    a. Nine unique carbon types.

    b. Carbon units.

        i. 1 C=O.

        ii. 3 CH$_3$.

        iii. 2 CH$_2$.

        iv. 1 CH (low field).

        v. 2 -C- (low field, 1 high field).

3. HMR.

    a. Hydrogen units.

        i. A 1H, br s, 3.05 δ.
        ii. B 1H, br d (J=15Hz), 2.65 δ.
        iii. C 1H, br d (J=15Hz); 2.10 δ.
        iv. D 1H, br dd (J=15, 2Hz), 1.80 δ.
        v. E 1H, br dd (J=15, 2Hz), 1.65 δ.
        vi. F 3H, s, 1.39 δ.
        vii. G 3H, s, 1.00 δ.
        viii. H 3H, s, 0.92 δ.

    b. Decoupled HMR: None.

II. DEDUCTION OF THE STRUCTURAL FORMULA

A. <u>Summarize the Data</u>

1. Formula $C_9H_{14}O_2$ (three unsaturation, one C=O, two rings).

2. Part structures.

   a. Three CH_3 groups not further coupled to any other hydrogens (two high field – probably on carbon not close to deshielding groups; one lower field [1.39 δ] probably close to a deshielding group).

   b. One non-strained ketone (acyclic, or six-ring ketone).

   c. One CH unit with only small (long-range?) coupling.

   d. Two geminal CH_2 AB units (the 15 Hz coupling constant is too large to be a vicinal coupling constant) with further small coupling.

B. <u>Determine the Symmetry</u>. None.

C. <u>Assemble the Structure</u>

1. $C_9H_{14}O_2$ (three unsaturations):

$$\underset{X}{\overset{O}{\underset{\|}{C}}}\diagdown Y \quad + \quad \begin{bmatrix} CH_3, CH_3, CH_3 & -C-, -C-, O \\ CH_2, CH_2, CH & + \text{ 2 unsaturations} \end{bmatrix}$$

   a. X or Y cannot be oxygen (IR and CMR show no ester).

   b. X or Y cannot be hydrogen (no C=O doublet in the off-resonance-decoupled CMR, IR shows no aldehyde).

   c. X or Y cannot be CH_3 (the calculated* HMR chemical shift of a CH_3 group α to ketone is 2.15 δ).

2. The subunits can be simplified.

   a. Since none of the CH_3 groups is attached to the "insulating" C=O group, they <u>must</u>, therefore, be attached to the "insulating" quaternary carbons. Since attaching all three CH_3 groups would generate a tert-butyl group where all three CH_3 groups would most likely be equivalent, two CH_3 groups are the most which can be attached to a single atom. Note that a $(CH_3)_2C\diagdown$ group will have non-equivalent methyls, provided there is no plane of symmetry bisecting the quaternary carbon.

   b. No OCH_3 group is possible since it would have a calculated* chemical shift of 3.35 δ.

---

* See Appendix III.

278

c. Therefore:

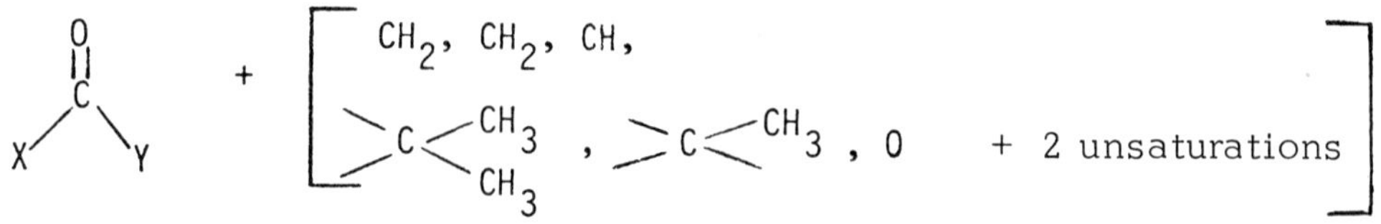

$$\underset{X}{\overset{O}{\underset{\|}{C}}}_{Y} + \left[CH_2, CH_2, CH, \underset{}{>}C\underset{CH_3}{\overset{CH_3}{<}}, \underset{}{>}C\underset{}{\overset{CH_3}{<}}, O + 2 \text{ unsaturations} \right]$$

3. The CH_2, CH_2, and CH groups cannot be adjacent to each other; otherwise larger coupling constants would have resulted. Since they are isolated, there must be an insulating subunit surrounding each of them at each of their valance points. Therefore, seven bonds to insulating subunits must be made. The C=O, $(CH_3)_2C<$, $CH_3\overset{|}{\underset{|}{C}}-$ and O units must serve as these insulators.

4. The oxygen unit cannot be connected to either CH_2 unit since this would generate *two* low field protons having an AB pattern.

   a. $-CH_2-O-R$ (calculation*).

   $\uparrow$

   1.20   base
   +2.30   α-OR

   3.50 δ   calculated *vs.* lowest observed AB pattern at 2.65 δ and 1.80

   b. This means that the oxygen must be connected to two of the three remaining groups. This generates three possibilities:

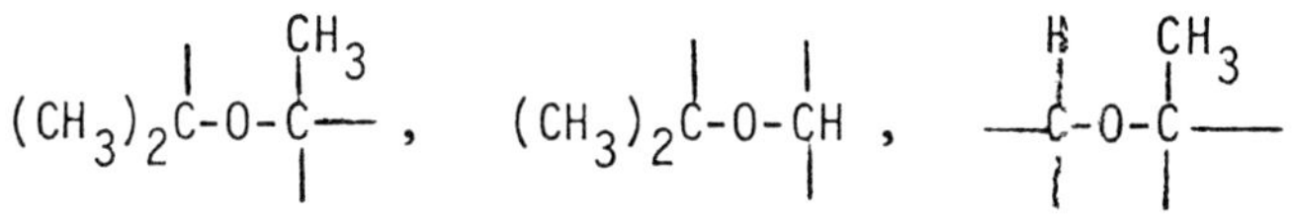

   $$(CH_3)_2\overset{|}{\underset{|}{C}}-O-\overset{CH_3}{\underset{|}{\overset{|}{C}}}- \quad , \quad (CH_3)_2\overset{|}{\underset{|}{C}}-O-\overset{|}{\underset{|}{C}}H \quad , \quad -\overset{H}{\underset{|}{\overset{|}{C}}}-O-\overset{CH_3}{\underset{|}{\overset{|}{C}}}-$$

   $\underline{1}$ $\qquad\qquad\quad$ $\underline{2}$ $\qquad\qquad\quad$ $\underline{3}$

5. Building further from the larger subunits $\underline{1}$, $\underline{2}$, and $\underline{3}$:

   a. 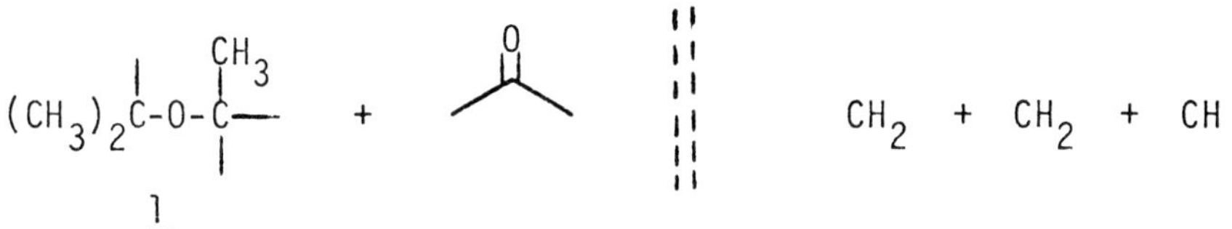

   $$(CH_3)_2\overset{|}{\underset{|}{C}}-O-\overset{CH_3}{\underset{|}{\overset{|}{C}}}- \quad + \quad \underset{}{\overset{O}{\underset{\|}{\bigwedge}}} \qquad\qquad CH_2 + CH_2 + CH$$

   $\underline{1}$

   insulation valances (I.V.'s)   insulation valances (I.V.'s)
      available 3 + 2   required 2 + 2 + 3

   Since structures with this unit would require the CH_2 or CH groups to be joined together, $\underline{1}$ can be eliminated.

---

*See Appendix III.

b.

$$-\underset{|}{\overset{|}{C}}-CH_3 \quad + \quad \overset{O}{\underset{}{\bigwedge}} \quad \| \quad (CH_3)_2\underset{|}{\overset{|}{C}}-O-\underset{|}{\overset{|}{C}}H \quad \| \quad CH_2 + CH_2$$

$$\underline{2}$$

5 I.V.'s available   1 I.V. available,   4 I.V.'s needed
                            2 needed

$\underline{2}$ is still a possible subunit. (6 I.V. available, 6 I.V. needed).

c.

4 I.V. available   2 I.V. available;   4 I.V. needed
                          2 I.V. needed

$\underline{3}$ is still a possible subunit.

6. Further elaboration with subunit $\underline{2}$:

   a. The $\overset{|}{\underset{|}{CH}}$ moiety of $\underline{2}$ requires two additional insulator units (one each $H_3C-\overset{|}{\underset{|}{C}}-$ and $C=O$).

   b. Therefore the $(CH_3)_2COR$ moiety of $\underline{2}$ must be bonded to a CH_2 unit or to itself.

   c. Bonding the $(CH_3)_2COR$ moiety to a CH_2 unit gives new subunit $\underline{4}$:

   $\underline{4}$

   3 I.V. available +   2 I.V. required
   1 I.V. required

   i. Connecting the remaining two pieces (a-c, b-d, c'-d') generates structure $\underline{5}$. (The alternate structure with the <u>trans</u>-fused bicyclo [3.2.0] system is far too strained to seriously consider.)

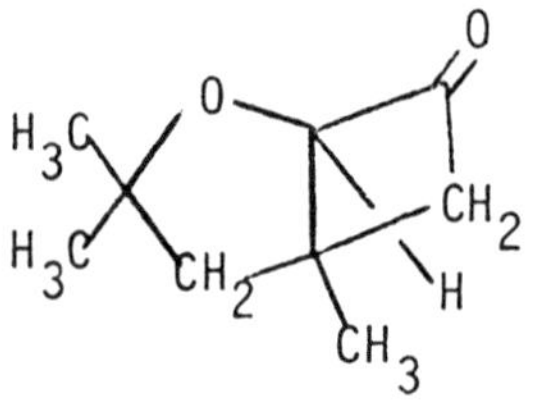

280

[Structure 5]

ii. Connecting a-b, c-d, c'-d' produces the structure 6.

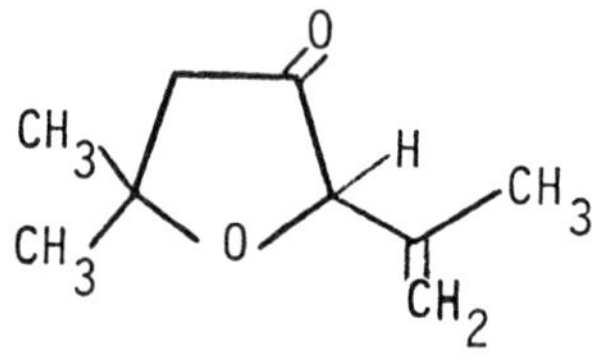

[Structure 6]

d. Bonding the (CH$_3$)$_2$CHOR moiety to itself generates new subunit 7:

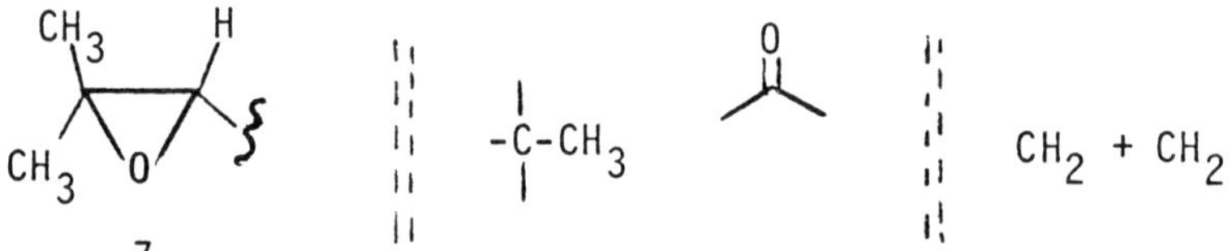

[Structure 7]   -C-CH$_3$ [C=O on isopropyl]   CH$_2$ + CH$_2$

1 I.V. needed   5 I.V.'s available   4 I.V.'s needed

i. Connecting 7 to the -C-CH$_3$ group eventually produces structure 8 since the remaining groups can only be insulated in this manner.

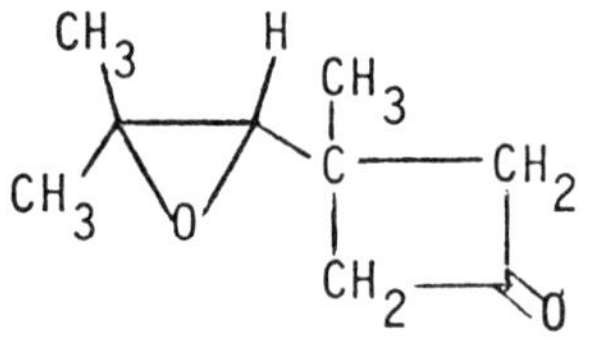

[Structure 8]

ii. Connecting 7 to the >C=O group eventually produces structure 9 as the only possibility.

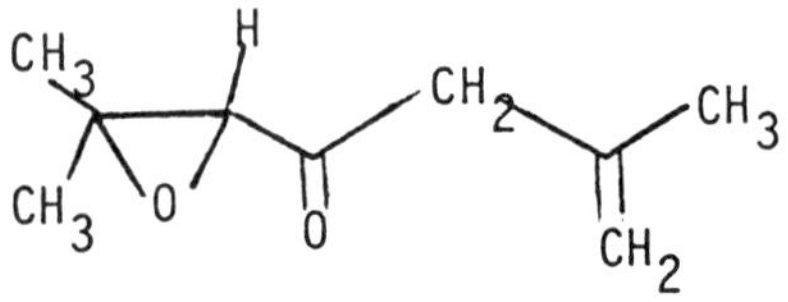

[Structure 9]

7. Further elaboration with subunit 3:

   a. The CH moiety of 3 requires the attachment of two additional insulator units. Three are available: the $(CH_3)_2C$, C=O, and the other end of 3.

   b. The $\begin{smallmatrix}CH_3\\RO\end{smallmatrix}\!\!>\!\!C\!\!<$ moiety of 3 must be attached to a CH_2 group.

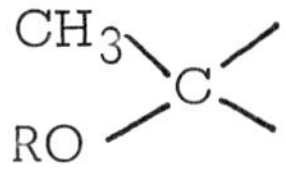

   c. This gives three new subunits (10, 11, 12); which in turn, generate four new possibilities (13, 14, 15, 16).

      i.

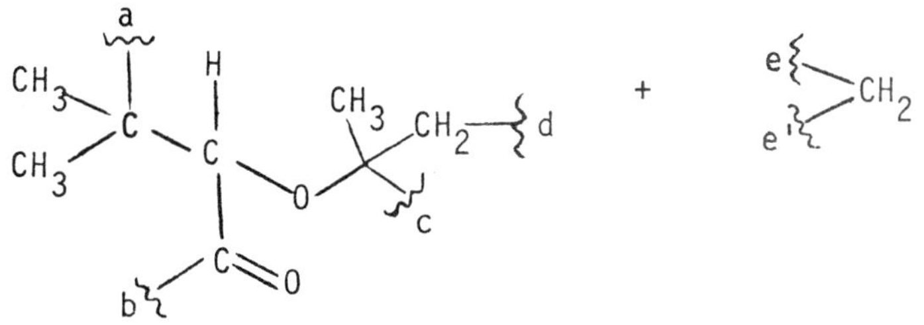

      10

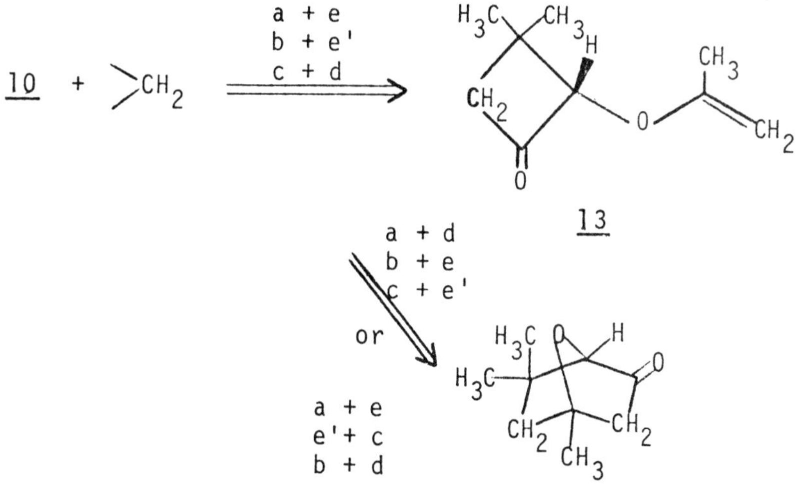

      10 + >CH₂ $\xrightarrow[\substack{a+e\\b+e'\\c+d}]{}$ 13

      $\searrow$ or $\substack{a+d\\b+e\\c+e'}$ 14

      $\substack{a+e\\e'+c\\b+d}$

      ii.

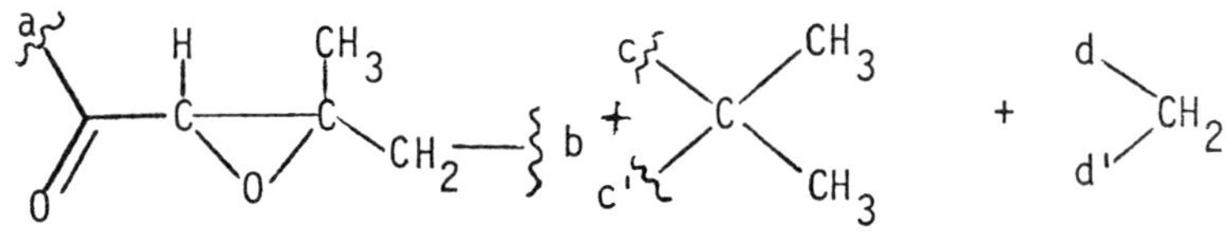

      11

$$\underline{11} \xrightarrow[c' + b]{\substack{a + d \\ d' + c}} \underline{15}$$

(structure 15 shown: bicyclic ketone-epoxide)

The epoxide moiety must have the indicated <u>cis</u> stereochemistry because the alternative <u>trans</u>-fused structure is far too strained to be realistically possible. All other combinations will not insulate the two CH$_2$ groups.

iii. (structure 12 shown with fragments a, b, c, c', d, d')

$$\underline{12} \xrightarrow[c' + b]{\substack{a + d \\ d' + c}} \underline{16}$$

(structure 16 shown: bicyclic ketone-epoxide)

As above, all other combinations will not insulate the CH$_2$ groups.

8. Evaluation of the structures <u>5</u>, <u>6</u>, <u>8</u>, <u>9</u>, <u>13</u>, <u>14</u>, <u>15</u>, and <u>16</u>.

   a. Structures <u>6</u>, <u>9</u>, and <u>13</u> can be eliminated since they have olefins and are not bicyclic (absolutely inconsistent with the observed chemical shifts in the CMR and HMR).

   b. Structures <u>5</u> and <u>8</u> have a strained cyclobutanone carbonyl and would be expected to have IR absorption about 5.5μ. (Structure <u>8</u> also has a plane of symmetry through the cyclobutanone ring and would have one fewer CH$_2$ group.)

   c. Structure <u>14</u> can be eliminated on two grounds: (1) The IR frequency of the carbonyl should be lower than 5.73μ (5-ring ketone) because of the additional strain of the fused bicyclic system; and (2) The methine α to both the ketone and the bridging ether is calculated* to have a shift of

*See Appendix III.

4.60 δ vs. the observed shift of 3.05 δ.

d. This leaves the two epoxides 15 and 16 as the reasonable candidates.

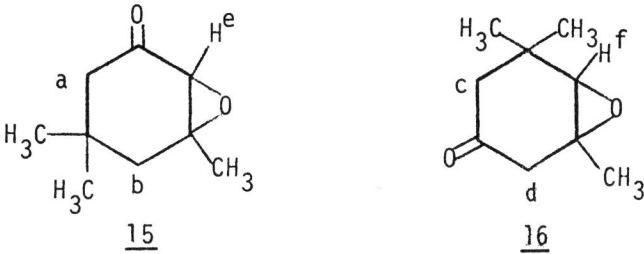

i. The calculated* shifts for the methylene groups of 15 and 16 are 2.30 (a), 1.35 (b), 2.30 (c), 2.45 (d) δ. This favors structure 15 because the greater spread in chemical shift is more in accord with the observed spectra. Note that each of the methylene groups would be expected to exist as an AB pattern due to the asymmetry created by the epoxide group.

ii. Since epoxides (like cyclopropanes) show a tendency to shield methine and methylene protons, care must be used in calculating the expected chemical shift of the epoxide methine. Utilization of the standard shift increment for an α-ether group will yield anomalously low field values. Examination of a number of simple epoxyalkanes and cycloalkanes suggests that the shift increment of ∼1.45 - 1.55 δ is appropriate for methine or methylene groups of epoxides. This generates an expected shift of 4.0 δ for He and 3.05 for Hf. This calculation would conversely tend to favor structure 16.

iii. The choice between 15 and 16 cannot confidently be made on the basis of the available data. In cases like this it is often wise to consult the literature for "model" compounds. The only evidence which does not support structure 15 is the estimated chemical shift of 4.0 δ for the methine hydrogen, He (as compared with the 3.05 δ observed shift). This concern vanishes upon examination of several closely related α-epoxyketones 17, 18, 19 which all have methine absorption in the 3.15-3.20 δ range!

*See Appendix III.

This example underscores the danger in placing too much faith in using the additivity constants to calculate chemical shifts in cyclic systems containing anisotropic groups such as the carbonyl moiety.

III. CHECK THE ASSIGNMENT

A. IR : Six-membered ring ketone.

B. CMR : Consistent.

C. HMR : Consistent.

D. MS

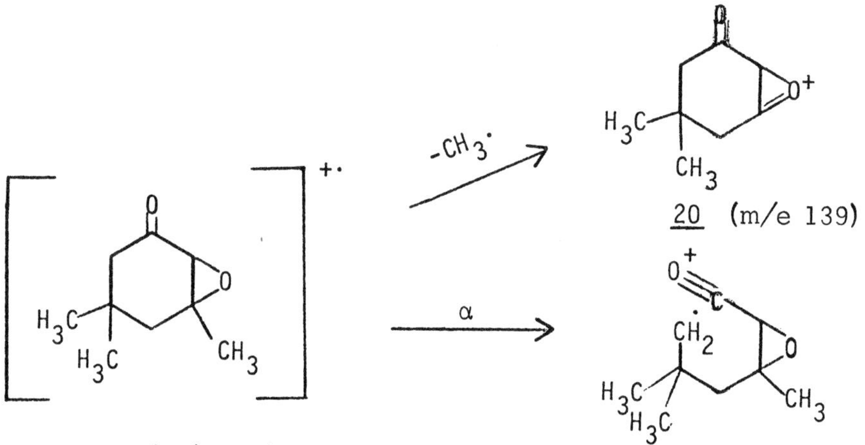

15 (m/e 154) 20 (m/e 139) 21 (m/e 154)

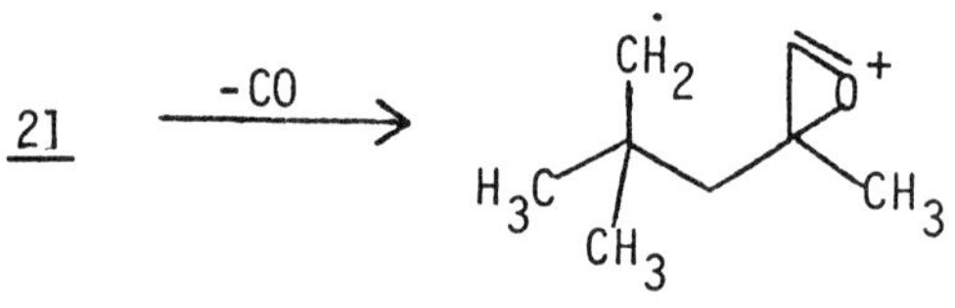

22 (m/e 126)

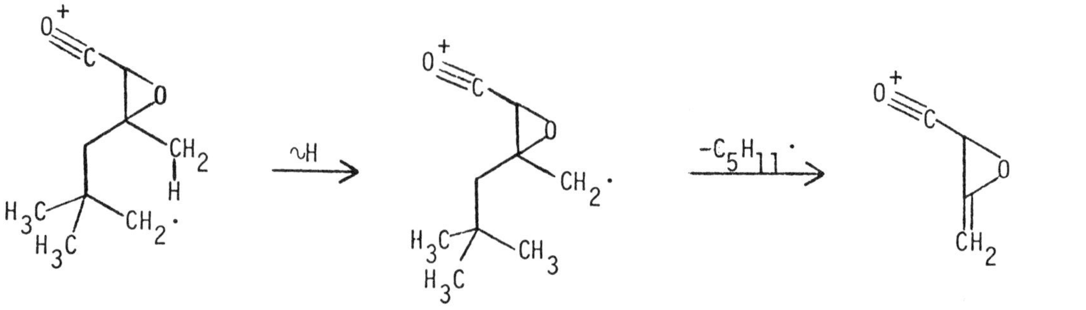

21 23 (m/e 154) 24 (m/e 83)

UNKNOWN 65

I. DATA COLLECTION (UNKNOWN 65)

A. Formula

1. Empirical formula: $C_{10}H_{14}N_2$.
2. Exact formula: Same.

B. Unsaturation

1. U.N. = 5.

2. Types of unsaturation.

   a. The IR shows aromatic absorptions but no C≡N.

   b. The CMR shows five aromatic carbons.

   c. The HMR shows four aromatic protons.

   d. Since there are five aromatic carbons there are at least three double bonds in this structure.

   e. Therefore, the two extra unsaturations must be rings.

C. Functional Groups and Part Structure

1. IR : aromatic absorption, no C≡N, no C=C=C.

2. CMR.

   a. Ten unique carbons.

   b. Carbon units.

   i. 4 CH (aromatic or sp^2).
   ii. 1 C (aromatic or sp^2).
   iii. 1 CH (aliphatic).
   iv. 3 CH_2 (aliphatic).
   v. 1 CH_3 (aliphatic).

3. HMR.

   a. Hydrogen units.

   i. A 2H, m, 8.5 δ.
   ii. B 1H, dt (10, 3Hz), 7.75 δ.
   iii. C 1H, dd (10, 10Hz), 7.30 δ.
   iv. D 2H, m, 3.26 δ.
   v. E 2H, m, 2.3 δ.
   vi F 3H, s, 2.2 δ.
   vii. G 3H, m, 1.9 δ.

b. Decoupled HMR: The B and C groupings indicate a partial aromatic structure.

$$J_{ac} = J_{bc} = 10 Hz$$
$$J_{ab} = 3 Hz.$$

II. DEDUCTION OF THE STRUCTURAL FORMULA

A. <u>Summarize the Data.</u> $C_{10}H_{14}N_2$ - five unsaturations; three double bonds, two rings.

B. <u>Determine the Symmetry.</u> None.

C. <u>Assemble the Structure</u>

1. Since there is no allene (C=C=C) in the IR, one of the double bonds must be of the C=N type.

2. This suggests the aromatic structure is a pyridine.

— 3H's
— $C_5H_{10}N$

3. Pyridine itself has the following chemical shifts:

H(7.60)
H(7.00)
H(8.60)

Therefore, the unknown is probably a 3-substituted pyridine, since <u>two</u> low field hydrogens are observed. (A 4-substituted pyridine would make two pairs of ring carbons equivalent by symmetry.)

$H_4 = \underline{B}$, $H_5 = \underline{C}$, and H_2 plus $H_6 = \underline{A}$. Note that the multiplet $\underline{A}$ is consistent with an overlapping dd (10, 3Hz) for H_6 and a d (3Hz) for H_2.

4. The $C_5H_{10}N$ group contains the following subunits.

   CH, 3CH$_2$, CH$_3$, N, one Ring

5. The point of attachment of the pyridine ring to the $C_5H_{10}N$ unit must by necessity either be via the CH moiety or the nitrogen atom. (A trivalent or tetravalent connector is necessary to generate a ring and an outside connection.)

6. The HMR chemical shift of the CH$_3$ group requires that it is on a deshielding atom and not a carbon. Therefore, the structure has been simplified to the following:

7. There are only two ways to put these subunits together.

8. As with the previous problem(s), the anisotropic nature of the pyridine ring makes calculation of the expected HMR chemical shifts for $\underline{1}$ and $\underline{2}$ very risky.

9. A better way to distinguish between these two possibilities is on the basis of the pyrrolidine ring carbon CMR shifts. Structure $\underline{1}$ should have a low field CH (pyridine and nitrogen deshielding) and a low field CH$_2$ (nitrogen deshielding), while structure $\underline{2}$ should have two low field CH$_2$ (nitrogen deshielding) and also a moderately deshielded CH (pyridine deshielding). The observed spectrum is clearly in better accord with structure $\underline{1}$.

III. CHECK THE ASSIGNMENT

   A. IR: Consistent.

   B. CMR: Consistent.

   C. HMR: Consistent.

   D. MS:

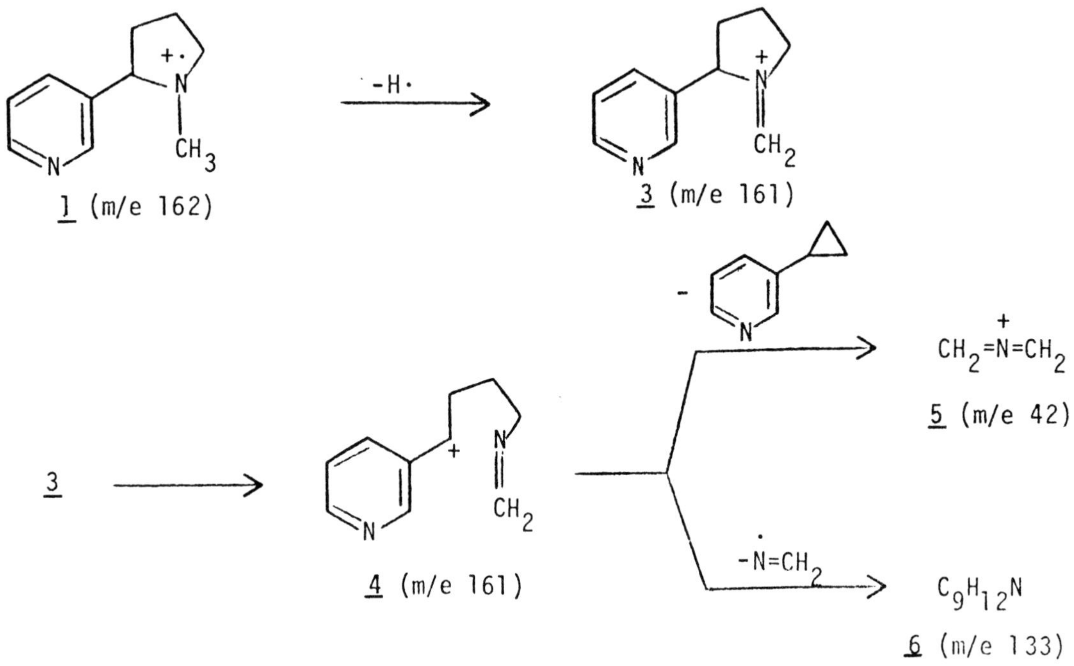

[Scheme: 1 (m/e 162) → −H· → 3 (m/e 161); 3 → 4 (m/e 161) → either 3-cyclopropylpyridine + CH₂=N⁺=CH₂ 5 (m/e 42), or −Ṅ=CH₂ → C₉H₁₂N 6 (m/e 133)]

UNKNOWN 73

I. DATA COLLECTION (UNKNOWN 73)

   A. Formula

      1. Empirical formula: $C_{11}H_{14}O_2$.
      2. Exact formula: Same.

   B. Unsaturation

      1. U.N. = 5.

2. Types of unsaturation.

   a. The IR shows two types of C=O 5.82μ (ketone) and 6.0μ (enone); note also the C=C (6.19μ) of the enone.

   b. The CMR shows four low field sp^2 carbons (δ 126, 166, 198, 211).

   c. The HMR shows a single olefinic hydrogen.

   d. Therefore, the unsaturations are: two C=O, one olefin, and two rings.

C. Functional Groups and Part Structure

   1. IR: See 2a above.

   2. CMR.

      a. Eleven unique carbon types.

      b. Carbon units.

         i. 2 C=O.

         ii. 5 CH$_2$.

         iii. 1 -C-.

         iv. 1 HC=.

         v. 1 >C=.

         vi. 1 CH$_3$.

   3. HMR.

      a. Hydrogen units.

         i. <u>A</u> 1H, br. s., 5.86 δ.
         ii. <u>B</u> 10 H, complex pattern, 1.3-2.9 δ.
         iii. <u>C</u> 3H, s., 1.45 δ.

      b. Decoupled HMR: None.

II. DEDUCTION OF THE STRUCTURAL FORMULA

A. Summarize the Data

   1. Formula: C$_{11}$H$_{14}$O$_2$ - five unsaturations (two carbonyl, one olefin, two rings).

   2. The following units must be present:

B. **Determine the Symmetry.** None.

C. **Assemble the Structure**

1. The point of attachment of the two rings must be at a union of higher connectivity value than two valences. Since the carbonyl unit and the five methylene units have only two unused valences, they cannot be the point at which two rings are joined. Units 2 and 3 have three valences available for the formation of rings.

2. There are three ways the ring-forming units can be put together.

   a. 4 can form two types of ring systems.

      i. Building between a + b and c + c'.

ii. Building between a + c and b + c'.

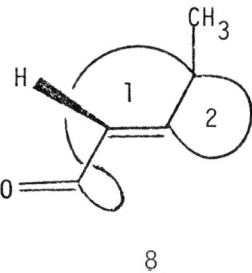

8

This possibility looks far too strained. There are only a total of six connecting carbons for both rings - the trans double bond makes this type of ring fusion very unlikely.

b. 5 can also form two types of ring systems.

i. Building between d + d' and e + f.

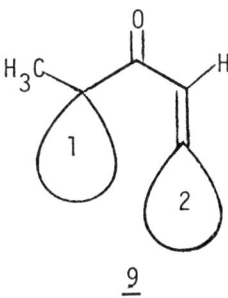

9

ii. Building between d + e and d' + f. This is even more strained than 8 (violates Bredt's Rule).

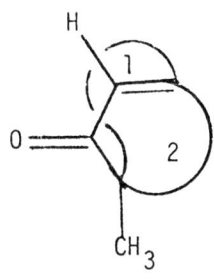

10

c. 6 will also yield a pair of ring systems.

i. Building between g + h and h' and i.

11

ii. Building between g + i and h + h'. Again, there are not enough carbons to prevent such a system from being too strained.

12

3. The choices *7*, *9*, and *11* can now be further narrowed down.

   a. For *7*: Since the additions to one of the new rings must be composed entirely of CH_2 units, the ring #2 must contain the carbonyl group (otherwise there would be a plane of symmetry through the ring #2 unit having only CH_2 groups). Therefore, since the isolated ketone has IR absorption of 5.82μ, it is probably a cyclohexanone (cyclopentanone = 5.72μ; cycloheptanone = 5.90μ). This requires that four CH_2 and one C=O units be added to ring #2 and this in turn, means there is only one CH_2 left for ring #1. The structures which are thus generated are *13* and *14* (The ketone could not be at the remote position of the cyclohexanone ring because that structure would have reduced symmetry.)

   13   *14*

Both of these structures would have a very strained enone IR absorption (definitely not expected at 6.0μ) and can be eliminated from further consideration.

b. For 9: The same considerations as above generate 15, 16, as the possible candidates in the 9 series. (Smaller ring sizes in the #1-ketone-containing ring are again not possible because of the IR carbonyl stretch frequency.)

15　　　　16

Both of these isomers are obviously inconsistent with the CMR data--too many olefinic carbons!

c. For 11: The choices from 11 are somewhat more numerous since the carbonyl group is not restricted by symmetry to be in only one ring. Since the ring which has the saturated ketone must be a six-membered ring, it is necessary to generate a bicyclo [4.4.0] system. Six of them are possible.

i.

17 ⇌ 17'

This α-dione would probably prefer to exist in the enolic form 17'; even if it did not, the IR would not be "normal" for the two carbonyls. This isomer can be eliminated.

ii.

18 ⇌ 18'

As in the previous example, the compound would probably exist in the enol form; even if it did not, the isolated CH$_2$ group would be expected to exhibit an AB pattern in the proton NMR. This isomer

can be rejected because the isolated CH$_2$ unit would be calculated*
to absorb at ~3.50 δ (2 α-C=O + 1 β olefin).

iii.

19

This is an enedione and would not have the IR, NMR, or CMR properties observed for the unknown under consideration; therefore, this isomer is eliminated.

iv. All three of these isomers seem reasonable at first glance.

20   21   22

4. The following values are the calculated* shift positions for the above three isomers:

20	21	22
CH$_2$(k) 3.05 δ	CH$_2$(m) 2.40 δ	CH$_3$(o) 1.25 δ
CH$_3$(l) 0.95 δ	CH$_3$(n) 0.95 δ	

Since the increment used for the β-olefin is that which would be afforded by a simple isolated olefin, the calculated shift of the methyl group for all three isomers is undoubtedly not representative. The more polarized enone olefin would be expected to be substantially more deshielding at the β carbon than would be a simple olefin (This effect is expressly seen in the CMR by the shifts of the specific olefinic carbons involved.) On this basis structure 22 (with the most deshielded methyl group) would seem to be the best choice. In order to firmly exclude structures 20 and 21 (22 is the correct answer), it

*See Appendix III.

would be necessary to resort to spectral comparison with rather similar models; alternatively, simple chemical reduction of the non-conjugated ketone would be expected to produce a substantial (~0.3 ppm) upfield shift of the angular methyl group for compound 22.

III. CHECK THE ASSIGNMENT

  A. **IR**: Enone carbonyl and olefin.

  B. **CMR**: The 166 δ singlet is the deshielded β carbon of the enone while the 126 δ doublet is the α-carbon.

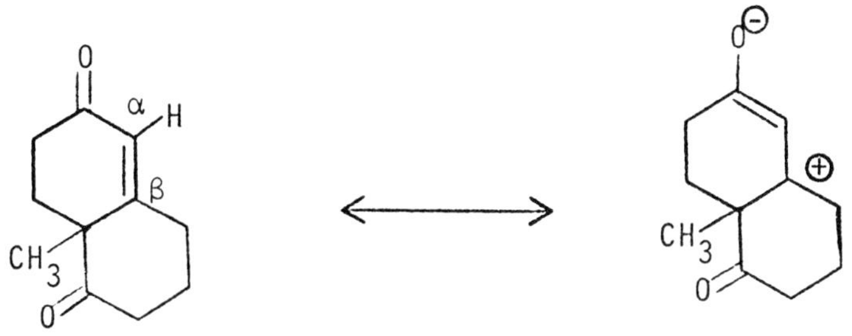

  C. **HMR**: Consistent.

  D. **MS**

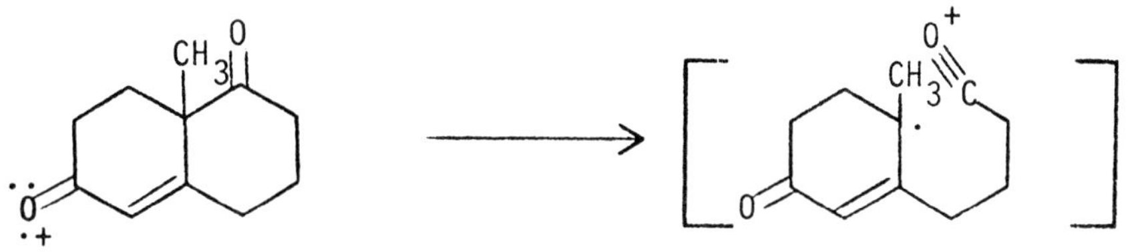

20 (m/e 178)

23 (m/e 178)

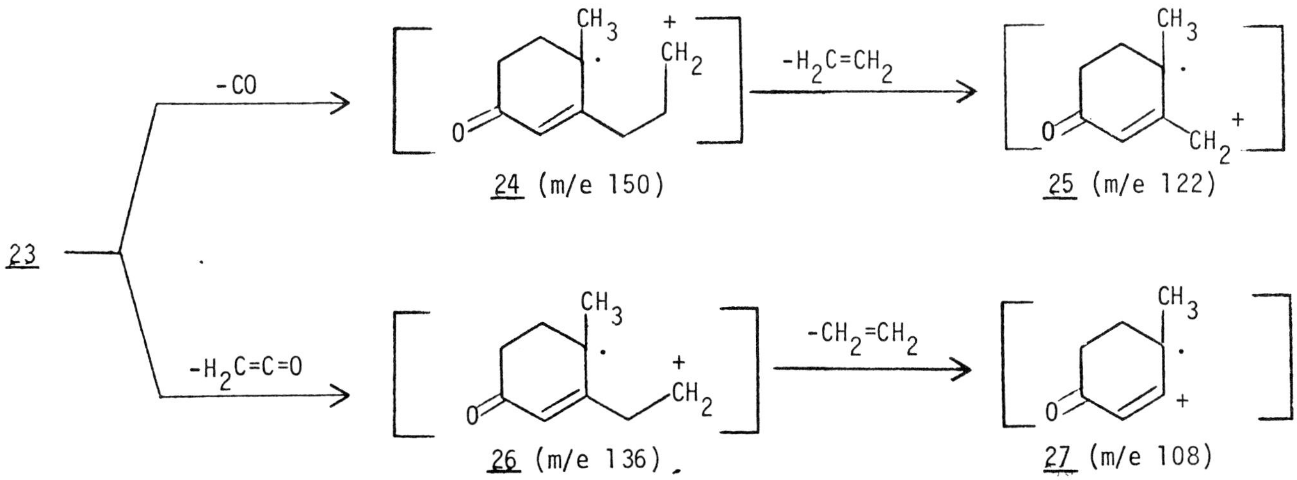

24 (m/e 150)

25 (m/e 122)

26 (m/e 136)

27 (m/e 108)

296

UNKNOWN 83

I. DATA COLLECTION (UNKNOWN 83)

A. Formula

1. Empirical formula: $C_{14}H_{18}N_2$.
2. Exact formula: Same.

B. Unsaturation

1. U.N. = 7.

2. Types of unsaturation.

a. The IR shows absorption consistent with aromatic hydrocarbons. (6-7µ, 12.2µ, 13.4µ).

b. The CMR shows six carbon types in the aromatic region; therefore, at least three C=C are of the aromatic type.

c. The HMR shows six protons in the aromatic region; therefore, at least three C=C are of the aromatic type.

C. Functional Groups and Part Structure

1. IR: No, NH, or C≡C-H.

2. CMR.

a. Seven unique carbon types.

b. Carbon units.

i. 3 CH (aromatic or sp^2).

ii. 3 -C- (aromatic or sp^2).

iii. 1 CH$_3$ (deshielded).

c. This accounts for only seven of the fourteen carbons present in the molecule; therefore, there is probably a plane of symmetry which bisects the molecule.

3. HMR.

a. Hydrogen units.

i. A 4H, m, 6.75 δ.
ii. B 2H, dd, (J=8, 2Hz), 6.40 δ.
iii. C 12H, s, 2.30 δ.

b. Decoupled HMR.

The 6.75 δ 4H m looks like an AB type pattern with further coupling. The 6.40 δ dd is consistent with a 1,2,3 trisubstituted aromatic structure where the ortho coupling is 8Hz and the meta coupling is 2Hz.

II. DEDUCTION OF THE STRUCTURAL FORMULA

A. <u>Summarize the Data</u>

1. $C_{14}H_{18}N_2$ - (seven unsaturations ≥ three of aromatic type).
2. Four equivalent CH_3 groups (CMR + HMR).

B. <u>Determine the Symmetry</u>. High symmetry from HMR and CMR--probably a plane of symmetry bisects the molecule.

C. <u>Assemble the Structure</u>

1. The chemical shift of the methyl groups is consistent with methyls bonded to either nitrogen or an aromatic ring.

2. Since CMR shows six aromatic C (3 CH + 3 -C-) but the HMR shows six aromatic protons (i.e. six aromatic CH), it is possible to account for at least nine aromatic carbons (4.5 unsaturations).

3. The proton NMR shows four additional carbons as equivalent CH_3 groups.

4. This must mean that the "missing" carbon is of the aromatic type and is a carbon bearing no protons.

5. That is, (six CH + four C + five unsaturations) + four CH_3.

6. Subtraction from $C_{14}H_{18}N_2$ leaves two rings (the remaining two unsaturations) and two nitrogen atoms for which to assign.

7. Since there are two rings and five unsaturations, the rings must be fused together to share an unsaturation!

8. There are two general possibilities: (1) nitrogen within the ring (fused pyridine or pyrrole rings) bearing aromatic methyl groups; <u>or</u> (2) a Naphthalene ring system bearing dimethylamino groups.

9. Inspection of the pyridine and pyrrole systems reveals no reasonable structures.

10. The <u>bis</u> dimethylaminonaphthalene isomers which have a symmetry element are:

1 (1,5) **2** (2,7) **3** (2,6) **4** (1,8) **5** (1,4) **6** (2,3)

11. Due to their symmetry (**1**,**3**, a C_2 axis; **5** and **6** a plane bisecting the dimethylamino groups), each of structures **1**,**3**,**5**,**6** have only five unique aromatic carbon types (3 CH + 2C).

    It is important to note that symmetry arguments can only be used to <u>unambiguously</u> eliminate isomers of higher symmetry; that is, if the CMR of this unknown had shown five carbon types instead of the six that were observed, it would have been <u>possible</u> that the unknown still <u>could have</u> been structures **2** and **4**, because an accidental equivalence could have reduced the number of observed carbon types. It is impossible to observe six carbon types in instances where only five can exist because of symmetry.

12. The choice between **2** and **4** can be made on the basis of the coupling constants in the proton NMR.

    a. Structure **2** would show the proton at C-1 as a small doublet (J=2-3Hz) coupled to the <u>meta</u> proton at C-3. The C-3 proton would be a doublet of doublets (J=2-3Hz, 8-9Hz) coupled to the <u>meta</u> proton at C-1 and the <u>ortho</u> proton at C-4. The C-4 proton would be a doublet (J=8-9Hz) coupled to the <u>ortho</u> proton at C-3; furthermore, the C-1, C-3 protons would be expected to be upfield of the C-4 proton resonance because of the shielding effect of the nitrogen lone pairs.

b. Structure 4 is in far better accord with the proton NMR. The upfield dd (J=2,8Hz) is from <u>meta</u> and <u>ortho</u> coupling to the protons at C-4 and C-3, respectively. The patterns for protons 2, 3, and 4 can be diagrammed as follows:

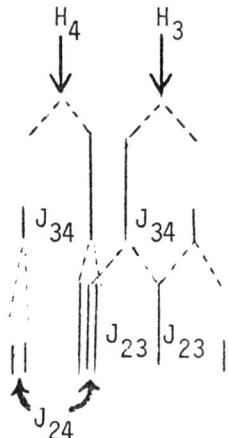

III. CHECK THE ASSIGNMENT

A. <u>IR</u>: Consistent.

B. <u>CMR</u>: Note that the furthest upfield carbons are a singlet and doublet. This is consistent with the expected shifts of carbons which are strongly shielded by the resonance interaction of the dimethylamino group.

C. <u>HMR</u>: Consistent.

D. <u>MS</u>: Strong parent ion typical of aromatics, loss of CH_3, $N(CH_3)_2$ at m/e 199 and 170, respectively.

UNKNOWN 84

I. DATA COLLECTION (UNKNOWN 84)

A. Formula

1. Empirical formula: $C_{14}H_{18}O_5$.
2. Exact formula: Same.

B. Unsaturation

1. U.N. = 6.

2. Types of unsaturation.

a. The IR shows two types of C=O absorption: ester and unsaturated ketone (5.75 and 6.0µ, respectively).

b. The CMR shows six carbons in the aromatic range (~110-160 ppm) and two carbons in the carbonyl region (1 ketone, 200 δ; 1 ester, 170 δ). This establishes all six unsaturations: one aromatic ring + two C=O groups.

c. The HMR shows two aromatic hydrogens and a CH_2 consistent with the OCH_2 of an ethyl ester.

C. Functional Groups and Part Structure

1. IR: Ester and conjugated ketone.

2. CMR.

   a. Fourteen unique carbon types.

   b. Carbon units.

   i. CO_2R (170 ppm).

   ii. C=O (200 ppm).

   iii. 4 -C- (aromatic or sp^2).

   iv. 2 CH (aromatic or sp^2).

   v. 2 CH_2 (deshielded).

   vi. 4 CH_3 (2 strongly deshielded).

3. HMR.

   a. Hydrogen units.

   i. A 1H, s, 7.35 δ.
   ii. B 1H, s, 6.75 δ.
   iii. C 2H, q (J=7Hz), 4.15 δ.
   iv. D 6H, s, 3.90 δ.
   v. E 2H, s, 3.85 δ.
   vi. F 3H, s, 2.55 δ.
   vii. G 3H, t(J=7Hz), 1.28 δ.

   b. Decoupled HMR: None.

II. DEDUCTION OF THE STRUCTURAL FORMULA

A. Summarize the Data

1. $C_{14}H_{18}O_5$ (six unsaturations; one aryl ring, one ketone, one ester).

2. IR shows that the ketone is conjugated (therefore, attached directly to the aryl ring), but the ester is not directly attached to the ring (CMR + IR).

3. There is an OCH₂CH₃ group in the molecule (either an ether or ester function).

B. <u>Determine the Symmetry</u>. None.

C. <u>Assemble the Structure</u>

1. The part structure X-CH₂-CH₃ is clearly due to Protons <u>C</u> + <u>G</u>.

2. C₁₄H₁₈O   (six unsaturations)
   -C₆H₂CO (the aromatic keto group = five unsaturations)
   ─────────
   C₇H₁₆O₄ + [aromatic ring with H, CO group, and H]

3. The two low field CH₃ groups must be directly attached to an oxygen atom.

4. The observed value of 3.90 δ can be compared to a calculated* value of 3.35 δ for a saturated OCH₃ group.

5. The deshielding increment of a beta aryl group 0.35 δ provides the means necessary to generate a chemical shift closer to the observed value.

This yields: C₅H₁₀O₂ + [aromatic ring with CO, H₂, (OCH₃)₂]

6. If the assignment of the OCH₃ group is correct, the OCH₂CH₃ group is on the ester moiety.

i.e.   [aromatic ring with -CO—, (OCH₃)₂, H₂] + $\overset{O}{\overset{\parallel}{C}}$-OC₂H₅ + CH₂ + CH₃

7. Since the ester is not directly attached to the aromatic ring, the CH₂ unit must be the means of connection. This places the methyl group on the ketone by default and provides this basic structure.

*See Appendix III.

8. The two hydrogens are not coupled to each other; therefore, they must be *para* to one another. This generates three possible structures:

1   **2**   **3**

9. In order to establish the correct isomer, it will be necessary to have some method of approximating the shielding and deshielding effects of the alkoxy and keto groups on the aromatic hydrogens. (The ester group is not directly attached to the ring and should have a minimal effect.) The following shift positions are known for simple "model" compounds:

	δo	δm	δp
R=H	7.25	7.25	7.25
R=COCH$_3$	7.90	7.60	7.60
R=OCH$_3$	6.80	7.20	6.80

We can generate therefore these approximate shielding and deshielding constants for each group.

ortho COCH$_3$ = +.65
meta, para COCH$_3$ = +.35
ortho, para OCH$_3$ = -.45
meta OCH$_3$ = -.05

10. The following shifts are calculated on the assumption that these crude increments are absolutely additive. (This assumption is clearly not going to be quantitatively valid; however, it should be in the correct direction.)

a.

[Structure 1: benzene ring with COCH₃, Hₐ, CH₂CO₂C₂H₅, OCH₃, H_b, CH₃O substituents]

1

$H_a \cong 7.25$ (base value) +
.65 (o-COCH₃) − 2 × .05
(2-m-OCH₃) = 7.80 δ

$H_b \cong 7.25 − (2 × .45)$
$+ .35 = 6.70 δ$

b.

[Structure 2: benzene ring with CH₃O, H_c, CH₂CO₂C₂H₅, OCH₃, H_d, CH₃CO substituents]

2

$H_c \cong 7.25 − .45 − .05 + .35$
$= 7.10 δ$

$H_d \cong 7.25 + .65 − .45 − .05$
$= 7.40 δ$

c.

[Structure 3: benzene ring with CH₃O, H_c, CH₂CO₂C₂H₅, COCH₃, H_d, CH₃O substituents]

3

Same as isomer **2**.

d. The absolute shifts of both calculations are not in good accord with any of the structures; but the net difference between the absorption position (**1** = 1.10 δ, **2,3** = 0.30 δ vs 0.60 observed) tends to favor structures **2** or **3**.

e. In order to absolutely identify the unknown, further spectral or chemical experiments would be necessary. (The correct answer is actually **3**.)

III. CHECK THE ASSIGNMENT

　A. <u>IR</u>: Consistent.

　B. <u>CMR</u>: Consistent.

　C. <u>HMR</u>: Calculation of chemical shift:

　　1. CH₂ (E) = 1.20 + 1.45
　　　(δ-aryl) + 1.00 (δ-CO₂R) = 3.65 <u>vs</u>. 3.88 observed.

2. CH_3 (F) = 0.90 + 1.25 (α-C=O) + .35 (β-aryl) = 2.50 vs. 2.56 observed.

D. MS : Loss of $CO_2C_2H_5$ group to generate the substituted tropylium ion (m/e 182).

UNKNOWN 90

I. DATA COLLECTION (UNKNOWN 90)

A. Formula

1. Empirical formula: $C_{18}H_{22}$.
2. Exact formula: Same.

B. Unsaturation

1. U.N. = 8.

2. Types of unsaturation.

a. The IR shows aromatic absorptions.

b. The CMR shows six aromatic type carbons, three CH + three C (plus some minor impurities); this means at least one trisubstituted aromatic ring.

c. The HMR shows a sharp multiplet for six aromatic hydrogens.

d. Therefore, there is a plane of symmetry which makes at least two groups of three CH carbons equivalent.

C. Functional Groups and Part Structure

1. IR : Aromatic (12.2μ, 2 adjacent H; 11.3μ, 1 isolated H?) no C≡C-H.

2. CMR.

a. Ten unique carbon types.

b. Carbon units.

i. 3 CH (aromatic).
ii. 3 C- (aromatic).
iii. 1 CH (aliphatic).
iv. 3 CH_3.

3. HMR.

a. Hydrogen units.

i. <u>A</u> 6H, m, 7.0 δ.
ii. <u>B</u> 1H, q (J=7Hz), 4.0 δ.
iii. <u>C</u> 12H, s, 2.2 δ.
iv. <u>D</u> 3H, d (J=7Hz), 1.6 δ.

b. Decoupled HMR: The coupling pattern shows that <u>B</u> + <u>D</u> are the following group where X has no adjacent hydrogens.

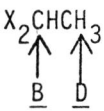

II. DEDUCTION OF THE STRUCTURAL FORMULA

A. Summarize the Data

1. Formula $C_{18}H_{22}$, eight unsaturations.

2. X_2CHCH_3 group is present.

3. Probably two aromatic rings having three CH + three C each (total = eight unsaturations) are present.

B. Determine the Symmetry

1. Two groups of $[(CH)_3 + (C)_3]$.
2. Two groups of $[CH_3 + CH_3]$.

C. Assemble the Structure

1. The data is in excellent accord with this structure. The question is: What is the methyl substitution pattern?

2. There are six possible trisubstituted aromatic structures.

      4   5   6

- a. Structures 4 and 6 can be eliminated because they are of higher symmetry than the rest and would only have seven CMR absorption signals. (The aromatic methyls and two of the ring carbons are equivalent.)

- b. Structure 1 would be expected to have IR absorption for three adjacent hydrogen atoms between 12.5–13.1 μ.

- c. This leaves the three 1,2,4 trisubstituted aromatic structures 2, 3, 5.

- d. These structures cannot be distinguished on the basis of the data available. A nuclear Overhauser experiment would show that the aromatic methyl groups are *not* near neighbors of the CHCH₃ group. This would eliminate structures 2 and 5 and prove the structure to be 3.

III. CHECK THE ASSIGNMENT

- A. **IR**: Aromatic absorption.
- B. **CMR**: Consistent with the assigned structure.
- C. **HMR**: Typical aromatic methyl chemical shift.
- D. **MS**

3 (m/e 240)

7 (m/e 239)

8 (m/e 225)

APPENDIX III

THE CURPHEY-MORRISON ADDITIVITY CONSTANTS FOR PROTON NMR CHEMICAL SHIFT APPROXIMATION*

Substituent Effects on R—C—C with $H\alpha$, $H\beta$

Standard Shift Positions:
Methyl 0.90δ; Methylene 1.20δ; Methine 1.55δ.

Functional Group R	Type of Hydrogen	Alpha Shift	Beta Shift
Chlorine	CH_3-	2.30	0.60
	$-CH_2-$	2.30	0.55
	$-CH-$	2.55	0.15
Bromine	CH_3-	1.80	0.80
	$-CH_2-$	2.15	0.60
	$-CH-$	2.20	0.25
Iodine	CH_3-	1.30	1.10
	$-CH_2-$	1.95	0.60
	$-CH-$	2.70	0.35
Aryl	CH_3-	1.45	0.35
	$-CH_2-$	1.45	0.55
	$-CH-$	1.35	----
$-C(=O)H$, $-C(=O)R'$	CH_3-	1.25	0.25
	$-CH_2-$	1.10	0.30
	$-CH-$	0.95	----
$-C(=O)OH$, $-C(=O)O-R'$	CH_3-	1.20	0.25
	$-CH_2-$	1.00	0.30
	$-CH-$	0.95	----
$-C=C-$	CH_3-	0.90	0.05
	$-CH_2-$	0.75	0.10
	$-CH-$	1.25	----
Hydroxyl	CH_3-	2.45	0.40
	$-CH_2-$	2.30	0.20
	$-CH-$	2.30	----

–O–alkyl	CH₃– –CH₂– –CH–	2.45 2.30 2.10	0.30 0.15 ----
–O–Aryl	CH₃– –CH₂– –CH–	2.95 3.00 3.30	0.40 0.45 ----
$-\text{O}-\underset{\parallel}{\overset{\text{O}}{\text{C}}}-\text{alkyl}$	CH₃– –CH₂– –CH–	2.90 2.95 3.45	0.40 0.45 ----
Amines (R=nitrogen)	CH₃– –CH₂– –CH–	1.25 1.40 1.35	0.20 0.15 ----
SH, S-Alkyl	CH₃– –CH₂– –CH–	1.20 1.30 1.30	0.40 0.30 ----
–C≡N	CH₃– –CH₂– –CH–	1.10 1.10 1.05	0.45 0.40 ----
–NO₂	CH₃– –CH₂– –CH–	3.50 3.15 3.05	0.65 0.85 ----
–C≡C–	CH₃– –CH₂– –CH–	0.90 0.80 0.35	0.15 0.05 ----

These additive increments were developed with acyclic systems where free rotation of the deshielding groups was possible. These values can also be used to approximate chemical shifts in cyclic systems. Configurational and conformational changes in these systems may produce significant anisotropic effects; therefore, the calculated and actual chemical shifts may have greater variance than what is commonly found in acyclic systems.

*Taken from the Ph.D. Dissertation of Thomas J. Curphey, Harvard University and modified by Professor H. A. Morrison at Purdue; by permission of T. J. Curphey and H. A. Morrison.

APPENDIX IV

SELECTED REFERENCE LIST

A. General Spectroscopy

1. "Spectrometric Identification of Organic Compounds," 3rd Ed., R. M. Silverstein, G. C. Bassler, T. C. Morrill, John Wiley, New York, 1974.

2. "Spectroscopic Methods in Organic Chemistry," 2nd Ed., D. H. Williams, I. Fleming, McGraw-Hill, New York, 1973.

3. "Organic Structural Analysis," J. B. Lambert, H. F. Shurvell, L. Verbit, R. G. Cooks, G. H. Stout, MacMillan, New York, 1976.

4. "Stereochemistry Fundamentals and Methods," H. B. Kagan, Ed., Volume 1, Determination of Configurations by Spectroscopic Methods, Thieme, Stuttgart 1977.

B. Infrared Spectroscopy

1. "Infrared Spectroscopy," M. Avram, Gh. D. Matteeseu, Wiley-Interscience, New York, 1972.

2. "Infrared Absorption Spectroscopy," 2nd Ed., K. Nakanishi, P. H. Solomon, Holden-Day, San Francisco, 1977.

C. Proton NMR Spectroscopy

1. "Applications of Nuclear Magnetic Resonance Spectroscopy in Organic Chemistry," 2nd Ed., L. M. Jackman, J. Sternhell, Pergamon Press, New York, 1969.

2. "Nuclear Magnetic Resonance," W. W. Paudler, Allyn and Bacon, Boston, 1971.

D. Carbon-13 NMR Spectroscopy

1. "Carbon-13 Nuclear Magnetic Resonance for Organic Chemists," G. C. Levy, G. L. Nelson, Wiley-Interscience, New York, 1972.

2. "Carbon-13 NMR Spectroscopy," J. B. Stothers, Academic Press, New York, 1972.

E. Mass Spectroscopy

1. "Interpretation of Mass Spectra," 2nd Ed., F. W. McLafferty, Benjamin, London, 1973.

2. "Interpretation of Mass Spectra of Organic Compounds," H. Budzikiewicz, C. Djerassi, D. H. Williams, Holden-Day, San Francisco, 1964.

ROBERT P. BORRIS